AF412835

THE RNA WORLD

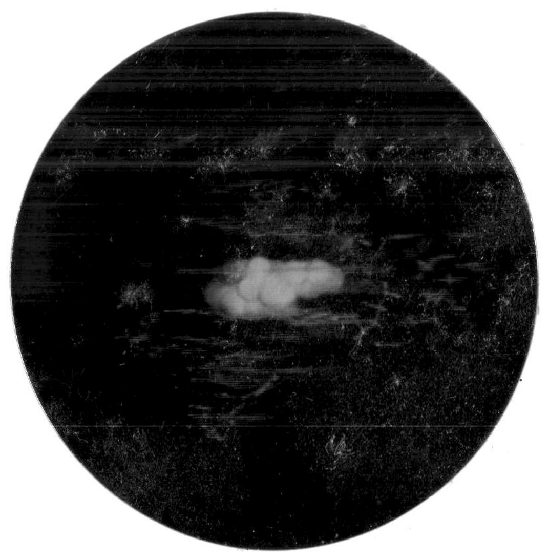

A binding pocket in RNA. Holographic view of a model of HIV TAR RNA with arginine bound at the Tat-protein-binding site. The model is consistent with NMR data (Puglisi et al., *Science 257:* 76 [1992]). The part of TAR shown consists of the three-nucleotide bulge and adjoining helical regions. The six-nucleotide loop of TAR is not shown. The arginine recognition site involves a base triple between a bulge nucleotide and a base pair in the upper helix. The arginine contacts a guanine base in the upper helix, as well as two phosphates below the bulge. We thank James Williamson (MIT) for the model coordinates, Ignacio Tinoco (Berkeley) for helpful advice, Marco Molinaro (Berkeley) and Paul Leo (Utah) for assistance with computer graphics, and Chromagem, Inc. (Youngstown, Ohio) for producing the hologram.

For optimum viewing, hold hologram under a point light source (e.g., a clear light bulb or halogen spotlight) or sunlight at a 30º angle. The image will appear slightly blurred if viewed under diffuse or fluorescent light.

THE RNA WORLD

The Nature of Modern RNA Suggests a Prebiotic RNA World

Edited by

Raymond F. Gesteland
University of Utah

John F. Atkins
University of Utah
University College Cork

COLD SPRING HARBOR LABORATORY PRESS
1993

COLD SPRING HARBOR MONOGRAPH SERIES

The Lactose Operon
The Bacteriophage Lambda
The Molecular Biology of Tumour Viruses
Ribosomes
RNA Phages
RNA Polymerase
The Operon
The Single-Stranded DNA Phages
Transfer RNA:
    Structure, Properties, and Recognition
    Biological Aspects
Molecular Biology of Tumor Viruses, 2nd Ed.
    DNA Tumor Viruses
    RNA Tumor Viruses
The Molecular Biology of the Yeast
    Saccharomyces:
    Life Cycle and Inheritance
    Metabolism and Gene Expression

Lambda II
Mitochondrial Genes
Nucleases
Gene Function in Prokaryotes
Microbial Development
The Nematode *Caenorhabditis elegans*
Oncogenes and the Molecular Origins
    of Cancer
Stress Proteins in Biology and Medicine
DNA Topology and Its Biological Implications
The Molecular and Cellular Biology of the
    Yeast *Saccharomyces:*
    Genome Dynamics, Protein Synthesis,
    and Energetics
    Gene Expression
Transcriptional Regulation
Reverse Transcriptase
The RNA World

THE RNA WORLD

Monograph 24
Copyright 1993 by Cold Spring Harbor Laboratory Press
All rights reserved
Printed in the United States of America
Book design by Emily Harste

Library of Congress Cataloging-in-Publication Data

The RNA world / edited by Raymond F. Gesteland and John F. Atkins.
        p. cm. -- (Monograph / Cold Spring Harbor Laboratory ; 24)
    Includes bibliographical references and index.
    ISBN 0-87969-380-0
    1. RNA--Evolution.   I. Gesteland, Raymond F.   II. Atkins, John F.
(John Fuller) III. Series: Cold Spring Harbor monograph series ;
24.
QP623.R6   1993
574.87'3283--dc20   93-2840
                                                       CIP

All Cold Spring Harbor Laboratory Press publications may be ordered directly from Cold Spring Harbor Laboratory Press, 10 Skyline Drive, Plainview, New York 11803-2500. Phone: 1-800-843-4388 in Continental U.S. and Canada. All other locations: (516) 349-1930. FAX: (516) 349-1946.

Contents

Preface

This book invokes a prebiotic RNA world that existed some 3.8 billion years ago, only 50 times more ancient than the dinosaur era. The essence of this RNA world was RNA-catalyzed replication, which ultimately gave rise to the predecessor of the DNA-based organisms of modern life.

The notion that prebiotic evolution absolutely depended on RNA replication has intrigued many since the early suggestions by Woese, Orgel, and Crick. Those of us with interests in the current function of RNA were especially fascinated as we, of course, hoped that RNA was the pivotal molecule. However, no one predicted the richness of apparent "relics" from the RNA world that have been discovered in present-day organisms. In fact, the overwhelming stimulus for the new wave of interest in the RNA world comes from discoveries of catalytic functions and structural diversity among contemporary RNAs. Experiments with catalytic RNAs reveal the potential for RNA to possess the plasticity of functions needed for complex evolution—functions which could have resulted in the "breakthrough organism" that was the predecessor of all life. The plasticity of RNA is critically dependent on 2′-hydroxyl groups; it is remarkable that this distinction between RNA and DNA can so dramatically influence versatility of function.

New technologies allow iterative selective examination of large numbers of RNA sequence variants with altered binding properties, novel catalytic activities, or perhaps even replicating functions, opening the door to real-time exploration of RNA molecular evolution. If the current rate of experimental progress continues, we anticipate, in the near future, creation of self-replicative elements perhaps typical of the primordial RNA world. Self-replicating RNA, by combining template and catalytic functions, could solve the evolutionary "chicken and egg" dilemma of how to generate proteins without nucleic acid templates and how to replicate nucleic acids without protein catalysis.

We believe that the time is right for this volume. It delves into the necessarily hypothetical description of the prebiotic RNA world and examines the current roles of RNA in order to reveal the "fossil record" and the versatility of RNA functions. Obviously, not all topics relevant to the RNA world are included in this book. The chapters comprise our subjective view of a book that presents the basic principles of the roles of RNA and that also communicates the current excitement about the world of RNA. We are not aware of other books on this topic, except perhaps the Cold Spring Harbor Symposium volume of 1987 that dealt with catalytic RNAs, and much has happened since then.

It is fitting that Francis Crick and Jim Watson, the canons of the canonical pair, should write the Foreword and Prologue for this volume. After all, they abandoned DNA shortly after discovery of its structure to devote their energies to RNA.

We are grateful to many people for making this book possible: in particular, all of the authors and the staff of Cold Spring Harbor Laboratory Press, especially Inez Sialiano, Patricia Barker, Nancy Ford, and John Inglis.

R.F. Gesteland

J.F. Atkins

Foreword

The term "RNA world" originally referred to a hypothetical time in the evolution of earthly life when there was no elaborate mechanism for protein synthesis such as we have today. As described by Joyce and Orgel (Chapter 1), there were speculations in the 1960s that RNA catalysts existed at that stage in evolution, that RNA was the sole genetic material, and that the standard Watson-Crick pairing was the basis of genetic replication. None of these early authors was smart enough to suggest that relics of these hypothetical catalytic RNAs might still be around today. Indeed, it was speculated that the original ribosomes might have been made solely of RNA but not that their main catalytic activity (the formation of the peptide bond) might still be performed today by RNA alone, as recent evidence seems to suggest.

This hypothesis of an RNA world without protein was largely forgotten but has now become fashionable again because of the remarkable discoveries by Altman and by Cech (described by Cech, Chapter 11) of RNA molecules that do indeed have catalytic activity of one sort or another. These discoveries have removed the chief objection to the RNA world—that RNA by itself cannot act catalytically.

This volume on the RNA world covers a somewhat wider field. It was realized in the 1960s that not all present-day RNA was messenger RNA or viral RNA. The ribosome was known to contain several distinct types of "structural" RNA molecules; tRNAs were familiar, and there were isolated examples of other small RNA molecules. What was not realized was the richness of the *present-day* RNA world, including snRNPs (as described by Baserga and Steitz, Chapter 14) and now many others. We have come to realize that RNA molecules can occur with many different shapes and can thus have many different functions. Several of the chapters in this book deal with what these structures are, what their functions may be, what makes them stable, and how to predict their structure. Just how many of these RNAs are, by themselves, truly catalytic remains to be seen. Some of them may use divalent cations to produce both their folding and their catalytic activities, as discussed by Pan et al. (Chapter 12).

Some of the properties of present-day RNA molecules were quite unexpected. One might perhaps have guessed that some limited form of messenger-RNA editing could occur very rarely as a freak, but who would have predicted that such editing would take so many distinct forms and occur as widely as Bass describes in Chapter 15?

That a retrovirus contains a sequence rather like part of a tRNA molecule was a surprise when it was first discovered. In 1987, Maizels and Weiner proposed that such structures tagged RNA genomes for replication in the early RNA world. In Chapter 23, they explore the many possible ramifications of this genome tag hypothesis. In particular, they propose that both tRNA and tRNA-aminoacylation activity were subject to selection *before* protein synthesis arrived on the scene. Whether one can safely extrapolate from a present-day property of rapidly evolving RNA viruses to as far back in time as the early RNA world remains to be seen.

A good case can be made (Gold et al., Chapter 19) that small lengths of RNA can often form relatively rigid structures more easily than can polypeptides of similar length. New techniques are now being used (Chapter 19 and Chapter 20 [Szostak and Ellington]) to explore this vast "space" of small RNA sequences for interesting properties such as the binding of one specific substrate or another and, in some cases, a particular type of catalytic activity.

The recombinant-DNA revolution of the 1970s was possible because molecular biologists could use the sophisticated products of billions of years of natural evolution — the replicases, restriction enzymes, and so forth — as precise and delicate chemical tools. These molecules were all proteins that acted on nucleic acids in one way or another. Now experimentalists (Chapters 19 and 20) are using a combination of biochemical processes, related to nucleic acid replication, which together embody the mechanism of natural selection. These powerful tools can provide them with new and desirable RNA molecules in the laboratory while they wait. Because these methods do not need whole cells, but only cellular components, they can handle many more individuals at one time and so explore the RNA "space" more quickly. The message is clear: Where possible, let Nature do the work, and if she proves a little slow, take your whip to her.

The exact nature of the early RNA world is now a matter of active debate. What can we learn from the present "relics" and, in particular, can we deduce the composition and nature of each particular type of early RNA from the detailed study of its many descendants today? Can we learn anything from small RNA-related molecules (such as coenzymes) about the general nature of the early RNA world? Early speculations had assumed that the RNA world was rather simple and rather inaccurate. Benner and his colleagues (Chapter 2) now believe it was much more complex and more accurate. As more sequence data accumulate, one would expect these suggestions to become less speculative.

It may be possible to deduce something about some of the activities in the RNA world by a detailed study of the present mechanisms of protein synthesis. For example, Weiss and Cherry (Chapter 3) have proposed an ingenious model for the origins of the smaller ribosomal subunit. They suggest that it was originally an RNA replicase that used oligonucleotides as a substrate, cannibalizing triplets from them for its own replication.

We may, in time, arrive at a rather plausible picture of this early stage in evolution, even if true molecular fossils (that is, actual specimens of the molecules that then existed) are forever unavailable to us because of the ceaseless battering of thermal motion over billions of years.

The details of the transition from the RNA world to the present protein-dominated world are also debatable. For example, was there some sort of protein synthesis before the existence of a primitive ribosome? Did these first ribosomes have some protein, or did the ribosomal proteins arrive a little later?

It may turn out that we will eventually be able to see how this RNA world got started. At present, the gap from the primal "soup" to the first RNA system capable of natural selection looks forbiddingly wide. Was there perhaps a pre-RNA world, with replication based on some even more primitive system?

Such a pre-RNA world might have been of several types. It would presumably have been based on some form of genetic polymer or sheet that was formed more easily than RNA under prebiotic conditions. It might eventually have transcribed its detailed genetic information directly onto RNA to form the RNA world. Alternatively, it might have acted solely as a midwife, setting up the basis of RNA synthesis for some use of its own, with RNA replication then taking over and replacing it. Cairns-Smith has already suggested that this pre-RNA system might have been based on clays, or on some organic polymer (Cairns-Smith, A.G., *Genetic Takeover and the Mineral Origins of Life.* Cambridge University Press [1982]). This time we should be smart enough to ask if there are any relics of this *pre*-RNA world still with us today.

Finally, a word of heresy: As Moore has pointed out (Chapter 5), the real fossil record suggests that our present form of protein-based life was already in existence 3.6 billion years ago and evolved rather slowly for a billion or so years after that. This leaves an astonishingly short time to get life started. Moreover, the three main lines of descent (see Fig. 3 in Woese and Pace, Chapter 4) seem a very long distance from their hypothetical common ancestor.

The rather far-fetched hypothesis of Directed Panspermia would predict that life was sent here in the form of "bacteria" suitable for growth in anaerobic conditions, *and that several somewhat different forms would probably have been sent at the same time,* in the hope that at least one would survive. All are likely to have evolved originally from a common ancestor (on another planet) that existed some billions of years before the formation of our solar system. Therefore, the final question about the RNA world and the pre-RNA world (if it existed) is: *Where* did it occur? Are we totally confident that our form of life started here, or did it perhaps originate elsewhere in the universe? It might have been easier to start elsewhere because, for example, the atmosphere there was more reducing than the Earth's early atmosphere appears to have been.

The lively and authoritative chapters of this book deal comprehensively both with the hypothetical RNA world and with the complexities of RNA structure and function we find around us today. I recommend it to all molecular biologists and especially to anyone fascinated by the baroque complexity of the nucleic acids and of RNA in particular.

Francis Crick

Prologue
Early Speculations and Facts about RNA Templates

RNA first came alive to me during the fall of 1947 at Indiana University when I took Salvador Luria's course on viruses. There I first learned that whereas the then-known phage, pox and papilloma viruses, contained DNA, this molecule was totally absent in several purified plant viruses as well as in the viruses that caused encephalitis and polio, which instead contained RNA. Apparently a given virus had either RNA or DNA, in contrast to cells which contained both. But whether it was the nucleic acid component that carried their genetic specificity was still unclear. At that time, most scientists wanted Avery, MacLeod, and McCarty's experiment on pneumococcal transformation by purified DNA to be extended to other life forms before jumping on the nucleic acid bandwagon. Then in the spring of 1952 came the report from Al Hershey that the DNA component of phage T2 carried genetic specificity. This immediately thrilled me, but I remember well the audience's indifference when in mid-April I read Hershey's letter at an Oxford meeting of The Society for General Microbiology.

When I had arrived at Cambridge in the fall of 1951, I started taking seriously the work of Brachet and his collaborators in Brussels, who emphasized the correlation between the RNA content and the protein-synthesizing capacity of cells. Those cells making large amounts of protein possessed large numbers of virus-sized ribonucleoprotein particles, known initially as microsomal particles but since 1958 as ribosomes.

Most importantly, these particles had been pinpointed as the actual sites of protein synthesis by means of the then just-developed cell-free systems for protein synthesis. Here the key lab was that of Paul Zamecnik at Massachusetts General Hospital. Equally important was Brachet and Chantrenne's demonstration that the nucleus, and hence DNA, had no direct participation in protein synthesis. To show this, they cut the giant alga *Acetabularia* in half and observed that the half without a nucleus could maintain almost normal protein synthesis for more than a month. Yet from the one gene–one enzyme (protein) results of Beadle and Tatum, the ultimate source of the genetic information that specifies the amino acid sequences of proteins had to be the genes found in the nucleus. I thus postulated a two-stage scheme for protein synthesis in which DNA first serves as a template for nucleus-located synthesis of RNA, and this RNA then in turn moves to the cytoplasm where it functions as the template for protein synthesis.

No one then had any compelling reason to take my hypothesis seriously, but by November 1952 I liked it well enough to print DNA→RNA→protein on a small piece of paper which I taped on the wall above my writing table in my rooms at Clare College. From the day of our first meeting, Francis Crick and I thought it highly likely that the genetic information of DNA is conveyed by the sequence of its four bases, but we knew it was premature to promote this idea before the structure of DNA was known. However, from the moment we first saw how to build a double helix out of the four base pairs, it was clear that the essential uniqueness of a gene must reside in its respective sequence of base pairs. Moreover, not only could base pairing provide the way for genes to be copied exactly during gene duplication, but it was also very likely to underlie the process by which the genetic information of a DNA molecule is transferred to its RNA product.

Still totally unclear, however, was how RNA might serve as the template for ordering the amino acids in their respective polypeptide products. Emboldened by our fantastic good luck in so simply finding the structural essence of the gene duplication process, I saw no reason not to take on the challenge of finding out what RNA molecules looked like in three dimensions. Such knowledge, I felt, would be indispensable to understanding how they functioned in protein synthesis. I took on this task when I moved to Pasadena in the fall of 1953. There I joined forces with Alex Rich, who had started working on DNA just before Francis and I found the double helix. There was no difficulty in getting him to move on to a better pasture, and soon I was collecting RNA samples and drawing them into fibers that Alex exposed to X-ray beams. But despite much travail, even those fibers displaying high birefringence never gave rise to ordered diffraction patterns like those of DNA. Although we thought we saw reflections that might have come from short sections of double helices, we could never be sure and saw no way to decide whether RNA was a one- or two-chain molecule. Here the base composition was a bad tease. Viral RNAs clearly did not show equivalence of A with U or G with C, but the RNA from cells had A/U and G/C ratios sometimes closely approaching 1/1. But we could see no difference in the general features of the X-ray diagrams from viral or cellular RNAs. To our annoyance, RNA, no matter from what source, showed identical X-ray diagrams characterized by strong reflections at 3.36 Å and 4.00 Å. Clearly, there was some ordered structure in RNA, but we saw no way to get to it. After some six months of such frustration, we gave up.

A very different approach to understanding protein synthesis came from the very clever Russian-born theoretical physicist George Gamow, who was struck by the fact that the 3.6-Å distance between adjacent amino acids in extended polypeptide chains was very similar to the 3.4-Å separation between adjacent base pairs in the B form of the double helix. Not then cognizant of RNA's primary role in protein synthesis, Gamow proposed that amino acids are directly ordered by contacts with the DNA base pairs, with the polypeptide products containing the same number of amino acids as there are base pairs in their DNA templates. To deal with the fact that DNA has only 4 letters in its alphabet, whereas 20 different letters (amino acids) are used to specify proteins, Gamow assumed that each amino acid must be coded by several adjacent base pairs. Because there are only 16 (4 x 4) combinations of the four bases, taken two at a time, Gamow made the assumption that each amino acid must be specified by groups of three adjacent base pairs (a codon) along DNA chains. To deal with the fact that there are 64 (4 x 4 x 4) such triplets, he assumed that many amino acids must be specified by more than one triplet (redundant triplets). In such an "overlapping" code, adjacent amino acid codons have two out of the three base pairs in common, thereby restricting which amino acids can lie adjacent to each other. Gamow's overlapping code was only the first of several that would later be devised, each leading to different combinations of forbidden amino acid neighbors.

When George first told me his scheme, I quickly dismissed it since DNA was not the template that ordered the amino acids. But he was having combinatorial fun, and besides I could not rule out the possibility that some RNA molecules were double helices. With time, moreover, I realized there was a real virtue to Gamow-like codes. In disproving them, the possibility of overlapping codes could be ruled out, pointing the way to codes in which adjacent groups of most likely three bases specified successive amino acids along a polypeptide chain. In fact, the first known amino acid sequence, that determined by Sanger for insulin, disproved the first code, although Gamow did not at first realize this because his initial list of the 20 amino acids had several embarrassing mistakes (e.g., he included both cystine and cysteine). Other such overlapping codes devised over the next year were also ruled out as more proteins became sequenced. It was during this first coding rush that Leslie Orgel and I, on a trip to Berkeley where Gamow was spending the spring of 1954, suggested that we form a club of 20 members whose purpose was to crack the RNA structure and in so doing reveal how the genetic code operated. Soon to be known as the "RNA Tie Club," with its members reflecting Gamow's eclectic taste, it never had a formal meeting, nor did all its members ever cough up the money to purchase their RNA ties and tiepins bearing their respective amino acid code letters. Gamow's tiepin sported ALA (alanine), and mine, PRO (proline). Much more important in the long run was the opportunity it provided to exchange ideas about the code through "Notes to the RNA Tie Club."

Several of these communications, the most important of which came from Francis Crick, Leslie Orgel, and Sydney Brenner, later became incorporated into published manuscripts. Among these was the definitive disproof by Brenner of any form of overlapping code, which he wrote in Johannesburg in the fall of 1956 just before he returned to England to join Francis Crick. Equally important was the communication by Crick, John Griffiths, and Orgel suggesting a comma-less code as a device to let non-overlapping triplets be read in the appropriate reading frame. But the most influential paper intellectually was the RNA Tie Club's first Note, written by Francis Crick and sent out early in 1955 under the title *On Degenerate Templates and the Adaptor Hypothesis*. After spending the previous August in Woods Hole batting about potential codes with Gamow and Brenner, Crick began questioning the basic assumption that a nucleic acid template provided specific cavities complementary in shape and charge to the amino acid side groups. Here he argued that the nucleic acid bases want to hydrogen-bond and are not at all suitable for forming cavities that could attract the hydrophobic side groups of amino acids such as valine, leucine, or isoleucine. Equally tricky to imagine was any structural basis for degenerative codes in which many amino acid side groups are specified by more than one set of triplets. Faced with what he considered insuperable obstacles, Crick made the radical proposal that prior to peptide bond formation, amino acids are enzymatically attached to small "adaptor" molecules that have surfaces specifically tailored to bond to nucleic acid triplets. Here Crick suggested that the adaptor molecules might be tiny polynucleotides that base pair to RNA templates.

My reaction to the adaptor hypothesis was initially very negative, even though I had spent the fall of 1954 in futile efforts trying to fold RNA chains into shapes bearing cavities appropriate for the amino acid side groups. The adaptor idea seemed much too complicated to me to have ever gotten started at the origin of life several billion years ago. Francis, too, had his moments of doubt; he even concluded his now famous Tie Club Note with the phrase, "In the comparative isolation of Cambridge I must confess there are times when I have no stomach for decoding." In fact, by the time I returned to Cambridge for the year beginning in June 1955, Francis and Alex Rich were immersed in building new three-dimensional models for collagen to compete with one earlier proposed by Linus Pauling, this despite the fact that Francis and I had often referred to collagen as the most boring of macromolecules.

Equally frustrated, I reverted to thinking about plant viruses, in particular tobacco mosaic virus (TMV), whose helical construction I had worked out in the spring of 1952 after Francis and I had been told by Sir Lawrence Bragg to stop trying to work out the structure of DNA through model building. Always troublesome to me was the apparent necessity to postulate both genetic and protein synthesis roles for RNA. Knowing that only a tiny fraction of TMV particles are actually infectious, I speculated whether in fact these rare particles contained DNA, not RNA, chains. But this was ruled out when it became possible to reconstitute infectious TMV particles from their purified RNA and protein components. This experiment, first successfully accomplished in the spring of 1955 in Berkeley by Heinz Fraenkel-Conrat and Robley Williams, generated much newspaper publicity which gave uninformed readers the idea that life itself had been created. Francis, however, put the matter in its proper light, being quoted in the English press as saying that this was a finding he had anticipated.

Reconstitution by itself, however, did not answer the question of whether the protein component played any more than a protective-coat role for the genetic-information-bearing RNA component. Less than a year later, however, Alfred Gierer, working in Gerhard Schramm's lab in Tübingen, clearly showed that the RNA alone was infectious. This primacy of nucleic acids as bearers of genetic information then lay at the heart of the way Francis and I thought about cells and viruses. But this was far from an acceptable paradigm for many of the attendees at the key, late March 1956, CIBA Foundation meeting on the Structure of Viruses. They were not at home with the concept that information flows unidirectionally from nucleic acids to proteins and never backward. This was awkwardly shown when André Lwoff and I passed on to Robley Williams a telegram message supposedly from Wendell Stanley reading "TMV protein infectious—be cautious!" To our amazement, Robley didn't question the result until we revealed the hoax.

At this gathering of some 30 scientists, Francis presented our ideas on why the protective coats of viruses are made up of protein subunits. In our view, it was a consequence of the fact that no viral nucleic acid had sufficient coding capacity to specify a single polypeptide chain large enough to surround a more centrally located core of nucleic acid. The 2-million-molecular-weight RNA of TMV, for example, contains only 6000 bases and, assuming a coding ratio of three bases per amino acid, is only capable of specifying a 2000-amino-acid polypeptide or about 230,000 molecular weight. To make up the 38-million-molecular-weight protein coat, at least 165 subunits would be needed. In fact, the TMV subunit contains only some 150 amino acids, suggesting that the TMV RNA codes for several proteins or that the coding ratio is very much larger than three. Initially, we thought that tomato bushy stunt virus (TBSV), with a much higher RNA content, might be more useful in giving a realistic value for the coding ratio. It contains only four bases for every amino acid in the crystallographic subunit. Later, direct analysis of its protein subunit size suggested that each crystallographic repeat contains some five protein subunits, very likely implying that TBSV RNA, like TMV RNA, also codes for several proteins.

After that 1956 meeting, I again had a go at the RNA structure, taking advantage of the newly discovered enzyme polynucleotide phosphorylase that Marianne Grunberg-Manago and Severo Ochoa found could be used to make synthetic RNA molecules. By then back at the National Institutes of Health, Alex Rich had shown the month before that random AU and AGCU co-polymers gave X-ray diffraction patterns similar to those we had obtained using purified cellular and viral RNAs. I focused instead on poly(A) (adenine) fibers drawn from material prepared in the Molteno Institute by Roy Markham and David Lipkin. To my delight, they generated clean helical X-ray diagrams that were best interpreted as base-paired double helices built up from two parallel poly(A) chains. Initially, I was disturbed by the fact that many of the key reflections over-lapped with those generated by purified RNA, which by then we had every reason to believe was single-chained. Later, this apparent paradox was possibly resolved by the finding that sections of hydrogen-bonded hairpins form along most RNA chains. Conceivably, it is these short sections of imperfect double helices that generate the DNA-like feature of the RNA X-ray patterns.

By then discouraged that the study of purified RNA would lead toward understanding how protein synthesis occurs, I decided to concentrate on the structure of ribosomal particles, at that time believing that they must carry the genetic information for ordering amino acids in proteins. In the spring of 1956 I convinced Alfred Tissières, then a Fellow at Kings College working in the Molteno Institute on oxidative phosphorylation, to join me at Harvard, where I would be moving in the fall of 1956. Alfred had in fact already done some preliminary experiments on the then-called microsomal particles and was keen to follow them up.

I arrived at Harvard some six months before Alfred but was preoccupied most of this time trying to be an effective teacher for the seniors and beginning graduate students who were taking my course on viruses. As soon as my lectures were under control, however, I went across the Charles River to see Paul Zamecnik whom I had first met the year before while briefly stopping at Harvard on my way back to England. There, in the building where Fritz Lipmann also had his lab, I first appreciated the importance of the discovery made there a year earlier by Mahlon Hoagland of the activated high-energy acyl-amino acid intermediates for protein synthesis. Even more important, I first learned of the more recent observation by Mahlon, Elizabeth Keller, and Paul of a soluble RNA (sRNA) fraction to which the activated amino acids are transferred prior to protein synthesis. Quickly, I realized that these sRNA molecules might be the polynucleotide adaptors postulated two years before by Francis in his first Note to the RNA Tie Club. Prior to my visit, Crick's ideas were unknown to the Massachusetts General group and they eagerly sought out Francis when they came together at the 1957 Gordon Conference on Nucleic Acids and Proteins.

Tissières and I commenced our molecular characterization of *Escherichia coli* ribosomes in the spring of 1957, suspecting that they might have structural plans like those of the small RNA viruses, whose isocahedral-shaped protein shells are formed by the regular aggregation of a single protein building block. We could not have been more wrong about how they are organized. To start with, we found that the *E. coli* ribosomes, like those from all other organisms, are formed by the aggregation of two RNA-containing subunits, the larger 50S subunit approximately twice the size of the smaller 30S subunit. Each subunit contains a single major RNA chain, with the 50S ribosomal subunits possessing 23S RNA chains and the 30S subunits possessing 16S RNA chains. Moreover, both subunits contain a large number of different small proteins that for the most part are subunit-specific. At low Mg^{++} levels, the 30S and 50S subunits do not associate with each other, but when the Mg^{++} concentration is raised, they come together to form the 70S complex that subsequent work has shown to be the ribosomal form that carries out protein synthesis.

Naively, at first, we assumed that either the 16S RNA, or the 23S RNA, or both, were the actual templates for protein synthesis. Puzzling to us, however, was why the templates existed in only two size classes while there was great variation in the sizes of their putative polypeptide products. Equally disturbing was why the base composition of ribosomal RNAs barely varied between bacterial species with highly different AT/GC ratios. A priori we had expected to find that the base compositions of the RNA templates would reflect those of their DNA templates. Luckily, there was one powerful exception. In 1956 the phage T2-specific RNA made after T2 infection of *E. coli* was shown by Volkan and Astrachan to have a T2 DNA-like base composition. Moreover, in contrast to the metabolically very stable ribosomal RNA chains, T2 RNA had been found to have a half-life of only several minutes.

In retrospect, Tissières and I should have gravitated early to T2 RNA, but in fact not until the fall of 1959 were its molecular properties investigated anywhere. Then, Masayasu Nomura and Ben Hall, working with Sol Spiegelman at the University of Illinois, provided tentative evidence for the incorporation of T2 RNA into abnormally small ribosomes. In trying to follow up this observation early in 1960, my graduate student Bob Risebrough came to a radically different conclusion. After T2 RNA is synthesized it does not become part of a ribosomal subunit by aggregating with newly made ribosomal proteins. What in fact happens is that in the presence of Mg^{++}, the T2 RNA becomes attached to the smaller 30S ribosomal subunit, which in turn binds the larger 50S subunit to form the >70S complex that actually carries out protein synthesis. This result instantly changed the way we visualized protein synthesis. Instead of serving template roles, ribosomes function as stable assemblage sites for protein synthesis. The true template had to be a new RNA class unknown until that moment both because it comprises such a small percentage of the total RNA and because it is heterogeneous in length. Later, these metabolically unstable templates, whose amounts respond to cellular needs, would be named messenger RNA (mRNA) by Jacques Monod and François Jacob.

The first lab outsider to whom I revealed this conceptual breakthrough was Leo Szilard, whom I had gone down to see in New York where he was successfully plotting the radiation therapy that would cure his bladder cancer. Leo's reaction was entirely negative, not being convinced that we had the right interpretation for Risebrough's experimental results. Predictably, he wanted us to show that mRNA existed in uninfected *E. coli* cells before he would change his mind-set. These were experiments that we had in fact already planned to start several weeks later, as soon as François Gros arrived from Paris to spend several months working with Alfred and me. Also soon to join this effort was Wally Gilbert, then still teaching theoretical physics to Harvard students, but who was increasingly tempted to move into molecular biology by our excitement about mRNA. By the time of the June Gordon Conference on Nucleic Acids we were virtually convinced that *E. coli* mRNA also existed, and by the summer's end we had data for a convincing publication. Already at the Gordon Conference we had heard rumors that Sydney Brenner and François Jacob, using very different arguments, had also postulated the existence of mRNA and that their idea was being tested by Sydney that week in Matt Meselson's lab at Caltech. Eventually, we were to publish our independent proofs for mRNA's existence early in 1961 in back-to-back articles in *Nature*.

With the basic scheme for how RNA participates in protein synthesis known, the path became open for definitive experiments on the exact nature of the genetic code. Using enzymatically synthesized RNA as messengers for in vitro protein synthesis, the correct assignments for all of the triplet codons were determined by early 1966. With this major goal achieved, the time had clearly come to ask how the DNA→RNA→protein flow of information had ever gotten started. Here, Francis was again far ahead of his time. In 1968 he argued that RNA must have been the first genetic molecule, further suggesting that RNA, besides acting as a template, might also act as an enzyme and, in so doing, catalyze its own self-replication. As the chapters in this book will show, how right he was!

J.D. Watson

THE RNA WORLD

1

Prospects for Understanding the Origin of the RNA World

Gerald F. Joyce
Departments of Chemistry and Molecular Biology
The Scripps Research Institute
La Jolla, California 92037

Leslie E. Orgel
The Salk Institute for Biological Studies
San Diego, California 92186

The general idea that, in the development of life on the earth, evolution based on RNA replication preceded the appearance of protein synthesis was first proposed more than 20 years ago (Woese 1967; Crick 1968; Orgel 1968). It was suggested that catalysts made entirely of RNA are likely to have been important at this early stage in the origins of life, but the possibility that RNA catalysts might still be present in contemporary organisms was overlooked. The unanticipated discovery of ribozymes (Kruger et al. 1982; Guerrier-Takada et al. 1983) initiated extensive discussion of the role of RNA in the origins of life (Pace and Marsh 1985; Sharp 1985; Lewin 1986) and led to the coining of the phrase "the RNA world" (Gilbert 1986).

The RNA world means different things to different authors, so it would be futile to attempt a restrictive definition. All RNA world hypotheses include three basic assumptions: (1) At some time in the evolution of life, genetic continuity was assured by the replication of RNA; (2) Watson-Crick base-pairing was the key to replication; (3) genetically encoded proteins were not involved as catalysts. RNA world hypotheses differ in what they assume about life that may have preceded the RNA world, about the metabolic complexity of the RNA world, and about the role of low-molecular-weight cofactors, possibly including polypeptides, in the chemistry of the RNA world.

It should be emphasized that the existence of an RNA world as a precursor of our DNA/protein world is a hypothesis. We find it an attractive hypothesis and believe that it derives some support from the results of experiments that it has inspired. The demonstration that the peptide-

The RNA World
© 1993 Cold Spring Harbor Laboratory Press 0-87969-380-0/93 $5 + .00

bond-forming step of protein synthesis is catalyzed by largely protein-free ribosomal RNA is particularly striking (Noller et al. 1992). We recognize, however, that not everyone will find the available evidence persuasive.

In our initial discussion of the RNA world we accept The Molecular Biologist's Dream: "Once upon a time there was a prebiotic pool full of β–D-nucleotides... ." We now consider what would have to have happened to make the dream come true. This discussion triggers The Prebiotic Chemist's Nightmare: how to make any kind of self-replicating system from the kind of intractable mixture that is formed in experiments designed to simulate the chemistry of the primitive earth.

THE MOLECULAR BIOLOGIST'S DREAM

Abiotic Synthesis of Polynucleotides

In this section, we discuss the synthesis of oligonucleotides from β–D-nucleoside-5′-phosphates. Two fundamentally different chemical reactions are involved. First, the nucleotide must be converted to an activated derivative, for example, a nucleoside 5′-polyphosphate. Next the 3′-hydroxyl group of one nucleotide molecule must be made to react with the activated phosphate group of another. Synthesis of oligonucleotides from nucleoside 3′-phosphates will not be discussed since activated nucleoside 2′- or 3′-phosphates in general react readily to form 2′,3′-cyclic phosphates. These cyclic phosphates are unlikely to oligomerize efficiently because the equilibrium constant for dimer formation is only of the order of 1.0 L mole^{-1} (Erman and Hammes 1966; Mohr and Thach 1969).

In enzymatic RNA and DNA synthesis, the nucleoside 5′-triphosphates are the substrates of polymerization. Polynucleotide phosphorylase, although it is a degradative enzyme in nature, can be used to synthesize oligonucleotides from nucleoside 5′-diphosphates. Nucleoside 5′-polyphosphates are, therefore, obvious candidates for the activated forms of nucleotides. Although nucleoside 5′-triphosphates are not formed readily, the synthesis of nucleoside 5′-tetraphosphates from nucleotides and inorganic trimetaphosphate provides a reasonably plausible prebiotic route to activated nucleotides (Lohrmann 1975). A number of other more or less plausible prebiotic syntheses of nucleoside 5′-polyphosphates have been reported (Handschuh et al. 1973; Osterberg et al. 1973). Nucleoside 5′-polyphosphates, although they are high-energy phosphate esters, are relatively unreactive in aqueous solution. Consequently, little is known about their nonenzymatic oligomerization reactions.

In a different approach to the activation of nucleotides, the isolation of an activated intermediate is avoided by employing a condensing agent such as a carbodiimide (Khorana 1961). This is a popular method in organic synthesis, but its application to prebiotic chemistry is problematical. Potentially prebiotic molecules such as cyanamide and cyanoacetylene activate nucleotides in aqueous solution, but the subsequent condensation reactions are inefficient (Lohrmann and Orgel 1973).

Most attempts to study nonenzymatic polymerization of nucleotides in the context of prebiotic chemistry have used nucleoside $5'$-phosphorimidazolides. Although phosphorimidazolides can be formed from imidazoles and nucleoside $5'$-polyphosphates (Lohrmann 1977), they are only marginally plausible as prebiotic molecules. They were chosen because they are prepared easily and react at a convenient rate in aqueous solution.

Nucleotides contain three principal nucleophilic groups, the $5'$-phosphate, the $2'$-hydroxyl, and the $3'$-hydroxyl groups, in order of decreasing reactivity. The reaction of a nucleotide or oligonucleotide with an activated nucleotide, therefore, normally yields $5',5'$-pyrophosphate-, $2',5'$-phosphodiester-, and $3',5'$-phosphodiester-linked adducts, in order of decreasing abundance (Fig. 1a) (Sulston et al. 1968). The condensation of several monomers, therefore, would likely yield an oligomer containing one pyrophosphate and a preponderance of $2',5'$ phosphodiester bonds (Fig. 1b). There is little chance of producing entirely $3',5'$-linked oligomers from activated nucleotides unless a catalyst can be found that increases the yield of $3',5'$ phosphodiester bonds. Several metal ions, particularly Pb^{++} and UO_2^{++}, catalyze the formation of oligomers from nucleoside $5'$-phosphorimidazolides (Sleeper and Orgel 1979; Sawai et al. 1988). However, the product oligomers always contain a large proportion of $2',5'$ linkages.

What kinds of prebiotically plausible catalysts might lead to the production of $3',5'$-linked oligonucleotides directly from nucleoside $5'$-phosphorimidazolides or other activated nucleotides? It is unlikely, but not impossible, that a metal ion or simple acid-base catalyst would provide sufficient stereospecificity. The most attractive of the other hypotheses is that adsorption to a specific surface of a mineral might orient activated nucleotides rigidly and thus catalyze a highly stereospecific reaction.

The potential role of minerals in prebiotic chemistry is difficult to assess. Polynucleotides adsorb strongly to positively charged surfaces of minerals such as hydroxylapatite. It seems almost certain that much prebiotic chemistry was the chemistry of adsorbed monomers and polymers. What is unclear is the extent to which mineral surfaces can modify

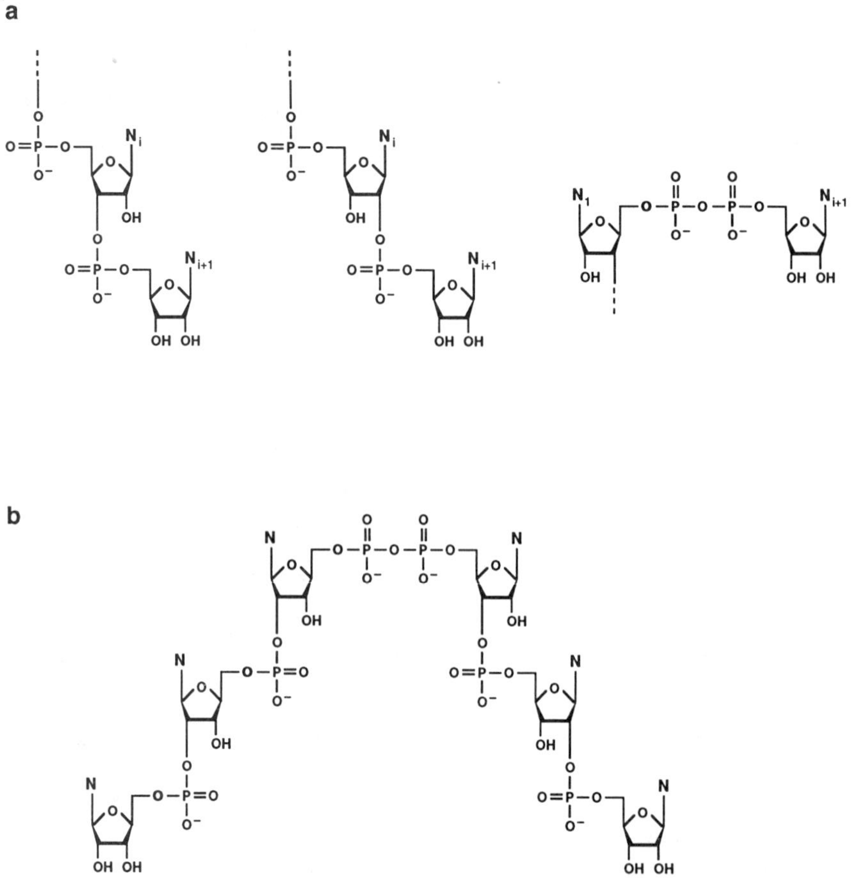

a

b

Figure 1   Phosphodiester linkages resulting from chemical condensation of nucleotides. (*a*) Reaction of an activated mononucleotide (N_{i+1}) with an oligonucleotide ($N_1 - N_i$) to produce a 3′,5′-phosphodiester (*left*), 2′,5′-phosphodiester (*middle*), or 5′,5′-pyrophosphate (*right*). (*b*) Typical oligomeric product resulting from chemical condensation of activated mononucleotides.

reactions that normally produce a complex mixture of products in such a way as to produce a single predominant product. This is equally unclear at the level of monomer synthesis and at the level of polymer formation.

Whenever a problem in prebiotic synthesis seems intractable, it is possible to postulate the existence of a mineral that catalyzes the reaction. Because there are a very large number of minerals, and most of them can tolerate a substantial range of metal-ion substitutions, such claims cannot easily be refuted. It would be nice to see a few relevant examples of stereospecific mineral catalysis, but at present we know of none. Clearly, this is an unresolved issue central to all discussions of the origins of life.

Nonenzymatic Replication of RNA

If a mechanism existed for the polymerization of activated nucleotides, it would have generated a complex mixture of product oligonucleotides, differing in both sequence and length. The next stage in the evolution of an RNA world would have been the replication of some of these molecules, so that a process equivalent to natural selection could begin. The reaction central to replication of nucleic acids is template-directed synthesis; that is, the synthesis of a complementary oligonucleotide under the direction of a preexisting oligonucleotide. A good deal of work has already been done on this aspect of nonenzymatic replication. Progress up to about 1987 has been reviewed previously (Joyce 1987), so only a summary of the results are given here.

The first major conclusion is that most activated nucleotides do not undergo efficient, regiospecific, template-directed reactions. In general, only a small proportion of template molecules succeed in directing the synthesis of a complement, and the complement contains a mixture of $2'$, $5'$ and $3'$, $5'$ phosphodiester bonds. After a considerable search, a set of activated nucleotides was found that undergo efficient and highly regiospecific template-directed reactions. Working with guanosine $5'$-phospho-2-methylimidazolide (2-MeImpG), it was shown that poly(C) can direct the synthesis of long oligo(G)s in a reaction that is highly efficient and highly regiospecific (Inoue and Orgel 1981). If poly(C) is incubated with an equimolar mixture of the four 2-MeImpNs (N = G, A, C, or U), less than 1% of the product consists of noncomplementary nucleotides (Inoue and Orgel 1982).

Random copolymers containing an excess of C residues can be used to direct the synthesis of products containing G and the complements of the other bases present in the template (Inoue and Orgel 1983). The reaction with a poly(C,G) template is especially interesting because the products, like the template, are composed entirely of C and G residues. If these products in turn could be used as templates, it might allow the emergence of a self-replicating sequence. Self-replication, however, is unlikely, mainly because poly(C,G) molecules that do not contain an excess of C residues tend to form stable self-structures that prevent them from acting as templates (Joyce and Orgel 1986). The self-structures are of two types: the standard Watson-Crick variety based on C·G pairs, and a quadrahelix structure that results from the association of four G-rich sequences (Fig. 2). As a consequence, any C-rich oligonucleotide that can serve as a good template will give rise to G-rich complementary products that tend to be locked in self-structure and so cannot act as templates. Overcoming the self-structure problem is very difficult because it requires the discovery of conditions that favor the binding of mononucleo-

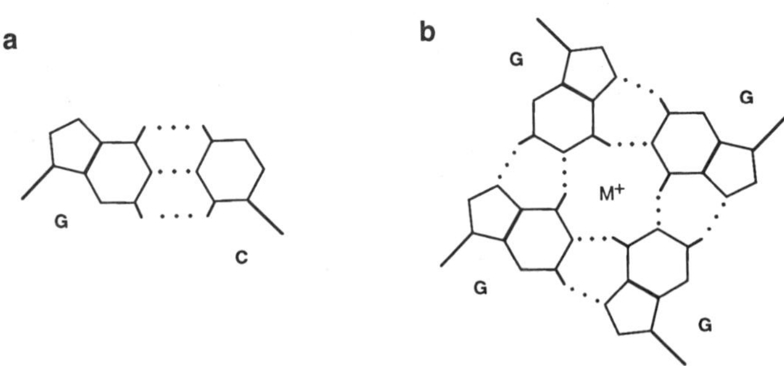

Figure 2   Two types of self-structure interactions expected for poly(C,G) templates. (*a*) Standard G•C Watson-Crick pair. (*b*) G quadrahelix pairing, stabilized by a monovalent cation (M+). Dotted lines indicate hydrogen-bonding interactions.

tides in order to allow template-directed synthesis to occur but that suppress the formation of long duplex regions which would exclude activated monomers from the template.

Some progress has been made in discovering defined-sequence templates that are copied faithfully to yield complementary products (Inoue et al. 1984; Acevedo and Orgel 1987; Wu and Orgel 1992a). Successful templates typically contain an excess of C residues, with A and U residues isolated from each other by at least three C residues. Runs of G residues are copied into runs of C residues, so long as the formation of self-structures by G residues can be avoided (Wu and Orgel 1992b). In light of the available evidence, it seems unlikely that a pair of complementary sequences can be found, each of which facilitates the synthesis of the other using nucleoside 5′-phospho-2-methylimidazolides as substrates. Some of the obstacles to self-replication may be attributable to the choice of reagents and reaction conditions, but others seem to be intrinsic to the template-directed condensation of activated mononucleotides.

Ribozymes catalyze transesterification reactions in which the 3′-terminal nucleotide of a preexisting oligonucleotide is added to the 3′ terminus of a growing polymer chain (Zaug and Cech 1986; Been and Cech 1988), e.g., the reaction

$$C_5 + C_i \rightarrow C_4 + C_{i-1}$$

Related chemistry could, in principle, be used to copy an oligonucleotide template. This suggestion, although perhaps applicable to an established RNA world, is not very plausible for the nonenzymatic replication of RNA. The synthesis of the 3′, 5′-linked oligonucleotide substrates, as we have seen, presents a very difficult problem. Furthermore, the use of an "energy-neutral" transesterification reaction would make it impossible to convert more than a small fraction of the substrate to high-molecular-weight products. The free energy of association of the template with its complement might alleviate this difficulty to some extent.

Another nonenzymatic replication scheme suggested by ribozyme-catalyzed reactions involves synthesis by the ligation of short 3', 5'-linked oligomers (Sharp 1985; Doudna and Szostak 1989). This is certainly an attractive hypothesis, but it faces two major obstacles. The first is again the difficulty of obtaining the substrates to begin with. The second is concerned with fidelity. Pairs of oligonucleotides containing a single base mismatch, particularly if the mismatch forms a G·U wobble pair, still hybridize as efficiently as fully complementary oligomers, except in a temperature range very close to the melting point of the perfectly paired structure. Maintaining fidelity would therefore be difficult under any plausible temperature regime.

The First RNA Replicase

The molecular biologist's notion of the RNA world places emphasis on an RNA molecule that catalyzes its own replication. Such a molecule must function as an RNA-dependent RNA polymerase, acting on itself (or copies of itself) to produce complementary RNAs, and acting on the complementary RNAs to produce additional copies of itself. The efficiency of this process must be sufficient to produce "progeny" RNA molecules at a rate that exceeds the rate of decomposition of the "parents." The fidelity of polymerization must be sufficient to allow the exact copies to compete favorably against the distribution of mutant copies. Beyond these requirements, the details of the replication process are not highly constrained.

In the Molecular Biologist's Dream, it is assumed that a supply of activated β–D-nucleotides was available by some as yet unrecognized abiotic process. Furthermore, it is assumed that a magic catalyst existed to convert the activated nucleotides to a random ensemble of polynucleotide sequences, a subset of which had the ability to replicate. It seems to be implicit in the model that such sequences replicate themselves but, for whatever reason, do not replicate unrelated neighbors. It is not clear whether replication is literally one molecule copying itself or whether it involves a molecule copying a second molecule with the same or very similar sequence. We ignore these questions for the moment and discuss, in a formal way, the question of whether an RNA molecule of reasonably short length can catalyze its own replication with reasonably high fidelity.

The Error Threshold

Consider a set of polynucleotides that catalyze their own replication, drawing on a fixed supply of activated mononucleotides to produce RNAs having the same sequence as themselves. If a particular RNA sequence is to survive, it must accumulate at a positive rate and its rate of reproduction must exceed the mean rate of production of other RNA sequences in its locale. Satisfying these requirements places restrictions on the fidelity of replication, which in turn places restrictions on the length of the RNA that can be replicated.

The rate of synthesis of new copies of a self-replicating RNA molecule, R_i, is proportional to the concentration of R_i

$$\mathrm{d}[R_i] / \mathrm{d}t = W_i [R_i] \tag{1}$$

The autocatalytic rate constant, W_i, is the difference between the rate of formation of error-free copies of R_i and the rate of decomposition of existing copies of R_i

$$W_i = A_i Q_i - D_i \tag{2}$$

In this formulation, first developed by Manfred Eigen (1971), A_i refers to the rate of amplification of R_i, Q_i refers to the proportion of copies of R_i that are error-free, and D_i refers to the rate of decomposition of R_i. For simplicity, the nondiagonal error terms are neglected in this treatment; that is, the contributions to $[R_i]$ resulting from error copies of some other $R_{k \neq i}$ are ignored. In subsequent formulations (Eigen and Schuster 1977; McCaskill 1984), the nondiagonal terms were included, leading to more precise results that are qualitatively similar to those described here.

In the simplified model, for an advantageous self-replicating RNA, R_m, to outgrow its competitors, $R_{k \neq i}$, it is necessary that

$$W_m > \overline{W}_{k \neq m}, \text{ where } \overline{W}_{k \neq m} = \Sigma W_k[R_k] / \Sigma[R_k] \tag{3a}$$

$$A_m Q_m - D_m > \overline{A}_{k \neq m} - \overline{D}_{k \neq m} \tag{3b}$$

Note that the term $\overline{Q}_{k \neq m}$ does not appear on the right side of Equation 3b because error copies of a particular $R_{k \neq m}$ give rise to some other $R_{k \neq m}$. Rearranging Equation 3b gives

$$Q_m > (\overline{A}_{k \neq m} - \overline{D}_{k \neq m} + D_m) / A_m \tag{4}$$

In other words, to maintain the genetic information contained within an advantageous individual, it is necessary that the fidelity of copying that information exceed the inverse of the relative advantage enjoyed by the advantageous individual compared to the rest of the population. This relative advantage, or "superiority," is expressed as

$$\sigma_m = A_m / (\overline{A}_{k \neq m} - \overline{D}_{k \neq m} + D_m) \tag{5}$$

Combining Equations 4 and 5 gives

$$Q_m > 1 / \sigma_m \tag{6}$$

The proportion of copies of R_m that are error-free is determined by the fidelity of the component condensation reactions that are required to produce a complete copy. For simplicity, consider a self-replicating RNA that is formed by v condensation reactions, each having mean fidelity q. The probability of obtaining an error-free copy is the product of the fidelity of the component condensation reactions

$$Q_m = q^v \tag{7}$$

Combining Equations 6 and 7 and solving for v

$$\ln Q_m = v \ln q > -\ln \sigma_m \tag{8a}$$

$$v < -\ln \sigma_m / \ln q \tag{8b}$$

Equation 8b, sometimes referred to as the "Eigen error threshold," defines the trade-off between the fidelity of replication (q) and the maximum allowable number of component condensation reactions (v). For a self-replicating RNA of length n that is formed by template-directed condensation of activated mononucleotides, a total of $2n - 2$ condensation reactions are required to produce a complete copy. This takes into account the synthesis of both a complementary strand and a complement of the complement.

Figure 3 depicts the relationship between q and n for two different values of σ_m. The maximum allowable genome length is highly sensitive to the fidelity of replication but depends only weakly on the superiority of the most advantageous self-replicating species. It should be recognized that a marked superiority of one sequence over all other sequences could not be maintained over evolutionary time because novel variants would soon arise to challenge the dominant species. However, a marked initial superiority may be important in allowing a self-replicating RNA to emerge in the first place. In the absence of appreciable competition, a primitive self-replicating RNA that operates with low efficiency and low fidelity may gain a foothold by taking advantage of a somewhat less stringent error threshold. For example, if $\sigma_m = 10$, an RNA that replicates with 90% fidelity could be no longer than 12 nucleotides, and one that replicates with 70% fidelity could be no longer than 4 nucleotides. It seems highly unlikely that *any* of the 17 million possible RNA dodecamers is able to catalyze its own replication with 90% fidelity and even less likely that a tetranucleotide could catalyze its own replication

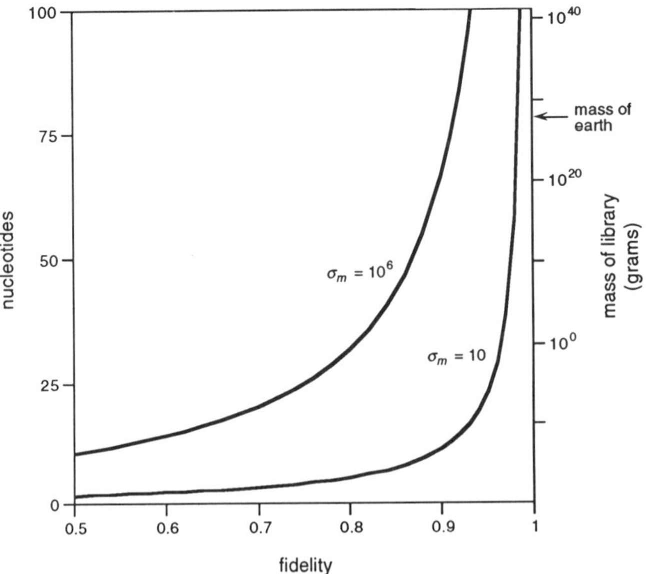

Figure 3 The Eigen error threshold, relating fidelity of replication to the maximum number of nucleotides, for two different values of superiority (σ_m). Fidelity (q) is that of the component condensation reactions in the replication cycle. The number of condensation reactions (v) is related to the number of nucleotides (n) by: $v = 2n - 2$. Scale at right indicates the mass of a combinatorial library containing one copy each of the 4^n possible sequences of length n.

with 70% fidelity. However, if $\sigma_m = 10^6$, an RNA that replicates with 90% fidelity could be as long as 67 nucleotides, and one that replicates with 70% fidelity could be as long as 20 nucleotides.

Once self-replication is established, there is strong selection pressure favoring improvement of the fidelity of the component template-directed condensation reactions. As fidelity improves, a lower superiority can be tolerated. More importantly, a larger genome can be maintained. This allows exploration of a larger number of possible sequences, some of which may lead to further improvement in fidelity, which in turn allows a still larger genome size, and so on. Once the evolving population has entered the steep part of the curve shown in Figure 3, RNA-based life can become firmly established. Until that time, it is a race between evolutionary improvement in the context of a sloppy self-replicating system and the risk of extinction due to either overstepping the error threshold or undergoing some fluctuation catastrophe that wipes out the population.

It is difficult to state with certainty the minimum possible size of an RNA replicase ribozyme. An RNA consisting of a single secondary structural element, that is, a small stem-loop containing 12–17 nucleotides, would not be expected to have replicase activity, whereas a double stem-loop, perhaps forming a "dumbbell" structure or a pseudoknot, might just be capable of a low level of activity. A triple stem-loop structure, containing 40–60 nucleotides, offers a reasonable hope of functioning as a replicase ribozyme (Fig. 4). One could, for example, imagine a molecule consisting of a pseudoknot and a pendant stem-loop that forms a cleft for template-dependent replication.

Suppose there is some 50-mer that enjoys a superiority (σ_m) of 200 and replicates with 90% fidelity. This should be regarded as a highly optimistic but still reasonable view of what is possible for a minimum replicase ribozyme. Would such a molecule be expected to occur within a population of random RNAs? A complete library consisting of one copy each of all 10^{30} possible 50-mers would weigh 10^{10} grams, roughly 10^{-9} the weight of the present biosphere. However, there may be many such 50-mers, encompassing both distinct structural motifs and, more importantly, a large number of equivalent representations of each motif. As a result, even a small fraction of the total library, consisting of perhaps 10^{24} sequences and weighing 10^4 grams, might be expected to contain at least one self-replicating RNA with the requisite properties. On the other hand, it is not sufficient that there be just one copy of a self-replicating RNA. Unless the molecule can literally copy itself, that is, act simultaneously as both template and catalyst, it must encounter another copy of itself that it can use as a template. If two or more copies of a particular 50-mer RNA are needed, then a much larger library, consisting of 10^{54} RNAs and weighing 10^{34} grams, would be required. This amount far exceeds the mass of the earth.

It should be recognized that concerns about the Eigen error threshold apply not to the problem of finding a self-replicating RNA within a pool of random RNAs, but to the need for stable propagation of the genetic information contained within a self-replicating species. If the error threshold is exceeded, the genetic information soon becomes delocalized and the population degenerates into a collection of random-sequence RNAs.

The Eigen error threshold would be eased somewhat if the requirement for high fidelity did not apply uniformly throughout the sequence of the replicase ribozyme. There may, for example, be positions within single-stranded regions of the replicase that can be occupied by any of the four nucleotides without any decrease in the molecule's ability to function as a replicase. There may also be places in base-paired regions where any of the four standard base pairs will do, or where a G·U wobble can substitute for a G·C Watson-Crick pair.

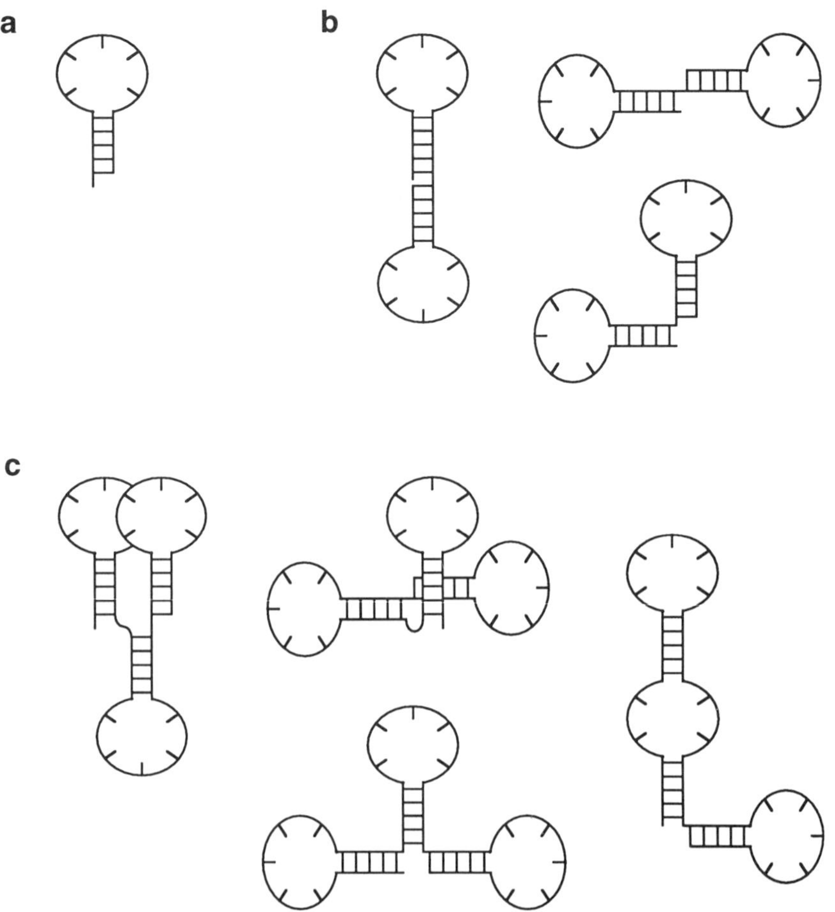

Figure 4 Examples of small RNA structures consisting of (*a*) one; (*b*) two; or (*c*) three stem-loop elements.

If insensitivity to base substitution is to result in significant relaxation of the error threshold, two requirements must be met. First, the unconstrained positions must be able to vary independently of one another; that is, if there are j unconstrained positions, the vast majority of the 4^j possible sequence variants resulting from base substitution at these positions must have comparable replicase activity. Second, each of the neutral variants must be surrounded by a comparable distribution of error copies. The latter requirement is more difficult to satisfy, particularly as the functionally significant positions in the molecule undergo evolutionary change over time.

At first sight, it might seem that another way to ease the error threshold would be for the replicase ribozyme to accept dinucleotide or trinucleotide substrates, so that a copy of an RNA of length n could be formed by $2(n/2 - 1)$ or $2(n/3 - 1)$ condensation reactions, respectively. Calculations show that, over a broad range of values for σ_m, RNAs that are required to replicate with 90% fidelity when using mononucleotide substrates would be required to replicate with roughly 80% fidelity when using dinucleotide substrates or roughly 70% fidelity when using trinucleotide substrates. This is not likely to offer a significant advantage because lessening of the error threshold would be outweighed by the greater difficulty in discriminating among the 16 possible dinucleotide or 64 possible trinucleotide substrates, rather than among the four mononucleotides.

Another Chicken-and-Egg Paradox

This discussion concerning the first RNA replicase ribozyme has, in a sense, focused on a straw man: the myth of a self-replicating RNA molecule that arose de novo from a soup of random polynucleotides. Not only is such a notion unrealistic in light of our current understanding of prebiotic chemistry, but it should strain the credulity of even an optimist's view of RNA's catalytic potential. If you doubt this, ask yourself whether you believe that a replicase ribozyme would arise in a solution containing nucleoside 5′-diphosphates and polynucleotide phosphorylase!

If one accepts the notion of an RNA world, then one is faced with the dilemma of how such a genetic system came into existence. To say that the RNA world hypothesis "solves the paradox of the chicken and the egg" is correct if one means that RNA can function both as a genetic molecule and as a catalyst that promotes its own replication. RNA-catalyzed RNA replication provides a chemical basis for Darwinian evolution based on natural selection. Darwinian evolution is a powerful way to search among vast numbers of potential solutions for those that best address a particular problem. Selection based on inefficient RNA replication, for example, could be used to search among a population of RNA molecules for those individuals that promote improved RNA replication; but here we encounter another chicken-and-egg paradox: Without evolution it appears unlikely that a self-replicating ribozyme could arise, but without some form of self-replication there is no way to conduct an evolutionary search for the first, primitive self-replicating ribozyme.

RNA evolution may have gotten started without the aid of an evolved catalyst by using nonenzymatic template-directed synthesis to permit some copying of RNA before the appearance of the first replicase ribozyme. Suppose that the initial ensemble of monomers was not produced by random copolymerization, but rather by a sequence of untemplated and templated reactions as illustrated in Figure 5. Also suppose that members of the initial ensemble of multiple stem-loop structures could be replicated, albeit inefficiently, by the template-directed process. This provides a way to generate stem-loop structures automatically. If it is true that ribozyme function is favored by stable self-structure, and if the base sequences of the stems in stem-loop structures are relatively unimportant for function, this model might provide an economical way of generating a relatively small ensemble of sequences that are enriched in catalytic sequences. Another suggestive feature of this model is that it would make it likely that any molecule with replicase function that appeared in the mixture would find in its neighborhood similar molecules, related by descent, thus eliminating the requirement for two unrelated replicases to meet.

How plausible is the assumption that replicases act on sequences

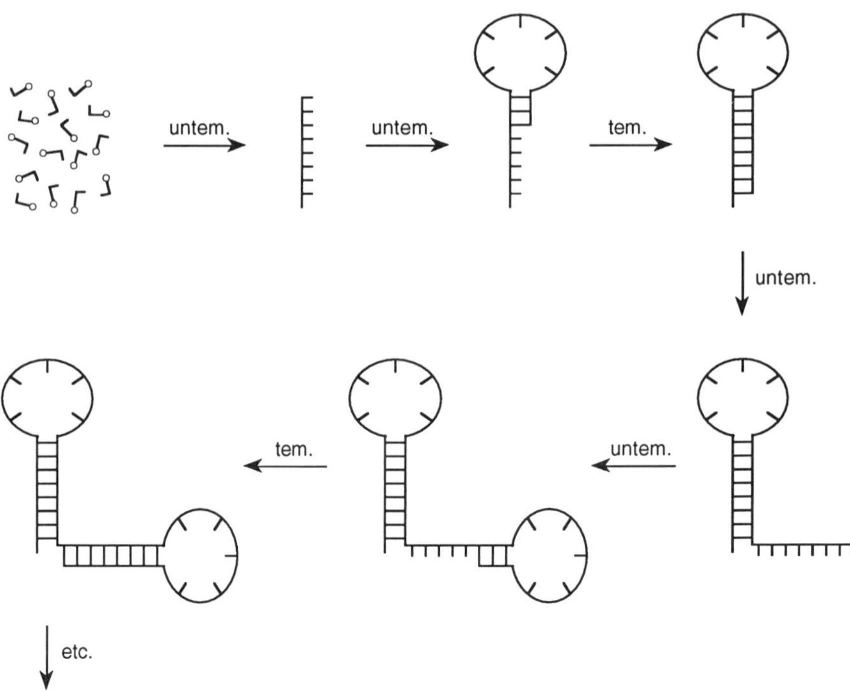

Figure 5 Nonenzymatic synthesis of multi-stem-loop structures as a result of untemplated (untem.) and templated (tem.) reactions. Template-directed synthesis is assumed to occur rapidly whenever a template and suitable primer are available. Once the complementary strand is completed, additional residues are added slowly, with random sequence.

similar to themselves, while ignoring unrelated sequences? This selectivity could be ensured by segregating individual molecules (or clonal lines) on the surface of mineral grains, on the surface of micelles, or within membranes. Closely related molecules might be segregated as a group through specific hydrogen-bonding interactions (the family that sticks together, replicates together). For any segregation mechanism, weak selection would result if the replicating molecules are sufficiently dispersed that diffusion over their intermolecular distance is slow compared to replication. Alternatively, the suggestions of Weiner and Maizels (1987) concerning "genomic tags" may be relevant. Among self-replicating sequences, it is plausible that some are restricted to copying molecules with a particular 3′-terminal subsequence. A replicator that happened by chance to carry a terminal sequence that matched the preference of its active site would replicate itself while ignoring its neighbors.

Another resolution of the paradox of how RNA evolution was initiated without the aid of an evolved ribozyme is to abandon the molecular biologist's dream and suppose that RNA was *not* the first genetic molecule (Shapiro 1984; Joyce 1987, 1989; Orgel 1989); perhaps RNA replication arose in the context of an evolving system based on something other than RNA (see below, Alternative Genetic Systems). Even if this is true, all the arguments concerning the relationship between the fidelity of replication and the maximum allowable genome length would still apply to this earlier genetic system. Of course, the challenge to those who advocate this approach is to demonstrate that there is an informational entity that is prebiotically plausible and is capable of initiating its own replication without the aid of a sophisticated catalyst.

Replicase Function in the Evolved RNA World

Although it is difficult to say how the first RNA replicase ribozyme arose, it is not difficult to imagine how such a molecule, once developed, would function. The chemistry of RNA replication would involve the template-directed polymerization of mononucleotides or short oligonucleotides, using chemistry in many ways similar to that employed by contemporary group I ribozymes (Cech 1986, 1987; Doudna and Szostak 1989). One important difference is that, unlike group I ribozymes, which rely on a nucleoside or oligonucleotide leaving group, an RNA replicase would more likely make use of a different leaving group that provides a substantial driving force for polymerization and that, after its release, does not become involved in some competing phosphoester transfer reaction.

The polymerization of activated nucleotides proceeds via nucleophilic attack by the 3'-hydroxyl of a template-bound oligonucleotide at the α-phosphorus of an adjacent template-bound nucleotide derivative (Fig. 6). The nucleotide is "activated" for attack by the presence of a phosphoryl substituent, for example, a phosphate, polyphosphate, alkoxide, or imidazole group. As discussed above, polyphosphates, such as inorganic pyrophosphate, are the most obvious candidates for the leaving group. The condensation reaction could be assisted by deprotonation of the nucleophilic 3'-hydroxyl, stabilization of the trigonal-bipyramidal intermediate, and protonation of the leaving group. All three of these tasks might be carried out by RNA, acting either alone or with the help of a suitably positioned metal cation or other cofactor.

Deprotonation of the 3'-hydroxyl may be facilitated by general base catalysis, perhaps utilizing N1 of adenine (pK = 3.5) or N3 of cytosine (pK = 4.2). Alternatively, Mg^{++}, Zn^{++}, or a similar metal could form a metal-hydroxide complex, which would then carry out the deprotonation. The metal could be positioned by binding to N7 of adenine or guanine, to a 2'-hydroxyl group, or to a nonbridging phosphoryl oxygen. An enzyme-bound metal cation would be especially well suited to coordinate the negative charge on the peripheral phosphoryl oxygens (Knowles 1980). This could promote the condensation reaction in three ways: (1) by stabilizing the developing negative charge on the phosphate as it undergoes nucleophilic attack (Breslow and Huang 1991); (2) by stabilizing the geometry of the trigonal-bipyramidal transition state relative to the

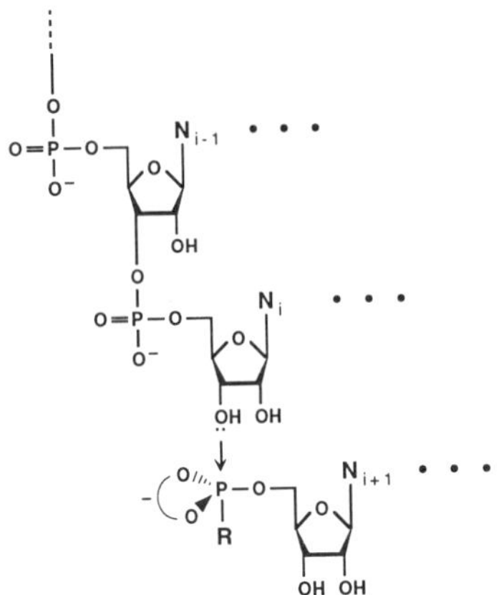

Figure 6 Nucleophilic attack by the 3'-hydroxyl of a template-bound oligonucleotide (N_1 – N_i) at the α-phosphorus of an adjacent template-bound mononucleotide (N_{i+1}). Dotted lines indicate base-pairing to a complementary template. R is the leaving group.

ground state; (3) by coordinating with both the phosphoryl oxygen and the nucleophilic 3'-oxygen in the transition state (Herschlag and Jencks 1990). Protonation of the leaving group could be accomplished by N1-H of guanine (pK = 9.4) or N3-H of uracil (pK = 9.3). An equivalent role could be played by a metal ion, in this case, by acting to stabilize the developing negative charge on the leaving group (Browne and Bruice 1992).

An RNA replicase ribozyme must ensure that the polymerization reaction works efficiently for a broad range of template sequences. Differences in the templating properties of the four bases and in the local structure of the template must be "smoothed out" without sacrificing the overall efficiency and fidelity of the reaction. This might prove difficult because factors that promote efficient template-directed incorporation of a nucleotide irrespective of its base might be expected to lower the fidelity. On the other hand, factors that lower duplex stability to prevent template self-structure might enhance the replicase's ability to discriminate against base mismatch.

RNA replicase activity is probably not the only catalytic behavior that was essential for the existence of the RNA world. Maintaining an adequate supply of the four activated nucleotides would have been a top priority. Even if the prebiotic environment contained a large reservoir of these compounds, the reservoir would eventually become depleted, and some capacity for nucleotide biosynthesis would have been required. This metabolic complexity necessitates a more complicated genome. Rather than consisting of a single "gene" that provided replicase activity, the genome must have contained several genes, specifying both metabolic and replicative functions. This would have required the evolution of some form of control of gene expression.

THE PREBIOTIC CHEMIST'S NIGHTMARE

The molecular biologist's pool contains pure β–D-nucleotides. How close to such a pool could one hope to get without magic (or enzymes) on the primitive earth? Could one hope to achieve replication in a pool containing a more realistic mixture of organic molecules, including, of course, β–D-ribonucleotides?

The synthesis of a nucleotide could occur in a number of ways. The simplest, conceptually, would be to synthesize a nucleoside base, couple it to ribose, and finally, phosphorylate the resulting nucleoside. However, a number of other routes are feasible; for example, the assembly of the base on a preformed ribose or ribose phosphate.

The classic prebiotic synthesis of sugars is by the polymerization of formaldehyde. It yields a very complex mixture of products including only a small proportion of ribose (Mizuno and Weiss 1974). This reaction does not provide a reasonable route to the ribonucleotides. An important discovery has been made by Eschenmoser and his colleagues. They have shown that the base-catalyzed aldomerization of glycoaldehyde phosphate in the presence of a half-equivalent of formaldehyde gives a relatively simple mixture of tetrose- and pentose-diphosphates and hexose-triphosphates, of which ribose 2,4-diphosphate is the major component (Müller et al. 1990).

More recently, it has been found that reactions of this kind, in particular the reaction of glycolaldehyde phosphate with glyceraldehyde-2-phosphate, proceed efficiently when 2 mM solutions of substrates are incubated at room temperature and pH 9.5 in the presence of layered hydroxides such as hydrotalcite (magnesium aluminum hydroxide) (Pitsch 1992; Pitsch et al. 1992). The phosphates are absorbed between the positively charged layers of the mineral. The reaction proceeds under mild conditions, presumably because of the high concentration of substrates in the interlayer and because the positive charge on the metal hydroxide layers favors enolization of glycolaldehyde phosphate.

Much remains to be demonstrated: for example, the plausibility of glycolaldehyde phosphate as a prebiotic precursor of sugar synthesis and the possibility of conversion of ribose 2,4-diphosphate to a phosphate from which nucleotides could be synthesized; e.g., ribose 5-phosphate or ribose 1,5-diphosphate. However, ribose synthesis, although still problematical, is no longer the intractable problem it seemed a few years ago.

The synthesis of the nucleoside bases is one of the success stories of prebiotic chemistry. Adenine is formed with remarkable ease from ammonia and hydrogen cyanide (Orò 1961). This synthesis has been described as "the rock of the faith" by Stanley Miller. Reasonably plausible syntheses of the other purine bases and of the pyrimidines have also been described (Sanchez et al. 1967). The coupling of the purine bases with ribose or its phosphates has been achieved under mild conditions, but in relatively low yield (Fuller et al. 1972). The formation of pyrimidine nucleosides in reasonable yield from the bases and ribose has not been achieved and is likely to prove difficult.

We conclude that the direct synthesis of the nucleosides or nucleotides from prebiotic precursors in reasonable yield and unaccompanied by larger amounts of related molecules could not be achieved by presently known chemical reactions. The only remotely plausible route to the molecular biologist's pool would involve a series of mineral-catalyzed reactions, coupled with a series of subtle fractionations of nucleotide-like materials based on charge, stereochemistry, etc.

Even minerals could not achieve on a macroscopic scale one desirable separation, the resolution of D-ribonucleotides from the L-enantiomers. This is a serious problem because experiments on template-directed synthesis using poly(C) and the imidazolides of G suggest that the polymerization of the D-enantiomer is often strongly inhibited by the L-enantiomer (Joyce et al. 1984). This difficulty may not be insuperable; perhaps with a different activator group inhibition would be less severe. However, enantiomeric cross-inhibition is certainly a serious problem.

Scientists interested in the origins of life seem to divide neatly into two classes. The first, usually but not always molecular biologists, believe that RNA must have been the first replicating molecule and that chemists are exaggerating the difficulties of nucleotide synthesis. They believe that a few more striking chemical "surprises" will establish that a reasonable approximation to a racemic version of the molecular biologist's pool could have formed on the primitive earth and that further experiments with different activating groups and minerals will solve the enantiomeric cross-inhibition problem. The second group of scientists are much more pessimistic. They believe that the de novo appearance of oligonucleotides on the primitive earth would have been a near miracle. (The authors subscribe to this latter view.) Time will tell which is correct.

ALTERNATIVE GENETIC SYSTEMS

The problems that arise when we try to understand how an RNA world could have arisen de novo on the primitive earth are sufficiently severe that we must explore other possibilities. What kind of alternative genetic systems might have preceded the RNA world? How could they have invented the RNA world? These topics have generated a good deal of speculative interest, but there are no relevant experimental data. Here we summarize some old ideas and present a few new possibilities.

One can conceive of two fundamentally different types of transitions from a simple genetic system to the RNA world. The first is a continuous transition from an informational macromolecule sufficiently similar to RNA that it could accommodate the substitution of nucleotides in place of its original monomers without necessitating major alterations in three-dimensional structure. Alternative monomers only become interesting if they are accessible by much simpler synthetic routes than are the nucleotides. This has not yet been demonstrated for any nucleotide analog that has been proposed as a "primitive" monomer.

The second type of transition can be described as a genetic takeover. A preexisting self-replicating system evolves, for its own selective advantage, the mechanism for synthesizing and polymerizing the components of a completely different genetic system and is taken over by it. Cairns-Smith (1982) proposes that the first genetic system was inorganic, perhaps a clay, and that it initiated a self-replicating system based on organic monomers. However, he clearly recognized the possibility of one organic genetic material replacing another (Cairns-Smith and Davies 1977).

The hypothesis of a genomic material completely different from a nucleic acid has one enormous advantage—it opens up the possibility of using very simple, easily synthesized prebiotic monomers in place of nucleotides. However, it also raises two new and difficult questions. Which prebiotic monomers are plausible candidates as the components of a replicating system? Why would an initial genetic system invent nucleic acids once it had evolved sufficient synthetic know-how to synthesize molecules as complex as nucleotides?

A number of prebiotic monomers that might have made up a simple genetic material have already been suggested. They include hydroxy acids (Weber 1987), amino acids (Orgel 1968; A. Rich and X. Zhang, pers. comm.), phosphomonoesters of polyhydric alcohols (Weber 1989), aminoaldehydes (Nelsestuen 1980), and molecules containing two sulfhydryl groups (Schwartz and Orgel 1985). The list could be expanded almost indefinitely. Here we discuss a small class of these monomers that we believe are particularly attractive in the light of recent work on enzyme mechanisms.

There is accumulating evidence that several enzymes that make or break phosphodiester bonds have two or three metal ions at their active sites (Cooperman et al. 1992). In the case of the editing site for phosphodiester hydrolysis in the Klenow fragment of *Escherichia coli* DNA polymerase I, no other functional groups of the enzyme come close to the phosphodiester bond that is cleaved. This has led to the suggestion that the major role of the enzyme is to act as scaffolding on which to hang metal ions in precisely determined positions (Beese and Steitz 1991). A similar suggestion has been made for ribozymes on the basis of indirect evidence (Freemont et al. 1988; M. Yarus, pers. comm.).

Perhaps these observations can be extended to suggest that if informational polymers preceded RNA, they may also have been dependent on metal ions for their catalytic activity. If so, the range of prebiotic monomers that needs to be considered is greatly reduced. In addition to the functional groups that react to form the backbone, the monomers must have carried metal-binding functional groups. If the metal ions involved were divalent ions such as Mg^{++} and Ca^{++}, the side groups are likely to have been carboxylate or phosphate groups. If transition-metal ions were involved, SH groups and possibly imidazole derivatives are likely to have been important.

Prebiotic monomers suitable for building polymers that bind Mg^{++} or Ca^{++}, for example, include aspartic acid, glutamic acid, and serine phosphate among biologically important amino acids. Hydroxy-dicarboxylic acids, such as α-hydroxysuccinic acid, and hydroxy-tricarboxylic acids, such as citric acid, are other possible candidates. A polymer containing D-aspartic acid, L-aspartic acid, and glycine as its subunits is typical of potentially informational copolymers that might, in the presence of divalent metal ions, both replicate and function as a catalyst. Transition-metal ions might play a corresponding role for polymers containing cysteine or homocysteine. The present challenge is to demonstrate replication experimentally in some such system.

What selective advantage could a simpler, metabolically competent system derive from the synthesis of oligonucleotides? This is a baffling question. Most arguments that come to mind do not stand up to detailed analysis. If, for example, one postulates that nucleotides were first synthesized as parts of cofactors such as DPN, one must explain why the particular heterocyclic bases and sugars were chosen. Even if we suppose that among the many "experiments" in metabolism carried out by early organisms one happened by accident on a pair of complementary nucleotides that could form a replicating polymer, one must still explain how polymerization subsequently contributed to the success of the "inventor." The only glimmer of an explanation that we can see is that oligonucleotides, by hybridization, functioned as selective "glues" to tie together appropriate macromolecules.

The discussion so far, even though highly speculative, is still conservative in overall outlook. It supposes that the original information-accumulating system that led to the evolution of life on earth was either RNA or some linear copolymer that replicated in an aqueous environment in much the same way as RNA. There remains a lingering doubt that we are on the right track at all; maybe the original system was not an organic copolymer (Cairns-Smith 1982), or maybe it replicated in a non-aqueous environment and RNA is an adaptation that permitted invasion of the oceans. Perhaps systems of high complexity can develop without any need for a genome in the usual sense (Kauffman 1986; Wächtershäuser 1988; De Duve 1991). Perhaps...

Laboratory simulations of prebiotic chemistry are dependent on organic chemistry and can only explore the kinds of reactions understood by organic chemists. A good deal is known about reactions in aqueous solution, but less is known about reactions at the interface between water and inorganic solids. Very little is known about reactions in systems where inorganic solids are depositing from aqueous solutions containing organic material. It is hard to see how speculative schemes involving heterogeneous aqueous systems can be tested until much more is known about the underlying branches of chemistry.

CONCLUSIONS AND PERSPECTIVES

After dreaming of self-replicating ribozymes emerging from pools of random polynucleotides and after having had nightmares over the difficulties that must have been overcome for RNA replication to occur in a realistic prebiotic soup, we awaken to the cold light of day. The constraints that must have been met in order to originate a self-sustained evolving system are now reasonably well understood. One can sketch out a logical order of events, beginning with prebiotic chemistry and ending with DNA/protein-based life. However, it must be said that the details of this process remain obscure and are not likely to be known in the near future.

The presumed RNA world should be viewed as a milestone, a plateau in the early history of life on earth. So too, the concept of an RNA world has been a milestone in the scientific study of life's origins. Although this concept does not explain how life originated, it has helped to guide scientific thinking and has served as a useful heuristic device to focus experimental efforts. Further progress will depend primarily on new experimental results, as chemists, biochemists, and molecular biologists work together to address problems concerning molecular replication, ribozyme enzymology, and RNA-based cellular processes.

ACKNOWLEDGMENTS

We are grateful to Steven Benner, Manfred Eigen, and Albert Eschenmoser for their helpful comments on the manuscript. This work was supported by NSCORT grant NAGW-2881 from the National Aeronautics and Space Administration.

REFERENCES

Acevedo, O.L. and L.E. Orgel. 1987. Non-enzymatic transcription of an oligodeoxynucleotide 14 residues long. *J. Mol. Biol.* **197:** 187–193.

Been, M.D. and T.R. Cech. 1988. RNA as an RNA polymerase: Net elongation of an RNA primer catalyzed by the *Tetrahymena* ribozyme. *Science* **239:** 1412–1416.

Beese, L.S. and T.A. Steitz. 1991. Structural basis for the 3′-5′ exonuclease activity of *Escherichia coli* DNA polymerase I: A two metal ion mechanism. *EMBO J.* **10:** 25–33.

Breslow, R. and D.-L. Huang. 1991. Effects of metal ions, including Mg^{2+} and lanthanides, on the cleavage of ribonucleotides and RNA model compounds. *J. Am. Chem. Soc.* **88**: 4080–4083.

Browne, K.A. and T.C. Bruice. 1992. Chemistry of phosphodiesters, DNA and models. 2. The hydrolysis of bis(8-hydroxyquinoline) phosphate in the absence and presence of metal ions. *J. Am. Chem. Soc.* **114**: 4951–4958.

Cairns-Smith, A.G. 1982. *Genetic takeover and the mineral origins of life.* Cambridge University Press, England.

Cairns-Smith, A.G. and C.J. Davies. 1977. The design of novel replicating polymers. In *Encyclopaedia of ignorance* (ed. R. Duncan and M. Weston-Smith), pp. 391–403. Pergamon Press, Oxford.

Cech, T.R. 1986. A model for the RNA-catalyzed replication of RNA. *Proc. Natl. Acad. Sci.* **83**: 4360–4363.

———. 1987. The chemistry of self-splicing RNA and RNA enzymes. *Science* **236**: 1532–1539.

Cooperman, B.S., A.A. Baykov, and R. Lahti. 1992. Evolutionary conservation of the active site of soluble inorganic pyrophosphatase. *Trends Biochem. Sci.* **17**: 262–266.

Crick, F.H.C. 1968. The origin of the genetic code. *J. Mol. Biol.* **38**: 367–379.

De Duve, C. 1991. *Blueprint for a cell: The nature and origin of life.* Neil Patterson, Burlington, North Carolina.

Doudna, J.A. and J.W. Szostak. 1989. RNA-catalysed synthesis of complementary-strand RNA. *Nature* **339**: 519–522.

Eigen, M. 1971. Selforganization of matter and the evolution of biological macromolecules. *Naturwissenschaften* **58**: 465–523.

Eigen, M. and P. Schuster. 1977. The hypercycle: A principle of natural self-organization. Part A: Emergence of the hypercycle. *Naturwissenschaften* **64**: 541–565.

Erman, J.E. and G.G. Hammes. 1966. Relaxation spectra of ribonuclease. IV. The interaction of ribonuclease with cytidine $2':3'$-cyclic phosphate. *J. Am. Chem. Soc.* **88**: 5607–5614.

Freemont, P.S., J.M. Friedman, L.S. Beese, M.R. Sanderson, and T.A. Steitz. 1988. Cocrystal structure of an editing complex of Klenow fragment with DNA. *Proc. Natl. Acad. Sci.* **85**: 8924–8928.

Fuller, W.D., R.A. Sanchez, and L.E. Orgel. 1972. Studies in prebiotic synthesis. VI. Synthesis of purine nucleosides. *J. Mol. Biol.* **67**: 25–33.

Gilbert, W. 1986. The RNA world. *Nature* **319**: 618.

Guerrier-Takada, C., K. Gardiner, T. Marsh, N. Pace, and S. Altman. 1983. The RNA moiety of ribonuclease P is the catalytic subunit of the enzyme. *Cell* **35**: 849–857.

Handschuh, G.J., R. Lohrmann, and L.E. Orgel. 1973. The effect of Mg^{2+} and Ca^{2+} on urea-catalyzed phosphorylation reactions. *J. Mol. Evol.* **2**: 251–262.

Herschlag, D. and W.P. Jencks. 1990. Catalysis of the hydrolysis of phosphorylated pyridines by $Mg(OH)^+$: A possible model for enzymatic phosphoryl transfer. *Biochemistry* **29**: 5172–5179.

Inoue, T. and L.E. Orgel. 1981. Substituent control of the poly(C)-directed oligomerization of guanosine $5'$-phosphoroimidazolide. *J. Am. Chem. Soc.* **103**: 7666–7667.

———. 1982. Oligomerization of (guanosine $5'$-phosphor)-2-methylimidazolide on poly(C). *J. Mol. Biol.* **162**: 204–217.

———. 1983. A nonenzymatic RNA polymerase model. *Science* **219**: 859–862.

Inoue, T., G.F. Joyce, K. Grzeskowiak, L.E. Orgel, J.M. Brown, and C.B. Reese. 1984. Template-directed synthesis on the pentanucleotide CpCpGpCpC. *J. Mol. Biol.* **178**: 669–676.

Joyce, G.F. 1987. Nonenzymatic template-directed synthesis of informational macromolecules. *Cold Spring Harbor Symp. Quant. Biol.* **52:** 41–51.

————. 1989. RNA evolution and the origins of life. *Nature* **338:** 217–224.

Joyce, G.F. and L.E. Orgel. 1986. Non-enzymic template-directed synthesis on RNA random copolymers: Poly(C,G) templates. *J. Mol. Biol.* **188:** 433–441.

Joyce, G.F., G.M. Visser, C.A.A. van Boeckel, J.H. van Boom, L.E. Orgel, and J. van Westrenen. 1984. Chiral selection in poly(C)-directed synthesis of oligo(G). *Nature* **310:** 602–604.

Kauffman, S.A. 1986. Autocatalytic sets of proteins. *J. Theor. Biol.* **119:** 1–24.

Khorana, H.G. 1961. A review of the synthetic applications of carbodiimides with reference to the possible mechanisms of reactions. In *Some recent developments in the chemistry of phosphate esters of biological interest*, pp. 126–141. Wiley, New York.

Knowles, J.R. 1980. Enzyme-catalyzed phosphoryl transfer reactions. *Annu. Rev. Biochem.* **49:** 877–919.

Kruger, K., P.J. Grabowski, A.J. Zaug, J. Sands, D.E. Gottschling, and T.R. Cech. 1982. Self-splicing RNA: Autoexcision and autocyclization of the ribosomal RNA intervening sequence of *Tetrahymena. Cell* **31:** 147–157.

Lewin, R. 1986. RNA catalysis gives fresh perspective on the origin of life. *Science* **231:** 545–546.

Lohrmann, R. 1975. Formation of nucleoside 5′-polyphosphates from nucleotides and trimetaphosphate. *J. Mol. Evol.* **6:** 237–252.

————. 1977. Formation of nucleoside 5′-phosphoramidates under potentially prebiological conditions. *J. Mol. Evol.* **10:** 137–154.

Lohrmann, R. and L.E. Orgel. 1973. Prebiotic activation processes. *Nature* **244:** 418–420.

McCaskill, J.S. 1984. A stochastic theory of molecular evolution. *Biol. Cybern.* **50:** 63–73.

Mizuno, T. and A.H. Weiss. 1974. Synthesis and utilization of formose sugars. *Adv. Carbohyd. Chem. Biochem.* **29:** 173–227.

Mohr, S.C. and R.E. Thach. 1969. Application of ribonuclease T_1 to the synthesis of oligoribonucleotides of defined base sequence. *J. Biol. Chem.* **244:** 6566–6576.

Müller, D., S. Pitsch, A. Kittaka, E. Wagner, C.E. Wintner, and A. Eschenmoser. 1990. Chemie von α-aminonitrilen. Aldomerisierung von Glykolaldehydphosphat zu *racemischen* hexose-2,4,6-triphosphaten und (in gegenwart von formaldehyd) *racemischen* pentose-2,4-diphosphaten: *rac.*-allose-2,4,6-triphosphat und *rac.*-ribose-2,4-diphosphat sind die reaktionshauptprodukte. *Helv. Chim. Acta* **73:** 1410–1468.

Nelsestuen, G.L. 1980. Origin of life: Consideration of alternatives to proteins and nucleic acids. *J. Mol. Evol.* **15:** 59–72.

Noller, H.F., V. Hoffarth, and L. Zimniak. 1992. Unusual resistance of peptidyl transferase to protein extraction procedures. *Science* **256:** 1416–1419.

Orgel, L.E. 1968. Evolution of the genetic apparatus. *J. Mol. Biol.* **38:** 381–393.

————. 1989. Was RNA the first genetic polymer? In *Evolutionary tinkering in gene expression* (ed. M. Grunberg-Manago et al.), pp. 215–224. Plenum Press, London.

Orò, J. 1961. Mechanism of synthesis of adenine from hydrogen cyanide under plausible primitive earth conditions. *Nature* **191:** 1193–1194.

Osterberg, R., L.E. Orgel, and R. Lohrmann. 1973. Further studies of urea-catalyzed phosphorylation reactions. *J. Mol. Evol.* **2:** 231–234.

Pace, N.R. and T.L. Marsh. 1985. RNA catalysis and the origin of life. *Origins Life* **16:** 97–116.

Pitsch, S. 1992. "Zur chemie von glykolaldehyd-phosphat: seine bildung aus oxirancarbonitril und seine aldomerisierumg zu den (racemischen) pentose–2,4-diphosphaten und hexose-2,4,6-triphosphaten." Ph.D. thesis, ETH, Zürich.

Pitsch, S., A. Eschenmoser, B. Gedulin, S. Hui, and G. Arrhenius. 1992. Mineral induced selective formation of sugar phosphates. *Helv. Chim. Acta* (in press).

Sanchez, R.A., J.P. Ferris, and L.E. Orgel. 1967. Studies in prebiotic synthesis. II. Synthesis of purine precursors and amino acids from aqueous hydrogen cyanide. *J. Mol. Biol.* **30:** 223–253.

Sawai, H., K. Kuroda, and T. Hojo. 1988. Efficient oligoadenylate synthesis catalyzed by uranyl ion complex in aqueous solution. *Nucleic Acids Symp. Ser.* **19:** 5–7.

Schwartz, A.W. and L.E. Orgel. 1985. Template-directed synthesis of novel, nucleic acid-like structures. *Science* **228:** 585–587.

Shapiro, R. 1984. The improbability of prebiotic nucleic acid synthesis. *Origins Life* **14:** 565–570.

Sharp, P.A. 1985. On the origin of RNA splicing and introns. *Cell* **42:** 397–400.

Sleeper, H.L. and L.E. Orgel. 1979. The catalysis of nucleotide polymerization by compounds of divalent lead. *J. Mol. Evol.* **12:** 357–364.

Sulston, J., R. Lohrmann, L.E. Orgel, and M.H. Todd. 1968. Nonenzymatic synthesis of oligoadenylates on a polyuridylic acid template. *Proc. Natl. Acad. Sci.* **59:** 726–733.

Wächtershäuser, G. 1988. Before enzymes and templates: Theory of surface metabolism. *Microbiol. Rev.* **52:** 452–484.

Weber, A.L. 1987. The triose model: Glyceraldehyde as a source of energy and monomers for prebiotic condensation reactions. *Origins Life* **17:** 107–119.

———. 1989. Model of early self-replication based on covalent complementarity for a copolymer of glycerate-3-phosphate and glycerol-3-phosphate. *Origins Life* **19:** 179–186.

Weiner, A.M. and N. Maizels. 1987. 3′ Terminal tRNA-like structures tag genomic RNA molecules for replication: Implications for the origin of protein synthesis. *Proc. Natl. Acad. Sci.* **84:** 7383–7387.

Woese, C. 1967. The evolution of the genetic code. In *The genetic code*, pp. 179–195. Harper & Row, New York.

Wu, T. and Orgel, L.E. 1992a. Nonenzymatic template-directed synthesis on oligodeoxy-cytidylate sequences in hairpin oligonucleotides. *J. Am. Chem. Soc.* **114:** 317–322.

———. 1992b. Nonenzymatic template-directed synthesis on hairpin oligonucleotides. II. Templates containing cytidine and guanosine residues. *J. Am. Chem. Soc.* **114:** 5496–5501.

Zaug, A.J. and T.R. Cech. 1986. The intervening sequence RNA of *Tetrahymena* is an enzyme. *Science* **231:** 470–475.

2

Reading the Palimpsest:* Contemporary Biochemical Data and the RNA World

Steven A. Benner,[1] Mark A. Cohen,[1] Gaston H. Gonnet,[2]
David B. Berkowitz,[3] and Kai P. Johnsson[1]
[1]Laboratory for Organic Chemistry
[2]Department of Computer Science
E.T.H. Zürich, CH-8092 Switzerland

The chemical behavior of contemporary living systems contains vestiges of the history of living chemistry on planet Earth. To read this "palimpsest" is a challenge of the first order (Benner et al. 1989). Some of the historical record has been written over by the demands of natural selection that have forced the evolution of new chemical structures to meet new biological challenges. Some has been lost in the noise arising from random events, the "neutral drift" that characterizes the structural divergence of biological molecules following the divergence of their host organisms (King and Jukes 1969; Kimura 1982). Some has undoubtedly been confused by lateral transfer of genetic information between phylogenetically distant organisms (Doolittle et al. 1990) and by "sequence convergence," the independent emergence of polypeptide sequences that offer unique chemical solutions to particular biological problems.

Three advances of the past decade have greatly improved our ability to deconvolute the information about earlier life forms written in the biological chemistry of contemporary organisms. First, substantial progress has been made toward an integration of structural theory from chemistry and evolutionary theory from biology (Benner and Ellington 1990b). The integrated theory allows us to proceed past the classic ques-

---

*A palimpsest is a parchment that has been inscribed two or more times, with the previous texts imperfectly erased and therefore still partially legible.

[3]Present address: Department of Chemistry, University of Nebraska, Lincoln, Nebraska 68588.

The RNA World
© 1993 Cold Spring Harbor Laboratory Press 0-87969-380-0/93 $5 + .00

tions in biological chemistry (what happens at a chemical level in a living system) to ask *why* it happens. A number of questions in biological chemistry have been addressed within the context of the integrated theory, and models are now available to understand the evolutionary status of many molecular aspects of living systems. Most of these hypotheses are directed to specific behaviors in biological macromolecules: Catalytic activity, stereospecificity, and thermal stability are three examples.

Second, there has been an explosion of sequence data (Bairoch and Boeckmann 1992), both from individual investigators reporting sequences of individually chosen proteins and from genome projects now beginning to yield results (Oliver et al. 1992; Sulston et al. 1992). The recently completed organization and exhaustive matching of the protein sequence database (Gonnet et al. 1992) make systematic analysis of these sequence data possible in their entirety for the first time. This analysis has in turn provided better methods for aligning homologous protein sequences, one of the more important tools for understanding the evolution of living systems.

Finally, substantial progress has been made in developing a manipulative understanding of how biological macromolecules work. It is now possible to predict many details of the folded structure of proteins de novo from sequence data (Crawford et al. 1987; Benner 1989b; Bazan 1990; Niermann and Kirschner 1990; Benner and Gerloff 1991; Knighton et al. 1991; Thornton et al. 1991; de Vos et al. 1992). Furthermore, the design of polypeptides that fold in solution and catalyze reactions, at least at some rate, is now possible, and the first examples of designed peptides whose structure in solution has been rigorously proven have recently appeared (Johnsson et al. 1990; Osterhout et al. 1992).

With these tools, a detailed reconstructed history of the evolution of biological macromolecules back to the protogenome, the genome from the most recent common ancestor of today's archaebacteria, eubacteria, and eukaryotes (Benner and Ellington 1987, 1990a; Benner et al. 1987, 1989), becomes a tangible research goal. The organism most closely corresponding to this reconstructed ancestor lived over a billion years ago. Using chemical assumptions, a somewhat less detailed history back to the "breakthrough organism," the last organism in the RNA world to use RNA as the sole genetically encoded component of biological catalysis (Benner et al. 1989), is available. The breakthrough organism lived perhaps 2.5 billion years ago (Benner et al. 1989). In this chapter, we outline contemporary tools for reconstructing models for ancient forms of life and provide examples of the use of these tools to solve specific problems in the history of life.

THE TOOLS

**Contemporary Methods for Analyzing
Sequence Data**

Central to historical reconstructions of earlier episodes in the history of
life are alignments of the sequences of homologous proteins and nucleic
acids. Although sequence alignments have been constructed for 30 years
(Edwards and Cavalli-Sforza 1963; Zuckerkandl and Pauling 1965; Fitch
and Margoliash 1967) and are routinely used in laboratories throughout
the world, alignments available to most biochemists are in fact quite
problematic (Thorne et al. 1992). In particular, the tools available for
constructing alignments, both the matrices that allow the scoring of
mutations and formulae for scoring gaps in alignments, remain primitive.
As a result, much research is handicapped by suboptimal alignments, and
the literature contains seemingly endless arguments over evolutionary issues that might be resolved by well-constructed alignments.

Recently, the protein sequence database was reconstructed using a
"patricia tree" data structure (Gonnet et al. 1992). This made possible an
exhaustive cross-matching of every subsequence in the database with
every other subsequence. From this came 1.7 million aligned pairs of sequences of potentially homologous proteins. This large collection of
aligned sequence pairs was the starting point for a broadly empirical
study of macromolecular evolution, providing the most advanced tools to
date for constructing alignments.

*Mutation Matrices, Gap Penalties, and Statistically
Rigorous Alignments*

A "log-odds" (or Dayhoff) matrix is a 20 x 20 table that shows the
logarithm of a probability (multiplied by 10) of each pair of the 20
proteinogenic amino acids being matched in an alignment (Table 1). The
matrix is defined for a particular evolutionary distance between the two
proteins. Thus, a 1% mutation matrix describes pairwise probabilities of
amino acids in two protein sequences that have undergone one point
mutation per 100 amino acid residues. In such a matrix, the sum of all the
off-diagonal terms is equal to 1%.

Such matrices are named in honor of Margaret Dayhoff, who proposed their use some 20 years ago (Dayhoff et al. 1978). Versions of the
1978 matrix are used by most commercially available computer alignment packages. However, two methodological problems make the 1978
matrix suboptimal for the alignments that are most interesting in the context of this article. First, and unavoidably in 1978, the amount of data

Table 1 Dayhoff "log-odds" matrix

	C	S	T	P	A	G	N	D	E	Q	H	R	K	M	I	L	V	F	Y	W
C	11.5																			
S	0.1	2.2																		
T	−0.5	1.5	2.5																	
P	−3.1	0.4	0.1	7.6																
A	0.5	1.1	0.6	0.3	2.4															
G	−2.0	0.4	−1.1	−1.6	0.5	6.6														
N	−1.8	0.9	0.5	−0.9	−0.3	0.4	3.8													
D	−3.2	0.5	−0.0	−0.7	−0.3	0.1	2.2	4.7												
E	−3.0	0.2	−0.1	−0.5	−0.0	−0.8	0.9	2.7	3.6											
Q	−2.4	0.2	0.0	−0.2	−0.2	−1.0	0.7	0.9	1.7	2.7										
H	−1.3	−0.2	−0.3	−1.1	−0.8	−1.4	1.2	0.4	0.4	1.2	6.0									
R	−2.2	−0.2	−0.2	−0.9	−0.6	−1.0	0.3	−0.3	0.4	1.5	0.6	4.7								
K	−2.8	0.1	0.1	−0.6	−0.4	−1.1	0.8	0.5	1.2	1.5	0.6	2.7	3.2							
M	−0.9	−1.4	−0.6	−2.4	−0.7	−3.5	−2.2	−3.0	−2.0	−1.0	−1.3	−1.7	−1.4	4.3						
I	−1.1	−1.8	−0.6	−2.6	−0.8	−4.5	−2.8	−3.8	−2.7	−1.9	−2.2	−2.4	−2.1	2.5	4.0					
L	−1.5	−2.1	−1.3	−2.3	−1.2	−4.4	−3.0	−4.0	−2.8	−1.6	−1.9	−2.2	−2.1	2.8	2.8	4.0				
V	−0.0	−1.0	0.0	−1.8	0.1	−3.3	−2.2	−2.9	−1.9	−1.5	−2.0	−2.0	−1.7	1.6	3.1	1.8	3.4			
F	−0.8	−2.8	−2.2	−3.8	−2.3	−5.2	−3.1	−4.5	−3.9	−2.6	−0.1	−3.2	−3.3	1.6	1.0	2.0	0.1	7.0		
Y	−0.5	−1.9	−1.9	−3.1	−2.2	−4.0	−1.4	−2.8	−2.7	−1.7	2.2	−1.8	−2.1	−0.2	−0.7	−0.0	−1.1	5.1	7.8	
W	−1.0	−3.3	−3.5	−5.0	−3.6	−4.0	−3.6	−5.2	−4.3	−2.7	−0.8	−1.6	−3.5	−1.0	−1.8	−0.7	−2.6	3.6	4.1	14.2

The terms represent ten times the logarithm of the probability of a pairwise matching of the indicated amino acids in two proteins separated by 250 PAM units. The terms were generated from data collected for all protein pairs separated by PAM distances between 6 and 100 PAM, extrapolated by exponential fitting to PAM 116.5, and then extrapolated directly to a PAM distance of 250 to conform to standard usage.

available to Dayhoff was small. Thus, the elements of the 1978 Dayhoff matrix have large variances.

Second, to ensure that the alignments yielding mutation data were of high quality, Dayhoff constructed matrices from pairs of proteins that were themselves quite similar (typically 90–95% identical) in sequence and then extrapolated the matrices to give a matrix suitable for aligning distant proteins. Data from the exhaustive matching of the contemporary sequence database (Gonnet et al. 1992) show that this extrapolation was problematic. The probability of certain substitution pairs turns out to be a strong function of evolutionary distance, apparently because constraints imposed by the genetic code influence the mutation matrix for closely related proteins but not distantly related proteins (Gonnet et al. 1992). A revised Dayhoff matrix suitable for aligning distant sequences is shown in Table 1.

Insertions and deletions also create problems in constructing alignments. If gaps are introduced into an alignment at no cost, any random sequences can be aligned. Classically, therefore, gaps have been penalized. The penalty most commonly used involves a gap cost calculated from a formula of the form $(ak+b)$, where k is the length of the gap and a and b are arbitrarily chosen parameters. At an intuitive level, such scoring is well known to be inadequate, because most biochemists "tweak" computer-produced alignments to achieve a more pleasing disposition of gaps. Data from the exhaustive matching show why such adjustment is necessary. The probability of a gap of length k declines as a function of $k^{-1.7}$ (Gonnet et al. 1992 and in prep.), not exponentially as implied by the gap cost calculated using the $(ak+b)$ formula. Thus, the classic formula is an inadequate approximation for constructing alignments containing gaps.

PAM Distances and Evolutionary Trees

Reconstructions of ancient protein sequences generally require alignments of several sequence pairs and arrangement of the sequences on an evolutionary tree constructed from a measure of evolutionary distance between sequence pairs. Often, evolutionary distance is defined by the fraction (or percent) sequence identity. It is well known that this measure is imperfect. Inspection of the revised Dayhoff matrix (Table 1) shows, for example, that conservation of a cysteine is more significant than conservation of an alanine. Furthermore, due to reverse mutations and multiple mutations at individual residues, evolutionary distance is not a simple function of percent identity after modest evolutionary divergence.

Evolutionary distance between two homologous protein sequences is best measured in PAM units, indicating the number of accepted point mutations per 100 amino acid residues separating the two sequences. Two protein sequences 1 PAM unit distant differ by 1 accepted point mutation per 100 amino acid residues. For two sequences separated by a PAM distance of x, transformation of the first sequence via the 1% mutation matrix (see above) x times gives the second sequence with the highest probability. PAM distances and trees constructed from them can both be subjected to a rigorous statistical analysis.

Given a set of homologous proteins, a PAM distance table (Fig. 1) can be constructed tabulating the PAM distances between each pair of aligned sequences. A tree is constructed from these distances, where vertices in the tree represent ancestral sequences from ancient organisms. If three sequences are available, the length of individual segments of the tree (in PAM distance units) connecting these vertices can be obtained by solving the set of linear simultaneous equations derived from the PAM distance data. With more than three sequences, the solution is overdetermined, and the best tree is obtained from a least-squares fit of the data (Gonnet et al. 1992). An idealized example of the process for constructing these trees is shown in Figure 1. A representative tree for the superfamily of proteins that includes aspartate aminotransferases (AAT), tyrosine aminotransferases (YAT), aminocyclopropanecarboxylate synthase, and histidinol phosphate aminotransferases (HPAT) is shown in Figure 2. A segment of the corresponding alignment is shown in Figure 3.

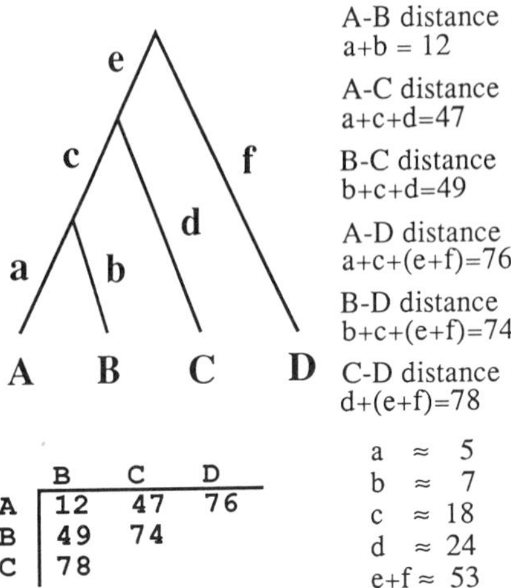

	B	C	D
A	12	47	76
B	49	74	
C	78		

Figure 1 A hypothetical tree reconstructed from a hypothetical PAM distance table (below). Note that in the simultaneous equations, ($e+f$) always appear as a pair, and the data do not permit assignment of the relative magnitudes of e and f (that is, the data do not permit the rooting of the tree). However, the remaining distances are calculated with known statistical parameters.

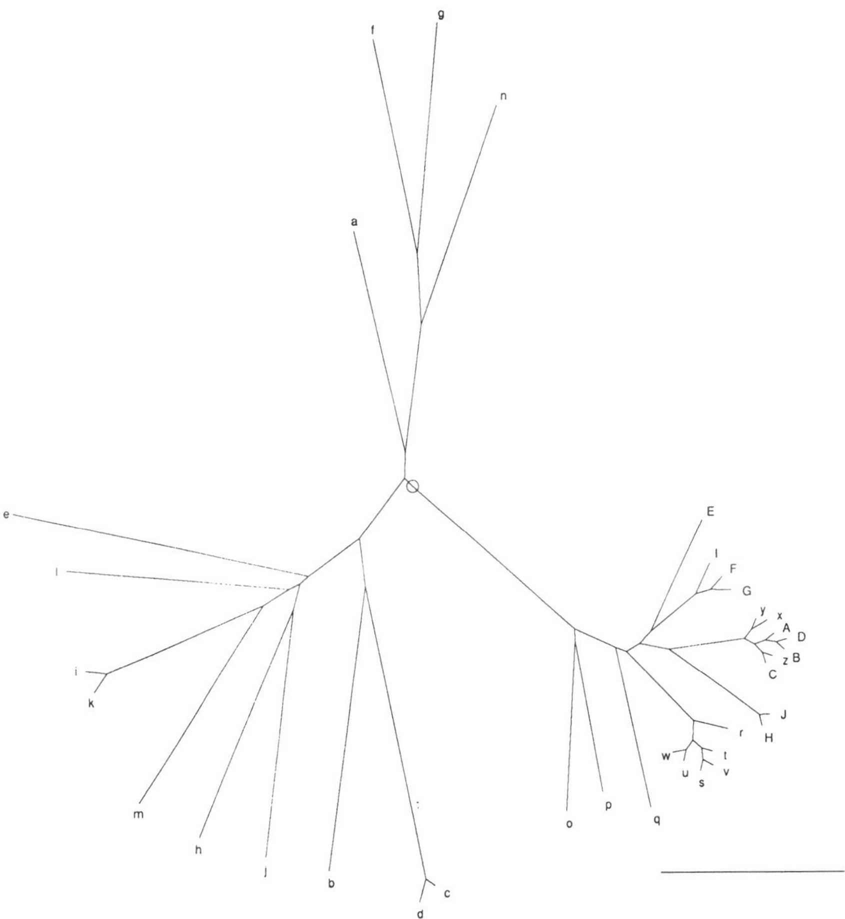

Figure 2 The most probable evolutionary tree relating HPAT, AAT, YAT, and a number of other enzymes. The bar represents an evolutionary distance of 50 PAM units. Key: (a) 1-Aminocyclopropanecarboxylate-1C synthetase, tomato; (b) AAT, *Sulfolobus solfataricus*; (c) YAT, human; (d) YAT, rat; (e) CobC protein, *Pseudomonas*; (f) lysine decarboxylase, *Hafnia alvei*; (g) ornithine decarboxylase, *E. coli*; (h) HPAT, *Bacillus subtilis*; (i) HPAT, *E. coli*; (j) HPAT, *Haloferax volcanii*; (k) HPAT, *Salmonella typhimurium*; (l) HPAT, *Streptomycete*; (m) HPAT, *Saccharomyces cerevisiae*; (n) hypothetical protein, *B. subtilis*; (o) aromatic amino acid aminotransferase, *E. coli*; (p) AAT, *E. coli*; (q) AAT, *S. cerevisiae*, cytoplasmic; (r) AAT, chicken, cytoplasmic; (s) AAT, horse, cytoplasmic; (t) AAT, human, cytoplasmic; (u) AAT, mouse, cytoplasmic; (v) AAT, pig, cytoplasmic; (w) AAT, rat, cytoplasmic; (x): AAT, chicken, mitochondrial; (y) AAT, turkey, mitochondrial; (z) AAT, horse, mitochondrial; (A) AAT, human, mitochondrial; (B) AAT, mouse, mitochondrial; (C) AAT, pig, mitochondrial; (D) AAT, rat, mitochondrial; (E) AAT, lupine, P2-root; (F) AAT, lupine, P1-root; (G) AAT, alfalfa, leaf; (H) AAT, *Panicum miliaceum*, isoenzyme 1; (I) AAT, *P. miliaceum*, isoenzyme 2; (J) AAT, *P. miliaceum*, isoenzyme 3.

```
      223 ..296
b - HNPTGTLFSPNDVKKIVDISR-DNKIILLSDEIYDNFVYEGKMRSTLE--DSDWRDF-LIYVNGFSKTFSMTGWRLG
c - SNPCGSVFSKRHLQKILAVAA-RQCVPILADEIYGDMVFSDCKYEPLA--TLSTDVP-ILSCGGLAKRWLVPGWRLG
d - SNPCGSVFSKRHLQKILAVAE-RQCVPILADEIYGDMVFSDCKYEPLA--NLSTNVP-ILSCGGLAKRWLVPGWRLG
e - NNPTGRALAPAELLAI-AARQKASGGLLLVDEAFGDLE-P---QLSVA--GHASGQGNLIVFRSFGKFFGLAGLRLG
h - NNPTGTYTSEGELLAFLE--RVPSRVLVVLDEAYYEYV-TAEDYPETV--PLLSKYSNLMILRTFSKAYGLAALRVG
j - HNPTGSVLPREELVELAE--SVEEHTLLVVDEAYGEFA----EEPSAI--DLLSEYDNVAALRTFSKAYGLAGLRIG
i - NNPTGQLINPQDFRTLLELTRGK--AIVVADEAYIEFC-PQA---SLA--GWLAEYPHLAILRTLSKAFALAGLRCG
k - NNPTGQLINPQDLRTLLELTRGK--AIVVADEAYIEFC-PQA---TLT--GWLVEYPHLVILRTLSKAFALAGLRRG
m - GNPTGAKIKTSLIEKVLQ-NWDN--GLVVVDEAYVDFC-GGS---T-A--PLVTKYPNLVTLQTLSKSFGLAGIRLG
l - NNPTGTAVPAETVLALYEAAQAAKPSMVVVDEAYIEFS-HGA---SLL--PLLDGRPNLVVSRTMSKAFGAAGLRLG
G - HNPTGVDPTLEQWEQIRQLIRSKSL-LPFFDSAYQGFASGSLDADAQPVRLFVADGGELLVAQSYAKNMGLYGERVG
F - HNPTGVDPTTEQWEQIRKLLRSKAL-LPFFDSAYQGFASGSLDIDAQAVRLFVADGGELLLAQSYAKNMGLYGERVG
I - HNPTGVDPTIDQWEQIRQLMRSKSL-LPFFDSAYQGFASGSLDKDAQPVRMFIADGGELLMAQSYAKNMGMYGERVG
E - HNPTGIDPTPEQWEKIADVIQEKNH-IPFFDVAYQGFASGSLDEDAASVRLFVARGLEVLVAQSYSKNLGLYAERIG
z - HNPTGVDPRPEQWKEIATLVKKNNL-FAFFDMAYQGFASGDGNKDAWAVRYFIEQGINVCLCQSYAKNMGLYGERVG
C - HNPTGVDPRPEQWKEMATLVKKNNL-FAFFDMAYQGFASGDGNKDAWAVRHFIEQGINVCLCQSYAKNMGLYGERVG
A - HNPTGVDPRPEQWKEIATVVKKRNL-FAFFDMAYQGFASGDGDKDAWAVRHFIEQGINVCLCQSYAKNMGLYGERVG
B - HNPTGVDPRPEQWKEIASVVKKKNL-FAFFDMAYQGFASGDGDKDAWAVRHFIEQGINVCLCQSYAKNMGLYGERVG
D - HNPTGVDPRPEQWKEMAAVVKKKNL-FAFFDMAYQGFASGDGDKDAWAVRHFIEQGINVCLCQSYAKNMGLYGERVG
x - HNPTGVDPRQEQWKELASVVKKRNL-LAYFDMAYQGFASGDINRDAWALRHFIEQGIDVVLSQSYAKNMGLYGERAG
y - HNPTGVDPRPEQWKEMATLVKKNNL-FAFFDMAYQGFASGDINRDAWAVRHFIEQGINVVLSQSYAKNMGLYGERAG
J - HNPTGVDPTEEQWREISHQFKVKKH-FPFFDMAYQGFASGDPERDAKAIRIFLEDGHQIGCAQSYAKNMGLYGQRVG
H - HNPTGVDPTEEQWREISHQFKVKKH-FPFFDMAYQGFASGDPERDAKAIRIFLEDGHQIGCAQSYAKNMGLYGQRVG
r - HNPTGTDPTPDEWKQIAAVMKRRCL-FPFFDSAYQGFASGSLDKDAWAVRYFVSEGFELFCAQSFSKNFGLYNERVG
s - HNPTGTDPTPEQWKQIASVMKRRFL-FPFFDSAYQGFASGNLDRDAWAVRYFVSEGFELFCAQSFSKNFGLYNERVG
v - HNPTGTDPTPEQWKQIASVMKRRFL-FPFFDSAYQGFASGNLEKDAWAIRYFVSEGFELFCAQSFSKNFGLYNERVG
t - HNPTGIDPTPEQWKQIASVMKRRFL-FPFFDSAYQGFASGNLERDAWAIRYFVSEGFEFFCAQSFSKNFGLYNERVG
u - HNPTGTDPTPEQWKQIAAVMQRRFL-FPFFDSAYQGFASGDLEKDAWAIRYFVSEGFELFCAQSFSKNFGLYNERVG
w - HNPTGTDPTEEEWKQIAAVMKRRFL-FPFFDSAYQGFASGDLEKDAWAIRYFVSEGFELFCPQSFSKNFGLYNERVG
q - HNPTGLDPTSEQWVQIVDAIASKNH-IALFDTAYQGFATGDLDKDAYAVRXXLSTVSPVFVCQSFAKNAGMYGERVG
p - HNPTGIDPTLEQWQTLAQLSVEKGW-LPLFDFAYQGFARG-LEEDAEGLRAFAAMHKELIVASSYSKNFGLYNERVG
o - HNPTGADLTNDQWDAVIEILKAREL-IPFLDIAYQGFGAG-MEEDAYAIRAIASAGLPALVSNSFSKIFSLYGERVG
    .**.*.   ..... ..... .... ...   *      ... ... ... .. ... ......*........*.*


αβ         aaaaaaaa      bbbbbbbbbbbb         aaaaaaaaabbbbbb   bbbbbbbbb  α₂ 1
αβ         aaaaaaaaaaaaaa      bbb         aaaaaa       bbbbbb            bb  α₂ 2
αβ         aaaaaaaaaaaaaaaabbbbbbbbbb   aaaaaaaaaaaaaa  bbbbbbbb         bbb  α₂ 3
```

Figure 3 Segment of a multiple alignment of the proteins in the tree in Fig. 2, excluding four (proteins a, f, g, and n) that do not give significant alignments in this region. The key is as in Fig. 2. A deletion is represented by a dash. Secondary structures predicted for proteins b, c, d, e, h, j, k, l, and m (the left branch of the tree in Fig. 2) are shown below the alignment (**1**), above secondary structures determined by crystallographic analysis for proteins p (**2**) and r (**3**). Especially noteworthy is the prediction of an αβ unit preceding, and an α₂ unit following, the segment where the alignment is statistically significant. The alignment in these regions is not shown.

Reconstructed Sequences of Ancient Proteins

Points in an evolutionary tree correspond to protein intermediates in the evolution of the protein family. Using the appropriate Dayhoff matrix, sequences for these ancient proteins can be reconstructed in a probabilistic form. In a reconstructed sequence, each position is represented by a vector of unit length in 20-dimension space, where the component of the vector in each of the 20 dimensions is the probability that each of the 20 amino acids was present in this protein at this position. Part of the probabilistic sequence for the reconstructed histidinol phosphate aminotransferase from the most recent common ancestor of *Bacillus* (a eubacterium), *Haloferax* (an archaebacterium), and *Saccharomyces* (a eukaryote) (the left branch of the tree in Fig. 2) is given in Table 2.[1]

With many contemporary sequences, a high branching order of the tree, and a slow rate of divergence, ancestral sequences can often be reconstructed with remarkable precision. For example, for the protein family that includes elongation factor 1α, more than 71% of the 478 positions in the reconstructed proto-eukaryotic sequence are reconstructed with more than 90% probability (Table 3). An ancient protein with a well-defined probabilistic sequence can be made and studied in the laboratory. The first example of an ancestral protein from an extinct organism to be made and studied was a ribonuclease from the most recent common ancestor of swamp buffalo, river buffalo, and ox, corresponding in the fossil record approximately to the fossil organism *Pachyportax latidens* (Stackhouse et al. 1990).

Predicting and Designing Protein Folds

Converting sequence data into conformational data has been among the most difficult challenges in structural biology. In fact, there are two challenges. The first is to design de novo proteins that fold in solution in a productive way (i.e., to form stable folded structures and catalyze reactions). The second is to predict de novo (i.e., without input of crystallographic data) the folded structure of natural proteins. Both problems have now been addressed in specific cases, and there is reason to believe that the approach that yielded the best solutions will apply, if not to all protein structures, at least to a large subset of them. For the purposes of this chapter, it is important only to know that solutions to these problems have improved substantially and to understand how they might be used in reconstructing ancient forms of life.

In protein design, the first polypeptides designed to fold in solution and catalyze reactions have been prepared, and their structure in solution has been proven (Johnsson et al. 1990). In one case, the catalytic mechanism of the designed enzyme has been explored in detail (Johnsson et al. 1990; K. Johnsson, unpubl.). These experiments provide a chemical basis for assumptions regarding how catalytically active proteins originated (see below). Likewise, the design of catalytic proteins yields

---

[1]Sequence data allow the calculation only of the most probable sequences for specified points in an evolutionary tree. These trees are, however, "unrooted"; the sequence data alone do not define a point on the tree that is geologically the most ancient protein sequence. Such points can only be specified given additional biological information and are less precisely defined for more ancient branch points than for more recent branch points. As a convention, we reconstruct here ancient sequences indicated by the circle on the tree, representing the center of gravity of the tree. This is a formalism; probabilistic sequences corresponding to all other points are also available. However, this point generally falls within the region of the tree where the root is likely to lie.

Table 2 Reconstructed probabilistic ancestral sequences

Position		Position	
223	N=59.4 H=35.0	179	N=98.3
224	N=98.9	180	N=99.9
225	P=99.8	181	P=100.0
226	T=98.3	182	T=99.8
227	G=99.7	183	G=100.0
228	T=64.2 V=17.6 S=7.4	184	T=68.1 S=23.8 A=5.2
229	A=39.5 V=17.0 D=14.1 E=5.9	185	V=35.8 L=24.1 A=14.8 I=6.3
230	I=32.2 L=27.6 P=17.8 V=11.2	186	I=64.1 V=25.8 L=7.4
231	S=57.9 T=31.2	187	P=80.0 S=9.6
232	P=99.3	188	A=24.2 T=15.4 E=14.1 P=13.1 S=9.2 R=8.7 K=6.5 Q=5.6
233	E=90.8 Q=5.1	189	E=74.2 Q=14.7
234	E=64.8 Q=24.2 D=6.4	190	E=78.3 D=12.7
235	L=65.7 W=21.5	191	L=82.1 I=12.0
236	Q=31.8 L=25.9 K=22.5 E=5.8 R=5.3	192	L=73.1 V=13.4 I=5.5
237	A=65.2 K=10.4 Q=8.2 T=6.6	193	A=66.8 T=9.9 E=8.8 S=5.8
238	I=61.6 L=33.0	194	L=95.3
239	L=66.5 A=14.5 V=8.1	195	L=93.7
240	E=86.2 Q=8.1	196	E=99.2
241	V=53.2 L=21.9 I=14.1 A=8.3	197	L=37.6 A=28.0 V=9.7 I=5.2
242	T=27.0 A=23.6 S=21.3 M=6.7 V=5.4 I=5.1	198	T=34.9 A=21.9 S=15.8 N=13.6
243	R=67.3 K=24.2 Q=6.5	199	R=88.7 Q=6.5
244	A=68.6 S=12.2 E=6.9	200	A=48.8 V=40.5
245	K=78.8 E=6.9	201	E=26.0 K=23.5 A=13.2 P=8.5 N=8.2 Q=6.1 S=5.7
246	N=57.1 S=16.9 E=7.8 K=6.4	202	E=30.7 K=30.2 S=13.9 Q=6.0
247	H=40.6 R=18.3 L=11.3 K=7.2 Q=5.5	203	R=28.1 H=25.8 P=18.0 K=7.7 Q=7.4
248	A=55.1 S=13.6 T=9.1 V=8.7	204	A=42.5 S=26.8 T=23.2
249	L=84.9 I=10.7	205	L=86.9 M=7.5 I=5.2

250   V=54.6 P=19.3 L=8.4 I=5.9
251   V=42.1 L=26.8 F=19.4 I=7.3
252   V=62.1 F=18.0 I=7.2 L=6.4
253   D=98.7
254   E=66.2
255   A=98.1
256   Y=99.3
257   G=53.1 Q=18.9 A=5.3
258   E=40.7 D=27.2 G=21.3
259   F=99.1
260   A=78.6 V=11.0 C=6.3
261   Y=26.3 S=24.2 F=11.9 A=8.3 T=5.6
262   P=35.1 G=30.0 S=16.4 A=6.6
263   G=72.4 D=16.2
264   A=58.4 L=18.9 V=5.2
265   E=67.6 D=27.1
266   E=46.2 Q=21.5 K=15.1 R=5.6
267   P=46.9 D=25.4 E=9.0 S=5.7
268   S=63.7 A=24.7 T=7.8
269   L=64.2 Y=12.2 F=7.8 W=5.7
270   A=97.6
271   V=39.5 I=26.3 L=14.5
272   R=64.7 K=10.2
273   P=60.0 A=24.0
274   L=70.4 F=24.4
275   A=44.3 L=33.7 V=11.1
276   S=92.2
277   E=68.6 K=13.2 D=6.2 Q=5.0
278   Y=47.2 G=18.8 H=6.2
279   P=70.2 L=15.0

206   V=95.6
207   V=99.2
208   V=93.9
209   D=99.8
210   E=99.7
211   A=99.8
212   Y=99.9
213   I=55.0 V=27.4
214   E=98.6
215   F=97.9
216   C=61.6 A=23.4 S=6.8 V=6.1

217   T=23.9 P=20.1 H=15.2 S=13.0 A=6.3 G=6.1
218   G=84.4 A=10.3
219   A=77.2 S=10.3 E=7.0
220   D=46.1 E=42.1
221   Y=28.6 E=23.4 H=6.9 D=6.0 Q=5.0
222   P=97.2
223   S=83.0 T=14.3
224   L=62.1 A=11.1 V=7.8 T=7.5
225   V=55.5 I=19.3 L=11.3 A=10.7

226   P=99.4
227   L=99.2
228   L=98.9
229   S=75.6 T=14.4 A=6.7
230   E=56.3 K=37.0
231   Y=99.9
232   P=86.9 S=6.0

Table 2 Reconstructed probabilistic ancestral sequences (*continued*)

Position		Position	
280	N=63.8 E=16.9 D=8.2	233	N=99.7
281	L=93.6	234	L=97.3
282	V=46.5 I=29.7 L=19.0	235	V=82.8 A=9.1
283	V=90.6 I=8.1	236	V=51.8 I=43.7
284	L=61.8 A=18.8 V=7.2	237	L=98.8
285	R=69.7 Q=14.2 K=5.6	238	R=99.6
286	T=55.6 S=39.3	239	T=99.8
287	F=94.7	240	F=58.8 L=32.7 M=7.3
288	S=96.2	241	S=99.6
289	K=98.2	242	K=99.7
290	A=56.5 N=17.0 S=9.7	243	A=99.1
291	F=98.8	244	F=64.2 Y=35.7
292	G=99.6	245	G=99.9
293	L=98.0	246	L=98.8
294	A=61.2 Y=17.0	247	A=99.8
295	G=99.7	248	G=99.6
296	L=61.7 E=13.4	249	L=98.5
297	R=98.9	250	R=99.9
298	L=65.4 V=21.8 I=9.8	251	L=51.7 I=26.3 V=19.6
299	G=99.7	252	G=100.0

The sequences presented correspond to the segment of the multiple alignment presented in Fig. 3. The column on the left is the probabilistic ancestral sequence for all the proteins in the multiple alignment (see footnote 1). The column on the right is the probabilistic ancestral sequence for only the histidinol phosphate aminotransferases (sequences h–m in Fig. 3). The numbering on the left corresponds to the position in the alignment in Fig. 3. The positions on the right indicate the equivalent positions in the multiple alignment of the histidinol phosphate aminotransferases alone. Only probabilities >5% are shown.

Table 3 Ancestral sequences of elongation factors 1α and Tu

Protein	No. of sequences in family	Length of ancestral sequence	Probability of most probable amino acid[a]		
			90–99%	99%	100%
Proto-eukaryotic EF-1α	26	478	36	243	65
Proto-eubacterial ET-Tu	19	475	85	115	0
Proto-archaeal EF-1α/Tu	6	442	35	226	2
Protogenomic EF-1α/Tu[b]	51	530	115	41	0

[a]Value indicates the percent probability of the most probable amino acid at this position in family ancestral sequence.

[b]See footnote 1.

an understanding of which reactions are easy to catalyze, which are difficult to catalyze, and what chemical structures are needed for catalysis. This brings at least some much-needed chemical rigor to speculations concerning primitive catalysts (see below).

Likewise, methods for predicting de novo the folded structure of proteins have advanced substantially over the past 5 years. To date, structures of at least three protein families have been predicted *before* crystallographic data were available where the predictions were shown to be remarkably accurate by subsequently determined crystal structures: the αβ-barrel domain of tryptophan synthase (Crawford et al. 1987; Hyde et al. 1988), the catalytic domain of protein kinase (Benner and Gerloff 1991; Knighton et al. 1991), and the extracellular domain of the human growth hormone receptor (Bazan 1990; de Vos et al. 1992). The only successful methods to date for predicting de novo the folded structure of proteins start with an alignment of homologous sequences. The most reliable of these methods extracts information from patterns of variation and conservation within a set of aligned homologous sequences (Benner 1989b; Benner and Gerloff 1991). Secondary structure can now be predicted with reasonable certainty for proteins for which a number of homologous sequences are known. Assignments of active-site residues, covariation analysis, and biochemical information (positions of S–S bonds and cross-linking studies) allow the predicted secondary structural units to be assembled to yield a model of the folded protein.

It is well known that secondary and tertiary structure in proteins diverges less rapidly than primary structure (Chothia and Lesk 1986). Motifs and consensus sequence elements also can indicate distant relationships that are sometimes not identified by sequence comparisons (Dever et al. 1987; Bairoch 1992), although these methods have proven on occasion to be quite unreliable (Bork 1992). If crystal structures are available for two families of proteins, alignments emphasizing structural similarities can establish homology in cases where sequence alignments alone might be unconvincing (Taylor and Orengo 1989; Valencia et al. 1991). Reliable structure predictions can replace crystal structures in this process.

For example, the evolutionary tree in Figure 2 joins proteins in two connected components[2] of the protein sequence database. Homology between these two sets of protein sequences is not strongly supported by sequence similarity, as evident from inspection of the segment of the multiple alignment (Table 2) produced by computational analysis. No crystallographic data are available for any protein in the left branch of the tree. It is possible, however, to predict de novo a secondary structure for proteins in this branch using methods developed in Zürich (Benner 1989b; Benner and Gerloff 1991). The predicted secondary structure for the aligned and flanking regions is shown beneath the alignment in Figure 3. The secondary structure predicted for the left branch of the tree (Fig. 2) corresponds well to the secondary structure obtained crystallographically for aspartate aminotransferases from eubacteria and eukaryotes from the right branch. The correspondence between a reliable predicted secondary structure in one branch of the tree with crystallographically determined secondary structures of representative proteins in a second branch secures what would otherwise be a marginal hypothesis regarding homology between the two branches.

Models for Understanding the Evolution of
Protein Structure and Behavior

Natural selection targets the properties of the folded sequence that contribute to survival and reproduction in the host organism. Issues related to function and survival are formulated in three ways (Benner and Ellington 1988). First and most classically, the sequence of a macromolecule determines its behavior, and much work in contemporary biochemistry focuses on the relationship between sequence and behavior. Second, behavior determines survival value. Finally, the survival value is itself determined by protein sequence. This final relationship has been the most difficult to probe, although much of the discussion of macromolecular evolution (e.g., molecular clocks) has been based on assumptions regarding this relationship (Lewin 1988).

It is now possible to distinguish (at least at the level of hypothesis) adaptive, neutral, and historical behaviors in biological macromolecules (Benner and Ellington 1990b). The stereospecificity of metabolic trans-

---

[2]Connected components are families of proteins within the database related by specific criteria, in this case a PAM distance of less than 200, a similarity score of >125, and a match length of >80 positions.

formations (Benner et al. 1990), the kinetic details of enzymatic reactions (Albery and Knowles 1976; Benner 1989a), and the form and nature of metabolic pathways (Benner et al. 1989) are three that have been examined in special detail. Adaptive differences influence survival. Neutral differences do not. Historical behaviors are those that, although not themselves adaptive in the contemporary world, are vestiges of ancient structures that were adaptive or that arose as a result of particular constraints in the evolution of earlier forms of life. The last are the most relevant here, as constraints imposed in a world where RNA was the sole genetically encoded macromolecule, or those imposed by nonbiological chemical reactions, can greatly influence models of the organisms that arose in these environments.

The Reactivity of Organic Molecules

Structure-reactivity theory from organic chemistry constrains speculations concerning what types of reactions are plausible in primitive catalytic systems. Such constraints are often lacking in the discussion of catalysis, especially by RNA molecules. A few comments are relevant in the present context.

First, most functionalized macromolecules catalyze reactions. The catalytic power depends on what kind of functional groups the macromolecule bears and what the reaction is. Some reactions are considerably more difficult to catalyze than others, and an understanding of structure-reactivity theory, in particular stereoelectronic theory, allows one to evaluate which reactions belong to which classes (Benner 1988). Proteins presumably came to be the predominant catalysts in the modern world because they carry more encoded functional groups than RNA molecules (Benner et al. 1987). Even with RNA, however, structure theory suggests that the problem is to control catalytic power in RNA, not to create it.

An understanding of chemical reactivity is also crucial for reconstructing ancient forms of life. For example, the most convincing case for an RNA world comes from the fact that many biological cofactors contained RNA fragments *that play no role* in the chemical reactivity of the cofactor in the contemporary world (White 1976; Visser and Kellogg 1978; Benner et al. 1989). Such behavior is presumably a vestige of a world where RNA was the primary biological macromolecule, because the absence of a function in the contemporary world (and, indeed, abundant examples of alternative structures in contemporary organisms that performed the same chemistry without the RNA fragment) precludes any functional reasons to create such structures in any other environment.

Finally, chemistry becomes a still more powerful tool by allowing scientists to ask "What if?" and "Why not?" Many of these questions are directed toward nucleic acid structure. For example, Figure 4 shows six isomorphic Watson-Crick base pairs constructed from 12 different purine and pyrimidine base analogs. In principle, oligonucleotides could incorporate all 12 bases, potentially providing an RNA molecule with much of the structural versatility of proteins (Switzer et al. 1989; Piccirilli et al. 1990). The only way to learn why Nature did not avail itself of this structural banquet is to make, by chemical synthesis, the bases themselves and to study their chemistry. Hexose DNA (Eschenmoser and Loewenthal 1992), floppy DNA (Schneider and Benner 1990), DNA with unnatural bases, DNA with altered linking groups (Huang et al. 1991), and a variety of other unusual structures (Van der Woerd et al. 1987) now grace the chemical literature, all synthesized to answer the question "Why not?"

GENERAL VIEWS CONCERNING DIVERGENT EVOLUTION IN PROTEINS

A systematic analysis of divergent evolution of protein macromolecules casts new light on a wide range of widely held beliefs concerning macromolecular evolution. We cannot review more than a few of these, but the discussion below should encourage caution throughout.

Molecular Clocks

If the rate of accumulation of point mutations is a constant function of time, it should be possible to assign a chronological date to the divergence of two lineages simply by a comparison of protein sequences. In the absence of dated fossils, this is one of the few approaches available for obtaining such chronology. Not surprisingly, many have hoped that such a "molecular clock" exists (Lewin 1988).

Although the divergence of nonfunctional (neutral) aspects of macromolecular sequences (e.g., codon use) may display clock-like behavior, it is clear from rigorously constructed evolutionary trees (e.g., Fig. 2) that this is not generally the case for functional sequences (see also Jukes and Holmquist 1972). In such trees, the lengths of the lines (representing the number of accepted point mutations) between the ends of the branches (representing contemporary sequences) and the common branch points are variable, even though the time interval in each case is identical. Thus, assigning chronological dates for branch points in an evolutionary tree remains problematic (see below).

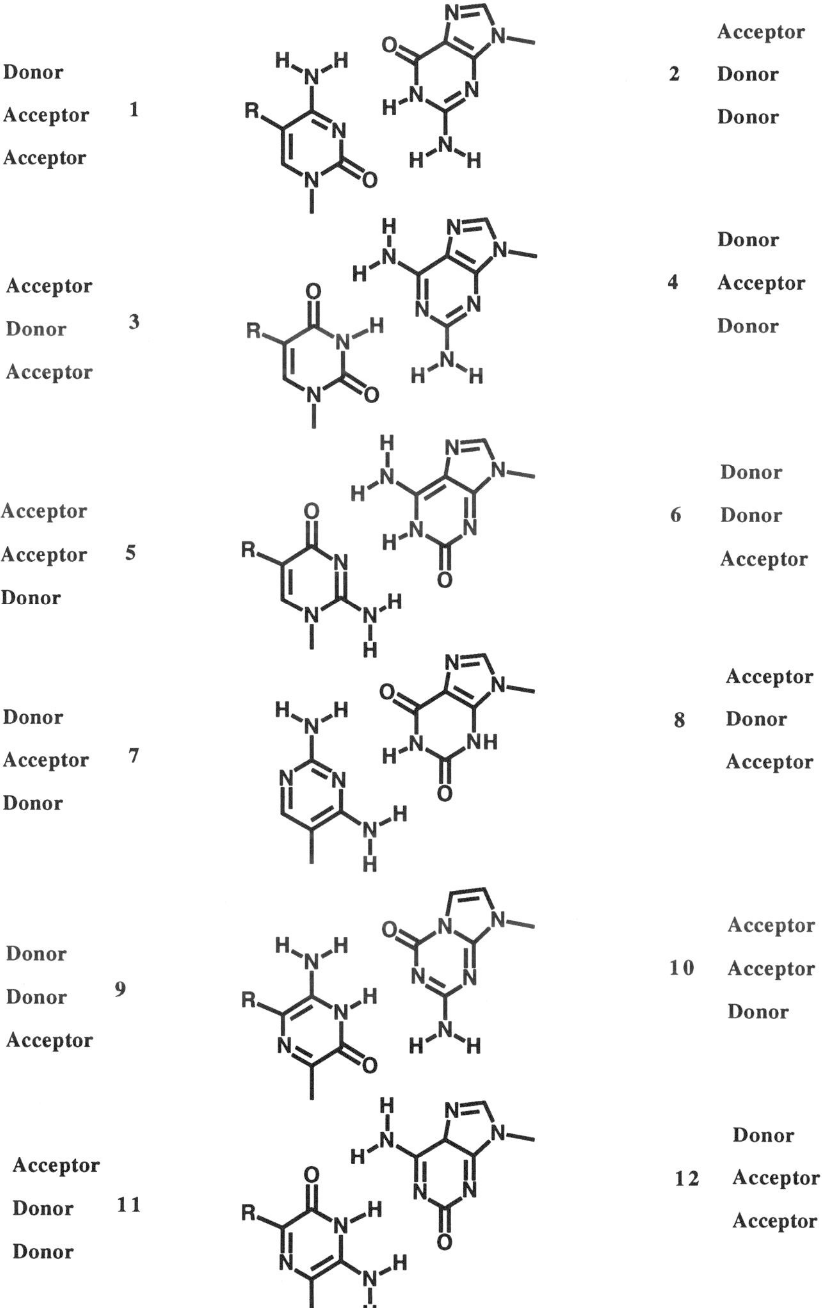

Figure 4 Six isomorphic Watson-Crick base pairs can be constructed from 12 purine and pyrimidine base analogs.

Exon Shuffling

In essentially all protein families that participate in central metabolism, divergence of function has been accomplished through the accumulation of point mutations, insertions, and deletions. Overall, these processes behave approximately statistically. The family of proteins represented by the tree in Figure 2 is an excellent example of this.

There is essentially no evidence in metabolic enzymes for modular behavior (e.g., domain shuffling), either for developing primitive catalysts or for altering the function of advanced enzymes. Rather, modular behavior is observed primarily in proteins involved in advanced regulatory systems in advanced organisms; in particular, proteins involved in the immune system, blood clotting, and other regulatory pathways that emerged only within the last 400 million years. The particular attention paid to these types of proteins by contemporary molecular biologists has, we believe, created the illusion that domain shuffling is more widespread (and more ancient) than it actually is.

Recruitment of Proteins and Deletion-replacement Events

Protein families containing two (or more) types of catalysts suggest a process of "recruitment," where an enzyme performing one function is recruited (often following gene duplication) to perform a second function. Recruitment presents special challenges in reconstructive efforts, because it is difficult to decide which of the two functions was performed by the common ancestral protein (i.e., which function is "primitive" and which is "derived"). Preparation of the ancestral proteins in the laboratory (see above) is possible only in special cases. Furthermore, whereas the derived function might have been created for the first time by recruitment, it is also possible that the ancestral organism had catalysts for both reaction types. If so, the recruitment was the second step of a "deletion-replacement" event, where the gene encoding the enzyme for one reaction was lost and a replacement was obtained by recruitment. Deletion-replacement events occurring on the laboratory time scale have been well studied (Li et al. 1983).

The frequency of deletion-replacement events is unknown, although inspection of the connected components obtained from the exhaustive matching (Gonnet et al. 1992) can provide a guess. Of the approximately 70 connected components that contain an archaebacterial sequence and a sequence from at least one other kingdom (Table 4, A–E), 16 (22%) contain enzymes catalyzing more than one type of reaction. It is difficult to know whether this overestimates or underestimates actual recruitment events. The criteria used to construct the connected components are quite liberal (footnote 2), and several of these trees almost certainly include proteins that are not homologs as a result (see notes to Table 4). This suggests that we will overestimate the frequency of deletion-replacement events by using this sample. However, the connected components may not include some homologous proteins that perform different functions where significant sequence similarity no longer remains. Furthermore, the sequence database is far from complete. These factors suggest that we have underestimated the frequency of deletion-replacement events.

Nevertheless, deletion-replacement events follow rules (Benner et al. 1989). For an existing enzyme performing a specific biological function to be replaced, it first must be deleted to yield an organism that lacks, at least for a time, this function. The more "lethal" the deletion is, the more difficult a deletion-replacement event. Furthermore, deletion-replacement events are more likely in an organism that already contains an enzyme that catalyzes a reaction chemically similar to that catalyzed by the deleted enzyme; this enzyme is ready to be recruited following a relatively small number of structural alterations. This implies that deletion-replacement events occur most rapidly in an organism that has a wide variety of enzyme types.

Although the literature occasionally focuses on conservation of binding sites (Yeh and Ornston 1980), conservation of catalytic mechanism is far more important during divergent evolution. For example, in the evolutionary tree indicated in Figure 2, at least three reaction types are represented. All require pyridoxal cofactors, however. The tree also suggests a solution to a particular biochemical puzzle in the literature: What is the enzymatic function of the cobC (entry e in Fig. 2), a genetic entity involved in the biosynthesis of vitamin B_{12} in *Pseudomonas*? (Crouzet et al. 1990). The literature suggests, on the basis of complementation and other biological data, that cobC encodes an amidase. However, there is clear homology between the cobC protein and several aminotransferases and decarboxylases dependent on pyridoxal. This evolutionary information coupled with chemical information concerning the biosynthesis of B_{12} makes it more likely that cobC protein is a threonine decarboxylase.

RECONSTRUCTING EARLY FORMS OF LIFE

With these tools in hand, we can begin to reconstruct detailed models for the biochemistry of ancient forms of life. This reconstruction proceeds stepwise; the metabolism of the "proto-mammal," the "proto-animal," the "proto-plant," the "proto-fungi," and the "proto-eukaryote" are reconstructed in this order. The prefix "proto-" is used exactly as in the field of

Table 4 Connected components containing archaebacterial sequences

		(A) Enzymes with at Least One Representative from Each Primary Kingdom
Component	Size	Comments
33[a]	16	pyridoxal enzymes (aminocyclopropanecarboxylate synthase, tyrosine aminotransferase, aspartate aminotransferase, histidinol phosphate aminotransferase, lysine decarboxylase, ornithine decarboxylase, CobC, and a hypothetical protein
332	17	ATP-citrate lyase, citrate synthase, succinyl CoA synthase
349[b]	126	oxidoreductases (acyl CoA desaturase, cytochrome b5, ferredoxin, flavodoxin, nitrate reductase, and others)
1183	8	argininosuccinate synthase
1277[c]	74	ATP synthase α,β and vacuolar subunits
1685	12	β-galactosidase, β-glucosidase, lactase-phlorizin hydrolase precursor
1735[a]	8	biotin sulfoxide reductase, dimethyl sulfoxide reductase, formate dehydrogenase, NAD-reducing hydrogenase, respiratory nitrate reductase, NAD-ubiquinone reductase
1859[a]	29	various oxidoreductases
2181[a]	51	carbamoyl-phosphate synthase, anthranilate synthase, D-ala-D-ala ligase, PAB synthase, isochorismate synthase, propionyl CoA carboxylase
4018	46	O-acetylserine sulfhydrylase, serine dehydratase, threonine dehydratase, threonine synthase, tryptophan synthase α and β subunits
4637	21	dihydrofolate reductase
5847	21	various hydrogenases
5942	51	glyceraldehyde 3-phosphate dehydrogenase
6465	43	glutamine synthase
13113	26	phosphoglycerate kinase
13288	5	DNA photolyase
13903	5	pyrroline-5-carboxylate reductase
14385	7	phosphoribosylaminoimidazole carboxylase
16521	23	superoxide dismutase Fe/Mn/Fe-Mn

(B) Ribosomal Proteins with at Least One Representative from Each Primary Kingdom

Component	Size	Comments
10487	5	S5 and putative LLREP3
12533	6	L6, L9, and outer membrane binding protein
14487	18	S14, S11, CRP-2, RP59
15018	6	L11, YL15
15028	16	L14, L17A
15089	19	L2, KD4, K37
15133	15	L23, L25
15180	10	L3
15250	10	39A, L5
15384	24	S12
15834	7	S10, S20
15904	15	S19, S15
15914	6	S16, S9
15976	12	S22, S8
16039	20	S7

(C) Ribosomal Proteins with at Least One Archaebacterial and One Eukaryotic Sequence and No Eubacterial Representative

Component	Size	Comments
26	4	S13, S15, and an unknown 17.4-kD protein
8352	4	HS3, S4E, S4
15005	6	L1, L1A, L1B, L2, L4
15047	6	L15, L29, L27A
15074	5	L18, L5, L5A, L5B
15078	5	L19, L19E, and probable ribosomal protein (OrfE)
15125	3	L22, L23

Table 4 Connected components containing archaebacterial sequences (*continued*)

(C) Ribosomal Proteins with at Least One Archaebacterial and One Eukaryotic Sequence and No Eubacterial Representative (continued)

Component	Size	Comments
15151	3	L24, L26
15192	7	L30, L7
15194	4	L30, L32, probable ribosomal protein (Orf1), hypothetical 11.5-kD protein
15204	5	L32, RP49, and probable ribosomal protein (OrfD)
15211	4	L32, L35A, and a hypothetical 9.7-kD protein
15258	4	YL4, L7A, HS6
15842	5	S11, S17
15865	4	HS12, S15A, S19
15958	4	S19, S24

(D) Enzymes with at Least One Archaebacterial and One Eubacterial Sequence and No Eukaryotic Representative

Component	Size	Comments
855	34	nitrogenase Fe-(Fe/Mo/V) protein; nitrogenase Fe-Mo cofactor biosynthesis protein NIFE
1616	2	bacterio-opsin activator/nitrogen fixation regulatory protein
4095	18	DNA-binding protein and integration host factor
5846	2	8-hydroxy-deazaflavin-reducing hydrogenase/hypothetical protein
8003	7	phosphoribosylformimino-5-aminoimidazole carboxamide ribotide isomerase; HISF protein (cyclase)
17808	6	indole-3-glycerol phosphate

(E) Archaebacterial Sequences from Connected Component 1[d]

Archaebacterial enzyme	Similarity	Related enzymes[e]
DNA gyrase subunit B	120	DNA topoisomerase II (E), DNA topoisomerase large subunit (V), DNA gyrase subunit B (B)
Acidic ribosomal protein	135	ribosomal protein L20/L12 (A), acidic ribosomal protein P0 (E), protein homolog P0
Ribosomal protein L20	120	ribosomal protein A/L12 (A), acidic ribosomal protein P1/P2 (E)
Ribosomal protein L12	120	ribosomal protein A/L20 (A), acidic ribosomal protein P1/P2 (E)

DNA-directed RNA polymerase A	120	DNA-directed RNA, polymerases (A,E,B and V)
DNA-directed RNA polymerase B	120	DNA-directed RNA, polymerases (A,E,B and V)
DNA-directed RNA polymerase C	130	DNA-directed RNA, polymerases (A,E,B and V)
Elongation factors[f]		elongation factors (A,E, and B)

(F) Archaebacterial Singletons (Unconnected Entries)[g]

22.6-kD Protein	*Desulfurococcus mobilis*
24-kD Flagellin (frag)	*Methanospirillum hungatei*
30S Ribosomal protein hs13	*Halobacterium marismortui*
30S Ribosomal protein s14	*Methanococcus vannielii*
31-kD Flagellin (frag)	*Methanococcus voltae*
50S Ribosomal protein hl5 (frag)	*Halobacterium cutirubrum*
50S Ribosomal protein hl5 (frags)	*H. marismortui*
50S Ribosomal protein hl9 (frag)	*H. cutirubrum*
50S Ribosomal protein hl10 (frag)	*H. cutirubrum*
50S Ribosomal protein hl16 (frag)	*H. cutirubrum*
50S Ribosomal protein hl21/hl22	*H. marismortui*
50S Ribosomal protein hl29 (l19)	*H. marismortui*
50S Ribosomal protein hl30 (frag)	*H. cutirubrum*
50S Ribosomal protein hl31 (frag)	*H. cutirubrum*
50S Ribosomal protein hl32	*H. marismortui*
50S Ribosomal protein l5 (hl19) (frag)	*H. cutirubrum*
50S Ribosomal protein l6 (hmal6) (hl10) (frag)	*H. marismortui*
50S Ribosomal protein l9 (frag)	*H. marismortui*
50S Ribosomal protein l12 (frag)	*H. marismortui*
50S Ribosomal protein l13 (frag)	*H. cutirubrum*
50S Ribosomal protein l22 (hl23) (frag)	*H. cutirubrum*
50S Ribosomal protein l29 (hmal29) (hl33)	*H. marismortui*
50S Ribosomal protein l29	*Halobacterium halobium*
50S Ribosomal protein l31	*H. marismortui*

Table 4 Connected components containing archaebacterial sequences *(continued)*

(F) Archaebacterial Singletons (Unconnected Entries)[g] (continued)

50S Ribosomal protein l34 (hl30)	*H. marismortui*
50S Ribosomal protein l46e	*Sulfolobus solfataricus*
50S Ribosomal protein lc12 (frag)	*H. marismortui*
50S Ribosomal proteins hl46e	*H. marismortui*
Cell surface glycoprotein precursor (csg)	*H. halobium*
Deoxyribodipyrimidine photolyase (E.C. 4.1.99.3) (frag)	*Methanobacterium thermoautotrophicum*
DNA binding protein hmf-2 (hmfb)	*Methanothermus fervidus*
DNA gyrase subunit a (E.C. 5.99.1.3) (gyra) (frag)	*Haloferax sp. (strain aa 2.2)*
DNA-binding protein 7e	*Sulfolobus acidocaldarius*
DNA-binding proteins 7a, 7b, and 7d	*S. acidocaldarius*
Ferredoxin	*Methanococcus thermolithotrophicus*
Ferredoxin	*Methanosarcina barkeri*
Ferredoxin	*S. acidocaldarius*
Formylmethanofuran--THMP formyltransferase (E.C. 2.3.1.101)	*M. thermoautotrophicum*
Gas vesicle protein, chromosomal (c-vac) (gvp) (gvpa)	*H. halobium*
Gas vesicle protein, plasmid (p-vac) (gvp) (gvpa)	*H. halobium*
Glucose dehydrogenase (E.C. 1.1.1.47) (frag)	*Thermoplasma acidophilum*
Gvpd protein (gvpd)	*H. halobium*
Gvpe protein (gvpe)	*H. halobium*
Hypothetical 7.1-kD protein in hmal3 5' region (Orf2)	*H. marismortui*
Hypothetical 8.5-kD protein (Orf75)	*H. halobium*
Hypothetical 9.9-kD protein in rpob 5' region (Orf88)	*S. acidocaldarius*
Hypothetical 10.3-kD protein in ribosomal proteins operon (Orf3)	*H. marismortui*
Hypothetical 10.9-kD protein in rs10 3' region (Orf4)	*M. vannielii*
Hypothetical 11.3-kD protein in gapdh 3' region (Orfb)	*Pyrococcus woesei*
Hypothetical 11.5-kD protein in rs17 5' region (Orfa)	*M. vannielii*
Hypothetical 11.6-kD protein in rs17 5' region (Orfb)	*M. vannielii*
Hypothetical 13.7-kD protein in transposable element ish50	*H. halobium*

Hypothetical 15.2-kD protein	*M. thermoautotrophicum*
Hypothetical 15.3-kD protein in rs10 3′ region (Orf3)	*M. vannielii*
Hypothetical 15.6-kD protein in phr 5′ region	*H. halobium*
Hypothetical 16.7-kD protein	*M. thermoautotrophicum*
Hypothetical 18.7-kD protein in ribosomal RNA operon	*Thermofilum pendens*
Hypothetical 23-kD protein in ribosomal protein gene cluster (nab)	*H. cutirubrum*
Hypothetical 23.1-kD protein in hmal3 5′ region (Orf1)	*H. marismortui*
Hypothetical 24.4-kD protein in lacs 3′ region	*S. solfataricus*
Hypothetical 24.7-kD protein in gapdh 5′ region (Orfa)	*P. woesei*
Hypothetical 28.5-kD protein in 7s RNA 5′ region (Orf260)	*M. fervidus*
Hypothetical 30.7-kD protein	*M. thermoautotrophicum*
Hypothetical 31-kD protein in transposable element ish50	*H. halobium*
Hypothetical 38-kD protein in 23s RNA operon	*Thermoproteus tenax*
Hypothetical 46.2-kD protein (Orfb)	*Methanobrevibacter smithii*
Hypothetical 60.5-kD protein	*M. thermoautotrophicum*
Hypothetical 80.2 kd protein (Orf4)	*Haloferax sp. (strain aa 2.2)*
Hypothetical protein in ftr 5′ region (frag)	*M. thermoautotrophicum*
Hypothetical protein in gapdh 3′ region (Orfx)	*P. woesei*
Hypothetical protein in gapdh 5′ region (Orfz) (frag)	*P. woesei*
Hypothetical protein in glna 3′ region (frag)	*M. voltae*
Hypothetical protein in hmal29 3′ region (frag)	*H. marismortui*
Hypothetical protein in nifh 3′ region (frag)	*M. thermolithotrophicus*
Hypothetical protein in nifh 5′ region (frag)	*M. thermolithotrophicus*
Hypothetical protein in ribosomal l1 protein 5′ region (frag)	*M. vannielii*
Hypothetical protein in ribosomal protein s11 3′ region (frag)	*H. marismortui*
Hypothetical protein in ribosomal proteins operon (Orf1) (frag)	*H. marismortui*
Insertion element ism1 Hypothetical 48.3-kD protein (Orfis)	*M. smithii*
Malate dehydrogenase (E.C. 1.1.1.37) (E.C. 1.1.1.82) (mdh)	*M. fervidus*
Malate dehydrogenase (E.C. 1.1.1.37) (frag)	*S. acidocaldarius*
Membrane-associated ATPase γ chain (E.C. 3.6.1.34) (atpd)	*S. acidocaldarius*

Table 4 Connected components containing archaebacterial sequences (*continued*)

(F) Archaebacterial Singletons (Unconnected Entries)[g] (continued)	
Membrane-associated ATPase γ chain (E.C. 3.6.1.34)	*S. acidocaldarius*
Methyl CoM methylreductase α subunit (frag)	*Methanosarcina thermophila*
Methyl CoM methylreductase β subunit (frag)	*M. thermophila*
Methyl CoM methylreductase γ subunit (frag)	*M. thermophila*
Methyl CoM reductase II α subunit (frag)	*M. thermoautotrophicum*
Methyl CoM reductase II β subunit (frag)	*M. thermoautotrophicum*
Methyl CoM reductase II γ subunit (frag)	*M. thermoautotrophicum*
N-(5′-phosphoribosyl)anthranilate isomerase (trpf) (frag)	*M. voltae*
Nitrogenase Mo-Fe protein β chain (E.C. 1.18.6.1) (frag)	*M. thermolithotrophicus*
Ribosomal protein "a" (frag)	*M. thermoautotrophicum*
Thermopsin precursor (E.C. 3.4.23.-)	*S. acidocaldarius*

From SwissProt Release 19 (November 1991). Connected components are defined as sets of sequences where every sequence in the set is connected to at least one other sequence in the set by a PAM distance less than 200 provided that the similarity of the match is greater than 125 and extends over at least 80 positions. These were generated using the DARWIN system (Gonnet and Benner 1991). The connected components are unedited and may include at this PAM distance grouping of nonhomologous families. In connected components that contain more than one functional class of enzyme, the name of the class that includes an archaebacterial sequence is underlined.

[a]Only the archaebacterial aminotransferase is in this connected component. Eubacterial and eukaryotic AATs are in a separate connected component.

[b]These connected components arise from matches that may be insignificant statistically. They are properly resolved into several probably nonhomologous protein families.

[c]The α and β subunits form a single connected component at this level. However, they are more properly treated as separate components because the duplication resulting in the two chains is believed to have occurred before the divergence of the 3 primary kingdoms (Gogarten et al. 1989).

[d]Connected Component 1 (the "black hole") contains a large number (>5527) of protein entries joined by repetitive sequences, possibly arising from convergent sequence evolution. In these repetitive sequences, a small number of amino acid types are overrepresented compared to the database as a whole, and independent variation at each position according to the standard Dayhoff matrix can no longer be plausibly assumed and statistical analysis is problematic. In the protein pairs reported here, the similarity criteria noted above were achieved outside the region of the repetitive sequence.

[e]Related enzymes are archaebacterial (A), eukaryotic (E), eubacterial (B), or viral (V) enzymes as indicated.

[f]A matching of archaebacterial elongation factor sequences against the database as a whole shows matches likely to be nonsignificant at a level that excludes some nonarchaebacterial elongation factors.

[g]Protein sequences unconnected with other proteins at PAM 200 units. Many of these sequences are fragments and therefore may be excluded from a relevant connected component because of the size constraint. Of these 93 sequences, 35 are labeled hypothetical.

historical linguistics (Lehmann 1973; Benner and Ellington 1990a), where it designates a language reconstructed from descendant languages using a principle of parsimony and rules for transformation (e.g., "proto-Indoeuropean").

Most interesting in the context of this volume is the organism containing the "protogenome," the genome in the most recent common ancestor of archaebacteria, eubacteria, and eukaryotes (Benner and Ellington 1990a). We therefore jump directly back to this organism, a leap that takes us back over 1 billion years and omits, unfortunately, much important discussion of evolution in the intervening period.

The Protogenome: The Most Recent Common Ancestor of Archaebacteria, Eubacteria, and Eukaryotes

Reconstruction of the protogenome (Benner and Ellington 1990a) begins with sequence data for proteins performing analogous functions in the three kingdoms of life (Balch et al. 1977; Woese and Fox 1977). Two situations can exist. First, proteins having analogous biological functions in archaebacteria, eubacteria, and eukaryotes might be found in homologous forms in all three kingdoms. In this case, the protogenomic sequence can be reconstructed (see above), and the function can be placed in the protogenome. The possibility that the protein appears in homologous form in all three kingdoms via lateral transfer of genetic information is generally discounted, unless the gene is plasmid-encoded or if lateral transfer is indicated on other grounds. For example, the gas vacuolar protein is homologous in the archaebacterium *Halobacterium halobium* and the eubacterium *Pseudoanabaena* (Horne and Pfeifer 1989). However, because a version of the gene is plasmid-borne (Dassarma et al. 1987), the contact between these organisms in brines is intimate, and the gene is unconnected with a metabolic pathway, the possibility of lateral transfer cannot be ignored.

Second, proteins performing analogous biological roles in archaebacteria, eubacteria, and eukaryotes might *not* be homologous in all three kingdoms. In this case, either the protogenome did not encode the function at all (implying that the function arose by independent convergent evolution in all three kingdoms or by lateral transfer between kingdoms), or it did encode the function, but the molecule encoded in the protogenome was replaced in one or more of the derived kingdoms.

Multiple deletion-replacement events in the three lineages derived from the protogenome are normally considered improbable, except in one particular case. If a reaction was catalyzed in the protogenome by a riboenzyme, the possibility of multiple deletion-replacement events is presumably larger. Protein enzymes are (again presumably) more efficient than riboenzymes because they have a larger repertoire of encoded functional groups (see above). Thus, if a metabolic pathway is clearly assigned to the protogenome on independent grounds, but is catalyzed in the descendant kingdoms by nonhomologous proteins, and if the function is critical and the reaction unusual (implying that deletion-replacement events are slow), the hypothesis that a riboenzyme encoded by the protogenome was replaced independently in the three kingdoms is considered plausible. This is the case, for example, with ribonucleotide reductases (Benner et al. 1989).

Table 4 lists connected components in the current database containing sequences from both archaebacteria and at least one other kingdom. These are therefore families of proteins that might have common ancestors encoded by the protogenome. The list suggests two conclusions of immediate significance. First, an analysis of the multiple alignments for these connected components permits the reconstruction of more than 6,000 amino acids derived from at least 18,000 base pairs in the protogenome.[3] The encoded proteins include at least 16 ribosomal proteins and 20 enzymes, including proteins involved in replication (DNA-gyrase), protein synthesis (DNA-directed RNA polymerases, elongation factors), amino acid biosynthesis (tryptophan synthase, histidinol phosphate aminotransferase), glycolysis (phosphoglycerate kinase), the urea cycle (argininosuccinate lyase, carbamoyl phosphate synthase), redox reactions (ferredoxins, NADH-ubiquinone oxidoreductase), and ATP synthesis (ATP synthase).

Table 4 also establishes that the metabolism encoded by the protogenome was rather complex (Fig. 5). Two enzymes from the glycolytic pathway leading to the citric acid cycle (phosphoglycerate kinase and citrate synthase) are rigorously reconstructed in the protogenome; a third (glyceraldehyde-3-phosphate dehydrogenase) is tentatively assigned. These reconstructions suggest that the protogenome coded for the full glycolytic pathway as phosphoglycerate (the product of the reconstructed phosphoglycerate kinase) is not likely to be an end product of metabolism, nor is acetyl CoA (one substrate for the reconstructed citrate synthase) likely to have been obtained from the diet. Such a hypothesis, of course, is experimentally testable. For example, archaebacterial lactate dehydrogenases and enolases should exist and should be homologous to their eubacterial and eukaryotic counterparts.

---

[3]To obtain this estimate, positions in the multiple alignments deleted in any of the proteins were ignored. It must be noted that the 6000 amino acids representing 36 proteins do not indicate an average protein size of 166 amino acids for these proteins in the last common ancestor of the three kingdoms, but rather that the amino acids can be reliably assigned in an average of 166 positions per protein.

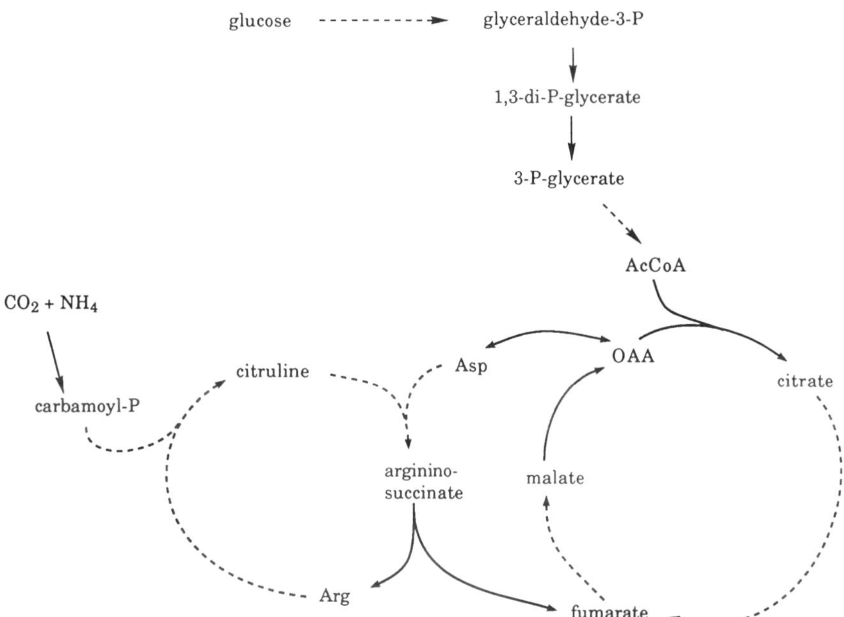

Figure 5 Diagram of the central metabolism encoded by the protogenome, the most recent common ancestor of the genome found in archaebacteria, eubacteria, and eukaryotes. Solid lines indicate metabolic reaction steps catalyzed by enzymes that can be reconstructed in the most recent common ancestor of archaebacteria, eubacteria, and eukaryotes. Broken lines indicate metabolic steps implied on chemical grounds from the reconstructed enzymes but catalyzed by enzymes yet to be reconstructed by sequence comparisons of homologous enzymes in all three kingdoms. The reconstruction of an ancestral malate dehydrogenase enzyme catalyzing the oxidation of malate to oxaloacetate is only poorly reconstructed.

Furthermore, the reconstructed tryptophan synthase, aspartate aminotransferase, and histidinol phosphate transaminase support the assumption that the protogenome encoded the biosynthesis of the full repertoire of amino acids. This assumption is, of course, also suggested by the fact that the reconstructed proteins encoded by the protogenome include all 20 proteinogenic amino acids found in contemporary organisms.

The reconstructed protogenome all but disproves the prevailing view that the most recent common ancestor of archaebacteria, eubacteria, and eukaryotes was a "progenote," an organism with highly imprecise mechanisms for replicating and translating genetic information (Woese and Fox 1977). Genome size and fidelity of handling of genetic information are directly correlated (Eigen and Schuster 1977). A genome of a size needed to encode such a metabolism is sustainable only by reasonably precise mechanisms for copying and translating genetic information.

Rather interesting evidence that the protogenome did not live in a progenote comes from the field of peptide design (see above). It has proven relatively easy to prepare relatively small polypeptides that catalyze the decarboxylation of oxaloacetate (Johnsson et al. 1990). It therefore appears that a substantial fraction of random functionalized polypeptides will catalyze destruction of oxaloacetate. An organism without reasonable control over what peptides it makes would therefore find oxaloacetate a problematic metabolic intermediate, because oxaloacetate would not survive in a cell where proteins with imprecise structures were being prepared. From the reconstruction, however, it is clear that the protogenome lived in a cell with oxaloacetate as a metabolic intermediate; at least two enzymes that use oxaloacetate as a substrate (citrate synthase and aspartate aminotransferase) can be reconstructed and placed in the protogenome. This implies that the protogenome encoded proteins that effected a metabolism involving oxaloacetate as a metabolic intermediate, suggesting in turn that the protogenome was not in a progenote.

Many pathways in contemporary organisms are missing from the reconstructed protogenome, however. Enzymes requiring biotin are absent in the reconstruction at this point. Pathways for the biosynthesis of straight-chain fatty acids are also absent; presumably, the protogenome resided in cells with terpenoid membranes, as do contemporary archaebacteria. In contrast, there is no evidence to reconstruct enzymes involved in methanogenesis, the classic metabolic pathway found in a major branch of archaebacteria. Thus, archaebacteria appear to display only some primitive traits, a statement that applies to each of the main kingdoms, suggesting a certain inappropriateness of the name "archaebacteria" (and, unfortunately, its more recently suggested alternative, "Archaea") (Woese et al. 1990).

Assigning a chronological date to the protogenome is, of course, extremely difficult in view of the absence of a reliable molecular clock (see above). It is possible, however, to reconstruct certain enzymes in the protogenome that suggest that it lived in an environment where molecular oxygen was available. Superoxide dismutase, gas vacuolar proteins, and the C5 pathway involved in the synthesis of chlorophyll (Friedmann et al. 1987; Kannangara et al. 1988) can all be assigned to the protogenome with varying degrees of reliability. Each of these is related in some way to oxygen in the atmosphere. Superoxide dismutase removes toxic products of oxygen metabolism. The gas vacuolar protein allows an organism to float at a position in water relative to an oxygen-containing atmosphere. Chlorophyll is essential (at least in the contemporary world) in the photosynthetic generation of molecular oxygen. Regardless of whether these arguments are definitive, making a connection between the protogenome and molecular oxygen is important in efforts to determine when the protogenome lived. Geological records suggest that molecular oxygen appeared on earth approximately 2.5 billion years ago. Fossil records make almost certain that the three kingdoms diverged before 1 billion years ago. Thus, if the protogenome lived in an oxygen atmosphere, it must have lived within these two limiting dates.

The Last Ribo-organism

In the metabolism encoded by the reconstructed protogenome, RNA is a more important component of catalysis than in contemporary organisms. RNA cofactors of all sorts are reliably placed in the protogenome, as is ribosomal RNA. Additional catalytic activities (e.g., ribonucleotide reductase) are also assigned to the protogenome in a riboenzymatic form (Benner et al. 1989). It is now well recognized that these facts are consistent with the presumption that the protogenome is itself derived from a more ancient genome that encoded RNA catalysts exclusively (Rich 1962; White 1976; Visser and Kellogg 1978; Gilbert 1986; Orgel 1986). It is useful to punctuate the episode in the history predating the protogenome by a breakthrough: the invention of a translation apparatus to translate an encoded mRNA (Benner et al. 1989). The breakthrough organism separates life that possessed an mRNA from life that had RNA as the sole genetically encoded component of biological catalysts.

Because information concerning only a single descendant lineage of the breakthrough organism (that leading to the protogenome) is available, rules of parsimony cannot assist reconstructions of the breakthrough organism. Rather, reconstructions must rely on rules of transformation alone. The simplest of these identifies components of the protogenomic metabolism that involve RNA performing roles where RNA is not an optimal chemical solution to the particular biochemical problem (Benner et al. 1989). RNA used in this capacity is presumed to be a vestige of an earlier time where RNA was the only available encoded molecule: the RNA world. Thus, these roles are assigned to a reconstructed metabolism for the breakthrough organism.

By applying this simple rule, one is forced to conclude that the last ribo-organism had a relatively complex metabolism that included oxidation and reduction reactions, aldol and Claisen condensations, transmethylations, porphyrin biosynthesis, and an energy metabolism based on nucleoside phosphates, all catalyzed by riboenzymes (Benner et al.

1989). It should be noted that this reconstruction cannot be weakened without losing much of the logical and explanatory force of the RNA world model. Nevertheless, the reconstructed metabolism of the breakthrough organism has proven to be the most controversial aspect of the "palimpsest" model (Maizels and Weiner 1987).

Many of the ribozymes in the breakthrough organism evidently were replaced by protein-based enzymes during the time separating the breakthrough from the protogenome. These are deletion-replacement events, governed by the rules noted above. The rate of deletion-replacement processes was determined by two factors: the degree to which a deletion of the riboenzyme would be lethal, and the possibility of finding a protein already participating in the metabolism that catalyzed a reaction chemically similar to the reaction of the deleted riboenzyme. In all cases, the rate of deletion replacement undoubtedly accelerated as the protein-based metabolism became more versatile. Some riboenzymes proved to be extremely difficult to replace; those involved in the ribosome-based synthesis of proteins apparently have survived until the present day (Noller et al. 1992). Others, such as the riboenzymes that encoded the conversion of ribonucleotides to 2′-deoxyribonucleotides (the ribonucleotide reductases) apparently resisted replacement until well after the divergence of the descendants of the protogenome (Benner et al. 1989).

It is impossible to estimate the time interval separating the protogenome from the breakthrough organism, other than to say that it must have been substantial. Each of the 36 or more proteins (with a total of 6000 encoded amino acids) apparently developed de novo in the interval separating the protogenome from the breakthrough organism. Thus, there is no reason to view the protogenome and the breakthrough genome as being close, either in time or metabolically.

Earlier in the RNA World

The reconstructions presented here bring us only to the end of the RNA world, the focus of this volume. Extrapolation further back in time is extremely difficult. As with the breakthrough organism, parsimony is unavailable to help us reconstruct events in the RNA world. Furthermore, transformation rules (e.g., mutation matrices) for reconstructing the evolution of catalytically functional nucleic acids are simply unavailable at present. Transformation rules might at best come from a better understanding of ribosomal RNA, RNAs from snRNPs, and other catalytic RNAs that are presumably vestiges of the RNA world. Even here, the process is risky, because many popular proposals that contemporary RNA molecules are vestiges of the RNA world, including the self-splicing introns from *Tetrahymena* (as a primitive vestige of the first self-replicating RNA molecule) and genomic tags (Weiner and Maizels 1987), are at best only weakly supported at this time by rigorous evolutionary analysis.

Nevertheless, the reconstructed metabolism of the breakthrough organism can influence models of earlier events and structures in the RNA world itself. For example, in most models, translation was assembled in an environment that is either prebiotic or primitive metabolically. The reconstructed metabolism of the breakthrough organism suggests that this view is incorrect. Rather, translation apparently arose in a relatively complex metabolic background provided by the breakthrough organism. It is, of course, much easier to design a new catalyst (including a catalyst for translating an mRNA) in a complex metabolic environment than in a simple one. Because none of the reactions involved in ribosome-based translation are exceptional (phosphate anhydride exchange, carboxylate ester formation, and aminolysis of a carboxylate ester being the only types), and because the reconstructed metabolism of the breakthrough organism contains several pathways that must have involved riboenzymes catalyzing similar reactions, there is no need to struggle to build models describing the origin of translation in the primitive world.

Indeed, specific RNA-catalyzed metabolic processes in the reconstructed breakthrough organism are plausible precursors for components in the first ribozyme. For example, the charging of amino acids is chemically similar to the first step in the pathway for preparing porphyrins reconstructed first in the protogenome on biochemical evidence (Friedmann et al. 1987), and from there to the breakthrough organism (Benner et al. 1989).[4]

THE PYRIDOXAL PARADOX

Reconstructions take on additional significance when directed toward specific problems in biological chemistry. Pyridoxal is perplexing in the context of the palimpsest model. It is present in all three kingdoms (Noll and Barber 1988) and therefore might be assigned to the protogenome according to a rule of parsimony. However, unlike nicotinamide cofactors, flavin cofactors, CoA, and ATP, for example, pyridoxal does not

---

[4]Sequence data suggest that the pyridoxal-dependent enzymes catalyzing the condensation of glycine with succinyl CoA in eubacteria and eukaryotes are not homologous, suggesting that this competing pathway for preparing aminolevulinate (without the need for an RNA cofactor) was not encoded by the protogenome (Li et al. 1989).

have an RNA substituent that identifies it as a cofactor originating in the RNA world (Orgel 1968; White 1976; Visser and Kellogg 1978). In this respect, it is similar to biotin. However, pyridoxal is unlike biotin in that its chemistry is easily modeled, a characteristic suggested by Visser and Kellogg (1978) for cofactors that originated in the RNA world. Although pyridoxal is formally accessible by combination of ribose and glyceraldehyde phosphate, its biosynthesis in contemporary organisms remains obscure (Hill et al. 1987), as do possible routes for nonbiological synthesis.

The enzymology of pyridoxal presents similar paradoxes. With ATPases, dehydrogenases (e.g., dihydrofolate reductase), enzymes that catalyze Claisen condensations (e.g., citrate synthase), and many other classes of enzymes (e.g., polymerases), analogous enzymes in the three kingdoms are generally homologous.[5] Building both trees and the reconstructed protogenomic sequences is direct. When enzymes catalyzing different reactions have been recruited from enzymes within these families, the derived function is generally apparent and not encoded by the protogenome (e.g., the sulfoxide reductases of connected component 1735 and the lactase-phlorizin hydrolase of connected component 1685; Table 5). There is little evidence for deletion-replacement events in central metabolic enzymes, and there is no evidence for large amounts of lateral transfer of genetic information between kingdoms.

This is not the case with pyridoxal-dependent enzymes. Consider just one family of pyridoxal-dependent enzymes, those in Figure 2. Aspartate aminotransferases (AAT) in this tree are found in all three kingdoms. However, the tyrosine aminotransferase (YAT) and AAT from two different kingdoms (archaebacteria and eukaryotes) are more closely related than YAT and AAT from the same kingdom (eukaryotes) (Cubellis et al. 1989). Furthermore, the AAT of *Escherichia coli* shares a common ancestor with the aromatic amino acid aminotransferase of *E. coli* (Fig. 2) that acts on tyrosine. Parsimony suggests that the protoeubacterial protein transaminated aspartate (Fig. 6). Therefore, the common ancestor of the entire family was, according to parsimony, also an AAT, and YATs

---

[5]Enzymes such as malate dehydrogenase and glyceraldehyde-3-phosphate dehydrogenase are found in all three kingdoms; sequence analysis indicates in one case that there may be no significant homology and in the other that the region of significant homology between all three kingdoms is very small. Although these activities might be tentatively assigned to the last common ancestor, complete reconstruction of ancestral malate and glyceraldehyde-3-phosphate dehydrogenases is not yet possible. Recent results from the *Caenorhabditis elegans* genome project indicate that an expressed protein in *C. elegans* is homologous to archaebacterial malate dehydrogenase (Waterston et al. 1992). This suggests that a new class of malate dehydrogenase is present in at least two of the kingdoms. Elucidation of the structure of this second class of malate dehydrogenases (either by crystallography or prediction after more homologous sequences have been obtained) should resolve the question of the relationship between the two classes of malate dehydrogenase.

were derived more than once from the ancestral AAT. Similarly, histidinol phosphate aminotransferases were also derived, in the most parsimonious model, from the ancestral AAT.

This implies either that there was no enzyme specifically catalyzing transamination reactions involving tyrosine or histidinol phosphate in the protogenome or that the enzymes catalyzing these transformations were deleted and replaced by an enzyme derived from the ancestral aspartate aminotransferase. Because the biosynthesis of tyrosine and histidine was almost certainly encoded by the protogenome (see above), the second case is more probable. This conclusion could, of course, be profoundly altered by additional sequence data; in particular, those for an archaebacterial YAT. However, we might predict that archaebacterial YATs were derived from archaebacterial AAT, if they are members of this family of proteins at all.

A small number of deletion-replacement events in a lineage are not, of course, excluded in the palimpsest model (Benner et al. 1989). However, we have invoked no fewer than three deletion-replacement events to account for the sequences in this tree. Still more must be invoked in the most parsimonious model describing other families of pyridoxal-dependent enzymes. For example, connected component 4018 (Table 5) contains enzymes essential for the biosynthesis of tryptophan, threonine, cysteine, and branched-chain amino acids. Again, multiple deletion-replacement events are required to account for these data, because enzymes involved in the biosynthesis of all these amino acids were presumably encoded by the protogenome.

The situation becomes still more complicated when other pyridoxal reactions are included. For example, the tree in Figure 2 includes some amino acid decarboxylases dependent on pyridoxal cofactors; other amino acid decarboxylases are found in other connected components (Table 5). The pattern has perplexed many authors (Martin et al. 1988; Mehta et al. 1989). Although decarboxylation of amino acids cannot as a pathway be reconstructed in the protogenome, recruitments were apparently still more prevalent within pyridoxal enzymes. To make matters worse, many decarboxylases do not use pyridoxal cofactors, even though the intrinsic chemistry involved would be well suited for pyridoxal chemistry. Instead, they use an enzyme-bound pyruvoyl residue (Recsei and Snell 1984). These include enzymes that decarboxylate aspartate, histidine, and *S*-adenosyl methionine and that are found in both eubacteria and eukaryotes. Furthermore, at least one enzyme, glycogen phosphorylase (from eubacteria and eukaryotes), uses pyridoxal phosphate as a Bronsted acid (Madsen and Withers 1986). In chemical terms, the function of this catalytically essential pyridoxal phosphate could be fulfilled

Table 5 Connected components containing pyridoxal-dependent enzymes

Connected component	Size	Representative entries and comments
1	5972	The "black hole";[a] glutamine decarboxylase
33	14	histidinol phosphate aminotransferase; lysine decarboxylase; ornithine decarboxylase, prokaryotic; 1-aminocyclopropane-1-carboxylate synthase; tyrosine aminotransferase; aspartate aminotransferase, archaebacterial; CobC protein; hypothetical protein, *B. subtilis*
216	14	aspartate aminotransferase, cytoplasmic and mitochondrial
223	1	aspartate aminotransferase, rabbit cytoplasmic (fragment)
225	1	aspartate aminotransferase, bovine mitochondrial (fragment)
231	1	aspartate aminotransferase, rabbit mitochondrial (fragment)
629	1	alanine aminotransferase, pig (fragment)
709	4	alanine racemase
713	1	amino acid racemase, *Pseudomonas* (fragment)
1076	13	acetylornithine aminotransferase, *E. coli*; 2,2-dialkylglycine decarboxylase, *Pseudomonas*; adenosylmethionine-8-amino-7-oxononanate aminotransferase; 4-aminobutyrate amino transferase; glutamate-1-semialdehyde (2,1)-aminomutase; ornithine aminotransferase, eukaryotic
1533	1	valine-pyruvate aminotransferase, *E. coli*
1717	1	adenosylmethionine-8-amino-7-oxononanate aminotransferase, *Salmonella*
1719	1	succinyldiaminopimelate aminotransferase, *Salmonella* (fragment)
1728	12	7-8-amino-7-oxononanate synthase; 5-aminolevulinic acid synthase; 2-amino-3-ketobutyrate CoA ligase
2515	6	cystathionine γ-lyase; methionine γ-lyase; cystathionine β-lyase, cystathionine γ-synthase
4018[b]	46	tryptophan synthase α and β; threonine synthase; L-serine dehydratase; threonine dehydratase; *O*-acetylserine sulfhydrylase

4074	3	D-alanine aminotransferase; branched-chain amino acid aminotransferase
4088	1	succinyldiaminopimelate aminotransferase, *E. coli* (fragment)
4131	13	aromatic amino acid decarboxylase; glutamate decarboxylase; histidine decarboxylase; α-methyl-DOPA hypersensitive protein; *Drosophila*
4137	11	diaminopimelate decarboxylase; arginine decarboxylase; ornithine decarboxylase, eukaryotic
4369	7	soluble hydrogenase, *Anabaena* and *Synechococcus*; NIFS protein *Anabaena, Synechococcus,* and *Azobacter;* serine-pyruvate aminotransferase, rat and human
5087	3	phosphoserine aminotransferase
6280	1	glycine dehydrogenase (decarboxylating), chicken (fragment)
6608	4	serine hydroxymethyltransferase
7741	3	5-aminolevulinic acid synthase
7747	1	5-aminolevulinic acid synthase, *Rhizobium* (fragment)
13296	11	glycogen phosphorylase, maltodextrin phosphorylase
16252	1	*S*-carboxymethylcysteine synthase, *E. coli* (fragment)
16330	1	D-serine dehydratase, *E. coli* (fragment)
16331	1	L-serine dehydratase, *E. coli* (fragment)
17353	1	threonine dehydratase, *Salmonella* (fragment)
17504	1	tryptophan synthase, *E. coli* (fragment)
17800	1	tryptophan synthase, *Klebsiella* (fragment)

SwissProt Version 19 identifies 153 enzymes identified as containing pyridoxal phosphate; SwissProt 22 identifies 179. The connected components generated at PAM <200, cost >125, and match length >80 place the 153 entries in SwissProt Version 19 in 33 connected components, as indicated below. Some sequences are singletons because they are only recorded as fragments and fail to fulfill the length requirement.

[a]See note d in Table 4.

[b]The two chains of tryptophan synthase fall in separate connected components. They are grouped together here by concatenation.

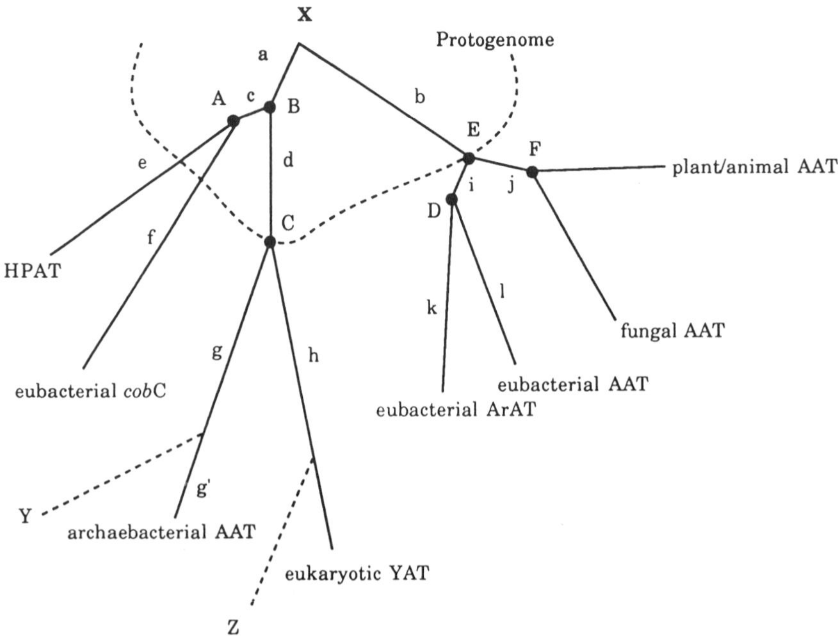

Figure 6 Schematic representation of the evolutionary tree in Fig. 2. Biological information suggests that the most recent common ancestor X lies above the curved dotted line. Functional parsimony indicates that X transaminated aspartate. Y and Z connected to the tree by broken lines indicate possible positions where an as yet not determined archaebacterial tyrosine aminotransferase might be treed. Introduction of an archaebacterial sequence at position Z does not change the most parsimonious interpretation of function on the tree, but it does indicate that the protogenome encoded a YAT. If an archaebacterial YAT sequence is treed at position Y, then the most functionally parsimonious reconstruction of X is as a tyrosine aminotransferase. However, no YAT can be assigned to the protogenome.

just as well by a simple phosphate ester such as a nucleoside phosphate or a phosphoserine residue.

More sequences, especially from archaebacteria, could of course alter the most parsimonious representation of the data presently available. Should this representation hold, however, there are three ways to account for the paradoxical behavior of pyridoxal enzymes. First, it may be intrinsically easy to delete and replace protein enzymes that use pyridoxal as cofactor. In this view, the protogenome encoded a full set of protein-based pyridoxal-dependent enzymes that were involved in multiple deletion-replacement events in the derived kingdoms. Supporting this view are several facts. Mutagenesis studies suggest that altering substrate specificity in pyridoxal enzymes is relatively easy (Cronin and Kirsch

1988). Pyridoxal-dependent decarboxylases could arise readily from a pyridoxal-dependent transaminase (and vice versa) (Smith et al. 1991). In fact, some enzymes (e.g., aspartate-β-decarboxylase; Novogrodsky and Meister 1964) have been shown to decarboxylate some amino acids while transaminating others. Amino acids might readily be obtained in the diet while recruitment is under way.

A second possibility derives from the hypothesis, noted above, that deletion-replacement events are likely to be especially facile if the enzyme being deleted and replaced is a riboenzyme. In this view, placing pyridoxal in the protogenome by a rule of parsimony is correct, but the protogenome-encoded aminotransferases dependent on pyridoxal cofactors were riboenzymes. The large number of deletion-replacement events in the derived lineages thus reflect this fact. In this view, RNA-based transaminases encoded by the protogenome were vestiges of the RNA world and were found in the breakthrough organism.

However, a third possibility should be considered. It is possible that pyridoxal was not in fact present in the most recent common ancestor, but rather was invented later and appeared in the three kingdoms followed by extensive lateral transfer once it was available and its unique chemical properties were appreciated. Just as with random peptides that destroy oxaloacetate (see above), considerations of the chemical reactivity of pyridoxal phosphate support this view. In the absence of any protein catalyst, pyridoxal catalyzes many irreversible reactions (Longenecker et al. 1957; Senda et al. 1977); this implies that sophisticated protein-binding sites (Visser and Kellogg 1978) are needed to control the reactivity of the cofactor. Furthermore, if unconstrained by a binding site, pyridoxal phosphate should lose the 5-phosphate group during its catalytic cycle (Benner 1988). Finally, the existence and evolutionary positions of pyruvoyl-dependent decarboxylases fit this view well, as do available data concerning the biosynthesis of pyridoxal.

We are unable to say at this time which model is correct, although the last seems to us to be the most plausible. If it is true, the suggestion of massive lateral transfer is most interesting, both for future efforts to use simple parsimony analysis for the reconstruction of models for ancient organisms and for evolutionary biology in general.

CONCLUSIONS AND PERSPECTIVES

A rigorous analysis of the divergent evolution of protein sequences and an integrated theory combining structural theory from chemistry and evolutionary theory from biology have provided the tools needed to address several important problems in structural biology. We can now construct high-quality alignments of protein sequences, predict reasonably well the conformation of proteins, design catalytic peptides de novo, and manipulate macromolecular structure in both proteins and nucleic acids. These and other tools sustain construction of preliminary models of the history of life on earth following the breakthrough that produced the first translation systems. The results of these reconstructions alter, often profoundly, our view of how life has evolved. If sequencing, enzymological, and design efforts continue at their present pace, and especially if they include an increased focus on archaebacterial systems, it should be possible at the end of the next decade to obtain definitive models for the protogenome and, perhaps, the breakthrough organism, tracing the molecular history of life back approximately 2.5 billion years before present.

ACKNOWLEDGMENTS

We are indebted to the Swiss National Science Foundation and Sandoz AG for support of the work described here. M.A.C. was supported by a fellowship from the Wellcome Trust. A portion of this work was presented at the meeting on Molecular Biology and the Origin of Life, organized by the Institute for Advanced Study in Biology, Berkeley, California, July 1990.

REFERENCES

Albery, W.J. and J.R. Knowles. 1976. Evolution of enzyme function and the development of catalytic efficiency. *Biochemistry* **15:** 5631–5640.

Bairoch, A. 1992. Prosite: A dictionary of sites and patterns in proteins. *Nucleic Acids Res.* **20:** 2013–2018.

Bairoch, A. and B. Boeckmann. 1992. The SWISS-PROT protein sequence data bank. *Nucleic Acids Res.* **20:** 2019–2022.

Balch, W.E., L.J. Magrum, G.E. Fox, R.S. Wolfe, and C.R. Woese. 1977. An ancient divergence among the bacteria. *J. Mol. Evol.* **9:** 305–311.

Bazan, J.F. 1990. Structural design and molecular evolution of a cytokine receptor superfamily. *Proc. Natl. Acad. Sci.* **87:** 6934–6938.

Benner, S.A. 1988. Stereoelectronic analysis of enzymatic reactions. *Mol. Struct. Energ.* **9:** 27–74.

———. 1989a. Enzyme kinetics and molecular evolution. *Chem. Rev.* **89:** 789–806.

———. 1989b. Patterns of divergence in homologous proteins as indicators of tertiary and quaternary structure. *Adv. Enzyme Regul.* **28:** 219–236.

Benner, S.A. and A.D. Ellington. 1987. Return of the "last ribo-organism." *Nature* **332:** 688–689.

———. 1988. Interpreting the behavior of enzymes: Purpose or pedigree? *CRC Crit. Rev. Biochem.* **23:** 369–426.

———. 1990a. "Progenote" or "Protogenote"? *Science* **248:** 943.

————. 1990b. Evolution and structural theory: The frontier between chemistry and biochemistry. *Bioorg. Chem. Front.* **1:** 1–70.

Benner, S.A. and D. Gerloff. 1991. Patterns of divergence in homologous proteins as indicators of secondary and tertiary structure: The catalytic domain of protein kinases. *Adv. Enzyme Regul.* **31:** 121–181.

Benner, S.A., A.D. Ellington, and A. Tauer. 1989. Modern metabolism as a palimpsest of the RNA world. *Proc. Natl. Acad. Sci.* **86:** 7054–7058.

Benner, S.A., A. Glasfeld, and J.A. Piccirilli. 1990. Stereospecificity in enzymology: Its place in evolution. *Top. Stereochem.* **19:** 127–207.

Benner, S.A., R.K. Allemann, A.D. Ellington, L. Ge, A. Glasfeld, G.F. Leanz, T. Krauch, L.J. MacPherson, S. Moroney, J.A. Piccirilli, and E. Weinhold. 1987. Natural selection, protein engineering, and the last riboorganism: Rational model building in biochemistry. *Cold Spring Harbor Symp. Quant. Biol.* **52:** 53–63.

Bork, P. 1992. Mobile modules and motifs. *Curr. Opin. Struct. Biol.* **2:** 413–421.

Chothia, C. and A. Lesk. 1986. The relation between the divergence of sequence and structure in proteins. *EMBO J.* **5:** 823–826.

Crawford, I.P., T. Niermann, and K. Kirschner. 1987. Prediction of secondary structure by evolutionary comparison: Application to the α subunit of tryptophan synthase. *Proteins* **2:** 118–129.

Cronin, C.N. and J.F. Kirsch. 1988. Role of arginine 292 in the substrate specificity of aspartate aminotransferase as examined by site-directed mutagenesis. *Biochemistry* **27:** 4572–4579.

Crouzet, J., L. Cauchois, F. Blanche, L. Debussche, D. Thabaut, M.-C. Rouyez, S. Rigault, J.-F. Mayaux, and B. Cameron. 1990. Nucleotide sequence of *Pseudomonas denitrificans* 5.5 kbase DNA fragment containing five cob genes and identification of structural genes encoding S-adenosyl-L-methionine: Uroprophyinogen III methyltransferase and cobrynic acid a,c-diamide synthase. *J. Bacteriol.* **172:** 5968–5979.

Cubellis, M.V., C. Rozzo, G. Nitti, M.I. Arnone, G. Marino, and G. Sannia. 1989. Cloning and sequencing of the gene coding for aspartate aminotransferase from the thermoacidophilic archaebacterium *Sulfolobus solfataricus. Eur. J. Biochem.* **186:** 375–381.

Dassarma S., T. Damerval, J.G. Jones, and N. Tandeau de Marsac. 1987. A plasmid-encoded gas vesicle protein gene in a halophilic archaebacterium. *Mol. Microbiol.* **1:** 365–370.

Dayhoff, M.O., R.M. Schwartz, and B.C. Orcutt. 1978. A model of evolutionary change in proteins. In *Atlas of protein sequence and structure* (ed. M.O. Dayhoff), vol. 5, suppl. 3, p. 345–352. National Biomedical Research Foundation, Washington, D.C.

Dever, T.E., M.J. Glynias, and W.C. Merrick. 1987. GTP-binding domain: Three consensus elements with distinct spacing. *Proc. Natl. Acad. Sci.* **84:** 1814–1818.

de Vos, A.M., M. Ultsch, and K.K. Kossiakoff. 1992. Human growth hormone and extracellular domain of its receptor: Crystal structure of the complex. *Science* **255:** 306–312.

Doolittle, R.F., D.F. Feng, K.L. Anderson, and M.R. Alberro. 1990. A naturally occurring horizontal gene transfer from a eukaryote to a prokaryote. *J. Mol. Evol.* **31:** 383–388.

Edwards, F.W. and L.L. Cavalli-Sforza. 1963. The reconstruction of evolution. *Ann. Hum. Genet.* **27:** 104.

Eigen, M. and P. Schuster. 1977. The hypercycle, a principle of natural self-organization. *Naturwissenschaften* **64:** 341–369.

Eschenmoser, A. and E. Loewenthal. 1992. Chemistry of potentially prebiological natural products. *Chem. Soc. Rev.* **21:** 1–16.

Fitch, W. and E. Margoliash. 1967. Construction of phylogenetic trees. *Science* **155:** 279–284.

Friedmann, H.C., R.K. Thauer, S.P. Gough, and C.G. Kannangara. 1987. Δ-Amino-levulinic acid formation in the archae bacterium *Methanobacterium thermo-autotrophicum* requires tRNA^{Glu}. *Carlsberg Res. Commun.* **52:** 363–371.

Gilbert, W. 1986. The RNA world. *Nature* **319:** 618.

Gogarten, J.P., H. Kibak, P. Diettrich, L. Taiz, E.J. Bowman, B.J. Bowman, M.F. Manolson, R.J. Poole, T. Date, T. Oshima, J. Konishi, K. Denda, and M. Yoshida. 1989. Evolution of the vacuolar H$^+$-ATPases: Implications for the origin of eukaryotes. *Proc. Natl. Acad. Sci.* **86:** 6661–6665.

Gonnet, G.H., and S.A. Benner. 1991. Computational biochemistry at ETH. Technical report 151. ETH, Zürich.

Gonnet, G.H., M.A. Cohen, and S.A. Benner. 1992. Exhaustive matching of the entire protein sequence database. *Science* **256:** 1443–1445.

Hill, R.E., A. Iwanow, B.G. Sayer, W. Wysocka, and I.D. Spenser. 1987. The regiochemistry and stereochemistry of the biosynthesis of vitamin B6 from triose units. *J. Biol. Chem.* **262:** 7463–7471.

Horne, M. and F. Pfeifer. 1989. Expression of two gas vacuole proteins in *Halobacterium halobium* and other related species. *Mol. Gen. Genet.* **218:** 437–444.

Huang, Z., K.C. Schneider, and S.A. Benner. 1991. Building bocks for analogs of ribo- and deoxyribonucleotides with dimethylene-sulfide, -sulfoxide, and -sulfone groups replacing phosphodiester linkages. *J. Org. Chem.* **56:** 3869–3882.

Hyde, C.C., S.A. Ahmed, E.A. Padlan, E.W. Miles, and D.R. Davies. 1988. Three-dimensional structure of the tryptophan synthase $\alpha_2\beta_2$ multienzyme complex from *Salmonella typhimurium*. *J. Biol. Chem.* **263:** 17857–17871.

Johnsson, K., R.K. Allemann, and S.A. Benner. 1990. Designed enzymes: New peptides that fold in aqueous solution and catalyze reactions. In *Molecular mechanisms in bioorganic processes* (ed. C. Bleasdale and B.T. Golding), pp. 166–187. Royal Society of Chemistry, Cambridge, England.

Jukes, T.H. and R. Holmquist. 1972. Evolutionary clock: Nonconstancy of rate in different species. *Science* **177:** 530–532.

Kannangara, C.G., S.P. Gough, P. Bruyant, J.K. Hoober, A. Kahn, and D. von Wettstein. 1988. tRNA^{Glu} as a cofactor in δ-aminolevulinate biosynthesis. *Trends Biochem. Sci.* **13:** 139–143.

Kimura, M. 1982. Molecular evolution, protein polymorphism, and the neutral theory. Springer-Verlag, Berlin.

King, J.L. and T.H. Jukes. 1969. Non-Darwinian evolution. *Science* **164:** 788–798.

Knighton, D.R., J. Zheng, L.F. Ten Eyck, V.A. Ashford, N.H. Xuong, S.S. Taylor, and J.M. Sowadski. 1991. Crystal structure of the catalytic subunit of cyclic adenosine monophosphate-dependent protein kinase. *Science* **253:** 407–414.

Lehmann, W.P. 1973. *Historical linguistics*. Holt, Rinehard and Winston, New York.

Lewin, R. 1988. Molecular clocks turn a quarter century. *Science* **239:** 561–563.

Li, B.F.L., S. Osborne, and M.L. Sinnott. 1983. Catalytic consequences of experimental evolution. Part 2. Rate-limiting degalactosylation in the hydrolysis of aryl beta-D-galactopyranosides by the experimental evolvants ebg(a) and egb(b). *J. Chem. Soc. Perkin Trans.* **II:** 365–369.

Li, J.-M., C.S. Russel, and S.D. Cosloy. 1989. Cloning and structure of HemA gene of *Escherichia coli* K-12. *Gene* **82:** 209–217.

Longenecker, J.B., M. Ikawa, and E.E. Snell. 1957. The cleavage of α-methylserine and α-methylolserine by pyridoxal and metal ions. *J. Biol. Chem.* **226:** 663–666.

Madsen, N.B. and S.G. Withers. 1986. Glycogen phosphorylase. In *Vitamin B6, pyridoxal phosphate*, Part B (ed. D. Dolphin et al.), pp. 355–389. Wiley, New York.

Maizels, N. and A.M. Weiner. 1987. The last riboorganism was no breakthrough. *Nature* **330:** 616.

Martin, C., B. Cami, P. Yeh, P. Stragier, C. Parsot, and J.-C. Patte. 1988. *Pseudomonas aeruginosa* diaminopimelate decarboxylase: Evolutionary relationship with other amino acid decarboxylases. *Mol. Biol. Evol.* **5:** 549–559.

Mehta, P.K., T.I. Hale, and P. Christen. 1989. Evolutionary relationships among aminotransferases. *Eur. J. Biochem.* **186:** 249–253.

Niermann, T. and K. Kirschner. 1990. Improving the prediction of secondary structure of "TIM-barrel" enzymes. *Protein Eng.* **4:** 137–147.

Noll, K.M. and T.S. Barber. 1988. Vitamin contents of archaebacteria. *J. Bacteriol.* **170:** 4315–4321.

Noller, H.F., V. Hottarth, and L. Zimniak. 1992. Unusual resistance of peptidyl transferase to protein extraction procedures. *Science* **256:** 1416–1419.

Novogrodsky, A. and A. Meister. 1964. Control of aspartate-b-decarboxylase activity by transamination. *J. Biol. Chem.* **239:** 879–888.

Oliver, S.G., Q.J.M. van der Aart, M.L. Agostoni-Carbone, M. Aigle, L. Alberghina, D. Alexandraki, G. Antoine, R. Anwar, J.P.G. Ballesta, P. Benit, G. Berben, E. Bergatino, N. Biteau, P.A. Bolle, M. Bolotin-Fukahara, A. Brown, A.J.P. Brown, J.M. Buhler, C. Carcano, G. Carignani, H. Cederberg, R. Chanet, R. Contreras, M. Crouzet, and B. Daignan-Fornier. 1992. The complete DNA sequence of yeast chromosome III. *Nature* **357:** 38–46.

Orgel, L.E. 1968. Evolution of the genetic apparatus. *J. Mol. Biol.* **38:** 381.

———. 1986. RNA catalysis and the origins of life. *J. Theor. Biol.* **123:** 127–149.

Osterhout, J.J., Jr., T. Handel, G. Na, A. Toumadje, R.C. Long, P.J. Connolly, J.C. Hoch, W.C. Johnson, Jr., D. Live, and W.F. DeGrado. 1992. Characterization of the structural properties of $\alpha_1\beta$, a peptide designed to form a four-helix bundle. *J. Am. Chem. Soc.* **114:** 331–337.

Piccirilli, J.A., T. Krauch, S.E. Moroney, and S.A. Benner. 1990. Extending the genetic alphabet: Enzymatic incorporation of a new base pair into DNA and RNA. *Nature* **343:** 33–37.

Recsei, P.A. and E.E. Snell. 1984. Pyruvoyl enzymes. *Annu. Rev. Biochem.* **53:** 357–387.

Rich, A. 1962. On the problems of evolution and biochemical information transfer. In *Horizons in biochemistry* (ed. M. Kasha and B. Pullman), pp. 103–126. Academic Press, New York.

Schneider, K.C. and S.A. Benner. 1990. Oligonucleotides containing flexible nucleoside analogs. *J. Am. Chem. Soc.* **112:** 453–455.

Senda, S., K. Hirota, T. Asao, and K. Maruhashi. 1977. A model for an intermediate in pyridoxal-catalyzed γ-elimination and γ-replacement reactions of amino acids. *J. Am. Chem. Soc.* **99:** 7356–7359.

Smith, D.M., N.R. Thomas, and D. Gani. 1991. A comparison of pyridoxal 5'-phosphate-dependent decarboxylase and transaminase enzymes at a molecular level. *Experientia* **47:** 1104–1118.

Stackhouse, J., S.R. Presnell, G. McGeehan, K.P. Nambiar, and S.A. Benner. 1990. The ribonuclease from an extinct bovid. *FEBS Lett.* **262:** 104–106.

Sulston, J., Z. Du, K. Thomas, R. Wilson, L. Hillier, R. Staden, N. Halloran, P. Green, J. Thierry-Mieg, L. Qiu, S. Dear, A. Coulson, M. Craxton, R. Durbin, M. Berks, M. Metzstein, T. Hawkins, R. Ainscough, and R. Waterston. 1992. The *C. elegans* genome sequencing project: A beginning. *Nature* **356:** 37–41.

Switzer, C.Y., S.E. Moroney, and S.A. Benner. 1989. Enzymatic incorporation of a new base pair into DNA and RNA. *J. Am. Chem. Soc.* **111:** 8322–8323.

Taylor, W.R. and C.A. Orengo. 1989. Protein structure alignment. *J. Mol. Biol.* **208:** 1–22.

Thorne, J.L., H. Kishino, and J. Felsenstein. 1992. Inching toward reality: An improved likelihood model of sequence evolution. *Mol. Biol. Evol.* **34:** 3–16.

Thornton, J.M., T.P. Flores, D.T. Jones, and M.B. Swindells. 1991. Prediction of progress at last. *Nature* **354:** 105–106.

Valencia, A., M. Kjeldgaard, E.F. Pai, and C. Sander. 1991. GTPase domains of ras p21 oncogene protein and elongation factor Tu: Analysis of three-dimensional structures, sequence families, and functional sites. *Proc. Natl. Acad. Sci.* **88:** 5443–5447.

Van der Woerd, R., C.G. Bakker, and A.W. Schwartz. 1987. Synthesis of P1,P2 dinucleotide pyrophosphates. *Tetrahedron Lett.* **28:** 2763–2766.

Visser, C.M. and R.M. Kellogg. 1978. Biotin. Its place in evolution. *J. Mol. Evol.* **11:** 171–178.

Waterston, R., C. Martin, M. Craxton, C. Huynh, A. Coulson, L. Hillier, R.K. Durbin, P. Green, R. Shownkeen, N. Halloran, T. Hawkins, R. Wilson, M. Berks, Z. Du, K. Thomas, J. Thierry-Mieg, and J. Sulston. 1992. A survey of expressed genes in *Caenorhabditis elegans*. *Nature Genet.* **1:** 114–123.

Weiner, A.M. and N. Maizels. 1987. tRNA-like structures tag the 3′ ends of genomic RNA molecules for replication: Implications for the origin of protein synthesis. *Proc. Natl. Acad. Sci.* **84:** 7383–7387.

White III, H.B. 1976. Coenzymes as fossils of an earlier metabolic state. *J. Mol. Evol.* **7:** 101–104.

Woese, C.R. and G.E. Fox. 1977. Phylogenetic structure of the prokaryotic domain: The primary kingdoms. *Proc. Natl. Acad. Sci.* **74:** 5088–5090.

Woese, C.R., O. Kandler, and M.L. Wheelis. 1990. Towards a natural system of organisms: Proposal for the domains *Achaea*, *Bacteria*, and *Eucarya*. *Proc. Natl. Acad. Sci.* **87:** 4576–4579.

Yeh, W.K. and L.N. Ornston. 1980. Origins of metabolic diversity: Substitution of homologous sequences into genes for enzymes with different catalytic activities. *Proc. Natl. Acad. Sci.* **77:** 5365–5369.

Zuckerkandl, E. and L. Pauling. 1965. Evolutionary divergence and convergence in proteins. In *Evolving genes and proteins* (ed. V. Bryson and H. J. Vogel), pp. 97–166. Academic Press, New York.

3

Speculations on the Origin of Ribosomal Translocation

Robert Weiss and Joshua Cherry
Department of Human Genetics
University of Utah
Salt Lake City, Utah 84112

The discovery of catalytic RNA suggested a way out of a troublesome problem concerning the origin of life (Gilbert 1986). Given the known roles of the various biological informational molecules, it had been unclear how a genetic system could ever have gotten started. The replication of a nucleic acid would require a proteinaceous polymerase, but without replication, nothing, protein synthesis included, could evolve. Put another way, there was no molecule known that could both serve as a template for replication and express its information without some sort of translation, in the loose sense of the word. The discovery of modern catalytic RNA demonstrated in dramatic fashion that RNA could be such a molecule. The first nucleic acid polymerase, it is hypothesized, was an RNA molecule that catalyzed its own replication. Such self-replicating RNAs were the first genetic systems and might well be called the first living things. Once there were replicators, there was selection. Whereas the first selection-driven changes probably were "improvements" to polymerase activity, eventually metabolic innovations arose, including template-directed protein synthesis. The precursor of the modern ribosome, then, was a purely, or mostly, RNA-based system, confirming early suspicions about the origins of rRNA, mRNA, and tRNA (Woese 1967; Crick 1968; Orgel 1968). The fact that the modern translational apparatus contains so much RNA is explained by such an origin of translation. In fact, it is tempting to think of ribosomal RNA as catalytic RNA, and experimental evidence indicates that at least some of the ribosome's activities are associated with ribosomal RNA rather than ribosomal protein (Noller 1991a). This discussion of the evolution of translation considers how translation might have arisen in the "RNA World."

Although the notion of an RNA world does much to demystify the origin of translation, much difficulty remains. The lack of a good theory for the evolution of translation reflects, in part, the limited understanding

The RNA World
© 1993 Cold Spring Harbor Laboratory Press 0-87969-380-0/93 $5 + .00

of the ways in which modern ribosomes work. A principal difficulty with
the origin of translation is that so many activities are required for protein
synthesis as we know it. The presence of RNAs that could function as
tRNAs, their aminoacylation, the peptidyl transferase, translocation, and
perhaps other activities of the ribosome, and, of course, some sort of
message, would all be necessary for translation. Even if we imagine that
we understand how a single, simple activity could (given a selection for
that activity) arise, the emergence of a complex system like translation,
whose constituent components seem useless on their own, remains a
problem. An obvious way out of such a problem is to suggest reasons for
the existence of the various constituents of translation, or of entities very
much like them, independent of their role in translation proper. Such
"partial theories" have been proposed for tRNA (Weiner and Maizels
1987; Maizels and Weiner, this volume) and for the decoding site of 16S
rRNA (Noller 1991b; von Ahsen et al. 1991; Davies et al.; Noller; both
this volume). It must be stressed that if a partial theory is to help explain
the evolution of a complex process, it must propose a selective advantage
for the simpler activity in question.

Recent speculators on the origins of translation have cast their nets
over modern RNA-only processes in search of functions that might relate
to translation. Although the discovery of catalytic RNA has been
responsible for a great change in our understanding of early evolution, it
is unclear what relationship these modern RNAs bear to the primordial
world. It may well be that modern catalytic RNAs do not trace their
origins back to the RNA world, but are of much more recent origin. If a
given catalytic RNA does in fact go back to the RNA world, it does not
follow that it shares a common ancestor, in any meaningful sense, with
any constituent of the ancient or modern translation system (or with the
ancient RNA polymerase). If such a common ancestor did indeed exist, it
may have had the modern RNA-only activity, the translation-related activity, or some other, very different activity. In any case, it is uncertain
whether specific mechanistic details, such as the involvement of a $5'G$
residue in catalysis, would have remained intact over the more than three
billion years since the glory days of RNA catalysis.

The origin and evolution of protein synthesis required solutions to
three fundamental problems regarding specificity: tRNA charging, triplet
decoding, and triplet translocation. Prior to the discovery of catalytic
RNA, the most complete theory of the origin of decoding and translocation was a model of *ribosome-free* template-directed protein synthesis
that contained inherent definition of the reading frame. Building on an
earlier proposal for a switch in the base-stacking configuration of the
anticodon loop, known as the ratchet (Woese 1970), it was theorized

(Crick et al. 1976) that a purine-purine-pyrimidine, or RRY, primitive code could form the basis for the readout of a triplet genetic code by two adjacent, interacting primitive tRNAs carrying anticodon loops of the form 3'-UG-YYR-UU-5', reminiscent of the sequence of modern anti-codon loops. This type of code provides inherent protection from framing errors by virtue of its repeating RRY structure. Furthermore, pepti-dyl-tRNA:mRNA complexes would be stabilized by the coaxial stacking of adjacent five-base tRNA:mRNA pairs with the two flanking anticodon stems (Fig. 1). The theory was generalized to a symmetric RNY code (Eigen and Schuster 1979) that would allow for internal complementarity of these hypothetical mRNAs. The features of enhanced stability via five base-pair tRNA:mRNA complexes, inherent translocation by ratcheting, and enforced frame maintenance by comma-free RNY coding make this model of primitive translation appealing. Related to such models are Eigen's view of tRNA as early adapter/mRNA whose phylogeny could still be detected in the sequence of modern tRNA (Eigen 1987) and the occurrence of RNY preferences in modern reading frames (Shepherd 1981), although claims that there is a remnant of an ancient RNY system observable in modern sequences are a subject of debate.

The ribosome-free ratcheting model leaves unanswered some important questions concerning the mechanism of primitive translation. The origin and nature of the peptidyl transferase activity is not specified by this model; perhaps the approximation of the two acceptor arms was the

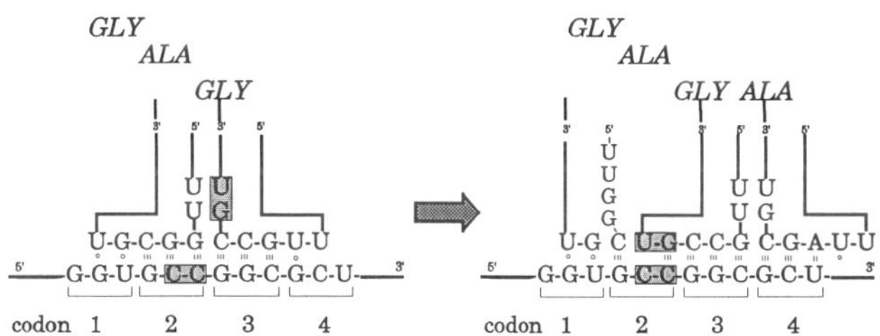

Pre-translocation   Post-translocation

Figure 1 Ribosome-free primitive translocation with an RNY code. Two tRNAs are shown, after transpeptidation, bound to adjacent codons with the P-site tRNA in a 3' stack, or "FH" configuration, which is forming two extra G-U base pairs 5' of the codon, and the A-site tRNA in a 5' stack, or "hf" configuration, which is forming one extra G-U base pair 3' of the codon. Translocation re-quires ratcheting of the peptidyl-tRNA's anticodon loop from "hf" configuration into the "FH" configuration, followed by binding of the next aminoacyl-tRNA in the "hf" configuration.

sole source of rate enhancement. The absence of a source of free energy to drive the molecular gymnastics of ratcheting sticks out like an amputated thumb. The RNA movements might well have been driven by the peptidyl transfer reaction, but some mechanism for coupling the two processes would be necessary. Some structure that interacted with both the mRNA:tRNA complex and the acceptor ends of the tRNAs could certainly have effected such a coupling and could have also contained a peptidyl transferase activity. However, translation involving such a structure could hardly be described as ribosome-free; rather, such a structure would be a primitive ribosome.

If credible evidence for a reciprocating ratchet mechanism (Woese 1980) at the decoding center of the modern ribosome existed, further conjecture would perhaps be unnecessary. Today's model of translocation postulates a mechanism in which P- and A-site tRNA:mRNA base-paired complexes move between separate locations on the 30S subunit. If translation had started in the manner suggested by Woese (1970), Crick (Crick et al. 1976), and Eigen (Eigen and Schuster 1979), it would seem to have been supplanted by the modern mechanism. Certainly, we must postulate that RNA versions of aminoacyl-tRNA synthetases were supplanted by more efficient protein versions. As for enzymes, nothing is left of the RNA world except for a handful of catalytic RNAs (which may or may not have ancient origins). Why proteins have not taken over completely is unclear. The RNA world solves the problem of proteins having to invent their own synthesis, but it does not answer to what extent they reinvented the process after it first got started. Are ribosomal proteins replacements, refinements, or additions to the components of the original process? Were there RNA predecessors of the now protein-aceous elongation, initiation, and termination factors? Why is rRNA still here? The approachable question, given an rRNA-based origin of translation, may be, What were the crucial activities that primitive rRNA carried in its sequence and structure that both made primitive translation function and made the primitive rRNA irreplaceable? In the spirit of partial theories accounting for the origins of constituent activities of the translation apparatus, what will be considered is whether there are attributes of the modern translocation mechanism that suggest functions which could plausibly have existed in the RNA world and which were initially (and are still) based within rRNA.

MODERN RIBOSOME TRANSLOCATION AND FRAME MAINTENANCE

Ribosome movement is a two-faceted problem. The first problem is how the tRNA:mRNA complex moves through the ribosome, and the second is how the reading frame is kept. Two experimental approaches allow investigation of the mechanisms of ribosome movement. Direct biochemical assays of the number of tRNA-binding sites and their occupancy during a single translation cycle have led to the current three-site model of the modern ribosome. In addition to the P and A sites, *Escherichia coli* ribosomes possess a third site, the "exit" or E site, that binds deacylated tRNA after it has left the P site (Rheinberger et al. 1990; Wintermeyer et al. 1990). The physical arrangement of tRNA and mRNA in the ribosome can be modeled using biochemical data, although a consensus view is still a subject of some controversy (Lim et al. 1992). Model builders attempt to coalesce the massive amounts of structural data on the ribosome with steric constraints imposed by the binding of two appropriately positioned tRNAs at adjacent codons. A second type of experimental probe into the workings of the translocation mechanism includes the study of extragenic suppressors of frameshift mutations, genetic assays that monitor the occurrence of reading frame errors, sequencing of peptides resulting from non-triplet translocations, and the analysis of naturally occurring mRNA sequences that promote efficient non-triplet translocation. Because a change of frame is, in a sense, a failure of the framing mechanism, the rules uncovered by such experiments may be viewed as design parameters for the frame monitoring system. Together, these areas of investigation have given us our current understanding of modern translocation.

Translocation: The Biochemical Trail

The original two-site model of tRNA-binding sites (Watson 1964) has recently matured into a three-site E-P-A model. How these sites are positioned within and between the ribosomal subunits, and, correspondingly, how the bound tRNAs shuttle between them, is a key area of current research. In vitro studies of the translocation reaction are performed with ribosome-mRNA complexes containing peptidyl-tRNA in the A site (operationally defined as nonreactive with puromycin) and deacylated tRNA in the P site. Upon addition of EF-G and GTP, peptidyl-tRNA is translocated to the P site, as monitored by the appearance of puromycin sensitivity. Results of RNase protection assays performed before and after EF-G/GTP addition are consistent with translocation of the mRNA by three nucleotides (Thach and Thach 1971).

More recently, precise site-directed mRNA cross-linking data have confirmed the movement of the mRNA after EF-G-GTP binding and have yielded a detailed view of the close contacts made between mRNA and 16S rRNA (Dontsova et al. 1992). A current view (Fig. 2) of tRNA-mRNA complexes proposed by Dontsova et al. is depicted in the elegant

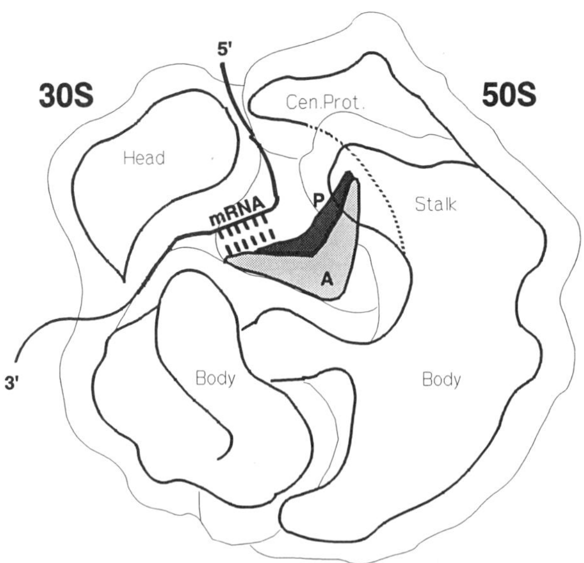

Figure 2 A current view of the tRNA-mRNA complex within the 70S ribosome. The sketch is based on reconstruction images of the *E. coli* ribosome by Frank et al. (1991), as interpreted by Dontsova et al. (1992).

reconstruction from electron micrographs of the 70S ribosome by Frank et al. (1991), which emphasizes the topography of the ribosomal surface. Here, the tRNAs are shown in the R configuration, in which the TpseudoUC loop of the A-site tRNA faces the P-site tRNA. An alternative placement of the tRNAs in the S configuration, with the tRNA elbows pointing toward the 30S surface, has also been proposed (Wower et al. 1989; Noller et al. 1990); the actual orientation is still controversial. The path of the message through the A and P sites, the nature of the "kink," and the orientation of the anticodons have been analyzed and reviewed in detail by Spirin (1985). The movement of the anticodon: mRNA complex from the A to the P site can be described as a right-handed helical motion, in which the complex rotates approximately 100° about a translocation axis. Cross-linking studies of tRNA, such as the cross-link of the 5′ base of the P-site anticodon to C-1400 of 16S rRNA (Prince et al. 1982), and footprinting of tRNA and antibiotics that effect decoding (Noller 1991a), all indicate a highly structured core in this conserved region of the 30S subunit that is in close contact with the tRNA:mRNA complexes. mRNA decoding seems to be primarily a 30S activity.

Recent structural studies have uncovered several hybrid states that occur during translocation arrest in vitro. Models involving hybrid states of tRNA binding, perhaps with one subunit moving relative to the other during translocation, had been originally proposed as explanations for the origin of a two-subunit ribosome (Bretscher 1968; Spirin 1970). Results of tRNA footprinting experiments on EF-G-deprived ribosomes (Noller 1991a) suggest that the movement of acceptor arms relative to the 50S is independent of the EF-G-dependent movement of codon:anticodon complexes relative to the 30S. Displacement of tRNA from the 50S P site during peptide bond formation is also supported by observations of changes in the fluorescence quantum yield and anisotropy of labeled tRNA probes during the peptidyl transferase reaction (Hardesty et al. 1990). This evidence supports the conclusion that after peptide bond formation, but before EF-G binding, there is a conformational change in the acceptor arm of deacylated tRNA, whereas there is relatively little movement of the anticodon loop relative to the 30S subunit. These results imply that translocation may be a two-step process, with the acceptor ends of the two adjacent tRNAs moving on the 50S subunit independent of translocation of the anticodon:mRNA complexes on the 30S subunit. The movement of the acceptor arms is somewhat independent of the movement along the message. The movement of the P-site acceptor may be a simple consequence of the tRNA's "letting go" of the peptide. The A site acceptor arm's motion may be equivalent to the motion of a processive peptidyl transferase along the growing peptide chain.

Translocation: The Genetic Trail

The triplet nature of the genetic code was deduced by clever examination of the combinatorial numerology of mutations that insert or delete nucleotides from a coding region. Intragenic frameshift suppressors can be described as mutations that are of sign opposite to that of the original mutation and thus are capable of restoring a translating ribosome to its starting reading frame after its brief excursion through an alternate frame. Frameshift mutations frequently exhibit residual levels of expression from the frameshifted open reading frame, and in many cases this residual activity represents the level of spontaneous frameshifting by translating ribosomes. Even after more than three billion years of refinement, triplet translocation is only so good.

The limitations on the modern ribosome's capacity for error-free translocation may provide some insight into the challenges faced by a primitive apparatus. The ability to make defined peptides, except for certain homopolypeptides such as polyphenylalanine from poly(U), is dependent on a successive linear readout in which one misstep ruins the whole product. Such lapses have a more severe consequence than a misincorporation event caused by misacylation or misreading (Gallant and Foley 1980; Kurland and Gallant 1986). Enhanced levels of non-triplet movement across homopolymeric sequences, especially but not exclusively on (A) or (U) repeats, are common (Weiss et al. 1990b) and can be interpreted as strand-slippage events between anticodons and the mRNA. Strand slippage on homopolymer or repeat sequences is a common problem observed not only with modern ribosomes, but in DNA and RNA synthesis as well. Binding of an incoming tRNA to a triplet other than the proper A-site codon can override the ribosome's definition of the decoding site and lead to reading-frame shifts in either direction (Atkins et al. 1979; Weiss et al. 1988). Disruption of codon:anticodon pairing can lead to drop-off events where the peptidyl-tRNA is lost from the ribosome (Menninger 1976). Codon:anticodon disruption can also result in tRNA hopping, in which the tRNA reanneals with the mRNA at a distant but synonymous site (Weiss et al. 1987). An awareness of the potential catastrophic effects of framing errors and drop-off events encouraged the early speculators in theories of ribosome-free translation to postulate comma-free codes, such as RRY, along with augmentation of codon: anticodon stability with extra base-pairing and coaxial base-stacking; such arguments are still valid for ribosome-based origin models.

This sequence specificity of non-triplet events involving the normal apparatus is mirrored in extragenic frameshift suppressors. Frameshift suppressors comprise a wide spectrum of alterations in tRNA, rRNA, ribosomal proteins, elongation factors, and release factors (Atkins et al. 1991). The canonical example is an addition of one nucleotide to the anticodon loop of the Gly_1-tRNA from *Salmonella typhimurium*, which gives this tRNA an enhanced capacity for translocating a glycine codon four nucleotides (Riddle and Carbon 1973). Thus, in an important sense, reading frame is determined by the location and character of codon: anticodon pairs. However, the view that a seven-nucleotide anticodon loop forms three-nucleotide codon:anticodon pairs and thus translocates triplets, whereas eight nucleotide loops form four nucleotide pairs and thus translocate quadruplets, is at best incomplete. The fact that reading frame maintenance is affected by the alteration of such a variety of translation components suggests that the link between tRNA:mRNA binding, translocation, and framing is indirect and complicated.

A common misinterpretation of certain findings is that the mRNA must be passively translocated by virtue of its base-pairing with tRNA, which must be the entity on which the translocation machinery acts. This is by no means certain. The extent and character of codon:anticodon pairing is certainly the paramount determinant of reading frame, but it does not follow from this which member of the pair, tRNA or mRNA, is the substrate for the machinery. Is the peptidyl-tRNA entering the higher affinity P site and dragging the mRNA along, or is a generic mRNA translocase moving tRNA from one ribosomal site to another by virtue of its pairing with the message, or is the mRNA:tRNA complex the substrate? Demonstrations of low levels of in vitro poly(U)-, (A)-, or (U,C)-dependent synthesis of polypeptides by *E. coli* ribosomes in the absence of elongation factors EF-Tu, EF-G, and GTP suggest that the ribosome contains the mechanism for catalyzing translocation (Rutkevitch and Gavrilova 1982). Further demonstration of EF-Tu-dependent *template-free* ribosomal synthesis of homopolymeric polypeptides suggests that peptidyl-tRNA is capable of translocation in the absence of codon: anticodon pairing and favors models in which the primary force moves tRNA and the mRNA moves as a consequence of its pairing with tRNA. However, in contrast to template-dependent synthesis, template-free synthesis is completely dependent on EF-G for translocation (Yusupova 1986). The implication is that translocation by the ribosomal (factor-independent) mechanism is dependent on codon:anticodon pairing, and therefore the tRNAs may move because the ribosome moves the mRNA or mRNA:tRNA complex.

mRNA sequence contexts have been found that elicit, in vivo, a wide range and degree of non-triplet translocations (Weiss et al. 1990b). Analysis of naturally occurring mRNA sequences that provoke high-level (>10% of the ribosomes alter their reading frame) ribosomal frameshifting has prompted the formulation of a new term for these phenomena: recoding, or reprogrammed genetic decoding. A general principle that defines two components of most recoding signals is a site of action delimited by the boundaries defining the passage of the ribosome from the zero-frame into the new frame and a stimulatory signal that enhances the level of these events (Gesteland et al. 1992). The sites of action for high-level frameshifts comprise both slippages and hops. The slippages are one-nucleotide frame changes and involve either a single tRNA, probably a P-site tRNA, slipping 3′ on the mRNA, as is the case with the *E. coli* gene for release factor 2 and the *Saccharomyces cerevisiae* Ty1 AB gene, or two tRNAs slipping 5′ on the mRNA within a heptameric sequence that allows simultaneous slippage of adjacent tRNAs, as is the case in many retroviral *gag-pol* genes (Fig. 3). The retroviral sequences elicit a similar frameshift when translated by *E. coli* ribosomes, implying that a common problem underlies this framing mishap.

A severe example of non-triplet translocation is observed in the bypass mechanism for translating the mRNA for bacteriophage T4 gene *60* (Huang et al. 1988; Weiss et al. 1990a). Here, the peptidyl-tRNA dissociates from the mRNA and reengages at an identical codon located 50

<u>A.</u> <u>E.coli</u> Release Factor 2 +1 recoding signals: (efficiency = 30%):

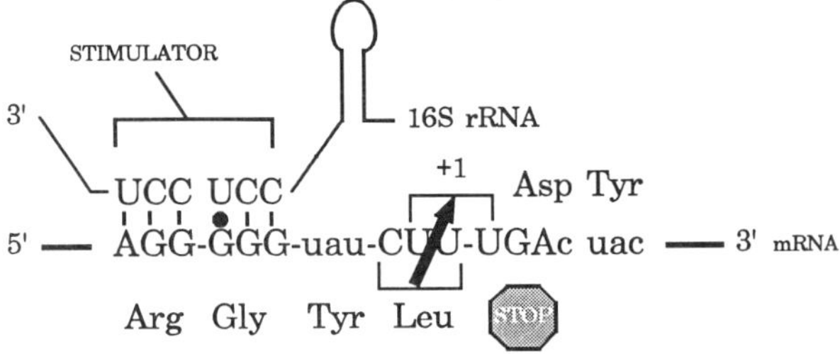

A. Coronaviral -1 recoding signals: (efficiency = 50%):

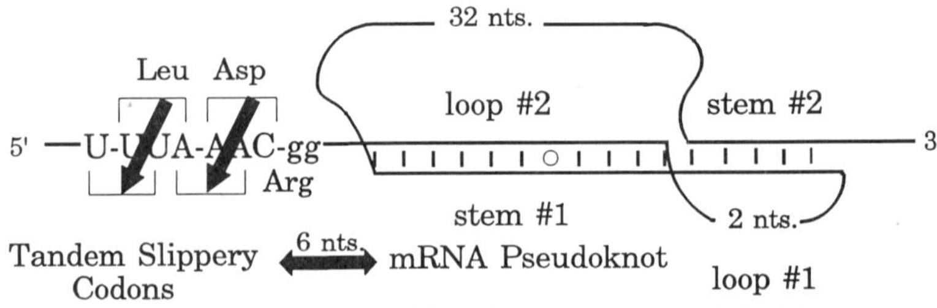

Figure 3 Modern reading frameshifts. The sequence and flanking structure requirements are shown for two examples of naturally occurring, high-level ribosomal frameshifts.

nucleotides 3′ of its original codon. The case of gene *60* bypass and the other cases of tRNA-hopping events suggest that there may be a mechanism in which elongating *E. coli* ribosomes propel themselves along mRNA independent of tRNA:mRNA pairing, perhaps analogous to the scanning mechanism used by eukaryotic 40S subunits during initiation.

The sequence contexts that stimulate the well-understood cases of recoding events provide some unique examples of how ribosomes currently contend with secondary structure in mRNA. That translating ribosomes melt mRNA stem-loop structures before such structures can occlude the decoding sites has been deduced from models of in vivo attenuation control of biosynthetic operons in enteric bacteria (Landick and Yanofsky 1987). In these cases, a dynamic coupling between the movement of RNA polymerase and a trailing ribosome is coordinated with the speed of ribosomal transit across a stretch of several codons. The ribosomal transit time controls the formation of alternate stem-loop structures simply by the position of the ribosome relative to the stems. One conclusion from these systems is that ribosomes are melting stem structures before they reach the decoding sites, and that the "melting" center may be positioned at a defined distance from the A site.

A different measure of the distance between the A site and the melting center is obtained from the effects of sequence context that stimulate the retroviral class of frameshifting (Fig. 4). The stimulator, in most cases, is a pseudoknot whose distance from the A site (measured as a 6- or 7-nucleotide spacer between the 3′nucleotide of the A-site codon and the first nucleotide of the pseudoknot stem) is essential for activity. The location and structure of the pseudoknot apparently cause the P- and A-site-bound tRNAs to slip one nucleotide 5′on the mRNA, and the sequence rules at the shift site suggest that some degree of base-pairing is required in both frames. How the encounter of the ribosome with the pseudoknot structure is linked to the rephasing event at the decoding site is an interesting, but unanswered, question. One possibility, shown in Figure 4, is that entry of the pseudoknot into the melting center on the 30S prevents the completion of a full triplet swing during an attempted translocation step. Repositioning of the mRNA, stabilized by the overlapping base-pairing sites for both tRNAs, would add one nucleotide to the spacer region, slackening the line between the codon:anticodon pairs and the pseudoknot, and thus permitting the tRNAs freedom to dock into their sites.

Another example of the effect of structure on framing is seen in an interaction that occurs 5′of the decoding site. In the release factor 2 gene

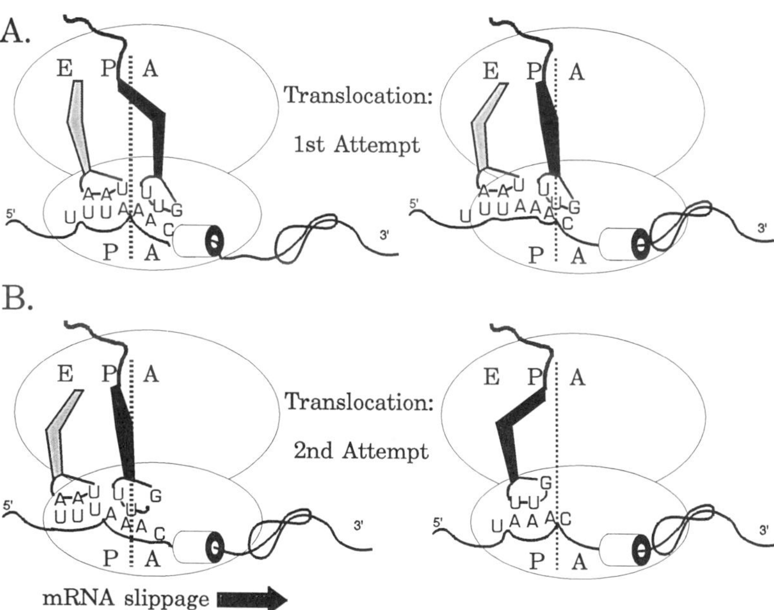

Figure 4 Postulated mechanism for pseudoknot-induced frameshifting. (*A*) First translocation attempt is jammed by pseudoknot structure binding in the 30S melting site. (*B*) Second attempt is successful because both tRNAs have slipped by one nucleotide toward the 5′end of the mRNA.

(Fig. 3), P-site tRNA slippage is stimulated by base-pairing between the mRNA and a sequence at the 3′ end of 16S rRNA, an interaction identical to the Shine-Dalgarno pairing involved in translation initiation. This base-pairing of the 3′ end of 16S rRNA occurs continuously during the elongation phase, forming and melting with different complementary sites on the mRNA. The formation of this short helix 5′ of the decoding sites can either enhance or suppress frameshifting at distal sites depending on the direction of the frameshift and the distance from the site (Weiss et al. 1990b). A final illustration of the effect of secondary structure on tRNA:mRNA pairs is seen in the requirement for a short, stable stem-loop at the 5′ junction of the gene *60* coding gap. The takeoff codon from which the ribosome begins the 50-nucleotide translocation event is flanked 3′ by a small, but stable, stem-loop that is essential for maximal bypass. It is not clear what phase of the bypass event requires this structure, but one model postulates that the short, compact structure can reform at the A site after having been melted initially, and that this structure plus the other sequence requirements (of which there are four distinct categories) conspire to move the tRNA by 50 nucleotides relative to the mRNA template.

PRIMITIVE TRANSLOCATION

The above discussion serves to illustrate the complexity of the functions necessary for modern framing and translocation as a starting point for a speculation about the origins of a primitive translocation mechanism. RRY codes solve some of the concerns about frame maintenance in the face of strand slippage and also reduce the need to melt template secondary structure, since RRY minimizes self-complementarity. However, it is a partial theory for the origin of translocation that is of immediate concern, regardless of the nature of the first codon:anticodon complexes. We start our speculation by proposing that the ancestor of small subunit RNA was an RNA replicase that used oligonucleotides as a substrate. More specifically, we imagine that this replicase was a predator living high on the food chain of the RNA world, with its prey being other oligoribonucleotides that it consumed during its self-replication. Such a replicase would have already evolved solutions to processive elongation along a template strand using sequence-specific, sequential binding of a substrate oligonucleotide (a.k.a. decoding). The unwinding of duplexed template RNA is a prerequisite for both replication and translation, and we imagine that the strand separation activity was first selected in a replicase. Additionally, we propose that variants of this replicase, defective in one portion of the polymerization cycle, were the progenitors of the small ribosomal subunit.

The first surprise from this conjecture is the direction of polymerization. All modern ribosomes translate mRNA in a 5′ to 3′ direction, just as all modern RNA and DNA polymerases elongate in a 5′ to 3′ direction. However, a ribosome moves from the *5′ to 3′* end of its template strand, whereas RNA and DNA polymerases elongate from the *3′ to 5′* end of their template strand. Consequently, our hypothetical replicase elongates in a direction opposite to that of all known template-directed polynucleotide polymerase, but in the same direction that ribosomes move relative to mRNA. The substrates and reaction cycle of this hypothetical replicase are shown in Figure 5.

There is no obvious chemical reason that polynucleotide synthesis must proceed in a 5′ to 3′ direction, or that 3′ ends involved in synthesis must be free hydroxyls and 5′ ends must be phosphorylated. Modern DNA polymerases, which utilize deoxyribonucleotide 5′ triphosphates, may synthesize in a 5′ to 3′ direction because, given this directionality, the exonucleolytic removal of an incorporated base (i.e., proofreading) regenerates the original (3′-OH) end. The base-catalyzed hydrolysis of

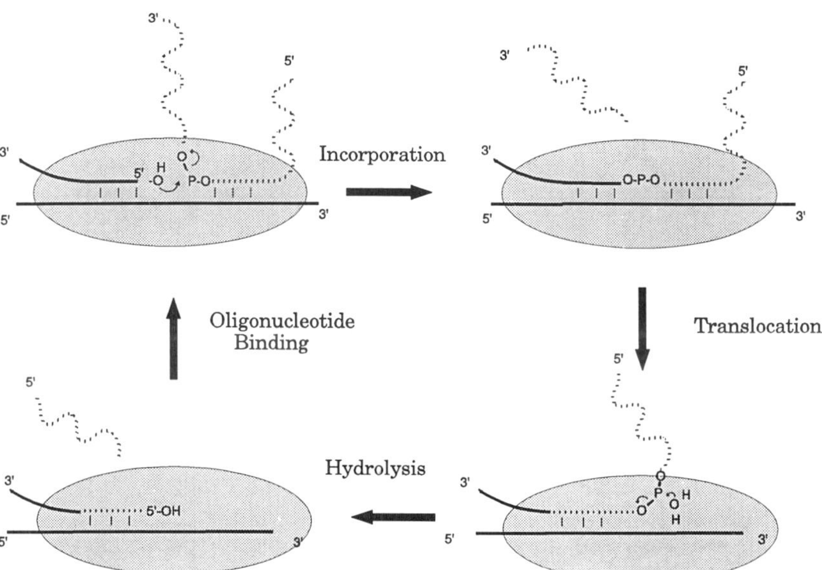

Figure 5 Proposed model of a processive RNA-dependent RNA polymerase that uses oligonucleotide substrates. Although the phosphoester transfer reactions are shown as single nucleophilic substitutions, they may have proceeded through a more complicated pathway, perhaps involving 2′,3′-cyclic phosphate intermediates. In each cycle, three nucleotides from the middle of an oligonucleotide substrate are added to the 5′ end of the growing chain. Two phosphodiester bonds are broken and one is formed.

RNA, as well as the hydrolysis catalyzed by many RNases, yields a free 5′ hydroxyl and a 2′ or 3′ phosphate. Use of the 5′-OH group rather than the 3′-OH group also solves the problem of suppression of reaction at the 2′-OH group (Orgel 1986). The nucleophilic attack of the 5′-OH group on the phosphodiester bond would incorporate the bound oligonucleotide into the growing chain. In a second reaction, hydrolysis of a phosphodiester bond generates the 5′-OH necessary for chain extension and provides some driving force for the process. Although the phosphoester transfers are shown as single nucleophilic replacements, either or both might proceed through a 2′,3′-cyclic phosphate intermediate, just as do base- and RNase-A-catalyzed hydrolysis. An oligonucleotide-consuming RNA replicase capable of processive template-dependent elongation, including unwinding of template secondary structure, would have solved a very difficult problem, and this solution may have been adapted for the decoding and translocation activities of a primitive translation apparatus.

Several authors have proposed that oligoribonucleotides, rather than mononucleotide derivatives such as NTPs, were the raw material for self-replicating RNAs (Sharp 1985; Orgel 1986; Doudna and Szostak 1989). There are at least two motivations for such proposals. Certain reactions catalyzed by derivatives of self-splicing introns are viewed as polymerase-like, and these reactions involve an oligonucleotide as the donor of nucleotide units. However, we see no reason to insist that a primitive replicase had an activity which is easily derivable from a modern catalytic RNA. A second reason for postulating the use of oligonucleotides is the difficulty in explaining the presence of NTPs in any abundance. The presence of random oligomers is also a problem, but it is one which has already been presumed to be "solved"; the pool of random oligomers from which the first replicators emerged could also have been the raw material for replication. Autocatalytic replicases using activated mononucleotides may have also existed (Pace and Marsh 1985), given a source of substrates, and such replicases may have coexisted with oligonucleotide replicases and may have, in some sense, been preyed upon by them.

A model for replication should be thermodynamically as well as kinetically sound. That is, the mechanism proposed should involve a net reaction which, under relevant conditions, will proceed in the "forward" direction. One view of the energetics of standard biological nucleic acid polymerization is that the formation of the new high-energy phosphoester bond is driven by the destruction of two such bonds. The α-β bond is broken upon addition of the (d)NTP to the growing chain, and the (former) β-γ bond is broken by the hydrolysis of the released pyrophosphate.

The two reactions are coupled by virtue of the fact that pyrophosphatase acts only on the released pyrophosphate, not on the unreacted (d)NTP. The models proposed for the use of oligonucleotides as substrates for replication involve the destruction of only one phosphodiester bond for each one formed. The reactions in question could certainly be driven forward by a vast excess of random oligomers. To assert that random oligomers were always present in vast excess to a self-replicator is to assert that the "replicator" never emerged in abundance from the random mix, i.e., that it was no self-replicator after all. One possible driving force for replication is the formation of extensive secondary and tertiary structure by the product of replication. Since the "intended" product would presumably be an enzyme, or at least the complement of an enzyme, extensive internal base-pairing and other interactions would be expected. Such interactions would stabilize the product relative to random oligomers and also would serve to displace the newly made strand from the template strand. The degree of stabilization would depend on temperature, ionic conditions, and, of course, the number and nature of the interactions. It is difficult to say whether such stabilization would be sufficiently large, or even how large would be "sufficient." The coupling of chain elongation to some favorable reaction would be an appealing, perhaps indispensable, part of any model of replication.

The model for replication shown in Figure 5 includes a driving force for chain elongation. The hydrolysis of the 5′ phosphodiester bond makes the "forward" reaction favorable; two phosphodiester bonds are broken for each one formed. Chain growth is coupled to the hydrolysis reaction because hydrolysis occurs only after translocation, which occurs only after the addition reaction. The block to "premature translocation" could be either kinetic or thermodynamic (or both). That is, there might be no low-energy path for translocation of unreacted complex, or binding of the unreacted complex in the translocated arrangement might be poor. Modern processive polymerases obviously perform a similar feat; they move relative to the template only after addition to the growing chain. Hydrolysis of the 5′ tail is dependent on translocation simply because of the location of a catalytic site. Translocation puts the relevant phosphate into a site that catalyzes hydrolysis. This catalytic center might be identical to the one that catalyzes chain elongation, as the two phosphoester transfer reactions are quite similar.

The first step in the conversion of such a polymerase to a 30S-like activity (Fig. 6) would be the loss of the addition activity. Loss of a catalytic activity is an easy change to get, although retention of the hydrolysis activity would complicate things if the two reactions were catalyzed by the same site. A low level of aberrant, non-incorporation-

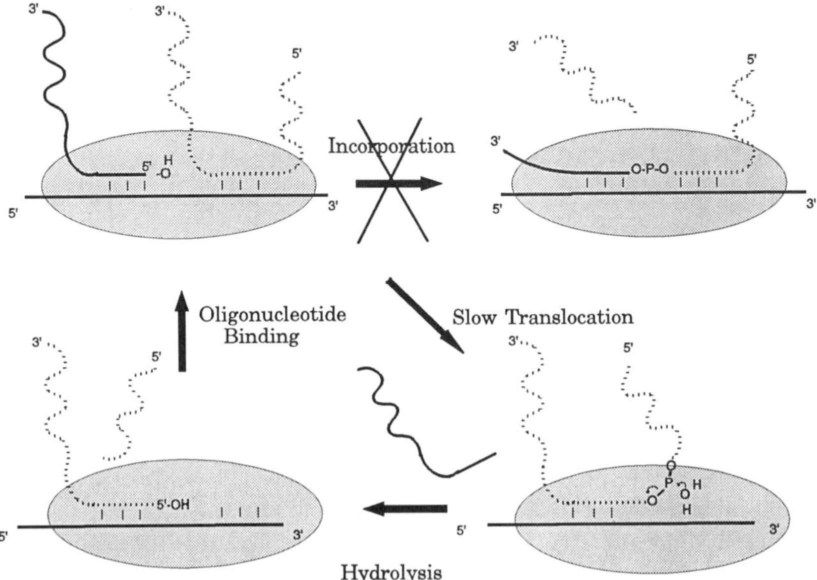

Figure 6   The "Highly Questionable" origin of the 30S subunit. A variant polymerase is disabled for the oligonucleotide incorporation event. Translocation proceeds, still driven by the hydrolysis event.

dependent translocation might be sufficient to get things started. Alternatively, a second mutation may have permitted such translocation, or one mutation could have had both effects. Translocation would be followed by the hydrolysis reaction, which would drive the whole process. Singly clipped oligonucleotide would leave following translocation. The net activity of such machinery would be the binding, in a base-pair-specific manner, of an incoming RNA, translocation along the template RNA, and binding of a new incoming RNA with retention of the previously bound RNA, all driven by the hydrolysis of a phosphodiester bond. In short, the mutant polymerase would perform the functions that seem to be associated with the modern 30S subunit. The RNA movements accomplished are similar to those accomplished by ratcheting, but the movements are given direction by coupling to a favorable reaction in a clearly defined way.

Nothing has been said concerning the origin of peptidyl transferase, which one might think of as the primordial large ribosomal subunit. It may be that some sort of template-independent synthesis of homopolymers or simple heteropolymers existed well before the advent of template-directed synthesis, although it is uncertain what use such polymers might have been. The primordial peptidyl transferase may have been anything from a free-floating, nonprocessive, fairly generic catalyst of acyl transfer to an elegant, multiactivity complex acting on covalently bound substrates in a manner analogous (but not homologous) to modern eukaryotic fatty acid synthesis. It may have acted directly on small activated amino acid derivatives (such as aminoacyl-AMP), or it may have acted on substrates covalently bound either to the enzyme complex itself or to carrier RNAs. Such carrier RNAs, if they existed, would have served a tRNA-like role in non-template-directed synthesis and could have been used in the first template-directed synthesis. Such RNAs might have, as part of their role in template-less synthesis, catalyzed their own aminoacylation. The presence of the synthetase on the tRNA would solve part of the problem of charging specificity. It may be that the "breakthrough" polymerase mutation merely altered the specificity of binding so that the aminoacylated RNAs, rather than random oligomers, were the preferred substrates. The aminoacylated RNAs may well have been poor substrates for the incorporation reaction, either as an incidental consequence of their structure, or, more appealingly, as a result of selection against the destruction of such metabolic intermediates.

CONCLUSIONS AND PERSPECTIVES

The origin of translation remains a notoriously difficult problem. In the turbulent year of 1968, it was realized that there should have been an RNA world, but these early speculators didn't have a leg to stand on, since "RNA and DNA are ... fit mainly for reproduction (with a little help from proteins) but of little use for much of the really demanding work" (Crick 1981). Now, with modern examples of catalytic RNA, they have several legs on which to stand. The only question that remains is how to walk out of the RNA world toward the light of protein synthesis without stepping in anything nasty. As we, and others, have shown, it is not an easy task.

REFERENCES

Atkins, J.F., R.F. Gesteland, B.R. Reid, and C.W. Anderson. 1979. Normal tRNAs promote ribosomal frameshifting. *Cell* **18:** 1119–1131.

Atkins, J.F., R.B. Weiss, S. Thompson, and R.F. Gesteland. 1991. Towards a genetic dissection of the basis of triplet decoding, and its natural subversion: Programmed reading frame shifts and hops. *Annu. Rev. Genet.* **25:** 210–228.

Bretscher, M.S. 1968. Translocation in protein synthesis: A hybrid structure model. *Nature* **218:** 675–677.

Crick, F.H.C. 1968. The origin of the genetic code. *J. Mol. Biol.* **38:** 367–379.

———. 1981. *Life itself, its origin and nature.* Simon and Schuster, New York.

Crick, F.H.C., S. Brenner, A. Klug, and G. Pieczenik. 1976. A speculation on the origin of protein synthesis. *Origins Life* **7:** 389–397.

Dontsova, O., S. Dokudovskaya, A. Kopylov, J. Rinke-Appel, N. Junke, and R. Brimacombe. 1992. Three widely separated positions in the 16S RNA lie in or close to the ribosomal decoding region; a site-directed cross-linking study with mRNA analogues. *EMBO J.* **11:** 3105–3116.

Doudna, J.A. and J.W. Szostak. 1989. RNA-catalyzed synthesis of complementary-strand RNA. *Nature* **339:** 519–522.

Eigen, M. 1987. New concepts for dealing with the evolution of nucleic acids. *Cold Spring Harbor Symp. Quant. Biol.* **52:** 307–320.

Eigen, M. and P. Schuster. 1979. *The hypercycle, a principle of natural self-organization.* Springer-Verlag, Berlin.

Frank, J., P. Penczek, R. Grassucci, and S. Srivastava. 1991. Three-dimensional reconstruction of the 70S *Escherichia coli* ribosome in ice: The distribution of ribosomal RNA. *J. Cell. Biol.* **115:** 597–605.

Gallant, J.A. and D. Foley. 1980. On the causes and prevention of mistranslation. In *Ribosomes: Structure, function and genetics* (ed. G. Chambliss et al.), pp. 615–638. University Park Press, Baltimore, Maryland.

Gesteland, R.F., R.B. Weiss, and J.F. Atkins. 1992. Recoding: Reprogrammed genetic decoding. *Science* **257:** 1640–1641.

Gilbert, W. 1986. Origin of life, the RNA World. *Nature* **319:** 618.

Hardesty, B., O.W. Odom, and J. Czworkowski. 1990. Movement of tRNA through ribosomes during peptide elongation. In *The ribosome: Structure, function and evolution* (ed. W.E. Hill et al.), pp. 366–372. American Society for Microbiology, Washington, D.C.

Huang, W.M., S. Ao, S. Casjens, R. Orlandi, R. Zeikus, R. Weiss, D. Winge, and M. Fang. 1988. A persistent untranslated sequence within bacteriophage T4 DNA topoisomerase gene 60. *Science* **239:** 1005–1012.

Kurland, C.G. and J.A. Gallant. 1986. The secret life of the ribosome. In *Accuracy in molecular processes* (ed. T.B.L. Kirkwood et al.), pp. 127–157. Chapman and Hall, London.

Landick, R. and C. Yanofsky. 1987. Transcription attenuation. In Escherichia coli *and* Salmonella typhimurium (ed. F.C. Neidhardt et al.), pp. 1276–1301. American Society for Microbiology, Washington, D.C.

Lim, V., C. Venclovas, A. Spirin, R. Brimacombe, P. Mitchell, and F. Muller. 1992. How are tRNAs and mRNA arranged in the ribosome? An attempt to correlate the stereochemistry of the tRNA-mRNA interaction with constraints imposed by the ribosome topography. *Nucleic Acids Res.* **20:** 2627–2637.

Menninger, J.R. 1976. Peptidyl transfer RNA dissociates during protein synthesis from ribosomes of *Escherichia coli. J. Biol. Chem.* **251:** 3392–3398.

Noller, H.F. 1991a. Ribosomal RNA and translation. *Annu. Rev. Biochem.* **60:** 191–227.

———. 1991b. Drugs and the RNA world. *Nature* **353:** 302–303.

Noller, H.F., D. Moazed, S. Stern, T. Powers, P.N. Allen, J.M. Robertson, B. Weiser, and K. Triman. 1990. Structure of rRNA and its functional interactions in translation. In *The ribosome: Structure, function and evolution* (ed. W.E. Hill et al.), pp.73–92. American Society for Microbiology, Washington, D.C.

Orgel, L.E. 1968. Evolution of the genetic apparatus. *J. Mol. Biol.* **38:** 381–393.

———. 1986. RNA catalysis and the origins of life. *J. Theor. Biol.* **123:** 127–149.

Pace, N. and T.L. Marsh. 1985. RNA catalysis and the origin of life. *Origins Life* **16:** 97–116.

Prince, J.B., B.H. Taylor, D.L. Thurow, J. Ofengand, and R.A. Zimmermann. 1982. Photochemical crosslinking of tRNA^{lys} and tRNA^{glu} to 16S RNA at the P site of *Escherichia coli* ribosomes. *J. Biol. Chem.* **254:** 4745–4749.

Rheinberger, H.J., U. Geigenmuller, A. Gnirke, T.P. Hauser, J. Remme, H. Saruyama, and K. Nierhaus. 1990. Allosteric three-site model for the ribosomal elongation cycle. In *The ribosome: Structure, function and evolution* (ed. W.E. Hill et al.), pp. 318–330. American Society for Microbiology, Washington, D.C.

Riddle, D.L. and J. Carbon. 1973. Frameshift suppression: A nucleotide addition in the anticodon of a glycine transfer RNA. *Nature* **242:** 230–234.

Rutkevitch, N.M. and L.P. Gavrilova. 1982. Factor-free and one-factor-promoted synthesis of polypeptides in cell-free systems from *Escherichia coli. FEBS Lett.* **143:** 115–118.

Sharp, P.A. 1985. On the origin of RNA splicing and introns. *Cell* **42:** 397–400.

Shepherd, J.C.W. 1981. Method to determine the reading frame of a protein from the purine/pyrimidine genome sequence and its possible evolutionary justification. *Proc. Natl. Acad. Sci.* **78:** 1596–1600.

Spirin, A.S. 1970. A model of the functioning ribosome: Locking and unlocking of the ribosome subparticles. *Cold Spring Harbor Symp. Quant. Biol.* **34:** 197–207.

———. 1985. Ribosomal translocation: Facts and models. *Prog. Nucleic Acid Res. Mol. Biol.* **32:** 75–114.

Thach, S.S. and R.E. Thach. 1971. Translocation of messenger RNA and "accommodation" of fmet-tRNA. *Proc. Natl. Acad. Sci.* **68:** 1791–1795.

von Ahsen, U., J. Davies, and R. Schroeder. 1991. Antibiotic inhibition of group I ribozyme function. *Nature* **353:** 368–370.

Watson, J.D. 1964. The synthesis of proteins upon ribosomes. *Bull. Soc. Chem. Biol.* **46:** 1399–1425.

Weiner, A.M. and N. Maizels. 1987. 3′ Terminal tRNA-like structures tag genomic RNA molecules for replication: Implications for the origin of protein synthesis. *Proc. Natl. Acad. Sci.* **84:** 7383–7387.

Weiss, R.B., W.M. Huang, and D.M. Dunn. 1990a. A nascent peptide is required for ribosomal bypass of the coding gap in bacteriophage T4 gene 60. *Cell* **62:** 117–126.

Weiss, R.B., D.M. Dunn, J.F. Atkins, and R.F. Gesteland. 1987. Slippery runs, shifty stops, backward steps and forward hops: −2, −1, +1, +2, +5 and +6 ribosomal frameshifting. *Cold Spring Harbor Symp. Quant. Biol.* **52:** 687–694.

———. 1990b. Ribosomal frameshifting from −2 to +50 nucleotides. *Prog. Nucleic Acid Res. Mol. Biol.* **39:** 159–183.

Weiss, R., D. Lindsley, B. Falahee, and J. Gallant. 1988. On the mechanism of ribosomal frameshifting at hungry codons. *J. Mol. Biol.* **203:** 403–410.

Wintermeyer, W., R. Lill, and J.M. Robertson. 1990. Role of the tRNA exit site in ribosomal translocation. In *The ribosome: Structure, function and evolution* (ed. W.E. Hill et al.), pp. 348–357. American Society for Microbiology, Washington, D.C.

Woese, C.R. 1967. *The genetic code.* Harper and Row, New York.

———. 1970. Molecular mechanics of translocation: A reciprocating ratchet mechanism. *Nature* 226: 817–820.

———. 1980. Just so stories and Rube Goldberg machines: Speculations on the origins of the protein synthetic machinery. In *Ribosomes: Structure, function and genetics* (ed. G. Chambliss et al.), pp. 357–373, University Park Press, Baltimore.

Wower, J., S.S. Hixson, and R.A. Zimmerman. 1989. Labeling the peptidyltransferase center of the *Escherichia coli* ribosome with photoreactive tRNA[Phe] derivatives containing azidoadenosine at the 3′ end of the acceptor arm: A model of the tTNA-ribosome complex. *Proc. Natl. Acad. Sci.* **86:** 5232–5236.

Yusupova, G.Z., N.V. Belitsina, and A.S. Spirin. 1986. Template-free ribosomal synthesis of polypeptides from aminoacyl-tRNA. *FEBS Lett.* **206:** 142–145.

4

Probing RNA Structure, Function, and History by Comparative Analysis

Carl R. Woese
Department of Microbiology
University of Illinois
Urbana, Illinois 61801

Norman R. Pace
Department of Biology
Indiana University
Bloomington, Indiana 47405

Life on this planet is a profusion of incredibly complex systems. From the biologist's perspective it is indeed fortunate that all these systems have sprung from a common ancestor. They have by their nature retained traces of their ancestries, and so are similar — homologous — to a greater or lesser extent. The science of biology has been built from its beginning upon cataloging similarities and differences among different living systems (and various states of the same system), the method that has become known as comparative analysis. It is only through this simple and sometimes tedious approach that biologists could begin to understand the complexity with which they are confronted, could begin to distinguish the important elements from the unimportant, and so reduce living systems to a set of understandable essentials. It is also through such comparisons, through measuring similarity and difference in degree and kind, that biologists have come to learn the genealogical relationships among all organisms.

Despite its compelling utility, comparative analysis has never been a commonly used tool in molecular biology. This is not because comparative analysis is without value at the molecular level: The secondary structures of the ribosomal RNAs, major accomplishments of modern molecular biology, are tribute to the comparative approach (Woese et al. 1980; Noller et al. 1981). The molecular biologist's aversion to comparative analysis would seem to lie in the molecular paradigm itself. Molecular biology arose from chemistry and physics. The entities with which these sciences deal, atoms and their relatively simple chemical combinations, do not have meaningful histories. Consequently, the con-

The RNA World
© 1993 Cold Spring Harbor Laboratory Press 0-87969-380-0/93 $5 + .00

ceptual and experimental outlook inherited from chemistry and physics effectively has no comparative, historical dimension. This may also explain why molecular biologists have tended to view evolution (apart from the origin of life issue) as a trivial part of biology, as merely a collection of relatively uninteresting historical accidents: No matter that evolution is the essence, the *sine qua non* of biology.

Our objective in this chapter is to review the various ways in which the comparative analysis of RNA at the molecular level has contributed and will contribute to biology. Comparative analysis contributes to molecular biology in four important ways: (1) in defining constraints and patterns from which molecular structure can be inferred; (2) as a necessary background for, and adjunct to, experimental analysis of molecular function and structure; (3) as the essence of genealogical analysis and classification; and (4) in providing a general conceptual and organizational framework for much of future biological research. The changes that have occurred in microbiology over the past decade bear witness to the purifying, sharpening effect that a phylogenetic/comparative framework has on experimental design and interpretation of results. There can be little doubt that the forthcoming deluge of genomic sequence information can only be handled effectively in a comparative framework.

COMPARATIVE ANALYSIS AND NUCLEIC ACID STRUCTURE

History: Molecular Biology's Flirtation with Comparative Analysis

The potential of a comparative approach to RNA higher-order structure became evident at the 1966 Cold Spring Harbor Symposium. By that time, the sequences of four tRNAs were known. The sequences reported had not been determined initially with a comparative approach in mind, but the fact that all four tRNA sequences could assume the same "clover leaf" configuration convinced most of those present at the meeting that all tRNAs had a common secondary structure. However, the general principle, that comparative analysis is an effective tool in the analysis of molecular structure, seems not to have been grasped, for the lesson had to be repeated when it came to 5S rRNA structure. Here, too, the initial attempts to determine secondary structure had not involved a systematic, comparative approach. Rather, structures were suggested on the basis of maximizing of base-pairing, for instance, or some other ingenious "first principle." The resulting structures tended to be awkwardly complex and unattractive (for review, see Erdmann 1976). A systematic comparative approach was eventually applied in this case, and the true structure of 5S rRNA then began to emerge (Fox and Woese 1973).

When the sequence of the small subunit rRNA was in the process of being determined, the comparative lesson seemed once more to have been forgotten. Again, the search for secondary structure initially consisted of merely looking for stretches of nucleotides with the potential to form base pairs. The credibility of the resulting structures did not seem to be an issue. There is even an unsubstantiated rumor (worth repeating) that when Brosius, Noller, and their colleagues submitted the first complete 16S rRNA sequence (that of *Escherichia coli*) for publication, one of the reviewers questioned why they had not included the molecule's secondary structure as well! By this time, however, the comparative lesson had to some extent been assimilated, and Noller's group soon teamed up with Woese's group to produce a second sequence, for the specific purpose of using comparisons to identify among the ten thousand or so possible helical elements in any given small subunit rRNA a manageable number of probable ones (Woese et al. 1980). By the time the large subunit rRNA was sequenced (Brosius et al. 1980), it was generally accepted that comparative analysis was essential to determining its secondary structure, and the appropriate experimental steps, determination and comparison of sequences from different organisms, were taken to realize this (Noller et al. 1981).

Using the Comparative Approach to Analyze RNA Structure

Today, comparative analysis has become the method of choice for establishing higher-order structure for large RNAs. Regardless of their other virtues, none of the more direct, physical-chemical approaches can provide so detailed a picture of the relationships among bases in large RNAs.

Comparative analysis starts with the alignment of sets of homologous sequences. In its most stringent definition, "homology" refers strictly to common ancestry. The simply stated objective of alignment is to juxtapose related sequences so that the homologous residues in each occupy the same column in the alignment. Although there have been significant advances in computer-generated alignment in recent years, the most precise method of alignment is still essentially a manual one (often computer-assisted). The reason for this is that a great deal more is involved in alignment than merely shifting nucleotides around until some maximum sequence similarity is achieved. Homology among individual sequence residues is best defined in the context of larger units of homologous structure and function. Therefore, the most useful and generally the most precise alignments are produced by iteratively invoking higher-order structure in the process. A practical condition imposed on sets of aligned sequences is that the sequences are arranged in an order that approximates their phylogenetic relationships to one another. This is the most useful context for comparative analysis and foreshadows a day when phylogenetic structuring will become an integral part of most if not all biological databases.

Given a phylogenetically ordered sequence alignment, one can begin to interpret Nature's evolutionary experimentation. Nature provides the results of those experiments that have worked, those in which molecular function remains optimized (within the limits defined by selective constraints); only rarely are we privy to natural experiments in which function has been adversely altered. Therefore, invariance in compositions of certain residues in a molecule identifies these as being important to the structure and function of the molecule. Conversely, areas of a molecule that frequently vary in composition from organism to organism, especially when length variation also occurs, mark themselves as probably of little or no direct functional significance. Rarely, a particular section of a molecule will become deleted, thereby identifying itself as some kind of functional or structural unit. Residues whose compositions covary (change in concert) must in some way be related, which in almost all cases means in direct physical contact.

Sequence comparisons reveal more than the simple existence of structural relationships, however. The pattern of a covariation may suggest the nature of an interaction. For example, if two positions in a sequence covary strictly according to the Watson-Crick rules (and cover all the allowable combinations), one can be fairly certain that they form a bona fide Watson-Crick base pair. On the other hand, a normal pairing geometry is unlikely in those cases exhibiting other than a canonical pattern of covariation. The pattern and frequency of variation at a given position is a far more subtle characteristic of that position than is generally appreciated. It is a refined measure of structural homology, for only if the full structural context, not merely the major contacts, is maintained from one group of organisms to the next, will the pattern (and frequency) of variation be similar in the groups.

Some Specific Examples Involving Ribosomal RNA

A principal use of comparative analysis in the study of rRNA has been to infer secondary structure. As used here, secondary structural elements are contiguous stretches of two or more nucleotides that form canonical (or G:U type) antiparallel pairs with one another. Comparative "proof" of such structures involves finding multiple phylogenetically independent instances in which the compositions of two potentially paired positions change in concert, according to the Watson-Crick pairing rules. Figure 1, taken from the recent report of Gutell et al. (1993), gives some idea of what comparative analysis has accomplished in terms of rRNA structure, and the extent of the comparative evidence that supports the various helices. About 60% of the nucleotides in the small subunit rRNA (Fig. 1) are involved in confirmed secondary structure, a proportion comparable to the 55% of nucleotides in tRNA that are known from its crystal structure to be involved in secondary structure (Kim 1979).

Comparative analysis shows that noncanonical pairs occur frequently in rRNAs, and that they occur in a variety of pairing geometries. The U:G type of pair is, of course, the most common of the noncanonical pairs, the next most common being the A:G type (Gutell et al. 1993). From their patterns of variation it is apparent that rRNA contains several different kinds of U:G pairs. One kind exhibits frequent compositional variation over the phylogenetic spectrum, but U:G remains its major composition. Perhaps the most striking example of this kind can be seen in the helix 829-40:846-857 shown in Figure 1. This structure, which typically comprises 10–12 pairs, almost always shows 5 or more U:G pairs in its central section. The helix is highly variable in composition, with U:G pairs often being replaced by other pairs (frequently G:U), and the exact location of the U:G pairs is somewhat variable (Gutell et al. 1993). It seems evident that at least some U:G pairs in rRNA have unique structural or functional significance.

A second kind of U:G pair is identified by the fact that its composition varies only slowly over evolutionary time, and when it does so, it almost always (or always) converts to a C:A pair (Gutell et al. 1993). This pattern of variation suggests a non-Watson-Crick pairing geometry.

Another interesting pattern of covariation is shown by the U:U pair that converts solely to C:C (and vice versa), again suggestive of an atypical pairing geometry (Gutell et al. 1993). Although the pattern of covariation does not allow us to infer the physical conformation of any pair with certainty, it does favor certain possibilities. For example, a U:U pairing that alternates (and is presumably isomorphic) with a C:C pairing could involve nucleotides in a syn-anti conformation. Direct physical measurement is needed to ascertain this.

One of the first unusual higher-order structures in the small subunit rRNA to be uncovered by comparative analysis was the so-called pseudoknot (Woese et al. 1983). Pseudoknot here describes a topology in which a stretch of nucleotides within a hairpin loop pairs with nucleotides external to that loop. Three such pseudoknots are now known in the small subunit rRNA. These involve the pairing of positions 17-

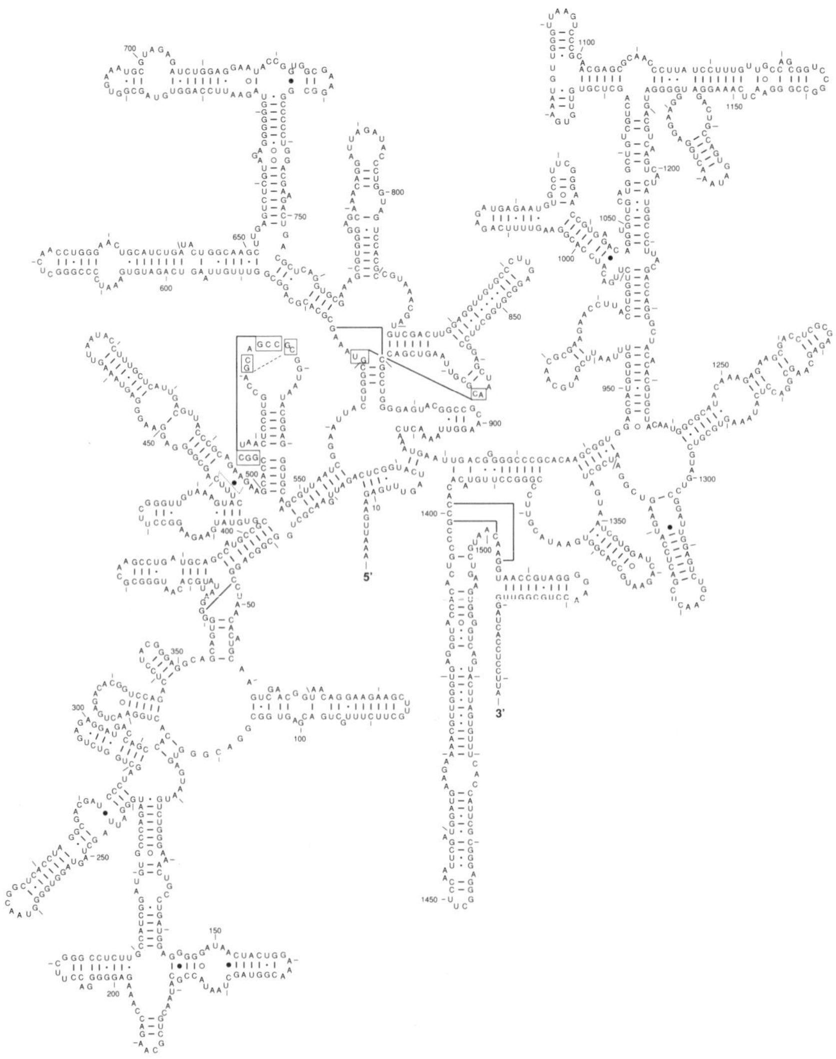

Figure 1 Higher-order structure of *E. coli* 16S rRNA (Gutell et al. 1993). As discussed in the text, this folded structure is based on comparative analysis. Watson-Crick pairs are connected with lines, G:U pairs are connected with filled circles, and A:G pairs are connected with open circles. Bases that are not juxtaposed as pairs, but connected, have yet to be specifically proven phylogenetically. Boxed regions pair as indicated by the connecting lines. Every 10th position is indicated with a line and each 50th position is numbered.

20:915-918, 505-507:524-526, and 570-571:865-866 (Fig. 1). In all three instances, the pseudoknot pairings tend to be nearly invariant in composition, suggesting that their three-dimensional context is stringently defined and that these structures are important to ribosome structure/function. Some of the proposed pseudoknot structures have been experimentally tested by mutagenesis: In all cases in which mispairings have been introduced, ribosome function was impaired, if not totally eliminated (Powers and Noller 1991).

Comparative analysis also suggested that the area of the small subunit rRNA between positions 500 and 545 (which contains the second of the three pseudoknot structures) is of major importance because of its highly conserved sequence and overall length (Woese et al. 1983). This particular structure is highly constrained. Of the 46 bases comprising it, only 14 are not known to be involved in secondary or tertiary interactions (see Fig. 1). Mutational analysis and rRNA probing studies show this region to be of functional significance (Noller et al. 1990; Powers and Noller 1991). The fact that the region's known structure is entirely self-contained (confined to these 46 nucleotides) makes it an excellent candidate for isolation and characterization by physical methods.

Another important RNA structural element identified by rRNA sequence comparisons is the so-called tetraloop (Woese et al. 1990b), a loop of four nucleotides that terminates (caps) a double-helical stem of two or more base pairs. Such structures account for the majority of all hairpin loops in rRNA (Gutell et al. 1993). The most interesting characteristic of tetraloops is that the sequence in the loop is highly constrained: The naturally occurring examples are confined to an extremely limited subset of the 256 possible permutations of the four nucleotides. The compositions of the first and last bases in the loop are strongly coordinated, and to a lesser extent, the compositions of the inner two bases are as well. Three motifs cover the overwhelming majority of the naturally occurring tetraloops in rRNA; UNCG, CUYG, and GMRA (Woese et al. 1990b). In the small subunit rRNA, the predominant compositions for each of the three major types are UUCG, CUUG, and GMAA (Woese et al. 1990b). The composition of the terminal base pair in the underlying stem is related to the sequence in the loop, especially for two of the three loop compositions: the UUCG loop strongly favors a C:G closure; CUUG loops are usually closed by a G:C pair, and the closing pair for the GCAA loop, although less constrained, tends to be R:Y (Woese et al. 1990b).

Some of the tetraloops in the small subunit rRNA are invariant or change only slowly in sequence over evolutionary time, whereas others change with remarkable frequency. In both cases, the variation is far from random, with the vast majority of variants tending to conform to one of the above three general types. The most frequently changing of the small subunit rRNA tetraloops is that located at positions 83-86 (Fig. 1). More than 50 phylogenetically independent alterations in loop sequence have been recorded among the (eu)Bacteria. The loop has a UUCG, CUUG, or GCAA sequence in 93% of cases (Woese et al. 1990b). Of the remaining 7%, most are themselves tetraloops whose compositions are related to one of the above three. In only 3% or so of cases does the size of the loop vary; and then, by addition or deletion of a single nucleotide (Woese et al. 1990b). The UUCG loop at position 83-86 shows a C:G closing pair 91% of the time, the CUUG loop here closes with a G:C pair in 95% of cases, and 86% of the GCAA loop closures have an R/Y composition (mainly A:U). Solution NMR spectroscopic analysis of two of the characteristic tetraloops, C(UUCG)G and C(GMAA)G (bases outside the parentheses form the closing pair), suggest structures in which the two terminal bases in the loop itself interact to form an unusual, noncanonical pair, which would seem to explain their rather strict covariation (Heus and Pardi 1991; Varani et al. 1991).

Tetraloops of the types described above are not confined to rRNAs; they clearly serve various other functions in the cell. They have been reported to occur in the context of controlling gene expression in T4 phage (Tuerk et al. 1988). What actual function(s) tetraloops serve in rRNA beyond favorable structures with which to cap helices remains a mystery. The fact that their compositions vary frequently at some locations in the molecule and infrequently, or not at all, at others suggests functional differentiation.

The Scope and Utility of Comparative Analysis

Comparative sequence analysis utilizes a highly abstract representation of a molecule, a mere string of symbols representing the four nucleotides. In its own right, comparative analysis reveals only patterns involving these symbols; it says nothing about actual molecular structure (or function). However, when combined with structural knowledge derived from the more direct physical methods of measurement, comparative analysis takes on physical meaning, and so, becomes a more powerful and useful approach to analysis of structure. The potential of this link between comparative analysis and actual structure determination should be fully appreciated and exploited. When the physical chemist pays attention to the results of comparative analysis, it helps him/her to uncover biologically important and physically interesting molecular structures. What is not generally recognized is that as the physicist's repertoire of biologically important structural motifs increases, the potential for comparative analysis of primary structure to reveal the existence of physical structure increases proportionately. Thus, comparative analysis is increasingly useful as an adjunct to physical and chemical determinations of molecular structure.

A fine example of the interplay between physical characterizations and comparative analysis is given by recent work on G:A-type pairs. There are many instances in which otherwise canonical helices contain GpA doublets paired (antiparallel) with GpA, but never ApG paired with ApG. Such a bias among natural RNAs indicates a profound difference between the two types of pairing. Based on these comparative data, Turner, Wilson, and their respective colleagues have examined thermodynamic and structural properties of RNA and DNA oligonucleotide duplexes containing the two types of paired doublets (SantaLucia et al. 1990; Li et al. 1991). They found that pairs formed from GpA are as energetically favorable as canonical pairs would be, but that the ApG pairing is unstable. Spectroscopic analysis revealed, moreover, that the more stable GpA pairing has a novel (non-Watson-Crick) geometry.

In a similar way, comparative analysis can serve as a guide in molecular genetic analyses. Much of the molecular biologist's approach to understanding and utilizing molecular structure and function turns upon manipulation of genetic sequences. Since sequence space is enormous, there is no way the molecular biologist can efficiently explore it without some kind of map. Taking random mutational "shots" at the ribosome, for example, is a most unproductive way to attempt to uncover the molecular basis for ribosome function. Comparative analysis can provide an initial guide, can give essential clues that turn an intractable task into a manageable one. The following are some examples.

A common need in the study of macromolecules is the identification of features responsible for an activity; for instance, the structural elements involved in catalysis by ribozymes. Sequences responsible for the catalytic activity of the so-called hammerhead self-cleaving RNA (see Pan et al., this volume) could be identified by their conservation throughout a collection of catalytic "satellite" RNAs (and their complements) with otherwise extremely variable sequences (Forster and Symons 1987a). Deletion analysis then demonstrated that the catalytic structure could be reduced to the only common structure in all the RNAs, about 50 nucleotides (Forster and Symons 1987b). Similar sequence comparisons coupled with deletion analyses have revealed functional aspects of self-splicing introns (see Cech, this volume).

RNase P is the enzyme that cleaves 5'-precursor sequences from precursor forms of tRNA (see Cech, this volume). The RNase P holoenzyme in vivo is a protein-RNA complex, but in vitro, at high ionic strength, the RNA alone catalyzes the reaction (Guerrier-Takada et al. 1983). Design of a simplified RNase P RNA required an extensive comparative study in order to define the key homologous residues, all of which were not immediately evident because the molecule is generally quite variable in length as well as in sequence, which makes alignment in certain areas problematic. Figure 2, which compares the RNase P RNAs from *E. coli* and *Bacillus megaterium*, illustrates the point. The two differ by a number of structural insertion/deletion events. However, a detailed comparative analysis of a number of such sequences permitted the definition of what appeared to be an essential core structure, common to all versions of the molecule (James et al. 1988). To test the theory that the catalytic activity of this ribozyme is contained in this homologous core, a simplified RNase P RNA, called Min 1 RNA, comprising just that core, was designed and synthesized (Waugh et al. 1989).

The *E. coli* RNase P RNA was used as the foundation of the design because its structure has fewer interruptions in the conserved core than do those of its counterparts from the gram-positive bacteria (i.e., *B. megaterium* in Fig. 2). There are four proposed helices in the *E. coli* RNase P RNA model that have no counterparts in the gram-positive versions of the structure, and therefore, are probably not essential for enzymatic activity. These were excluded from the "minimal" RNase P RNA design in one of two ways: complete omission, or replacement of *E. coli* sequence segments with their homologs from the gram-positive version of the molecule (*B. megaterium*) that lacks the *E.coli*-specific helical elements. A few other changes in the *E. coli* version were introduced at the ends of the synthetic RNA to optimize it for production by in vitro transcription (Waugh et al. 1989).

The strategy of intergeneric replacement for some structures was used because of concern that in the absence of a more detailed knowledge of the structure, tinkering with the molecular length of the *E. coli* sequence might not properly restore the remnant secondary and tertiary structure. Consequently, as indicated in Figure 2, the regions of the *E. coli* RNA

Figure 2 Ribonuclease P RNAs: design of Min 1 RNA and sites of cross-linking with substrate phosphate. As described in the text, secondary structures of the RNase P RNAs of *E. coli* and *B. megaterium* are based on phylogenetic comparisons (James et al. 1988; Haas et al. 1991). Watson-Crick pairs are shown with lines, non-Watson-Crick pairs with filled circles. Arcs connect long-range pairs. In the design of Min 1 RNA, the regions of the *E. coli* RNA indicated by italics were replaced with the corresponding, but shorter, sequences from the *B. megaterium* RNA, indicated by italics in that folded structure. An additional, indicated region of the *E. coli* sequence was deleted from Min 1 RNA. Arrows indicate nucleotides that are cross-linked to an arylazide associated with the 5′-phosphate of tRNA (Burgin and Pace 1990), as discussed in the text. Arrows on the *B. megaterium* folded structure are based on analyses with the RNase P RNA of *B. subtilis*.

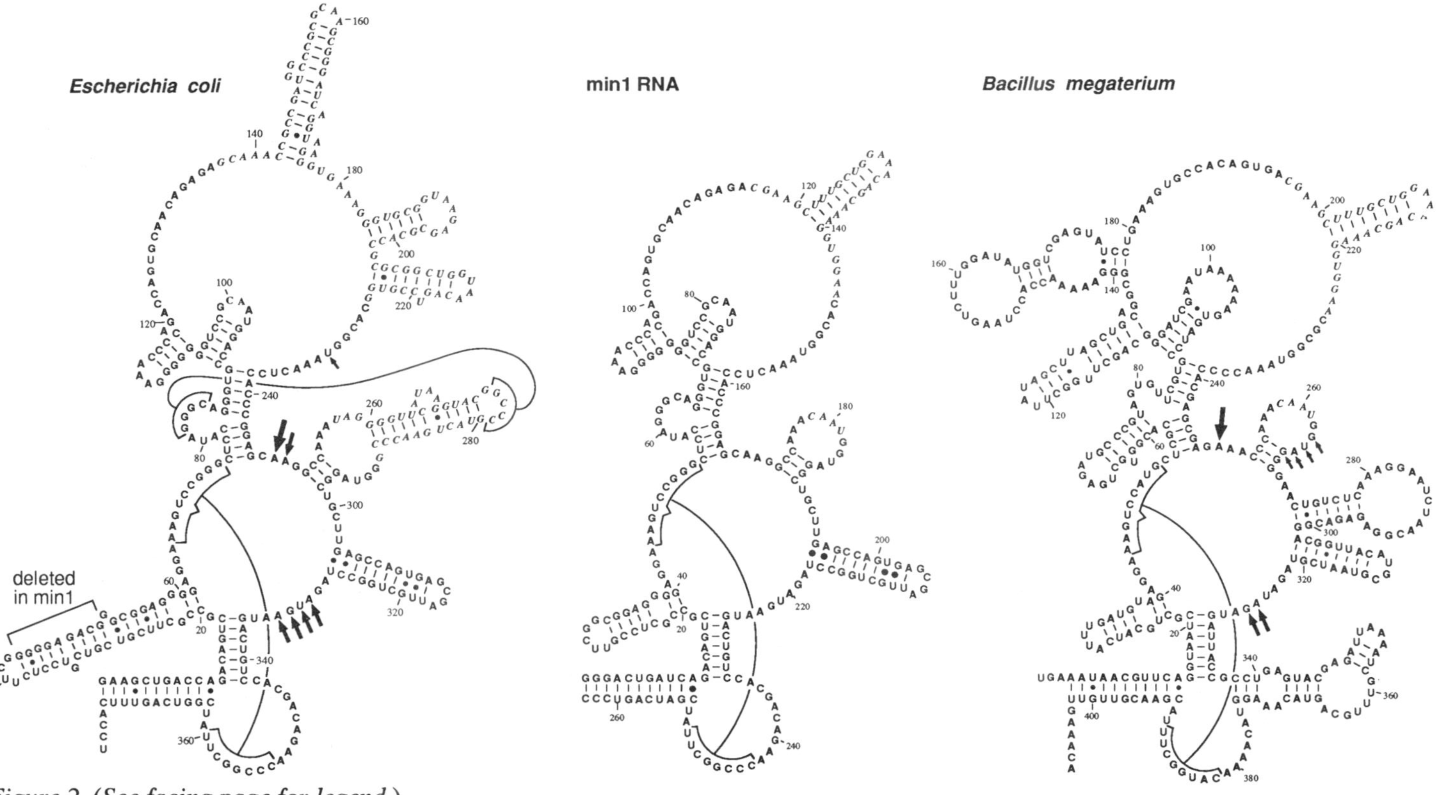

Figure 2 (*See facing page for legend.*)

that were significantly longer than the gram-positive versions were deleted or, alternatively, replaced with shorter *B. megaterium* sequences that in the context of the rest of the structure model would occupy the same positions. This approach presumed that the superstructure of the synthetic RNA would accommodate the smaller putative homolog without steric distortion of the catalytically active core of the molecule, which would be expected of truly homologous structures and which provides a stringent test of the RNase P RNA structure model. The synthetic RNA, produced by in vitro transcription, proved to be highly active, with a catalytic efficiency (k_{cat}/K_m) 20% that of the native *E. coli* RNase P RNA and a turnover rate (k_{cat}) 20-fold higher than native RNA (Waugh et al. 1989). It is unlikely that a successful, piecemeal removal of 30% of the molecular length of the native RNA could have been accomplished without approaching the problem from a comparative perspective.

Comparative Analysis as Interpretive Perspective

The interpretations of experimental results in molecular biology are often ambiguous because of the complexity of the systems and the consequent difficulty of differentiating meaningful from trivial information. Results derived from chemical cross-linking, footprinting, and nuclease or chemical structure-mapping experiments, for example, contain a wealth of information. However, it is often impossible to distinguish completely the significant data from the background of trivial data (idiosyncratic for the particular organism or an artifact of the particular experimental method) that invariably accompanies them. Such data are commonly presented with little attempt to evaluate their meaning and significance. Comparative studies with homologous molecules from different organisms add depth and precision to the interpretation of these kinds of data. Comparisons identify which elements in the data set correspond to general characteristics, universal properties, at the same time pointing out the possibly trivial results, the idiosyncratic ones.

One example of using the comparative approach in this manner involves a cross-linking analysis of the active site of RNase P RNA (Burgin and Pace 1990). In the study, an arylazide photoaffinity cross-linking agent was attached to the 5′-terminal phosphate in tRNA, the phosphate that is acted upon by RNase P. Ultraviolet irradiation cross-links the substrate to RNase P RNA, and sites of cross-linking can be identified by primer extension analysis. Three different RNase P RNAs, from *E. coli*, *B. subtilis*, and *Chromatium vinosum*, were used in the study. The three RNAs differ extensively in sequence and in the presence or absence of some structural elements (as indicated in part in Fig. 2).

Several nucleotides in each RNA formed cross-links with the photo-agent-containing substrate. However, only a subset of these cross-linked nucleotides was found to be common to all three tRNAs (Fig. 2). These are likely to be the most fruitful candidates for future study.

RNA AS HISTORICAL RECORD

The availability of rRNA sequences has allowed quantitative analysis of evolutionary relationships. Tapping the phylogenetic information in rRNA has had a revolutionary effect on microbiology, adding an evolutionary dimension where there had been none before. The prior lack of an evolutionary framework in microbiology was the unavoidable consequence of the fact that morphologies and physiologies of microorganisms are too simple and/or unpredictably variable to be of significant value in developing a natural classification. It is no wonder that, using those properties, microbiologists had been unable to produce a valid, natural (phylogenetic) classification for the bacteria. However, as Zuckerkandl and Pauling (1965) pointed out, at the molecular level there is no such problem; molecular sequences are historical records rich in readily interpretable genealogical information. Thus, it was possible, by comparison of partial sequences of 16S rRNAs (oligonucleotide catalogs), to make initial sense of microbial genealogical relationships (Fox et al. 1980; Woese et al. 1985). The resulting phylogenetic framework, the natural classification, brings clarity and order to the day-to-day conduct of microbiology. Experimental progress is enhanced; deeper interpretations of results are now possible; new directions of research are indicated. Microbial ecology has come of age: the capacity to identify microorganisms phylogenetically using rRNA sequence-based techniques, even without cultivation (Pace et al. 1985) and in situ (DeLong et al. 1989), gives microbial ecologists a power they have always lacked for comprehensive analysis of various niches. Not only can organisms be identified with greater precision and characterized in greater depth, but unculturable or previously unrecognized organisms in a particular niche can now be definitively related to those in other niches.

The early studies of microbial phylogeny based on rRNA sequences yielded the remarkable finding that the world of prokaryotes, which all biologists had taken to be phylogenetically unified (monophyletic), was indeed not so. There exist two distinct kinds of prokaryotes, now formally called the domains Archaea and Bacteria (Woese et al. 1990a), that are no more closely related to one another than either is to the eukaryotes. Although the early oligonucleotide cataloging approach could readily define and distinguish three primary groupings of organisms (Archaea,

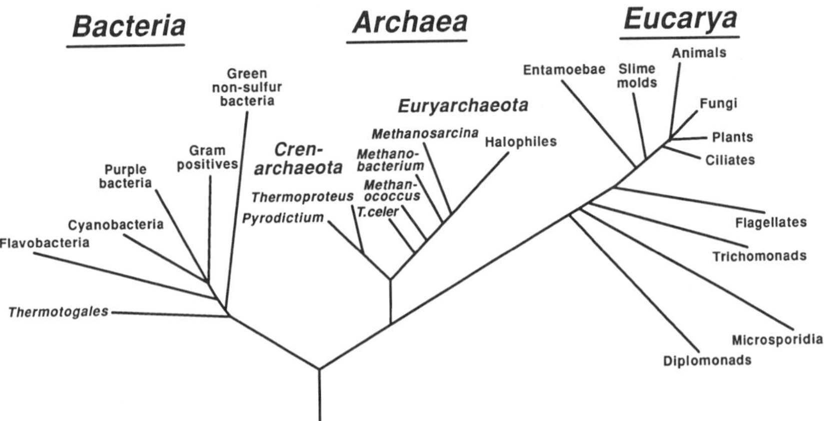

Figure 3 Rooted universal phylogenetic tree. The root of the tree was established as discussed in the text by comparing pairs of paralogous genes whose common ancestor predates the most recent universal ancestor (Gogarten et al. 1989; Iwabe et al. 1989).

Bacteria, and Eucarya), it could not relate them to one another in any precise way. This joining had to await the technology that permitted determination of complete sequences of rRNAs (or their genes), which in turn allowed construction of a universal phylogenetic tree. This tree is shown in Figure 3 in rooted form; its root has been inferred by the so-called Dayhoff strategy. In this method, an unrooted tree generated from a set of (homologous) sequences is rooted using a related (or paralogous) sequence. To root a tree that spans all extant life, it is necessary that the gene duplication that originally produced the paralogous genes occurred in the ancestral stem, prior to the initial phylogenetic radiations. The root of the universal tree is seen to lie between the Bacteria on the one hand and the common lineage that the Archaea and Eucarya share, on the other (Gogarten et al. 1989; Iwabe et al. 1989). The universal tree, in other words, predicts that the Archaea are specific relatives of the Eucarya (albeit at a deep level), to the exclusion of the Bacteria.

rRNA Definition of the Three Domains

rRNAs unequivocally define and distinguish the three primary phylogenetic domains. The rRNA in each domain has a characteristic sequence and higher-order structure. In terms of the small subunit rRNA, any sequence within one domain is more similar to all others in that domain than it is to any sequence from another domain (Woese 1987). A small subunit rRNA sequence signature that permits ready definition and distinction of the three domains is shown in Table 1 (Winker and Woese

1991). One example of a higher-order structural feature used to distinguish among the domains is the 500-545 region of Figure 1, mentioned above. The noteworthy feature is the bulge loop (nucleotides 505-510) that protrudes from the upstream side of the helical stem, whose overall length (number of base pairs) is the same in all three domains. Among the Bacteria, the loop in question arises after the fifth pair in the stalk and comprises six nucleotides (of a characteristic composition). In the Archaea and Eucarya the loop contains seven nucleotides (again of a characteristic composition) and arises after the sixth, not the fifth, pair of the stalk (Woese et al. 1983; Woese and Gutell 1989). Since there exist characteristic archaeal, bacterial, and eukaryal versions for just about every universal macromolecular function so far characterized in the cell, there can be no doubt that all life on this planet is organized into three, very distinctive groupings (Wheelis et al. 1992).

The Archaea

From a phenotypic perspective, the Archaea are a rather strange and disparate collection. Unlike the Bacteria, they show only a few major phenotypes: the methanogenic; the extremely halophilic; the sulfate reducing; and the sulfur-"dependent," extremely thermophilic phenotype (Woese 1987; Winker and Woese 1991). These four phenotypes are sufficiently dissimilar that, although they were known (except for the sulfate reducing phenotype) before the Archaea were recognized as a phylogenetic unit, their relationship to one another was not recognized. (This despite the existence of some molecular evidence suggestive of the relationship; e.g., the unusual archaeal ether-linked lipids [Winker and Woese 1991]). The four main archaeal phenotypes do not define four major taxa of equivalent taxonomic rank, however. As shown in Figure 4, the Archaea comprise two major branches, two "kingdoms," the Euryarchaeota and the Crenarchaeota (Woese et al. 1990a). The latter taxon is phenotypically uniform; its species are all of the sulfur-dependent, thermophilic type. In contrast, the former encompasses examples of all four archaeal phenotypes, with methanogens dominating the group. The methanogens comprise three major groups in addition to a very deeply branching lineage represented by the genus *Methanopyrus* (Burggraf et al. 1991). Two of the three main methanogen lineages are phenotypically uniform, but the third, the *Methanomicrobiales* lineage, has spawned other phenotypes as well: the extreme halophiles, the sulfate reducers, and perhaps *Thermoplasma*. The remaining euryarchaeal lineage is the phylogenetically compact cluster of species that constitute the *Thermococcales*, a group phenotypically resembling the crenarchaeotes.

Table 1 Small subunit rRNA sequence signatures defining the three domains, Archaea, Bacteria, and Eucarya

Position(s) of base or pair[a]	Bacteria		Archaea		Eucarya	
	composition	percent[b] of total	composition	percent of total	composition	percent of total
8	A	99	U	95	Y[c,d](C)	98
9:25	G:C*	99	C:G	100[e]	C:G	100
10:24	R(A):U	100	Y:R	100	U:A	96
33:551	A:U	100	Y(C):R(G)*	100	A:U	100
52:359	Y:R	96	G:C	98	G:C	96
53:358	A:U*	97	C:G	100	C:G	100
113:314	G:C*	99	C:G	100	C:G	100
121	Y(C)	97	C	100	A	98
292:308	G:C	99	G:C	100	R:U	94
307	H	99	G	95	Y(U)	100
335	C	99	C	100	A*	96
338	A	100	G*	100	A	100
339:350	C:G*	100	G:Y(C)	98	C:G	98
341:348	C:G	99	C:G	100	U:A	96
361	R(G)*	100	C	100	C	98
365	U	100	A	98	A	100
367:393	U:A	100	C:G*	93	U:A	98
377:386	R(G):Y(C)	99	Y(C):G	100	Y(C):R(G)	96
500:545	G:C	100	G:C	100	U:A*	94
514:537	Y(C):R(G)*	100	G:C	100	G:C	98
549	C	100	U	98	C	94
558	G	99	Y	95	A	96
569:881	Y(C):R(G)	100	Y:R	100	G:C*	100
585:756	R(G):Y(C)*	100	C:G*	98	U:A	100

675	A	100	U	100	U	98
716	A*	100	C	100	Y(C)	100
867R(G)	98	Y(C)	95	Y	94	
880	C	98	C	100	U	100
884	U	99	U	100	G	98
923:1393	A:U	99	G:C	100	A:U	100
928:1389	G:C	96	G:C	100	A:U	98
930:1387	Y(C):R(G)*	100	A:U	98	G:C	96
931:1386	C:G*	98	G:C	100	G:C	98
933:1384	G:C*	99	A:U	100	A:U	100
962:973	C:G	96	G:C	100	U:G	98
966	G*	98	U	100	U	100
974	M	97	G	95	G	96
1098	Y(C)*	97	G	100	G	98
1109	C*	99	A	100	A	100
1110	A*	100	G	100	G	100
1194	U*	100	R(G)	100	R(A)	100
1201	M(A)	99	M(A)	100	U	96
1211	U	99	G*	100	Y(U)	100
1212	U*	99	A	100	A	94
1381	U*	99	C	100	C	100
1487	G	100	G	100	A*	96
1516	R(G)	98	G	100	U	98

Except where indicated, the compositions shown are invariant in each domain. Analysis based upon approximately 380 bacterial, 40 archaeal, and 50 eukaryal sequences. Those cases in which a signature composition is "pure," i.e., is not seen at all in the other two domains, are marked with an asterisk.

[a]Numbering follows *E. coli* 16S rRNA standard (Brosius et al. 1980). A few exceptions (among bacteria) are, therefore, designated as 99%.

[b]100% applies only to cases showing complete invariance; one or a few exceptions among the Bacteria is scored as 99%.

[c]Abbreviations used are: Y = U or C, R = A or G, M = A or C, H = C, U, or A.

[d]R:Y as used herein does not include A:C pairs.

[e]A, Archaea; B, Bacteria; E, Eucarya.

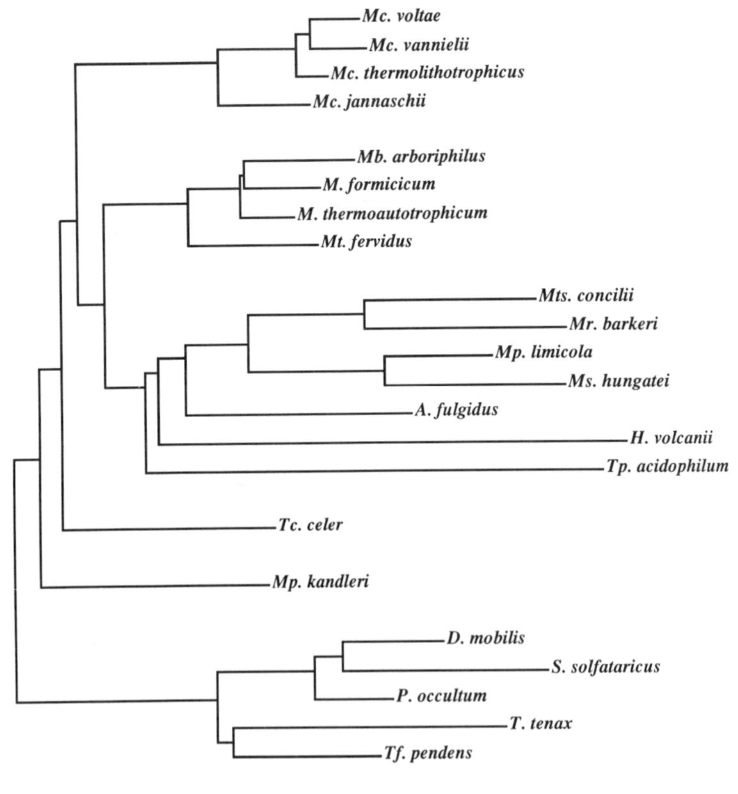

Figure 4 Evolutionary tree for the Archaea. The tree is based on transversion distances among small subunit rRNA sequences (Woese et al. 1991). Scale bar indicates five changes per 100 nucleotides.

The distribution of phenotypes on the tree of Figure 4 strongly suggests the Archaea to be of thermophilic origin. All known crenarchaeal species are thermophilic, some growing optimally at temperatures above 100°C; the deepest branchings on the euryarchaeal side, the *Thermococcales* and the genus *Methanopyrus*, are thermophilic as well. This is true for the deepest branchings within all the major euryarchaeal sublineages. For the most part, Archaea are anaerobic; ability to grow aerobically, when it occurs, is generally facultative. Similarly, the Archaea are most often chemoautotrophic, particularly the deeply branching lineages. Thermophily, anaerobic growth, and chemoautotrophy, then, can be considered ancestral characteristics of the Archaea.

The surprising and scientifically inviting relationship of the Archaea to the Eucarya (Fig. 3) deserves comment. The sequences of many, but not all, archaeal genes resemble their eukaryotic homologs decidedly more than their bacterial homologs. Ribosomal protein sequences are this way (Auer et al. 1989; Ramirez et al. 1989), as are the major subunits of the archaeal RNA polymerase (Pühler et al. 1989), as well as an archaeal histone (Sandman et al. 1990) and a chaperone (Trent et al. 1991). Although the archaeal rRNA overall is an exception to this rule (in sequence and secondary structure it most resembles the bacterial type), the most highly conserved positions in the archaeal sequence do show a slight eukaryotic bias in composition (see Table 1) (Winker and Woese 1991).

Two examples of the above relationship between archaeal and eukaryal proteins are of particular interest, namely, RNA polymerase and histone. The archaeal RNA polymerase is more similar in sequence to both of the eukaryal polymerases II and III than these two are to each other (Pühler et al. 1989). The case of the archaeal histone is even more striking. Its sequence is reported to be closer to each of the sequences of the four eukaryal histones, H2a, H2b, H3, and H4, than any of these four is to one another (Sandman et al. 1990). In other words, in these two examples, the archaeal version of a molecular type appears to resemble the inferred common ancestor of a eukaryal family of proteins more closely than does any of the extant members of that eukaryal family. Is this perhaps the tip of an iceberg? Do many families of eukaryotic genes have an archaeal homolog that resembles the (inferred) ancestor of that family more than its extant representatives do?

The Bacteria

Figure 5 shows the (eu)bacterial phylogenetic tree inferred from small subunit rRNA sequences. Anyone familiar with classic bacterial taxonomy will see immediately that the groupings defined by molecular sequence analysis bear little relationship to the groupings defined by classic methods (Woese 1987). Microbiologists previously grouped all phototrophs in a single taxon, which contained few if any nonphotosynthetic species. In actuality, photosynthetic and nonphotosynthetic phenotypes are often intimately intermixed: Note in particular the phylum of purple bacteria and relatives in Figure 5. Microbiologists formerly used morphology as a primary determinant of classification. Morphology turns out to be an extremely poor indicator of bacterial phylogenetic relationships, although there are a few notable exceptions, such as the spirochetes and the endospore-formers. Properties such as gliding motility,

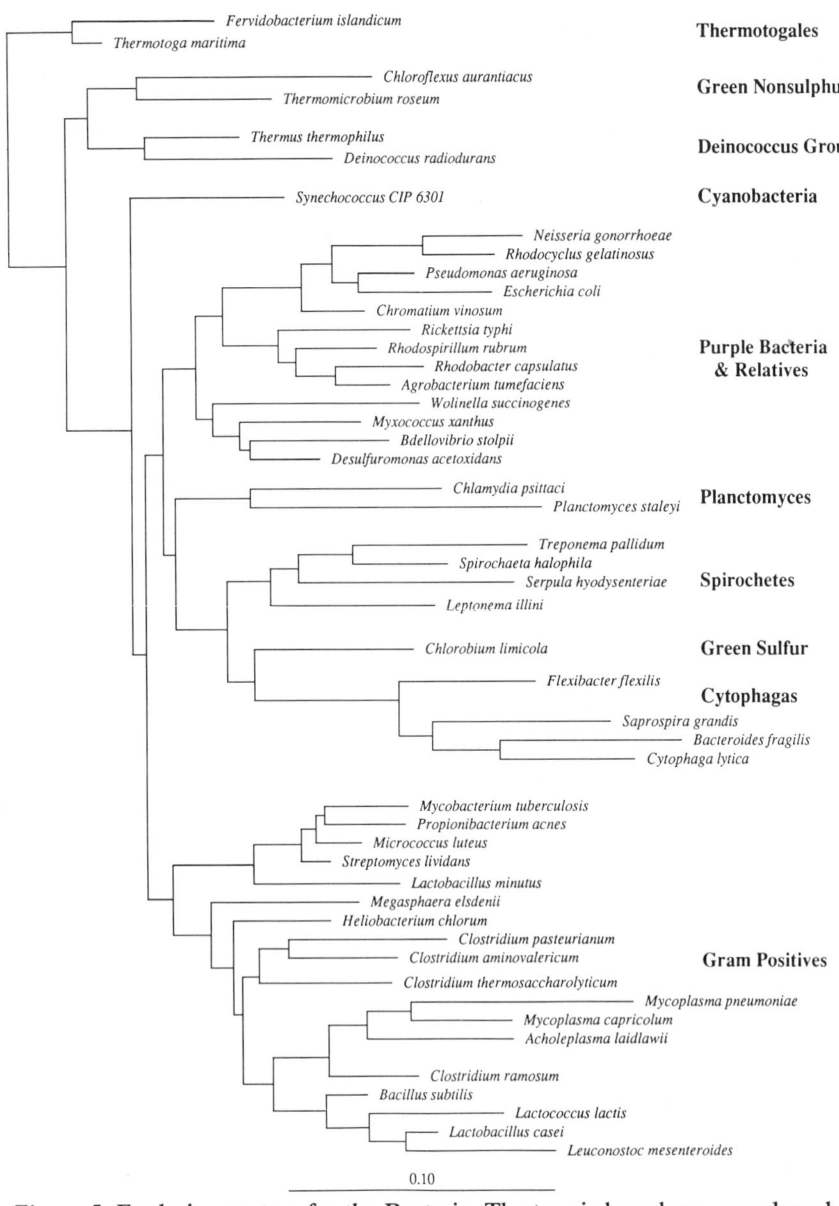

Figure 5 Evolutionary tree for the Bacteria. The tree is based on normal evolutionary distances among small subunit rRNA sequences (Weisburg et al. 1989). It is derived from a ~400-organism tree constructed by maximum likelihood (Olsen and Woese 1993). Scale bar indicates ten changes per 100 nucleotides.

formerly used to group organisms into a few high-level taxa, also are phylogenetically widely dispersed. The mycoplasmas, rickettsias, and a variety of other pathogens do not warrant the high-level taxonomic distinction previously accorded them on the basis of their parasitic character.

The distribution of phenotypes in the (eu)bacterial tree suggests the following. (1) The Bacteria arose from a thermophilic ancestor: Thermophily is widespread on the tree, dominant in its deeper branches (Achenbach-Richter et al. 1987; Woese 1987). (2) Unless there has been lateral transfer of the system, photosynthesis evolved at an early stage in the evolution of the Bacteria (Woese 1987): Note the deeply branching phototrophic genus *Chloroflexus* (Fig. 5). No evidence exists to support the old notion that life began heterotrophically and that, consequently, the first organisms were heterotrophs; if anything, bacterial phylogeny is more consistent with an aboriginal autotrophy (Woese 1987).

The Eucarya

The textbook picture of eukaryotes emphasizes four great "kingdoms": animals, plants, fungi, and protists. rRNA sequences tell a rather different story; see Figure 3. The animal, green plant, and fungal kingdoms are seen from that perspective to constitute some of the more superficial branchings on the tree, whereas the protist kingdom is a polyphyletic collection of (little-related) lineages, which together cover the full span of the eukaryal tree. Since mitochondria appear only among the higher branchings in the eukaryal tree, it is clear that the aboriginal eukaryotic cell possessed none, a conclusion consistent with the fact that the earlier branchings on the tree must have arisen at a time when there was little or no free oxygen on this planet. Moreover, the eukaryal lineage originated before the ancestors of mitochondria, the α-purple Bacteria and relatives (Yang et al. 1985).

So far, the lower eukaryal branchings are defined by parasitic species only. These offer few clues as to the conditions under which their presumed free-living ancestors lived, during the early, anaerobic phase in Earth history. Thus, the question of whether the eukaryotes, like the prokaryotes, arose from thermophilic ancestry cannot be addressed.

The Universal Ancestor

The universal ancestor of all extant life, sitting at the base of the universal tree, represents the limit of how far back in the evolutionary course rRNA sequences (or any other individual gene) can take us. The best that this "written record" can do is to underscore the importance of the problem of the universal ancestor and to leave us with the general question of its nature. The tree of Figure 3 (and the sequences underlying it) provide only a few clues here, which, unfortunately, cannot be interpreted with any certainty at this stage.

The fossil record, according to conventional interpretation, suggests that photosynthetic bacteria have existed for at least 3.5 billion years. If this is true, then the common ancestor of all Bacteria, and by implication the ancestors of the Archaea and Eucarya as well, existed at an even earlier time (see Fig. 3). In other words, life as we know it has existed, and the three primary branchings have been evolving, for perhaps 4 billion years, 90% of the earth's existence! The fact that each domain has a characteristic rRNA, and that the amount of divergence in rRNA sequence within any domain is slight by comparison to that between any two domains, means that the evolutionary process by which the rRNA of the universal ancestor was transformed into the rRNAs characteristic of each of the three domains was of a more drastic nature than the rRNA evolution that occurred subsequently within each of the three domains. Yet, this drastic type of evolution must have occurred over a far shorter time period than the subsequent, 3- or more billion year evolution within each domain. If other universally distributed genes also are more similar within the domains than between them, then it would be difficult to avoid concluding that the evolutionary events that transformed the universal ancestor into its primary descendant lineages occurred at a far more rapid pace than the subsequent evolution in each of the three primary lineages. It seems likely that cellular life as we know it could not survive an evolution of this pace and quality (although simpler entities such as viruses appear to do so). This, then, is the question with which the rRNA sequence record leaves us: Was the universal ancestor of all extant life a more rudimentary (simpler, less well-defined, and/or less tightly integrated) entity than its descendants, the common ancestors of each of the three primary lineages? This question is one of the most compelling rationales for the sequencing and comparison of a large number of prokaryotic and lower eukaryotic genomes. This will allow a definition of the aboriginal genome in terms of both the approximate number of genes and their types.

RIBOSOMAL RNA AS RELIC

What is perhaps the most interesting part of the grand evolutionary progression, the succession of ancestral states that culminated in the most recent universal ancestor, is left untouched by rRNA sequences. Although these earlier evolutionary stages are not accessible through rRNA sequences, the "written record," the "relic" information in rRNA may help us ultimately to explore those early events. Like all biological entities, the translation apparatus embodies, reflects, its evolutionary history and is not intelligible apart from that history. Thus, buried in the form and function of the translation apparatus are the relics of the complex and protracted history that led to the modern mechanism, which existed already in the ancestors of the three primary lineages. Reading this relic record is difficult, however; at present impossible, in good part because we do not yet understand how this marvelous molecular machine, the translation apparatus, carries out protein synthesis. Nonetheless, we make a few general comments pertaining to this problem.

The chicken-and-egg nature of the evolution of translation essentially demands the conclusion that the mechanism initially was RNA-based, and probably remains so today (Woese 1967). This conclusion has been strongly reinforced by the recent demonstration that the all-important peptidyl transfer reaction is mediated by rRNA, rather than protein (Noller et al. 1992). Thus, it seems inescapable that when the molecular mechanics of translation become known, they will be seen fundamentally to involve conformational transitions in RNA (Woese 1970, 1973). This expectation is not reflected in our current concept of translation, the essence of which is embodied in the so-called "A-site-P-site" theory—a conceptualization we believe to be flawed; one that has impeded rather than fostered understanding.

Our dissatisfaction with the A-site-P-site theory stems from several concerns. First, this pictorial representation of the translation apparatus is ill-defined in any molecular sense, and as a consequence, lacks the two main characteristics of a good theory, predictive power and explanatory power (Popper 1959). In other words, the theory is largely untestable (it is not falsifiable) and so does not "explain" translation in terms that would either help us to understand the process or to design new experiments that would do so. The theory is compatible with almost any fact, and in cases where it is not, biologists have made it so by adding assumptions (e.g., the assumption that the initiator tRNA enters the P site, not the A site).

Second, we believe that the imagery projected by the A-site-P-site theory overly constrains the ways in which we conceptualize translation. The tRNA pictured by this theory is a conformationally static, passive entity, something moved ("translocated") by an undefined process, from one site to another. The role of tRNA is only to recognize, to pair passively with, its corresponding codon(s). The ribosome (with its "translation factors"), on the other hand, becomes a black-box machine, which harbors the A and P sites and somehow supplies the mechanical movement the process requires. The A and P "sites," of course, have functional definitions, mainly in terms of the action of certain antibiotics that block translation at various stages. However, these defining phenomena are related to *stages* in the translation process, to *states* of a dynamic system, not merely to physical sites. When experimental findings are interpreted solely in terms of static sites, the data are deprived of their full meaning, their richness. Our point is, we do not know that any of the imagery, the relationships, and so on that are inherent in the A-site-P-site model are valid. We do not know that the tRNA molecule is conformationally static in the process; we do not know whether it is the ribosome or the tRNA that initiates or defines the movement and the state changes that must occur. It is counterproductive to be locked into imagery that automatically precludes interesting possibilities, that is neither a guide to experimentation nor a useful framework for understanding.

Our final criticism of the A-site-P-site theory is that it leads to an evolutionary impasse. Even if the proteins associated with translation are taken to be dispensable in the primitive process, the ribosome/tRNA mechanism that remains is still too complicated to have arisen as such. Yet, according to the A-site-P-site theory, some sort of ribosome must have been present to effect "translocation," to serve as the "molecular machine." A view of translation is needed in which this strict division of labor between dynamic ribosome and passive tRNA is not an essential feature. We need to entertain the idea that tRNA may function dynamically in the translation process. If indeed a dynamic tRNA were to be defining of major steps in translation, then one could consider a primitive process based solely on tRNAs or some abbreviated predecessors thereof (Woese 1970, 1973).

The ultimate theory of translation must be couched in terms of the molecular mechanics of the process, must be a physical-chemical model. Our task at present is to work toward such a theory, at the very least by ceasing to work within the stultifying confines of A-site-P-site. One simple means by which to start such a conceptual retrenchment is to stop using the word "site" and, so, thinking of translation in static terms. Translation is a process, a passage through a series of states. It is relatively simple to replace the static term "site" by the dynamic term "state," which is far more forgiving. We find a description such as "during peptidyl transfer the complex of the incoming tRNA with the 30S subunit remains in the A state, but that with the 50S subunit shifts to the P state," inviting to the imagination, whereas an attempt to frame these facts (Moazed and Noller 1989) in terms of "sites" is at very best uninspiring.

ACKNOWLEDGMENTS

The authors' research activities are supported by grants from the National Science Foundation and the National Aeronautics and Space Agency

(C.R.W.), and the National Institutes of Health and Department of Energy (N.R.P.). We are most grateful to Dr. Robin R. Gutell for providing Figure 1, and to Dr. James W. Brown for assistance with Figure 2.

REFERENCES

Achenbach-Richter, L., R. Gupta, K.O. Stetter, and C.R. Woese. 1987. Were the original eubacteria thermophiles? *Syst. Appl. Microbiol.* **9:** 34–39.

Auer, J., K. Lecher, and A. Böck. 1989. Gene organization and structure of two transcriptional units from *Methanococcus* coding for ribosomal proteins and elongation factors. *Can. J. Microbiol.* **35:** 200–204.

Brosius, J., T.J. Dull, and H.F. Noller. 1980. Complete nucleotide sequence of a 23S ribosomal RNA gene from *Escherichia coli. Proc. Natl. Acad. Sci.* **77:** 201–204.

Brosius, J., J.L. Palmer, J.P. Kennedy, and H.F. Noller. 1978. Complete nucleotide sequence of a 16S ribosomal RNA gene from *Escherichia coli. Proc. Natl. Acad. Sci.* **75:** 4801–4805.

Burggraf, S., K.O. Stetter, P. Rouviere, and C.R. Woese. 1991. *Methanopyrus kandleri*: An archaeal methanogen unrelated to all other known methanogens. *Syst. Appl. Microbiol.* **14:** 346–351.

Burgin, A. and N.R. Pace. 1990. Mapping the active site of ribonuclease P RNA using a substrate containing a photoaffinity agent. *EMBO J.* **9:** 4111–4118.

DeLong, E.F., G.S. Wickham, and N.R. Pace. 1989. Phylogenetic stains: Ribosomal RNA-based probes for the identification of single cells. *Science* **243:** 1360–1363.

Erdmann, V. 1976. Structure and function of 5S and 5.8S RNA. *Prog. Nucleic Acid Res. Mol. Biol.* **18:** 45–90.

Forster, A.C. and R.H. Symons. 1987a. Self-cleavage of plus and minus RNAs of a virusoid and a structural model for the active sites. *Cell* **49:** 211–220.

———. 1987b. Self-cleavage of virusoid RNA is performed by the proposed 5S-nucleotide active site. *Cell* **50:** 9–16.

Fox, G.E. and C.R. Woese. 1973. 5S RNA secondary structure. *Nature* **256:** 505–507.

Fox, G.E., E. Stackenbrandt, R.B. Hespell, J. Gibson, J. Maniloff, T.A. Dyer, R.S. Wolfe, W.E. Balch, R. Tanner, L. Magrum, L.B. Zablen, R. Blakemore, R. Gupta, L. Bonen, B.J. Lewis, D.A. Stahl, K.R. Luehrsen, K.N. Chen, and C.R. Woese. 1980. The phylogeny of prokaryotes. *Science* **209:** 457–463.

Gogarten, J.P., H. Kibak, P. Dittrich, L. Taiz, E.J. Bowman, B.J. Bowman, M.F. Manolson, R.J. Poole, T.E. Date, T. Oshima, J. Konishi, K. Denda, and M. Yoshida. 1989. Evolution of the vacuolar H^+-ATPase: Implication for the origin of eukaryotes. *Proc. Natl. Acad. Sci.* **86:** 9355–9359.

Guerrier-Takada, C., K. Gardiner, T. Marsh, N.R. Pace, and S. Altman. 1983. The RNA moiety of ribonuclease P is the catalytic subunit of the enzyme. *Cell* **35:** 849–857.

Gutell, R.R., N. Larsen, and C.R. Woese. 1993. Lessons from an evolving ribosomal RNA: 16S and 23S rRNA structure from a comparative perspective. In *Ribosomal RNA: Structure, evolution, gene expression and function in protein synthesis* (ed. R.A. Zimmerman and A.E. Dahlberg). Telford Press, Caldwell, New Jersey (in press).

Haas, E.S., D. Morse, J.W. Brown, F.J. Schmidt, and N.R. Pace. 1991. Sequence covariation reveals new long-range structure in ribonuclease P RNA. *Science* **254:** 853–856.

Heus, H.A. and A. Pardi. 1991. Structural features that give rise to the unusual stability of RNA hairpins containing GNRA loops. *Science* **253:** 191–194.

Iwabe, N., K. Kuma, M. Hasegawa, S. Osawa, and T. Miyata. 1989. Evolutionary relationship of archaebacteria, eubacteria and eukaryotes inferred from phylogenetic trees of duplicated genes. *Proc. Natl. Acad. Sci.* **86:** 9355–9359.

James, B.D., G.J. Olsen, J. Liu, and N.R. Pace. 1988. The secondary structure of ribonuclease P RNA, the catalytic element of a ribonucleoprotein enzyme. *Cell* **52:** 19–26.

Kim, S.-H. 1979. Crystal structure of yeast tRNA-phe and general structural features of other tRNAs. In *Transfer RNA: Structure, properties, and recognition* (ed. P.R. Schimmel et al.), pp. 83–100. Cold Spring Harbor Laboratory, Cold Spring Harbor, New York.

Li, Y., G. Zon, and W.D. Wilson. 1991. Thermodynamics of DNA duplexes with adjacent G•A mismatches. *Biochemistry* **31:** 7566–7572.

Moazed, D. and H.F. Noller. 1989. Intermediate states in the movement of transfer RNA in the ribosome. *Nature* **342:** 142–148.

Noller, H.F., V. Hoffarth, and L. Zimniak. 1992. Unusual resistance of peptidyl transferase to protein extraction procedures. *Science* **256:** 1416–1419.

Noller, H.F., D. Moazed, S. Stern, T. Powers, P.N. Allen, J.M. Robertson, B. Weiser, and K. Triman. 1990. Structure of rRNA and its functional interaction in translation. In *The ribosome: Structure, function and evolution* (ed. W.E. Hill et al.), pp. 73–92. American Society for Microbiology, Washington, D.C.

Noller, H.F., J. Kop, V. Wheaton, J. Brosius, R. Gutell, A.M. Kopylov, F. Dohme, W. Herr, D.A. Stahl, R. Gupta, and C.R. Woese. 1981. Secondary structure model for 23S ribosomal RNA. *Nucleic Acids Res.* **9:** 6167–6189.

Olsen, G.J. and C.R. Woese. 1993. Ribosomal RNA: A key to phylogeny. *FASEB J.* (in press).

Pace, N.R., D.A. Stahl, D.J. Lane, and G.J. Olsen. 1985. Analyzing natural microbial populations by rRNA sequences. *Am. Soc. Microbiol. News* **51:** 4–12.

Popper, K.R. 1959. *The logic of scientific discovery*. Hutchinson, London.

Powers, T. and H.F. Noller. 1991. A functional pseudoknot in 16S ribosomal RNA. *EMBO J.* **10:** 2203–2214.

Pühler, G., H. Leffers, F. Gropp, P. Palm, H.-P. Klenk, F. Lottspeich, R.A. Garrett, and W. Zillig. 1989. Archaebacterial DNA-dependent RNA polymerases testify to the evolution of the eucaryotic nuclear genome. *Proc. Natl. Acad. Sci.* **86:** 4569–4573.

Ramirez, C., L.C. Shimmin, C.H. Newton, A.T. Matheson, and P.P. Dennis. 1989. Structure and evolution of the L11, L1, L10, and L12 equivalent ribosomal proteins in eubacteria, archaebacteria and eukaryotes. *Can. J. Microbiol.* **35:** 234–244.

Sandman, K., J.A. Krzycki, B. Dobrinski, R. Lurz, and J.N. Reeve. 1990. DNA binding protein HMf, isolated from the hyperthermophilic archaea *Methanothermus fervidus*, is most closely related to histones. *Proc. Natl. Acad. Sci.* **87:** 5788–5791.

SantaLucia, J., Jr., R. Kierzek, and D.H. Turner. 1990. Effects of GA mismatches on the structure and thermodynamics of RNA internal loops. *Biochemistry* **29:** 8813–8819.

Trent, J.D., E. Nimmersgern, J.S. Wall, F.-U. Hartl, and A.L. Horwich. 1991. A molecular chaperone from a thermophilic archaebacterium is related to the eukaryotic protein t-complex polypeptide-1. *Nature* **354:** 490–493.

Tuerk, C., P. Gauss, C. Thermes, D.R. Groebe, M. Gayle, N. Guild, G. Stormo, Y. D'Aubenton-Carafa, O.C. Uhlenbeck, I. Tinoco, E.N. Brody, and L. Gold. 1988. CUUCGG hairpins: Extraordinarily stable RNA secondary structures associated with various biochemical processes. *Proc. Natl. Acad. Sci.* **85:** 1364–1368.

Varani G., C. Cheong, and I. Tinoco, Jr. 1991. Structure of an unusually stable RNA hairpin. *Biochemistry* **30:** 3280–3289.

Waugh, D.S., C.J. Green, and N.R. Pace. 1989. The design and catalytic properties of a simplified ribonuclease P RNA. *Science* **244:** 1569–1572.

Weisburg, W.G., J.G. Tully, D.L. Rose, D.L. Petzel, J.P. Oyaizu, H. Yang, D. Mandelco, L. Sechrest, J. Lawrence, T.G. Van Etten, J. Maniloff, and C.R. Woese. 1989. Phylogenetic analysis of the mycoplasmas: Basis for their classification. *J. Bacteriol.* **171:** 6455–6467.

Wheelis, M.L., O. Kandler, and C.R. Woese. 1992. On the nature of global classification. *Proc. Natl. Acad. Sci.* **89:** 2930–2934.

Winker, S. and C.R. Woese. 1991. A definition of the domains archaea, bacteria and eucarya in terms of small subunit ribosomal RNA characteristics. *Syst. Appl. Microbiol.* **14:** 305–310.

Woese, C.R. 1967. *The genetic code: The molecular basis of genetic expression.* Harper and Row, New York.

———. 1970. Molecular mechanics of translation: A reciprocating ratchet mechanism. *Nature* **226:** 817–820.

———. 1973. Evolution of the genetic code. *Naturwissenschaften* **60:** 447–459.

———. 1987. Bacterial evolution. *Microbiol. Rev.* **51:** 221–271.

Woese, C.R. and R.R. Gutell. 1989. Evidence for several higher order structural elements in ribosomal RNA. *Proc. Natl. Acad. Sci.* **86:** 3119–3122.

Woese, C.R., O. Kandler, and M.L. Wheelis. 1990a. Toward a natural system of organisms: Proposal for the domains Archaea, Bacteria, and Eucarya. *Proc. Natl. Acad. Sci.* **87:** 4576–4579.

Woese, C.R, S. Winker, and R.R. Gutell. 1990b. Architecture of ribosomal RNA: Constraints on the sequence of tetra-loops. *Proc. Natl. Acad. Sci.* **87:** 8467–8471.

Woese, C.R., L. Achenbach, P. Rouviere, and L. Mandelco. 1991. Archael phylogeny: Examination of the phylogenetic position of *Archaeoglobus fulgidus* in light of certain composition-induced artifacts. *Syst. Appl. Microbiol.* **14:** 364–371.

Woese, C.R., R. Gutell, R. Gupta, and H.F. Noller. 1983. Detailed analysis of the higher-order structure of 16S-like ribosomal ribonucleic acids. *Microbiol. Rev.* **47:** 621–669.

Woese, C.R., E. Stackenbrandt, T.J. Macke, and G.E. Fox. 1985. A phylogenetic definition of the major eubacterial taxa. *Syst. Appl. Microbiol.* **6:** 143–151.

Woese, C.R., L.J. Magrum, R. Gupta, R.B. Siegel, D.A. Stahl, J. Kop, N. Crawford, J. Brosius, R. Gutell, J.J. Hogan, and H.F. Noller. 1980. Secondary structure model for bacterial 16S ribosomal RNA: Phylogenetic, enzymatic and chemical evidence. *Nucleic Acids Res.* **8:** 2275–2293.

Yang, D., Y. Oyaizu, H. Oyaizu, G.J. Olsen, and C.R. Woese. 1985. Mitochondrial origins. *Proc. Natl. Acad. Sci.* **82:** 4443–4447.

Zuckerkandl, E. and L. Pauling. 1965. Molecules as documents of evolutionary history. *J. Theoret. Biol.* **8:** 357–366.

5

Ribosomes and the RNA World

Peter B. Moore
Departments of Chemistry and of Molecular
Biophysics and Biochemistry
Yale University
New Haven, Connecticut 06511

The hypothesis that all modern organisms are descendants of an ancient, self-replicating entity that depended on RNA both for its genetic material and for catalysis has gained a large following over the past decade, as the existence of this volume attests. This idea, also known as the RNA world hypothesis, has been around in various forms for almost 30 years, and it has been known for most of that time that RNA can serve a DNA-like function. The reason there is so much interest in the RNA world hypothesis today is that we now know something our predecessors did not, namely, that RNAs are capable of catalyzing biochemical reactions.

Boiled down to its essence, the RNA world hypothesis is a proposal for the origin of life that explains why RNA is so important for gene expression in modern organisms but has no role in any other aspect of metabolism. Ribosomes are important for two reasons in this context. First, because ribosomal RNA accounts for about 80% of the RNA in most modern cells, the RNA world hypothesis either makes sense of them or fails entirely. Second, again because the ribosome contains so much RNA, information about its structure and function has to be an important part of the input used for formulating all such theories.

This paper is intended to provide the reader with a broadbrush picture of the ribosome field that emphasizes the findings that bear most directly on evolutionary issues. Readers wishing to find out more about the topics covered here, as well as about the ribosome field in general, should consult the symposium volumes that have appeared at irregular intervals since the late 1960s (Nomura et al. 1974; Chambliss et al. 1980; Hardesty and Kramer 1986; Hill et al. 1990), which are excellent summaries of the field's progress.

The information we now have about ribosomes points to three conclusions that are important for theories about the origin of life. In order of increasing controversiality they are (1) that the Last Common Ancestor (LCA) of all modern organisms had ribosomes, (2) that the ribosomes of the LCA were nucleoprotein complexes, and (3) that the concept that the progenitors of the LCA had protoribosomes made entirely of RNA is supportable, but irrelevant to the RNA World Hypothesis.

A SHORT SCIENTIFIC HISTORY OF THE RIBOSOME

The Ribosome Is Discovered

The ribosome field is rooted in observations made long before it was recognized as a distinct biochemical subdiscipline. The word "ribosome" was coined in 1958 (Roberts 1958), about 20 years after the first observations that led to the development of the field (see Tissières 1974).

The origins of the ribosome field can be traced back to histological studies made before World War II that established several important facts about eukaryotic cells: (1) DNA is a nuclear substance (Feulgen et al. 1937), (2) most of the RNA in both plant and animal cells is cytoplasmic, and (3) the amount of RNA a cell contains is correlated with its activity in protein synthesis (Brachet 1941; Caspersson 1941, 1950). Cell fractionation techniques, which were developed in the late 1930s, enabled Claude to identify an RNA-rich, particulate fraction from cytoplasmic extracts called "microsomes" (see, e.g., Claude 1941), and by 1942, Jeener and Brachet were convinced that microsomes were universal in eukaryotic cells and important for protein synthesis (Jeener and Brachet 1941, 1942).

Ultracentrifugation and electron microscopy, technologies that entered common use in the years immediately after the war, had a revolutionary impact on cytology. Through their use, it was soon demonstrated that the RNA in eukaryotic microsomes is associated with the discrete ribonucleoprotein particles we now call ribosomes (Palade 1955; Palade and Siekevitz 1956) and that bacterial cells also contain ribosomes (Luria et al. 1943; Schachman et al. 1952).

Physical studies soon demonstrated that the ribosomes in any given organism are all the same size and are roughly half protein and half RNA. They are more or less spherical, with diameters in the neighborhood of 250 Å and molecular weights of several million. Furthermore, all ribosomes are 1:1 complexes of two nonequivalent subunits that associate in a magnesium-dependent way (Chao 1957).

Radioisotopes also became available to the biochemical world in the early postwar years, and they were used to test the hypothesis that microsomes are involved in protein synthesis. Whole-animal experiments done around 1950 supported that theory (see Tissières 1974), but definitive proof was not obtained until several years later when Zamecnik and his collaborators published the results of their classic pulse-chase experiments which demonstrated that nascent polypeptides form on ribosomes (Littlefield et al. 1955).

Nothing that has happened since has shaken the general conclusions drawn in those early years. All cellular life forms that have been examined so far contain ribosomes, and the only known function of ribosomes is protein synthesis. Both because ribosomes are universal and because the role they play in metabolism is so basic, the conclusion that the LCA must also have made proteins using ribosomes is all but inescapable and is widely accepted.

The Function of the Ribosome Is Defined

From the mid-1950s until the late 1960s, elucidation of the role of the ribosome in protein synthesis was a central issue in biochemistry. Most of the work done in that period depended on cell-free systems that synthesize proteins in vitro. A perusal of Zamecnik's 1969 review of those years will remind the reader just how difficult it was to develop such systems (Zamecnik 1970). It could never have been done without radioisotopes, since cell-free systems efficient enough to make more than traces of product were not available then or for many years thereafter. The knowledge gleaned from these systems was of the utmost importance; we note only those findings that are important here.

One of the first fruits of the cell-free work was the demonstration that aminoacyl-tRNAs are the activated amino acid intermediates used by the protein-synthesizing system (Hoagland et al. 1957, 1958). These were the molecules called for by Crick's adapter hypothesis (Crick 1958), and the proof of their existence provided by Zamecnik's group and others launched the tRNA/aminoacyl-tRNA synthetase field, which is as active today as ever. The significance of the synthetases is that they are the entities that actually "read" the code; once an amino acid is attached to a tRNA, its fate is sealed.

Although it was clear from the outset that ribosomes have to be included in a cell-free system, it was not clear precisely what they contribute. Since the Central Dogma was well established by the late 1950s (see Crick 1958), the presence of RNA in ribosomes was not a source of anxiety. After all, DNA makes RNA makes protein. However, up until 1960 or so, many thought that the RNA in ribosomes was informational (see Zamecnik 1970). The ribosome was envisioned as a segment of sequence-specifying RNA around which a set of proteins clustered busily translating RNA sequence information into amino acid sequences. There even was a phase where the notion that ribosomes are built like little viruses was entertained (Crick 1958).

Even then there were reasons for feeling uncomfortable with the idea that rRNA is informational. The rRNAs in an organism do not vary from one tissue type to the next. They are uniform in size, and base compositional data alone suggested that their sequences do not vary from tissue to tissue and certainly indicated that their sequences are not representative of the genomes with which they are associated (see Spirin 1964). Both characteristics were worrying, since it was already known that proteins vary enormously in chain length and sequence. One's expectation, therefore, was that there ought to be a corresponding variability in rRNA size and base composition. The theory that rRNA is informational was put to rest for good in 1961 by the announcement of the discovery of messenger RNA (Brenner et al. 1961; Gros et al. 1961; Jacob and Monod 1961).

By the close of play in 1961, one knew that the protein-synthesizing apparatus is programmed by mRNA and that the interactions between mRNA and tRNAs that lead to protein synthesis occur on the ribosome. The final chapter of the basic story was put in place a few years later by Monro and co-workers (Monro 1967; Maden et al. 1968), who demonstrated that the activity responsible for peptide bond formation, i.e. peptidyl transferase, is an intrinsic part of the large ribosomal subunit.

The Ribosome Is a Polymerase

The pace of discovery in the protein synthesis field was so rapid that by 1964 or so, one could confidently write a reaction scheme for the ribosomal steps of protein synthesis

$$n(\text{aminoacyl-tRNA}) + 2n\text{GTP} \xrightarrow[\text{ribosome + protein factors}]{\text{template}} (\text{amino acid})_n + n\text{tRNA} + 2n\text{P}_i + 2n\text{GDP}$$

When the corresponding reaction scheme is written for DNA (or RNA) polymerase the result is the following

$$n\text{dXTP} \xrightarrow[\text{DNA polymerase + protein factors}]{\text{template}} (\text{dXMP})_n + n\text{P}_i\text{P}_i$$

The parallels are obvious. DNA polymerase is an enzyme that catalyzes the formation of heteropolymers, starting with activated monomers, and the sequences of the polymers it makes are determined by polynucleotide templates. If you replace "DNA polymerase" with "the ribosome," the sentence still works.

The ribosome is not an organelle, a term one still sees attached to it in the literature; it is not comparable in any way to intracellular entities like the mitochondrion or the lysozome, which are organelles. It is an enzyme. It would make sense to rename the ribosome to reflect that fact; "polypeptide polymerase" would be appropriate (see Moore 1986). However, the weight of tradition is such that it would be folly to agitate for reform at this late date — even a reform as sensible as this.

Ribosomes Contain Many Proteins

By 1966 or so, a lot of the biochemistry connected with the ribosomal phase of gene expression had been worked out, but not as thoroughly, perhaps, as many thought at the time. Important new observations continue to be made. Our understanding of the elongation phase of protein synthesis — which biochemists who do not work on protein synthesis have regarded as a closed book for 25 years — has changed significantly (see Nierhaus 1990).

Nevertheless, starting in the late 1960s, the focus of the ribosome field began to shift from pathway elucidation toward structure determination and the study of the relationship between ribosome structure and function. From about 1965 to 1980, most questions of this kind were addressed using the particles from *Escherichia coli*, and by 1980, their chemical structure was fully understood. The ribosomes from this organism contain 3 RNA molecules: 16S rRNA, which is part of the small ribosomal subunit, and 23S and 5S rRNAs, which are found only in the large ribosomal subunit. All three are present in 1 mole per mole of ribosomes (see Van Holde and Hill 1974).

5S rRNA was sequenced in 1968 (Brownlee et al. 1968), and although progress was made with the large rRNAs by direct sequencing methods, their sequences were not completed until their genes were analyzed in the late 1970s (Brosius et al. 1978, 1980; Carbon et al. 1979). Even though *E. coli* has several genes for each of its rRNAs, the sequence differences between their products are slight, one or two residues per hundred, and have no known physiological significance. From the point of view of the RNAs they contain, all ribosomes in *E. coli* are equivalent.

Starting in the early 1960s, methods were gradually worked for isolating the components of the protein mixture that is found in ribosome preparations. The invention of acrylamide gel electrophoresis, which occurred at about that time, provided a convenient means for identifying individual proteins and assessing the purity of ribosomal protein preparations. Despite some confusion, it was evident quite early that ribosomal protein mixtures are very complicated (Waller and Harris 1961; Waller 1964). Approximately 56 different proteins associate strongly with the ribosomes from *E. coli* (see Wittmann-Liebold 1986). Their average molecular weight is about 15,000, and they range in molecular weight from about 8,000 to about 65,000.

By 1984, sequences were available for all of the ribosomal proteins from *E. coli*. Sequence comparisons proved what the fingerprinting studies done almost 20 years earlier had indicated (Traut et al. 1967); every one of the ribosomal proteins is a distinct species. The differences between these proteins are so large that the extent of homologies between them is still uncertain (see Wittmann-Liebold 1986).

The stoichiometry of the proteins in the ribosome was a bone of contention in the late 1960s and early 1970s, but when the dust cleared, the consensus was that almost all ribosomal proteins occur in single copies in the ribosome in vivo (Hardy 1975). The exception is the acidic, large subunit protein, L7/L12, which occurs in four copies per ribosome (Subramanian 1975; Marquis and Fahnestock 1978). Normal preparations contain substoichiometric amounts of many proteins because ribosomes shed proteins as they are purified.

Bacterial Ribosomes Self-assemble

In the late 1960s, Nomura and colleagues discovered that the small subunit from bacterial ribosomes will reassemble from its dissociated macromolecular components in vitro (Traub and Nomura 1968). Techniques were worked out subsequently for reconstituting the large subunit as well (Nomura and Erdmann 1970; Nierhaus and Dohme 1974). Reconstitution provided the field with a powerful experimental tool that is still bearing fruit. Shortly after its discovery, a series of experiments was done that answered an important question about the distribution of proteins in ribosome populations.

The stoichiometric data require that every small subunit have one copy of 16S rRNA and that every large subunit have a single 23S rRNA molecule, because both molecules account for more than half the mass of the particles to which they belong. Although the stoichiometric data also tell us that there is one copy of every protein per particle, they do not require that every particle have one copy of every protein. Some particles could carry two copies of protein x and none of protein y, provided others had two copies of y and none of x, etc.

The single protein omission experiments done by Nomura's group on the small subunit settled the issue for a significant number of proteins (Mizushima and Nomura 1970; Held et al. 1974). They found that the omission of any one of many ribosomal proteins led to complete failure of reconstitution; no normally sedimenting particles would form. This proved that every normal subunit had to contain (at least) one copy of each protein whose omission has this effect on reconstitution. Everything that has been learned since supports the idea that each subunit in vivo carries one copy of every protein (except L7/L12), not just the proteins essential for normal reconstitution.

The Ribosome Is a Single-site Enzyme

Any macromolecular complex that contains single copies of effectively all of its components is asymmetric and can have no equivalent sites, formally speaking. Consistent with this inference, all the experiments that have ever been done to measure the number of active sites of a given kind on the ribosome have produced one as their answer. The ribosome has one peptidyl transferase site, one mRNA-binding site, one A-type site for tRNA binding, etc. It follows that when one is thinking of all-protein enzymes to compare to the ribosome, the molecules to think about are small enzymes like RNase A or lysozyme, which have single active sites, not large, multi-enzyme complexes like pyruvate dehydro-genase, which are comparable to the ribosome in size, but which have multiple copies of every site.

There is every reason to believe that *E. coli* ribosomes are typical. The qualitative conclusions we have just discussed would have emerged had the ribosomes from any other organism been subjected to com-parable scrutiny.

ARE RIBOSOMES RIBOZYMES?

As soon as one realizes that the ribosome is an enzyme, the following question arises. Why does it contain RNA? After all, RNA and DNA polymerase, which have analogous functions, get along quite well without it.

Some have attempted to explain the presence of RNA in ribosomes by arguing that base-pairing is the most economical way to obtain the substrate-enzyme interactions required during the ribosomal phase of protein synthesis, because all the substrates involved are nucleic acids. Although this argument has some force, it does not carry the day. Given the versatility of proteins, there is no reason to believe that an all-protein catalyst could not have evolved to perform the same function. Since there is no compelling chemical explanation for rRNA, the reason it exists must have an evolutionary explanation.

Do Proteins Do It or Does the RNA?

Two extreme hypotheses can be entertained for the function of the rRNAs. On the one hand, all the important catalytic functions of the modern ribosome could be mediated by its proteins, and its RNA might serve only as a scaffold on which proteins are hung (see, e.g., Fellner 1974). Alternatively, rRNA could be responsible for all important ribosomal functions, with the proteins serving only auxiliary functions.

Someone evaluating the merits of these two models in 1970 would have noted with satisfaction that the first model ascribes catalytic activity to molecules of a class well known for it, but its defect would have been equally obvious. What purpose is served by attaching catalytic proteins to large pieces of RNA? If the primordial ribosome was an all-protein assembly, how did it ever acquire RNA components (Crick 1968; Woese 1980)?

The ribocentric model is easier to rationalize from an evolutionary point of view. If the primordial protein-synthesizing system depended on a catalytic RNA, the retention of RNA in the modern ribosome reflects nothing more than the conservatism of evolution. Furthermore, no difficulties are created by proposing that proteins were added to the ribosome as time went on to improve the functional properties of the RNA that is its core. The only difficulty the ribocentric picture posed for the biochemist of 1970 was its implication that RNAs are catalytic, a concept for which there was no precedent.

The collective response of the ribosome field to this conundrum was not as perverse as some who have written about it since enjoy making it sound. Yes, the field did concentrate on ribosomal proteins from about 1968 until 1978. This does not imply, however, that the field had collectively decided in favor of the proteocentric ribosome. Biochemistry is a technique-driven science. One deals today with those problems that today's technology gives one a chance of solving, not necessarily with those problems one would like most to solve.

There were sound, practical reasons for working on ribosomal proteins in that era: (1) Ribosomal proteins could be purified, (2) proteins could be sequenced, and (3) protein activities could be examined one by one using reconstitution techniques. Nothing half so interesting was possible with rRNA. If the members of the field had been polled on the RNA-protein question in that era, many (myself included) would have claimed to be agnostics.

Today, as a result of the revolution in molecular biology, RNA is much easier to work with than protein, and most ribosome workers are using those technologies to study the involvement of rRNA in protein synthesis to the exclusion of all else. Furthermore, since it is now clear that some RNAs have catalytic properties, the conceptual block that earlier held the ribocentric model at bay has disappeared. Now that the pendulum has swung decisively in favor of the ribocentric ribosome, it would be wise to review the evidence on which those who profess that doctrine base their beliefs.

The Last Common Ancestor's Ribosomes Were Ribonucleoproteins

I can remember wondering in the mid-1970s whether ribosome structure really is conserved across major phylogenetic boundaries. On the one hand, we all knew that every ribosome has two unequal subunits, that the subunits played distinctive roles that are the same from one species to the next, and that the protein synthesis pathway is basically the same in all species, granted, of course, the substantial mechanistic differences between prokaryotes and eukaryotes. We also were pretty sure that the three-dimensional structure of 5S RNAs does not vary from one taxonomic group to the next (Fox and Woese 1975). All these observations argued for conservation. On the other hand, eukaryotic ribosomes are almost twice as big as their prokaryotic counterparts, and the few experiments that had been done suggested that parts were not interchangeable between eukaryotic and prokaryotic ribosomes (see, e.g., Wrede and Erdmann 1973). Moreover, the protein sequence comparisons one could do (which were not very numerous) suggested that there might be substantial differences between major phylogenetic groups in this regard, too.

The nucleic acid sequencing revolution of the late 1970s immediately resolved this problem. Sequences began pouring in for the large rRNAs from species belonging to all kingdoms. In addition, the number of different 5S rRNA sequences available skyrocketed (see Specht et al. 1990). Cross-species comparisons led to several important conclusions. First, it was confirmed that all 5S rRNAs are homologous. Second, it was established that the two large rRNAs from *E. coli* have homologs in the ribosomes of all other species, which had not been quite so obvious. We know the latter is true because all of the sequences available can be folded into similar secondary structures that place conserved sequences at corresponding locations (see Noller 1984). Thus, there is no doubt that the ribosomes in the LCA had rRNAs whose secondary structures resembled those of their descendants. Because this is so, rRNA sequence comparisons have become extremely important for determining phylogenetic relationships (see Woese and Pace, this volume).

The existence of strong homologies at the RNA level and their apparent absence at the protein level was taken by some (myself included) as evidence for the paramount importance of RNA in ribosome function. Despite the obscurity that had descended over their activities, the ribosomal protein-sequencing laboratories continued beavering away, and as a result of their efforts, the picture has recently become more interesting than that. In the last few years, enough protein sequences from eubacteria, archaebacteria, and eukaryotes have become available to make meaningful searches for sequence homologies possible across kingdom boundaries. They exist, and there are quite a lot of them (Wool et al. 1990; I.G. Wool, pers. comm.).

This discovery has many interesting implications. If ribosomal proteins were "invented" after the three kingdoms diverged, there should be no homologies. Every kingdom should have its own set whose members would bind to rRNA in their own ways at their own special locations. The existence of homologies thus argues powerfully that ribosomal proteins appeared before the great divergence; the LCA had ribosomal proteins. Furthermore, because we expect that homologous proteins will bind to rRNAs at homologous locations, another implication of the existence of protein homologies is that ribosome structure should be conserved at the quaternary level, as well as at the primary, secondary, and tertiary levels. Thus, the LCA's ribosomes must have had a three-dimensional structure we would recognize as ribosomal today—if only we knew what any modern ribosome's structure looks like in three dimensions.

The reasons the field has been so slow coming to these conclusions are both technological and scientific. The technological reason is the difficulty connected with sequencing proteins; proteins are harder to sequence than RNAs. The scientific problem has to do with the fact that the rules that relate sequence to structure in proteins are much less well understood than they are in nucleic acids. Thus, the sequence database for ribosomal proteins was, and remains, thin compared to that for rRNAs, and its meaning is much harder to read. Let me add that until we have determined the structures of ribosomes from organisms belonging to at least two different kingdoms, it is unlikely that we will fully understand the extent of the homology that exists between the ribosomal proteins from different kingdoms.

Ribosome Function Does Not Depend Absolutely on Its Proteins

Clearly, evidence other than sequence comparisons must be appealed to in order to decide whether RNA is functionally more important than protein in the ribosome. Fortunately, there is a lot of non-sequence data that speaks to this point, and it all points in the same direction—toward RNA. Only the most decisive data are summarized here due to limitations of space; a more complete analysis may be found elsewhere (Moore 1986).

The genetic studies of Dabbs have provided the strongest evidence we have on the nonessentiality of the ribosomal proteins for ribosome function. He has constructed deletion mutations for 16 of the 56 ribosomal proteins in *E. coli* and found that none of them are lethal (see Dabbs 1991). Cells that lack specific ribosomal proteins usually grow slowly, but they do grow, and that means that their ribosomes make protein.

To the list of proteins Dabbs has identified as nonessential for activity mutationally, we can add the species identified as nonessential by reconstitution techniques (Held et al. 1973; Held and Nomura 1975; Ramakrishnan et al. 1986). These experiments, as well as many others, indicate that there are no ribosomal proteins that make all-or-none contributions to ribosome activity. There are no serine hydroxyl groups or histidine imidazole groups that are absolutely required by the catalytic mechanism of the ribosome.

Ribosomal proteins do have a function, however. Modern rRNAs are incapable of assuming the conformations they display in the ribosome in the absence of ribosomal proteins, and many of the proteins identified as nonessential by experiments of the types just described catalyze assembly. Thus, efficient ribosome assembly depends on ribosomal proteins. It is also certain that ribosomal proteins are the means by which the structure of the ribosome is tuned for optimal function, and the selective pressure for optimization is overwhelming (see Kurland et al. 1990).

RNA Is Directly Involved in Most Ribosomal Functions

The only aspect of protein synthesis that we know for certain depends exclusively on RNA-RNA interactions is the one identified by Shine and Dalgarno (1974) many years ago. The pairing between the 3′ end of 16S rRNA and upstream sequences in bacterial mRNAs which they postulated is responsible for aligning mRNAs on ribosomes during initiation has been amply documented (see Steitz 1980; Hui and de Boer 1987; Jacob et al. 1987). The impact this discovery made on the RNA-protein controversy was modest, however, because the process it explained is peripheral to decoding and peptide transfer, which are the essence of protein synthesis.

Although proof of exclusively RNA-based mechanisms is lacking for other aspects of protein synthesis, the list of processes in which RNA is known to be intimately involved is a long one. It includes: (1) peptidyl transferase (Barta et al. 1984), (2) the decoding center on the 30S ribosomal subunit (Prince et al. 1982), (3) the thiostrepton site (Cundliffe 1986), (4) the α-sarcin–ricin site (Endo and Wool 1982; Endo et al. 1987), and (5) the colicin E3 site (Bowman et al. 1971; Senior and Holland 1971), to name a few. Either RNA is a part of each of these sites, or the properties of the site are altered by RNA mutation, or damage to the RNA in the region in question blocks ribosome activity. Structure-function studies directed at the rRNAs have often led to the sort of all-or-none results that those studying ribosomal proteins in the 1970s wanted but never obtained.

The most ambitious effort to demonstrate rRNA involvement in ribosome activity to date has been undertaken by the Noller group at Santa Cruz, which has been trying to show that the large ribosomal subunit RNA alone will catalyze the fragment reaction of peptidyl transferase (Noller 1991). They have managed to strip almost all the proteins from the large ribosomal subunit of *Thermus aquaticus* by a combination of detergent and protease treatments without destroying its fragment reaction activity. Unfortunately, some protein resists removal, and as long as any is present, absolute proof that the peptidyl transferase is RNA will be lacking. Nevertheless, the finding that a particle that has lost most of its protein retains transferase activity strongly suggests that the peptidyl transferase is indeed made of RNA.

The answer to the question raised at the beginning of this section is thus clear. Ribosomes contain RNA because RNA is directly involved in many of the interactions on which ribosomal activity depends. "Many" may ultimately have to be replaced by "all."

CONCLUDING COMMENTS

Three basic messages that advocates of the RNA world should be receiving from their friends and colleagues who work on ribosomes are: (1) Ribosomes are enzymes, (2) ribosomal RNA is vital for the enzymatic activity of ribosomes, and (3) the LCA had ribonucleoprotein ribosomes.

How these nuggets of wisdom get incorporated into one's personal theory about the origin of life is a matter of taste, both because the constraints placed on theories about the early history of this planet by the evidence available are few, and because hypotheses about the origin of life are impossible to prove. In addition, origin theories have always had a disconcerting tendency to ebb and flow with intellectual fashion. It is wise to regard all of them with skepticism — even the entirely sound and well-supported theory that is the subject of this volume.

Of one thing we can be certain: The RNA world — if it ever existed — was short-lived. The earth came into existence about 4.5×10^9 years ago, and fossil evidence suggests that cellular organisms resembling modern bacteria existed by 3.6×10^9 years before the present (Schopf 1983). There are even hints that those early organisms engaged in oxygenic photosynthesis, which is likely to have been a protein-dependent process then, as it is now. Thus, it appears likely that organisms with sophisticated, protein-based metabolisms existed only 0.9×10^9 years after the planet's birth.

The "window of opportunity" for the RNA world was much shorter than 0.9×10^9 years. The earth's surface was uninhabitable at the beginning due to the heat generated by meteoric bombardment and its geological differentiation. The oldest rocks discovered so far, which date to only 3.9×10^9 years before the present, could be remnants of the planet's original crust. Moreover, we have to allow some time for cellular forms to evolve the photosynthetic apparatus that they may have had by 0.9×10^9 years after the beginning. Thus, the interval in which the biosphere could have been dominated by RNA-based life forms may be less than 100 million years. Incidentally, when one starts thinking along these lines, one must consider the unthinkable, i.e., that the length of time that RNA-based organisms bestrode the earth might actually be zero.

Did the denizens of the fleeting, transient RNA world have ribosomes? Absolutely not. Any entity that contains structures that can be identified as protoribosomes must be making proteins, and if so, it is not a bona fide member of the RNA world. It is instead a participant in the brave new Protein world, as indeed the LCA certainly was. Thus, the question of what the first ribosomes were like, although of the utmost interest, has not got much to do with the RNA world per se.

In fact, it is quite easy to account for the presence of RNA in modern ribosomes. Simply postulate that the first useful peptidyl transferase to evolve was made of RNA, not protein. In addition, it is easy to understand how such a thing might have arisen in a biosphere where RNA was much more important than it is today, i.e., in the RNA world. Nevertheless, there is no need to propose that the first ribosomes were made entirely of RNA. Proteins must have come before ribosomes. After all, why would a device for making polypeptides evolve in an organism that had no use for protein? A protoribosome that contained both RNA and protein is thus entirely plausible.

ACKNOWLEDGMENT

This work was supported by a grant from the National Institutes of Health (AI-09167).

REFERENCES

Barta, A., G. Steiner, J. Brosius, H.F. Noller, and E. Kuechler. 1984. Identification of a site on the 23S ribosomal RNA located at the peptidyl transferase center. *Proc. Natl. Acad. Sci.* **81:** 3607–3611.

Bowman, C.M., J.E. Dahlberg, T. Ikemura, T., J. Konisky, and M. Nomura. 1971. Specific inactivation of 16S ribosomal RNA induced by colicin E3 in vitro. *Proc. Natl. Acad. Sci.* **68:** 964–968.

Brachet, J. 1941. La detection histochimique et le microdosage des acides pentosenucleique. *Enzymologia* **10:** 87–96.

Brenner, S., F. Jacob, and M. Meselson. 1961. An unstable intermediate carrying information from genes to ribosomes for protein synthesis. *Nature* **190:** 567–581.

Brosius, J., T.J. Dull, and H.F. Noller. 1980. Complete nucleotide sequence of a 23S ribosomal RNA gene from *Escherichia coli. Proc. Natl. Acad. Sci.* **77:** 201–204.

Brosius, J., M.L. Palmer, P.J. Kennedy, and H.F. Noller. 1978. Complete nucleotide sequence of a 16S ribosomal RNA gene from *Escherichia coli. Proc. Natl. Acad. Sci.* **75:** 4801–4805.

Brownlee, G.G., F. Sanger, and B.G. Barrell. 1968. The sequence of 5S ribosomal ribonucleic acid. *J. Mol. Biol.* **34:** 379–412.

Carbon, P., C. Ehresmann, B. Ehresmann, and J.-P. Ebel. 1979. The complete nucleotid sequence of the ribosomal 16S RNA from *Escherichia coli. Eur. J. Biochem.* **100:** 399–410.

Caspersson, T. 1941. Studien uber den Eiweissumsatz der Zelle. *Naturwissenschaften* **29:** 33–43.

———. 1950. *Cell growth and cell function: A cyto-chemical study.* Norton, New York.

Chambliss, G., G.R. Craven, J. Davies, K. Davis, L. Kahan, and M. Nomura, eds. 1980. *Ribosomes: Structure, function and genetics.* University Park Press, Baltimore, Maryland.

Chao, F.C. 1957. Dissociation of macromolecular ribonucleoprotein of yeast. *Arch. Biochem. Biophys.* **70:** 426–431.

Claude, A. 1941. Particulate components of cytoplasm. *Cold Spring Harbor Symp. Quant. Biol.* **9:** 263–271.

Crick, F.H.C. 1958. On protein synthesis. *Symp. Soc. Exp. Biol.* **12:** 138–163.

———. 1968. The origin of the genetic code. *J. Mol. Biol.* **38:** 367–379.

Cundliffe, E. 1986. Involvement of specific portions of ribosomal RNA in defined ribosomal functions: A study utilizing antibiotics. In *Structure, function, and genetics of ribosomes* (ed. B. Hardesty and G. Kramer), pp. 586–604. Springer-Verlag, New York.

Dabbs, E.R. 1991. Mutants lacking individual ribosomal proteins as tools to investigate ribosomal properties. *Biochimie* **73:** 639–645.

Endo, Y. and I.G. Wool. 1982. The site of action of alpha sarcin on eukaryotic ribosomes. *J. Biol. Chem.* **257:** 9054–9060.

Endo, Y., M. Mitsui, M. Motizuki, and K. Tsurugi. 1987. The mechanism of action of ricin and related toxic lectins on eukaryotic ribosomes. The site and the characteristics of the modification in 28S ribosomal RNA caused by the toxins. *J. Biol. Chem.* **262:** 5908–5912.

Fellner, P. 1974. Structure of the 16S and 23S ribosomal RNAs. In *Ribosomes* (ed. M. Nomura et al.), pp. 169–192. Cold Spring Harbor Laboratory, Cold Spring Harbor, New York.

Feulgen, R., M. Behrens, and S. Mahdihassan. 1937. Darstellung und Identifizierung der in den pflanzlichen Zellkernen vorkommenden Nucleinsaure. *Hoppe-Seyler's Z. Physiol. Chem.* **246:** 203–211.

Fox, G.E. and C.R. Woese. 1975. 5S RNA secondary structure. *Nature* **256:** 505–507.

Gros, F., H. Hiatt, G. Gilbert, C.G. Kurland, R.W. Risebrough, and J.D. Watson. 1961. Unstable ribonucleic acid revealed by pulse labelling of *E. coli. Nature* **190:** 581–585.

Hardesty, B. and G. Kramer, eds. 1986. *Structure, function, and genetics of ribosomes.* Springer-Verlag, New York.

Hardy, S.J.S. 1975. The stoichiometry of the ribosomal proteins of *Escherichia coli. Mol. Gen. Genet.* **140:** 253–274.

Held, W.A. and M. Nomura. 1975. *Escherichia coli* 30S ribosomal proteins uniquely required for assembly. *J. Biol. Chem.* **250:** 3179–3184.

Held, W.A., S. Mizushima, and M. Nomura. 1973. Reconstitution of *Escherichia coli* 30S ribosomal subunits from purified molecular components. *J. Biol. Chem.* **248:** 5720–5730.

Held, W.A., B. Ballou, S. Mizushima, and M. Nomura. 1974. Assembly mapping of 30S ribosomal proteins from *Escherichia coli. J. Biol. Chem.* **249:** 3103–3111.

Hill, W.E., A. Dahlberg, R.A. Garrett, P.B. Moore, D. Schlessinger, and J.R. Warner, eds. 1990. *The ribosome: Structure, function and evolution.* American Society for Microbiology, Washington, D.C.

Hoagland, M.B., P.C. Zamecnik, and M.F. Stephenson. 1957. Intermediate reactions in protein biosynthesis. *Biochim. Biophys. Acta* **24:** 215–216.

Hoagland, M.B., M.L. Stephenson, J.F. Scott, L.I. Hecht, and P.C. Zamecnik. 1958. A soluble ribonucleic acid intermediate in protein synthesis. *J. Biol. Chem.* **231:** 241–257.

Hui, A. and H.A. de Boer. 1987. Specializer ribosome system: Preferential translation of a single mRNA species by a subpopulation of mutated ribosomes in *Escherichia coli. Proc. Natl. Acad. Sci.* **84:** 4762–4766.

Jacob, F. and J. Monod. 1961. Genetic regulatory mechanisms in the synthesis of proteins. *J. Mol. Biol.* **3:** 318–356.

Jacob, W.F., M. Santer, and A.E. Dahlberg. 1987. A single base change in the Shine-Dalgarno region of 16S rRNA of *Escherichia coli* affects translation of many proteins. *Proc. Natl. Acad. Sci.* **84:** 4757–4761.

Jeener, R. and J. Brachet. 1941. Association, dans le meme granule, de ferments et des pentosenucleoproteides cytoplasmique. *Acta Biol. Belg.* **1:** 476–480.

—————. 1942. Sur la presence d'hormones proteiques et d'hemoglobine dans les granules a pentosenucleoproteides. *Acta Biol. Belg.* **2:** 447–450.

Kurland, C.G., F. Jorgensen, A. Richter, M. Ehrenberg, N. Bilgin, and A.-M. Rojas. 1990. Through the accuracy window. In *The ribosome: Structure, function and evolution* (ed. W.E. Hill et al.), pp. 513–526. American Society for Microbiology, Washington, D.C.

Littlefield, J.W., E.B. Keller, J. Gross, and P.C. Zamecnik. 1955. Studies on cytoplasmic ribonucleoprotein particles from the liver of the rat. *J. Biol. Chem.* **217:** 111–124.

Luria, S.E., M. Delbruck, and T.F. Anderson. 1943. Electron microscope studies of bacterial viruses. *J. Bacteriol.* **46:** 57–78.

Maden, B.E.H., R.R. Traut, and R.E. Monro. 1968. Ribosome catalyzed peptidyl transfer: The polyphenylalanine system. *J. Mol. Biol.* **35:** 333–345.

Marquis, D. and S.R. Fahnestock. 1978. A complex of acidic ribosomal proteins. Evidence of a four-to-one complex of proteins in the *Bacillus stearothermophilus* ribosome. *J. Mol. Biol.* **119:** 557–567.

Mizushima, S. and M. Nomura. 1970. Assembly mapping of 30S ribosomal proteins from *E. coli. Nature* **226:** 1214–1218.

Monro, R.E. 1967. Catalysis of peptide bond formation by 50S ribosomal subunits from *Escherichia coli. J. Mol. Biol.* **26:** 147–151.

Moore, P.B. 1986. Polypeptide polymerase: The structure and function of the ribosome in 1985. In *Proceedings of the Robert A. Welch Conferences on Chemical Research XXIX. Genetic Chemistry: The Molecular Basis of Heredity*, pp. 185–214. Robert A. Welch Foundation, Houston, Texas.

Nierhaus, K.H. 1990. The allosteric three-site model for the ribosomal elongation cycle: Features and future. *Biochemistry* **29:** 4997–5008.

Nierhaus, K.H. and F. Dohme. 1974. Total reconstitution of functionally active 50S ribosomal subunits from *Escherichia coli. Proc. Natl. Acad. Sci.* **71:** 4713–4717.

Noller, H.F. 1984. Structure of ribosomal RNA. *Annu. Rev. Biochem.* **53:** 119–162.

———. 1991. Ribosomal RNA and translation. *Annu. Rev. Biochem.* **60:** 191–227.

Nomura, M. and V.A. Erdmann. 1970. Reconstitution of 50S ribosomal subunits from dissociated molecular components. *Nature* **228:** 744–748.

Nomura, M., A. Tissières, and P. Lengyel, eds. 1974. *Ribosomes.* Cold Spring Harbor Laboratory, Cold Spring Harbor, New York.

Palade, G.E. 1955. A small particulate component of the cytoplasm. *J. Biophys. Biochem. Cytol.* **1:** 59–68.

Palade, G.E. and P. Siekevitz. 1956. Liver microsomes, an integrated morphological and biochemical study. *J. Biophys. Biochem. Cytol.* **2:** 171–200.

Prince, J.B., B.H. Taylor, D.L. Thurlow, J. Ofengand, and R.A. Zimmerman. 1982. Covalent crosslinking of transfer RNA val to 16S RNA in the ribosomal A site. Identification of crosslinked residues. *Proc. Natl. Acad. Sci.* **79:** 5450–5454.

Ramakrishnan, V., V. Graziano, and M.S. Capel. 1986. A role for proteins S3 and S14 in the 30S ribosomal subunit. *J. Biol. Chem.* **261:** 15049–15052.

Roberts, R.B. 1958. Preface. In *Microsomal particles and protein synthesis*, p. viii. Pergamon Press, New York.

Schachman, H.K., A.B. Pardee, and R.Y. Stanier. 1952. Studies on the macromolecular organization of microbial cells. *Arch. Biochem. Biophys.* **38:** 245–260.

Schopf, J.W., ed. 1983. *Earth's earliest biosphere: Its origins and evolution.* Princeton University Press, Princeton, New Jersey.

Senior, B.W. and I.B. Holland. 1971. Effect of colicin upon the 30S ribosomal subunit of *Escherichia coli. Proc. Natl. Acad. Sci.* **68:** 959–963.

Shine, J. and L. Dalgarno. 1974. The 3′-terminal sequence of *Escherichia coli* 16S ribosomal RNA: Complementarity to nonsense triplets and ribosome binding sites. *Proc. Natl. Acad. Sci.* **71:** 1342–1346.

Specht, T., J. Wolters, and V.A. Erdmann. 1990. Compilation of 5S ribosomal RNA and 5S ribosomal RNA gene sequences. *Nucleic Acids Res.* **19(S):** 2189–2191.

Spirin, A.S. 1964. *Macromolecular structure of ribonucleic acids.* Reinhold Publishing, New York.

Steitz, J.A. 1980. RNA-RNA interactions during polypeptide chain initiation. In *Ribosomes: Structure, function and genetics* (ed. G. Chambliss et al.), pp. 479–496. University Park Press, Baltimore, Maryland.

Subramanian, A.R. 1975. Copies of proteins L7 and L12 and the heterogeneity of the large subunit of the *Escherichia coli* ribosome. *J. Mol. Biol.* **95:** 1–8.

Tissières, A. 1974. Ribosome research: Historical background. In *Ribosomes* (ed. M. Nomura et al.), pp. 3–12. Cold Spring Harbor Laboratory, Cold Spring Harbor, New York.

Traub, P. and M. Nomura. 1968. Structure and function of *E. coli* ribosomes. V. Reconstitution of functionally active 30S ribosomal particles from RNA and proteins.

Proc. Natl. Acad. Sci. **59:** 777–784.

Traut, R.R., P.B. Moore, H. Delius, H. Noller, and A. Tissières. 1967. Ribosomal proteins of *E. coli.* Demonstration of different primary structures. *Proc. Natl. Acad. Sci.* **57:** 1294–1301.

Van Holde, K.E. and W.E. Hill. 1974. General physical properties of ribosomes. In *Ribosomes* (ed. M. Nomura et al.), pp. 53–91. Cold Spring Harbor Laboratory, Cold Spring Harbor, New York.

Waller, J.P. 1964. Fractionation of the ribosomal protein from *Escherichia coli. J. Mol. Biol.* **10:** 319–336.

Waller, J.P. and J.H. Harris. 1961. Studies on the composition of the protein from *Escherichia coli* ribosomes. *Proc. Natl. Acad. Sci.* **47:** 18–23.

Wittmann-Liebold, B. 1986. Ribosomal proteins: Their structure and evolution. In *Structure, function, and genetics of ribosomes* (ed. B. Hardesty and G. Kramer), pp. 326–361. Springer-Verlag, New York.

Woese, C.R. 1980. Just so stories and Rube Goldberg machines: Speculations on the origin of the protein synthetic machinery. In *Ribosomes: Structure, function and genetics* (ed. G. Chambliss et al.), pp. 357–376. University Park Press, Baltimore, Maryland.

Wool, I.G., Y. Endo, Y.-L. Chan, and A. Gluck. 1990. Structure, function and evolution of mammalian ribosomes. In *The ribosome: Structure, function and evolution* (ed. W.E. Hill et al.), pp. 203–214. American Society for Microbiology, Washington, D.C.

Wrede, P. and V.A. Erdmann. 1973. Activities of *B. stearothermophilus* 50S ribosomes reconstituted with procaryotic and eucaryotic 5S RNSA. *FEBS Lett.* **33:** 315–319.

Zamecnik, P.C. 1970. An historical account of protein synthesis, with current overtones—A personalized view. *Cold Spring Harbor Symp. Quant. Biol.* **34:** 1–16.

6

On the Origin of the Ribosome: Coevolution of Subdomains of tRNA and rRNA

Harry F. Noller
Sinsheimer Laboratories
University of California
Santa Cruz, California 95064

Translation is one of the most complicated of biological processes, involving literally hundreds of specific macromolecules. Not the least part of this complexity is the structure of the ribosome itself, which even in the relatively simple *Escherichia coli* version consists of more than 50 different proteins and 3 RNA molecules, giving an aggregate mass of around 2.5 million daltons (Hill et al. 1990). The difficulty of imagining how such a structure evolved is eased somewhat by accepting the notion that the original ribosome was made solely of RNA, as tentatively suggested by Crick (1968) more than two decades ago and asserted with increasing force (Woese 1980) and enthusiasm (Woese and Pace, this volume) since then. Adherents of RNA world scenarios imagine, to varying degrees, that something resembling protein synthesis was carried out by ribozyme-like protoribosomes prior to the advent of ribosomal proteins and translation factors (not to mention aminoacyl-tRNA synthetases). Such scenarios solve the chicken-or-the-egg problem of the molecular evolution of ribosomes but raise some difficult new questions. The most obvious problem is that rRNA itself has a vast and intricate structure containing thousands of nucleotides (see, e.g., Fig. 1) and is not likely to have evolved by chance in a few simple evolutionary events. Compounding this problem is that, prior to the existence of protein synthesis, it is difficult to imagine what selective pressures could have driven its evolution; in other words, the RNA world could not have anticipated the invention of protein synthesis. More likely, it evolved from some preexisting function. The question is, what was this function?

In this chapter, I begin by summarizing the evidence that supports the view that rRNA plays a central role in translation. This is followed by a simplified view of translation that is centered on interactions between

The RNA World
© 1993 Cold Spring Harbor Laboratory Press 0-87969-380-0/93 $5 + .00

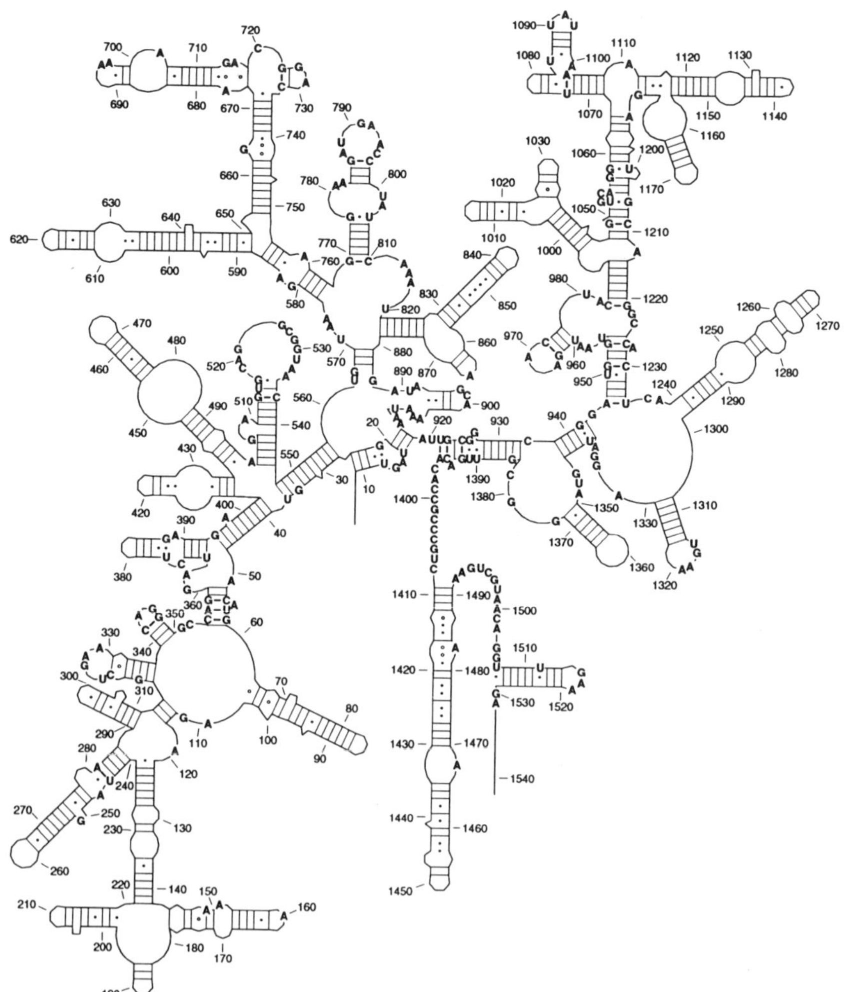

Figure 1 Secondary structure of 16S rRNA, showing (in capital letters) the bases that are universally conserved among the 16S-like rRNAs in all three primary kingdoms.

tRNA and rRNA. From this view emerges the idea of three prototypical functional domains: two "RNA domains" that define essential aspects of the mechanism and an "RNP domain" that is introduced later as a major refinement of the preexisting RNA mechanism. I then argue that each of the two RNA domains emerged from preexisting ribozymes and, specifically, that these ribozymes each contained both proto-tRNA and proto-rRNA elements. Thus, different parts of tRNA and rRNA coevolved separately in the RNA world, eventually merging into larger, more complex, multifunctional, multidomain molecules resembling the ones that we now find in ribosomes. Finally, some speculations are presented concerning details of structure, function, and mechanism.

EVIDENCE FOR THE FUNCTIONAL ROLE OF rRNA

There is now extensive evidence, both biochemical and genetic, to support the involvement of rRNA in the fundamental mechanism of translation. This subject has been reviewed elsewhere (Noller 1991a), so only a few of the many results are summarized here. In retrospect, the results of early in vitro reconstitution studies should have raised suspicions about the correctness of the ribosomal protein paradigm. With few exceptions, omission of single proteins from 30S subunits resulted in particles that retained at least partial activity in in vitro translational assays (Nomura et al. 1970). Most of the exceptions were proteins whose omission caused major assembly defects. The finding that cleavage of a single covalent bond in 16S rRNA by colicin E3 abolished the activity of 30S subunits (Bowman et al. 1971; Senior and Holland 1971) also had surprisingly little impact on mainstream ribosomology at that time.

The colicin result, along with the inactivation of tRNA binding by kethoxal modification of 16S rRNA (Noller and Chaires 1972) and direct cross-linking of the tRNA anticodon to 16S rRNA (Prince et al. 1982), gradually led to a narrowly held perception that 16S rRNA was somehow important for tRNA-ribosome interaction. Affinity-labeling studies using reactive groups coupled to the aminoacyl end of tRNA showed that 23S rRNA was in the immediate environment of the peptidyl transferase catalytic center (Breitmeyer and Noller 1976; Barta et al. 1984, 1990). Cleavage of a single bond in 23S rRNA by α-sarcin or ricin inactivated ribosomes, pointing again to the functional importance of the large subunit rRNA (Endo and Wool 1982; Endo et al. 1987). More often, however, the results of these and many other biochemical experiments were rationalized in terms of the protein paradigm (Fellner 1974).

Following the discovery of catalytic RNA (Kruger et al. 1982; Guerrier-Takada et al. 1983), the rRNA view gained wider acceptance, to the extent that at least one reviewer has cautioned that the pendulum may have swung too far (Moore 1990). Recently, the application of site-directed mutagenesis and other genetic approaches has led to the study of a large number of mutations in 16S and 23S rRNA. Although analysis of the effects of such mutants is still in its infancy, the results obtained so far are consistent with the functional importance of rRNA. This includes a mutation that impairs EF-Tu-dependent binding of aminoacyl-tRNA (Powers and Noller 1990 and unpubl.) and mutations that affect subunit association (Santer et al. 1990), initiation (Tapprich et al. 1989), antibiotic sensitivity (for review, see Cundliffe 1990), and translational accuracy (Allen and Noller 1991). In vivo mutagenesis, on the other hand, has led to isolation of mutants lacking individual ribosomal proteins that are nevertheless viable (Dabbs 1986). Consistent with this is the rather unremarkable sequence conservation found between ribosomal proteins from different species (Mian et al. 1993).

Recently, it has been shown that peptidyl transferase activity survives vigorous extraction with phenol following treatment with detergents and proteinase K (Noller et al. 1992). The activity of the protein-depleted particles is inhibited by established peptidyl transferase inhibitors such as chloramphenicol and carbomycin and is highly sensitive to ribonuclease. The nearly fully active particles are highly depleted in their protein content. These results suggest that even peptide bond formation, the single chemical reaction that is unambiguously catalyzed by the ribosome itself, may be a function carried out by rRNA.

FUNCTIONAL DOMAINS IN rRNA

tRNA-16S rRNA Interactions: The Decoding Site

Early studies showed that limited chemical modification of 30S ribosomal subunits with kethoxal caused loss of tRNA binding (most likely P-site binding), even though interactions between the ribosome and poly(U) message were largely unimpaired (Noller and Chaires 1972). Reconstitution experiments showed that the site of inactivation was 16S rRNA; proteins from the modified subunits were fully active when reconstituted with unmodified 16S rRNA. The subunits were protected from inactivation by bound tRNA, implying that modification of a small number of guanines in the tRNA-binding site was responsible for loss of function. We now know the identity of the protected guanines, as well as bases that are protected from other chemical probes by A- and P-site tRNA (Moazed and Noller 1986, 1990); Figure 2 shows their location in the secondary structure of 16S rRNA.

The bases protected by tRNA have the characteristics expected for functional sites: Most of them are completely invariant in the nearly 1000 16S-like rRNA sequences now available, including examples spanning the extremes of all three primary kingdoms (Ribosomal Database Project, University of Illinois). Among these protected bases, the site of anticodon-codon interaction is almost certainly in the vicinity of the cluster of conserved nucleotides around positions 1400 and 1500 (Fig. 1). The site of direct cross-linking of the wobble base of the P-site anticodon, at C1400 (Prince et al. 1982), is among the protected P-site bases and is near several others. The site of colicin E3 cleavage, A1493

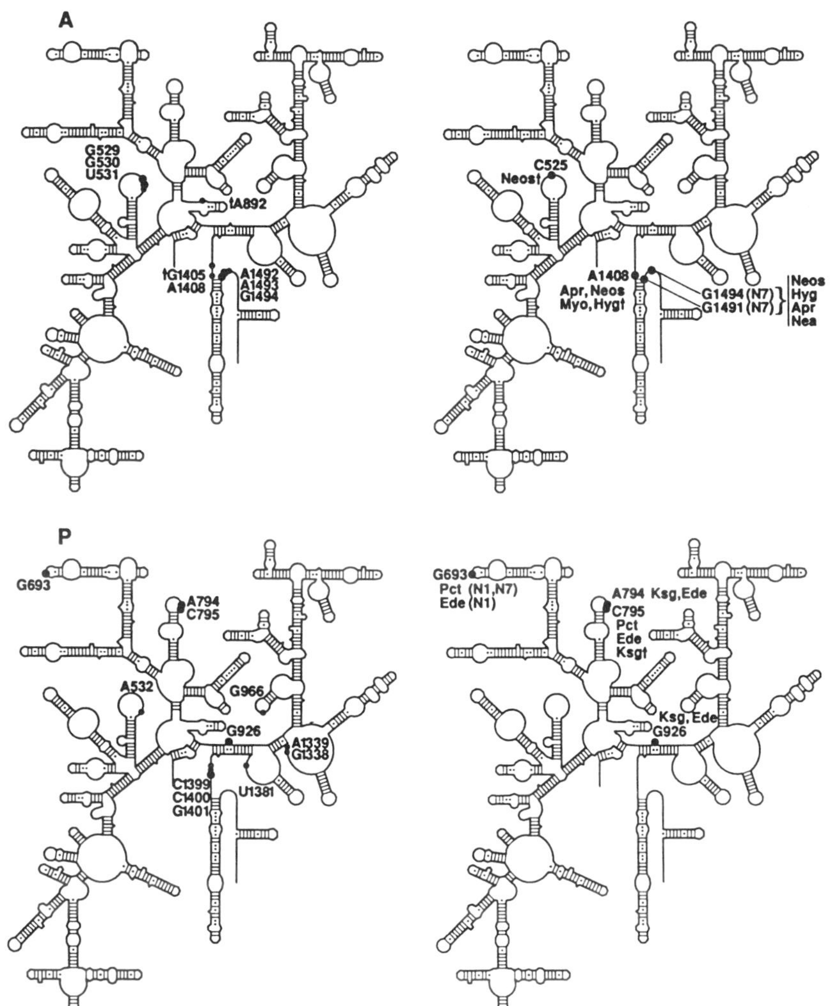

Figure 2 Bases in 16S rRNA that are protected from chemical modification by A-site (*upper left*) or P-site (*lower left*) tRNA (Moazed and Noller 1986, 1990). On the right are shown the bases that are protected by A-site-specific (*upper right*) and P-site-specific (*lower right*) antibiotics (Moazed and Noller 1987a; Woodcock et al. 1991).

(Bowman et al. 1971; Senior and Holland 1971), is adjacent to a cluster of strong A-site protections. In addition, many drugs that cause miscoding protect bases in this same region (Moazed and Noller 1987a), and mutations that confer resistance to these drugs are also found here (Cundliffe 1990).

There is good evidence that interactions between tRNA and the small ribosomal subunit involve exclusively (or nearly so) the anticodon stem-loop region of tRNA. First, the binding constant of a 15-nucleotide anticodon stem-loop fragment for the 30S P site does not differ significantly from that observed for the intact tRNA (Rose et al. 1983). Second, this same fragment protects all the same A- and P-site bases in 16S rRNA that are protected by intact tRNA (Moazed and Noller 1986). Finally, when bound to the 30S P site, the region of tRNA protected from hydroxyl radical attack corresponds almost precisely to the anticodon stem-loop (Fig. 3) (Hüttenhofer and Noller 1992). This indicates that the small subunit must contain a binding pocket that accommodates this stem-loop structure rather specifically and tightly, such that it becomes inaccessible even to solvent.

These results are consistent with what is known about the structure of the decoding site of the small subunit from model-building studies (Stern et al. 1988). According to this model, the P-site tRNA-protected nucleotides, although distributed widely in the 16S rRNA secondary structure, are localized in a compact neighborhood in three dimensions (Fig. 4). This is the same location (the "cleft") that has been shown experimentally to be the site of interaction of the anticodon end of P-site tRNA with the 30S subunit by difference electron microscopy (EM) (Wagenknecht et al. 1988). It has also been shown to be the location of the 1400 region of 16S rRNA (to which the anticodon has been cross-linked) by EM localization of a bound complementary oligonucleotide (Oakes et al. 1986).

By virtue of the fact that A- and P-site codons are contiguous, the A-site codon-anticodon interaction must also take place in, or very close to, the cleft. Indeed, all but one of the main A-site tRNA protections are at positions (1408, 1492, and 1493) very close to position 1400 and are contained in two of the longest stretches of universally conserved se-

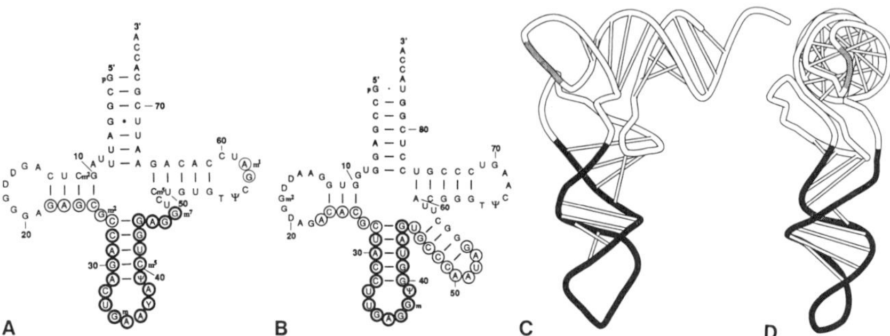

Figure 3 Protection by 30S ribosomal subunits of backbone ribose residues in tRNA from hydroxyl radical attack (Hüttenhofer and Noller 1992). Protected residues are circled (*A,B*) or shaded (*C,D*).

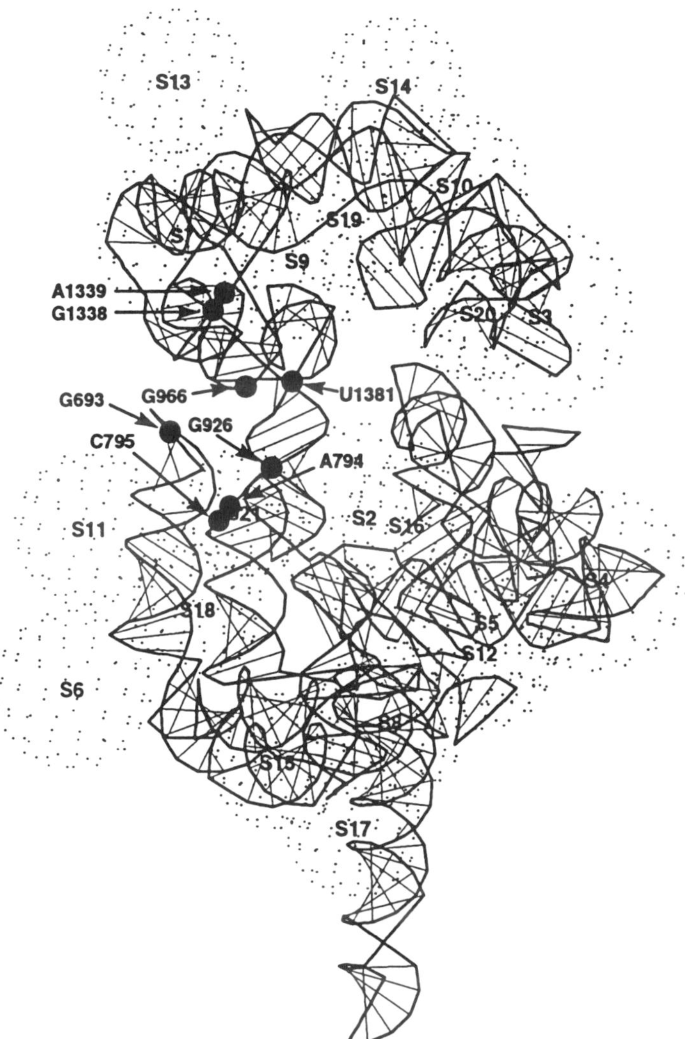

Figure 4 Model for the folding of 16S rRNA in the 30S subunit (Stern et al. 1988). Filled circles show bases protected from chemical probes by tRNA bound to the P site (Moazed and Noller 1990).

quence in the ribosome. This region, which contains these A-site tRNA protections, the sites of protection by many miscoding antibiotics, as well as sites of mutation conferring resistance to these drugs, the site of cleavage by colicin E3, and the site of a frameshift antisuppressor mutation (Weiss-Brummer and Hüttenhofer 1989), is, by all indications, the heart of the decoding site. It represents one of the primary functional domains of the ribosome.

tRNA-23S rRNA Interactions: The Peptidyl Transferase Site

Correspondingly, the other end (the CCA, or acceptor end) of tRNA interacts with the large ribosomal subunit; bases in 23S rRNA that are protected by tRNA in the A, P, and E sites are shown in Figure 5 (Moazed and Noller 1989). Almost all of the protected bases are located in domain V, mainly in the vicinity of the highly conserved central loop. Stepwise deletion of the aminoacyl moiety or the terminal A and C show that protection of specific bases depends on different structural components of the CCA terminus. Conversely, oligonucleotides such as CAACCA(f-Met), bound in the presence of sparsomycin and ethanol to the P site, protect almost exactly the same bases in 23S rRNA as in intact tRNA (Moazed and Noller 1991). These results are convincing evidence that the interaction of tRNA with the 23S rRNA P site involves almost exclusively its CCA terminus.

These protection studies are in close agreement with cross-linking results, where the aminoacyl group of P/P- and A/P-state tRNA has been cross-linked to bases 2451 and 2584, respectively (Fig. 6) (Barta et al. 1984, 1990). In addition, antibiotics that inhibit peptidyl transferase protect bases in this region, and mutations that confer resistance to these drugs are also found here (Fig. 6) (Moazed and Noller 1987b; Vester and Garrett 1988; Cundliffe 1990). These results, in addition to the peptidyl transferase activity of protein-depleted 23S rRNA (Noller et al. 1992), support the likelihood that the central loop of domain V of 23S rRNA,

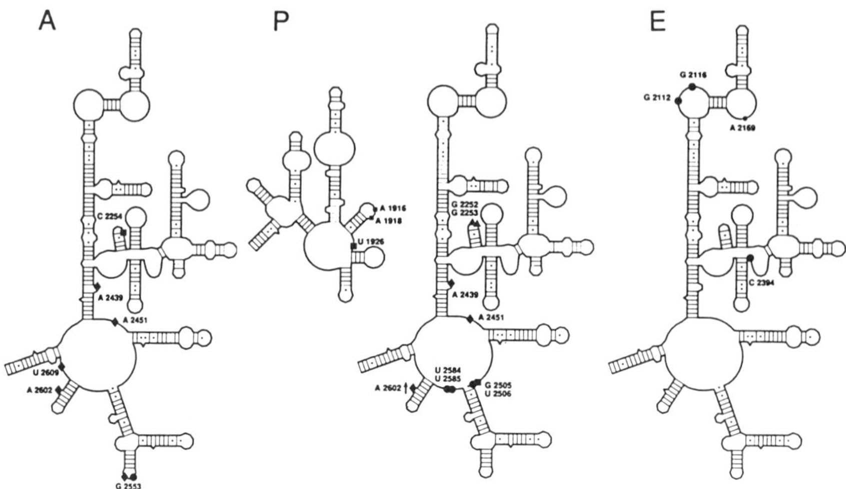

Figure 5 Protection of bases in 23S rRNA from chemical modification by binding tRNA to ribosomes in the A, P, and E sites (Moazed and Noller 1989).

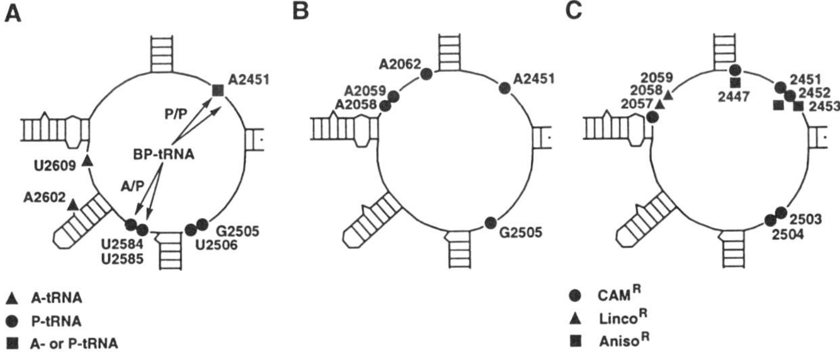

Figure 6 The central (peptidyl transferase) loop of domain V of 23S rRNA. (*A*) Bases protected from chemical probes by A-site (triangles), P-site (circles), or A- or P-site tRNA (square) (Moazed and Noller 1989). Bases affinity-labeled by the aminoacyl end of tRNA in the A/P or P/P states via benzophenone (Barta et al. 1984, 1990) are indicated by arrows. (*B*) Bases protected by chloramphenicol or carbomycin, two peptidyl transferase inhibitors (Moazed and Noller 1987b). (*C*) Sites of mutations conferring resistance to chloramphenicol (circles), lincomycin (triangles), or anisomycin (squares), three antibiotics that inhibit peptidyl transferase (for review, see Cundliffe 1990).

and peripheral elements, form a second primary functional domain, the peptidyl transferase site.

Elongation Factor-rRNA Interactions: An Intersubunit Regulatory Domain

According to a strict RNA paradigm, the GTP-dependent elongation factors EF-Tu and EF-G would be late evolutionary additions to the translation apparatus. Indeed, under the appropriate in vitro conditions, aminoacyl-tRNA can be bound to ribosomes in the absence of EF-Tu, and translocation occurs in the absence of EF-G. It is likely that the basic mechanism is, therefore, embodied in the ribosome itself and that the factors refine the process by enhancing its speed and accuracy.

Both EF-Tu and EF-G footprint the highly conserved loop around position 2660 of 23S rRNA (Fig. 7) (Moazed et al. 1988). This loop is commonly known as the α-sarcin loop because it is the site of attack by the cytotoxin α-sarcin (Endo and Wool 1982). The importance of this loop for factor-related functions has been reviewed recently (Wool et al. 1992). If GTP hydrolysis is blocked, the aminoacyl-tRNA.EF-Tu.GTP ternary complex is bound to the ribosome in a way that allows interaction of EF-Tu with the α-sarcin loop but prevents access of tRNA to the 23S rRNA peptidyl transferase site (Fig. 7) (Moazed et al. 1988). This is

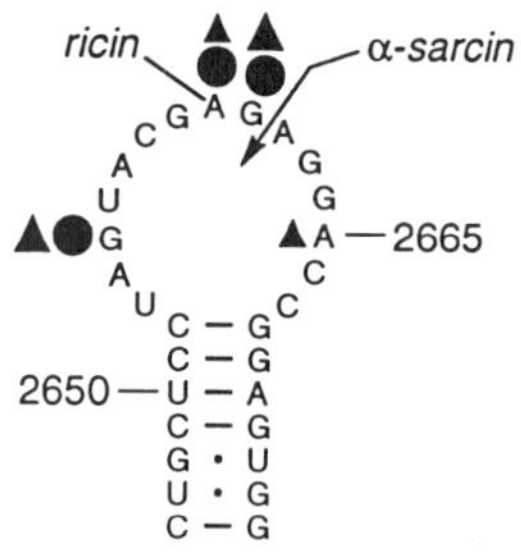

Figure 7 The α-sarcin loop of 23S rRNA, showing the sites of bond cleavage by α-sarcin and ricin (Endo and Wool 1982; Endo et al. 1987), and bases protected from chemical probes by EF-G (circles) and EF-Tu (triangles) (Moazed et al. 1988).

probably the manifestation of a proofreading mechanism by which the accuracy of tRNA selection is increased (Thompson 1988; Kurland et al. 1990). EF-G is known to interact with another site, around position 1067 in domain II of 23S rRNA, which has been shown to be the site of action of the EF-G-dependent GTPase inhibitor thiostrepton (Thompson et al. 1982).

Another conserved feature, the 530 loop of 16S rRNA (Fig. 8), has recently been implicated in EF-Tu-related function. Mutations of G530 carry a dominant lethal phenotype (Powers and Noller 1990). This is due to specific impairment of EF-Tu-dependent binding of aminoacyl-tRNA (T. Powers and H.F. Noller, unpubl.); nonenzymatic A-site binding of aminoacyl-tRNA is unaffected by such mutations, as are initiation, P-site binding, translocation, and subunit association. Furthermore, mutations in the α-sarcin loop appear to interfere with release of EF-Tu, but only in S12 restrictive strains (Tapprich and Dahlberg 1990). Since S12 is known to modulate the conformation of the 530 loop (Stern et al. 1989), these results provide further evidence for communication between the α-sarcin loop and the 530 loop, mediated by EF-Tu.

Structurally, the α-sarcin and 530 loops have some common features. Both are large, highly conserved hairpin loops (among the most highly conserved elements in rRNA) containing, according to chemical probing data, additional higher-order structure that is not yet completely understood. The 530 loop is involved in at least one pseudoknot interaction (Woese and Gutell 1989; Powers and Noller 1991), and the α-sarcin loop appears to exist as a tetraloop structure in at least one of its physiological states (Wool et al. 1992). Both loops, although themselves conserved, are embedded in domains that otherwise show rather low sequence conservation, in contrast to the decoding and peptidyl transferase sites. This, in

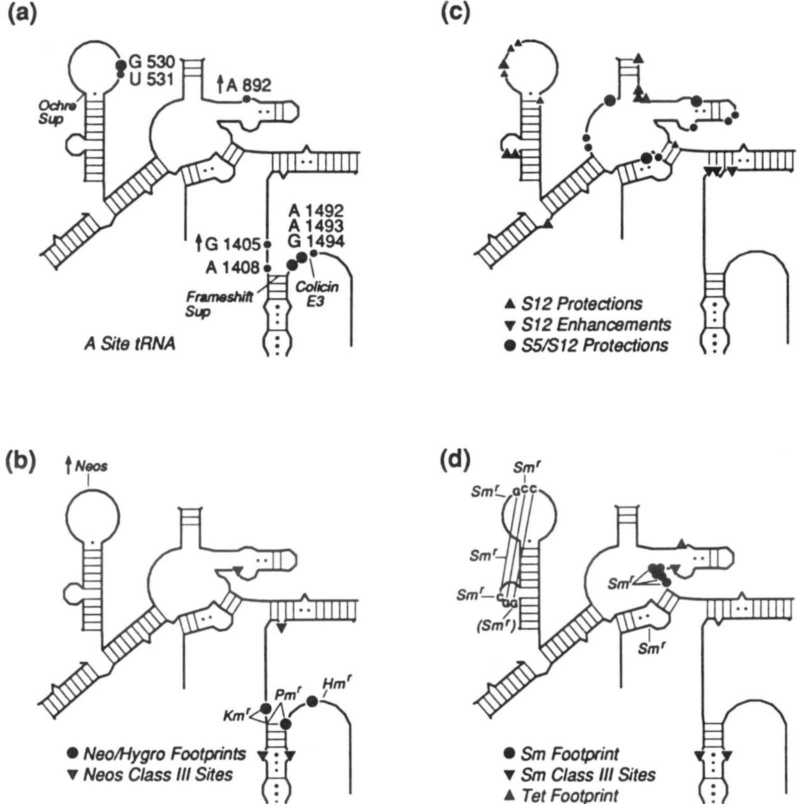

Figure 8   A-site-related regions of 16S-rRNA. (*a*) Bases protected by A-site tRNA (circles), locations of ochre and frameshift suppressor mutations, and site of cleavage by colicin E3. (*b*) Bases protected by neomycin-related antibiotics (circles), and class III sites: bases that are protected by antibiotics, tRNA, or 50S subunits (inverted triangles). (*c*) Bases affected by the assembly of ribosomal protein S12. (*d*) Sites of protection by streptomycin and locations of mutations in 16S rRNA that confer streptomycin resistance. Also shown is the pseudoknot interaction between the 530 loop and the side bulge around position 510. For details, see text, and Powers and Noller (1991).

addition to their structural and functional association with some of the most highly conserved proteins that are involved in protein synthesis, including S12, L7/L12, and EF-Tu, suggests a somewhat different character for these two RNA features: that of a ribonucleoprotein domain. This view is compatible with the suggestion that the factor-dependent functions evolved after establishment of the basic translational mechanisms of codon recognition and peptidyl transferase.

Although these two loops are on different RNA molecules, in different subunits, there is reason to believe that they are near each other in the structure of the 70S ribosome. The location of the site of interaction of EF-Tu and EF-G with the 50S subunit, established mainly by cross-linking studies (Maassen and Möller 1974), places the factors near protein L11 at the base of the L7/L12 stalk on the right side of the subunit (Fig. 9). EF-Tu has also been found to interact with the small subunit, on the right side, according to EM studies (Fig. 9) (Girshovich et al. 1986; Langer and Lake 1986). Thus, the sites of interaction of EF-Tu with the 30S and 50S subunits are juxtaposed in the 70S ribosome, as one might well expect. In the large subunit, EF-Tu interacts with the α-sarcin loop, which has not yet been localized, but by interpolation must lie somewhere between the peptidyl transferase site and L11. The site of interaction of EF-Tu with the 30S subunit is very near the position of the 530 loop, consistent with the genetic results. The RNP domain is, therefore, physically distinct from the two primary RNA functional domains, which are located on the left side of the ribosome (Fig. 9).

EVOLUTION OF TRANSLATIONAL PROTODOMAINS
FROM THE RNA WORLD

One of the main problems in thinking about the evolution of the ribosome is its size and structural complexity. This is especially crucial in any RNA world scenario, because even modern RNA genomes are notorious for their low fidelity of replication. Because of this, it is reasonable to imagine that the RNA genes, and therefore the corresponding

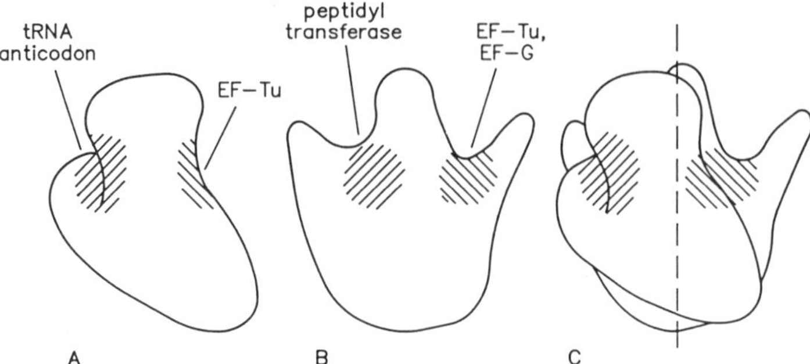

Figure 9 Location of the sites of interaction of the tRNA anticodon, its CCA end (peptidyl transferase), EF-Tu, and EF-G with (*A*) 30S subunits, (*B*) 50S subunits, and (*C*) 70S ribosomes, in the EM model (Girshovich et al. 1986; Langer and Lake 1986; Oakes et al. 1986; Wagenknecht et al. 1988). The proposed RNA protodomains are located to the left of the dashed line, and the proposed RNP protodomains are to the right (see text for details).

ribozymes, of the RNA world were rather small—a few hundred nucleotides at the largest. Even so, it is likely that their sequences showed much higher mutational variation than seen in DNA genes. The challenge, then, is to understand how the large rRNA structures could have evolved from a world where the largest RNAs were most likely an order of magnitude smaller.

Although the ribosome is large and complex, it seems increasingly evident that, for at least two primary translational functions, the regions of tRNA and rRNA that interact with each other are relatively compact. Codon recognition appears to involve no more than about 15 nucleotides of the tRNA structure and a small subdomain of 16S rRNA. The peptidyl transferase reaction involves only three nucleotides (the conserved CCA terminus) of tRNA and, apparently, only a limited region of one domain of 23S rRNA. Thus, the two functional domains, including their tRNA subfragment "substrates," are each within RNA world ribozyme size limits.

Here, a possible evolutionary advantage of RNA over proteins becomes evident. It is unlikely that the active site of a protein enzyme could exist on its own; this is because structures are usually stabilized by an elaborate hydrophobic core, typically involving a much larger proportion of the total molecular mass than the active site itself. RNA structure, in contrast, is largely determined locally. Nucleic acids are not believed to have hydrophobic interiors; rather, stability is provided by local hydrogen-bonded and base-stacking interactions and by coordination with magnesium and other multivalent cations. The existence of very small functional RNAs, or even RNA "naked active sites" is, therefore, a real possibility. An example is the "hammerhead" ribozyme, which catalyzes a specific RNA cleavage reaction with as few as 19 nucleotides (Uhlenbeck 1987). Thus, the existence of compact, functional protodomains provides a plausible way out of the dilemma of the spontaneous evolution of large, complex rRNAs. Furthermore, it may actually be easier to realize such structures with RNA than with protein.

Starting with small RNAs carries the further advantage that the daunting problem of correctly folding a very large RNA is avoided (the difficulty of correctly folding an RNA probably increases with something like the second power of its length). Given the existence of compact protodomains, larger complex assemblies resembling the ribosome could then arise by noncovalent association of protodomains, which could eventually be ligated into large, contiguous RNA molecules. Because gene products are also templates for replication in the RNA world, any such ligation products become potential genes, which can then be carried forward in evolution.

A central question underlying the evolution of protodomains is, What was the driving force (selective advantage) for their evolution? This leads to the further question, What were the preexisting RNA molecules in the RNA world from which the protodomains evolved? As mentioned above, the RNA world could not have "known" that it was in the process of evolving protein synthesis until the whole process became functional. Most likely, the functions of protodomains evolved as improvements or "spinoffs" from existing ribozyme-catalyzed functions. What were they? An RNA world genetic system would most likely be based on two well-known functions: replication and recombination. The latter would seem to be important in order to exploit to their full potential any catalytic RNAs that might emerge. Fortunately, there is no need to speculate on whether such functions could have been RNA-catalyzed, since both have been shown to be carried out by present-day ribozymes. Recombination in an RNA world amounts to what is currently referred to as trans-splicing. Both RNA replication and trans-splicing can be catalyzed by group I introns (Cech and Bass 1986).

Whether group I introns are ancient or modern is not yet clear; however, their existence alone is sufficient for arguing for the plausibility of such functions as starting points for the evolution of translational protodomains. The discovery that certain ribosome-directed antibiotics such as neomycin, kanamycin, and gentamycin also inhibit the function of group I introns raises the intriguing possibility that there may actually be an evolutionary link between present-day ribozymes and ribosomes (Von Ahsen et al. 1991).

A possible mechanistic basis for such a link has been suggested recently (Noller 1991b). Recognition of splice sites usually involves base-pairing via short complementary sequences. An interesting consequence of this is that for a generalized splicing reaction, in which the splice junctions are recognized in this way, the complex, at the instant of splicing, bears a striking resemblance to two tRNAs bound to mRNA at adjacent A- and P-site codons (Fig. 10). If the transesterification activity of the ribozyme could be adapted to catalysis of the chemically analogous acyl-transfer reaction of peptide bond formation, the stage would be set for the evolution of a primitive ribosome. Recent experiments suggest that such a possibility may not be out of the question. Piccirilli et al. (1992) engineered a group I intron so that its internal guide sequence could base-pair with the P-site peptidyl transferase substrate CAACCA(f-Met). This places the formyl-methionyl carbonyl group at the position of the phosphate group of the normal transesterification reaction. The modified ribozyme was shown to catalyze the hydrolysis of the model aminoacyl ester substrate. If the ribozyme can now be altered to

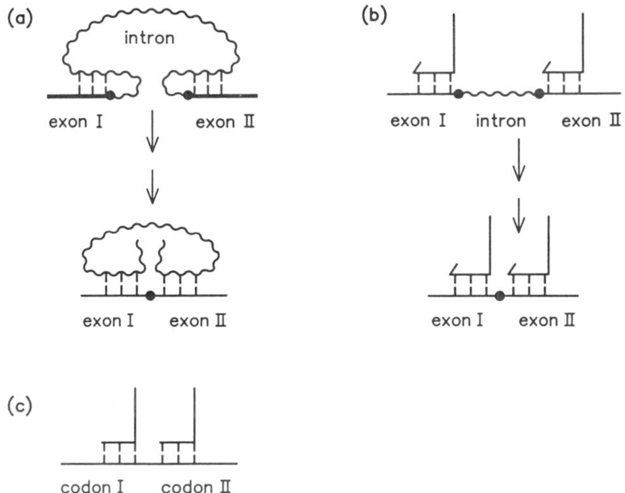

Figure 10 Base-pairing of short sequences flanking splice junctions in putative ancestral RNAs related to (*a*) group I introns, (*b*) small nuclear RNAs, which could have given rise to complexes that are structurally analogous to two tRNAs bound to adjacent codons on mRNA (*c*) (after Noller 1991b).

donate an aminoacyl group, rather than water, to the acyl group of methionine, catalysis of peptide bond formation would be achieved.

Conversely, it has been shown that ribosomal peptidyl transferase can catalyze transfer of aminoacyl groups to a phosphate center. Tarussova et al. (1981) have reported the formation of Ac-Met.GlyP-Phe from Ac-Met.GlyP and Phe-tRNA, where GlyP is a glycine analog in which the carboxyl group is replaced by phosphate. The reaction results in a phosphinoamide, in which a P-N bond is formed. This result is consistent with the evolution of peptidyl transferase from nucleic acid-based chemistry.

Independent evolution of the two extremities of tRNA (proto-tRNAs) in separate protodomains introduces a new paradox: uncoupling of the evolution of coding and peptidyl transferase. Coding without polypeptide synthesis or polypeptide synthesis without coding both seem pointless from an evolutionary standpoint. One solution to this paradox is that each of the proto-tRNAs had the ability to participate in both coding and polymerization, but that each became specialized for one of the two functions as the modern tRNA-ribosome system emerged. In this scenario, the anticodon stem-loop proto-tRNA would have had amino acid acceptor activity, and the acceptor-end proto-tRNA would have had recognition capability, in addition to their present-day properties. Certain emerging characteristics of tRNA identity and charging could be relics of proto-tRNA origins. Here, I take the extreme but parsimonious view that each proto-tRNA was directly involved in recognition of its own amino acid. This constrains the site of aminoacylation to the immediate vicinity of its amino acid recognition elements. In the first case, where the anticodon serves as a recognition element, esterification is only possible at 2′-OH groups, because of the remoteness of the 3′ end from the anticodon. In the second case, where the acceptor end contains the recognition element, esterification is possible at the chemically more active 2′,3′-OH group. Accumulating evidence suggests that aminoacyl-tRNA synthetases fall essentially into two classes; one class charges its tRNAs at the 2′-OH and the other almost exclusively at the 3′-OH (Eriani et al. 1990). Interestingly, three tRNAs that have been shown to have major identity elements in their acceptor ends, alanine, histidine, and glycine (Francklyn et al. 1992), belong to the class that is charged at the 3′-OH. Furthermore, two tRNAs whose anticodons are known to be crucial for recognition, glutamine and methionine, are charged at their 2′-OH groups. Unfortunately, there are exceptions; for example, tRNA aspartic acid has identity elements in its anticodon, but is charged at its 3′-OH. Nevertheless, the correlation is suggestive.

Recent studies have demonstrated the ability of RNA to recognize amino acids directly. Yarus (1988) has shown that arginine binds specifically to the guanosine cofactor-binding site of the group I intron. Remarkably, he has noted that, in over 100 different group I intron sequences, the binding site always contains an arginine codon. Yarus has suggested that the origin of the arginine codons might thus have occurred via direct amino acid-RNA recognition, possibly involving the codon-anticodon duplex. Although the guanidino side chain of arginine well suits this amino acid for specific hydrogen-bonded recognition by RNA sequences, it is not so easy to imagine specific recognition of some of the other amino acids, in particular the hydrophobics. However, the recent discovery of in-vitro-selected RNAs that bind specifically to hydrophobic amino acids (Famulok and Szostak 1992) suggests that even this may not be beyond the evolutionary scope of an RNA world.

CONCLUSION

The point of departure for this discussion is the rapidly unfolding picture of ribosome structure and function, and of ribosomal RNA in particular. Our current understanding is already beginning to place significant constraints on models for the evolution of translation and, as asserted in this chapter, may provide substantial clues to the nature of early versions of the translational apparatus. It seems almost inevitable that ribosomes evolved from pieces that were functionally independent, or nearly so. The absence of any detectable structural or sequence homology within or between 16S and 23S rRNA is consistent with independent evolutionary origins, as is the ability of the two extremities of tRNA to interact independently with their respective ribosomal subunits. However, some serious conceptual gaps remain in the scheme suggested here. Not the least of these is the merging of the two ends of tRNA. For example, in the case of a proto-tRNA destined to become an anticodon stem-loop, it is not clear how the recognition of an amino acid by its anticodon could eventually result in esterification of the amino acid to the CCA end. This and other difficulties left unaddressed here emphasize the importance of another long-standing question, How did tRNA identity evolve? Continuing progress in comparative molecular biology and in newly discovered in vitro RNA selection methods provide hope for answers to questions that have, until recently, been considered to be inaccessible.

ACKNOWLEDGMENTS

This work was supported by National Institutes of Health grant GM-17129 and by a grant to the Center for the Molecular Biology of RNA from the Lucille P. Markey Charitable Trust. This chapter was motivated by a Symposium on Molecular Biology and the Origin of Life, held at Berkeley, California in July 1990, sponsored by the Institute for Advanced Studies in Biology. I thank J. Abelson, H. Hartman, T. Powers, J. Sampson, M. Saks, and M. Yarus for discussions.

REFERENCES

Allen, P.N. and H.F. Noller. 1991. A single base substitution in 16S ribosomal RNA suppresses streptomycin dependence and increases the frequency of translational errors. *Cell* **66:** 141–148.

Barta, A., E. Kuechler, and G. Steiner. 1990. Photoaffinity labelling of the peptidyltransferase region. In *The ribosome: Structure, function and evolution* (ed. W.E. Hill et al.), pp. 358–365. American Society for Microbiology, Washington, D.C.

Barta, A., G. Steiner, J. Brosius, H.F. Noller, and E. Kuechler. 1984. Identification of a site on 23S ribosomal RNA located at the peptidyl transferase center. *Proc. Natl. Acad. Sci.* **81:** 3607–3611.

Bowman, C.M., J.E. Dahlberg, T. Ikemura, J. Konisky, and M. Nomura. 1971. Specific inactivation of 16S ribosomal RNA induced by colicin E3 in vivo. *Proc. Natl. Acad. Sci.* **68:** 964–968.

Breitmeyer, J.B. and H.F. Noller. 1976. Affinity labeling of specific regions of 23S rRNA by reaction of N-bromoacetyl-phenylalanyl-transfer RNA with *Escherichia coli* ribosomes. *J. Mol. Biol.* **101:** 297–306.

Cech, T.R. and B.L. Bass. 1986. Biological catalysis by RNA. *Annu. Rev. Biochem.* **55:** 599–629.

Crick, F.H.C. 1968. The origin of the genetic code. *J. Mol. Biol.* **38:** 367–379.

Cundliffe, E. 1990. Recognition sites for antibiotics within rRNA. In *The ribosome: Structure, function and evolution* (ed. W.E. Hill et al.), pp. 479–490. American Society for Microbiology, Washington, D.C.

Dabbs, E.R. 1986. Mutant studies on the prokaryotic ribosome. In *Structure, function and genetics of ribosomes* (ed. B. Hardesty and G. Kramer), pp. 733–748. Springer-Verlag, New York.

Endo, Y. and I.G. Wool. 1982. The site of action of α-sarcin on eukaryotic ribosomes. The sequence at the α-sarcin cleavage site in 28S ribosomal ribonucleic acid. *J. Biol. Chem.* **257:** 9054-9060.

Endo, Y., M. Mitsui, M. Motizuki, and K. Tsurugi. 1987. The mechanism of action of ricin and related toxic lectins on eukaryotic ribosomes. The site and the characteristics of the modification in 28S ribosomal RNA caused by the toxins. *J. Biol. Chem.* **262:** 5908–5912.

Eriani, G., M. Delarue, O. Poch, J. Gangloff, and D. Moras. 1990. Partition of tRNA synthetases into two classes based on mutually exclusive sets of sequence motifs. *Nature* **347:** 203–206.

Famulok, M. and J. Szostak. 1992. Stereospecific recognition of tryptophan agarose by in vitro selected RNA. *J. Am. Chem. Soc.* **114:** 3990–3991.

Fellner, P. 1974. Structure of the 16S and 23S ribosomal RNAs. In *Ribosomes* (ed. M. Nomura et al.), pp. 169–191. Cold Spring Harbor Laboratory, Cold Spring Harbor, New York.

Francklyn, C., J.P. Shi, and P. Schimmel. 1992. Overlapping nucleotide determinants for specific aminoacylation of RNA microhelices. *Science* **255:** 1121–1125.

Girshovich, A.S., E.S. Bochkareva, and V.D. Vasiliev. 1986. Localization of elongation factor Tu on the ribosome. *FEBS Lett.* **197:** 192–198.

Guerrier-Takada, C., K. Gardiner, T. Marsh, N. Pace, and S. Altman. 1983. The RNA moiety of RNase P is the catalytic subunit of the enzyme. *Cell* **35:** 849–857.

Hill, W.E., A.E. Dahlberg, R.A. Garrett, P.B. Moore, D. Schlessinger, and J.R. Warner. eds. 1990. *The ribosome: Structure, function and evolution.* American Society for Microbiology, Washington, D.C.

Hüttenhofer, A. and H.F. Noller. 1992. Hydroxyl radical cleavage of tRNA in the ribosomal P site. *Proc. Natl. Acad. Sci.* **89:** 7851–7855.

Kruger, K., P.J. Grabowski, A. Zaug, J. Sands, D.E. Gottschling, and T.R. Cech. 1982. Self-splicing RNA: Autoexcision and autocyclization of the ribosomal RNA intervening sequence of *Tetrahymena. Cell* **31:** 147–157.

Kurland, C.G., F. Jorgensen, A. Richter, M. Ehrenberg, N. Bilgin, and A.M. Rojas. 1990. Through the accuracy window. In *The ribosome: Structure, function and evolution* (ed. W.E. Hill et al.), pp. 513–526. American Society for Microbiology, Washington, D.C.

Langer, J.A. and J.A. Lake. 1986. Elongation factor Tu localized on the exterior surface of the small ribosomal subunit. *J. Mol. Biol.* **187:** 617–621.

Maassen, J.A. and W. Möller. 1974. Identification by photo-affinity labelling of the proteins in *Escherichia coli* ribosomes involved in EF-G-dependent GDP binding. *Proc. Natl. Acad. Sci.* **71:** 1277–1280.

Mian, I.S., B. Weiser, and H.F. Noller. 1993. Small subunit ribosomal proteins: Sequence and structure. *Prog. Nucleic Acids Res. Mol. Biol.* (in press).

Moazed, D. and H.F. Noller. 1986. Transfer RNA shields specific nucleotides in 16S ribosomal RNA from attack by chemical probes. *Cell* **47:** 985–994.

———. 1987a. Interaction of antibiotics with functional sites in 16S ribosomal RNA. *Nature* **327:** 389–394.

————. 1987b. Chloramphenicol, erythromycin, carbomycin and vernamycin B protect overlapping sites in the peptidyl transferase region of 23S ribosomal RNA. *Biochimie* **69:** 879–884.

————. 1989. Interaction of tRNA with 23S rRNA in the ribosomal A, P and E sites. *Cell* **57:** 585–597.

————. 1990. Binding of tRNA to the ribosomal A and P site protects two distinct sets of nucleotides in 16S rRNA. *J. Mol. Biol.* **211:** 135–145.

————. 1991. Sites of interaction of the CCA end of peptidyl-tRNA with 23S rRNA. *Proc. Natl. Acad. Sci.* **88:** 3725–3728.

Moazed, D., J.M. Robertson, and H.F. Noller. 1988. Interaction of elongation factors EF-G and EF-Tu with a conserved loop in 23S ribosomal RNA. *Nature* **334:** 362–364.

Moore, P.B. 1990. Comments on the 1989 International Conference on Ribosomes. In *The ribosome: Structure, function and evolution* (ed. W.E. Hill et al.), pp. xxi–xxiii. American Society for Microbiology, Washington, D.C.

Noller, H.F. 1991a. Ribosomal RNA and translation. *Annu. Rev. Biochem.* **60:** 191–227.

————. 1991b. Drugs and the RNA world. *Nature* **353:** 302–303.

Noller, H.F. and J.B. Chaires. 1972. Functional modification of 16S ribosomal RNA by kethoxal. *Proc. Natl. Acad. Sci.* **69:** 3115–3118.

Noller, H.F., V. Hoffarth, and L. Zimniak. 1992. Unusual resistance of peptidyl transferase to protein extraction methods. *Science* **256:** 1416–1419.

Nomura, M., S. Mizushima, M. Ozaki, P. Traub, and C.V. Lowry. 1969. Structure and function of ribosomes and their molecular components. *Cold Spring Harbor Symp. Quant. Biol.* **34:** 49–61.

Oakes, M.I., M.W. Clark, E. Henderson, and J.A. Lake. 1986. DNA hybridization electron microscopy: Ribosomal RNA nucleotides 1392–1407 are exposed in the cleft of the small subunit. *Proc. Natl. Acad. Sci.* **83:** 275–279.

Piccirilli, J.A., T.S. McConnell, A.J. Zaug, H.F. Noller, and T.R. Cech. 1992. Aminoacyl esterase activity of the *Tetrahymena* ribozyme. *Science* **256:** 1420–1424.

Powers, T. and H.F. Noller. 1990. Dominant lethal mutations in a conserved loop in 16S rRNA. *Proc. Natl. Acad. Sci.* **87:** 1042–1046.

————. 1991. A functional pseudoknot in 16S ribosomal RNA. *EMBO J.* **10:** 2203–2214.

Prince, J.B., G.N. Taylor, D. Thurlow, J. Ofengand, and R.A. Zimmermann. 1982. Covalent crosslinking of tRNAVal to 16S RNA at the ribosomal P site: Identification of crosslinked residues. *Proc. Natl. Acad. Sci.* **79:** 5450–5454.

Rose, S.J., P.T. Lowry, and O.C. Uhlenbeck. 1983. Binding of yeast tRNAPhe anticodon arm to *Escherichia coli* 30S ribosomes. *J. Mol. Biol.* **167:** 103-117.

Santer, M., E. Bennett-Guerrero, S. Byahatti, S. Czarnecki, D. O'Connell, M. Meyer, J. Khoury, X. Cheng, I. Schwartz, and J. McLaughlin. 1990. Base changes at position 792 of *Escherichia coli* 16S rRNA affect assembly of 70S ribosomes. *Proc. Natl. Acad. Sci.* **87:** 3700–3704.

Senior, B.W. and I.B. Holland. 1971. Effect of colicin E3 upon the 30S ribosomal subunit of *Escherichia coli. Proc. Natl. Acad. Sci.* **68:** 959–963.

Stern, S., B. Weiser, and H.F. Noller. 1988. Model for the three-dimensional folding of 16S ribosomal RNA. *J. Mol. Biol.* **204:** 447–481.

Stern, S., T. Powers, L.-M. Changchien, and H.F. Noller. 1989. RNA-protein interactions in 30S ribosomal subunits: Folding and function of 16S rRNA. *Science* **244:** 783–790.

Tapprich, W.E. and A.E. Dahlberg. 1990. A single base mutation at position 2661 in *E. coli* 23S RNA affects the binding of ternary complex to the ribosome. *EMBO J.* **9:** 2649–2655.

Tapprich, W.E., D.J. Goss, and A.E. Dahlberg. 1989. Mutation at position 791 in *Escherichia coli* 16S ribosomal RNA affects processes involved in the initiation of protein synthesis. *Proc. Natl. Acad. Sci.* **86:** 4927–4931.

Tarussova, N.B., G.M. Jacovleva, L.S. Victorova, M.K. Kukhanova, and R.M. Khomutov. 1981. Synthesis of an unnatural P-N bond catalyzed with *Escherichia coli* ribosomes. *FEBS Lett.* **130:** 85–87.

Thompson, J., F. Schmidt, and E. Cundliffe. 1982. Site of action of a ribosomal RNA methylase conferring resistance to thiostrepton. *J. Biol. Chem.* **257:** 7915–7917.

Thompson, R.C. 1988. EF-Tu provides an internal kinetic standard for translational accuracy. *Trends Biochem. Sci.* **13:** 91–93.

Uhlenbeck, O.C. 1987. A small catalytic oligoribonucleotide. *Nature* **328:** 596–600.

Vester, B. and R.A. Garrett. 1988. The importance of highly conserved nucleotides in the binding region of chloramphenicol at the peptidyl transferase centre of *Escherichia coli* 23S ribosomal RNA. *EMBO J.* **7:** 3577–3587.

Von Ahsen, U., J. Davies, and R. Schroeder. 1991. Antibiotic inhibition of group 1 ribozyme function. *Nature* **353:** 368–370.

Wagenknecht, T., J. Frank, M. Boublik, K. Nurse, and J. Ofengand. 1988. Direct localization of the tRNA-anticodon interaction site on the *Escherichia coli* 30S ribosomal subunit by electron microscopy and computerized image averaging. *J. Mol. Biol.* **203:** 753–760.

Weiss-Brummer, B. and A. Hüttenhofer. 1989. The paromomycin resistance mutation (par r-454) in the 15S rRNA gene of the yeast *Saccharomyces cerevisiae* is involved in ribosomal frameshifting. *Mol. Gen. Genet.* **217:** 362–369.

Woese, C.R. 1980. Just so stories and Rube Goldberg machines: Speculations on the origin of the protein synthetic machinery. In *Ribosomes: Structure, function and genetics* (ed. G. Chambliss et al.), pp. 357-373. University Park Press, Baltimore, Maryland.

Woese, C.R. and R.R. Gutell. 1989. Evidence for several higher order structural elements in ribosomal RNA. *Proc. Natl. Acad. Sci.* **86:** 3119–3122.

Woodcock, J., D. Moazed, M. Cannon, J. Davies, and H.F. Noller. 1991. Interaction of antibiotics with A-and P-site-specific bases in 16S ribosomal RNA. *EMBO J.* **10:** 3099–3103.

Wool, I.G., A. Glück, and Y. Endo. 1992. Ribotoxin recognition of ribosomal RNA and a proposal for the mechanism of translocation. *Trends Biochem. Sci.* **17:** 266–269.

Yarus, M. 1988. A specific amino acid binding site composed of RNA. *Science* **240:** 1751–1758.

7

Transfer RNA: An RNA for All Seasons

Dieter Söll
Department of Molecular Biophysics and Biochemistry
Yale University
New Haven, Connecticut 06511

Long before the existence of an RNA world was accepted, Francis Crick attributed to tRNA enzymatic properties (like a protein) because of its versatility and ability to recognize so many proteins with exquisite specificity. Presumably, this was meant to reflect a well-known property of proteins: conformational flexibility, which is a prerequisite for the manifold recognition characteristics of this small RNA molecule. Even though tRNA alone has been shown to have only marginal enzymatic properties (as discussed below), the many roles of tRNA demand that this molecule assume different conformations to provide diverse points of contact for the selective recognition of many different proteins.

Over the years, tRNA has provided complex challenges in different experimental areas. For biophysicists it provided the first high-resolution X-ray crystal structure of an RNA (Kim et al. 1974; Robertus et al. 1974) and of an RNA:protein complex (Rould et al. 1989), whereas it challenged biochemists to achieve the first sequence determination of an RNA molecule (Holley et al. 1965). For chemists it afforded the first illustration of synthesis of a gene that was biologically active (Ryan et al. 1979), whereas informational suppression acquainted geneticists with tRNA and led them to discover basic principles of protein biosynthesis and the genetic code (for review, see Garen 1968; Steege and Söll 1979). Molecular biologists used the highly specific interaction of tRNAs with aminoacyl-tRNA synthetases to define principles of protein:RNA interaction (for review, see Pallanck and Schulman 1992), whereas molecular archaeologists found tRNA molecules useful in establishing evolutionary relationships of organisms (Cedergren et al. 1991), in determining the age of the genetic code (Eigen et al. 1989), and in searching for fossils of the RNA world (Weiner and Maizels 1987; Maizels and Weiner, this volume). In this chapter, I first summarize some general properties of tRNA and then describe briefly the diverse roles of this versatile RNA molecule.

The RNA World
© 1993 Cold Spring Harbor Laboratory Press 0-87969-380-0/93 $5 + .00

General Characteristics of tRNA

tRNA molecules are ubiquitous in every cell. The tRNA population varies in complexity from 29 species in *Mycoplasma* (Muto et al. 1990), the smallest autonomously self-replicating organism, to presumably more than 150 species in a mammalian cell. The major and best-known function of this class of RNA is to activate amino acids for protein synthesis (Lapointe and Giege 1991). For this function there are usually several tRNA species (isoacceptors) for each amino acid. Two critical regions of the tRNA are the anticodon (interacting with the messenger RNA codon) and the acceptor end, where the adenosine of the universal 3'-terminal CCA-sequence is esterified at the 2' or 3' position with the amino acid specific (cognate) to the particular tRNA species. Different cell compartments (e.g., cytoplasm, chloroplast, mitochondrion) contain distinct protein-synthesizing systems; the chloroplast and mammalian mitochondria contain their own set of tRNAs different from those in the cytoplasm, whereas some plant mitochondrial tRNAs are of nuclear origin.

Structure

Currently there are almost 2000 tRNA or tRNA gene sequences known from more than 200 organisms and organelles (Steinberg et al. 1993). Most of these sequences can be folded into the familiar cloverleaf model of secondary structure (Fig. 1) that was predicted in 1965 from yeast tRNA[Ala], the first tRNA sequence to be determined (Holley et al. 1965). Although the chain length of the known tRNAs varies from 72 to 95 nucleotides, the different sequences can fit into the cloverleaf structure as the additional nucleotides are accommodated in the "extra arm." Notable exceptions to this rule are mitochondrial tRNAs from mammals, flagellates, and nematodes, which show gross deviations in the D-stem/loop or T-stem/loop structure or lack the D-loop altogether. Because these structures represent functional tRNAs, they may embody interesting remnants in the evolution of tRNA structure (Okimoto and Wolstenholme 1990) that have preserved some semblance with the minimal adapter. Such an RNA must have existed in early protein biosynthesis (Eigen and Winkler-Oswatitsch 1981) and is still manifested in the ability of the acceptor helix alone, part of the more ancient top half of tRNA (Maizels and Weiner, this volume), to be aminoacylated (Imura et al. 1969; Francklyn and Schimmel 1989).

Although the crystal structures are known only for tRNAs with short extra arms, the three-dimensional structure of tRNAs appears quite conserved (Basavappa and Sigler 1991). It is this structural similarity of different tRNAs that demands rigorous discrimination for specific protein-

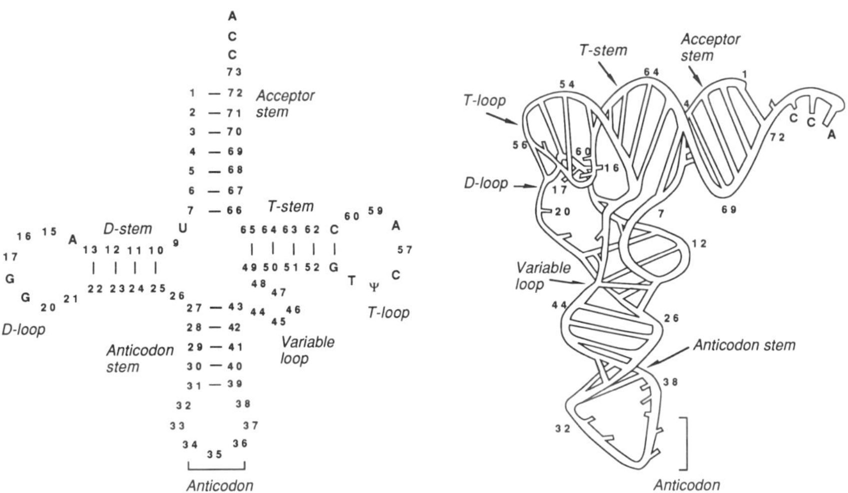

Figure 1 Secondary and tertiary structure of tRNA. (*Left*) Invariant nucleotides in the cloverleaf structure and base-pairing (dashes) are indicated. (*Right*) The L-shaped model of tertiary structure.

tRNA recognition in the diverse functions of this class of RNA molecules.

Another interesting facet of tRNAs is their high content of modified nucleosides, which display a dazzling structural variety (Yokoyama and Nishimura 1993). The degree of modification varies with the source of the tRNA and appears to increase in amount with the evolutionary development of the organism. For example, in human tRNAs, up to 25% of the bases may be modified, whereas there is much less modification in bacterial tRNAs (Steinberg et al. 1993). Currently, the structures of almost 80 modified nucleotides are known (Edmonds et al. 1991), with additional ones isolated but not yet determined. The role of modified nucleosides is not clear, although a number of them have been implicated in processes ranging from restriction (Ohashi et al. 1970; Yoshida et al. 1970; Björk 1992) and expansion (Crick 1966; Söll et al. 1966; Osawa et al. 1992; Atkins and Gesteland 1993) of codon recognition to their necessity as antideterminants or identity elements for aminoacylation by some aminoacyl-tRNA synthetases (Muramatsu et al. 1988; Sylvers et al. 1993).

Biosynthesis

tRNA genes are transcribed into larger precursors and then processed to mature tRNA. The initial transcript may include several tRNAs (up to 21 in *Bacillus subtilis* [Green and Vold 1983]), whereas eukaryotic transcripts are mostly monomeric, since RNA polymerase III promoters are internal in the tRNA gene. In some cases, the initial transcript also contains introns. A number of exo- and endonucleases are required for properly trimming away the extra nucleotides (Deutscher 1990; Altman et al. 1993). At some stages in this maturation process, the modified bases are formed either by additional modifications (e.g., methylation) of the bases in the transcript (see, e.g., Persson et al. 1992) or by excision of the original base and transglycosylase-mediated insertion of a preformed modified base (Shindo-Okada et al. 1980).

Most interesting is the fact that tRNA genes contain two different types of introns and thus have acquired different mechanisms for removal of these sequences. In eukaryotes the conventional protein-mediated splicing process (Zillmann et al. 1991) removes introns of limited length (8–60 nucleotides in yeast tRNA genes [Gamulin et al. 1983; Ogden et al. 1984]) from tRNA precursors derived from nuclear-encoded tRNA genes. However, some chloroplast tRNA genes with very long introns (458 nucleotides) do self-splice, as was predicted from the structural similarity of their intervening sequence with group I introns (Steinmetz et al. 1982; Dinter-Gottlieb 1986). Recently it was shown that shorter (205–237 nucleotides) group I introns also occur in tRNA genes of widely divergent bacteria, e.g., *Agrobacterium tumefaciens* and *Azoarcus* (Reinhold-Hurek and Shub 1992). If eubacteria are phylogenetically older than archaebacteria or eukaryotes, these findings may support the view that self-splicing tRNA introns are older than the protein-spliced introns of archaebacterial and eukaryotic tRNA genes (Cavalier-Smith 1991).

RNA editing, a fascinating mechanism of RNA formation, may also play a role in tRNA biosynthesis. The rat genome contains at least 10 tRNA[Asp] genes with the sequence $C_{32}T_{33}$ just preceding the anticodon. Single-strand conformation polymorphism analysis and sequence analysis of PCR fragments demonstrated that all tRNA[Asp] genes possess this sequence. However, the major cytoplasmic aspartate tRNA in rat liver and rat ascites contains a $U_{32}C_{33}$ sequence (Beier et al. 1992). These results strongly suggest that the major rat tRNA[Asp] is generated by post-transcriptional pyrimidine transitions in positions 32 and 33.

Aminoacyl-tRNA Formation

Aminoacyl-tRNA is synthesized by the cognate aminoacyl-tRNA synthetase, which forms an aminoacyl ester linkage with the 2′ or 3′

hydroxyl of the ribose of the terminal adenosine of the acceptor stem (Moras 1992). The activated amino acid can then participate in various acylation reactions. The formation of aminoacyl-tRNA is a highly specific process, as attachment of the wrong amino acid will cause an error in protein synthesis. Only recently have some of the principles assuring specificity in this reaction been understood (for review, see Pallanck and Schulman 1992). In most tRNA molecules, the bases making specific contacts with the cognate aminoacyl-tRNA synthetase are located in two regions, the anticodon loop and the acceptor stem. This fact was already accurately demonstrated and predicted in the 1960s (Hayashi and Miura 1966; Imura et al. 1969). The direct attachment of the correct amino acid to the cognate tRNA also implies that at least one aminoacyl-tRNA synthetase is present for each amino acid. However, there is a notable exception, as some organisms and organelles lack a glutaminyl-tRNA synthetase activity. In these cases, Gln-tRNA^{Gln} is formed instead by transamidation of glutamate mischarged to tRNA^{Gln} (see below) (Schön et al. 1988b).

With What Macromolecules Does tRNA Interact?

The average tRNA molecule interacts with probably 30–40 proteins during its life cycle. The components of the transcription machinery are numerous, and specific interaction of *Escherichia coli* RNA polymerase with fMet-tRNA$^{Met}_i$ has been observed (Pongs and Ulbrich 1976). Even greater are the possibilities of interaction during tRNA biosynthesis where, in addition to the processing nucleases (Deutscher 1990), a complex enzymatic machinery exists for the formation of modified nucleotides (Björk et al. 1987). For example, the formation of 5-methylaminomethyl-2-thiouridine, a nucleotide important for the identity of *E. coli* tRNA^{Glu} (Sylvers et al. 1993), requires at least five enzymes for its formation (Björk et al. 1987). During protein synthesis, the ribosome with its 55 proteins provides a complex interface of interactions; elongation, initiation, and termination factors also exhibit specific tRNA interactions. In addition, other proteins (see below) facilitate the function of tRNAs in various cellular reactions.

Other RNAs are also partners in interactions with tRNA. Most important in protein synthesis is the well-known codon-anticodon interaction with mRNA. There is also a complex interaction on the ribosome where the anticodon is in the vicinity of nucleotide 1400 of 16S rRNA (Ciesiolka et al. 1985) and the acceptor end close to the peptidyl transferase center, which is partly made up of RNA (Noller et al. 1992).

tRNA AS CARRIER OF ACTIVATED AMINO ACIDS

Many functions of tRNA rely on its role as carrier of activated amino acids. In some cases the activated amino acid is used intact, whereas for other functions it is further modified.

Activated Amino Acids Are Used Intact

Protein Biosynthesis

The role of aminoacyl-tRNA in ribosomal protein synthesis is well known (Hill et al. 1990). However, interest in the mechanism of protein synthesis has undergone a renaissance since it became clear that the protein-synthesizing apparatus and certain tRNAs are quite "flexible" in bringing about frameshifting, hopping, or selenocysteine insertion (Böck et al. 1991). These reprogrammed genetic decoding (Gesteland et al. 1992) events have stretched the common notion of a simple trinucleotide as a universal codon.

Regulation of Protein Synthesis by Codon Restriction

Early fractionation studies of tRNA revealed significant differences in the abundance of individual isoacceptor RNA species in organisms. Later it was shown that a correlation exists between codon usage and the amount of tRNA with the corresponding anticodon (Yamao et al. 1991). As one would expect, low concentrations of a particular tRNA slow down the rate of in vitro translation of the cognate codon (Anderson 1969). A beautiful example of how an organism may make use of such a "codon restriction" is seen in the developmental control of antibiotic production in *Streptomyces coelicolor* and *Streptomyces lividans* (Fernandez-Moreno et al. 1991; Leskiw et al. 1991). In this organism, aerial mycelium formation and the formation of the antibiotic actinorhodin are blocked by mutations in the *bldA* gene, which encodes a tRNA^{Leu} species capable of recognizing the codon UUA. This extremely rare codon in *Streptomyces* is present in the gene for the transcriptional activator of the actinorhodin biosynthetic genes and in the gene for a putative transmembrane export protein. Changing the rare TTA codon to TTG restores the normal synthesis of actinorhodin. The characterized *bldA* mutations all map within the mature tRNA coding region (Lawlor et al. 1987). Compared to other tRNAs, the *S. lividans* tRNA$^{Leu}_{UUA}$ has a less stable structure of the D-loop, as it lacks the conserved $G_{17b}G_{18a}$ sequence (Ueda et al. 1992). This may augment the deleterious effect of the *bldA* mutations.

A similar explanation may hold for the *E. coli argU* gene (Chen et al.

1991a), which encodes a minor tRNA^{Arg} capable of decoding the rare AGA codon (Anderson 1969; Yamao et al. 1991). The *argU10* (previously *dnaY*) mutation effects a G→A change in the first nucleotide of the mature tRNA and causes temperature sensitivity of growth. The mutation limits the expression of inducible genes with multiple AGA codons, whereas genes lacking this arginine codon are less affected.

Cell Wall Biosynthesis

The involvement of tRNA in cell wall biosynthesis was discovered in the 1960s and actively investigated for a number of years thereafter; however, there has been no new work in this area (Bumsted et al. 1968; Petit et al. 1968; Roberts et al. 1968). The cell wall peptidoglycan of *Staphylococcus* contains interpeptide bridges. It is known that glycine, serine, and threonine are transferred by their respective tRNAs via a lipid intermediate into the bridge peptides. The process is catalyzed by a particulate enzyme fraction and is specific for the amino acid. The tRNAs are charged by the normal aminoacyl-tRNA synthetase. Because one of the *S. aureus* and also *S. epidermidis* tRNA^{Gly} species did not support in vitro protein synthesis (Bumsted et al. 1968; Stewart et al. 1971), it is not clear whether a special tRNA species is involved in this process. Sequence analysis of the tRNA involved in peptidoglycan synthesis revealed two closely related *S. epidermidis* tRNA^{Gly} species (Roberts 1972), which were also seen in *S. aureus* (R.J. Roberts, pers. comm.). This makes it likely that this tRNA is sequestering glycine for use in cell wall biosynthesis.

Targeting Proteins for Degradation

In early studies on cell-free protein synthesis, a "soluble" amino acid incorporation system was observed where an activity in ribosome-free *E. coli* extracts transferred leucine and phenylalanine from their respective tRNAs onto proteins with amino-terminal arginine or lysine (Kaji et al. 1965; Leibowitz and Soffer 1969). In eukaryotes a similar enzymatic activity, arginyl-tRNA-protein transferase, was observed. This enzyme transfers arginyl residues onto proteins with acidic amino termini (Kaji 1968; Soffer 1980). More recent work establishes in *E. coli* and in eukaryotic systems that these aminoacyl-tRNA-protein transferases are involved in protein turnover.

Degradation of some proteins in the ubiquitin-mediated proteolytic pathway (Hershko and Ciechanover 1992) depends crucially on the nature of the amino-terminal amino acid. Proteins with acidic amino-terminal amino acids are degraded by the ubiquitin system only after conversion of the amino terminus to a basic moiety by the covalent addition of arginine (Ferber and Ciechanover 1987; Gonda et al. 1989). The reason for this is the inability of proteins with acidic amino termini to bind to the ubiquitin-protein ligase (*E3*), which is the required substrate recognition component of the ubiquitin pathway. However, proteins with amino-terminal arginine bind readily to *E3* (Elias and Ciechanover 1990). Partial purification of the rabbit reticulocyte arginyl-tRNA-protein transferase yielded a native enzyme of molecular mass ≈ 360 kD composed of a complex of arginyl-tRNA synthetase and arginyl-tRNA-protein transferase (Ciechanover et al. 1988). The gene (*ATE1*) encoding this enzyme has been cloned from yeast (Balzi et al. 1990), and the deduced amino acid sequence predicts a 503-residue protein.

Recently it has been shown that amino-terminal leucine and phenylalanine (and a few other amino acids) confer short half-lives to a test protein in *E. coli* extracts (Tobias et al. 1991). Although bacteria do not possess the ubiquitin pathway, it is believed that a similar recognition process of proteins involving their amino-terminal amino acids is a part of the protein degradation pathway involving the ATP-dependent protease Clp (Ti). Proteins having amino-terminal arginine and lysine are targeted for degradation by this pathway after posttranslational addition of leucine or phenylalanine onto their amino termini. Thus, the *E. coli* leucyl- and phenylalanyl-tRNA-protein transferases described earlier (Kaji et al. 1965; Leibowitz and Soffer 1969) play a role in Clp-dependent proteolysis in *E. coli* that is analogous to that played by the arginyl-tRNA-protein transferase in ubiquitin-dependent proteolysis in eukaryotes.

Experimental indications that amino acids other than arginine, leucine, and phenylalanine may also be transferred in this fashion may portend the discovery of additional aminoacyl-tRNA-protein transferases as the proteolytic pathways become better understood. Although the pathways of protein degradation differ in prokaryotes and eukaryotes, the principle of amino-terminal addition of amino acids activated by ligation to tRNA was probably a biochemically smart strategy which was kept, even though the pathways evolved differently.

tRNA as a Barometer of Metabolism

A sufficient supply of aminoacyl-tRNA is crucial for the well-being of a bacterial cell that normally contains a functional excess of charged tRNA to meet all requirements of metabolism and balanced synthesis of macromolecular components. Low concentration of charged tRNA (e.g., as brought about by insufficient activity of an aminoacyl-tRNA synthetase due to nutritional conditions) causes amino acid deprivation in two ways.

One is an amino-acid-specific response via attenuation, where the role of aminoacyl-tRNA in governing transcriptional regulation is well understood (for review, see Yanofsky and Crawford 1987). The second is a general response via the *relA*-mediated production of the magic spot compounds ppGpp and pppGpp, alarmones, which signal modestly increased general transcription of the biosynthetic enzymes for several amino acids (Xiao et al. 1991). Other consequences of variations in these alarmones can be induction of amino acid auxotrophy and increased sensitivity to oxidation.

Transcriptional regulation appears to play a role in the well-known phenomenon that amino acid transport and biosynthesis are associated with intracellular availability of aminoacyl-tRNA. Best-studied among the transport systems is that for the branched-chain amino acids leucine, isoleucine, and valine (LIV) in *E. coli*. Control of expression of the genes of the LIV-I permease system for the high-affinity transport of leucine, isoleucine, and valine involves rho-dependent transcription termination of *livJ* mRNA in which antitermination is associated with the availability of leucyl-tRNA (Williamson and Oxender 1992). This transcription termination mechanism is most reminiscent of amino acid biosynthetic operons (Yanofsky and Crawford 1987). It is suggested that leucyl-tRNA levels may alter the activity or expression of a *trans*-acting regulatory factor that prevents premature termination, although there is no information on the mechanism of this process. However, it has been observed in structural modeling studies that the leucine-specific binding protein, a component of the LIV-I transport system, contains a 121-amino-acid sequence with sequence and structural similarities to a region in the putative nucleotide-binding domain of leucyl-tRNA synthetase. Possibly, leucine transport and leucyl-tRNA formation are performed by evolutionarily related proteins (Williamson and Oxender 1990).

Activated Amino Acids Are Further Modified

tRNA-mediated amino acid transformations have been known for some time but only recently have attracted more attention. In some cases, the modifications require the activation of the amino acid's carboxyl group, whereas in other cases, activation by tRNA is not important for the subsequent modification.

Methionine → Formylmethionine

Initiation of protein synthesis in eubacteria (e.g., *E. coli*) requires the presence of formylated initiator tRNA^{Met} (Marcker and Sanger 1964).

E. coli K12 has two initiator tRNA species differing by a single nucleotide. Gene disruption studies showed that either one or the other is required for cell viability (Kenri et al.. 1992). *E. coli* methionyl-tRNA transformylase, the enzyme catalyzing the formylation of Met-tRNA$^{Met}_i$ with N_{10}-formyltetrahydrofolate, has been purified (Kahn et al. 1980), and its gene has been cloned (Guillon et al. 1992a). The importance of formylation is illustrated by the following experiments. The use of the cloned gene permitted the construction of an *E. coli* strain with a disrupted transformylase gene. Growth of the strain lacking any transformylase was severely impaired at 37°C, and the strain failed to grow at 42°C (Guillon et al. 1992a). In addition, studies on mutants of initiator tRNAs have shown that there is a strong correlation between formylation of initiator tRNA and its activity in initiation (Varshney et al. 1991).

Because of the central importance of this tRNA for initiation of protein synthesis, this RNA molecule has been studied extensively to define how two enzymes, methionyl-tRNA synthetase and transformylase, can each recognize this molecule so specifically among many tRNAs. The recognition elements for the transformylase are clustered in the acceptor stem of tRNA$^{Met}_i$ (Lee et al. 1991; Guillon et al. 1992b), whereas the major recognition elements for the synthetase are in the anticodon region (Pallanck and Schulman 1992).

Serine → Selenocysteine

One of the recent excitements in protein biosynthesis was the finding that selenocysteine, a nonstandard amino acid found in proteins in prokaryotic and eukaryotic cells, is cotranslationally inserted into polypeptides under the direction of the codon UGA assisted by a new elongation factor and a structural signal in the mRNA (for review, see Böck et al. 1991). The formation of selenocysteine from serine represents a very interesting tRNA-mediated amino acid transformation. Most of our knowledge stems from studies with *E. coli*.

The key to selenocysteine modification and insertion is a special tRNA species, tRNA^{Sec}. This tRNA (or the gene encoding it) has been found in many organisms (Hatfield et al. 1992; Steinberg et al. 1993). In all cases, it has the anticodon UCA capable of recognizing UGA. The RNA was first characterized in mammals (Hatfield et al. 1982) and more recently in *E. coli*, where it is a minor tRNA^{Ser} species (Baron and Böck 1991). The tRNA^{Sec} gene (*selC*) is one of six *E. coli* tRNA^{Ser} genes (Komine et al. 1990). The gene (Leinfelder et al. 1988) and the derived tRNA sequence (Schön et al. 1989) show unusual structure; it is the longest tRNA (95 nucleotides), it lacks the universal U_8 (replaced by G), and it possesses an acceptor stem with an additional (eighth) base pair, which results from abnormal cleavage by RNase P (Burkard and Söll 1988).

The two-step conversion of seryl-tRNA^{Sec} into selenocysteyl-tRNA^{Sec} is shown in Figure 2. Selenocysteine synthase, the product of the *E. coli selA* gene, catalyzes the formation of aminoacrylyl-tRNA^{Sec} with concomitant elimination of water (Forchhammer and Böck 1991; Forchhammer et al. 1991). This homo-dodecameric enzyme of 600 kD with a pyridoxal phosphate cofactor reacts specifically with tRNA^{Sec} and not with the other tRNA^{Ser} isoacceptors. Thus, like the methionyl-tRNA transformylase, this enzyme shows high discrimination among the different tRNA species. The recognition of the enzyme is based both on elements in the tRNA structure and on the amino acid (Böck et al. 1991). The second step is still incompletely understood. Formally, this is the addition of hydrogen selenide to the aminoacrylyl residue. This ATP-dependent reaction is catalyzed by the *selD* gene product, a 37-kD protein (Ehrenreich et al. 1992). The nature of the immediate selenium donor is not completely known. Because free hydrogen selenide does not react, it is suggested that the reactive aminoacrylyl group is made inaccessible by the enzyme (Forchhammer and Böck 1991; Forchhammer et al. 1991). Double-label experiments employing [^{75}Se]selenite and [γ-^{32}P]ATP, and NMR studies, indicate that the product of the SELD protein is phosphoselenoate (Ehrenreich et al. 1992; Veres et al. 1992). A preliminary investigation of the properties of partially purified murine selenocysteine synthase suggests that mammals use the same mechanism of selenocysteine formation as *E. coli* (Mizutani et al. 1992).

Studies with tRNA^{Sec} mutants revealed facets of the recognition of this molecule by the selenocysteine synthase, the translation factors SELB and EF-Tu, and by seryl-tRNA synthetase (Baron and Böck 1991). Because of its unusual structure, the wild-type tRNA^{Sec} shows in aminoacylation with seryl-tRNA synthetase a 100-fold reduction in K_{cat}/K_M compared to a tRNA^{Ser} species. The additional base pair in the

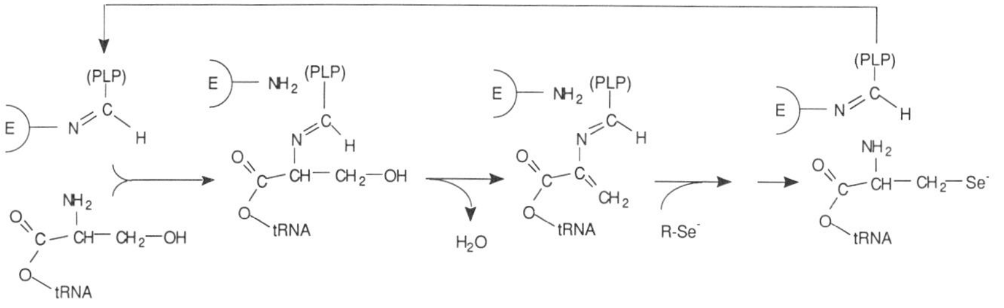

Figure 2 Proposed mechanism of selenocysteyl-tRNA formation.

acceptor stem is important for serine→selenocysteine conversion by purified selenocysteine synthase. In addition, the deletion of a base pair in tRNA^{Sec} to generate the "normal" length 7-bp acceptor stem permitted binding of EF-Tu but prevented interaction with the translation factor SELB. Thus, the longer acceptor stem is crucial for proper selenocysteine insertion during protein synthesis.

Glutamate → Glutamine

The notion that each amino acid has its own activating enzyme, which therefore indicates the existence of at least 20 aminoacyl-tRNA synthetases in every cell, is widespread. However, it was observed in the 1960s that the formation of Gln-tRNA^{Gln} in some *Bacilli* requires the presence of two enzymes, a glutamyl-tRNA synthetase that misacylates tRNA^{Gln} with glutamate and a tRNA-dependent amidotransferase that forms the "correct" aminoacyl-tRNA (Wilcox and Nirenberg 1968). The significance of this reaction and some unexplained observations in the literature became clear when it was shown that this misacylation/transamidation route (Fig. 3) appears to be the normal pathway of Gln-tRNA^{Gln} formation in archaebacteria, gram-positive eubacteria, mitochondria, and chloroplasts (Schön et al. 1988a,b). Because the presence of a functional glutaminyl-tRNA synthetase, the enzyme that directly ligates glutamine to tRNA^{Gln}, could not be demonstrated in these organisms and organelles, this pathway provides the sole source of glutamine for protein biosynthesis. This situation contrasts with that found in the cytoplasm of eukaryotic cells and in gram-negative eubacteria, where free glutamine is acylated directly onto tRNA^{Gln} by glutaminyl-tRNA synthetase (Hoben et al. 1982).

This pathway rests on the existence of the natural misacylating enzyme, glutamyl-tRNA synthetase, which is able to misacylate tRNA^{Gln} with glutamate (Lapointe et al. 1986; Chen et al. 1990b). Studies with partially purified *B. subtilis* Glu-tRNA^{Gln} amidotransferase (Wilcox 1969; Strauch et al. 1988) and the pure *Chlamydomonas* enzyme (Jahn et al. 1990) have shed light on the reaction mechanism. The enzyme shows strict tRNA selectivity, since it utilizes only Glu-tRNA^{Gln} but not Glu-tRNA^{Glu}. ATP, Mg^{++}, and an amide donor are necessary for the conver-

$$\text{tRNA}^{Gln} + \text{Glu} + \text{ATP} \underset{\textit{synthetase}}{\overset{\textit{Glu-tRNA}}{\rightleftharpoons}} \text{Glu-tRNA}^{Gln} + \text{AMP} + \text{PP}_i$$

$$\text{Glu-tRNA}^{Gln} + \text{Gln} + \text{ATP} \underset{\textit{amidotransferase}}{\overset{\textit{Glu-tRNA}^{Gln}}{\rightleftharpoons}} \text{Gln-tRNA}^{Gln} + \text{ADP} + \text{Glu} + \text{P}_i$$

Figure 3 The transamidation pathway of Gln-tRNA formation.

sion. Acceptable amide donors are glutamine, asparagine, and ammonia. The amidation reaction may involve glutamyl-phosphate formation, since ATP is cleaved to ADP when the enzyme is incubated with Glu-tRNA^{Gln} and ATP. The Glu-tRNA^{Gln} amidotransferase has many aspects of other well-known tRNA-independent glutamine amidotransferases: It utilizes ammonia as amide donor independently of glutamine binding; it can be inhibited by 6-diazo-5-oxonorleucine; it has low glutaminase activity; and it uses ATP as an energy source. The enzyme may be a hybrid between an aminoacyl-tRNA synthetase with its exquisite tRNA discrimination (Perona et al. 1989; Rould et al. 1989) and a glutamine amidotransferase with its conserved glutamine amide transfer domain (see, e.g., Mei and Zalkin 1989). Furthermore, the ATP cleavage by this enzyme to give ADP and P_i resembles the amidotransferases and not the aminoacyl-tRNA synthetases that produce AMP and PP_i (Moras 1992). The high substrate specificity of the *C. reinhardtii* aminotransferase enzyme for Glu-tRNA^{Gln} species from different organisms confirms the widespread evolutionary conservation of this mechanism.

How does the cell maintain accuracy in protein synthesis while coping with the mischarged tRNA? Clearly, the Glu-tRNA^{Gln} must not be among the pool of aminoacyl-tRNAs utilized on the ribosome, because this would lead to errors in protein synthesis. A channeling mechanism (Srivastava and Bernhard 1986) might be possible where the mischarged tRNA would be transferred directly to the amidotransferase from the glutamyl-tRNA synthetase and only released when the "correct" aminoacyl-tRNA is made. This may involve the formation of a tight complex between both enzymes.

Given the occurrence of the different pathways in the various phyla, it is plausible that they evolved independently, possibly to assure a greater degree of fidelity in protein synthesis. Conversely, the transamidation pathway may be the precursor of the direct aminoacylation pathway, in which case Glu-tRNA^{Gln} amidotransferase or the glutamyl-tRNA synthetase is the evolutionary precursor of the cytoplasmic glutaminyl-tRNA synthetase. Interestingly, the amidotransferase is in some ways similar to glutaminyl-tRNA synthetase; both enzymes bind in a specific manner the same substrates, glutamine, ATP, and the tRNA^{Gln} molecule.

Glutamate → Glutamate-1-semialdehyde

Porphyrin derivatives such as hemes and chlorophylls are important in respiratory and photosynthetic metabolic reactions. Eight molecules of the early precursor 5-aminolevulinic acid (δ-aminolevulinic acid, ALA) provide all the carbon and nitrogen atoms for the porphyrin skeleton of these molecules. In the chloroplasts of plants and green algae, in cyanobacteria, in some eubacteria (e.g., *E. coli* and *B. subtilis*) and archaebacteria, ALA is formed via the C_5 pathway from the carbon skeleton of glutamate (for review, see Jahn et al. 1992). A second route of ALA formation, the single-step condensation of succinyl-coenzyme A and glycine catalyzed by ALA synthase (May et al. 1986), operates in a limited number of eubacteria, in yeast, and in avian and mammalian cells.

The formation of ALA is known to be a key regulatory step in chlorophyll and heme biosynthesis in many organisms, since both feedback inhibition and light regulation are known (for review, see Beale and Weinstein 1990; Jordan 1990; O'Neill et al. 1991). During the past few years, the C_5 pathway has attracted much interest because of the involvement of a tRNA (tRNA^{Glu}) (Huang and Wang 1986; Schön et al. 1986), required for the biochemical transformation of glutamate to ALA. Our current understanding of this pathway is summarized in Figure 4. The initial metabolite for the two-step C_5 pathway is the normal Glu-tRNA^{Glu}, which is converted by the action of a unique enzyme, Glu-tRNA reductase (GluTR), to glutamate-1-semialdehyde (GSA) with the concomitant release of tRNA^{Glu}. GSA, whose chemical structure in solution is not clear (Jordan 1990), is converted to ALA by a specific transaminase, GSA-2,1-aminomutase. Thus, GSA is the first committed precursor of porphyrin synthesis in organisms/organelles that use the C_5 pathway. Studies indicate that there is sufficient charged tRNA^{Glu} in the cell/organelle, because tRNA^{Glu} gene expression does not appear to be regulated by light or other factors in order to satisfy the demand for porphyrin precursors, e.g., during greening of etioplasts (Berry-Lowe 1987; O'Neill and Söll 1990).

GluTR is a very low abundance enzyme in the cell. Most of our knowledge derives from the deduced amino acid sequences of prokaryotic GluTR genes, which display significant similarity (Verkamp et al.

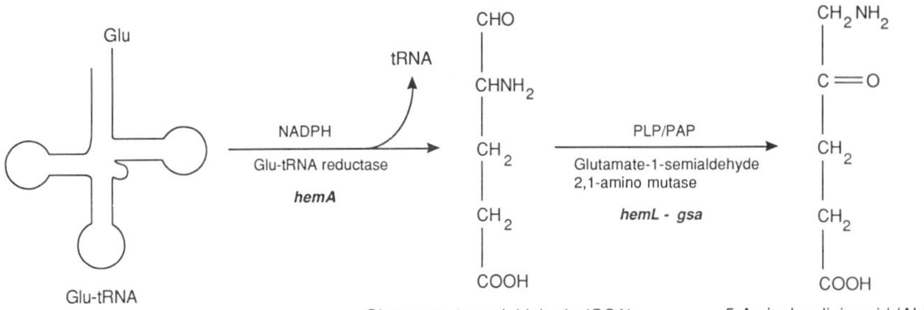

Figure 4 Scheme of the C_5 pathway: ALA formation from Glu-tRNA^{Glu}.

1992) and have molecular masses of 46–52 kD. NADPH is the only required cofactor in the Glu-tRNA^{Glu} → GSA conversion catalyzed by the purified GluTR enzymes. All GluTR enzymes tested so far have shown the ability to discriminate between different tRNA species. The clearest example is the barley GluTR, which can use charged chloroplast tRNA^{Glu} as substrate, but not *E. coli* or barley cytoplasmic Glu-tRNA^{Glu} (Peterson et al. 1988). It appears that GluTR can discriminate between different tRNA molecules even if they are aminoacylated with glutamate. Thus, if there is any recognition of the amino acid by GluTR, it must be secondary to that of the tRNA. The proposed mechanism for the GluTR-catalyzed reduction (Chen et al. 1990a) is formally analogous to the back reaction of glyceraldehyde-3-phosphate dehydrogenase, where the activated (by phosphorylation) carboxyl group of 3-phosphoglycerate is reduced to its aldehyde in the presence of NADH via an acyl-enzyme intermediate. Thus, intermediary metabolism makes use of the well-known property of activating an amino acid by ligation to tRNA.

This example shows that Glu-tRNA^{Glu} provides glutamate for protein biosynthesis and GSA for porphyrin synthesis. This is a novel role for aminoacyl-tRNA, where the activated amino acid is converted to another low-molecular-weight metabolite that is not incorporated into proteins.

ROLES OF UNCHARGED tRNA

Self-cleavage of tRNA

More than two decades ago, it was shown that several purified tRNAs can be cleaved at specific sites by metal ions (Dirheimer et al. 1972). A detailed study showed that yeast tRNA^{Phe} is cleaved by lead acetate in micromolar concentration at neutral pH between D_{17} and G_{18}, giving rise to 2′,3′-cyclic phosphate and 5′-hydroxyl termini (Werner et al. 1976). The authors concluded that the cleavages depend more on the conformation than on sequence of the RNA. Later, a crystallographic study showed the Pb^{++} ion is coordinated between eight residues in the T- and D-loops of the tRNA, so that the metal ion can remove a proton from the 2′hydroxyl of D_{17}, initiating the RNA chain cleavage (Brown et al. 1983). The same cleavage reaction also occurs with tRNA^{Phe} devoid of modified nucleotides (Behlen et al. 1990; Pan et al., this volume). This result then allowed a study with mutant tRNA^{Phe} transcripts that confirmed the identity of the two lead-coordinating pyrimidines (U_{59} and C_{60}) identified by X-ray crystallography (Brown et al. 1983). The understanding of the structural requirements for the Pb^{++} cleavage at U_{17} was confirmed by the production of mutant *E. coli* tRNA^{Phe} and yeast tRNA^{Arg} species that had acquired the specific cleavage (Behlen et al.

1990). The conserved G_{18} G_{19} sequence is not required for cleavage, as shown by the properties of in vitro selected tRNA^{Phe} variants with deletions in the D-loop (Pan and Uhlenbeck 1992).

Primer for Reverse Transcriptase

Initiation of retrovirus replication normally requires a tRNA, which is used as primer for the reverse transcriptase (Harada et al. 1979). This phenomenon, first described for mammalian and avian viruses, is also present in retroviral elements, *copia* in *Drosophila* (Kikuchi et al. 1986), and TY1 in *Saccharomyces cerevisiae* (Chapman et al. 1992). Different enzymes prefer different host tRNAs as primers. For instance, tRNA$^{Lys}_3$ is the primer for the HIV reverse transcriptase (Wain-Hobson et al. 1985), whereas *Drosophila* initiator methionine tRNA (tRNA$^{Met}_i$) primes transcription by the *copia*-encoded enzyme (Kikuchi et al. 1986). To function as a primer, the 3′ end of the tRNA base-pairs with the primer-binding site of the retroviral RNA. Presumably, the different priming tRNAs reflect the different sequences of the primer-binding site. Remarkably, when the primer-binding site sequence of the HIV template is made complementary to the 3′ end of *E. coli* tRNA$^{Gln}_2$, then transcription can be primed with this bacterial tRNA instead of the normal human tRNA$^{Lys}_3$ (Kohlstaedt and Steitz 1992). Thus, proper base-pairing and tRNA-like structure are important for tRNA specificity rather than sequence-specific recognition of tRNA$^{Lys}_3$. What parts of the tRNA are bound to the enzyme? Although a crystal structure of the HIV reverse transcriptase is available (Kohlstaedt et al. 1992), the RNA:protein complex is missing. Apart from the 3′ end of the tRNA, the anticodon also interacts with the HIV enzyme as shown in cross-linking experiments with *trans*-diamine-dichloro-platinum (Barat et al. 1989). A similar conclusion was reached from in vitro enzyme inhibition studies with oligonucleotides corresponding to the anticodon regions of tRNA$^{Lys}_3$ (Sarih-Cottin et al. 1992). It is interesting to see that two different forms of tRNA primers are found. In all the mammalian and avian cases examined, the intact tRNA is used as primer. However, in the case of the *Drosophila copia* reverse transcriptase, the first DNA extension does not start from the 3′ end of the tRNA, but from an internal site, two nucleotides after the anticodon loop of the *Drosophila* tRNA$^{Met}_i$ (Kikuchi et al. 1986). It is unclear what cellular enzyme produces this 39-nucleotide-long primer fragment from the 5′ region of the tRNA, although RNase P may assume this role, since a synthetic *Drosophila* tRNA$^{Met}_i$ precursor could be correctly cleaved by M1RNA, the catalytic RNA of *E. coli* RNase P (Kikuchi et al. 1990). The implications of the primer role of tRNA in reverse transcription for ancient strategies of genomic replication are addressed in the genomic tag hypothesis (Maizels and Weiner, this volume).

tRNA as Transcription Factor for RNA Polymerase III

Eukaryotic tRNA genes are transcribed by RNA polymerase III in concert with a number of transcription factors. Nuclear extracts from the silk gland of *Bombyx mori* were shown to direct tRNA synthesis from the well-studied silkworm tRNA^{Ala} gene. Partial purification of these extracts yielded preparations of transcription factors that had lost the ability to transcribe the tRNA^{Ala} gene and led to the discovery of a new transcription factor (TFIIIR). TFIIIR displayed sensitivity to micrococcal nuclease and was shown to be an RNA (Young et al. 1991). Complete purification of TFIIIR and sequence analysis revealed a tRNA whose amino acid specificity (isoleucine) was deduced from the anticodon sequence IAU (K.U. Sprague et al., in prep.). The specific activity of an unmodified T7 RNA polymerase transcript of the corresponding tRNA^{Ile} gene was approximately 50-fold less than that of native tRNA. It is not clear whether this difference in efficiency indicates a role for the modified bases in the native TFIIIR molecule. For comparison, the T7 transcript of an unrelated tRNA is more than 500-fold less active than native TFIII in vitro. There is no indication that aminoacylation is necessary for function. Two yeast isoleucine tRNAs with different anticodons have high TFIIIR activity. Of these, tRNA$^{Ile}_{IAU}$ has 20-fold higher specific activity than tRNA$^{Ile}_{UAU}$. The mechanism of TFIIIR action is not clear at present. However, single round transcription experiments demonstrate conclusively that the factor does not influence the number of active transcription factor complexes formed, but rather affects the efficiency with the complexes' direct repetitive transcription cycles (K.U. Sprague et al., in prep.). The fact that specific heterologous yeast tRNAs are efficient in transcription with *B. mori* factor preparations suggests that the tRNA nature of a transcription factor may be a general phenomenon also existing in other organisms.

tRNA Genes as Integration Sites for Prokaryotic Genetic Elements

Plasmids, phages, or retroelements integrate by site-specific recombination into the host chromosome of various bacteria. The attachment sites of the host (*attB*) and the corresponding attachment sites in the plasmid or phage (*attP*) include common sequences of approximately 40–112 bp. Examination of these sequences has shown that tRNA genes constitute the host attachment site in the genome of various organisms. These include the thermophilic archaebacterium *Sulfolobus* strain B12 (Reiter et al. 1989), *E. coli* and *Salmonella typhimurium* (for review, see Reiter et al. 1989; Sun et al. 1991), *Haemophilus influenzae* (Hauser and Scocca 1990), and *Streptomyces* strains (Mozodier et al. 1990; Martin et al. 1991; Vögtli and Cohen 1992). Despite the widespread occurrence of tRNA genes as target sites, tRNA genes per se are not required for the integration process, as integration sites for a number of eubacterial phages and plasmids do not contain tRNA genes (for review, see Reiter et al. 1989). The genetic element SLP1 integrates into an essential tRNA gene in *Streptomyces coelicolor*, revealing a novel mechanism of biological adaptation (Vögtli and Cohen 1992).

tRNA and Disease

A number of human diseases have been linked to mutations in mitochondrial tRNA genes (for review, see Lestienne 1992). Myoclonic epilepsy with ragged red fibers (MERRF) is associated with a G→A mutation in position 50 of the tRNA^{Lys} gene (Shoffner et al. 1990). A base substitution in the first base of the D-loop of tRNA$^{Leu}_{UUR}$ was often found associated with MELAS (*m*itochondrial myopathy, *e*ncephalopathy, *l*actic *a*cidosis, and *s*troke-like episodes) (Goto et al. 1990). Additional mutations in mitochondrial tRNA genes have been found in tRNA^{Gly}, tRNA^{Ile}, tRNA$^{Leu}_{CUN}$, tRNA$^{Leu}_{UAG}$, tRNA$^{Ser}_{GCU}$, and tRNA^{Thr} that have been related to CIPO (*c*hronic *i*ntestinal *p*seudo-obstruction with myopathy and *o*phthalmoplegia), with CPEO (*c*hronic *p*rogressive *e*xternal *o*phthalmoplegia), infantile cardiomyopathy, or infantile myopathy (Lauber et al. 1991; Yoon et al. 1991; Lestienne 1992; van den Ouweland et al. 1992).

It appears that mitochondrial tRNA genes represent hot spots for point mutations causing neuromuscular diseases (Lauber et al. 1991). The pathogenic mechanism is not yet understood. Mutations in different tRNAs render different disease phenotypes. These genetic changes exert their effect in the co-presence of the wild-type tRNA genes in a heteroplasmic situation. The changes lead to respiratory-chain impairment and to decreased mitochondrial protein synthesis (Kobayashi et al. 1991; Chomyn et al. 1992; Moraes et al. 1992) without specific alterations of mRNA:rRNA ratios. However, the exact nature of the tRNA involvement is not clear. The effect could be mediated by the tRNA itself such that the mutant tRNAs could cause reduced synthesis of particular proteins. On the other hand, because of the tight gene arrangement on the mitochondrial genome, tRNA mutations may affect other functions. For instance, it was shown that the tRNA$^{\text{Leu}}_{\text{UUR}}$ mutation causing MELAS occurs within the binding site for a protein that promotes transcription termination at the 16S rRNA/tRNA^{Leu} gene boundary. Further studies on the linkage of specific tRNA gene mutations to particular diseases may clarify the pathogenic mechanism.

OUTLOOK

It is clear from this discussion that tRNA is a ubiquitous molecule which is involved in many more reactions than protein biosynthesis. With the increased sensitivity of detection of unusual amino acids attached to tRNA, and with our ever-increasing knowledge of the interconnections of the various reactions in cellular metabolism, new roles of tRNA will be discovered and others will be clarified in mechanistic detail. A number of observations suggesting additional roles of tRNA have appeared in the literature.

Activation of amino acids by tRNA could be important in other processes. For instance, the observed pyrrolidone carboxylate tRNA^{Gln} (Bernfield and Nestor 1968) or *N*-acetyl-cysteinyl-tRNA (Pawelek et al. 1971) species are still in search of a role. Another example of amino acid transfer from aminoacyl-tRNA is the recent finding of specific glycine incorporation into bacterial lipopolysaccharide (Gamian et al. 1991). This reaction may have some similarity to the old observation that aminoacyl-phosphatidyl-glycerols are formed in a very specific reaction from aminoacyl-tRNA (Gould et al. 1968). The function of *E. coli* mutants, which indicate a tRNA involvement in cell division (*divE*, Tamura et al. 1984; and *feeB*, Chen et al. 1991b), also needs to be clarified. Another case of tRNA's being a required "cofactor" for an enzyme was recently uncovered. A murine endoribonuclease with sequence specificity for CCCCCGGUUUGU, a sequence found in the spacer of mouse pre-rRNA, requires tRNA for activity (Nashimoto 1992). The tRNA lacks the 3′ terminus and consists of either tRNA^{Gln}, tRNA^{Gly}, or tRNA^{Met}. It is suggested that the tRNA may have evolved from a ribozyme that possessed ribonucleolytic activity.

The wide variety of uses for tRNA in cell metabolism may indicate that in the past, tRNA or tRNA-like molecules did assume various enzymatic roles which in the course of evolution were usurped by the proteins that today interact with tRNA.

ACKNOWLEDGMENTS

I am grateful to A. Böck, J. Davies, D. Gonda, U.L. RajBhandary, M.J. Rogers, R. LaRossa, and K. Sprague for helpful discussions and communication of unpublished results. Work in the author's laboratory was supported by grants from the National Institutes of Health and the Department of Energy.

REFERENCES

Altman, S., L. Kirsebom, and S. Talbot. 1993. Recent studies on RNase P. *FASEB J.* **7:** 7–14.

Anderson, W.F. 1969. The effect of tRNA concentration on the rate of protein synthesis. *Proc. Natl. Acad. Sci.* **62:** 566–573.

Atkins, J.F. and R.F. Gesteland. Breaking an initial engagement and re-pairing to a new triplet codon as the basis for natural high-level frameshifting and hopping. In *Transfer RNA* (ed. D. Söll and U. RajBhandary). ASM Publications, Washington, D.C. (In press.)

Balzi, E., M. Choder, W.N. Chen, A. Varshavsky, and A. Goffeau. 1990. Cloning and functional analysis of the arginyl-tRNA-protein transferase gene *ATE1* of *Saccharomyces cerevisiae. J. Biol. Chem.* **265:** 7464–7471.

Barat, C., V. Lullien, O. Schatz, G. Keith, M.T. Nugeyre, F. Gruninger-Leitch, F. Barre - Sinoussi, S.F. LeGrice, and J.L. Darlix. 1989. HIV-1 reverse transcriptase specifically interacts with the anticodon domain of its cognate primer tRNA. *EMBO J.* **8:** 3279–3285.

Baron, C. and A. Böck. 1991. The length of the aminoacyl-acceptor stem of the selenocysteine-specific tRNA^{Sec} of *Escherichia coli* is the determinant for binding to elongation factors SELB or Tu. *J. Biol. Chem.* **266:** 20375–20379.

Basavappa, R. and P.B. Sigler. 1991. The 3 Å crystal structure of yeast initiator tRNA: Functional implications in initiator/elongator discrimination. *EMBO J.* **10:** 3105–3111.

Beale, S.I. and J.D. Weinstein. 1990. Tetrapyrrole metabolism in photosynthetic organisms. In *Biosynthesis of heme and chlorophyll* (ed. H.A. Dailey), pp. 287–391. McGraw-Hill, New York.

Behlen, L.S., J.R. Sampson, A.B. DiRenzo, and O.C. Uhlenbeck. 1990. Lead-catalyzed cleavage of yeast tRNA^{Phe} mutants. *Biochemistry* **29:** 2515–2523.

Beier, H., M.C. Lee, T. Sekiya, Y. Kuchino, and S. Nishimura. 1992. Two nucleotides next to the anticodon of cytoplasmic rat tRNA^{Asp} are likely generated by RNA editing. *Nucleic Acids Res.* **20:** 2679–2683.

Bernfield, M.R. and L. Nestor. 1968. The enzymatic conversion of glutaminyl-tRNA to pyrrolidone carboxylate-tRNA. *Biochem. Biophys. Res. Commun.* **33:** 843–849.

Berry-Lowe, S. 1987. The chloroplast tRNA glutamate gene required for δ-aminolevulinate synthesis. *Carlsberg Res. Comm.* **52:** 197–210.

Björk, G.R. 1992. The role of modified nucleosides in transfer RNA interactions. In *Transfer RNA in protein synthesis* (ed. D.L. Hatfield et al.), pp. 23–85. CRC Press, Boca Raton, Florida.

Böck, A., K. Forchhammer, J. Heider, and C. Baron. 1991. Selenoprotein synthesis: An expansion of the genetic code. *Trends Biochem. Sci.* **16:** 463–467.

Brown, R.S., B.E. Hingerty, J.C. Dewan, and A. Klug. 1983. Pb(II)-catalysed cleavage of the sugar-phosphate backbone of yeast tRNA^{Phe}—Implications for lead toxicity and self-splicing RNA. *Nature* **303:** 543–546.

Bumsted, R.M., J.L. Dahl, D. Söll, and J.L. Strominger. 1968. Biosynthesis of the peptidoglycan of bacterial cell walls. X. Further study of the glycyl transfer ribonucleic acid active in peptidoglycan synthesis in *Staphylococcus aureus. J. Biol. Chem.* **243:** 779–782.

Burkard, U. and D. Söll. 1988. The unusually long amino acid acceptor stem of *E. coli* selenocysteine tRNA results from abnormal cleavage by RNase P. *Nucleic Acids Res.* **16:** 11617–11624.

Cavalier-Smith, T. 1991. Intron phylogeny: A new hypothesis. *Trends Genet.* **7:** 145–148.

Cedergren, R., Y. Abel, and D. Sankoff. 1991. Evaluating gene *versus* genome evolution. In *Molecular techniques in taxonomy* (ed. G.M. Hewitt et al.), pp. 87–99. Springer Verlag, Berlin.

Chapman, K.B, A.S. Bystrom, and J.D. Boeke. 1992. Initiator methionine tRNA is essential for Ty1 transposition in yeast. *Proc. Natl. Acad. Sci.* **89:** 3236–3240.

Chen, K.S., T.C. Peters, and J.R. Walker. 1991a. A minor arginine tRNA mutant limits translation preferentially of a protein dependent on the cognate codon. *J. Bacteriol.* **172:** 2504–2510.

Chen, M.W., D. Jahn, G. O'Neill, and D. Söll. 1990a. Purification of the glutamyl-tRNA reductase from *Chlamydomonas reinhardtii* involved in δ-aminolevulinic acid formation during chlorophyll biosynthesis.*J. Biol. Chem.* **265:** 4058–4063.

Chen, M.W., D. Jahn, A. Schön, G.P. O'Neill, and D. Söll. 1990b. Purification and characterization of *Chlamydomonas reinhardtii* chloroplast glutamyl-tRNA synthetase, a natural misacylating enzyme.*J. Biol. Chem.* **265:** 4054–4057.

Chen, M.X., N. Bouquin, V. Norris, S. Casaregola, S.J. Seror, and I.B. Holland. 1991b. A single base change in the acceptor stem of tRNA$^{Leu}_3$ confers resistance upon *Escherichia coli* to the calmodulin inhibitor, 48/80. *EMBO J.* **10:** 3113–3122.

Chomyn, A., A. Martinuzzi, M. Yoneda, A. Daga, O. Hurko, D. Johns, S.T. Lai, I. Nonaka, C. Angelini, and G. Attardi. 1992. MELAS mutation in mtDNA binding site for transcription termination factor causes defects in protein synthesis and in respiration but no change in levels of upstream and downstream mature transcripts. *Proc. Natl. Acad. Sci.* **89:** 4221–4225.

Ciechanover, A., S. Ferber, D. Ganoth, S. Elias, A. Hershko, and S. Arfin. 1988. Purification and characterization of arginyl-tRNA-protein transferase from rabbit reticulocytes. Its involvement in post-translational modification and degradation of acidic NH$_2$ termini substrates of the ubiquitin pathway.*J. Biol. Chem.* **263:** 11155–11167.

Ciesiolka, J., P. Gornicki, and J. Ofengand. 1985. Identification of the site of cross-linking in 16S rRNA of an aromatic azide photoaffinity probe attached to the 5′-anticodon base of A site bound tRNA. *Biochemistry* **24:** 4931–4938.

Crick, F.H.C. 1966. Codon-anticodon pairing: The wobble hypothesis. *J. Mol. Biol.* **19:** 548–555.

Deutscher, M.P. 1990. Ribonucleases, tRNA nucleotidyltransferase, and the 3′ processing of tRNA. *Prog. Nucleic Acid Res. Mol. Biol.* **39:** 209–240.

Dinter-Gottlieb, G. 1986. Viroids and virusoids are related to group I introns. *Proc. Natl. Acad. Sci.* **83:** 6250–6254.

Dirheimer, G., J.P. Ebel, J. Bonnet, J. Gangloff, G. Keith, B. Krebs, B. Kuntzel, A. Roy, J. Weissenbach, and C. Werner. 1972. Structure primaire des tRN. *Biochimie* **54:** 127–144.

Edmonds, C.G., P.F. Crain, R. Gupta, T. Hashizume, C.H. Hocart, J.A. Kowalak, S.C. Pomerantz, K.O. Stetter, and J.A. McCloskey. 1991. Posttranscriptional modification of tRNA in thermophilic archaea (Archaebacteria).*J. Bacteriol.* **173:** 3138–3148.

Ehrenreich, A., K. Forchhammer, P. Tormay, B. Veprek, and A. Böck. 1992. Selenoprotein synthesis in *E. coli.* Purification and characterisation of the enzyme catalysing selenium activation. *Eur. J. Biochem.* **206:** 767–773.

Eigen, M. and R. Winkler-Oswatitsch. 1981. Transfer RNA, an early gene? *Naturwissenschaften* **68:** 282–292.

Eigen, M., B.F. Lindemann, M. Tietze, R. Winkler-Oswatitsch, A. Dress, and A. von Haeseler. 1989. How old is the genetic code? Statistical geometry of tRNA provides an answer. *Science* **244:** 673–679.

Elias, S. and A. Ciechanover. 1990. Post-translational addition of an arginine moiety to acidic NH_2 termini of proteins is required for their recognition by ubiquitin-protein ligase. *J. Biol. Chem.* **265:** 15511–15517.

Ferber, S. and A. Ciechanover. 1987. Role of arginine-tRNA in protein degradation by the ubiquitin pathway. *Nature* **326:** 808–811.

Fernandez-Moreno, M.A., J.L. Caballero, D.A. Hopwood, and F. Malpartida. 1991. The act cluster contains regulatory and antibiotic export genes, direct targets for translational control by the *bldA* tRNA gene of *Streptomyces. Cell* **66:** 769–780.

Forchhammer, K. and A. Böck. 1991. Selenocysteine synthase from *Escherichia coli.* Analysis of the reaction sequence. *J. Biol. Chem.* **266:** 6324–6328.

Forchhammer, K., W. Leinfelder, K. Boesmiller, B. Veprek, and A. Böck. 1991. Selenocysteine synthase from *Escherichia coli.* Nucleotide sequence of the gene (*selA*) and purification of the protein. *J. Biol. Chem.* **266:** 6318–6323.

Francklyn, C. and P. Schimmel. 1989. Aminoacylation of RNA minihelices with alanine. *Nature* **337:** 478–481.

Gamian, A., A. Krzyzaniak, M.Z. Barciszewska, I. Gawronska, and J. Barciszewski. 1991. Specific incorporation of glycine into bacterial lipopolysaccharide. Novel function of specific transfer ribonucleic acids. *Nucleic Acids Res.* **19:** 6021–6025.

Gamulin, V., J. Mao, B. Appel, M. Sumner-Smith, F. Yamao, and D. Söll. 1983. Six *Schizosaccharomyces pombe* tRNA genes including one with an eight base intervening sequence. *Nucleic Acids Res.* **11:** 8537–8546.

Garen, A. 1968. Sense and nonsense in the genetic code. *Science* **160:** 149–159.

Gesteland, R.F., R.B. Weiss, and J.F. Atkins. 1992. Recoding: Reprogrammed genetic decoding. *Science* **257:** 1640–1641.

Gonda, D.K., A. Bachmair, I. Wunning, J. W. Tobias, W.S. Lane, and A. Varshavsky. 1989. Universality and structure of the N-end rule. *J. Biol. Chem.* **264:** 16700–16712.

Goto, Y., I. Nonaka, and S. Horai. 1990. A mutation in the tRNA$^{Leu}_{UUR}$ gene associated with the MELAS subgroup of mitochondrial encephalomyopathies. *Nature* **348:** 651–653.

Gould, R.M., M.P. Thornton, V. Leipkalns, and W.J. Lennarz. 1968. Participation of aminoacyl transfer ribonucleic acid in aminoacyl phosphatidylglycerol synthesis. II. Specificity of alanyl phosphatidylglycerol synthetase. *J. Biol. Chem.* **243:** 3096–3104.

Green, C.J. and B.S. Vold. 1983. Sequence analysis of a cluster of twenty-one tRNA genes in *Bacillus subtilis. Nucleic Acids Res.* **11:** 5763–5774.

Guillon, J.M., Y. Mechulam, J.M. Schmitter, S. Blanquet, and G. Fayat. 1992a. Disruption of the gene for Met-tRNA^{fMet} formyltransferase severely impairs growth of *Escherichia coli. J. Bacteriol.* **174:** 4294–4301.

Guillon, J.M., T. Meinnel, Y. Mechulam, C. Lazennec, S. Blanquet, and G. Fayat. 1992b. Nucleotides of tRNA governing the specificity of *Escherichia coli* methionyl-tRNA^{fMet} formyltransferase. *J. Mol. Biol.* **224:** 359–367.

Harada, F., G.G. Peters, and J.E. Dahlberg. 1979. The primer tRNA for Moloney murine leukemia virus DNA synthesis. Nucleotide sequence and aminoacylation of tRNA^{Pro}. *J. Biol. Chem.* **254:** 10979–10985.

Hatfield, D., A. Diamond, and B. Dudock. 1982. Opal suppressor serine tRNAs from bovine liver form phosphoseryl-tRNA. *Proc. Natl. Acad. Sci.* **79**: 6215–6219.

Hatfield, D., I.S. Choi, S. Mischke, and L.D. Owens. 1992. Selenocysteyl-tRNAs recognize UGA in *Beta vulgaris*, a higher plant, and in *Gliocladium virens*, a filamentous fungus. *Biochem. Biophys. Res. Commun.* **184**: 254–259.

Hauser, M.A. and J.J. Scocca. 1990. Location of the host attachment site for phage HP1 within a cluster of *Haemophilus influenzae* tRNA genes. *Nucleic Acids Res.* **18**: 5305.

Hayashi, H. and K.I. Miura. 1966. Functional sites in transfer ribonucleic acid. *Nature* **209**: 376–378.

Hershko, A. and A. Ciechanover. 1992. The ubiquitin system for protein degradation. *Annu. Rev. Biochem.* **61**: 761–807.

Hill, W.E., A. Dahlberg, R.A. Garrett, P.B. Moore, D. Schlessinger, and J.R. Warner, eds. 1990. *The ribosome: Structure, function and evolution.* American Society for Microbiology, Washington, D.C.

Hoben, P., N. Royal, A. Cheung, F. Yamao, K. Biemann, and D. Söll. 1982. *Escherichia coli* glutaminyl-tRNA synthetase II. Characterization of the gene product. *J. Biol. Chem.* **257**: 11644–11650.

Holley, R.W., J. Apgar, G.A. Everett, J.T. Madison, M. Marquisee, S.H. Merrill, J.R. Penswick, and A. Zamir. 1965. Structure of a ribonucleic acid. *Science* **147**: 1462–1465.

Huang, D.D. and W.Y. Wang. 1986. Chlorophyll biosynthesis in *Chlamydomonas* starts with the formation of glutamyl-tRNA. *J. Biol. Chem.* **261**: 13451–13455.

Imura, N., G.B. Weiss, and R.W. Chambers. 1969. Reconstitution of alanine acceptor activity from fragments of yeast tRNA^{Ala}II. *Nature* **222**: 1147–1148.

Jahn, D., E. Verkamp, and D. Söll. 1992. Glutamyl-transfer RNA: A precursor of heme and chlorophyll biosynthesis. *Trends Biochem. Sci.* **17**: 215–218.

Jahn, D., Y.C. Kim, Y. Ishino, M.W. Chen, and D. Söll. 1990. Purification and functional characterization of the glu-tRNA^{Gln} amidotransferase from *Chlamydomonas reinhardtii. J. Biol. Chem.* **265**: 8059–8064.

Jordan, P.M. 1990. Biosynthesis of 5-aminolevulinic acid and its transformation into coproporphyrinogen in animals and bacteria. In *Biosynthesis of heme and chlorophylls* (ed. H.A. Dailey), pp. 55–121. McGraw-Hill, New York.

Kahn, D., M. Fromant, G. Fayat, P. Dessen, and S. Blanquet. 1980. Methionyl-transfer-RNA transformylase from *Escherichia coli*. Purification and characterisation. *Eur. J. Biochem.* **105**: 489–497.

Kaji, H. 1968. Further studies on the soluble amino acid incorporating system from rat liver. *Biochemistry* **7**: 3844–3850.

Kaji, A., H. Kaji, and G.D. Novelli. 1965. Soluble amino acid-incorporating system. Preparation of the system and nature of the reaction. *J. Biol. Chem.* **240**: 1192–1197.

Kenri, T., F. Imamoto, and Y. Kano. 1992. Construction and characterization of an *Escherichia coli* mutant deficient in the *metY* gene encoding tRNA$^{fMet}_2$: Either tRNA$^{fMet}_1$ or tRNA$^{fMet}_2$ is required for cell growth. *Gene* **114**: 109–114.

Kikuchi, Y., Y. Ando, and T. Shiba. 1986. Unusual priming mechanism of RNA-directed DNA synthesis in *copia* retrovirus-like particles of *Drosophila. Nature* **323**: 824–826.

Kikuchi, Y., M. Sasaki, and Y. Ando-Yamagami. 1990. Cleavage of tRNA within the mature tRNA sequence by the catalytic RNA of RNase P: Implication for the formation of the primer tRNA fragment for reverse transcription in *copia* retrovirus-like particles. *Proc. Natl. Acad. Sci.* **87**: 8105–8109.

Kim, S.H., F.L. Suddath, G.J. Quigley, A. McPherson, J.L. Sussman, A.H.J. Wang, N.C. Seeman, and A. Rich. 1974. Three-dimensional tertiary structure of yeast phenylalanine transfer RNA. *Science* **185**: 435–440.

Kobayashi, Y., M.Y. Momoi, K. Tominaga, H. Shimoizumi, K. Nihei, M. Yanagisawa, Y. Kagawa, and S. Ohta. 1991. Respiration-deficient cells are caused by a single point mutation in the mitochondrial tRNA$^{Leu}_{UUR}$ gene in mitochondrial myopathy, encephalopathy, lactic acidosis, and stroke-like episodes (MELAS). *Am. J. Hum. Genet.* **49**: 590–599.

Kohlstaedt, L.A. and T.A. Steitz. 1992. Reverse transcriptase of human immunodeficiency virus can use either human tRNA$^{Lys}_3$ or *Escherichia coli* tRNA$^{Gln}_2$ as a primer in an *in vitro* primer-utilization assay. *Proc. Natl. Acad. Sci.* **89**: 9652–9656.

Kohlstaedt, L.A., J. Wang, J.M. Friedman, P.A. Rice, and T.A. Steitz. 1992. Crystal structure at 3.5 Å resolution of HIV-1 reverse transcriptase complexed with an inhibitor. *Science* **256**: 1783–1790.

Komine, Y., T. Adachi, H. Inokuchi, and H. Ozeki. 1990. Genomic organization and physical mapping of the transfer RNA genes in *Escherichia coli* K12. *J. Mol. Biol.* **212**: 579–598.

Lapointe, J. and R. Giege. 1991. Transfer RNAs and aminoacyl-tRNA synthetases. In *Translation in eukaryotes* (ed. H. Trachsel), pp. 35–68. CRC Press, Boca Raton, Florida.

Lapointe, J., L. Duplain, and M. Proulx. 1986. A single glutamyl-tRNA synthetase aminoacylates tRNA^{Glu} and tRNA^{Gln} in *Bacillus subtilis* and efficiently misacylates *Escherichia coli* tRNA$^{Gln}_1$ *in vitro. J. Bacteriol.* **165**: 88–93.

Lauber, J., C. Marsac, B. Kadenbach, and P. Seibel. 1991. Mutations in mitochondrial tRNA genes: A frequent cause of neuromuscular diseases. *Nucleic Acids Res.* **19**: 1393–1397.

Lawlor, E.J., H.A. Baylis, and K.F. Chater. 1987. Pleiotropic morphological and antibiotic deficiencies result from mutations in a gene encoding a tRNA-like product in *Streptomyces coelicolor* A3(2). *Genes Dev.* **1**: 1305–1310.

Lee, C.P., B.L. Seong, and U.L. RajBhandary. 1991. Structural and sequence elements important for recognition of *Escherichia coli* formylmethionine tRNA by methionyl-tRNA transformylase are clustered in the acceptor stem. *J. Biol. Chem.* **266**: 18012–18017.

Leibowitz, M.J. and R.L. Soffer. 1969. A soluble enzyme from *Escherichia coli* which catalyzes the transfer of leucine and phenylalanine from tRNA to acceptor proteins. *Biochem. Biophys. Res. Commun.* **36**: 47–53.

Leinfelder, W., E. Zehelein, M.A. Mandrand-Berthelot, and A. Böck. 1988. Gene for a novel tRNA species that accepts L-serine and cotranslationally inserts selenocysteine. *Nature* **331**: 723–725.

Leskiw, B., E.J. Lawlor, J.M Fernandez-Abalos, and K.F. Chater. 1991. TTA codons in some genes prevent their expression in a class of developmental, antibiotic-negative, *Streptomyces* mutants. *Proc. Natl. Acad. Sci.* **88**: 2461–2465.

Lestienne, P. 1992. Mitochondrial DNA mutations in human diseases: A review. *Biochimie* **74**: 123–130.

Marcker, K.A. and F. Sanger. 1964. N-formylmethionyl-sRNA. *J. Mol. Biol.* **8**: 835–840.

Martin, C., P. Mazodier, M.V. Mediola, B. Gicquel, T. Smokvina, C.J. Thompson, and J. Davies. 1991. Site-specific integration of the *Streptomyces* plasmid pSAM2 in *Mycobacterium smegmatis. Mol. Microbiol.* **5**: 2499–2502.

May, B.K., I.A. Borthwick, G. Srivastava, B.A. Pirola, and W.H. Elliott. 1986. Control of 5-aminolevulinate synthases in animals. *Curr. Top. Cell. Regul.* **28**: 233–262.

Mazodier, P., C. Thompson, and F. Boccard. 1990. The chromosomal intergration site of the *Streptomyces* element pSAM2 overlaps a putative tRNA gene conserved among actinomycetes. *Mol. Gen. Genet.* **222**: 431–434.

Mei, B. and H. Zalkin. 1989. A Cys-His-Asp catalytic triad is involved in glutamine amidotransferase function in *purF*-type glutamine amidotransferase. *J. Biol. Chem.* **264:** 16613–16619.

Mizutani, T., H. Kurata, K. Yamada, and T. Totsuka. 1992. Some properties of murine selenocysteine synthase. *Biochem. J.* **284:** 827–834.

Moraes, C.T., E. Ricci, E. Bonilla, S. DiMauro, and E.A. Schon. 1992. The mitochondrial tRNA$^{Leu}_{UUR}$ mutation in mitochondrial encephalomyopathy, lactic acidosis, and strokelike episodes (MELAS): Genetic, biochemical, and morphological correlations in skeletal muscle. *Am. J. Hum. Genet.* **50:** 934–949.

Moras, D. 1992. Structural and functional relationships between aminoacyl-tRNA synthetases. *Trends Biochem. Sci.* **17:** 159–164.

Muramatsu, T., K. Nishikawa, F. Nemoto, Y. Kuchino, S. Nishimura, T. Miyazawa, and S. Yokoyama. 1988. Codon and amino-acid specificities of a transfer RNA are both converted by a single post-transcriptional modification. *Nature* **336:** 179–181.

Muto, A., Y. Andachi, H. Yuzawa, F. Yamao, and S. Osawa. 1990. The organization and evolution of transfer RNA genes in *Mycoplasma capricolum*. *Nucleic Acids Res.* **18:** 5037–5043.

Nashimoto, M. 1992. Characterization of the spermidine-dependent, sequence-specific endoribonuclease that requires transfer RNA for its activity. *Nucleic Acids Res.* **20:** 3737–3742.

Noller, H.F., V. Hoffarth, and L. Zimniak. 1992. Unusual resistance of peptidyl transferase to protein extraction procedures. *Science* **256:** 1416–1419.

Ogden, R.C., M.C. Lee, and G. Knapp. 1984. Transfer RNA splicing in *Saccharomyces cerevisiae*: Defining the substrates. *Nucleic Acids Res.* **12:** 9367–9382.

Ohashi, Z., M. Saneyoshi, F. Harada, H. Hara, and S. Nishimura. 1970. Presumed anticodon structure of glutamic acid tRNA from *E. coli*: A possible location of a 2-thiouridine derivative in the first position of the anticodon. *Biochem. Biophys. Res. Comm.* **40:** 866–872.

Okimoto, R. and D.R. Wolstenholme. 1990. A set of tRNAs that lack either the TΨC arm or the dihydrouridine arm: Towards a minimal tRNA adaptor. *EMBO J.* **9:** 3405–3411.

O'Neill, G.P. and D. Söll. 1990. Expression of the synechocystis 6803 tRNA^{Glu} gene provides tRNA for both protein and chlorophyll biosynthesis. *J. Bacteriol.* **172:** 6363–6371.

O'Neill, G., D. Jahn, and D. Söll. 1991. Transfer RNA involvement in chlorophyll biosynthesis. *Subcell. Biochem.* **17:** 235–264.

Osawa, S., T.H. Jukes, K. Watanabe, and A. Muto. 1992. Recent evidence for evolution of the genetic code. *Microbiol Rev.* **56:** 229–264.

Pallanck, L. and L.H. Schulman. 1992. tRNA discrimination in aminoacylation. In *Transfer RNA in protein bioynthesis* (ed. D. Hatfield et al.), pp. 279–318. CRC Press, Boca Raton, Florida.

Pan, T. and O.C. Uhlenbeck. 1992. In vitro selection of RNAs that undergo autolytic cleavage with Pb^{2+}. *Biochemistry* **31:** 3887–3895.

Pawelek, J., W. Godchaux III, J. Grasso, A.I. Skoultchi, J. Eisenstadt, and P. Lengyel. 1971. Occurrence of cysteinyl transfer ribonucleic acid with a blocked α-amino group in rabbit reticulocytes. *Biochim. Biophys. Acta* **232:** 289–305.

Perona, J.J., R.N. Swanson, M.A. Rould, T.A. Steitz, and D. Söll. 1989. Structural basis for misaminoacylation by mutant *E. coli* glutaminyl-tRNA synthetase enzymes. *Science* **246:** 1152–1154.

Persson, B.C., C. Gustafsson, D.E. Berg, and G.R. Björk. 1992. The gene for a tRNA modifying enzyme, m^{5}U$_{54}$-methyltransferase, is essential for viability in *Escherichia coli. Proc. Natl. Acad. Sci.* **89:** 3995–3998.

Peterson, D.M., A. Schön, and D. Söll. 1988. The nucleotide sequences of barley cytoplasmic glutamate transfer RNAs and structural features essential for formation of δ-aminolevulinic acid. *Plant Mol. Biol.* **11:** 293–299.

Petit, J.F., J.L. Strominger, and D. Söll. 1968. Biosynthesis of the peptidoglycan of bacterial cell walls. VII. Incorporation of serine and glycine into interpeptide bridges in *Staphylococcus epidermidis. J. Biol. Chem.* **243:** 757–767.

Pongs, O. and N. Ulbrich. 1976. Specific binding of formylated initiator-tRNA to *Escherichia coli* RNA polymerase. *Proc. Natl. Acad. Sci.* **73:** 3064–3067.

Reinhold-Hurek, B. and D.A. Shub. 1992. Self-splicing introns in tRNA genes of widely divergent bacteria. *Nature* **357:** 173–176.

Reiter, W.-D., P. Palm, and S. Yeats. 1989. Transfer RNA genes frequently serve as integration sites prokaryotic genetic elements. *Nucleic Acids Res.* **17:** 1907–1914.

Roberts, R.J. 1972. Structures of two glycyl-tRNAs from *Staphylococcus epidermidis. Nat. New Biol.* **237:** 44–45.

Roberts, W.S.L., J.L. Strominger, and D. Söll. 1968. Biosynthesis of the peptidoglycan of bacterial cell walls. VI. Incorporation of L-threonine into interpeptide bridges in *Micrococcus roseus. J. Biol. Chem.* **243:** 749–756.

Robertus, J.D., J.E. Ladner, J.R. Finch, D. Rhodes, R.S. Brown, B.F.C. Clark, and A. Klug. 1974. Structure of yeast phenylalanine tRNA at 3 Å resolution. *Nature* **250:** 546–551.

Rould, M.A., J. Perona, D. Söll, and T. Steitz. 1989. Structure of *E. coli* glutaminyl-tRNA synthetase complexed with tRNA^{Gln} and ATP at 2.8 Å resolution. *Science* **246:** 1135–1142.

Ryan, M.J., E.L. Brown, T. Sekiya, H. Kupper, and H.G. Khorana. 1979. Total synthesis of a tyrosine suppressor tRNA gene. XVIII. Biological activity and transcription, *in vitro*, of the cloned gene. *J. Biol. Chem.* **254:** 5817–5826.

Sarih-Cottin, L., B. Bordier, K. Musier-Forsyth, M.L. Andreola, P.J. Barr, and S. Litvak. 1992. Preferential interaction of human immunodeficiency virus reverse transcriptase with two regions of primer tRNA^{Lys} as evidenced by footprinting studies and inhibition with synthetic oligoribonucleotides. *J. Mol. Biol.* **226:** 1–6.

Schön, A., H. Hottinger, and D. Söll. 1988a. Misaminoacylation and transamidation are required for protein biosynthesis in *Lactobacillus bulgaricus. Biochimie* **70:** 391–394.

Schön, A., C.G. Kannangara, S. Gough, and D. Söll. 1988b. Protein biosynthesis in organelles requires misaminoacylation of transfer RNA. *Nature* **331:** 187–190.

Schön, A., A. Böck, G. Ott, M. Sprinzl, and D. Söll. 1989. The selenocysteine-inserting opal suppressor serine tRNA from *E. coli* is highly unusual in structure and modification. *Nucleic Acids Res.* **17:** 7159–7165.

Schön, A., G. Krupp, S. Berry-Lowe, C.G. Kannangara, and D. Söll. 1986. The RNA required in the first step of chlorophyll biosynthesis is a chloroplast glutamate tRNA. *Nature* **322:** 281–284.

Shindo-Okada, N., N. Okada, T. Ohgi, T. Goto, and S. Nishimura. 1980. Transfer ribonucleic acid guanine transglycosylase isolated from rat liver. *Biochemistry* **19:** 395–400.

Shoffner, J.M., M.T. Lott, A.M.S. Lezza, P. Seibel, S.W. Ballinger, and D.C. Wallace. 1990. Myoclonic epilepsy and ragged-red fiber disease (MERRF) is associated with a mitochondrial DNA tRNA^{Lys} mutation. *Cell* **61:** 931–937.

Soffer, R.L. 1980. Biochemistry and biology of aminoacyl-tRNA-protein transferases. In *Transfer RNA: Biological aspects* (ed. D. Söll et al.), pp. 493–505. Cold Spring Harbor Laboratory, Cold Spring Harbor, New York.

Söll, D., D.S. Jones, E. Ohtsuka, R.D. Faulkner, R. Lohrmann, H. Hayatsu, H.G. Khorana, J.D. Cherayil, A. Hampel, and R.M. Bock. 1966. Specificity of sRNA for recognition of codons as studied by the ribosomal binding technique. *J. Mol. Biol.* **19:** 556–575.

Srivastava, D.K. and S.A. Bernhard. 1986. Metabolite transfer via enzyme-enzyme complexes. *Science* **234:** 1081–1086.

Steege, D. and D. Söll. 1979. Suppression. In *Biological regulation and control* (ed. R.F. Goldberger), pp. 433–485. Plenum Publishing, New York.

Steinberg, S., A. Misch, and M. Sprinzl. 1993. Compilation of tRNA sequences and sequences of tRNA genes. *Nucleic Acids Res.* **21:** (in press).

Steinmetz, A. E.J. Gubbins, and L. Bogorad. 1982. The anticodon of the maize chloroplast gene for tRNA$^{Leu}_{UAA}$ is split by a large intron. *Nucleic Acids Res.* **10:** 3027–3037.

Stewart, T.S., R.J. Roberts, and J.L. Strominger. 1971. Novel species of tRNA. *Nature* **230:** 36–38.

Strauch, M.A., H. Zalkin, and A.I. Aronson. 1988. Characterization of the glutamyl-tRNA^{Gln}-to-glutaminyl-tRNA^{Gln} amidotransferase reaction of *Bacillus subtilis*. *J. Bacteriol.* **170:** 916–920.

Sun, J., M. Inouye, and S. Inouye. 1991. Association of a retroelement with a P4-like cryptic phophage (retronphage ϕR73) integrated into the selenocystyl tRNA gene of *Escherichia coli*. *J. Bacteriol.* **173:** 4171–4181.

Sylvers, L.A., K.C. Rogers, M. Shimizu, E. Ohtsuka, and D. Söll. 1993. A 2-thiouridine derivative in tRNA^{Glu} is a positive determinant for aminoacylation by *Escherichia coli* glutamyl-tRNA synthetase. *Biochemistry* **32:** (in press).

Tamura, F., S. Nishimura, and M. Ohki. 1984. The *E. coli divE* mutation, which differentially inhibits synthesis of certain proteins, is in tRNA^{Ser1}. *EMBO J.* **3:** 1103–1107.

Tobias, J.W., T.E. Shrader, G. Rocap, and A. Varshavsky. 1991. The N-end rule in bacteria. *Science* **254:** 1374–1377.

Ueda, Y., I. Kumegai, and K.I. Miura. 1992. The effects of a unique D-loop structure of a minor tRNA$^{Leu}_{UUA}$ from *Streptomyces* on its structural stability and amino acid accepting activity. *Nucleic Acids Res.* **20:** 3911–3917.

van den Ouweland, J.M., G.J. Bruining, D. Lindhout, J.M. Wit, B.F. Veldhuyzen, and J.A. Maassen. 1992. Mutations in mitochondrial tRNA genes: Non-linkage with syndromes of Wolfram and chronic progressive external ophthalmoplegia. *Nucleic Acids Res.* **20:** 679–682.

Varshney, U., C.P. Lee, B.L. Seong, and U.L. RajBhandary. 1991. Mutants of initiator tRNA that function both as initiators and elongators. *J. Biol. Chem.* **266:** 18018–18024.

Veres, Z., L. Tsai, T.D. Scholz, M. Politino, R.S. Balaban, and T.C. Stadtman. 1992. Synthesis of 5-methylaminomethyl-2-selenouridine in tRNAs: ^{31}P NMR studies show the labile selenium donor synthesized by the *selD* gene product contains selenium bonded to phosphorus. *Proc. Natl. Acad. Sci.* **89:** 2975–2979.

Verkamp, E., M. Jahn, D. Jahn, A.M. Kumar, and D. Söll. 1992. Glutamyl-tRNA reductase from *E. coli* and *Synechocystis* 6803: Gene structure and expression. *J. Biol. Chem.* **267:** 8275–8280.

Vögtli, M. and S.N. Cohen. 1992. The chromosomal integration site for the *Streptomyces* plasmid SLP1 is a functional tRNA^{Tyr} gene essential for cell viability. *Mol. Microbiol.* **6:** 3041–3050.

Wain-Hobson, S., P. Sonigo, O. Danos, S. Cole, and M. Alizon. 1985. Nucleotide sequence of the AIDS virus, LAV. *Cell* **40:** 9–17.

Weiner, A.M. and N. Maizels. 1987. tRNA-like structures tag the 3′ ends of genomic

RNA molecules for replication: Implications for the origin of protein synthesis. *Proc. Natl. Acad. Sci.* **84:** 7383–7387.

Werner, C., B. Krebs, G. Keith, and G. Dirheimer. 1976. Specific cleavages of pure tRNAs by plumbous ions. *Biochim. Biophys. Acta* **432:** 161–175.

Wilcox, M. 1969. Gamma-glutamyl phosphate attached to glutamine-specific tRNA. A precursor of glutaminyl-tRNA in *Bacillus subtilis. Eur. J. Biochem.* **11:** 405–412.

Wilcox, M. and M. Nirenberg. 1968. Transfer RNA as a cofactor coupling amino acid synthesis with that of protein. *Proc. Natl. Acad. Sci.* **61:** 229–236.

Williamson, R.M. and D.L. Oxender. 1990. Sequence and structural similarities between the leucine-specific binding protein and leucyl-tRNA synthetase of *Escherichia coli. Proc. Natl. Acad. Sci.* **87:** 4561–4565.

————. 1992. Premature termination of in vivo transcription of a gene encoding a branched-chain amino acid transport protein in *Escherichia coli. J. Bacteriol.* **174:** 1777–1782.

Yamao, F., Y. Andachi, A. Muto, T. Ikemura, and S. Osawa. 1991. Levels of tRNAs in bacterial cells as affected by amino acid usage in proteins. *Nucleic Acids Res.* **19:** 6119–6122.

Yanofsky, C. and I.P. Crawford. 1987. The tryptophan operon. In Escherichia coli *and* Salmonella typhimurium (ed. F. Neidhardt), pp. 1453–1472. American Society for Microbiology, Washington, D.C.

Yokoyama, S. and S. Nishimura. 1993. Stuctures of modified nucleosides found in tRNA. In *Transfer RNA* (ed. D. Söll and U.L. RajBhandary) American Society for Microbiology, Washington, D.C. (In press.)

Yoon, K.L., J.R. Aprille, and S.G. Ernst. 1991. Mitochondrial tRNA[Thr] mutation in fatal infantile respiratory enzyme deficiency. *Biochem. Biophys. Res. Commun.* **176:** 1112–1115.

Yoshida, M., K. Takeishi, and T. Ukita. 1970. Anticodon structure of GAA-specific glutamic acid tRNA from yeast. *Biochem. Biophys. Res. Commun.* **39:** 852–857.

Young, L.S., H.M. Dunstan, P.R. White, T.P. Smith, S. Ottonello, and K.U. Sprague. 1991. A class III transcription factor composed of RNA. *Science* **252:** 506–507.

Xiao, H., M. Kalman, K. Ikehara, S. Zemel, G. Glase, and M. Cashel. 1991. Residual guanosine 3′,5′-bispyrophosphate synthetic activity of *relA* null mutants can be eliminated by *spoT* null mutations. *J. Biol. Chem.* **266:** 5958–5990.

Zillmann, M., M.A. Gorovsky, and E.M. Phizicky. 1991. Conserved mechanism of tRNA splicing in eukaryotes. *Mol. Cell. Biol.* **11:** 5410–5416.

8

Antibiotics and the RNA World: A Role for Low-molecular-weight Effectors in Biochemical Evolution?

Julian Davies
Department of Microbiology
University of British Columbia
Vancouver, British Columbia, Canada B6T 1Z3

Uwe von Ahsen[1] **and Renée Schroeder**
Institut für Mikrobiologie und Genetik der Universität Wien
A-1030 Vienna, Austria

There is much more to molecular and cell biology than DNA, RNA, and proteins! A very large number of secondary metabolites produced by microbes and plants have been identified; these natural products include some of the most potent inhibitors of cellular metabolic reactions. The best-known inhibitors are antibiotics, used in the therapy of infectious disease, which have a number of cellular targets. Not surprisingly, antibiotics may interact with different macromolecules to inhibit cellular processes.

The protein synthesis inhibitors have been studied for many years, beginning with the pioneering work of the laboratory of E.F. Gale in Cambridge on the inhibition of bacterial cells, which could be interpreted as translation inhibition. The development of cell-free polypeptide synthesis in the early 1960s permitted the testing of many putative protein synthesis inhibitors. Probably the first conclusive experiments to focus on cellular targets for translation inhibitors were those of Erdös and Ullmann (1959) with streptomycin. Subsequently, the use of fractionated cell-free extracts with separated ribosomal subunits allowed the identification of many antibiotics as specific inhibitors of ribosome function. Within a few years, the mechanism of action of most inhibitors of translation had been worked out in general terms, although not in any great detail at the chemical level. An important concept, first elaborated by Vazquez (1964), was the species specificity of the interaction between

---

[1]Present address: University of California at Santa Cruz, Sinsheimer Labs, Santa Cruz, California 95064.

The RNA World
© 1993 Cold Spring Harbor Laboratory Press 0-87969-380-0/93 $5 + .00

inhibitors of protein synthesis and their targets (for an excellent review, see Cundliffe 1981). The narrow range of activity of many of these compounds is well known and is critical to the use of the antibiotics as therapeutic agents.

One surprising conclusion of these studies was that natural products with molecular weights less than 1000 can interact specifically (often in a 1:1 ratio) with macromolecules more than 3000 times larger (e.g., ribosomes). Research focused on the nature of these interactions, and combined genetic and biochemical studies led first to the conclusion that antibiotic inhibitors of ribosomal function might interact directly with ribosomal proteins. Detailed studies of the binding of these inhibitors to ribosomes were carried out in many laboratories, and kinetic analyses confirmed that the binding was highly specific. Indications that some antibiotics (e.g., erythromycin) might bind to ribosomal proteins have been described (Teraoka and Nierhaus 1978), but this has not proved to be a general phenomenon and the result cannot be considered of significance. However, in view of increasing evidence for the key role of RNA in ribosome function, it became apparent that the earlier conclusions with respect to putative interactions between antibiotics and ribosomal proteins were no longer tenable.

Perhaps the first indication of a functional interaction between an antibiotic (streptomycin) and rRNA came from Gorini's laboratory; it was found in equilibrium dialysis studies that radioactively labeled streptomycin would bind stoichiometrically with 16S rRNA (1–2 molecules of streptomycin per rRNA) (Garvin et al. 1974). Apparently, these experiments have never been repeated. More persuasive proof of antibiotic/rRNA interaction was soon to follow from four different types of studies:

1. The analysis of "self-resistance" in the bacteria producing antibiotic inhibitors of protein biosynthesis. In a majority of these cases, one mechanism of avoiding suicide is methylation of a ribose hydroxyl group on a specific nucleoside in rRNA, which engenders a resistant ribosome (Cundliffe 1989).
2. The isolation of antibiotic-resistant mutants of lower eukaryotes (yeast, fungi, algae) with alterations in rRNA, and not ribosomal proteins. Genetic mapping studies located the mutations to base changes in rRNA genes (Spangler and Blackburn 1985).
3. The isolation of antibiotic-resistant mutants of *Escherichia coli* by cloning mutagenized rRNA genes onto a multicopy plasmid and introducing them into a drug-sensitive background (Sigmund et al. 1984). Sequencing studies confirmed that resistance was due to single base alterations in the rRNA.

4. "Protection" studies, which revealed that the binding of antibiotic inhibitors of translation to ribosomes protects single bases in the rRNAs from subsequent chemical modification (Moazed and Noller 1987b).

Detailed maps are now available that show the nucleoside sites for interaction of antibiotic inhibitors of ribosome function in rRNA (Cundliffe 1990). A new class of receptors has been identified. These findings not only changed our perspectives with respect to how these antibiotics work, but also suggested that there may be evolutionary significance in the antibiotic/RNA interactions.

In all schemes of biochemical and cellular evolution, the origin of the mechanism of protein biosynthesis is inadequately described. Although there have been serious attempts to consider this problem, it is far too common to see flow charts for biochemical evolution with a box marked simply "evolution of the translation system"! Given the complexity of this process (multiple RNA species, more than 50 proteins, numerous protein cofactors, etc.), an evasive description is not surprising, but there have been relatively few attempts to provide plausible scenarios. The situation is more enigmatic than the "chicken-and-egg" problem, since in the modern protein biosynthetic system, proteins are needed to make proteins. How did the translation system evolve? Now that we have convincing evidence that (essentially) protein-free 23S rRNA has peptidyl transferase activity, and may in fact be a ribozyme (Noller et al. 1992), we can begin to formulate reasonable notions based on the prebiotic formation of a catalytic RNA that was capable of generating peptides from the amino acids present in primordial soups or being delivered to the primitive earth atmosphere from extraterrestrial sources (the earliest ETs were likely to have been simple organic molecules!). One can imagine that some of the primitive peptides made by RNA catalysis or coming from prebiotic sources would have properties permitting them to remain associated with the RNA and perhaps promote conformational changes or stability in the pre-rRNA, thereby providing more efficient (and accurate?) polypeptide formation.

A ROLE FOR LOW-MOLECULAR-WEIGHT EFFECTORS IN EVOLUTION

We propose a modification of this scenario: that reactions taking place on the earliest forms of catalytic RNA could have been "effected" or modulated by other low-molecular-weight compounds in the prebiotic environment. If nucleic acid bases and amino acids were present, surely condensation products and oligomers of these precursors must also have existed. Among such molecules may have been simple peptides that resembled some present-day antibiotics, a number of which are known to influence translation. Table 1 lists a number of the prebiotic amino acids (Miller 1987) and some of the antibiotics known to be derived from them. Although this does not constitute proof of the presence of such molecules at an early stage of biochemical evolution, it is consistent with the notion.

Could such compounds have played an intermediary role in the evolution of the modern ribosome? We have suggested that low-molecular-weight compounds could have been functional precursors of some ribosomal proteins in "early" translation reactions (Davies 1990). These molecules were subsequently displaced by more effective ribosomal proteins. In considering this question, we return to our earlier discussion of antibiotic sites in ribosomes. It is known that many of the antibiotic inhibitors of ribosome function have specific sites in rRNA, as identified by studies of resistance mechanisms and chemical protection analyses. Analysis of the sites concerned indicates that the antibiotics interact with bases in conformationally important regions of conserved sequence, in all probability in the most evolutionarily significant regions of rRNAs.

Since the low-molecular-weight compounds interact with rRNA to influence catalytic functions, presumably as a consequence of their effector role in the evolution of ribosomes, we wanted to examine whether com-

Table 1 Antibiotics formed from "primordial" amino acids

Amino acid	Antibiotic
Alanine	gramicidin
β-Alanine	destruxin
α-Amino-*n*-butyric acid	lysergic acid
α-Aminoisobutyric acid	alamethicin
β-Amino-*n*-butyric acid	brevistin
γ-Aminobutyric acid	subtilin
Diaminobutyric acid	polymixin
α,β-Diaminopropionic acid	edeine, viomycin
Glutamic acid	bacitracin
Glycine	nisin
α-Hydroxy-γ-aminobutyric acid	butirosin
Isoserine	edeine
N-Methylalanine	verticillin, tentoxin
Norvaline	cyclosporin
Pipecolic acid	actinomycin, ostreogrycin
Proline	tyrocidin
Sarcosine	cyclosporin, actinomycin
Valine	gramicidin

"Primoridal" amino acids are those that have been detected in electric discharge reactions or in meteorites.

pounds of the antibiotic class would interact actively with other types of catalytic RNA. Accordingly, we have tested the activity of a large number of antibiotic inhibitors of ribosome function on intron self-splicing.

A number of protein synthesis inhibitors have been found to interact with group I introns and to influence reactions of these ribozymes; some were inhibitors, some were potentiators, and some had no effect. A summary of the results obtained to date with the compounds tested for their activity is given in Table 2; these results are considered in more detail in the ensuing discussion. The immediate conclusion is that group I introns and rRNA (ribosomes) must have common features that specify binding sites for antibiotics of the aminoglycoside and peptide classes, raising the possibility of similarities between translation and splicing at the mechanistic level (Davies et al. 1992).

THE ACTION OF ANTIBIOTICS ON SPLICING REACTIONS

Streptomycin

This antibiotic was among the first therapeutically useful secondary metabolites to be discovered and was employed successfully for some time in the treatment of tuberculosis; its use is limited nowadays because more active and less toxic molecules are available. The mode of action of streptomycin as an inhibitor of protein synthesis has been the subject of much research activity. It was the first antibiotic shown to act on a specific ribosomal subunit; investigation of the mode of action of streptomycin led to the early demonstration of the role of the ribosome in determining the fidelity of translation. The genetics of streptomycin resistance (and dependence) has been studied in detail, and mutations affecting cellular response to streptomycin were the first known to affect ribosome structure and function. Although spontaneous mutations to streptomycin resistance alter the ribosomal protein genes (primarily S12), subsequent chemical protection analyses have identified a 16S rRNA region from U-911 to A-915, and to a lesser extent A-1413 to G-1494, as being crucial for binding of the antibiotic, thus identifying the parameters of the rRNA receptor. These conclusions have been confirmed by in vitro mutagenesis of *E. coli* rRNA genes; the studies also identified an A to C change at position 523 as a determinant of streptomycin resistance. Streptomycin dependence is an unusual phenomenon that may have a number of different biochemical explanations. One mechanism of streptomycin dependence would suggest that the ribosome cannot be assembled correctly in mutants unless streptomycin is present (Hummel et al. 1983). Dependence is also associated with amino acid substitutions in ribosomal protein S12.

It is not surprising, therefore, that streptomycin was the first aminoglycoside to be tested and found to inhibit group I intron splicing at high concentrations. The initial step of splicing (guanosine binding followed by cleavage of the 5′ exon-intron boundary; Fig. 1) was most markedly affected by the antibiotic; excision of the intron was influenced (apparently) to a lesser extent (von Ahsen and Schroeder 1990). Inhibition was found to be competitive with guanosine, which is consistent with the fact that both guanosine and (probably) streptomycin interact with the G-site in the intron RNA through their guanidino moieties (von Ahsen and Schroeder 1991); molecules of similar structure also inhibited the reaction. Kinetic analyses of the effect of streptomycin at different guanosine levels confirmed the competitive nature of inhibition by the antibiotic. A number of different molecules with guanidino residues have now been shown to inhibit the first step of splicing, presumably because of competition with the binding of guanosine at the G-site in the intron (Fig. 1).

Another group of antibiotics that contain guanidine residues are the viomycins. These compounds, also, inhibited self-splicing in a competitive reaction with guanosine. However, the viomycins can influence intron activity in a different way, as discussed below.

The guanidino-containing competitive inhibitors of splicing affected all group I introns but did not inhibit the splicing of group II introns, as might be predicted. Confirmation of the mechanism of streptomycin action was obtained by studies with a mutant of the td intron that has a higher K_m for guanosine without affecting the interaction between the G-binding site and guanidino groups. This mutant ribozyme was about tenfold more sensitive to inhibition by streptomycin, presumably because of easier access to the binding site (von Ahsen and Schroeder 1991). Thus we can be confident that streptomycin acts on the intron at a specific site and does not cause inhibition as a result of generalized nucleic acid binding. It is known that arginine is a competitive inhibitor of the splicing of group I introns, acting by the binding of its guanidino group to the G-binding site on the intron RNA. Interestingly, the conserved G of the G-binding site plus the two neighboring bases always constitute an arginine codon (AGA, CGA, CGG, or AGG) (Yarus, this volume). It is not known if streptomycin has a binding site that overlaps with the G-binding site, or if it interacts at an independent site(s) in the intron RNA.

Other Aminoglycoside Antibiotics

It has been well established that the aminoglycoside antibiotics are potent inhibitors of ribosome function (Cundliffe 1981). This class of secondary

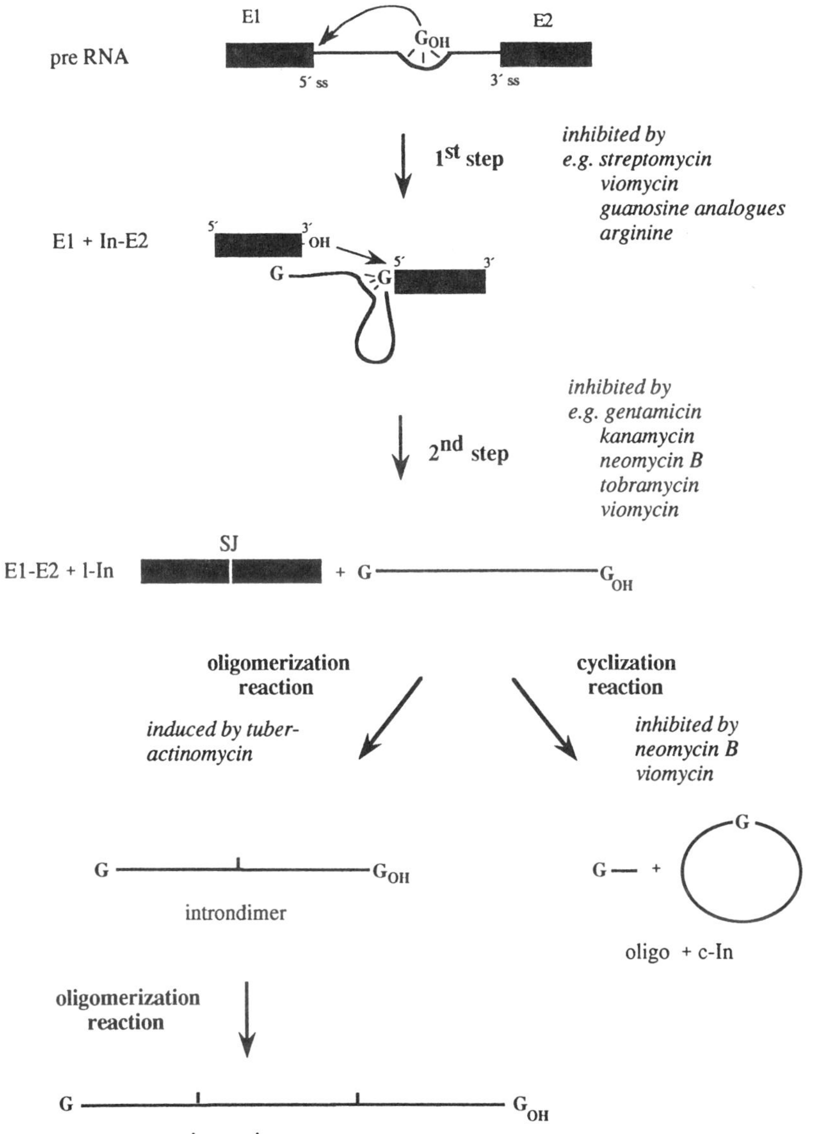

Figure 1 Scheme for splicing of group I introns, indicating steps affected by antibiotics.

metabolites causes (like streptomycin) extensive misreading of the genetic code, and the "receptor" sites have been mapped to the 16S rRNA. Several discrete base alterations have been identified in the ribosomes of resistant strains; these rRNA sites have been confirmed by chemical protection studies.

An obvious next step was to examine the effects of a variety of different aminoglycosides on splicing reactions. The results were strikingly different from those obtained with streptomycin, in both a quantitative and a qualitative sense.

The major group of aminoglycoside antibiotics is based on 2-deoxystreptamine (DOS) and includes a number of the most important agents used for the treatment of infectious diseases in hospitals: kanamycin, gentamicin, and tobramycin. The DOS-aminoglycosides are strongly basic water-soluble compounds of low molecular weight (<1000). Because of their basic nature, they are known to bind to many anions, including nucleic acids, and this binding can influence the biological properties of the anion. The mode of action of these compounds has been studied extensively, but full details of their action on ribosomes have not been worked out. These antibiotics may influence several steps in the process of translation, which can be identified with interference with ribosome A-site function (Hausner et al. 1988). Unlike streptomycin, no spontaneous mutations in ribosomal protein genes are known to confer resistance to gentamicin, kanamycin, tobramycin, neomycin, etc. Confirmation of the ribosomal site of action of the aminoglycosides comes from studies of "self-protection" in antibiotic-producing organisms. Specific rRNA methylases modify single bases in 16S rRNA, leading to high-level resistance to the DOS-aminoglycosides. As an aside, these studies, which confirm the ribosomal target site, also imply that the lethal effects of gentamicin, etc., in sensitive microbes must be related uniquely to an interaction with the small ribosomal subunit. Resistance is conferred by enzymatic methylation of G-1405 or A-1408 in 16S rRNA. That this region of 16S rRNA was implicated in aminoglycoside binding was confirmed by chemical protection studies, which identified bases in the same region. Other regions of 16S rRNA are probably involved in aminoglycoside binding, as shown by protection of A-790, G-791, and A-909 in 50S ribosomes in the presence of aminoglycosides and tRNA, implying that a highly ordered structure is required for binding (Moazed and Noller 1987b).

The 2-deoxystreptamine-based aminoglycosides had inhibitory activities that were as much as 3 logs more potent than streptomycin, acting mainly (but not exclusively) at the second step of self-splicing, and inhibition could not be reversed by the addition of excess guanosine (see Fig. 2) (von Ahsen et al. 1991). That is, these antibiotics were non-competitive inhibitors and independent of the guanosine cofactor. This was confirmed by kinetic studies of the inhibition of self-splicing by the gentamicin and neomycin-type antibiotics. The use of a mutant intron that responded to A but not G was especially revealing, since the inhibi-

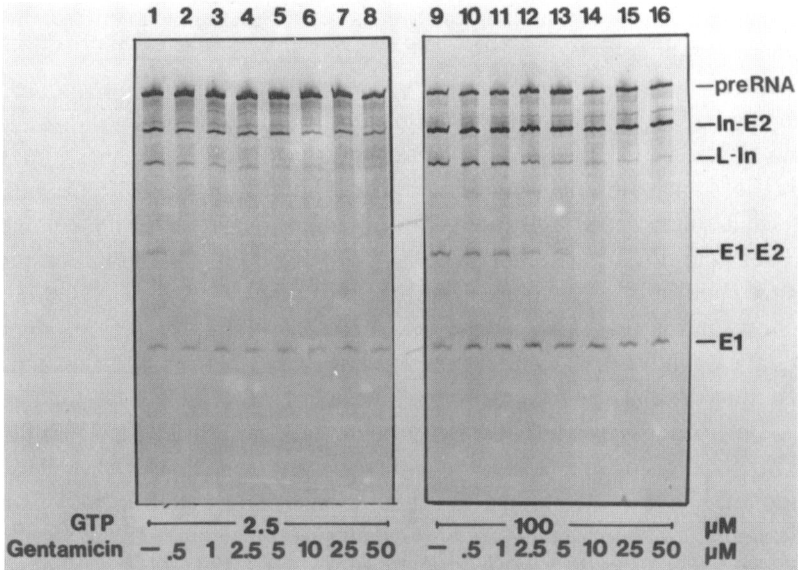

Figure 2 Effect of the aminoglycoside gentamicin on the splicing of the td intron. (Reprinted, with permission, from von Ahsen et al. 1991.)

tion by aminoglycosides shows that the antibiotics bind to a site on the intron RNA distinct from G (von Ahsen et al. 1992).

A number of other antibiotic inhibitors of ribosome function were examined for effects on the self-splicing of group introns (Table 2). Among the compounds tested were those that act on the 30S ribosome (kasugamycin, spectinomycin, hygromycin) and on the 50S ribosome (thiostrepton, chloramphenicol, erythromycin, pristinamycin, clindamycin). None of these compounds was active. Thiostrepton is known to bind with high affinity to 50S ribosomes and to 23S RNA; its inactivity in the self-splicing reaction clearly emphasizes the specificity of action of the aminoglycosides (Cundliffe 1990). As described below, only the peptide antibiotics of the viomycin class have demonstrable effects on group I intron self-splicing in vitro; these compounds have multiple effects on translation through interactions at several ribosome sites, including the A-site. The aminoglycosides act on 30S-specific functions in translation, preferentially at the A-site. Thus, we conclude that only antibiotics that interfere with ribosomal A-site function have activity against the group I introns.

Viomycin and Tuberactinomycin

As mentioned above, one other class of antibiotic inhibitors of ribosome function interacts with both rRNA and intron RNA. The viomycin antibiotics are extremely potent inhibitors of protein biosynthesis in bacteria; these are pentapeptide antibiotics. They are known to be synthesized by the peptide synthetase mechanism, which does not involve ribosomes. Like many peptide antibiotics of this class, they contain non-protein amino acids as D-isomers. We have mentioned the presence of such amino acids in prebiotic synthetic mixtures. Interestingly, an RNA molecule that binds specifically to a D-tryptophan agarose affinity column has been identified in a random pool of RNA oligomers (Famulok and Szostak 1992).

These antibiotics have a complex mode of inhibitory action, since they have been shown to affect functions of both 30S and 50S ribosome, which may explain their potency (Hausner et al. 1988). The binding site(s) in 16S rRNA has not been recognized, but Moazed and Noller (1987a) have mapped the viomycin binding sites on 23S rRNA. These compounds have been studied for their effects on translation in *Myco-bacterium sp.*, but since they are only weakly active against *E. coli* due to poor uptake, they have not been examined in detail in this latter, more tractable organism. Mutants of *E. coli* supersensitive to viomycin and tuberactinomycin are now available, so appropriate studies should be possible. When viomycin and its analogs were tested for activity on group I intron splicing, they gave an expected result (H. Wank et al., in prep.). Viomycin possesses a guanidino residue and was found to be a competitive inhibitor of self-splicing at high concentrations. However, anomalous RNA species detected by electrophoresis encouraged us to examine other viomycin-like antibiotics (e.g., tuberactinomycin) in more detail, with unexpected results. The primary products of the self-splicing of group I introns are the spliced exons and the free linear intron. The released linear intron can undergo a transesterification reaction by attack of the terminal G of the intron at the cyclization site, resulting in the formation of a circular form of the intron RNA. Under certain conditions, oligomers of the intron can be detected in low amounts. The antibiotic tuberactinomycin induced (by an unknown mechanism) extensive and efficient oligomerization of introns, which implies another primitive function for antibiotic-like molecules: involvement in the oligomerization of catalytic RNA to produce higher-molecular-weight RNAs. This could be the means by which more complex RNA molecules were formed in prebiotic reactions. (It is worth noting that tandem duplications of introns have been identified in certain classes of rRNA.)

Are Splicing and Decoding Related?

We propose that the processes of splice-site selection during splicing and decoding during translation are mechanistically related. Both processes

Table 2 Effects of various antibiotics on group I (td) intron splicing and translational misreading

Compounds	Self-splicing[a]		Protein synthesis[a] (in vitro misreading)
	1st step	2nd step	
Gentamicins (4,6-disubstituted DOS)			
gentamicin complex	–	2.0 μM	1.0 μM
gentamicin C1	–	2.0 μM	1.0 μM
gentamicin C2	–	2.0 μM	1.0 μM
gentamicin C1a	–	2.0 μM	1.0 μM
gentamicin B	–	20.0 μM	20.0 μM
G418	–	–	20.0 μM
5-episisomicin	20.0 μM	0.5 μM	1.0 μM
sisomicin	1.0 μM	50.0 μM	1.0 μM
dihydrosisomycin	–	100.0 μM	weak
garamine	100.0 μM	200.0 μM	–
2-phosphogentamicin	–	–	–
1-acetylsisomicin	–	100.0 μM	weak
5-deoxygentamicin	50.0 μM	0.5 μM	n.d.
Neomycins (4,5-disubstituted DOS)			
ribostamycin	–	1.0 mM	5.0 μM
neomycin B	100.0 μM	0.5 μM	0.5 μM
paromomycin	1.0 μM	100.0 μM	10.0 μM
lividomycin	–	1.0 mM	10.0 μM
neamine	–	200.0 μM	50.0 μM
hybrimycin C1	50.0 μM	0.5 μM	n.d.
Kanamycins (4,6-disubstituted DOS)			
tobramycin	–	0.5 μM	1.0 μM
kanamycin A	–	–	5.0 μM
kanamycin B	–	10.0 μM	1.0 μM
kanamycin C	–	–	100.0 μM
amikacin	–	–	10.0 μM
6′-*N*-ethyltobramycin	–	100.0 μM	100.0 μM
Other aminocyclitols			
apramycin	–	200.0 μM	10.0 μM
kasugamycin	–	–	–
hygromycin B	–	–	50.0 μM
spectinomycin	–	–	–
sorbistin	–	–	weak
streptomycin	4.0 μM	n.d.	1.0 μM
2-deoxystreptamine	–	–	–
Peptides			
viomycin	500.0 μM	100.0 μM	weak[xx]
tuberactinomycin	500.0 μM	xxx	n.d.
Antibiotics acting on the large ribosome unit			
thiostrepton, clindamycin erythromycin, pristinamycin chloramphenicol	good inhibitors of translation but all inactive in self-splicing inhibition assay		

Dash indicates no effect (at >1 mM). n.d. indicates not done. (xx) Strong inhibitor of translation (<1 μm); (xxx) slight stimulation.

[a]Concentrations required for approximately 50% inhibition.

are entirely RNA-based. Splice-site selection in group I introns occurs via Watson-Crick base-pairing between the 5'splice site and the internal guide sequence (IGS), which results in the formation of the intron P1 stem. Binding of the substrate (5'splice site) to the intron core RNA involves the interaction of the 2'OH groups of the substrate with bases in the catalytic core (junction 8/7 = J8/7). This enables the core of the intron, which is highly sequence-conserved, to interact and catalyze the cleavage of substrates that are not conserved in sequence (Michel and Westhof 1990). The nature of the substrate bases is recognized by the IGS, whereas the correctly base-paired stem P1 is recognized by the junction J8/7, which interacts with the 2'OH groups of the substrate; these are not "base specific." A similar mechanistic scenario can be seen for the decoding process; the anticodon interacts with mRNA by Watson-Crick pairing. The correctly base-paired triplet (not sequence-conserved) interacts with the 16S rRNA A-site "region," which is highly sequence-conserved. This interaction cannot occur by the usual base-base recognition and might take place through the interaction of the 2'OH groups of the anticodon with nucleotides in 16S rRNA. In this model, the 5'exon sequences would correspond to the anticodon, the IGS to the mRNA, and the sequence-conserved core of the intron (J8/7) to the 16S rRNA decoding site. That the two processes share common features is demonstrated by their sensitivity to aminoglycoside antibiotics. The self-splicing group I intron of *Anabena sp.* presents an intriguing overlap of the two processes (Xu et al. 1990), in which the intron is inserted in a tRNA anticodon between the wobble base and the second base of the anticodon. The P1 helix of the *Anabena* intron is shorter than usual, with only 3 base pairs preceding the 5'splice site. In this case, an anticodon (5' splice site) base-paired with the IGS interacts with the J8/7 junction via its 2'OH groups. What is interesting from the point of view of RNA chemistry is that the AG antibiotics have very specific interactions with two types of RNA molecules: 16S ribosomal RNA and group I intron RNA. Thus we suggest that an AG-recognizing conformation is present in both types of RNA. High-resolution nuclear magnetic resonance (NMR) studies are likely to be informative in future research, in the absence of three-dimensional X-ray information.

We have shown a clear relationship between group I introns and 16S rRNA through their common interaction with certain antibiotics; complementary studies have confirmed this relationship. It is likely that group I introns, although they self-splice in vitro, do require the mediation of protein factors in vivo. Belfort and her collaborators (T. Coetzee and M. Belfort, pers. comm.) have shown this to be the case for the T4 introns and have evidence that ribosomal proteins are involved! This exciting demonstration makes the existence of an evolutionary pathway linking group I introns and rRNA even more convincing.

Apart from viomycin, which affects both ribosome subunits, all of the ribosome inhibitors that interfere with the splicing of group I introns act at the ribosomal A-site. The latter has been identified as the decoding site in mRNA translation, and the aminoglycosides as a class are thought to influence the proofreading process during the selection of the correct aminoacyl-tRNA by the mRNA codons on the ribosome. The binding sites for these molecules are presumed to be at specific regions in rRNA, which are determined to a large extent by RNA conformation, as dictated by the binding of ribosomal proteins. The actual roles of the RNA-binding sites in the process of translation are unknown; they could provide "points" for attachment of tRNA or protein factors, or they could equally well be required for the binding of specific Mg^{++} ions in ribosome-catalyzed reactions. Unfortunately, our knowledge of the detailed chemistry of peptide formation on ribosomes is seriously wanting in these areas. Should we expect to find conformations related to the 16S rRNA decoding site in intron RNA? We believe this to be the case, since no nucleotide sequence similarities can be detected by comparing group I intron and rRNAs, implying that a three-dimensional structure is recognized by the antibiotics.

A Role for Magnesium?

The role of Mg^{++} in translation has been known for a long time. Apart from being required to maintain the integrity of ribosome structure, variations in Mg^{++} concentration have been shown to affect all identified steps in peptide bond formation. In the case of the A-site functions, tRNA binding and the fidelity of codon/anticodon recognition are Mg^{++}-dependent. However, as shown by Davies et al. (1968), the effects of aminoglycosides are generally over and above those resulting from changes in magnesium concentration. Y. Bao and D.L. Herrin (pers. comm.) have shown that self-splicing of the chloroplast group I ribozymes is actually promoted by some of the aminoglycoside antibiotics at suboptimal concentrations of Mg^{++}. The result might be explained as a direct replacement of the magnesium requirement by the cationic aminoglycosides, but this cannot be established without more extensive comparisons with other cations. Nonetheless, we wish to point out that interactions of aminoglycosides with specific, functional, magnesium-binding sites within macromolecules can be considered a key aspect of their mode of action. After all, the aminoglycosides have been shown to cross the outer membranes of gram-negative bacteria by a mechanism that involves competitive displacement of magnesium ions bound within the membrane.

Structure/Activity Relationships

In principle, studies of the relationship between the structures of the aminoglycosides and their activities should give some clues as to the sites of interaction with the intron. Limited information is available on the functions important for the effects of the aminoglycosides on translation. For example, the 2-deoxystreptamine ring is an important determinant, and only minor modifications can be tolerated; 5-epi-derivatives are usually equally active. A trisaccharide is generally more active than a disaccharide, and a tetrasaccharide is more potent when additional amino groups are added. The glucosamine ring attached to the 4-position of 2-DOS is the most important, and activity is related to the number and position of amino groups; typically, 2,6-diamino > 6-amino > 2-amino. The inactivation of aminoglycosides by resistance enzymes tends to confirm these general observations. Figure 3 summarizes the structure/function information that has been obtained to date and compares effects on translation with those on splicing inhibition. It is clear that the structural requirements for the action of aminoglycosides on splicing and translation reactions are similar, but not identical. For one thing, the intron site must have conformational or size limitations, in the

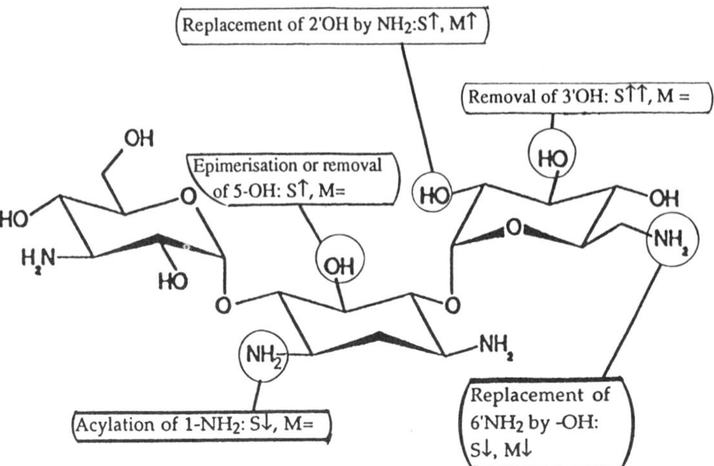

Figure 3   Structure/activity relationships for effects of aminoglycoside antibiotics on translation and on the splicing of group I introns. (↑) Increased sensitivity; (↓) decreased sensitivity; (=) no change.

sense that amikacin (1-HABA-kanamycin A) is a good inhibitor of translation but is much less active as an inhibitor of splicing. As would be expected, butirosin also failed to interact with the intron. The properties of structural variants of the kanamycins indicate the differences in structure/activity that arise from changes in amino and hydroxyl residues.

Much less is known concerning structure/activity relationships among the viomycin antibiotics. These antibiotics are essentially variants on a central pentapeptide structure, tuberactinamine, and it will be of interest to see whether tuberactinamine has activity in either assay. There is a striking difference between viomycin and tuberactinomycin; the latter promotes intron oligomerization and is a weak inhibitor of the first step of splicing. The only obvious difference in structure between the two compounds is the lack of a hydroxyl group on the diamino acid substituent of viomycin, and no hydroxyl on the reduced pyrimidine ring of tuberactinomycin. Capreomycin, the only other member of the group, has activity similar to viomycin. The information available is not sufficiently detailed to permit any confident predictions of antibiotic/intron complex formation. Three-dimensional NMR studies will likely be useful in giving clues to the nature of this process, which in turn will be valuable in interpreting the functional interactions between antibiotics and rRNA.

Attempts to establish a group I splicing assay in vivo, in which the antibiotic effects on splicing can be separated from the inhibition of ribosome function, are worth while, since the obtention of an appropriate phenotypic selection dependent on splicing would permit the isolation of mutants resistant to antibiotic inhibition. Will these map to the RNA or in protein genes? And at what sites in the intron RNA? The antibiotic inhibitors of self-splicing reactions affect primarily (although not exclusively) the steps involved in tRNA selection on the 30S subunit, and all cause mistranslation of the genetic code. How do the aminoglycoside and viomycin-like antibiotics affect ribozymes in vivo? Is it possible that they interfere with or replace the function of proteins in some cases?

CONCLUSIONS AND PERSPECTIVES

It is important to remember that prebiotic reaction conditions are cell-free extract conditions. If we want to know anything about prebiotic reactions, it is not possible to investigate them using whole cells. In the recent past, experiments with cell-free systems have often given results that were not completely consistent with expectations from in vivo observations. Rather than rejecting them as in vitro anomalies, we should profit from such experiments by considering them as potential clues to prebiotic events of biochemical importance.

Prebiotic soups must have contained tens of thousands of molecules, all of which could have participated, at some time, in the ultimate shotgun experiment—that of prebiotic evolution. Many of the molecules found as secondary metabolites, such as unusual amino acids or nucleic acid bases, may have participated in early reactions. For example, the simple RNA molecules produced may well have contained unusual nucleic acid bases before the adventitious incorporation of the modern quartet. The same is true of amino acids and their incorporation into peptides; we believe there is much to be learned from the properties of simple oligopeptides containing non-protein amino acids that may have contributed to the evolutionary process leading to the sophisticated macromolecular complexes found in cells.

It is all too easy to criticize these notions using anthropomorphic arguments—that millions of secondary metabolites "must have a function" because they are made. Obviously, many of these compounds do play important roles in the ecology of microbial and plant cells; however, until we have a better conception of when and how biosynthetic pathways of the precursors of macromolecules evolved and of the selective pressures leading to their evolution, we must keep open the notion that prebiotic evolution involved many low-molecular-weight compounds that are still products of metabolism and may have roles in the life-styles of the organisms that produce them. These roles are a matter for speculation, and various functions have been proposed (Davies et al. 1992). It is commonly believed that they participate in a form of biological warfare with competing organisms in the environment, although this has been shown experimentally in only a few cases. Every producing organism so far examined (microbe or plant) possesses at least one biochemical mechanism of resistance against its toxin, thus preventing suicide of the producer during production phase. In some instances, the resistance mechanism has been proposed to play a role in the regulation of the biosynthesis of the antibiotic (Chater and Hopwood 1989).

It had been shown previously that catalytic RNA interacts specifically and functionally with molecules such as nucleic acid bases (e.g., guanosine) and with amino acids (arginine). Our studies on the effects of antibiotics on self-splicing reactions have expanded the range of simple organic molecules that interact directly with introns. We believe that there are numerous aspects of these RNA receptor/organic ligand reactions that warrant further study. First of all, are there other low-molecular-weight natural products that can effect catalytic RNA reactions? There are tens of thousands of different molecules to screen! Such molecules may provide the means for controlling ribozyme reactions in vivo, by the design of specific organic ligands. Turning off a particular gene by preventing splicing is one such scenario. From another point of view, the antibiotics and related molecules are likely to be useful tools for the exploration of RNA structure. This may permit the identification of structural transitions taking place during RNA catalysis and aid in the three-dimensional modeling of ribozyme structures. New classes of inhibitors with potential as pharmaceutical agents could be developed, based on the antibiotic inhibition studies described. For example, might it be possible to obtain a new class of antimicrobial agents that inhibit the splicing of fungal group I introns required in the propagation of the pathogen? There is evidence that the aminoglycosides cause inhibition of fungal growth under aerobic conditions because they interfere with the translation on mitochondrial ribosomes. However, it has not been demonstrated in all cases that the effect is on the ribosome rather than on the splicing of the rRNA. In preliminary studies, we have found that *Saccharomyces cerevisiae* lacking mitochondrial introns are less sensitive to inhibition by certain antibiotics (M. Wögerbauer and R. Schroeder, unpubl.). These studies also suggest that the mode of action of some antibiotic inhibitors might need reinvestigation. In addition, aminoglycosides may influence RNA reactions involved in virus replication, etc. Certain aminoglycosides and their derivatives have been found to be inhibitors of viral nucleic acid replication (Matsunaga et al. 1986); could this be due to interference with the function of an RNA component involved in the replication process?

We have discussed the evolutionary aspects of the interaction of ribozymes with low-molecular-weight compounds above. The relationship between 16S rRNA and the group I introns, suggested by their interaction with the same groups of antibiotics, is obvious, and we believe that the two RNAs must have a common evolutionary precursor. The fact that the aminoglycosides have such specific interactions with ribosome and ribozyme would also suggest that there is a strong relationship between group I introns and 16S rRNA function. Furthermore, the fact that the viomycin antibiotics interfere with the properties of both 16S and 23S rRNA indicates that there may be a related site within these two rRNAs. Are both rRNAs capable of catalytic function, and might they act cooperatively in the ribosome? Did the two molecules evolve separately or as a unit from prebiotic precursor RNA (Fig. 4)? Finally, our investigations emphasize the value of low-molecular-weight effectors and inhibitors in the analysis of the biological properties of macromolecules. Given the large number of secondary metabolites produced by microbes, plants, invertebrates, and vertebrates, it can be anticipated that this rich store of chemical diversity will provide many applications to studies of cell biology—and many surprises!

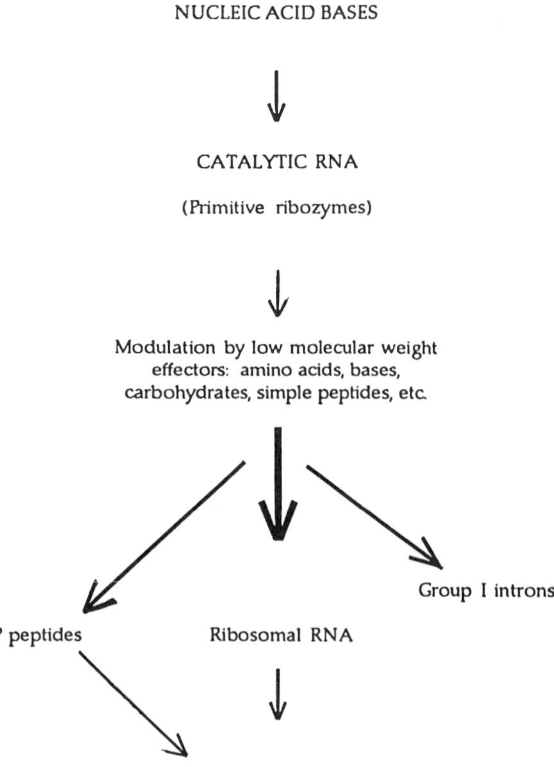

Figure 4 Scenario for an evolutionary relationship between rRNA and group I introns.

ACKNOWLEDGMENTS

We thank many colleagues for their instructive and constructive criticism, especially H. Wank for help in preparing figures, and R. Bauzon and D. Davies for preparing the manuscript. Financial support was provided by the Austrian Fonds zur Förderung der Wissenschaftlichen Forschung (R.S.), an EMBO fellowship (U.v-A.), and the National Science and Engineering Research Council of Canada (J.D.).

REFERENCES

Chater, K.F. and D.A. Hopwood. 1989. Antibiotic biosynthesis in *Streptomyces*. In *Genetics of bacterial diversity* (ed. D.A. Hopwood et al.), pp. 129–150. Academic Press, London.

Cundliffe, E. 1981. Antibiotic inhibitors of ribosome function. In *The molecular basis of antibiotic action*, 2nd ed. (ed. E.F. Gale et al.), pp. 401–457. Wiley, Chichester, England.

————. 1989. How antibiotic-producing organisms avoid suicide. *Annu. Rev. Microbiol.* **43:** 207–233.

————. 1990. Recognition sites for antibiotics within rRNA. In *The ribosome: Structure, function and evolution* (ed. W.E. Hill et al.), pp. 479–490. American Society for Microbiology, Washington, D.C.

Davies, J. 1990. What are antibiotics? Archaic functions for modern activities. *Mol. Microbiol.* **4:** 1227–1232.

Davies, J.E., W. Gilbert, and L. Gorini. 1968. Streptomycin, suppression, and the code. *Proc. Natl. Acad. Sci.* **51:** 883–890.

Davies, J., U. von Ahsen, H. Wank, and R. Schroeder. 1992. Evolution of secondary metabolite production; potential roles for antibiotics as prebiotic effectors of catalytic RNA reactions. *CIBA Found. Symp.* **171:** 24–44.

Erdös, T. and A. Ullmann. 1959. Effect of streptomycin on the incorporation of amino acids labelled with carbon-14 into ribonucleic acid and protein in a cell-free system of a mycobacterium. *Nature* **183:** 618–619.

Famulok, M. and J.W. Szostak. 1992. Stereospecific recognition of tryptophan agarose by in vitro selected RNA. *J. Am. Chem. Soc.* **114:** 3990–3991.

Garvin, R.T., D.K. Biswas, and L. Gorini. 1974. The effects of streptomycin or dihydrostreptomycin binding to 16S RNA or to 30S ribosomal subunits. *Proc. Natl. Acad. Sci.* **71:** 3814–3818.

Hausner, T.-P., U. Geigenmüller, and K.H. Nierhaus. 1988. The allosteric three-site model for the ribosomal elongation cycle. *J. Biol. Chem.* **263:** 13103–13111.

Hummel, H., M.H. Ahmad, and A. Böck. 1983. On the basis of aminoglycoside dependent growth of mutants of *Escherichia coli: In vitro* studies and the model. *Mol. Gen. Genet.* **191:** 176–181.

Matsunaga, K., H. Yamaki, T. Nishimura, and N. Tanaka. 1986. Inhibition of DNA replication initiation by aminoglycoside antibiotics. *Antimicrob. Agents Chemother.* **30:** 468–474.

Michel, F. and E. Westhof. 1990. Modelling of the three-dimensional architecture of group I catalytic introns based on comparative sequence analysis. *J. Mol. Biol.* **216:** 585–610.

Miller, S.L. 1987. Which organic compounds could have occurred on the prebiotic earth? *Cold Spring Harbor Symp. Quant. Biol.* **52:** 17–27.

Moazed, D. and H.F. Noller. 1987a. Chloramphenicol, erythromycin, carbomycin and vernamycin B protect overlapping sites in the peptidyl transferase region of 23S ribosomal RNA. *Biochimie* **69:** 879–884.

————. 1987b. Interaction of antibiotics with functional sites in 16S ribosomal RNA. *Nature* **327:** 389–394.

Noller, H.F., V. Hoffarth, and L. Zimniak. 1992. Unusual resistance of peptidyl transferase to protein extraction procedures. *Science* **256:** 1416–1424.

Sigmund, C.D., M. Ettayebi, and E.A. Morgan. 1984. Antibiotic resistance mutations in 16S and 23S ribosomal RNA genes of *Escherichia coli. Nucleic Acids Res.* **12:** 4653–4663.

Spangler, E.A. and E.H. Blackburn. 1985. The nucleotide sequence of the 17S ribosomal RNA gene of *Tetrahymena thermophila* and the identification of point mutations resulting in resistance to the antibiotics paramomycin and hygromycin. *J. Biol. Chem.* **260:** 6334–6340.

Teraoka, H. and K.H. Nierhaus. 1978. Proteins from *Escherichia coli* ribosomes involved in the binding of erythromycin. *J. Mol. Biol.* **126:** 185–193.

Vazquez, D. 1964. Uptake and binding of chloramphenicol by sensitive and resistant organisms. *Nature* **203:** 257–260.

von Ahsen, U. and R. Schroeder. 1990. Streptomycin and self-splicing. *Nature* **346:** 801.

————. 1991. Streptomycin inhibits splicing of group I introns by competition with the guanosine substrate. *Nucleic Acids Res.* **19:** 2261–2265.

von Ahsen, U., J. Davies, and R. Schroeder. 1991. Antibiotic inhibition of group I ribozyme function. *Nature* **353:** 368–370.

————. 1992. Non-competitive inhibition of group I intron RNA self-splicing by antibiotics. *J. Mol. Biol.* **226:** 935–941.

Xu, M.-Q., S.D. Kathe, H. Goodrich-Blair, S.A. Nierzwick-Bauer, and D.A. Shub. 1990. Bacterial origin of a chloroplast intron: Conserved self-splicing group I introns in cyanobacteria. *Science* **250:** 1566–1570.

9

An RNA-Amino Acid Affinity

Michael Yarus

Department of Molecular, Cellular, and Developmental Biology
University of Colorado
Boulder, Colorado 80309-0347

The origin of the genetic code is obscure, but this event draws continuing comment because it was a critical early informational innovation. However, the first such evolutionary innovation was the appearance of replication. Reasonably accurate replication made possible the transmission of information. Consequently, selection among genetic alternatives, and therefore Darwinian evolution, became possible. Before replication, history was determined by the most probable simple chemical events. Afterward, at an increasing tempo, first replication itself, then other protobiological qualities, could be refined by selection. The previously improbable processes and molecules that characterize biology could emerge from the vast manifold of the possible, ultimately dominating the planet. Among these more complex, accelerating outcomes was the genetic code, allowing early replicators access to a versatile new class of peptide catalysts.

The molecules that participated in the code's origin can be loosely specified. The most appealing hypothesis is that the informational molecule was an ancient analog of modern RNA (Gilbert 1986). An evolutionary precursor that resembled RNA, but was more resistant to the disruptive effects of chemically heterogeneous monomers (Joyce et al. 1987), may be plausible. However, it seems likely that the first peptidyl transferase was RNA-like. Several highly deproteinized modern 50S ribosomal RNAs can carry out a model peptide synthesis reaction at a rate virtually the same as the ribosomal particle (Noller et al. 1992). An essential activity like the peptidyl transferase is strongly constrained by the principle of continuity (Orgel 1968). Thus, the present peptidyl transferase has evolved from the primordial one by transformations that do not disrupt catalysis. Assuming that the present-day catalyst is ribosomal RNA, not the 5% peptide that survives protease and protein denaturants (Noller et al. 1992), it is likely that RNA at the time of the appearance of

The RNA World
© 1993 Cold Spring Harbor Laboratory Press 0-87969-380-0/93 $5 + .00

peptide synthesis was not too different from the modern molecule, because it possessed a similar active site.

In addition, it is likely that a large set of α-amino acids (including many modern ones) was present at the code's origin, because of their facile prebiotic synthesis from abundant precursors (Weber and Miller 1981).

The code's origin can be dated. Because the code is similar in all modern organisms as well as their organelles, its origin most likely predates the common ancestor of all modern biota. This time can be fixed unexpectedly precisely, at least in the sense that the likely error is small by comparison with the magnitude of the number itself: $3.7 \pm 0.3 \times 10^9$ years before the present. Organisms that resemble modern cyanobacteria are fossilized in rock isotopically dated to 3.4×10^9 years ago (Schopf and Packer 1987). However, the cyanobacteria were a late radiation among the prokaryotes, as judged from the rRNA phylogenetic tree (Woese 1987). Therefore, at 3.4×10^9 years ago, life was old enough to have undergone the early evolution of the prokaryotes, and its origins must be yet older. This conclusion is not substantially changed even if the fossilized cells and colonies are not true cyanobacteria. On the other hand, life is unlikely to have evolved until the early bombardment of the planet subsided. Until then, the interval between impacting planetesimals large enough to sterilize the planet was too short. Calculating from the cratering record on the earth and moon, this bombardment is believed to have sufficiently abated around 4×10^9 years ago (Maher and Stevenson 1988; for review, see Joyce 1991).

What can be known of this dauntingly ancient event? Two favorable possibilities exist. The primordial code may have relied on chemistry that can be reproduced, or alternatively, recognizable modern remnants of the origin may survive. What follows is a short review of an interaction that may fit these descriptions.

SUMMARY OF THE GROUP I RNA-ARGININE COMPLEX

Any catalytic RNA may be usable as a reporter for the binding of ligands within the catalytic domain. Thus, to elucidate the binding of amino acids to RNA, the standard amino acids were tested for effects on the rate of self-splicing by the *Tetrahymena* pre-rRNA (Yarus 1988). Arginine was found to be a uniquely effective inhibitor, with a dissociation constant of 4 mM. Inhibition was stereoselective, with L-arginine preferred by about 2-fold. Inhibition was side-chain-selective, and only arginine and lysine detectably inhibited the reaction of guanosine with the intron, although lysine had an apparent dissociation constant 12-fold higher than arginine. The structure of the *Tetrahymena* site now appears atypical. Using a site that better resembles the majority of group I self-splicing RNA introns, arginine's dissociation constant is 400 μM, the preference for L-arginine is 10-fold, and the dissociation constant for L-arginine is 160-fold smaller than for L-lysine (Yarus and Majerfeld 1992). Thus, arginine binding can be relatively strong, stereoselective, and side-chain-selective.

Inhibition is kinetically competitive with the G (guanosine) splicing substrate (Yarus 1988). In addition, RNA mutations that impair interaction with G also make inhibition by L-arginine undetectable (Michel et al. 1989; Yarus et al. 1991a). These observations suggest that arginine inhibits by obstructing a part of the G-binding region of the group I active site and does not cause inhibition by binding elsewhere. The binding site has been extensively modeled (Yarus et al. 1991b), catalysis by *Tetrahymena* has been reviewed recently (Cech et al. 1992), and the role of active-site divalents has also been recently reviewed (Yarus 1993; Pan et al., this volume). L-Arginine and G can occupy this same site because of a molecular resemblance between the two molecules (Yarus 1989).

Mutational studies and comparison of inhibitory arginine analogs suggest a four-part model for the function of the arginine site, illustrated in Figure 1 (Yarus 1989; Yarus and Majerfeld 1992). First, the side-chain guanidinium group engages part of the RNA's H-bond pattern also used by G; the positive charge on guanidinium also makes a small contribution. By comparison to lysine, the specific guanidinium contribution can be estimated as −1.6 kcal/mole.

Second, both G and the arginine guanidinium appear to stack beneath nucleotide 262, which caps the end of the binding site. These two interactions with guanidinium explain the side-chain selectivity of the site. By using the span of free energy of binding for different nucleotides 262, this interaction may be roughly approximated as −1.5 kcal/mole, although this value may include an effect of structural rearrangement after mutation.

Third, the aliphatic part of the arginine side chain makes no direct contribution to binding free energy, but must be of the correct length. Binding is weaker for side chains that are either too long or too short. For example, L-homoarginine, one methylene longer than L-arginine, sacrifices 1.3 kcal/mole in binding free energy (Yarus 1989). The arginine site therefore has a definite size.

Fourth, the α-amino group of the amino acid, if spaced properly from the guanidinium, engages the RNA backbone, adding to binding free energy (not shown in Fig. 1). Comparing arginine derivatives with and without the α-amino, this interaction appears to provide −1.9 kcal/mole.

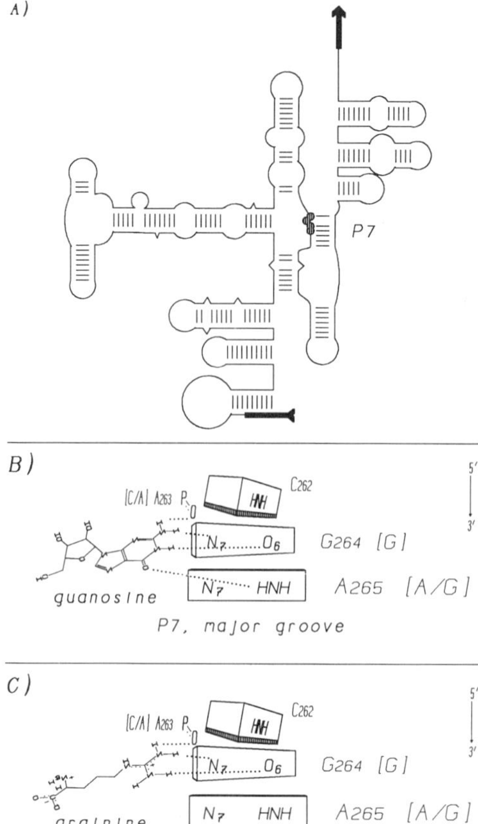

Figure 1 (A) Outline of the standard secondary structure of the *Tetrahymena* rRNA precursor, with intronic sequences shown as a thin line and exonic sequences as a bold line. The catalytic core of the molecule is in the center of the drawing. The P7 helix where the G/arginine site resides is labeled, and the four binding site nucleotides are portrayed as filled circles, in the same orientation as in *B* and *C*. (B) Schematic docking of guanosine to its RNA-binding site. Rectangular blocks represent the bases of three of the four nucleotides shown as filled circles in *A*. Potential H-bonding groups appear on the major groove edges of these bases (HNH, N_7, O_6). The standard numbering of the *Tetrahymena* sequence is shown as A_{263},... Only the 5′ phosphate of A_{263} is shown; its unstacked base is behind the plane of the drawing. Beside the names of the particular nucleotides found in *Tetrahymena* are the conserved nucleotides in that position in all group I sequences (A_{265} [A/G]). Dotted lines show the logic of a specific H-bonding scheme, but both G and arginine have been rotated out of the site to the left in order to clarify the site's function. C_{262} caps the site; the shaded lower surface of the base overhangs the bottom of the major helical groove and stacks on ligands. (Adapted from Yarus et al. 1991b.) (C) Schematic docking of arginine to its RNA-binding site.

Simultaneous binding of both the side chain and α-carbon groups explains the stereoselectivity of the site.

One interaction is notable for its absence: Although large repulsions can exist between ion pairs, the negative charge on the arginine carboxylate is brought into the negative RNA site at an apparent cost of only 0.6 kcal/mole (Yarus 1989). This moderate charge penalty suggests that even negatively charged free amino acids might be bound by RNA.

The groups that participate in specific H-bonds to G, a subset of which also contact arginine (Fig. 1) (Yarus et al. 1991b), reside on a contiguous triplet of nucleotides (263, 264, and 265 in Fig. 1). In each of the more than 120 known group I sequences, this triplet is identical to one of four of the modern codons for arginine: AGA, CGA, AGG, and CGG (Yarus and Christian 1989; Yarus 1991). Thus, arginine binds within arginine triplets, and no other sequences occur at this site. In addition, the arginine site, the stereoselectivity, and the side-chain selectivity are all conserved in five disparate group I RNAs (Hicke et al. 1989). These four conservations taken together suggest that an L-stereoselective, side-chain-selective complex between arginine and four coding triplets is ancient, dating from the ancestor of all modern group I RNAs.

I have suggested that this complex may be of evolutionary significance (Yarus 1991). The "restrained form" of the evolutionary argument is that the arginine complex may be taken as a model, showing that direct, specific interaction between an amino acid and coding sequences is possible. It therefore strengthens any argument for such interaction during the origin of the code. The "exuberant form" of the evolutionary argument is more assertive, holding that the arginine complex is a surviving remnant of the interaction that gave rise to four of six codons for arginine.

What follows is an attempt to combine the above with other data on RNA affinities for small ligands. These findings are compared to frequent assumptions about the origin of the genetic code in an attempt to define an origin for coding that would be consistent with RNA biochemistry.

ARGININE BINDS TO AN RNA SITE

The *Tetrahymena* arginine complex, most evidently, involves a coding RNA sequence; its existence supports the notion that RNA, and not another informational entity, originated the code. Such an implication requires only the restrained form of the evolutionary argument. Thus, the complex weighs in favor of Woese's argument (Woese et al. 1967) that specific affinity underlies the relationship between amino acids and codons. The alternative position is Crick's (1968), which states that the coding table may be a "frozen accident." However, this oversimplifies Crick's argument, which allowed for a specific linking of a subset of amino acids to their codons. The possibility that the origins of coding are different for different amino acids is a recurring theme in what follows.

The Crick and Woese positions are not mutually exclusive. There is another well-characterized arginine site: on TAR RNA, with a dissociation constant of 4 mM (Tao and Frankel 1992). The structure and specificity of this complex differ from the group I example; the TAR site may emphasize the arginine's guanidinium (Pugilisi et al. 1992). When more than one sequence-specific amino acid-RNA complex exists, a code may be based on specific affinity and may simultaneously involve an arbitrary choice among complexes.

It has often been assumed that if an interaction with RNA was responsible for the code, *anti*codons were responsible; the most expansive argument is reviewed by Lacey et al. (1985). This implies a tRNA-like function at the code's origin. However, arginine binds in the middle of sequences that correspond to its codons. Strength of binding depends on the ionic conditions (Yarus 1988) and on the nucleotide adjacent to the site, but the dissociation constant of the free amino acid ranges down to 400 µM (–4.7 kcal/mole; Yarus and Majerfeld 1992) under our standard conditions. This suggests that arginine could be recruited by a codon directly from dilute solution. In fact, carboxyl-activated arginine would likely lack a negative carboxyl charge and would bind even better than free arginine. Therefore, a peptide synthesis scheme in which a template RNA acts directly to order amino acids or activated derivatives is conceivable; tRNA and the rest of the ribosome could be later innovations (Yarus 1991). This is appealingly simple. In fact, no simpler scheme than one involving a template RNA, activated amino acids, and a peptidyl transferase is possible, unless the template and the transferase should be merged. If the transferase is a ribozyme (Noller at al. 1992), this final simplification is conceivable. Reliance on only a few molecules is not just a logical virtue. A small-power-dependence on reactant concentration and a small target for disruption seem adaptive qualities for a mechanism that might have had to tolerate dilute, inaccurately made participants and poor homeostasis.

The arginine complex forms in the major groove of an RNA helix, although at a specially adapted site (Fig. 1). A helical groove necessarily contains both coding sequences and their complements in close apposition on the two strands. The arginine site therefore exemplifies a way to simultaneously use proto-codons and proto-anticodons, in an era before they could be meaningfully distinguished (Yarus 1991).

DEFT HETEROGENEITY IS MORE PLAUSIBLE THAN EXQUISITE UNIFORMITY

It has been frequently supposed that a uniform RNA motif interacted with all or most of the amino acids during the code's origin: for example, a pocket formed of five nucleotides (Balasubramanian and Seetharamulu 1985), or four nucleotides (Shimizu 1982), or three nucleotides (Pelc and Welton 1966; but see Crick 1967). The scope and symmetry of such schemes are seductive. However, the *Tetrahymena* arginine site suggests particularity instead. Consider the unique set of bonds to the side-chain guanidinium, which are unlikely to be generalized to make a site for all or most other amino acids. Although common α-carbon groups on amino acids can be treated alike, it seems more plausible that different side chains will demand different RNA solutions. For example, an RNA has been selected that binds to a covalently immobilized tryptophan residue (Famulok and Szostak 1992). This affinity, unlike the arginine site, might rely on intercalation between bases, as reported when short tryptophan-containing oligopeptides interact with poly(U) and poly(A) (Rajeswari et al. 1987).

Primordial use of different RNA motifs to associate different amino acids with their proto-codons and proto-anticodons therefore seems likely. The motifs actually chosen to constitute the primordial code, however, must permit mutual assembly into a template surface that brings amino acid carboxyl and amino functions together for peptide formation. When a larger variety of motifs is available for comparison, this stereochemical criterion may be useful in choosing among them. As one example, the *Tetrahymena* motif (Fig. 1) might template for oligoarginine by forming an extended helical groove.

BINDING BENEFITS FROM RNA STRUCTURE

The *Tetrahymena* site engages G, according to the most detailed model (Yarus et al. 1991b), using a pocket composed of the three nucleotides of the arginine triplets, capped by the terminal base pair of the P7 helix (Fig. 1). Stable unstacking of the semi-conserved A_{263}, and capping by the terminal base C_{262}, are unusual features of the site, probably stabilized by the tertiary structure of the *Tetrahymena* core sequence. Thus, a free trinucleotide may interact more weakly, or not detectably, with arginine. However, there is a related difficulty even for an amino acid binding to an undistorted oligonucleotide. Because a free nucleotide has seven rotatable bonds, the entropic penalty for fixing a short, flexible oligonucleotide to fit it to the amino acid would significantly weaken binding. Therefore, it is likely that other RNA sites with similar affinity for amino acids will be supported within a stable structure; that is, will be a part of a larger RNA. In one exuberant form of the evolutionary hypothesis for arginine, structured arginine-binding RNAs preexist because of their role as RNA-RNA replicases (Yarus 1991).

A parallel may be drawn to the binding of a codon trinucleotide (viewed as a small ligand) to its anticodon trinucleotide. When the anticodon is incorporated into the structure of a tRNA's anticodon hairpin, the dissociation constant for its trinucleotide codon decreases more than three orders of magnitude (> 4.2 kcal/mole; for review, see Grosjean and Chantrenne 1980). For similar reasons, amino acid sites are most likely to be found in RNAs and under conditions that favor a stable higher-order RNA structure.

CODING MAY HAVE SCAVENGED EXISTING ACTIVITIES

It is not required that the arginine complex have intrinsic selective value. It may have arisen as a side effect of formation of a G-site, and later may have been captured during the origin of coding. That is, the binding of arginine and G derivatives in the same site in *Tetrahymena* RNA is based on a similarity in G and arginine H-bonding patterns (Yarus 1988), a similarity in size, a similarity in stacking (Yarus and Majerfeld 1992), and a similarity in interaction with the RNA backbone (Yarus 1989). In fact, for the two positions within the *Tetrahymena* site whose mutations alter the K_i for arginine, the K_m for GTP is found to be comparably altered (nucleotide 264, Yarus et al. 1991a; nucleotide 262, Yarus and Majerfeld 1992). This situation strongly supports the idea that arginine exploits a subset of G's contacts with the RNA, making analogous use of them. In independent experiments, G. Connell and M. Yarus (1992; unpubl.) have selected RNAs of different structure that also bind both G and arginine.

There is therefore no need to devise a separate rationale for arginine binding. Any RNA selected for use of guanine nucleotide energy may bind arginine in its G-site. Perhaps primordial group I catalysts, acting as replicators (Yarus 1991), evolved the G-site, its conserved arginine triplets, and an affinity for arginine without direct selection for any function of the amino acid. Not only does an amino acid site exist within the G-site, but the site most reactive with nucleotides also binds arginine most tightly. Selection for group I catalysis with the nucleotide would therefore have optimized the arginine site (Yarus et al. 1991a; Yarus and Majerfeld 1992). This suggests that other such preadaptations may have played a part in the origin of the code. For example, even a very simple

RNA metabolism probably would require binding sites for adenine nucleotides, and adenine presents H-bonding groups that are isosteric with the glutamine side chain.

AMINO ACIDS INTERACTING WITH RNA MAY HAVE BEEN ENCODED FIRST

The nature of the arginine complex suggests a set of qualities required for effective interaction with RNA. That is, the rigid guanidinium group in the arginine side chain offers multiple interactions to an RNA: ionic, H-bonding, and possibly stacking. These multiple contacts radiate from a rigid group whose shape matches its potential ligands in the RNA and which therefore can bind without an entropically unfavorable rearrangement. Histidine, asparagine, glutamine, aspartic acid, glutamic acid, tryptophan, tyrosine, and lysine also may fit this prescription. The glutamic, aspartic, and histidine side chains have the particular option of inner-sphere coordination to an RNA-bound metal ion (Yarus 1993).

This list differs dramatically from the amino acids prevalent in spark tubes and the Murchison meteorite (Weber and Miller 1981). Spark tube amino acids are stable and easily synthesized and therefore are often taken as self-evidently primordial. However, the arginine complex, or the acceptance of any important interaction with RNA, is congenial to a different postulate. Amino acids that interact well with RNA may have been the first ones accurately encoded. This idea does not deny a simultaneous role to glycine, alanine, valine, leucine, and the other prevalent chemically synthesized amino acids. An interesting possibility is that unencoded oligomers were useful. Oligopeptides are readily synthesized under prebiotic conditions (Fox 1960). Very simple peptides, even as short as hexa- to decapeptides (Brack and Barbier 1990), can show enzymatic activity. For example, alternating hydrophobic-cationic peptides combine to form a 2′(3′)-nucleotide-generating ribonuclease (Barbier and Brack 1988). One intriguing possibility is that such uncoded ribonucleolytic peptides actually functioned to stabilize a similar transition state, but for the *reverse* reaction: That is, they accelerated the primordial *synthesis* of RNA.

Of course, all modern amino acids were added to the accurately encoded repertoire eventually, by means that are unknown. However, RNA should not be intuitively denied a role, even for aliphatic hydrophobic side chains like valine's, whose interaction with RNA is the most difficult to envision. An RNA has been selected with measurable affinity and selectivity for the valine side chain (I. Majerfeld and M. Yarus, unpubl.).

PRIMORDIAL CODING NEED NOT HAVE BEEN UNSELECTIVE

It is frequently postulated that the primordial code was highly inaccurate, progressing from coding for groups of similar amino acids to a more resolved state (Woese 1965). The arginine complex demonstrates that this should not be axiomatic, at least for a subset of amino acids. The only other amino acid with a measurable affinity for the *Tetrahymena* site is L- or D-lysine (Yarus 1988), which has roughly the same charge distribution as arginine. Under different reaction conditions, the wild-type *Tetrahymena* site makes a 12-fold (Yarus 1988) or 44-fold (Yarus 1989) discrimination against L-lysine. A nearby mutation that enhances a proposed stacking interaction with the arginine guanidinium increases this arginine/lysine distinction to 160-fold (3 kcal/mole; Yarus and Majerfeld 1992). No other normal amino acid interacts detectably ($K \leq$ 250 mM; $\geq$ −0.8 kcal/mole). Thus, side-chain-selective RNA sites for some amino acids are worth consideration as evolutionary intermediates.

D-RIBOSE AND L-AMINO ACIDS HAVE A SELECTED, NOT AN INTRINSIC, RELATIONSHIP

Previous theorizing has sometimes asserted an intrinsic chiral complementarity between D-ribose and L-amino acids (see, e.g., Carter and Traut 1974). An intrinsic preference does not solve the chirality problem but displaces the question of the chiral preference of modern coding to the point at which D-ribose was selected. The chemical acylation of the internal 2′ OH of IpI with activated DL-alanine gives slight enantiomeric excesses of either L- or D-product, depending on whether the α-amino is protected or not (Profy and Usher 1984). This weighs against a strong intrinsic selection. The conserved preference of the *Tetrahymena* (Yarus 1988) and other group I sites (Hicke et al. 1989) for L-arginine again raises the question of an intrinsic enantiomeric preference. This preference for L- over D-arginine is 10-fold when the side-chain guanidinium is bound most effectively (Yarus and Majerfeld 1992).

However, mutation of a nucleotide near the *Tetrahymena* binding site removes L-stereoselectivity (Yarus et al. 1989). Neither catalysis nor binding of a small analog with an H-bonding pattern like G is greatly perturbed in the mutant, so the guanidinium end of the G/arginine site may be dislocated as a unit as a result of the mutation. In addition, G. Connell and M. Yarus (unpubl.) have selected D-stereoselective RNA sites for arginine. Thus, if an ancient relative of the group I active center was the progenitor of the code for arginine (the exuberant hypothesis; see above), a preference for L-arginine is explained, because this preference is completely conserved in modern group I RNAs (Hicke et al. 1989).

However, under the restrained interpretation, there is no consistent tendency for D-ribonucleic acids to prefer L-arginine.

THE IMMEDIATE FUTURE OF THE REMOTE PAST

Relatively quickly it has been shown that RNA can fold to form a site for guanosine (Bass and Cech 1984), two sites for arginine (Yarus 1988; Pugilisi et al. 1992), and sites for several complex polycyclic dyes (Ellington and Szostak 1990), for the guanidinium (von Ahsen and Schroeder 1991) and amino sugars in antibiotics (von Ahsen et al. 1992), and for α-amino-linked tryptophan agarose (Famulok and Szostak 1992). The expansion of knowledge about the interaction of RNA with small molecules will accelerate, thanks to means of selecting such affinity from pools of random oligomers (Ellington and Szostak 1990). This information will elucidate the versatility of RNA. In addition, increased knowledge about the ribosome will add to this discussion. For example, it seems likely that the mechanisms and structure of the coding sites will be as suggestive for molecular evolution as is the peptidyl transferase (Noller et al. 1992). Of course, such information is not literal history. These new findings, even if unexpectedly copious and provocative, can only yield a plausible scheme for the distant origin of coding. Nevertheless, such plausibility may be justified with measurements of phylogeny, free energy, and selectivity. This is more than we could have hoped a short time ago.

ACKNOWLEDGMENTS

The author was supported by National Institutes of Health research grant GM-30881. He also thanks the W.M. Keck Foundation for support of RNA science in Boulder.

REFERENCES

Balasubramanian, R. and P. Seetharamulu. 1985. Origins of life: Conformational energy calculations on primitive tRNA nestling an amino acid. *J. Theor. Biol.* **113:** 15–28.

Barbier, B. and A. Brack. 1988. Basic polypeptides accelerate the hydrolysis of ribonucleic acids. *J. Am. Chem. Soc.* **110:** 6880–6882.

Bass, B.L. and T.R. Cech. 1984. Specific interaction between the self-splicing RNA of *Tetrahymena* and its guanosine substrate: Implications for biological catalysis by RNA. *Nature* **308:** 820–826.

Brack, A. and B. Barbier. 1990. Chemical activity of simple basic peptides. *Origins Life Evol. Biosphere* **20:** 139–144.

Carter, C.W., Jr. and J. Traut. 1974. A proposed model for interaction of polypeptides with RNA. *Proc. Natl. Acad. Sci.* **71:** 283–287.

Cech, T.R., D. Herschlag, J.A. Piccirilli, and A.M. Pyle. 1992. RNA catalysis by a group I ribozyme: Developing a model for transition state stabilization. *J. Biol. Chem.* **267:** 17479–17482.

Crick, F.H.C. 1967. An error in model building. *Nature* **213:** 798.

———. 1968. The origin of the genetic code. *J. Mol. Biol.* **38:** 367–379.

Ellington, A.D. and J.W. Szostak. 1990. *In vitro* selection of RNA molecules that bind specific ligands. *Nature* **346:** 818–822.

Famulok, M. and J.W. Szostak. 1992. Stereospecific recognition of tryptophan agarose by *in vitro* selected RNA. *J. Am. Chem. Soc.* **114:** 3990–3991.

Fox, S.W. 1960. How did life begin? *Science* **132:** 200–208.

Gilbert, W. 1986. The RNA world. *Nature* **319:** 618.

Grosjean, H. and H. Chantrenne. 1980. On codon-anticodon interactions. In *Chemical recognition in biology* (ed. F. Chapville and A.-L. Haenni), pp. 347–367. Springer-Verlag, New York.

Hicke, B.J., E.L. Christian, and M. Yarus. 1989. Stereoselective arginine binding is a phylogenetically conserved property of group I self-splicing RNAs. *EMBO J.* **8:** 3843–3851.

Joyce, G.F. 1991. The rise and fall of the RNA world. *New Biol.* **3:** 399–407.

Joyce, G.F., A.W. Schwartz, S.L. Miller, and L.E. Orgel. 1987. The case for an ancestral genetic system involving simple analogues of the nucleotides. *Proc. Natl. Acad. Sci.* **84:** 4398–4402.

Lacey, J.C., Jr., L.M. Hall, and D.W. Mullins, Jr. 1985. Rationalization of some genetic anticodonic assignments. *Origins Life* **16:** 69–79.

Maher, K.A. and D.J. Stevenson. 1988. Impact frustration of the origin of life. *Nature* **331:** 612–614.

Michel, F., M. Hanna, R. Green, D.P. Bartel, and J.W. Szostak. 1989. The guanosine binding site of the *Tetrahymena* intron. *Nature* **342:** 391–395.

Noller, H.F., V. Hoffarth, and L. Zimniak. 1992. Unusual resistance of peptidyl transferase to protein extraction procedures. *Science* **256:** 1416–1419.

Orgel, L.E. 1968. Evolution of the genetic apparatus. *J. Mol. Biol.* **38:** 381–393.

Pelc, S.R. and M.G.E. Welton. 1966. Stereochemical relationships between coding triplets and amino acids. *Nature* **209:** 868–870.

Profy, A.T. and D.A. Usher. 1984. Stereoselective aminoacylation of a dinucleoside monophosphate by the imidazolides of DL-alanine and N-(tert-butoxycarbonyl)-DL-alanine. *J. Mol. Evol.* **20:** 147–156.

Pugilisi, J.D., R. Tan, B.J. Calnan, A.D. Frankel, and J.R. Williamson. 1992. Conformation of the TAR RNA-arginine complex by NMR spectroscopy. *Science* **257:** 76–80.

Rajeswari, M.R., T. Montenay-Garestier, and C. Helene. 1987. Does tryptophan intercalate in DNA? A comparative study of peptide binding to alternating and non-alternating A•T sequences. *Biochemistry* **26:** 6825–6831.

Schopf, J.W. and B.M. Packer. 1987. Early archean (3.3 billion to 3.5 billion-year-old) microfossils from Warrawoona group, Australia. *Science* **237:** 70–73.

Shimizu, M. 1982. Molecular basis for the genetic code. *J. Mol. Evol.* **18:** 297–303.

Tao, J. and A.D. Frankel. 1992. Specific binding of arginine to TAR RNA. *Proc. Natl. Acad. Sci.* **89:** 2723–2726.

von Ahsen, U. and R. Schroeder. 1991. Streptomycin inhibits splicing of group I introns by competition with the guanosine substrate. *Nucleic Acids Res.* **19:** 2261–2265.

von Ahsen, U., J. Davies, and R. Schroeder. 1992. Non-competitive inhibition of group I intron RNA self-splicing by aminoglycoside antibiotics. *J. Mol. Biol.* **226:** 935–941.

Weber, A.L. and S.L. Miller. 1981. Reasons for the occurrence of the twenty coded protein amino acids. *J. Mol. Evol.* **17:** 273–284.

Woese, C.R. 1965. On the evolution of the genetic code. *Proc. Natl. Acad. Sci.* **54:** 1546–1552.

————. 1987. Bacterial evolution. *Microbiol. Rev.* **51:** 221–271.

Woese, C.R., D.H. Dugre, S.A. Dugre, M. Kondo, and W.C. Saxinger. 1967. On the fundamental nature and evolution of the genetic code. *Cold Spring Harbor Symp. Quant. Biol.* **31:** 723–736.

Yarus, M. 1988. A specific amino acid binding site composed of RNA. *Science* **240:** 1751–1758.

————. 1989. Specificity of arginine binding by the *Tetrahymena* intron. *Biochemistry* **28:** 980–988.

————. 1991. An RNA-amino acid complex and the origin of the genetic code. *New Biol.* **3:** 183–189.

————. 1993. How many catalytic RNAs? Ions and the Cheshire cat conjecture. *FASEB J.* **7:** 31–39.

Yarus, M. and E. Christian. 1989. Genetic code origins. *Nature* **342:** 349–350.

Yarus, M. and I. Majerfeld. 1992. Co-optimization of ribozyme substrate stacking and L-arginine binding. *J. Mol. Biol.* **225:** 945–949.

Yarus, M., M. Illangesekare, and E. Christian. 1991a. Selection of small molecules by the *Tetrahymena* catalytic center. *Nucleic Acids Res.* **19:** 1297–1304.

————. 1991b. An axial binding site in the *Tetrahymena* precursor RNA. *J. Mol. Biol.* **222:** 995–1012.

Yarus, M., J. Levine, G.B. Morin, and T.R. Cech. 1989. A *Tetrahymena* intron nucleotide connected to the GTP/arginine site. *Nucleic Acids Res.* **17:** 6969–6981.

10

Similarities and Differences between RNA and DNA Recognition by Proteins

Thomas A. Steitz
Department of Molecular Biophysics and Biochemistry
and Department of Chemistry
and Howard Hughes Medical Institute
Yale University
New Haven, Connecticut 06511

Many DNA and RNA molecules are recognized by proteins that interact preferentially with a specific DNA sequence or a particular RNA molecule. I address here the structural basis by which these proteins recognize their target nucleic acid and show in what ways recognition of RNA and DNA is both similar and different. Sequence-specific DNA-binding proteins interact with duplex DNA that is in B-form. RNA molecules, on the other hand, invariably consist of duplex regions, often stacked one on another, that are A-form, as well as regions of single-stranded loops and bulges, making possible a more complex and richly varied three-dimensional shape than can be assumed by duplex DNA. Presently, the crystallographic and nuclear magnetic resonance (NMR) structural database of proteins complexed with DNA is very large, revealing some patterns and general conclusions about the source of sequence-specific DNA recognition (for reviews, see Steitz 1990; Harrison 1991; Pabo and Sauer 1992). On the other hand, the structural database for RNA-binding proteins, particularly in complex with RNA, is very meager indeed, so that any generalizations made may soon be overturned by the next structure determination of an RNA-protein complex. Nevertheless, some patterns of similarity and difference in the structural basis of nucleic acid recognition by proteins can be seen at this time.

Structural, biochemical, and molecular genetic studies of protein nucleic acid complexes have established at least three important sources of sequence specificity in protein-nucleic acid interactions: (1) Direct hydrogen bonding and van der Waals interaction between protein side chains and the exposed edges of base pairs provide structural complementarity to the correct, but not to the incorrect, sequences. The interactions are primarily, but not exclusively, in the major groove of B-DNA

The RNA World
© 1993 Cold Spring Harbor Laboratory Press 0-87969-380-0/93 $5 + .00

and to both the minor groove and the major groove at the end of a helix or at a bulge in RNA structures. (2) The sequence-dependent bendability or deformability of duplex DNA or RNA molecules provides sequence selectivity by virtue of the ability of some nucleic acid sequences to take up a particular structure required for binding to a protein at a lower free energy cost than other sequences. (3) Bases of RNA that are in single-stranded regions or in bulges can be directly recognized by pockets on the protein that are complementary to these bases in shape and hydrogen-bonding capabilities.

THE PROBLEM THAT IS SET: WHAT IS BEING RECOGNIZED?

Let us first consider the problem confronting proteins interacting with either duplex DNA or the duplex portion of an RNA molecule. The three-dimensional structure of double-stranded DNA is highly polymorphic (Kennard and Hunter 1989), but variations of two forms, A-form and B-form, are of relevance to the proteins of interest here. Figure 1 shows an important difference between A-form and B-form DNA. In B-DNA, the major groove is wide enough to accommodate either an α-helix or an antiparallel β-ribbon, and the functional groups on the exposed edges of the base pairs can be directly contacted by side chains of the protein. The minor groove, on the other hand, is deep and narrow (5.8 Å wide) and thus less accessible to secondary structures such as an α-helix. For RNA, which is always A-form, the opposite is true. The minor groove is shallow and broad (10–11 Å wide), whereas the major groove is very deep and narrow (4 Å) (Delarue and Moras 1989). The width of the minor groove in B-DNA varies depending on its base composition. AT-rich sequences have a narrower minor groove (3.5 Å) than GC-rich sequences (Yoon et al. 1988). Where adequate information is available it appears (as might be expected) that in general most DNA-binding proteins directly decode DNA sequences via interactions in the major groove, although some important exceptions are known. *Escherichia coli* integration host factor and eukaryotic TFIID appear to recognize sequences by interactions in the minor groove, although co-crystal structures of these proteins interacting with DNA are not yet available.

Whereas on the basis of RNA structure alone one might expect proteins to discriminate among duplex RNA molecules by interactions with sequences via the minor groove, examples of interaction between protein and RNA in both the major and minor groove are now known (Rould et al. 1989; Ruff et al. 1991). Although it is true that the edges of base pairs are inaccessible in the major groove of A-form RNA in the central portion of a long duplex, most naturally occurring RNA mole-

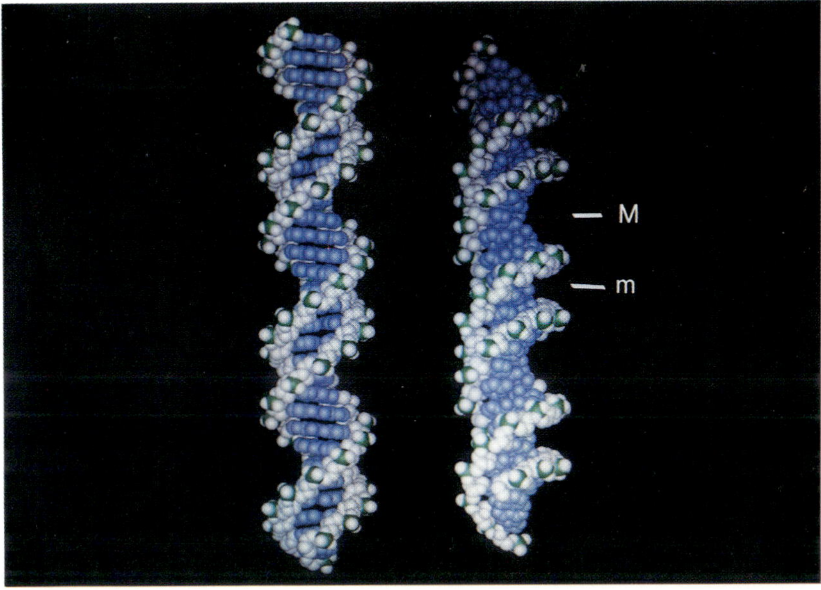

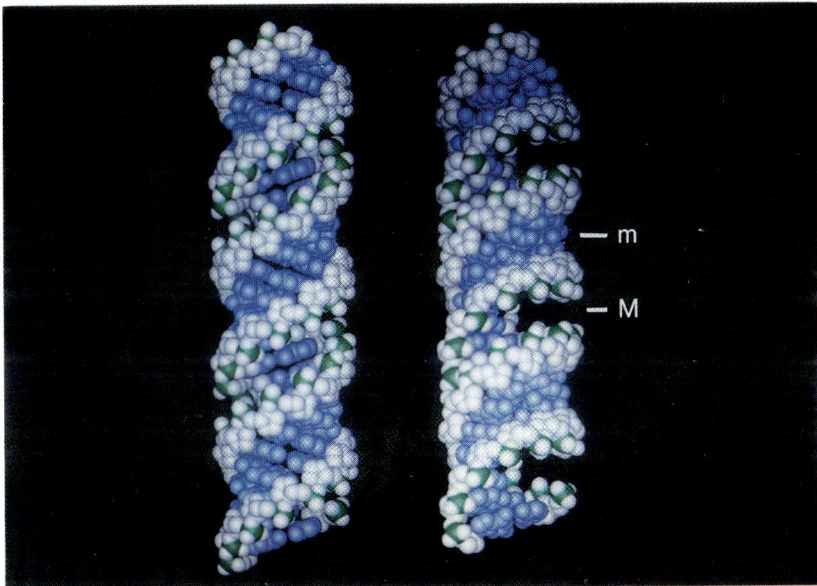

Figure 1 Structures of A-form (*top*) and B-form (*bottom*) DNA in space-filling representation showing differences in major and minor groove widths and shapes. In the models on the left, the helix axes are parallel to the page; on the right, the helix axes have been tilted up by 32° to show the groove shapes. Bases are colored blue, phosphorus atoms are green, and all other atoms are white. The edges of the bases are easily accessible from the major groove of B-DNA and the minor or shallow groove of A-DNA (or RNA). (*m*) Minor groove; (*M*) major groove. (Reprinted, with permission, from Steitz 1990).

cules contain relatively short duplex regions interrupted by bulges or loops. Although these short duplex regions may be expected to stack as occurs in tRNA, the edges of base pairs exposed in the major grooves are accessible at the ends of these RNA helices.

A second important consideration in the suitability of the major and minor grooves for direct sequence recognition is the degree of structural variation of the four base pairs as viewed from the two grooves. Seeman et al. (1976) pointed out that the base pairs presented a more richly varied set of hydrogen-bond donors to the major groove as compared to the minor groove. Figure 2 shows that the minor groove side of base pairs is a veritable recognition desert with only the N2 of guanine distinguishing AT from GC. The patterns of donors and acceptors on the major groove side, however, can distinguish all four base pairs. In duplex regions, RNA has an opportunity available that does not exist in duplex DNA sequences: Non-Watson-Crick base pairs can exist within RNA helices (see, e.g., Fig. 8) and thus present to both the major and the minor groove hydrogen-bonding and shape differences not seen in the four orientations of the two Watson-Crick base pairs. Although GU base pairs

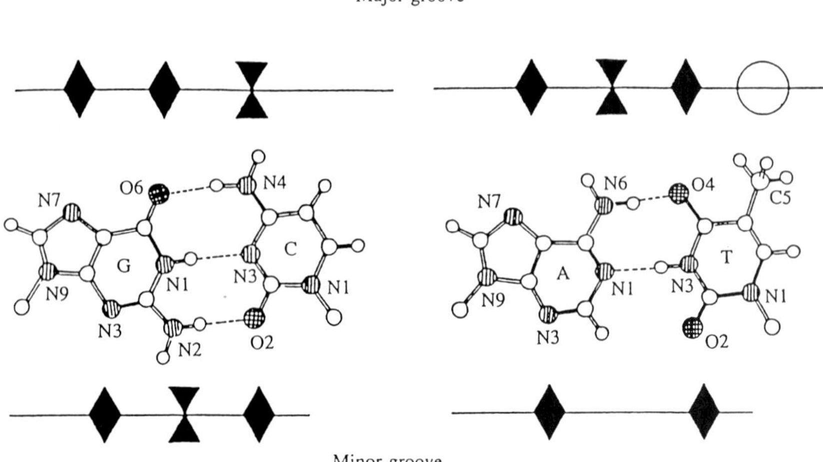

Figure 2 Hydrogen-bond donors and acceptors presented by Watson-Crick pairs to the major groove and the minor groove (adapted from Lewis et al. 1985). The symbols for hydrogen-bond donors (*hourglasses*) and acceptors (*diamonds*) (Woodbury et al. 1980) show a varied pattern presented by the base pairs into the major groove and a poor information array presented into the minor groove. Although it is possible to distinguish among AT, TA, GC, and CB in the major groove, functional groups in the minor groove allow easy discrimination only between AT- and GC-containing base pairs. (*Open circle*) Methyl group. (Reprinted, with permission, from Steitz 1990.)

are perhaps the most common non-Watson-Crick base pairs seen in RNA, AG and various kinds of UU base pairs have been seen, and others may exist.

Three other recognition opportunities that occur in RNA and not in duplex DNA, and that appear to be utilized by proteins binding specifically to RNA, are the single-stranded loop regions at the ends of helices, single-stranded bulges within helices, and modified bases.

ROLE OF THE MAJOR GROOVE IN DNA AND RNA RECOGNITION

The extensive hydrogen bonding in shape complementarity between the major groove of B-DNA and the surfaces of many of the sequence-specific DNA-binding proteins as a source of recognition has been extensively documented from high-resolution crystal structures and a few NMR structures of DNA complexes (for reviews, see Steitz 1990; Harrison 1991; Pabo and Sauer 1992). In general, structural complementarity between a protein and a specific DNA sequence is achieved in idiosyncratic manners: There does not appear to be a code for nucleic acid sequence recognition (Pabo 1983; Matthews 1988). Although particular amino acid side chains do not always recognize the same base pair, there are some apparent preferences, as suggested by Seeman et al. (1976). The guanidinium group of arginine very often makes a bidentate interaction with the N7 and O6 of guanine, although other interactions are also seen. Similarly, the hydrogen-bond donors and acceptors of the glutamine side chains are observed frequently to interact with the corresponding hydrogen-bond donors and acceptors of adenine. The ability of these side chains to make bidentate interactions with DNA greatly enhances their suitability for sequence-specific recognition (Seeman et al. 1976). The van der Waals interactions between the protein and the 5-methyl group of thymine appear also to contribute to specificity. Presumably, the close packing of a protein against the GC base pair would in many cases sterically exclude its replacement by an AT base pair with its accompanying bulky 5-methyl group.

Information concerning protein interaction in the major groove of RNA is sparse. Although details of the interaction have not yet been published, a protein loop at the end of a β-hairpin in aspartyl-tRNA synthetase is observed to interact with at least the terminal base pair via the major groove of the acceptor stem of tRNA[Asp] (Ruff et al. 1991). Interactions between human immunodeficiency virus (HIV) *tat* and its target RNA TAR are hypothesized to occur in the region of a 3-nucleotide bulge and on the major groove side (Weeks et al. 1990, 1991).

The potential accessibility of an RNA major groove to protein side chains has been probed by K.M. Weeks and D.M. Crothers (in prep.), using diethyl pyrocarbonate (DEPC). DEPC carbethoxylates purines primarily at the N7 position in a reaction that is sensitive to the solvent exposure of the base (Vincze et al. 1973; Peattie and Gilbert 1980). The reagent is comparable in size to those protein side chains such as arginine that mediate RNA-protein interactions, suggesting that the rates of reactivity of this probe are likely to reflect the steric accessibility of purines to protein interaction. Although the major groove of an uninterrupted RNA duplex is relatively inaccessible to this reagent, as expected, the major groove at helix termini is accessible to modification, with the effect extending further on the 3' strand (Fig. 3). Furthermore, bulges in RNA helices larger than one nucleotide greatly increase the accessibility of flanking duplexes to reaction with DEPC.

The structure of a portion of HIV TAR RNA containing a 3-nucleotide bulge and bound to arginine has been deduced from NMR data, showing one example of how a bulge can make the major groove of RNA accessible to a protein side chain (Puglisi et al. 1992). A cytosine from the 3-nucleotide bulge makes a triple base pair with an adjacent GC forming a binding site for the guanidinium group of arginine and opening the major groove.

ROLE OF NUCLEIC ACID BENDABILITY

The sequence-dependent nucleic acid distortability is a very important source of specificity in many protein-RNA, as well as protein-DNA, interactions. Nucleic acid distortability as a more indirect source of sequence specificity arises from two facts: (1) Proteins often bind a con-

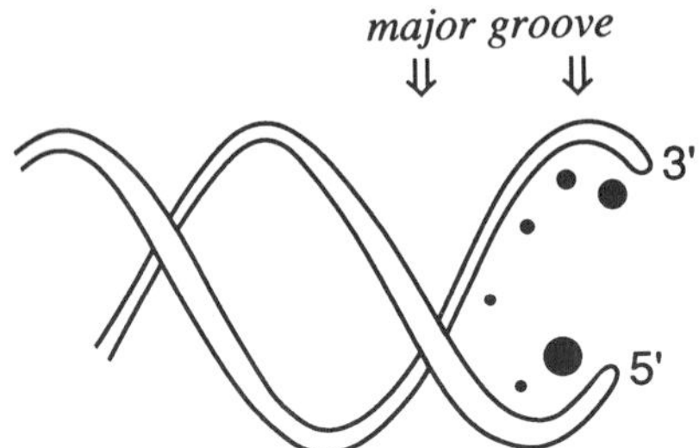

Figure 3 Schematic representation of an RNA duplex with the reactivity of purines to DEPC shown by filled circles whose diameter is proportional to reactivity (accessibility). Although the 5' base is most accessible, the accessibility to DEPC extends further into the duplex on the 3' strand (K.M. Weeks and D.M. Crothers, in prep.).

formation of a nucleic acid that is altered from its uncomplexed solution conformation. (2) The free energy cost for various nucleic acid sequences to assume the conformation that is required for its binding to the protein is not the same for different sequences.

Evidence for significant distortion of DNA upon binding to proteins now abounds, and in a few cases this protein-induced DNA distortion has been experimentally correlated with the ability of a protein to bind a specificity sequence. DNA distortion is seen in the crystal structures of DNA complexes with *Eco*RI (Frederick et al. 1984), 434 repressor (Aggarwal et al. 1988), trp repressor (Otwinowski et al. 1988), DNase I (Suck et al. 1988), Klenow fragment (Freemont et al. 1988), CAP (Schultz et al. 1991), met repressor, and a growing roster of other proteins. The distortions of duplex DNA structure that have been observed in complexes include changes in twist, groove width, and kinks (Steitz 1990).

Perhaps the most completely documented example of the correlations among DNA sequence, bendability, and affinity for proteins is in the case of *E. coli* catabolite gene activator protein (CAP). Gartenberg and Crothers (1988) found that CAP-binding sites containing AT bases at base-pair positions 10 and 11 from the center of the binding site bend more than those containing GC bases when bound to CAP (as assessed by polyacrylamide gel electrophoresis) and also bind CAP 14-fold more tightly. Thus, sequence, bending, and binding are correlated. The crystal structure of the CAP-DNA complex shows an 8° bend near base pair 10 and a very narrow minor groove that allows better interaction with the protein (Schultz et al. 1991). AT-rich sequences favor bending into the minor groove (as occurs) (Drew and Travers 1984) and also favor a narrow minor groove (Yoon et al. 1988). Experiments with 434 repressor also show a sequence dependence to DNA binding that most likely is a result of DNA distortability (Koudelka et al. 1987). Replacement of AT by GC base pairs at the dyad axis reduces binding of intact repressor by as much as 50-fold, despite the lack of base-specific contacts in this region.

The sequence-dependent deformability of duplex DNA or RNA that provides specificity for sequences being recognized by a protein can include the melting of base pairs. If binding to a protein requires melting of one or more base pairs, then the binding of mismatched base pairs should be favored over AT pairs, which in turn should bind better than GC base pairs. The order of binding should reflect the thermodynamic stability of base pairs. Two examples of the role of duplex meltability in sequence specificity can be cited—one in RNA and one in DNA. Binding of tRNA[Gln] to its cognate synthetase results in the breaking of the terminal base pair of the acceptor stem between nucleotides U1 and A72 (Rould et al. 1989). For glutaminyl-tRNA synthetase (GlnRS) recognition in charging of tRNA, it is important that this base-pair not be GC (Yarus et al. 1977). The added free energy cost of breaking the GC base pair makes tRNAs containing a GC at 1-72 less suitable for proper binding to the enzyme, reducing k_{cat}/K_m by about 10-fold (Jahn et al. 1991). In a second example, the 3′,5′-exonuclease active site of *E. coli* DNA polymerase I is observed to denature duplex DNA and bind four single-stranded nucleotides at the 3′ terminus (Freemont et al. 1988). In a competition between the duplex-binding polymerase active site and the single-strand-binding exonuclease active site for the 3′ end of the primer strand, duplex DNA containing a mismatch base pair will bind to the exonuclease site with greater frequency than a correctly matched duplex, thus enhancing the editing out of mismatch base pairs (Joyce and Steitz 1987; Freemont et al. 1988).

Sequence recognition in RNA also arises from the sequence-dependent ability of single-stranded RNA to take up the conformation required for protein binding, as occurs in the single-stranded acceptor end of tRNA^{Gln} (Fig. 4). The observed interaction between the N2 of G73 and the backbone phosphate of A72 is not possible for the other three bases (Rould et al. 1989), consistent with the observation that changing G73 to A, C, or U reduces the k_{cat}/K_m for charging by one, three, and four orders of magnitude, respectively (Jahn et al. 1991). Furthermore, two non-Watson-Crick base pairs are formed at the end of the anticodon stem in tRNA^{Gln} (see Fig. 8), producing a structure that is recognized by the synthetase (Rould et al. 1991). Other bases unable to make these non-Watson-Crick base pairs would not allow formation of the structure being recognized and bound to this enzyme. Although binding of tRNA^{Asp} to its cognate synthetase results in a very major change in the conformation of the anticodon loop (Ruff et al. 1991), it is not yet published whether or not any part of this structural change involves alterations in RNA-RNA interactions that are dependent on the RNA sequence, as occurs with GlnRS.

ROLE OF WATER MOLECULES IN SEQUENCE RECOGNITION

Buried water molecules appear to play a very important but underrecognized role in both DNA and RNA sequence recognition. Ascertaining the role of water molecules in sequence recognition requires crystal structures at sufficiently high resolution (usually 2.5 Å or better) and refinement that water molecules can be reliably located. Water (or a protein hydroxyl group) can only make a base-specific hydrogen bond if

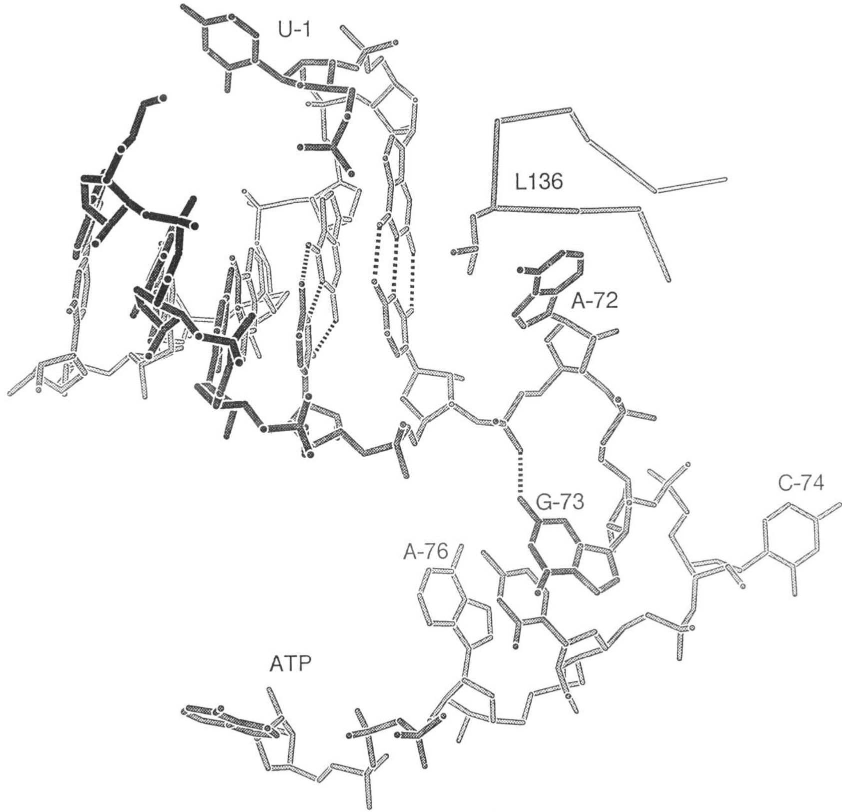

Figure 4 Conformation of the end of the acceptor stem and the 3'strand in tRNA^{Gln} bound to GlnRS (from Rould and Steitz 1992). The expected base pair between U1 and A72 is broken by Leu-136, which packs against the guanine of the G2-C71 base pair. The 2-amino group of guanine 73 hydrogen-bonds to the phosphate backbone, stabilizing the hairpin conformation of the 3'strand into the active site. Cytosine 74 binds into a tight pocket in the protein, allowing the bases of nucleotides 73, 75, and 76 to stack.

it is also making at least two other hydrogen bonds with obligate donors or acceptors on the protein and is sequestered from bulk solvent. In this circumstance, the two unsatisfied water H-bond donors/acceptors directed toward the nucleic acid become obligate donors/acceptors and consequently become part of the H-bonding template surface of the protein to which the nucleic acid must be complementary for optimal binding (Fig. 5). In trp repressor-DNA complex, there are three water molecules per half operator bound in the major groove between the protein and the DNA bases; at least two of them appear to be making

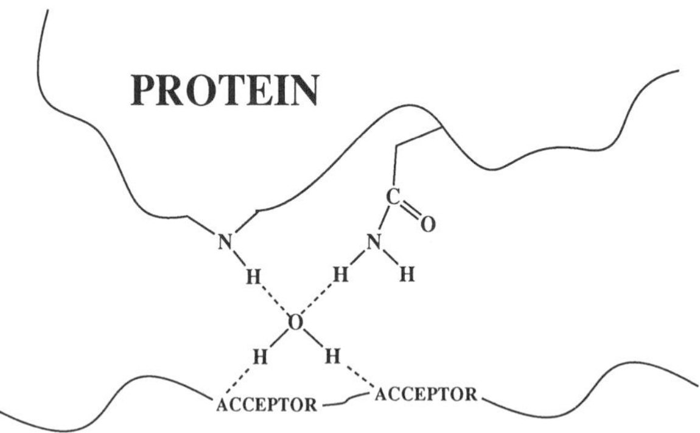

Figure 5 Schematic drawing showing how a water molecule can be specifically oriented by interactions with the protein turning it into a surrogate side chain. For example, here two obligate proton donors from the protein bind a water molecule such that it requires H-bond acceptors on the nucleic acid.

hydrogen bonds that specify base pairs 5, 6, and 7 from the dyad axis (Otwinowski et al. 1988; Steitz 1990). In this case, water molecules are playing the role of "honorary" protein side chains. In the GlnRS complex with tRNA, two buried water molecules are an integral part of the hydrogen-bonding matrix presented in the shallow groove of the tRNA acceptor stem (Rould et al. 1989; Rould and Steitz 1992). Hydrogen bonds between these two water molecules, as well as both a buried carboxylate of Asp-235 and a backbone amide of residue 183, serve to orient one hydrogen-bond donor of water toward the O2 of cytosine 71 and one acceptor toward the N2 of guanine (Fig. 6).

ROLE OF THE MINOR GROOVE IN DNA AND RNA RECOGNITION

As pointed out by Seeman et al. (1976), there are fewer features presented by base pairs in the minor groove that allow discrimination among the two base pairs in their two orientations (Fig. 2). The hydrogen-bond acceptors (N3 on guanine and adenine and O2 on cytosine and thymine) occur in almost the identical place in the minor groove for all four bases. Only the exocyclic of N2 of guanine distinguishes AT from GC and perhaps GC from CG. Furthermore, the minor groove of B-DNA is in general too narrow to accommodate an α-helix or too deep for bases to be reached by side chains alone.

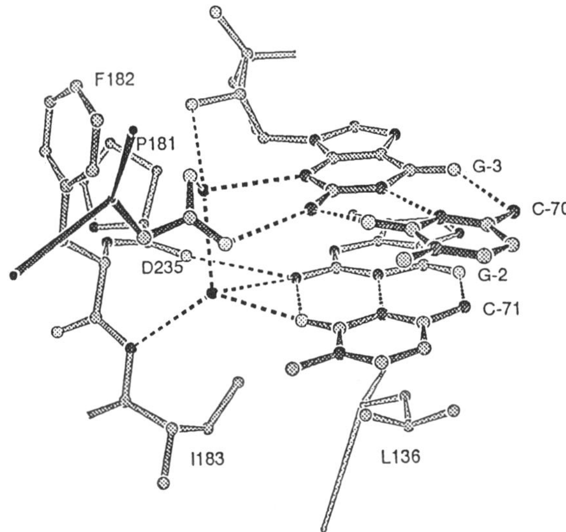

Figure 6 View of the recognition interface between GlnRS and base pairs G2-C71 and G3-C70 of tRNA^{Gln} (from Rould and Steitz 1992). Asp-235 directly bonds to the 2-amino group of guanine 3 via the minor groove. The backbone carbonyl of Pro-181 is rigidly directed to hydrogen-bond to the 2-amino group of guanine 2. A network of water molecules between the proteins and minor groove of the tRNA, only two of which are shown here, appear to enforce a requirement for GC base pairs at these positions. The hydrophobic environment formed by the proline, phenylalanine, isoleucine, and the underside of the ribose sugars enhances the strength and specificity of these direct and water-mediated hydrogen bonds.

DNA Interaction in the Minor Groove

There are ways, however, in which the interactions in the minor groove can be made sequence-specific. For example, the sequence preferences exhibited in the DNase I cleavage of DNA arise from its interactions in the minor groove (Suck et al. 1988). This side chain of a tyrosine observed to bind in the minor groove will fit into the normal-width minor groove but not into the narrower minor groove that characterizes AT-rich sequences.

Biochemical evidence for several sequence-specific DNA-binding proteins implies that they interact with DNA via the minor groove, although direct structural visualization of such an interaction has not yet been achieved. Yang and Nash (1989) have argued on the basis of methylation protection studies that *E. coli* integration host factor (IHF) interacts in the minor groove. IHF has significant sequence similarity with *E. coli* Hu protein, whose crystal structure (Tanaka et al. 1984) shows two long antiparallel β-loops, one from each subunit of the dimer, which form outstretched arms that create a large cleft sufficient in size to accommodate duplex DNA. The model for an IHF-DNA complex (Yang and Nash 1989) is based on one for Hu-DNA (Tanaka et al. 1984) and places the antiparallel β-loops in the minor groove in a manner proposed earlier for antiparallel β-strands (Carter and Kraut 1974; Church et al. 1977). The recently determined crystal structure of *Arabidopsis thaliana* TFIID similarly portrays a protein with pseudo-dyad symmetry and a twisted, antiparallel β-sheet forming a cleft of size sufficient to accommodate B-DNA (Nikolov et al. 1992). Biochemical data likewise point to minor groove interaction by this protein (Lee et al. 1991; Starr and Hawley 1991), although the structural basis of this interaction is not yet established.

RNA Interaction in the Minor Groove

There are now well-established examples of specific recognition of RNA in the minor groove (Rould et al. 1989; Musier-Forsyth and Schimmel 1992). With duplex RNA, which is A-form, the minor groove is shallow, wide, and accessible. Several sequence-specific interactions between GlnRS in the minor groove of tRNA^{Gln} have been observed (Rould et al. 1989, 1991; Rould and Steitz 1992). Base pairs G2-C71 and G3-C70 are recognized in a base-specific manner by two protein "fingers," one an α-helix and the other a turn of an antiparallel β-loop. In both cases, recognition involves contact between the 2-amino group of the guanine and hydrogen-bond acceptors of the protein. The carboxylate of an aspartic acid side chain 235 emanating from the amino end of α-helix H interacts with both the N2 of guanine 3 and a buried water molecule (Fig. 6). The peptide carboxyl group of Pro-181 interacts with the N2 of guanine 2. Substantiating the hypothesis that these two base pairs are among the recognition elements of tRNA^{Gln}, replacement of either by AU reduces the k_{cat}/K_m for charging by two to three orders of magnitude (Jahn et al. 1991). Furthermore, mutations in GlnRS that have increased rates of mis-charging of noncognate tRNAs are changes of Asp-235 to asparagine or glycine (Conley et al. 1988; Perona et al. 1989), showing the importance of this interaction for discrimination. An additional interaction in the minor groove that is important for discrimination is between the carboxylate of Glu-323 and the N2 of G10.

The importance of protein interaction with the 2-amino group of guanine in the minor groove of RNA has also been established in the case of alanine tRNA synthetase recognition of tRNA^{Ala} (Hou and Schimmel 1988; McClain and Foss 1988; Hou et al. 1989; Musier-

Forsyth and Schimmel 1992). The alanine synthetase has been clearly shown to recognize base pair 3-70, which is GU in tRNA[Ala]. Replacing G3 by an inosine, which lacks the N2, dramatically reduces charging of a minihelix (Musier-Forsyth and Schimmel 1992).

ROLE OF SINGLE-STRANDED REGIONS IN RECOGNITION

Since RNA molecules have single-stranded regions in loops and bulges and between helical stems, these regions are potential targets for recognition by proteins that are not available in duplex DNA. The recognition of anticodon loops in tRNAs by cognate synthetases provides the best-characterized examples of protein recognition of single-stranded regions. Molecular genetic and biochemical studies have shown that the anticodon bases of tRNA serve as recognition elements for many of the aminoacyl-tRNA synthetases (Schulman and Pelka 1985; Normanly and Abelson 1989; Sampson et al. 1989). The co-crystal structures of glutaminyl-tRNA synthetase and aspartyl-tRNA synthetase complexed with their cognate tRNAs show that, upon forming a complex, the anticodon bases become unstacked so that they may bind into separate base recognition pockets. The energy required to unstack the anticodon bases (as they exist in the uncomplexed tRNA) is provided by interactions with the protein. Since bases in loop regions of uncomplexed RNAs tend to be stacked on each other and since optimal recognition of bases by a protein requires their unstacking in order for them to interact in separate recognition pockets, it may be the case more often than not that protein recognition of an RNA single-stranded region is accompanied by a significant conformational change in the RNA.

Although details of the anticodon base interactions are not yet published for the aspartic acid enzyme (Ruff et al. 1991), Figure 7 shows how the three anticodon bases of tRNA[Gln] are interacting with GlnRS (Rould et al. 1991; Rould and Steitz 1992). Each anticodon nucleotide is recognized primarily by a polypeptide segment of five or six amino acids. In all three cases, at least one positively charged amino acid from this segment forms a salt link with an adjacent negatively charged phosphate. The aliphatic portion of this residue generally packs against either the base or the hydrophobic "underside" of ribose. With all three anticodon bases, recognition is achieved by direct hydrogen bonding between the backbone and side chains of a short recognition peptide and the Watson-Crick hydrogen-bonding groups of the bases. Furthermore, several of the interactions presumed to be discriminating involve hydrogen bonds with charged side chains that are buried from solvent in the complex. Although there is no conserved sequence or structural sim-

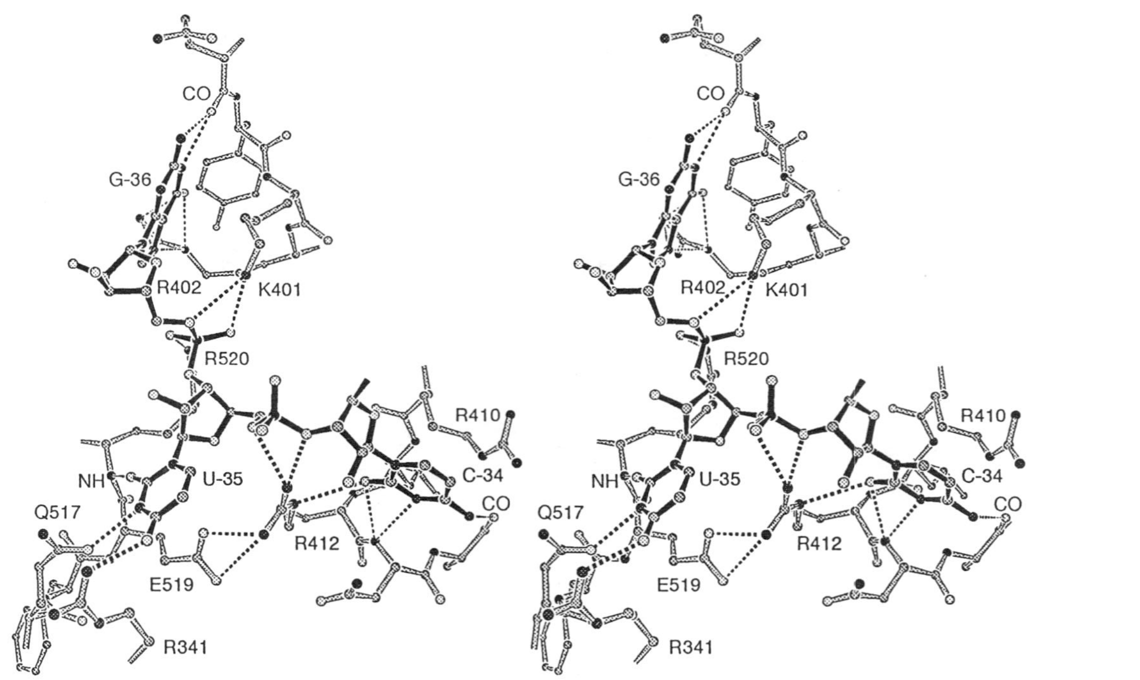

Figure 7 Stereo view of the binding pockets for the anticodon bases C34, U35, and G36 (*dark*) in the GlnRS-tRNA^{Gln} complex. Each nucleotide is recognized primarily by a single short polypeptide segment in the enzyme (*light*). In each case, an arginine or lysine from the polypeptide anchors the nucleotide by its phosphate group, allowing peptide backbone and side chains of the segment to specifically recognize the base (from Rould and Steitz 1992).

ilarity among these segments, they are predominantly in extended β-type conformation.

That the three anticodon bases of tRNA^{Gln} serve as recognition elements is confirmed by kinetic studies of mutant tRNAs (Jahn et al. 1991). Changes of the anticodon bases reduce $k_{\text{cat}}/K_{\text{m}}$ by 3–4 orders of magnitude. Interestingly, it is k_{cat} that changes the most, not K_{m}. A structural mechanism involving an anticodon-induced conformational change in the protein transmitted to the ATP-binding site has been hypothesized to account for the importance of anticodon base identity for catalysis (Rould et al. 1991).

ROLE OF MODIFIED BASES

Many of the RNA molecules that are recognized by proteins contain modified bases, whose role in specific recognition remains largely unknown. There are at least two ways that modified bases might enhance recognition: by stabilizing RNA conformations that otherwise would be less favored and by changing the shape of a recognition site. The N at position 5 of pseudo-uridine is observed in tRNA^{Gln} to interact with a water molecule, which in turn is interacting with the phosphate of the pseudo-uridine and the preceding phosphate, an interaction and conformation seen in all but one of the pseudo-uridines in tRNAs of known structure (J. Arnez and T.A. Steitz, unpubl.). This structure so stabilized may be significant in protein recognition.

Base modifications of noncognate tRNAs at nucleotides 34 and 37 may act as negative determinants of aminoacylation by GlnRS. The tightly packed interface between A37 and the protein (Fig. 8) suggests that bases with bulky modifications of A37, for example, 6-carbamoyl-threonyl adenine or 2-methylthio-N6-isopentenyl adenine, may provide an additional source of discrimination against the many noncognate tRNA molecules bearing these modifications. A role for modified bases at position 37 in tRNA discrimination has already been suggested by biochemical studies showing that ArgRS misacylates tRNA^{Asp} lacking modified bases (Perret et al. 1990). Likewise, C34 is tightly packed into a pocket that is covered by a loop of protein; this pocket may not accommodate certain modified bases at position 34, such as queuosine.

The 2$'$-ribosylated adenosine 64 in the initiator methionine tRNA from yeast appears to play an important role in assuring that this tRNA is only used in the initiation of protein synthesis and not in elongation (Kiesewetter et al. 1990). Again, the modification appears to function as a negative effector, since the modified tRNA$_{\text{i}}^{\text{Met}}$ will not bind to an EF-Tu or get inserted in elongation, whereas the tRNA$_{\text{i}}^{\text{Met}}$ that is un-

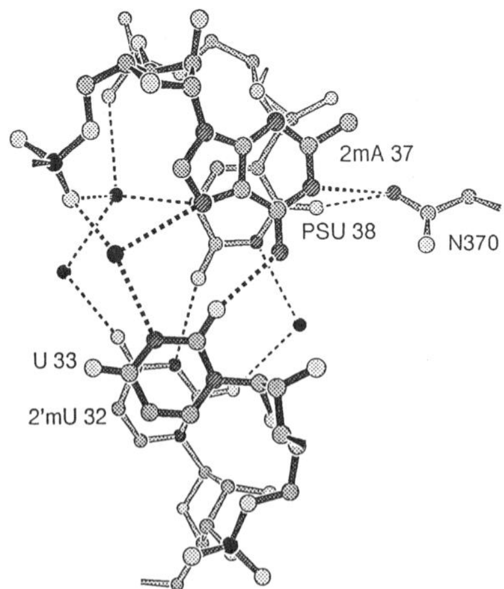

Figure 8 Stereo view of the two novel non-Watson-Crick base pairs that extend the anticodon stem of tRNA^{Gln} when complexed to GlnRS, showing the water network between these bases and the sugar-phosphate backbone. Asp-370 directly contacts both base pairs via the minor groove.

modified at position 64 will both bind to EF-Tu and participate in elongation.

SUMMARY

Many aspects of protein recognition of RNA and of DNA are very similar, such as the importance of sequence-dependent distortability of the nucleic acid and the role of specific water-mediated interaction. Although recognition of both nucleic acids can be achieved through true direct protein interaction with the exposed edges of base pairs in either the major or minor grooves of duplex, interactions via the major groove appear to dominate in DNA recognition, whereas the opposite preference may occur with RNA (although many more examples are required to establish this point). Interactions with single-stranded bases in RNA may prove to be the most significant in RNA recognition and are not at all characteristic of DNA recognition.

Whether there are simple, recurring protein motifs involved in RNA recognition, as has been found for DNA, is not yet known. Direct recognition of bases in DNA is achieved by various simple motifs that present an α-helix, antiparallel β-loop, or polypeptide chain end into the major groove of DNA. Although RNA recognition domains such as the RNP motif are known, it is too early to tell whether there are simple and general ways in which protein secondary structures — helix, antiparallel β-strands, or loops — interact with RNA.

REFERENCES

Aggarwal, A.K., D.W. Rodgers, M. Drottar, M. Ptashne, and S.C. Harrison. 1988. Recognition of a DNA operator by the repressor of phage 434: A view at high resolution. *Science* **242:** 99–107.

Carter, C.W. and J. Kraut. 1974. A proposed model for interaction of polypeptides with RNA. *Proc. Natl. Acad. Sci.* **71:** 283–287.

Church, G.M., J.L. Sussman, and S.-H. Kim. 1977. Secondary structure complementarity between DNA and proteins. *Proc. Natl. Acad. Sci.* **74:** 1458–1462.

Conley, J., H. Uemura, F. Yamao, J. Rogers, and D. Söll. 1988. *E. coli* glutaminyl tRNA synthetase: A single amino acid replacement relaxes tRNA specificity. *Protein Sequences Data Anal.* **1:** 479–485.

Delarue, M. and D. Moras. 1989. RNA structure. *Nucleic Acids Mol. Biol.* **3:** 182–196.

Drew, H.R. and A.A. Travers. 1984. DNA structural variations in the *E. coli tyr*T promoter. *Cell* **37:** 491–502.

Frederick, C.A., J. Grable, M. Melia, C. Samudzi, L. Jen-Jacobsen, B.-C. Wang, P. Greene, H.W. Boyer, and J.M. Rosenberg. 1984. Kinked DNA in crystalline complex with EcoRI endonuclease. *Nature* **309:** 327–331.

Freemont, P.S., J.M. Friedman, L.S. Beese, M.R. Sanderson, and T.A. Steitz. 1988. Co-crystal structure of an editing complex of Klenow fragment with DNA. *Proc. Natl. Acad. Sci.* **85:** 8924–8928.

Gartenberg, M.R. and D.M. Crothers. 1988. DNA sequence determinants of CAP-induced bending and protein binding affinity. *Nature* **333:** 824–829.

Harrison, S.C. 1991. A structural taxonomy of DNA-binding domains. *Nature* **353:** 715–719.

Hou, Y.-M. and P. Schimmel. 1988. A simple structural feature is a major determinant of the identity of a transfer RNA. *Nature* **333:** 140–145.

Hou, Y.-M., C. Francklyn, and P. Schimmel. 1989. Molecular dissection of a transfer RNA and the basis for its identity. *Trends Biochem. Sci.* **14:** 233–237.

Jahn, M., J. Rogers, and D. Söll. 1991. Anticodon and acceptor stem nucleotides in tRNA^{Gln} are major recognition elements for *E. coli* glutaminyl-tRNA synthetase. *Nature* **352:** 258–260.

Joyce, C.M. and T.A. Steitz. 1987. DNA polymerase I. From crystal structure to function via genetics. *Trends Biochem. Sci.* **12:** 288–292.

Kennard, O. and W.N. Hunter. 1989. Oligonucleotide structure: A decade of results from single crystal X-ray diffraction studies. *Q. Rev. Biophys.* **22:** 327–379.

Kiesewetter, S., G. Ott, and M. Sprinzl. 1990. The role of modified purine 64 in initiator/elongator discrimination of tRNA$_i^{Met}$ from yeast and wheat germ. *Nucleic Acids Res.* **18:** 4677–4682.

Koudelka, G.B., S.C. Harrison, and M. Ptashne. 1987. Effect of non-contacted bases on the affinity of 434 operator for 434 repressor and *cro*. *Nature* **326:** 886–888.

Lee, D.K., M. Horikoshi, and R.G. Roeder. 1991. Interaction of TFIID in the minor groove of the TATA element. *Cell* **67:** 1241–1250.

Lewis, M., J. Wang, and C. Pabo. 1985. Structure of the operator binding domain of lambda repressor. In *Biological macromolecules and assemblies*, vol. 2 (ed. F.A. Jurnak and A. McPherson), pp. 266–287. Wiley, New York.

Matthews, B.W. 1988. No code for recognition. *Nature* **335:** 294–295.

McClain, W.C. and K. Foss. 1988. Changing the identity of a tRNA by introducing a G-U wobble pair near the 3′ acceptor end. *Science* **240:** 793–796.

Musier-Forsyth, K. and P. Schimmel. 1992. Functional contact of a transfer RNA synthetase with 2'-hydroxyl groups in the RNA minor groove. *Nature* **357:** 513–515.

Nikolov, D.B., S.-H. Hu, J. Lin, A. Gasch, A. Hoffmann, M. Horikoshi, H.-H. Chua, R.G. Roeder, and S.K. Burley. 1992. Crystal structure of TFIID TATA-box binding protein. *Nature* **360:** 40–46.

Normanly, J. and J. Abelson. 1989. tRNA identity. *Annu. Rev. Biochem.* **58:** 1029–1049.

Otwinowski, Z., R.W. Schevitz, R.-G. Zhang, C.L. Lawson, A. Joachimiak, R.Q. Marmostein, B.F. Luisi, and P.B. Sigler. 1988. Crystal structure of trp repressor/operator complex at atomic resolution. *Nature* **335:** 321–329.

Pabo, C.O. 1983. DNA-protein interactions. In *Proceedings of the 27th Robert A. Welch Foundation Conference on Chemical Research: Stereospecificity in chemistry and biochemistry*, pp. 223–255. Houston, Texas.

Pabo, C.O. and R.T. Sauer. 1992. Transcription factors: Structural families and principles of DNA recognition. *Annu. Rev. Biochem.* **61:** 1053–1095.

Peattie, D.A. and W. Gilbert. 1980. Chemical probes of higher-order structure in RNA. *Proc. Natl. Acad. Sci.* **77:** 4679–4682.

Perona, J.J., R.N. Swanson, M.A. Rould, T.A. Steitz, and D. Söll. 1989. Structural basis for misaminoacylation by mutant *E. coli* glutaminyl-tRNA synthetase enzymes. *Science* **246:** 1152–1154.

Perret, V., A. Garcia, H. Grosjean, J.-P. Ebel, C. Florentz, and R. Giegé. 1990. Relaxation of a transfer RNA specificity by removal of modified nucleotides. *Nature* **344:** 783–789.

Puglisi, J.D., R. Tan, B.J. Calnan, A.D. Frankel, and J.R. Williamson. 1992. Conformation of the TAR RNA-arginine complex by NMR spectroscopy. *Science* **257:** 76–80.

Rould, M.A. and T.A. Steitz. 1992. Structure of the glutaminyl-tRNA synthetase-tRNA^{Glu}-ATP complex. *Nucleic Acids Mol. Biol.* **6:** 225–245.

Rould, M.A., J.J. Perona, and T.A. Steitz. 1991. Structural basis of anticodon loop discrimination by glutaminyl-tRNA synthetase. *Nature* **352:** 213–218.

Rould, M.A., J.J. Perona, D. Söll, and T.A. Steitz. 1989. Structure of *E. coli* glutaminyl-tRNA synthetase complexed with tRNA^{Gl} and ATP at 2.8 Å resolution: Implications for tRNA discrimination. *Science* **246:** 1135–1142.

Ruff, M., S. Krishaswarmy, M. Baeglin, A. Poterszman, A. Mitschler, A. Podjarny, B. Rees, J.C. Thierry, and D. Morase. 1991. Class II aminoacyl tRNA synthetase: Crystal structure of yeast aspartyl-tRNA synthetase complexed with tRNA^{Asp}. *Science* **252:** 1682–1689.

Sampson, J.R., A.B. Direnzo, L.S. Behlen, and O.C. Uhlenbeck. 1989. Nucleotides in yeast tRNA^{Phe} required for the specific recognition by its cognate synthetase. *Science* **243:** 1363–1366.

Schulman, L.H. and H. Pelka. 1985. *In vitro* conversion of a methionine to a glutamine-acceptor tRNA. *Biochemistry* **24:** 7309–7314.

Schultz, S.C., G.C. Shields, and T.A. Steitz. 1991. Crystal structure of a CAP-DNA complex: The DNA is bent by 90°. *Science* **253:** 1001–1007.

Seeman, N.C., J.M. Rosenberg, and A. Rich. 1976. Sequence-specific recognition of double helical nucleic acids by proteins. *Proc. Natl. Acad. Sci.* **73:** 804–808.

Starr, D.B. and D.K. Hawley. 1991. TFIID binds in the minor groove of the TAT box. *Cell* **67:** 1231–1240.

Steitz, T.A. 1990. Structural studies of protein-nucleic acid interaction: The sources of sequence specific binding. *Q. Rev. Biophys.* **23:** 205–280.

Suck, D., A. Lahm, and C. Oefner. 1988. Structure refined to 2 Å of a nicked DNA octanucleotide complex with DNase I. *Nature* **332:** 465–468.

Tanaka, I., K. Appelt, K.L. Dij, S.W. White. and K.S. Wilson. 1984. 3 Å resolution structure of a protein with histone-like properties in prokaryotes. *Nature* **310:** 376–381.

Vincze, A., R.E.L. Henderson, J.J. McDonald, and N.J. Leonard. 1973. Reaction of diethyl pyrocarbonate with nucleic acid components. Bases and nucleosides derived from guanine, cytosine, and uracil. *J. Am. Chem. Soc.* **95:** 2677–2682.

Weeks, K.M. and D.M. Crothers. 1991. RNA recognition by tat-derived peptides: Interaction in the major groove? *Cell* **66:** 577–588.

Weeks, K.M., C. Ampe, S.C. Schultz, T.A. Steitz, and D.M. Crothers. 1990. Fragments of the HIV-1 *tat* protein specifically bind TAR RNA: Peptide recognition of bulged RNA. *Science* **249:** 1281–1285.

Woodbury, C.P., O. Hagenbüchle and P.H. von Hippel. 1980. DNA site recognition and reduced specificity of the *Ecor* I endonuclease. *J. Biol. Chem.* **255:** 11534–11546.

Yang, C.-C. and H.W. Nash. 1989. The interaction of *E. coli* IHF protein with its specific-binding sites. *Cell* **57:** 869–880.

Yarus, M., R. Knowlton, and L. Soll. 1977. Aminoacylation of the ambivalent Su+7 amber suppressor tRNA. In *Nucleic acids protein recognition* (ed. H.J. Vogel), pp. 391–409. Academic Press, New York.

Yoon, C., G.G. Prive, D.S. Goodsell, and R.E. Dickerson. 1988. Structure of an alternating-B DNA helix and its relationship to A-tract DNA. *Proc. Natl. Acad. Sci.* **85:** 6332–6336.

11

Structure and Mechanism of the Large Catalytic RNAs: Group I and Group II Introns and Ribonuclease P

Thomas R. Cech
Howard Hughes Medical Institute
Department of Chemistry and Biochemistry
University of Colorado
Boulder, Colorado 80309-0215

Living cells require thousands of different chemical reactions to grow, store energy, move, respond to external stimuli, and reproduce. These reactions may take place spontaneously, but they rarely occur at a rate fast enough to support life. Biological catalysts accelerate these reactions, often by a factor of a billion or even a trillion, and also impart on them exquisite specificity. Biological catalysis is mostly the realm of protein enzymes, at least in modern cells. During the past decade, however, we have come to see that proteins do not have a monopoly on biological catalysis. Some RNA molecules are able to fold up to form active sites that catalyze chemical reactions, perhaps including a key reaction of protein synthesis. Furthermore, there is much speculation that RNA catalysis may have made an essential contribution to the establishment of self-replication and information-directed protein synthesis in early evolution.

The purpose of this chapter is to describe what we know and what we would like to know about some of these catalytic RNAs, or "ribozymes" (Kruger et al. 1982). The specific questions are those of structure (How do these RNAs fold to form an active site?) and mechanism (What chemical reactions occur and how are they catalyzed?). The decision to cover the group I and group II introns and the catalytic RNA subunit of RNase P in a single review is justifiable on two grounds. First, these are all large ribozymes compared, for example, to the hammerhead (see Pan et al., this volume). Although size could turn out to be an arbitrary distinction, it seems reasonable that some minimum size is required for an RNA to maintain a stable, catalytically active structure in the absence of its substrates. The large ribozymes might have such a preformed struc-

The RNA World
© 1993 Cold Spring Harbor Laboratory Press 0-87969-380-0/93 $5 + .00

ture, whereas the small ribozymes might attain their active structure only after the substrate has bound by base-pairing. Second, all three of the large ribozymes cleave RNA to generate 3′ hydroxyl termini, which distinguishes them from the small ribozymes, which cleave RNA to generate 2′,3′-cyclic phosphate termini. The similar chemical reaction promoted by the large ribozymes makes it of interest to compare their catalytic mechanism.

Although I focus on only three ribozymes, many of the considerations about RNA structure and function are expected to be applicable to other folded RNA molecules that participate actively in cellular metabolism, including ribosomal RNAs and the small nuclear RNAs involved in mRNA splicing. Catalytic RNAs have the special feature that the intactness of their structure can be directly assessed by monitoring the chemical reactions they catalyze. Thus, they provide powerful systems for studying the general problem of RNA folding.

The audience that I envision as I write this chapter consists of students who are already solidly grounded in biochemistry but have little knowledge in the specific area of catalytic RNA. (Among "students" I include everyone from advanced undergraduates to independent scientists working in other fields.) Keeping this audience in mind, I give more background than is customary in review articles, and I cite the current literature in a selective rather than a comprehensive manner. For those who desire more references, there have been a number of recent reviews (Cech 1988, 1990, 1992; Altman 1989; Michel et al. 1989a; Jacquier 1990; Lambowitz and Perlman 1990; Pace and Smith 1990; Brown and Pace 1992; Cech et al. 1992).

OVERVIEW

Especially in eukaryotes, many genes have their coding regions (exons) interrupted by stretches of noncoding DNA called introns or intervening sequences. RNA directly transcribed from such a "split gene" still contains introns, which are subsequently removed by a process termed RNA splicing. Introns have been categorized into at least four groups based on the presence of conserved nucleotide sequences and /or RNA structures. Of these, only the group I and group II introns provide the active site for their own splicing reaction. This does not mean that RNA catalysis is restricted to these groups. For example, there is much speculation that the small nuclear RNAs may assemble to provide a ribozyme active site for nuclear mRNA splicing, which does not involve group I or group II introns (see Baserga and Steitz; Moore et al.; both this volume).

Group I Introns

How do we recognize a group I intron when we see one? All members of the group share several short patches of conserved nucleotides, as shown in Figure 1a. Furthermore, as first proposed by Michel et al. (1982) and Davies et al. (1982), all group I introns can be folded into a secondary structure with a common "core" region (centered on paired regions P3, P4, P6, and P7). The conserved nucleotides always occupy the same position in this secondary structure. The 5′ splice site resides within paired region P1, which consists of the last few nucleotides of the 5′ exon base-paired to an internal guide sequence (IGS) within the intron. In some introns, an adjacent portion of the IGS pairs with the first nucleotides of the 3′ exon, thereby helping to bring the exons together for ligation (Davies et al. 1982). Additional nucleotides, often comprising long open reading frames (ORFs) that encode a protein, form stem-loop structures peripheral to the core.

Group I introns can also be recognized by their common splicing pathway. They all undergo splicing by a two-step transesterification mechanism, using an exogenous molecule of guanosine as the attacking group (nucleophile) for the first step (Fig. 2a). Furthermore, the active site for splicing, which includes the guanosine-binding site and other elements of the catalytic apparatus, is composed of RNA. Thus, many of the group I introns are self-splicing in vitro, which means they accurately and efficiently excise themselves from the precursor RNA and ligate the flanking exon sequences in the complete absence of protein (Cech et al. 1981; Kruger et al. 1982; Garriga and Lambowitz 1984; Van der Horst and Tabak 1985).

Proteins are known to be required for the splicing of some group I introns in vivo. For example, intron 4 of the cytochrome *b* pre-mRNA of yeast mitochondria does not self-splice in vitro, and genetic evidence has revealed the need for several proteins for its splicing in vivo. One of these is a "maturase" encoded by an ORF within the intron itself (Lazowska et al. 1989), and another is a mitochondrial leucyl-tRNA synthetase encoded in the yeast nucleus and imported into mitochondria (Herbert et al. 1988). In *Neurospora crassa*, mutations in the nuclear cyt-18 gene have been found to affect splicing in vivo. The cyt-18 protein is the mitochondrial tyrosyl-tRNA synthetase (Akins and Lambowitz 1987). It functions directly in splicing by binding to the catalytic core of the intron, and this protein-RNA complex is necessary and sufficient for splicing (Kittle et al. 1991; Guo and Lambowitz 1992). The involvement of tRNA synthetases in group I intron splicing might indicate that the intron RNA contains a tRNA-like structure (Guo and Lambowitz 1992). In these cases of protein-dependent splicing, the splicing pathway is still that described in Figure 2, and the RNA structure continues to be important, e.g., by providing the G-binding site (Gampel and Cech 1991). Thus, these systems show that proteins can facilitate an intrinsically RNA-catalyzed reaction.

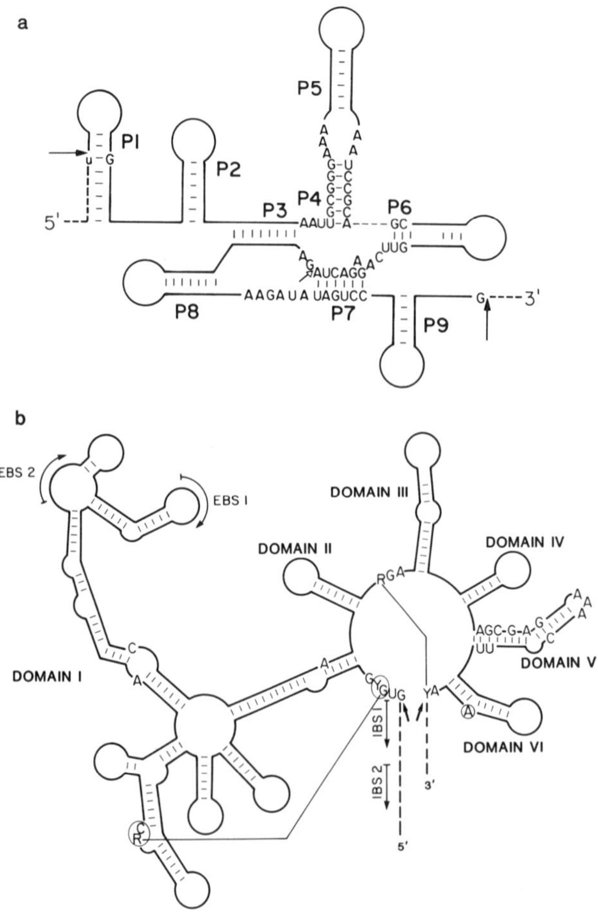

Figure 1 Secondary structures of (*a*) group I and (*b*) group II introns as proven by comparative sequence analysis. Dashed lines and lowercase *u* indicate exons. Solid lines and uppercase letters indicate introns. Filled arrows indicate 5′ and 3′ splice sites. (*a*) Group I structure showing most common nucleotide at each position in core region. Dashed line between P4 and P6 was added to make the diagram less crowded. Open arrow indicates site of insertion of extra stem-loops in a subgroup of these introns. (Reprinted, with permission, from Cech 1988.) (*b*) Group II secondary structure for a subgroup that includes two self-splicing introns from yeast mitochondria. Indicated nucleotides are present in all introns with at most one exception; R = purine; Y = pyrimidine. Each exon binding site (EBS) pairs with the corresponding intron binding site (IBS). (Redrawn, with permission, from Michel et al. 1989a.)

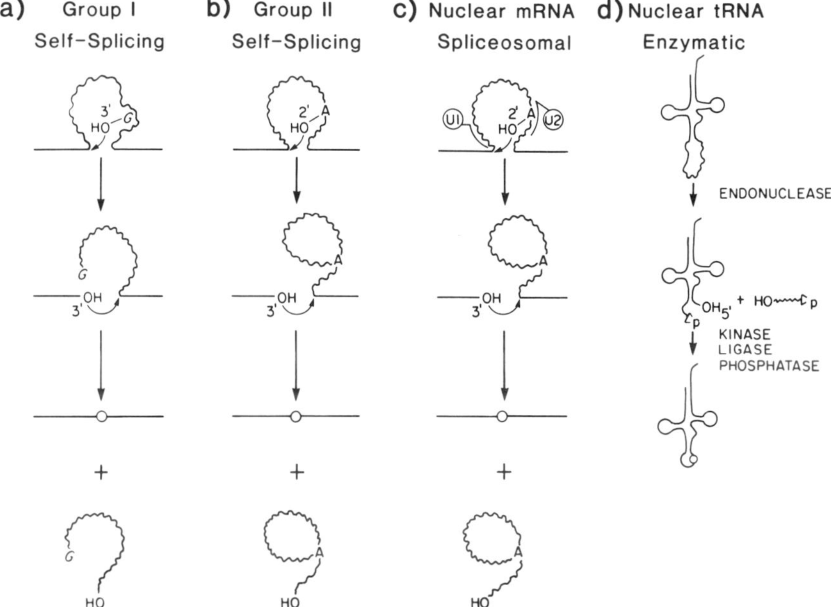

Figure 2 Splicing mechanisms of the four major groups of precursor RNAs. Wavy lines indicate introns. Smooth lines indicate flanking exons. For nuclear mRNA splicing, only two of the small nuclear RNPs and other factors that assemble with the pre-mRNA are shown. (Reprinted, with permission, from Cech 1990 [Annual Review of Biochemistry, vol. 59], copyright by Annual Reviews.)

Group I introns interrupt genes for mRNA, rRNA, and tRNA and are extremely widespread among species. They have been found in the mitochondrial, chloroplast, and nuclear genomes of diverse eukaryotes, although not yet in vertebrates. Their discovery in bacteriophage T4 marked the first time that RNA splicing had been found in a prokaryote (Chu et al. 1986; Gott et al. 1986). Subsequently, group I introns were also found to inhabit eubacterial genomes (Xu et al. 1990; Kuhsel et al. 1990; Reinhold-Hurek and Shub 1992).

The distribution of group I introns is irregular. For example, some species of *Tetrahymena* have group I introns in their genes for the large subunit rRNA, whereas closely related species do not. Considering those species that have the intron, the intron has a phylogenetic tree different from that inferred from rRNA sequences and other biochemical features (Sogin et al. 1986). This irregular distribution suggests that these introns are not restricted to normal genetic inheritance but can be acquired independently in different species. Such horizontal transmittance has been shown in some cases to occur by intron "mobility," a known mechanism involving recombination at the DNA level (Dujon 1989). It has been speculated that introns might also be able to relocate by a mechanism involving reverse splicing (Woodson and Cech 1989; Mohr and Lambowitz 1991).

Group II Introns

Like group I introns, group II introns were first defined by their common structural features. Michel et al. (1982) proposed the secondary structure in which six helical domains emanate from a central wheel (Fig. 1b). Two major subgroups (IIA and IIB) have similar secondary structures but differ in their quasi-universal nucleotides (those present at the same location in all examples of a subgroup, with at most one exception). The seven known fungal mitochondrial group IIA introns contain ORFs that potentially encode a reverse transcriptase-like protein (Michel and Lang 1985). Most other group II introns do not contain ORFs.

Several group II introns have been shown to undergo self-splicing in vitro, demonstrating that the folded RNA structure is sufficient for formation of an active site for this splicing reaction (Peebles et al. 1986; Schmelzer and Schweyen 1986; Van der Veen et al. 1986). However, self-splicing requires extreme conditions that are not physiological (e.g., 100 mM $MgCl_2$ and 500 mM $(NH_4)_2SO_4$ at 45°C) and even under these conditions proceeds at a rate ten times slower than that of group I RNA self-splicing. This observation is consistent with genetic evidence showing that splicing of these introns in vivo requires specific proteins, either maturases encoded by intronic ORFs or nuclear-encoded proteins (Lambowitz and Perlman 1990).

Group II introns undergo splicing by a two-step transesterification mechanism (Fig. 2). The attacking group for the first step is the 2' hydroxyl of an adenosine in domain VI (circled in Fig. 1b). Attack at the phosphorus atom of the 5' splice site generates a branched "lariat" intermediate, the same sort of intermediate found in nuclear mRNA splicing (Padgett et al. 1984; Ruskin et al. 1984). The second step involves attack of the 3' hydroxyl of the 5' exon at the 3' splice site, ligating the exons and releasing the lariat intron.

All known group II introns are found in eukaryotic organelles. They exist in plant and fungal mtDNAs and comprise the majority of introns in plant chloroplasts. The *Euglena* chloroplast genome contains numerous group II introns that have the typical splice-site sequences and counterparts of all six helical domains, with domains V and VI being particularly well conserved; however, they tend to be unusually short with domain I highly abbreviated (Michel et al. 1989a; Copertino and Hallick 1991). These do not undergo self-splicing under a variety of conditions tested in vitro (R.B. Hallick, pers. comm.), suggesting that their splicing requires some *trans*-acting component in vivo.

Ribonuclease P

The only catalytic RNA known to act in vivo as an enzyme, in the sense that it turns over, is RNase P. This enzyme acts as a ribonucleoprotein in cells, but at least in the case of several eubacterial RNase P enzymes, it is clear that its RNA subunit is the catalytic subunit.

The reaction catalyzed by RNase P is the removal of a block of nucleotides from the 5′end of a tRNA precursor to produce the mature 5′end of the tRNA (Fig. 3). This is one of a number of RNA processing reactions required to convert a primary transcript to a functional tRNA; others include 3′end-processing, nucleoside modification, and for some transcripts, RNA splicing (Fig. 2d). RNase P cleaves by hydrolysis, generating 5′phosphate and 3′hydroxyl termini. This is the same specificity apparent in reactions of the group I and II introns but opposite to that of many common ribonucleases. A single RNase P processes the precursors to all the diverse cellular tRNAs, so it must be recognizing structures and nucleotides common to all these substrates.

The *Escherichia coli* holoenzyme consists of a 377-nucleotide RNA (M1 RNA) and a 13.8-kD polypeptide (C5 protein) (for review, see Altman 1989). The secondary structure of the RNA subunit, determined by comparative sequence analysis with homologous RNAs from other eubacteria, is shown in Figure 3 (James et al. 1988; Brown et al. 1991). Regions shown as unpaired might in fact be single-stranded or might be engaged in secondary or tertiary structure interactions not yet identified. The C5 protein component is basic. It binds the M1 RNA with high affinity and other RNAs with 100-fold lower affinity. The protein subunit is required for viability of *E. coli*, so it is an essential component in vivo (Schedl and Primakoff 1973; Ikemura et al. 1975).

At first it was thought that the C5 protein subunit was also essential for RNase P activity in vitro (Altman 1989). However, when assayed under salt concentrations higher than those that had been used for the holoenzyme, the RNA subunit alone proved capable of carrying out the tRNA processing reaction (Guerrier-Takada et al. 1983; Guerrier-Takada and Altman 1984). The pre-tRNA was cleaved at the correct site, with multiple turnover. Thus emerged our current picture that RNase P has a catalytic subunit composed of RNA and an accessory polypeptide subunit. The polypeptide allows the reaction to proceed efficiently at physiological ionic conditions. Furthermore, the maximum velocity of cleavage (k_{cat}) is about 20 times higher for the holoenzyme than for the RNA alone, when each is assayed under its own optimal ionic conditions

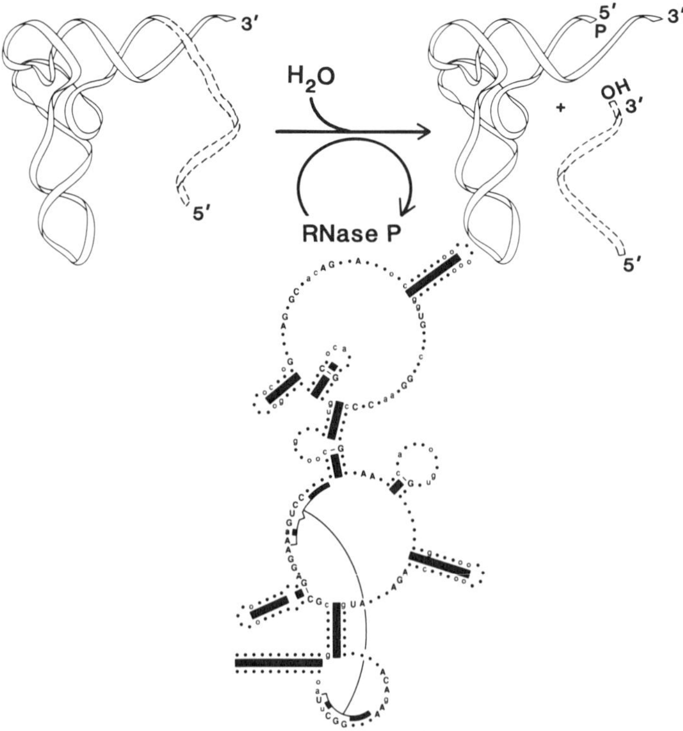

Figure 3 Maturation of tRNA catalyzed by RNase P, an enzyme with a catalytic RNA subunit. 5′ leader sequence removed by processing may have variable structure, so it is shown as a dashed ribbon. The eubacterial consensus structure of RNase P RNA with phylogenetically proven helices designated by filled rectangles, invariant nucleotides in uppercase letters, >90% conserved nucleotides in lowercase letters, and nucleotides less conserved in sequence shown as dots. (Reprinted, with permission, from Brown and Pace 1992 by permission of Oxford University Press.)

(Guerrier-Takada et al. 1983; Reich et al. 1988). The protein may be changing the rate-limiting step of the multiple-turnover reaction: The RNA-only reaction shows an initial burst of cleavage indicative of rate-limiting product release, whereas the holoenzyme reaction does not (Reich et al. 1988). The protein subtly alters the recognition of certain pre-tRNA substrates and is necessary for the cleavage of pre-4.5 S RNA, the only non-tRNA substrate identified in *E. coli* (Guerrier-Takada et al. 1983).

Eukaryotic RNase P enzymes from the nuclei and mitochondria of vertebrates and fungi have been studied (for review, see Brown and Pace 1992). These enzymes, like those of eubacteria, function as ribonucleoprotein complexes. Their activity is sensitive to micrococcal nuclease digestion, indicative of an essential RNA component, but none of the isolated RNAs appears to have catalytic activity by itself. Two rather different explanations come to mind. Perhaps in these enzymes the polypeptide has taken over part of the RNA function, with amino acids participating directly in catalysis. Alternatively, the protein may be stabilizing the correct fold of the RNA, the same sort of model that has been suggested for the protein-dependent group I introns (see above).

In evaluating these models, it would be very useful to know how the structures of the eukaryotic RNase P RNAs compare to the eubacterial consensus structure. At the moment, however, their relationship is unclear. It is possible to fold the eukaryotic RNAs into structures resembling the lower half of the eubacterial structure shown in Figure 3 (Forster and Altman 1990a). On the other hand, it has been argued that the variability of sequences among the small number of eukaryotic RNase P RNAs available is so great as to make sequence alignment and structure modeling premature (Brown and Pace 1992).

CATALYTIC RNA STRUCTURE

Mind-set

To understand RNA catalysis, we must understand the detailed three-dimensional shape of the catalytic entity, its dynamics, and the specific contributions that individual components of the molecule make to transition-state stabilization.

What picture should we have in mind when we think about the structure of one of the large ribozymes? Let us first consider the structure at low resolution. We can imagine that we are viewing the molecule from a considerable distance or observing it through an imperfect lens that prevents us from seeing atomic detail. At this resolution, these catalytic RNAs resemble protein enzymes. Each has its own specific shape, roughly globular, with an active-site cleft or pocket on its surface. Although there is certainly molecular motion, including large-scale movement of one part of the molecule relative to other parts, it is still useful to speak of the molecule as having "a structure" or a defined set of structures.

Evidence for this low-resolution view comes from diverse sources. First, what is the evidence for a defined three-dimensional structure, rather than a loose assemblage of secondary structure elements that samples the active conformation only occasionally? When group I introns are bombarded with free radicals produced by Fe(II)-EDTA present in the solution, different nucleotides along the RNA chain are cleaved with quite different efficiency (Latham and Cech 1989; Celander and Cech 1991; Heuer et al. 1991). The pattern of cleavage of tRNA correlates with solvent accessibility, so the pattern of cleavage of the catalytic RNA is thought to indicate the inside and outside surfaces of the folded molecule (Latham and Cech 1989; Celander and Cech 1990). The observation of high-efficiency UV cross-links between distant sequence elements of a ribozyme (Downs and Cech 1990; J.F. Wang et al., in prep.) or between a ribozyme and a bound RNA substrate (Guerrier-Takada et al. 1989; Burgin and Pace 1990) also argues that these large RNAs exist in the form of discrete three-dimensional structures. Second, what is the evidence for a surface location of the active site? In the *Tetrahymena* ribozyme, A301 and A302 are both highly accessible to methylation in the free ribozyme, but become protected when the RNA substrate binds and the resulting P1 helix docks into the catalytic center (Pyle et al. 1992). Third, what is the evidence for a compact, globular shape? Three-dimensional structure models of two group I introns, proposed by Michel and Westhof (1990), have been confirmed at least at low resolution by independent physical data (see below). Their models portray globular molecules with the P1 helix nestled into an active-site cleft. In addition, electron microscopic examination of the *Tetrahymena* ribozyme reveals an overall globular shape (Y.-H. Wang et al., unpubl.).

Having squinted at a ribozyme structure from afar, we now zoom in for a closer look. When we get close enough to recognize the individual secondary structure elements that comprise the molecule, we see that the "globular protein" analogy has major shortcomings.

Protein secondary structure involves hydrogen bonding between portions of the backbone of the polypeptide chain, which leaves the amino acid side chains displayed on the outside. Nucleic acid secondary structure involves hydrogen bonding between the bases to form double-helical regions, leaving the repetitious backbone on the outside. Thus, the secondary structure of nucleic acids is inside-out compared to that of proteins (Sigler 1975). This has a major implication for formation of higher-order structure. Protein tertiary structure involves the packing together of the exposed amino acid side chains, stabilized by hydrophobic interactions and, to some extent, other interactions such as hydrogen bonding. In order to form tertiary structure, RNA must pack together its helical elements in a specific way. Such packing seems problematic. First, the helices are polyanions, so packing seems unlikely to be driven by hydrophobic interactions and must overcome electrostatic repulsion. Second, backbones are information-poor compared to the bases, but the bases are tucked away in the helix and only partially accessible for tertiary interactions. The most available portion of an extended RNA helix is the minor groove, and the functional groups protruding into a minor groove are not very different from one base pair to the next (Seeman et al. 1976). Solutions to this RNA structure problem involve hydrogen bonding (base triples, backbone interactions) and, as described next, metal ions.

The importance of metal ions distinguishes RNA folding from protein folding. Most proteins fold without metal ions, although there are notable exceptions such as "zinc fingers." The *Tetrahymena* ribozyme requires divalent cations for folding into its active tertiary structure. This requirement can be met by either Mg^{++} or Mn^{++} (about 2 mM at 42°C) and at least partially fulfilled by Ca^{++} and Sr^{++} (Grosshans and Cech 1989; Celander and Cech 1991). The cations are presumably acting at the level of tertiary rather than secondary structure, because monovalent cations that stabilize secondary structure cannot replace the divalent cation requirement even at very high concentration. What might these metal ions be doing? Considering the X-ray crystallographic analysis of tRNA/metal ion complexes, it seems likely that the divalent cations stabilize structures in which two or more segments of the phosphodiester backbone come into close proximity (Holbrook et al. 1977; Jack et al. 1977). There is also the "Mg^{++} glue" hypothesis: Arrays of Mg^{++} ions could be precisely positioned to bridge phosphates on two separate helical elements, promoting their packing together (Celander and Cech 1991).

Another difference between protein and RNA folding concerns the autonomy of individual secondary structure elements. Most often an α-helix found in a protein structure unfolds if excised from its environment in the protein. In contrast, a stem-loop found in an RNA structure can usually be trusted to form the same structure as a separate entity. This in part accounts for the ability of fragments of an RNA molecule, representing separate secondary structure elements, to self-assemble into an active whole. Dramatic examples have been reported for group I and group II introns (see, e.g., Jarrell et al. 1988; Doudna and Szostak 1989; Van der Horst et al. 1991). One consequence is that in RNA, it may often be possible to attribute discrete elements of the RNA function to individual secondary structure features. Conversely, one might be able to build an RNA machine by properly organizing a set of secondary structure elements, each of which makes a distinct contribution to catalysis. For example, a metal ion-binding loop plus a guanosine-binding duplex held in the correct orientation might serve to orient and activate a guanosine nucleophile. Although this "cassette model" for ribozyme construction is undoubtedly naive, it may provide a useful way of decomposing structure-function relationships in RNAs, including ribozymes, ribosomal RNAs, and snRNAs.

Secondary Structure

The first step toward understanding RNA structure is to determine its base-paired secondary structure. Because the individual elements (e.g., double helices) that make up a complex folded RNA are typically stable by themselves, they can be considered to be the building blocks of the three-dimensional RNA structure in a simpler way than α-helices or β-sheets can be considered to be building blocks of protein structure. The most common secondary structure elements found in RNA have been reviewed recently (Chastain and Tinoco 1991) and are not repeated here. Instead, I review the methods that have been most useful for determining the secondary structures of the large catalytic RNAs (Figs. 1 and 3).

Phylogenetic Sequence Comparisons

This method, first used to determine the cloverleaf secondary structure of tRNA, requires the availability of sequences of homologous RNAs from different organisms. The sequences need to be related enough to allow the corresponding regions to be properly aligned, but also divergent enough to be informationally rich. Norman Pace calls this "finding the correct phylogenetic depth." Two portions of a given RNA could be complementary either by coincidence or because they are base-paired to one another. If in a second homologous RNA the same two portions have a somewhat different sequence, with base changes in both partners serving to maintain complementarity, then this constitutes strong evidence for base-pairing. This concept and its application are discussed by Woese and Pace (this volume).

Proof by Mutagenesis and Suppression

A candidate base pair can be tested by mutation of the two bases. If the single-base changes are deleterious to function, but the combination of two deleterious single-base changes that restores complementarity restores function, the base pair is considered to be proven. As with the phylogenetic method, proof of at least two base pairs is considered to prove the helix in which they reside. This approach is particularly well suited to catalytic RNAs which, in the words of Michael Yarus, "report their structure through their activity." In the example given in Figure 4, the existence of the RNA duplex was established by analysis of the splicing activity of RNA containing single-base changes and compensatory double mutations at multiple positions.

The U·G base pair at the 5´ splice site (positions –1 and 22 in Fig. 4) is an informative special case. Standard Watson-Crick combinations cannot substitute for the U·G (Barfod and Cech 1989; Doudna et al. 1989), which by itself would leave one wondering whether these bases were paired. However, C·A substitutes to some extent, leading to the model that a wobble base pair which can be formed by both U·G and C·A is important at the reaction site (Doudna et al. 1989). More generally, if substitution of one base pair for another in a proposed RNA duplex reduces activity, it does not necessarily disprove base-pairing; pairing might be necessary but not sufficient for activity. The base pair could simultaneously pair with a third base to form a base triple, one partner of the pair might need to alternate between two secondary structure interactions, or the detailed structure of the base pair, which depends on sequence, might be important for activity (Cech 1988; Flor et al. 1989; Michel et al. 1989b, 1990).

Tertiary Structure of Group I Introns

RNA catalysis does not take place in two dimensions. Although secondary structure determination is important, it is only an intermediate step in understanding RNA folding and active-site interactions in three dimensions. Bridging the gap from two to three dimensions should ultimately be feasible by determining distances using the nuclear Overhauser effect of NMR spectrometry, which has already proven useful for determining hairpin loop structures in RNA (Cheong et al. 1990; Heus and Pardi 1991). X-ray crystallography, which proved so successful in cracking the structure of tRNA (Kim et al. 1973; Robertus et al. 1974),

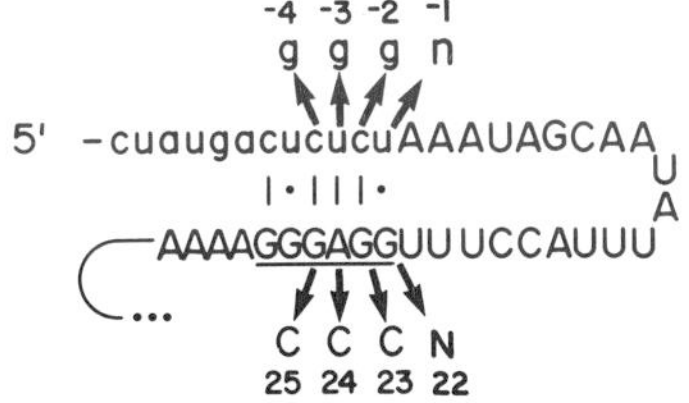

Figure 4 Testing a putative RNA duplex by mutagenesis in vitro. Paired region P1 of the self-splicing pre-rRNA of *Tetrahymena* consists of the last few nucleotides of the 5′ exon (lowercase letters) paired to the IGS (bottom strand). Intron nucleotides are in uppercase letters. Single-base changes at positions −2, −3, −4, 23, 24, or 25 result in inefficient or inaccurate splicing. Double mutations that restore complementarity (such as −2→g and 23→C) suppress the effect of the single-base changes. Watson-Crick combinations at positions −1 and 22 (n and N) do not substitute for u·G, although c·A has some activity (see text). (Adapted, with permission, from Been and Cech 1986, copyright by Cell Press.)

has not been widely attempted for other RNAs because until recently it was difficult to obtain the requisite quantity of pure material. Now that the material is available, there is much hope that X-ray structures of other biologically important RNAs will be forthcoming. In the absence of such direct structural information, it is still possible to obtain a reasonably high-resolution picture of a catalytic RNA molecule, as evidenced by the stunning success of the Michel-Westhof model.

Michel and Westhof (1990) modeled the architecture of the catalytic center of the group I intron using (1) the known secondary structure described above; (2) tertiary structure constraints provided by the localization of the G-binding site (Michel et al. 1989b), the proposal of a triple-helical scaffold involving P4 and P6 (Michel et al. 1990), and additional interactions that emerged from their careful analysis of phylogenetic covariation of bases; and (3) stereochemical constraints based on a consideration of reasonable bond angles for RNA. The resulting model (Fig. 5a) has an active-site cleft at the junction of two helical domains. One of these domains comprises coaxially stacked P4 and P6 helices (see also Kim and Cech 1987); this positions the conserved A residues in the internal loop above P4 in a position where they can make specific contacts with the reaction site in the P1 helix. The second domain comprises the highly conserved P7 duplex (which provides the G-binding site; see Fig. 5b), the P3 duplex, and the P8 duplex. Paired region P1, which contains the 5′ splice site that is cleaved by guanosine, is nested in the cleft formed by these two major domains. P1 is held in place in part by contacts between its backbone and the catalytic core.

The general architecture of the group I catalytic core proposed by Michel and Westhof (1990) has been confirmed by three independent tests. Two of these have given direct physical information: Attachment of a destructive reagent to the guanosine substrate has revealed that A207 is close to the 5′ end of bound GMP (Wang and Cech 1992), and site-specific cross-linking with derivatized G22 has shown that the 5′ end of the IGS in P1 is in proximity to A114 and A115 (J.F. Wang et al., in prep.). (The positions of these nucleotides are indicated on the secondary structure diagram of Fig. 5b.) In the third study, a combination of biochemical and genetic analysis has identified A302 as the residue in the catalytic core responsible for interacting with a 2′ OH group on P1 at position −3 relative to the cleavage site (Pyle et al. 1992). This identification agrees well with the general architecture proposed by Michel and Westhof (1990), although the actual contact point is different. Confirmation of a different type has come with the finding of a tertiary interaction involving two peripheral elements of the *sunY* group I intron, which is fully compatible with the model of the catalytic core (Michel et al. 1992).

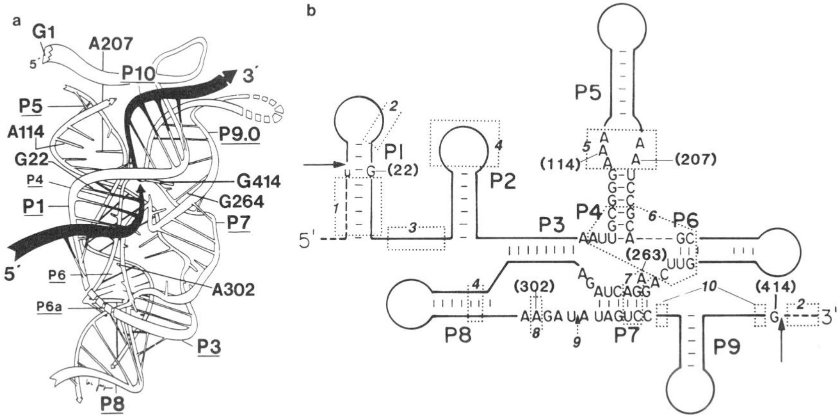

Figure 5 (*a*) Three-dimensional model of the core of the *Tetrahymena* group I intron. The stage shown is just prior to the exon ligation step. Filled backbone indicates exons. (Reprinted, with permission, from Michel and Westhof 1990.) (*b*) Conserved nucleotides and structures implicated in specific aspects of the catalytic mechanism are indicated on the secondary structure of Fig. 1a. Nucleotides mentioned in the text are numbered in parentheses, using the numbering system of the *Tetrahymena* pre-rRNA intron. (*1*) Length of P1 duplex preceding the conserved u•G pair helps determine reaction site (horizontal arrow) (Doudna et al. 1989). (*2*) Pairing between this portion of IGS and complementary sequence in 3' exon contributes to 3' splice site reaction in some group I introns (Davies et al. 1982). (*3*) Joining region contributes to positioning of P1 in *Tetrahymena* ribozyme (Young et al. 1991) and contributes in a different manner in another subgroup (Michel and Westhof 1990). (*4*) Hairpin loop docks into P8 to position 5' splice site in one subgroup of group I introns (Michel and Westhof 1990). (*5*) Conserved A residues in P4-P5 loop proposed to organize P7 relative to P1 (Michel and Westhof 1990; Wang and Cech 1992; J.F. Wang et al., in prep.). (*6*) Triple-helical scaffold proposed as a major structural element of the core (Michel et al. 1990). (*7*) Binding site for the guanosine substrate (Michel et al. 1989b; Yarus et al. 1991). (*8*) A302 stabilizes binding of 5' exon (Pyle et al. 1992). (*9*) Phosphate preceding A306 may help coordinate an active-site Mg^{++} ion (Piccirilli et al. 1992a; E. Christian and M. Yarus, in prep.). (*10*) Base-pairing contributes to positioning the 3'-terminal G of the intron in the G-binding site.

Thus, it has been confirmed that the Michel-Westhof model provides a remarkably insightful view of the three-dimensional structure of the active site of group I introns. It appears to be correct at least to a resolution of 10 Å, which is the resolution of some of the physical approaches that have been used to test it. It is likely that some of the atomic-resolution interactions proposed in the model will be correct, whereas others will require some adjustment, as was the case with A302 discussed above.

One critical element of the group I RNA structure not modeled by Michel and Westhof is the divalent metal ion component. Group I ribozyme activity requires either Mg^{++} or Mn^{++}. Ca^{++}, Sr^{++}, and Ba^{++} can mediate the correct overall folding of the RNA, but cannot by themselves confer catalytic activity, suggesting that two classes of metal ion-binding sites (structural and catalytic) must be occupied for catalysis (Grosshans and Cech 1989; Celander and Cech 1991; see also Guerrier-Takada et al. 1986). Positions of metal ions required for activity have been inferred from a phosphorothioate interference study (E. Christian and M. Yarus, in prep.). Mg^{++} has a very large preference for coordinating to the oxygen atom of phosphate relative to the sulfur atom of phosphorothioate, whereas Mn^{++} has much less of a preference. Thus, sites where phosphorothioate interferes with ribozyme activity in Mg^{++} but where the activity is restored in Mn^{++} are excellent candidates for phosphates that coordinate functionally important divalent cations. These would include metal ions involved in RNA folding and directly in active-site chemistry (see below, Transition State Stabilization).

How Broad Is RNA Catalysis?

RNA-catalyzed reactions that have been found in nature all involve transesterification or hydrolysis of phosphodiester linkages in RNA (Table 1). The details of the chemistry can vary. Group I and group II ribozymes and RNase P leave 3'-OH groups at the cleavage site, analogous to DNases and exonuclease activities associated with DNA synthesis. The 5'-phosphorus atom is transferred to a 3'-oxygen (group I), to a 2'-oxygen (group II), or to H_2O (RNase P) (Fig. 6a). In contrast, the small ribozymes (hammerhead, hairpin, and hepatitis δ agent) form 2',3'-cyclic phosphates, using the 2'-hydroxyl group adjacent to the cleavage site as the nucleophile to cleave the phosphodiester linkage (Fig. 6b). This is the chemistry employed by common ribonucleases such as RNase A (Richards and Wycoff 1971), as well as random and non-random cleavage of RNA by metal ions (see Pan et al., this volume).

Phosphate monoesters at the 3' end of an RNA molecule are also substrates for reactions catalyzed by the group I ribozyme from *Tetrahymena* (Fig. 6c) (Zaug and Cech 1986). The substrate phosphate can be transferred to the 3' end of the ribozyme itself, forming a covalent phosphoenzyme (E-p). It can subsequently be transferred to the 3'O of an acceptor oligonucleotide or can undergo hydrolysis, the latter reaction being enhanced at acid pH. The activity of the ribozyme toward phos-

Table 1 RNA-catalyzed phosphotransfer reactions

Ribozyme	Reaction[a]	Nucleophile	Chemical signature
Large			
group I	transesterification	3'OH of G	3'-5'linkage/3'OH
	hydrolysis	water	5'P/3'OH
group II	transesterification	2'OH of A	2'-5'linkage/3'OH
	hydrolysis	water	5'P/3'OH
RNase P	hydrolysis	water	5'P/3'OH
Small			
hammerhead, hairpin, and δ	transesterification	2'OH (adjacent)	2',3'-cyclic P/5'OH

[a]For group I and group II ribozymes, the biologically relevant reaction is listed first, followed by the hydrolysis reaction observed in vitro.

phate monoesters was unexpected, because cleavages of mono- and diesters are chemically distinct; furthermore, it is unusual for a protein enzyme to react with both of these substrates, although not unprecedented (see Zaug and Cech 1986).

Ribozymes can also catalyze the attack of a 3'-OH group on the α-phosphate of a 5'-triphosphate group, forming a 3',5'-phosphodiester bond with release of pyrophosphate (Fig. 6d). This reaction is of special interest because it represents the way in which all contemporary RNA and DNA polymerases and replicases catalyze polynucleotide chain elongation. A group II ribozyme can carry out this reaction (Mörl et al. 1992), even though it is not part of normal group II RNA splicing. Ribozymes have also been selected in vitro for the ability to seal a nick in double-stranded RNA, where the nick is preceded by a 3'OH and followed by a 5'triphosphate. A random pool of 2 x 10^{15} different RNA sequences (each 210 nt) was subjected to six rounds of selection and amplification; the selected RNA population accelerated the reaction 10^5-fold above the uncatalyzed rate (Szostak and Bartel 1992).

All the reactions described above involve reaction at phosphorus centers (Fig. 6a–d). Recent results show that ribozymes can also catalyze reactions at carbon centers. An oligoribonucleotide that can pair to the IGS of the *Tetrahymena* ribozyme and that terminates in a 3'(2') aminoacyl ester is deacylated at a rate up to 15 times faster than the uncatalyzed rate (Fig. 6e) (Piccirilli et al. 1992b). Like the RNA cleavage reaction, this aminoacyl esterase reaction requires Mg^{++} and sequence elements of the ribozyme catalytic core. Although the observed catalytic rate enhancement is modest, one would not expect an active site that has been selected in nature for phosphoryl transfer to be set up well to catalyze carboxylate ester hydrolysis (Piccirilli et al. 1992b); thus, there is every reason to believe that a ribozyme selected to catalyze reactions

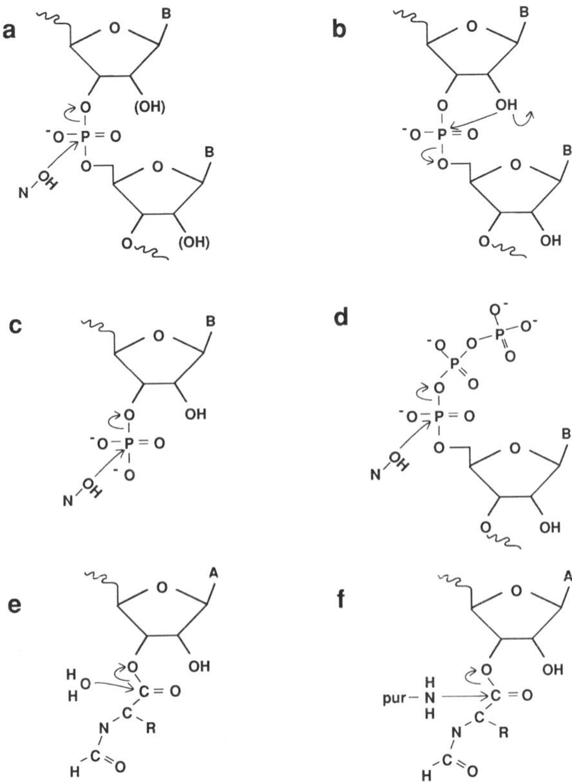

Figure 6 Ribozyme-catalyzed reactions described in the text. (*a*) Cleavage of RNA or DNA generating 3′-OH terminus at cleavage site, reactions catalyzed by group I and group II introns and RNase P. Nucleophiles (N-OH) vary as listed in Table 1. (*b*) Cleavage of RNA generating 2′,3′-cyclic phosphate end, the reaction catalyzed by hammerhead, hairpin, and hepatitis δ ribozymes as well as many common protein ribonucleases. (*c*) Phosphotransferase and phosphatase activities on a 3′-phosphomonoester by a group I ribozyme. (*d*) Chain extension by attack on α-phosphate releasing inorganic pyrophosphate, as catalyzed by a group II intron and RNAs selected by in vitro evolution as well as protein polymerases and replicases. (*e*) Hydrolysis of *N*-formylmethionine (R = side chain of met) from a hexanucleotide comprising the 3′ end of f-met tRNA[f-met] by a group I ribozyme. (*f*) Same substrate as *e*; amino acid is transferred to puromycin (pur-NH₂) by extensively deproteinized *T. aquaticus* ribosomes.

at carbon would be more efficient. In fact, the 23S ribosomal RNA of *Thermus aquaticus* may be an example.

Extensively deproteinized *T. aquaticus* ribosomes catalyze the puromycin reaction (Fig. 6f) with high efficiency; the activity is attributable to peptidyl transferase based on specific antibiotic inhibition (Noller et al. 1992). Thus 23S rRNA may be directly involved in the catalysis of peptide bond formation (see Noller; Moore; both this volume).

How Good Is RNA Catalysis?

One test of the quality of a catalyst is to compare the rate constant for the catalyzed reaction to that for an uncatalyzed reaction that gives the same reaction products. The isolation of the chemical step has allowed this comparison to be made for the *Tetrahymena* ribozyme acting as a sequence-specific endoribonuclease. The calculated rate enhancement of ~10^{11} (Herschlag and Cech 1990a) is comparable to rate advantages achieved by protein enzymes. Another criterion is the ability of a catalyst to perform its chemical step fast enough that the overall reaction is limited by diffusion of the substrate into its active site. For an RNA enzyme that base-pairs with its nucleic acid substrate, the equivalent of a "diffusion controlled rate" is the rate of formation of a base-paired duplex. The *Tetrahymena* ribozyme cleaves RNA with $k_{cat}/K_m = 10^8$ $\text{M}^{-1}\text{min}^{-1}$, within the range measured for duplex formation between oligonucleotides (Herschlag and Cech 1990a; Bevilacqua et al. 1992). Thus, it can perform the chemical step fast enough that the diffusive step is rate-limiting for cleavage.

In contrast to its high efficiency in catalyzing the first cleavage event, the *Tetrahymena* ribozyme has a limitation as a multiple-turnover enzyme. The ribozyme releases one of the cleavage products so slowly that it is easily saturated. Of course, the *Tetrahymena* intron was selected for the single-turnover self-splicing reaction, where tight binding of the cleaved 5′ exon is not a deficit but an advantage; it promotes exon ligation rather than nonproductive dissociation of the exon after 5′ splice site cleavage (Fig. 2a).

A number of site-specific deletions in the ribozyme greatly increase its turnover number by weakening the binding of the RNA substrate, thereby facilitating the rate-limiting product-release step (Young et al. 1991). The RNA subunit of RNase P also shows increased turnover number upon deletion mutagenesis, which has also been explained by acceleration of product release (Waugh et al. 1989).

Specificity: Choosing the Correct Reaction Site

A good biological catalyst is not only fast, but also highly specific. The *Tetrahymena* ribozyme chooses the site of guanosine addition (5′ splice site) at three levels: The substrate must have a sequence complementary to that of the IGS, it must be in a single-stranded region so that it is available for pairing to the IGS, and it must have a U residue at position –1 to pair with G22 of the IGS to give efficient cleavage (Waring et al. 1986; Zaug et al. 1986; Barfod and Cech 1989; Doudna et al. 1989; Woodson and Cech 1991). These considerations hold for both the intramolecular self-splicing reaction and the intermolecular ("*trans*") reaction. (Presumably, the choice of the 5' splice site in self-splicing is also facilitated by the proximity of the correct cleavage site to the catalytic core.) The sequence specificity for the ribozyme-catalyzed cleavage of target RNA molecules can be re-engineered by changing the sequence of the IGS (Zaug et al. 1986; Murphy and Cech 1989). The specificity for cleavage of a matched versus a slightly mismatched substrate RNA sequence also depends on kinetic factors, and these are subject to manipulation by site-specific mutagenesis (Herschlag and Cech 1990b; Herschlag 1991; Young et al. 1991).

Group II introns also catalyze a *trans* reaction in which a free 5' exon is specifically bound by the intron and proceeds to undergo exon ligation (Jacquier and Rosbash 1986). The interaction is bipartite: Two exon binding sites (EBS) located in two loops of domain I base-pair with two sequences (IBS) near the 3' end of the 5' exon (Fig. 1b) (Jacquier and Michel 1987). The sequence just downstream from the 5' splice site (GU<u>GY</u>G) is another obvious candidate for splice site recognition because it is conserved among different group II introns. (It also bears an intriguing similarity to the GURAGU sequence at the 5' splice site of nuclear mRNA introns, which is recognized by base-pairing to U1 snRNA; see Baserga and Steitz, this volume.) The underlined GY of the GU<u>GY</u>G sequence of the group II intron is recognized by base-pairing to yet another loop of domain I (Jacquier and Michel 1990). Components of the catalytic apparatus that recognize the conserved GU immediately following the 5' splice site have not yet been identified. Nevertheless, it is already apparent that 5' splice site recognition involves the collaboration of a number of interactions (Fig. 1b). The 3' splice site is always preceded by the dinucleotide AY, the pyrimidine (Y) always being complementary to a purine (R) in the "central wheel" of the group II core (Fig. 1b). Base-pairing between these nucleotides is not required for step 1 of splicing but is required for the efficiency of step 2 (Jacquier 1990), thereby identifying one element of 3' splice site recognition in group II RNA splicing.

A single RNase P enzyme can process the cell's entire set of tRNA precursors. These have only a small degree of nucleotide sequence conservation in the mature tRNA domain (such as the T-loop sequence

GTψCR) and no apparent conservation in the 5′ leader sequence which is cut off. This strongly suggests that the three-dimensional structure of the tRNA is a major recognition element, a proposal that has been supported by mutational studies: Base substitutions that disrupt tRNA conformation reduce RNase P cleavage (McClain and Seidman 1975). However, an RNA need not possess the full L-shaped tertiary structure of tRNA to be a substrate for RNase P. The *E. coli* RNase P RNA cleaves an artificial small substrate containing the amino acid acceptor stem (with 3′ terminal CCA) and the T stem and loop; cleavage is accurate, and the K_m and k_{cat} values are very similar to those obtained with the standard pre-tRNA^{Phe} substrate (McClain et al. 1987). In fact, the catalytic RNA subunit will cleave a single-stranded 5′ leader sequence adjacent to any double-stranded RNA segment, as long as there is an unpaired CCA sequence on the 3′ side (Forster and Altman 1990b). Consistent with these observations, chemical modifications of the substrate that inhibit cleavage by the catalytic RNA lie mostly along the amino acid acceptor and T stems (Kahle et al. 1990).

Insights from DNA Cleavage Reactions

It has recently been found that single-stranded DNA is also cleaved by all the large ribozymes. Perhaps this should have been anticipated earlier, because none of these ribozymes directly uses the substrate 2′-OH group in the chemical reaction. Furthermore, some protein enzymes that produce 3′-OH/5′-phosphate termini can cleave both RNA and DNA (e.g., S1 nuclease, P1 nuclease).

The *Tetrahymena* group I ribozyme cleaves DNA much more slowly than RNA (Robertson and Joyce 1990; Herschlag and Cech 1990c). The DNA substrate is bound 10^4-fold more weakly than RNA of the same sequence and, once bound, is cleaved 10^4-fold more slowly (Sugimoto et al. 1989; Herschlag and Cech 1990c; Pyle et al. 1990). The molecular basis for each of these deficiencies has been dissected in considerable detail and is most easily discussed in terms of the positive contributions made by the RNA 2′OHs that are missing in the DNA substrate (Fig. 7). The base-paired duplex formed between the substrate and ribozyme is more stable if it is an RNA·RNA duplex than if it is DNA·RNA; each 2′OH contributes to this difference, although not equally. The contribution of 2′OHs at positions −6, −5, and −4 appears to be explained by helical stability alone. The contributions of 2′-OH groups at positions −3 and −2 are also fully expressed at the binding step and retained in the chemical step, but their larger magnitude (1–2 kcal/mole each) suggests specific bonding to some site on the ribozyme (Bevilacqua and Turner

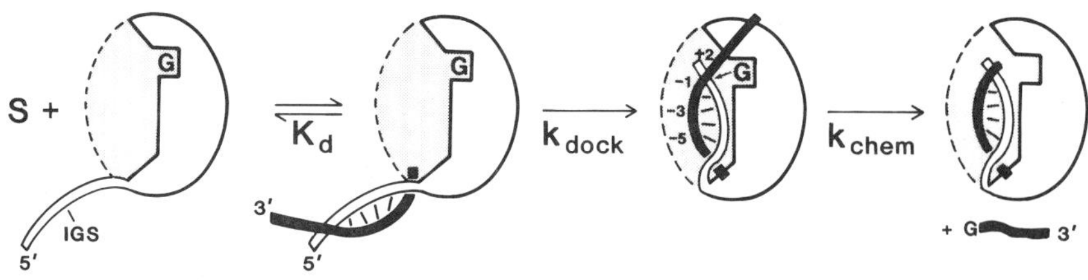

IMPORTANT 2'-OHs: all   -2,-3   -1 (-2,-3)

Figure 7 Dynamics of RNA cleavage by the L-21 *Sca*I form of the *Tetrahymena* ribozyme. The separate base-pairing, docking, and chemical steps required for cleavage of an RNA substrate provide a minimal model; there may very well be additional steps. Different 2'-OH groups along the substrate RNA contribute to one or more of these steps (see text). The parenthetical (–2, –3) indicates that these 2'-OH groups contribute to stabilization of the transition state and of the ground state E•S complex by the same amount. 2'-OH groups are numbered with the 2'OH adjacent to the site of G-addition = –1 (see D. Herschlag et al., in prep.).

1991; Pyle and Cech 1991; D. Herschlag et al., in prep.). The –3 position contact has in fact been localized to conserved nucleotide A302, and a 2'-OH–N_1 (A302) hydrogen bond has been proposed (Pyle et al. 1992). Finally, the 2'OH adjacent to the cleavage site at position –1 makes only a small contribution to binding but speeds the chemical step by a factor of 1700 (D. Herschlag et al., in prep.). The rate of the chemical step varies with the pK_a of the leaving group for a series of substrates with different 2'substituents at position –1, suggesting that the ribozyme does not use general acid catalysis to protonate the 2'O in the transition state and that RNA is cleaved faster than DNA simply because it has a more stable leaving group (D. Herschlag et al., in prep.).

The cleavage of a nucleic acid substrate (S) by the L-21 *Sca*I form of the *Tetrahymena* ribozyme has been divided into several steps: (1) binding of S to the IGS of the ribozyme by base-pairing, (2) docking of the resulting "P1" duplex to the active site, and (3) chemical cleavage by guanosine (Bevilacqua et al. 1992; Herschlag 1992; J.F. Wang et al., in prep.). Different 2'-hydroxyl groups facilitate each of the binding steps and the chemical step (Fig. 7). Thus, the natural ribozyme is highly tuned to react with RNA rather than DNA.

Random mutagenesis and in vitro evolution have been used to improve the DNA cleavage activity of this ribozyme. A small number of single-base changes in the ribozyme are sufficient to confer a 100-fold increase in k_{cat}/K_m for DNA cleavage (Beaudry and Joyce 1992). This kinetic parameter reflects both the binding and chemical steps (Herschlag and Cech 1990c; D. Herschlag et al., in prep.), so it will be interesting to see which of these steps is improved by the base changes.

RNase P also cleaves nucleic acid substrates that have a 2′ H replacing the 2′OH adjacent to the cleavage site. The ribonucleoprotein "holoenzyme" from yeast cleaves an artificial substrate consisting of an oligodeoxyribonucleotide joined to the 5′ end of a tRNA (J.-Y. Lee and D.R. Engelke, pers. comm.). The *E. coli* holoenzyme cleaves an RNA substrate with a single 2′ H at the cleavage site, although at a substantially reduced rate (Forster and Altman 1990b). The group II intron bI1 (yeast cytochrome *b* pre-mRNA intron 1) cleaves a single-stranded DNA version of the ligated exons with concomitant attachment of the 3′ exon to the intron (Mörl et al. 1992). The reaction is analogous to the first step of the reverse of RNA splicing. Once again, the efficiency of the DNA reaction is markedly less than that of the equivalent RNA reaction.

Transition State Stabilization by the Group I Ribozyme

The primary motivation for tertiary structure analysis of ribozymes is to determine how an active site composed of RNA can stabilize the transition state for RNA cleavage or RNA splicing. In the case of the group I intron catalytic center, results of the past few years have led to a partial understanding of transition state stabilization. Because this body of work has been reviewed recently (Cech et al. 1992), only the briefest summary is given here. In the model in Figure 8: (1) The guanosine nucleophile and the 3′-O leaving group have an in-line orientation, based on the determination of the stereochemical course of the reaction (inversion of configuration at phosphorus; McSwiggen and Cech 1989; Rajagopal et al. 1989). (2) An active-site metal ion is coordinated to the 3′-oxygen atom, thereby stabilizing the developing negative charge on the leaving group; this was indicated by a metal ion specificity switch with a 3′-thio-substrate (Piccirilli et al. 1992a; see Pan et al., this volume). The possibility of a second metal ion activating the 3′OH of G for nucleophilic attack (not shown in Fig. 8) has not been directly tested. If present, the two-metal-ion center would be remarkably similar to that found in a number of polymerases and in alkaline phosphatase. In fact, it has been proposed that a two-metal-ion mechanism would be equally well suited to phosphotransfer catalyzed by ribozymes or by protein enzymes (Beese and Steitz 1991). (3) The ribozyme does not provide general acid catalysis, as the leaving group oxyanion does not appear to be protonated in the transition state (D. Herschlag et al., in prep.). (4) Substitution of the pro-S_P oxygen with sulfur dramatically reduces the rate of the cleavage step, whereas thio-substitution of the pro-R_P oxygen (shaded in Fig. 8) is

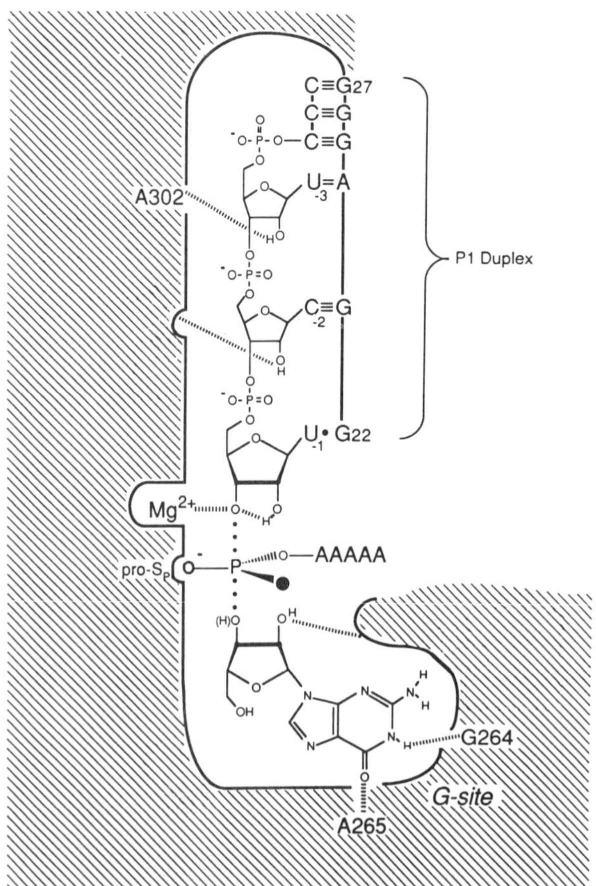

Figure 8 Transition-state stabilization by the group I catalytic center. Shaded area represents the three-dimensional surface of the ribozyme. Dashed lines indicate hydrogen bonds or metal ion-oxygen coordination. Dotted P-O bonds indicate bonds partially formed or partially broken in the transition state. (Reprinted, with permission, from Cech et al. 1992.)

tolerated (Herschlag et al. 1991). This argues against a "fluid" or "sloppy" model for the RNA active site and shows that an active site composed of RNA can achieve high stereoselectivity.

By 1982, the secondary structure common to the group I RNAs was largely determined (Davies et al. 1982; Michel et al. 1982). The next 6 years brought confirmation of that structure, along with a bit of fine-tuning (Cech 1988; Michel and Westhof 1990). In looking at the secondary structure in 1988, one saw the conserved nucleotides occupying fixed positions within a framework of helices of fixed length (Fig. 1a) and imagined that some day one might be able to assign specific catalytic functions to each of those nucleotides and structural elements. The following 3 years produced a flurry of structure-function relationships (Fig. 5b), along with a viable three-dimensional structure model (Fig. 5a). We can now provide answers (although certainly incomplete answers) to many questions that were complete mysteries in 1988, including the following: Why does the P7 helix have a conserved base sequence? Because several of these bases interact with the guanosine substrate. Why are there conserved As in the internal loop between P4 and P5? Because they interact with the P1 helix which contains the 5' splice site and help position it relative to the G-site in P7. Why is P4 always six base pairs? Because it has one end involved in the triple-helical scaffold that helps organize the structure and the other end supporting the conserved As in the P4-P5 loop, and these must be held at a precise distance and orientation in order to juxtapose the G-site with the 5' splice site. Although such answers are extremely satisfying at one level, they can also be considered to be merely the starting point for future studies aimed at understanding how a chemical reaction is so greatly facilitated by this active site composed of RNA.

ACKNOWLEDGMENTS

I thank Dan Herschlag for many helpful discussions; Luc Jaeger, François Michel, and Norm Pace for providing illustrations; Brenda Bass and Tom Steitz for critique of the manuscript; and Alice Sirimarco and Diane Lorenz for preparation of the manuscript and illustrations. I am an Investigator of the Howard Hughes Medical Institute and an American Cancer Society Professor.

REFERENCES

Akins, R.A. and A.M. Lambowitz. 1987. A protein required for splicing group I introns in *Neurospora* mitochondria is mitochondrial tyrosyl-tRNA synthetase or a derivative thereof. *Cell* **50:** 331–345.

Altman, S. 1989. Ribonuclease P: An enzyme with a catalytic RNA subunit. *Adv. Enzymol. Relat. Areas Mol. Biol.* **62:** 1–36.

Barfod, E.T. and T.R. Cech. 1989. The conserved U•G pair in the 5' splice site duplex of a group I intron is required in the first but not the second step of self-splicing. *Mol. Cell. Biol.* **9:** 3657–3666.

Beaudry, A.A. and G.F. Joyce. 1992. Directed evolution of an RNA enzyme. *Science* **257:** 635–641.

Been, M.D. and T.R. Cech. 1986. One binding site determines sequence specificity of *Tetrahymena* pre-rRNA self-splicing, trans-splicing and RNA enzyme activity. *Cell* **47:** 207–216.

Beese, L.S. and T.A. Steitz. 1991. Structural basis for the 3'-5' exonuclease activity of

Escherichia coli DNA polymerase I: A two metal ion mechanism. *EMBO J.* **10:** 25–33.

Bevilacqua, P.C. and D.H. Turner. 1991. Comparison of binding of mixed ribose-deoxyribose analogues of CUCU to a ribozyme and to GGAGAA by equilibrium dialysis: Evidence for ribozyme specific interactions with 2'OH groups. *Biochemistry* **30:** 10632–10640.

Bevilacqua, P.C., R. Kierzek, K.A. Johnson, and D.H. Turner. 1992. Dynamics of ribozyme binding of substrate revealed by fluorescence detected stopped-flow. *Science* **258:** 1355–1358.

Brown, J.W. and N.R. Pace. 1992. Ribonuclease P RNA and protein subunits from bacteria. *Nucleic Acids Res.* **20:** 1451–1456.

Brown, J.W., E.S. Haas, B.D. James, D.A. Hunt, and N.R. Pace. 1991. Phylogenetic analysis and evolution of RNase P RNA in proteobacteria. *J. Bacteriol.* **173:** 3855–3863.

Burgin, A.B. and N.R. Pace. 1990. Mapping the active site of ribonuclease P RNA using a substrate containing a photoaffinity agent. *EMBO J.* **9:** 4111–4118.

Cech, T.R. 1988. Conserved sequences and structures of group I introns: Building an active site for RNA catalysis. *Gene* **73:** 259–271.

————. 1990. Self-splicing of group I introns. *Annu. Rev. Biochem.* **59:** 543–568.

————. 1992. Ribozyme engineering. *Curr. Opin. Struct. Biol.* **2:** 605–609.

Cech, T.R., A.J. Zaug, and P.J. Grabowski. 1981. *In vitro* splicing of the ribosomal RNA precursor of *Tetrahymena*: Involvement of a guanosine nucleotide in the excision of the intervening sequence. *Cell* **27:** 487–496.

Cech, T.R., D. Herschlag, J.A. Piccirilli, and A.M. Pyle. 1992. RNA catalysis by a group I ribozyme: Developing a model for transition state stabilization. *J. Biol. Chem.* **267:** 17479–17482.

Celander, D.W. and T.R. Cech. 1990. Iron(II)-ethylenediaminetetraacetic acid-catalyzed cleavage of RNA and DNA oligonucleotides: Similar reactivity toward single- and double-stranded forms. *Biochemistry* **29:** 1355–1361.

————. 1991. Visualizing the higher order folding of a catalytic RNA molecule. *Science* **251:** 401–407.

Chastain, M. and I. Tinoco, Jr. 1991. Structural elements in RNA. *Prog. Nucleic Acid Res. Mol. Biol.* **41:** 131-177.

Cheong, C., G. Varani, and I. Tinoco, Jr. 1990. Solution structure of an unusually stable RNA hairpin. *Nature* **346:** 680–682.

Chu, F., G.F. Maley, D.K. West, M. Belfort, and F. Maley. 1986. Characterization of the intron in the phage T4 thymidylate synthase gene and evidence for its self-excision from the primary transcript. *Cell* **45:** 157–166.

Copertino, D.W. and R.B. Hallick. 1991. Group II twintron: An intron within an intron in a chloroplast cytochrome b-559 gene. *EMBO J.* **10:** 433–442.

Davies, R.W., R.B. Waring, J.A. Ray, T.A. Brown, and C. Scazzocchio. 1982. Making ends meet: A model for RNA splicing in fungal mitochondria. *Nature* **300:** 719–724.

Doudna, J.A. and J.W. Szostak. 1989. RNA-catalysed synthesis of complementary-strand RNA. *Nature* **339:** 519–522.

Doudna, J.A., B.P. Cormack, and J.W. Szostak. 1989. RNA structure, not sequence, determines the 5'splice-site specificity of a group I intron. *Proc. Natl. Acad. Sci.* **86:** 7402–7406.

Downs, W.D. and T.R. Cech. 1990. An ultraviolet-inducible adenosine-adenosine crosslink reflects the catalytic structure of the *Tetrahymena* ribozyme. *Biochemistry* **29:** 5605–5613.

Dujon, B. 1989. Group I introns as mobile genetic elements: Facts and speculations. *Gene* **82:** 91–114.

Flor, P., J.B. Flanegan, and T.R. Cech. 1989. A conserved base-pair within helix P4 of the *Tetrahymena* ribozyme helps to form the tertiary structure required for self-splicing. *EMBO J.* **8:** 3391–3399.

Forster, A.C. and S. Altman. 1990a. Similar cage-shaped structures for the RNA components of all ribonuclease P and ribonuclease MRP enzymes. *Cell* **62:** 407–409.

———. 1990b. External guide sequences for an RNA enzyme. *Science* **249:** 783–786.

Gampel, A. and T.R. Cech. 1991. Binding of the CBP2 protein to a yeast mitochondrial group I intron requires the catalytic core of the RNA. *Genes Dev.* **5:** 1870–1880.

Garriga, G. and A.M. Lambowitz. 1984. RNA splicing in *Neurospora* mitochondria: Self-splicing of a mitochondrial intron *in vitro*. *Cell* **39:** 631–641.

Gott, J.M., D.A. Shub, and M. Belfort. 1986. Multiple self-splicing introns in bacteriophage T4: Evidence from autocatalytic GTP labeling of RNA in vitro. *Cell* **47:** 81–87.

Grosshans, C.A. and T.R. Cech. 1989. Metal ion requirements for sequence-specific endoribonuclease activity of the *Tetrahymena* ribozyme. *Biochemistry* **28:** 6888–6894.

Guerrier-Takada, C. and S. Altman. 1984. Catalytic activity of an RNA molecule prepared by transcription *in vitro*. *Science* **223:** 285–286.

Guerrier-Takada, C., N. Lumelsky, and S. Altman. 1989. Specific interactions in RNA enzyme-substrate complexes. *Science* **246:** 1578–1584.

Guerrier-Takada, C., K. Haydock, L. Allen, and S. Altman. 1986. Metal ion requirements and other aspects of the reaction catalyzed by M1 RNA, the RNA subunit of ribonuclease P from *Escherichia coli*. *Biochemistry* **25:** 1509–1515.

Guerrier-Takada, C., K. Gardiner, T. Marsh, N. Pace, and S. Altman. 1983. The RNA moiety of ribonuclease P is the catalytic subunit of the enzyme. *Cell* **35:** 849–857.

Guo, Q. and A.M. Lambowitz. 1992. A tyrosyl-tRNA synthetase binds specifically to the group I intron catalytic core. *Genes Dev.* **6:** 1357–1372.

Herbert, C.J., M. Labouesse, G. Dujardin, and P.P. Slonimski. 1988. The NAM2 proteins from *S. cerevisiae* and *S. douglasii* are mitochondrial leucyl-tRNA synthetases, and are involved in mRNA splicing. *EMBO J.* **7:** 473–83.

Herschlag, D. 1991. Implications of ribozyme kinetics for targeting the cleavage of specific RNA molecules *in vivo*. More isn't always better. *Proc. Natl. Acad. Sci.* **88:** 6921–6925.

———. 1992. Evidence for processivity and two-step binding of the RNA substrate from studies of J1/2 mutants of the *Tetrahymena* ribozyme. *Biochemistry* **31:** 1386–1399.

Herschlag, D. and T.R. Cech. 1990a. Catalysis of RNA cleavage by the *Tetrahymena thermophila* ribozyme. 1. Kinetic description of the reaction of an RNA substrate complementary to the active site. *Biochemistry* **29:** 10159–10171.

———. 1990b. Catalysis of RNA cleavage by the *Tetrahymena thermophila* ribozyme. 2. Kinetic description of the reaction of an RNA substrate that forms a mismatch at the active site. *Biochemistry* **29:** 10172–10180.

———. 1990c. DNA cleavage catalysed by the ribozyme from *Tetrahymena*. *Nature* **344:** 405–409.

Herschlag, D., J.A. Piccirilli, and T.R. Cech. 1991. Ribozyme-catalyzed and non-enzymatic reactions of phosphate diesters: Rate effects upon substitution of sulfur for a non-bridging phosphoryl oxygen atom. *Biochemistry* **30:** 4844–4854.

Heuer, T.S., P.S. Chandry, M. Belfort, D.W. Celander, and T.R. Cech. 1991. Folding of group I introns from bacteriophage T4 involves internalization of the catalytic core. *Proc. Natl. Acad. Sci.* **88:** 11105–11109.

Heus, H.A. and A. Pardi. 1991. Structural features that give rise to the unusual stability of RNA hairpins containing GNRA loops. *Science* **253:** 191–194.

Holbrook, S.R., J.L. Sussman, R.W. Warrant, G.M. Church, and S.-H. Kim. 1977. RNA-ligand interactions: (1) Magnesium binding sites in yeast tRNA^{Phe}. *Nucleic Acids Res.* **4:** 2811–2820.

Ikemura, T., Y. Shimura, H. Sakano, and H. Ozeki. 1975. Precursor molecules of *Escherichia coli* transfer RNAs accumulated in a temperature-sensitive mutant. *J. Mol. Biol.* **96:** 69–86.

Jack, A., J.E. Ladner, D. Rhodes, R.S. Brown, and A. Klug. 1977. A crystallographic study of metal-binding to yeast phenylalanine transfer RNA. *J. Mol. Biol.* **111:** 315–328.

Jacquier, A. 1990. Self-splicing group II and nuclear pre-mRNA introns: How similar are they? *Trends Biochem. Sci.* **15:** 351–354.

Jacquier, A. and F. Michel. 1987. Multiple exon-binding sites in class II self-splicing introns. *Cell* **50:** 17–29.

———. 1990. Base-pairing interactions involving the 5′ and 3′-terminal nucleotides of group II self-splicing introns. *J. Mol. Biol.* **213:** 437–447.

Jacquier, A. and M. Rosbash. 1986. Efficient trans-splicing of a yeast mitochondrial RNA group II intron implicates a strong 5′ exon-intron interaction. *Science* **234:** 1099–1104.

James, B.D., F.J. Olsen, J. Liu, and N.R. Pace. 1988. The secondary structure of ribonuclease P RNA, the catalytic element of a ribonucleoprotein enzyme. *Cell* **52:** 19–26.

Jarrell, K.A., R.C. Dietrich, and P.S. Perlman. 1988. Group II intron domain 5 facilitates a *trans*-splicing reaction. *Mol. Cell. Biol.* **8:** 2361–2366.

Kahle, D., U. Wehmeyer, and G. Krupp. 1990. Substrate recognition by RNase P and by the catalytic M1 RNA: Identification of possible contact points in pre-tRNAs. *EMBO J.* **9:** 1929–1937.

Kim, S.-H. and T.R. Cech. 1987. Three-dimensional model of the active site of the *Tetrahymena* ribozyme. *Proc. Natl. Acad. Sci.* **84:** 8788–8792.

Kim, S.-H., G.J. Quigley, F.L. Suddath, A. McPherson, D. Sneden, J.J. Kim, J. Weinzierl, and A. Rich. 1973. Three-dimensional structure of yeast phenylalanine transfer RNA: Folding of the polynucleotide chain. *Science* **179:** 285–288.

Kittle, J.D., Jr., G. Mohr, J.A. Gianelos, H. Wang, and A.M. Lambowitz. 1991. The *Neurospora* mitochondrial tyrosyl-tRNA synthetase is sufficient for group I intron splicing *in vitro* and uses the carboxy-terminal tRNA-binding domain along with other regions. *Genes Dev.* **5:** 1009–1021.

Kruger, K., P.J. Grabowski, A.J. Zaug, J. Sands, D.E. Gottschling, and T.R. Cech. 1982. Self-splicing RNA: Autoexcision and autocyclization of the ribosomal RNA intervening sequence of *Tetrahymena*. *Cell* **31:** 147–157.

Kuhsel, M.G., R. Strickland, and J.D. Palmer. 1990. An ancient group I intron shared by eubacteria and chloroplasts. *Science* **250:** 1570–1573.

Lambowitz, A.M. and P.S. Perlman. 1990. Involvement of aminoacyl-tRNA synthetases and other proteins in group I and group II intron splicing. *Trends Biochem. Sci.* **15:** 440–444.

Latham, J.A. and T.R. Cech. 1989. Defining the inside and outside of a catalytic RNA molecule. *Science* **245:** 276–282.

Lazowska, J., M. Claisse, A. Gargouri, Z. Kotylak, A. Spyridakis, and P.P. Slonimski. 1989. Protein encoded by the third intron of cytochrome *b* gene in *Saccharomyces cerevisiae* is an mRNA maturase. Analysis of mitochondrial mutants, RNA transcripts, proteins and evolutionary relationships. *J. Mol. Biol.* **205:** 275–290.

McClain, W.H. and J.G. Seidman. 1975. Genetic perturbations that reveal tertiary conformation of tRNA precursor molecules. *Nature* **257**: 106–110.

McClain, W.H., C. Guerrier-Takada, and S. Altman. 1987. Model substrates for an RNA enzyme. *Science* **238**: 527–530.

McSwiggen, J.A. and T.R. Cech. 1989. Stereochemistry of RNA cleavage by the *Tetrahymena* ribozyme and evidence that the chemical step is not rate-limiting. *Science* **244**: 679–683.

Michel, F. and B.F. Lang. 1985. Mitochondrial class II introns encode proteins related to the reverse transcriptases of retroviruses. *Nature* **316**: 641–643.

Michel, F. and E. Westhof. 1990. Modelling of the three-dimensional architecture of group I catalytic introns based on comparative sequence analysis. *J. Mol. Biol.* **216**: 585–610.

Michel, F., A. Jacquier, and B. Dujon. 1982. Comparison of fungal mitochondrial introns reveals extensive homologies in RNA secondary structure. *Biochimie* **64**: 867–881.

Michel, F., K. Umesono, and H. Ozeki. 1989a. Comparative and functional anatomy of group II catalytic introns. *Gene* **82**: 5–30.

Michel, F., A.D. Ellington, S. Couture, and J.W. Szostak. 1990. Phylogenetic and genetic evidence for base-triples in the catalytic domain of group I introns. *Nature* **347**: 578–580.

Michel, F., M. Hanna, R. Green, D.P. Bartel, and J.W. Szostak. 1989b. The guanosine binding site of the *Tetrahymena* ribozyme. *Nature* **342**: 391–395.

Michel, F., L. Jaeger, E. Westhof, R. Kuras, F. Tihy, M.Q. Xu, and D.A. Shub. 1992. Activation of the catalytic core of a group I intron by a remote 3′ splice junction. *Genes Dev.* **6**: 1373–1385.

Mohr, G. and A.M. Lambowitz. 1991. Integration of a group I intron into a ribosomal RNA sequence promoted by a tyrosyl-tRNA synthetase. *Nature* **354**: 164–167.

Mörl, M., I. Niemer, and C. Schmelzer. 1992. New reactions catalyzed by a group II intron ribozyme with RNA and DNA substrates. *Cell* **70**: 803–810.

Murphy, F.L. and T.R. Cech. 1989. Alteration of substrate specificity for the endoribonucleolytic cleavage of RNA by the *Tetrahymena* ribozyme. *Proc. Natl. Acad. Sci.* **86**: 9218–9222.

Noller, H.F., V. Hoffarth, and L. Zimniak. 1992. Unusual resistance of peptidyl transferase to protein extraction procedures. *Science* **256**: 1416–1419.

Pace, N.R. and D. Smith. 1990. Ribonuclease P: Function and variation. *J. Biol. Chem.* **265**: 3587–3590.

Padgett, R.A., M.M. Konarska, P.J. Grabowski, S.F. Hardy, and P.A. Sharp. 1984. Lariat RNAs as intermediates and products in the splicing of messenger RNA precursors. *Science* **225**: 898–903.

Peebles, C.L., P.S. Perlman, K.L. Mecklenburg, M.L. Petrillo, J.H. Tabor, K.A. Jarrell, and H.-L. Cheng. 1986. A self-splicing RNA excises an intron lariat. *Cell* **44**: 213–223.

Piccirilli, J.A., J.S. Vyle, M.H. Caruthers, and T.R. Cech. 1992a. Metal ion catalysis by the *Tetrahymena* ribozyme. *Nature* (in press).

Piccirilli, J.A., T.S. McConnell, A.J. Zaug, H.F. Noller, and T.R. Cech. 1992b. Aminoacyl esterase activity of the *Tetrahymena* ribozyme. *Science* **256**: 1420–1424.

Pyle, A. M. and T.R. Cech. 1991. Ribozyme recognition of RNA by tertiary interactions with specific ribose 2′-OH groups. *Nature* **350**: 628–631.

Pyle, A.M., J.A. McSwiggen, and T.R. Cech. 1990. Direct measurement of oligonucleotide substrate binding to wild type and mutant ribozymes from *Tetrahymena. Proc. Natl. Acad. Sci.* **87**: 8187–8191.

Pyle, A.M., F.L. Murphy, and T.R. Cech. 1992. RNA substrate binding site in the catalytic core of the *Tetrahymena* ribozyme. *Nature* **358**: 123–128.

Rajagopal, J., J.A. Doudna, and J.W. Szostak. 1989. Stereochemical course of catalysis by the *Tetrahymena* ribozyme. *Science* **244:** 692–694.

Reich, C., G.J. Olsen, B. Pace, and N.R. Pace. 1988. Role of the protein moiety of ribonuclease P, a ribonucleoprotein enzyme. *Science* **239:** 178–181.

Reinhold-Hurek, B. and D.A. Shub. 1992. Self-splicing introns in tRNA genes of widely divergent bacteria. *Nature* **357:** 173–176.

Richards, F.M. and H.W. Wycoff. 1971. Bovine pancreatic ribonuclease. In *The enzymes* (ed. P.D. Boyer), vol. 4, pp. 647–806. Academic Press, New York.

Robertson, D.L. and G.F. Joyce. 1990. Selection *in vitro* of an RNA enzyme that specifically cleaves single-stranded DNA. *Nature* **344:** 467–468.

Robertus, J.D., J.E. Ladner, J.T. Finch, D. Rhodes, R.S. Brown, B.F. Clark, and A. Klug. 1974. Structure of yeast phenylalanine tRNA at 3 Å resolution. *Nature* **250:** 546–551.

Ruskin, B., A.R. Krainer, T. Maniatis, and M.R. Green. 1984. Excision of an intact intron as a novel lariat structure during pre-mRNA splicing in vitro. *Cell* **38:** 317–331.

Schedl, P. and P. Primakoff. 1973. Mutants of *Escherichia coli* thermosensitive for the synthesis of transfer RNA. *Proc. Natl. Acad. Sci.* **70:** 2091–2095.

Schmelzer, C. and R.J. Schweyen. 1986. Self-splicing of group II introns *in vitro*: Mapping of the branch point and mutational inhibition of lariat formation. *Cell* **46:** 557–565.

Seeman, N.C., J.M. Rosenberg, and A. Rich. 1976. Sequence-specific recognition of double helical nucleic acids by proteins. *Proc. Natl. Acad. Sci.* **73:** 804–808.

Sigler, P.B. 1975. An analysis of the structure of tRNA. *Annu. Rev. Biophys. Bioeng.* **4:** 447–527.

Sogin, M.L., A. Ingold, M. Karlok, H. Nielsen, and J. Engberg. 1986. Phylogenetic evidence for the acquisition of ribosomal RNA introns subsequent to the divergence of some of the major *Tetrahymena* groups. *EMBO J.* **5:** 3625–3630.

Sugimoto, N., M. Tomka, R. Kierzek, P.C. Bevilacqua, and D.H. Turner. 1989. Effects of substrate structure on the kinetics of circle opening reactions of the self-splicing intervening sequence from *Tetrahymena thermophila*: Evidence for substrate and Mg^{2+} binding interactions. *Nucleic Acids Res.* **17:** 355–371.

Szostak, J.W. and D.P. Bartel. 1992. Selection of new ribozymes from a large pool of random sequences. In *Abstracts from the 18th EMBO Symposium*, p. 68. European Molecular Biology Laboratory, Heidelberg.

Van der Horst, G. and H. Tabak. 1985. Self-splicing of yeast mitochondrial ribosomal and messenger RNA precursors. *Cell* **40:** 759–766.

Van der Horst, G., A. Christian, and T. Inoue. 1991. Reconstitution of a group I intron self-splicing reaction with an activator RNA. *Proc. Natl. Acad. Sci.* **88:** 184–188.

Van der Veen, R., A.C. Arnberg, G. Van der Horst, L. Bonen, H.F. Tabak, and L.A. Grivell. 1986. Excised group II introns in yeast mitochondria are lariats and can be formed by self-splicing in vitro. *Cell* **44:** 225–234.

Wang, J.-F. and T.R. Cech. 1992. Tertiary structure around the guanosine-binding site of the *Tetrahymena* ribozyme. *Science* **256:** 526–529.

Waring, R.B., P. Towner, S.J. Minter, and Davies, R.W. 1986. Splice-site selection by a self-splicing RNA of *Tetrahymena*. *Nature* **321:** 133–139.

Waugh, D.S., C.J. Green, and N.R. Pace. 1989. The design and catalytic properties of a simplified ribonuclease P RNA. *Science* **244:** 1569–1571.

Woodson, S.A. and T.R. Cech. 1989. Reverse self-splicing of the *Tetrahymena* group I intron: Implication for the directionality of splicing and for intron transposition. *Cell* **57:** 335–345.

———. 1991. Alternative secondary structures in the 5′ exon affect both forward and reverse self-splicing of the *Tetrahymena* intervening sequence RNA. *Biochemistry* **30:** 2042–2050.

Xu, M.Q., S.D. Kathe, H. Goodrich-Blair, S.A. Nierzwicki, and D.A. Shub. 1990. Bacterial origin of a chloroplast intron: Conserved self-splicing group I introns in cyanobacteria. *Science* **250:** 1566–1570.

Yarus, M., M. Illangesekare, and E. Christian. 1991. An axial binding site in the *Tetrahymena* precursor RNA. *J. Mol. Biol.* **222:** 995–1012.

Young, B., D. Herschlag, and T.R. Cech. 1991. Mutations in a nonconserved sequence of the *Tetrahymena* ribozyme increase activity and specificity. *Cell* **67:** 1007-1019.

Zaug, A.J. and T.R. Cech. 1986. The *Tetrahymena* intervening sequence ribonucleic acid enzyme is a phosphotransferase and an acid phosphatase. *Biochemistry* **25:** 4478–4482.

Zaug, A.J., M.D. Been, and T.R. Cech. 1986. The *Tetrahymena* ribozyme acts like an RNA restriction endonuclease. *Nature* **324:** 429–433.

12

Divalent Metal Ions in RNA Folding and Catalysis

Tao Pan, David M. Long, and Olke C. Uhlenbeck
Department of Chemistry and Biochemistry
University of Colorado, Boulder, Colorado 80309

Approximately three-quarters of all chemical elements are metals. Metal ions associate with RNA in solution in a number of interesting ways. The highly water-soluble ionic forms of certain metal ions are abundant in seawater and inside cells and play vital roles in RNA folding and catalysis. In this chapter, we examine the role of divalent ions in RNA folding. The properties of a number of divalent metal ions are summarized, and experiments examining their binding to RNA in solution and crystals are evaluated. We also review the evidence that precisely placed metal ions can promote catalysis by RNA. Since the mechanisms of group I introns and RNase P are the subject of another chapter (Cech, this volume), we focus here on biological and nonbiological examples of catalytic cleavage of the RNA that give products containing $2',3'$-cyclic phosphate and $5'$-hydroxyl termini.

PROPERTIES OF METAL IONS

The concentrations of a number of metal ions in seawater and cells are given in Table 1. Intracellular metal ion concentrations vary widely and are maintained and adjusted by a variety of active and passive transport mechanisms. Inside cells, free metal ion concentrations are much lower than total metal ion concentrations, since intracellular macromolecules bind substantial amounts of metal ions.

Table 2 lists certain chemical properties of selected metal ions that are relevant to their interaction with RNA. In aqueous solutions, metal ions form coordination complexes with water or solutes. The coordination number (number of ligands that bind) and coordination geometry (geometric arrangement of ligands) depend on characteristics of both the metal ion and the ligand(s). Most aqueous metal ions considered here are frequently present as hexacoordinate complexes with six ligand molecules

The RNA World
© 1993 Cold Spring Harbor Laboratory Press 0-87969-380-0/93 $5 + .00

Table 1 Metal content of water and bacterial cells

Metal	Median concentration in fresh water (μg/l)	Mean concentration in seawater (μg/l)	Concentration in bacterial cells (mg/kg)
Al	300	2	210
Ba	10	13	–
Be	0.3	0.0056	–
Ca	15,000	412,000	5.1×10^3
Cd	0.1	0.11	0.31
Co	0.2	0.02	7.9
Cr	1	0.3	4
Cu	3	0.25	150
Fe	500	2	170
K	2200	399,000	115×10^3
Li	2	180	–
Mg	4000	1.29×10^6	7×10^3
Mn	8	0.2	260
Mo	0.5	10	9.8
Na	6000	10.77×10^6	4.6×10^3
Ni	0.5	0.56	–
Pb	3	0.03	–
Sr	70	7900	–
Zn	15	0.03	83

arranged at the apexes of an octahedron. Less frequently, coordination complexes of four ligands in a tetrahedral arrangement occur.

Metal ions show differences in preferred partners in coordination complexes. On the basis of these preferences, both metal ions and ligands can be classified as "hard" or "soft." In solutions, hard metal ions (e.g., Li^+, Na^+, K^+, Mg^{++}, and Ca^{++}) preferentially bind hard ligands (e.g., H_2O, NH_3, OH^-, phosphates, ethers, and alcohols), and soft metal ions (e.g., Pb^{++}, Zn^{++}, Ni^{++}, Co^{++}, and Mn^{++}) preferentially bind soft

Table 2 Selected properties of divalent and trivalent metal ions

Metal	Preferred coordination	H_2O exchange rate (sec^{-1})	pK_a	pH at 10 mM solubility
Ba^{++}	6, 7	10^9	13.5	–
Ca^{++}	6, 8	10^8	12.9	12.4
Mg^{++}	6	10^5	11.4	9.6
Mn^{++}	6	10^7	10.6	7.9
Co^{++}	6	10^6	10.2	7.2
Ni^{++}	6	10^4	9.9	8.1
Zn^{++}	4, 6	10^7	9.0	6.6
Cd^{++}	4–7	10^8	9.6	8.1
Pb^{++}	6	10^9	7.7	7.2
Cr^{+++}	6	10^{-6}	–	–
Co^{+++}	6	1	–	–

ligands (e.g., cyanide, phosphorothioates, nitrogen-containing aromatic heterocycles, thioethers, and thiols). The hard/soft designation is explainable on the basis of several chemical theories and is not always definitive (Huheey 1983). For the purpose of RNA studies, ligands that coordinate through sulfur or nitrogen atoms are considered soft, and ligands that coordinate through oxygen are considered hard.

Hydrated metal ions exchange water ligands at rates that vary by nearly 18 orders of magnitude (Porterfield 1984). Ligands other than water are also likely to show large differences in exchange rate. Whereas most "common" divalent ions such as Mg^{++}, Ca^{++}, and Mn^{++} exchange very rapidly (>1 μ sec^{-1}), certain metal ions, notably Co^{+++} and Cr^{+++}, exchange so slowly that coordination compounds formed with these metals can be considered kinetically inert (Table 2). The kinetics and thermodynamics of ligand binding by different metal ions do not correlate in an intuitive manner. Metal ions that have the slowest exchange rates are not necessarily those that bind ligands the strongest.

Water molecules bound by metal ions can be substantially more acidic than free water. Indeed, the pK_a values of some hydrated metal ions (Table 2) are within the physiological range. Hydroxide ion bound to a metal ion is a potent nucleophile (Chaffee et al. 1973). Accordingly, hydrated metal ions can participate in general and/or specific acid/base chemistry. If a metal ion is coordinated to other ligands, the pK_a of a bound water can be affected. Conditions of high pH, where deprotonation of hydrated metal ion results in charge neutrality, cause many metal ions to aggregate and form insoluble hydrous oxides.

Although the oxidation or reduction of metal ions bound to RNA under physiological conditions has not been well studied, it is important to remember that many metal ions can easily gain or lose electrons. A practical consequence of this is that a number of metal ions often used in RNA studies can lose electrons at higher pH and react with dissolved oxygen to form insoluble metal oxides. For example, solutions of Mn^{++} rapidly convert to MnO_2 above pH 8.

BINDING OF DIVALENT METAL IONS TO RNA

Divalent metal ions can interact electrostatically with the phosphate backbone of RNA. Experiments measuring the binding of divalent metal ions to single-stranded RNA homopolymers (Krakaver 1971) reach saturation close to one metal ion per two phosphates. In the absence of other cations, dissociation constants of divalent metal ions to such unstructured RNAs are in the neighborhood of 0.1 to 10 mM. This value is lower than the binding of a metal ion to a nucleotide monophosphate

(K_d >10 mM; Belaich and Sari 1969), suggesting that each metal ion binds two (presumably adjacent) phosphates on the homopolymer chain. However, "bridging" the two phosphates with a single divalent metal ion requires that the conformation of the phosphodiester backbone be constrained, with an accompanying reduction in entropy and cost in energy. As would be expected, divalent metal ions bind more tightly to homopolymer helices where the phosphates are structurally constrained and two phosphates on opposite strands are sufficiently close in A-form RNA to bind a single metal ion. Since the divalent ions reduce the electrostatic repulsion between the negatively charged phosphates, this difference in divalent ion-binding affinity between single- and double-stranded RNA means that divalent metal ions are very effective in stabilizing the RNA double helix. Because of this bridging effect, a relatively low concentration of divalent metal ions (<10 mM) will increase the T_m of an RNA helix by more than much higher concentrations of monovalent ions.

The binding of divalent metal ions to heteromeric RNAs with complex, folded tertiary structures is more complicated. As is the case with homopolymers, divalent metal ions interact electrostatically with the phosphates in helical regions and thereby stabilize the RNA secondary structure. In addition, there are likely to be positions within the structure where single-stranded phosphates are constrained in such a way that they are relatively close in space and thereby form a "pocket" in the RNA where a metal ion can bind even more tightly than to an RNA helix. Formation of such metal ion-binding "pockets" is expected to be strictly dependent on complete folding of the RNA. Thus, for an experiment to measure these "tight" divalent metal ion-binding sites in a folded RNA structure, it is necessary to include another counterion, such as Na^+ or a polyamine, to promote RNA folding. This strategy assumes, of course, that the RNA is properly folded in the presence of the alternate counterion and no aspect of the overall structure is critically dependent on the presence of divalent metal ions. In addition, since the alternate counterion is likely to compete with the divalent metal ion, the apparent affinity constant of divalent metal ions could appear weaker.

Finally, it is important to point out that although the above discussion focuses on the electrostatic interactions of metal ions with phosphates, metal ions can bind to RNA in other ways. This is especially true for "soft" metal ions where interactions with nucleotide bases are more common. Thus, metal ion-binding pockets can potentially be formed involving both ribose phosphates and bases.

Experiments measuring divalent metal ion-binding to RNA are carried out by varying concentrations of metal ions in the presence of RNA, and the amount of bound metal ion can be determined either directly by measuring concentrations of metal ions bound, or indirectly by observing changes in the structure or activity of the RNA. In direct experiments, equilibrium dialysis has been very successful where free and bound metal ion concentrations are measured either by atomic absorption spectroscopy (Stein and Crothers 1976), scintillation counting of a radioactive metal ion (Danchin 1972), or fluorescence spectroscopy using a metal ion chelator (Römer and Hach 1975). An estimate of the number of bound Mg^{++} ions has also been obtained from the line broadening of the ^{25}Mg nuclear magnetic resonance (NMR) spectrum (Reid and Cowan 1990). Finally, the thermodynamics of Mg^{++} binding to RNA can also be analyzed calorimetrically (Rialdi et al. 1972). In indirect experiments, the folding of RNA is monitored by spectroscopic methods such as UV absorption spectroscopy, circular dichroism, fluorescence spectroscopy (Ehrenberg et al. 1979; Labuda and Pörschke 1982), or NMR (Cohn et al. 1969; Flanagan and Jacobson 1988). Chemical modification experiments that detect the formation of RNA secondary or tertiary structure have also been applied. The results obtained by these methods are most easily interpreted if an alternate counterion is used to fold the RNA so any changes that are observed represent only specific divalent metal ion binding. It is important to remember that indirect methods measure consequences of metal ion binding, not the binding of metal ions. Thus, considerable caution is needed to deduce metal ion-binding information from indirect experiments.

Table 3 summarizes the results of a number of direct experiments analyzing the binding of divalent metal ions to tRNA. Measurements carried out at low concentrations of monovalent ions (Bina-Stein and Stein 1976) indicate that divalent metal ion binding is cooperative, with the first divalent metal ion bound less tightly than subsequent ones. This cooperativity reflects the cooperative folding of RNA as divalent metal ion is added. When experiments are performed in higher monovalent ion concentrations, no cooperativity of divalent ion binding is observed (Stein and Crothers 1976; Crothers and Cole 1978). In these latter experiments, the data could be fit by assuming two classes of binding sites. Approximately 1–4 Mg^{++} ions bound with K_d values of 10–100 μM and were therefore termed "strong" metal ion sites. A much larger number (between 20 and 50) of "weak" sites with K_d values of 2–5 mM were also observed. Variation in the number of metal-binding sites and K_d values is a reflection of both the low accuracy of the experimental methods and the fact that different tRNA species were used. Since the strong sites are lost under conditions where the tertiary structure of tRNA is disrupted (Bina-Stein and Stein 1976), they are likely to be associated with the proper folding of the molecule. The weak sites have K_d values within the

Table 3 Mg^{++} binding to tRNAs in solution

tRNA	Method	Conditions	K_d (M)[a]		References
			strong sites	weak sites	
E. coli tRNA^{Met}	equilibrium dialysis	0.17 M Na$^+$; pH 7.0, 4°C	3.4 x 10^{-5} (1)[b]	2.4 x 10^{-3} (26)	Stein and Crothers (1976)
E. coli tRNA^{Glu}	equilibrium dialysis	0.1 M Na$^+$, pH 7.0, 4°C	1.3 x 10^{-5} (1)	1.2 x 10^{-3} (36)	Bina-Stein and Stein (1976)
Yeast tRNA^{Phe}	fluorescent titration	0.032 M Na$^+$, pH 6.0 10°C	1.1 x 10^{-5} (6.5 ± 3)	1.7 x 10^{-4} (17 ± 5)	Römer and Hach (1975)
Yeast tRNA^{Phe}	calorimetry	0.01 M Na$^+$, pH 7.2	1 x 10^{-6} (4)	9.1 x 10^{-5} (20)	Rialdi et al. (1972)
Yeast tRNA^{Phe}	^{25}Mg NMR	0.17 M Na$^+$, pH 7.0, 25°C	<10^{-4} (3–4)	4.5 x 10^{-3} (50 ± 8)	Reid and Cowan (1990)

[a]Determined under conditions when noncooperative binding is observed.
[b]Numbers in parentheses indicate the number of sites.

range expected for metal ions binding to single-stranded or double-helical RNA and, therefore, presumably reflect binding to other parts of the tRNA.

Perhaps the most significant difference between direct and indirect methods is that the latter cannot in general accurately determine the number of metal ions bound. Since indirect methods give information on the consequences of metal ion binding to RNA, they are useful for examining conformational changes and/or catalytic properties of RNA. For example, Mg^{++} binding to *Tetrahymena* group I intron as assayed by hydroxyl radical footprinting showed that the catalytically active tertiary structure is promoted by cooperative binding of at least three Mg^{++} ions as indicated by the Hill coefficient (Celander and Cech 1991). Half of the RNA molecules are properly folded at Mg^{++} concentrations of about 0.8 mM. The most useful information on metal ion binding to RNA by monitoring RNA folding can be obtained by keeping the RNA concentration much higher than K_d, so that all Mg^{++} ions present at low stoichiometric amounts will be bound to RNA. When all "strong" Mg^{++} binding sites are filled, addition of more Mg^{++} will not affect folding. By assuming that Mg^{++} binding is solely responsible for the observed changes, the number of binding sites can be deduced. This strategy has been successfully applied recently to show that an oligomer that has the sequence of the yeast tRNA[Phe] anticodon stem-loop undergoes a structural transition observed by circular dichroism when one Mg^{++} is bound per oligomer (Dao et al. 1992).

METAL ION BINDING SITES FROM CRYSTAL STRUCTURES

The best-characterized metal ion-binding sites in RNA have been identified in the crystal structures of yeast tRNA[Phe]. Crystallization of yeast tRNA[Phe] in the presence of relatively high concentrations of Mg^{++} has yielded orthorhombic and monoclinic crystal forms that were solved to 2.7 Å and 2.5 Å resolutions, respectively (Holbrook et al. 1977; Jack et al. 1977; Hingerty et al. 1978). At these resolutions, the relatively low electron density of Mg^{++} ions makes them difficult to locate unless they are in sites where their position is highly constrained. Table 4 summarizes the four Mg^{++} ion-binding sites in the orthorhombic crystal form (Holbrook et al. 1977; Quigley et al. 1978) and three sites in the monoclinic form (Jack et al. 1977; Hingerty et al. 1978). At the high Mg^{++} concentrations used, there are likely to be many other bound Mg^{++} ions in the crystal lattice that could not be located because their position was not identical for each molecule in the lattice. Although the resolution of the crystals may not be high enough to allow identification of the

Table 4 Mg^{++} binding sites in the crystal structures of yeast tRNA^{Phe}

		Distance (Å)	
Site[a]	Contacts	orthorhombic[b]	monoclinic[c]
1 (3)	G19-O2p	1.9	2.4
	G19-O1p	3.5	4.3
	G19-O5$'$	3.8	4.2
	G20-N7	4.1	3.7
	G18-O3$'$	4.4	>4.5
	U59-O4	>4.5	3.8
	G20-O6	>4.5	4.2
2 (2)	A21-O1p	1.8	2.3
	G20-O2p	2.3	2.7
	A21-O5$'$	3.7	>4.5
	A21-O2p	3.8	4.0
	G20-O3$'$	4.3	4.4
	G20-O5$'$	4.4	>4.5
	G20-O1p	4.4	4.3
3 (1)	U12-O1p	3.3	3.8
	U8-O1p	3.5	3.9
	U8-O2p	3.9	4.1
	C11-O1p	4.3	3.9
	A9-O1p	>4.5	3.9
	A9-O2p	>4.5	4.4
4	Y37-O1p	2.3	
	Y37-N7	3.8	
	C32-O2	3.9	
	A38-N7	4.1	
	U39-O2	4.2	
	Y37-O2p	4.4	
	C32-N3	4.4	
	A36-O3$'$	4.4	

[a]The designated numbers of the orthorhombic crystal form are given. The number of the monoclinic form is presented in parentheses.

[b]From Holbrook et al. (1977).

[c]From Hingerty et al. (1978). There is a fourth Mg^{++} found in the monoclinic form which is located at the interface of two tRNA^{Phe} molecules in the crystal.

coordinated ligands of each metal ion accurately, RNA ligands within 2.5 Å are assumed to be in direct coordination. Coordination sites not occupied by RNA are assumed to be occupied by water, which could form hydrogen bonds with neighboring oxygen and nitrogen atoms within 3.5–4.5 Å.

Three Mg^{++} binding sites are in excellent agreement between the two crystal forms, suggesting that crystal packing did not have a significant effect on the ligands which these Mg^{++} ions bind. Two Mg^{++} ions bind in the corner of the L-shaped structure of tRNA and may stabilize the tertiary interactions between the D loop and the T loop. Mg(2) is directly ligated by the oxygens of phosphates 20 and 21 with water as the remaining four ligands. Mg(3) is positioned about 9.9 Å from Mg(2) and is directly coordinated to phosphate 19. Four of the five water ligands of Mg(3) can form hydrogen bonds to several neighboring nucleotide bases as suggested by the Mg^{++}–O and Mg^{++}–N distances (Table 4). Mg(1) is found in the tight turn formed by residues 8–12 (Fig. 1). This Mg^{++} ion binds as hexahydrate with four water molecules forming hydrogen bonds with phosphate oxygens 8, 9, 11, and 12. A fourth Mg^{++}, located in the anticodon loop, is apparent in the orthorhombic form. In a recent refinement of the structure from monoclinic form, a Mg^{++} ion is identified in the anticodon loop as well (Westhof and Sundaralingam 1986). In all cases, Mg^{++} ions are located in constrained regions of the phosphate backbone and thus can be considered to occupy "pockets."

It is tempting to assume that Mg^{++} ions identified in crystal structures of tRNA[Phe] are the same 3–5 "strong" Mg^{++} ions identified in solution measurements (Table 4). However, it is important to realize that there is no evidence for this assumption. As discussed above, only a limited number of Mg^{++} ions bound to tRNA in crystals can be located with the available resolution, but higher resolution crystals would undoubtedly

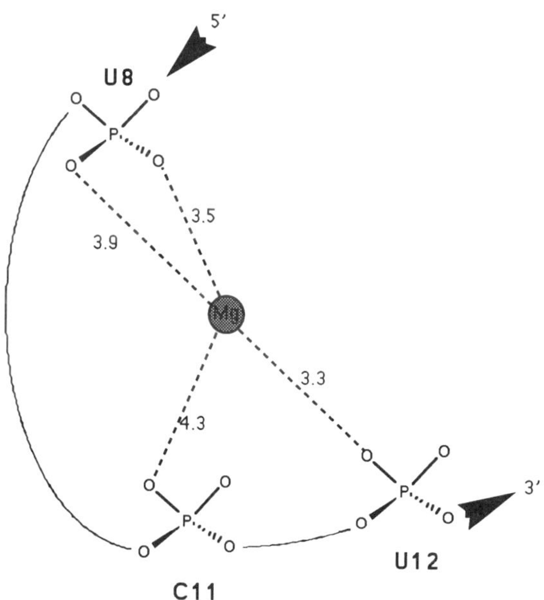

Figure 1 Mg(1) binding site of yeast tRNA[Phe] in the orthorhombic crystal form (Holbrook et al. 1977). This site is very similar to the Mg(3) site in the monoclinic form (Hingerty et al. 1978). Only the phosphates that are within 4.5 Å of the bound Mg^{++} are presented. The distances are given next to the dashed lines. The ribose phosphate backbone chain from 8 to 12 is shown as a solid line.

reveal additional Mg^{++} ions. To gain information about the affinity of these ions in the crystal, an experiment measuring the crystallographic occupancy of each site as a function of Mg^{++} ion concentration would be required. Alternatively, if one wanted to test whether a solution site was identical to a crystal site, appropriate phosphates could be substituted with phosphorothioate, thereby significantly decreasing Mg^{++} binding (see below, RNA Self-cleavage Reactions). Solution and crystallographic studies could be used to confirm reductions in metal ion binding that result from phosphorothioate substitution.

A large amount of data locating other metal ions in the tRNA[Phe] crystals has been collected during the course of obtaining isomorphic heavy-atom derivatives (Jack et al. 1977). tRNA[Phe] crystals initially obtained in the presence of Mg^{++} were soaked in dilute solutions of other metal ions. Consequently, any metal ion that binds specifically must either bind to another site or have a sufficiently higher affinity to displace a previously bound Mg^{++}. Due to their much higher electron densities, the location of these metal ions in the crystal structure is generally much more reliable than for Mg^{++} ions.

A general view of the placement of different heavy metals in the tRNA[Phe] structure is summarized in Figure 2. Many different binding sites are observed. The precise coordination of each metal ion presumably depends on their coordination preferences. Three types of binding sites are of particular interest to this review:

1. Lanthanide ions such as Sm^{+++} and Lu^{+++} tend to be coordinated by oxygen ligands with octahedral geometry, similar to Mg^{++} ions. In fact, among all metal ion-binding sites identified within tRNA[Phe] crystal structures, the only sites identical to Mg^{++} binding sites are two of five samarium sites (Jack et al. 1977). At Sm^{+++} concentrations 20–50-fold lower than Mg^{++}, Sm^{+++} replaces Mg(1) and Mg(3) precisely.

2. Solution studies have suggested several strong Co^{++} and Mn^{++} binding sites in yeast tRNA[Phe] (Danchin 1972; Schreier and Schimmel 1974). Crystallographic characterization of Co^{++} and Mn^{++} binding sites can be compared with solution data. Unfortunately, the resolutions of crystal structures containing these ions are low (3.5–4.0 Å) and only one major binding site of each metal ion could be identified (Jack et al. 1977). Both ions are directly ligated by N^7 of guanosine bases. Mn^{++} is bound 2 Å away from Mg(3) and results in displacement of this Mg^{++} ion.

3. As discussed below, both Pb^{++} and Zn^{++} are effective in RNA cleavage. Since Pb^{++} causes specific autolytic cleavage of yeast tRNA[Phe] at neutral pH, crystals containing intact tRNA with Pb^{++}

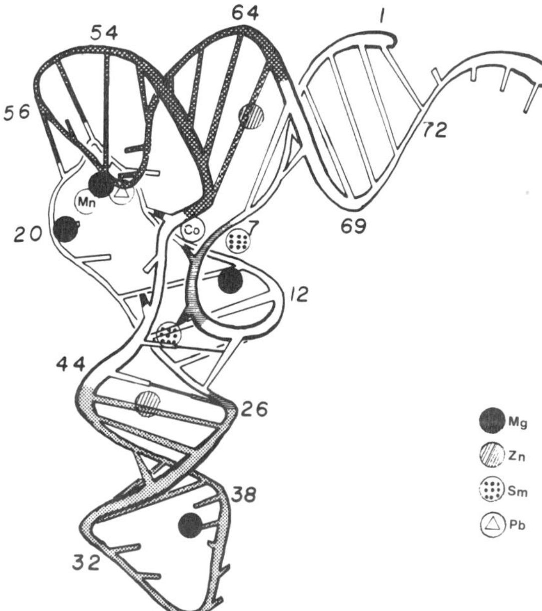

Figure 2 Metal ions found in the crystal structures of yeast tRNA[Phe]. Only the metal ions that are described in the text are shown. All sites illustrated are unimolecular. In several instances, bound metal ions are found at identical locations. They include two Sm^{+++} ions at Mg(1) and Mg(2) sites; three Zn^{++} ions at Mn^{++}, Co^{++}, and Mg(1) sites; two Pb^{++} ions at Sm(4) and Mg(4) sites.

bound were obtained at pH 5.0, under which the cleavage rate was greatly reduced (Brown et al. 1985). Three tightly bound Pb^{++} ions are present in the crystal structures. Pb(1) is located at the corner of the L-structure of tRNA and is directly coordinated by O^4 of U59 and N^3 of C60 in the T loop (Fig. 3). Hydrogen bonding of a water molecule ligated to this Pb^{++} can also form with a nonbridging oxygen of G19. The Pb(1) is 2.1 Å away from the Mg(3) site, and binding of Pb(1) causes Mg(3) to be displaced. Pb(2) is bound at the variable loop region with direct coordination to O^6 of G45. The Pb(2) site is also identical to one of the five lanthanide sites, Sm(4) (Jack et al. 1977; Brown et al. 1985). Pb(3) is located within the anticodon loop by direct ligation to N^7 of Y37 and can displace Mg(4) in the orthorhombic crystals (Rubin and Sundaralingam 1983).

In the 4 Å resolution structure of yeast tRNA[Phe] soaked with Zn^{++}, five Zn^{++} sites are found (Rubin et al. 1983). All but one of the Zn^{++} ions are directly coordinated to N^7 of guanosine bases, as observed for the major Co^{++} and Mn^{++} sites. In fact, the electron densities of Zn(1) and Zn(3) match precisely the Mn^{++} and Co^{++} sites,

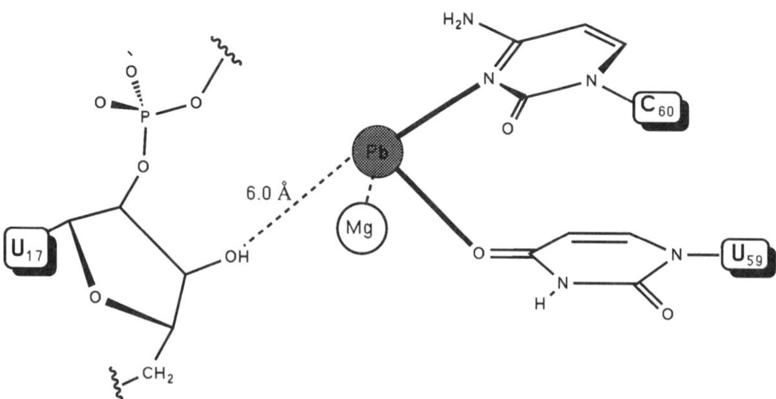

Figure 3 Pb(1) site in the crystal structure of yeast tRNA[Phe] (Brown et al. 1985). Direct coordination of Pb(1) to nucleotide bases U59 and C60 is shown as thick lines. The Mg^{++} ion in the figure is 2.1 Å away from Pb(1) and is likely to be absent when Pb^{++} is bound and vice versa.

respectively. The Zn(2) site is close enough to Mg(3) to cause replacement of this bound Mg^{++} ion.

RNA CLEAVAGE BY METAL IONS

RNA chain cleavage can occur via two different reaction pathways, depending on the nucleophile which attacks the phosphodiester bond. If an external nucleophile such as water or hydroxide ion attacks phosphorus directly, the reaction products will contain 5´ or 3´ monophosphate and free hydroxyl termini. For example, the reactions catalyzed by RNase P, group I, and group II introns generate 5´-phosphate and 2´, 3´-hydroxyl termini and are the subjects of another chapter (Cech, this volume). In this section, we focus on reactions where the 2´-hydroxyl group of the adjacent ribose is the nucleophile, resulting in products containing 2´,3´-cyclic phosphate and 5´-hydroxyl termini. This pathway is prevalent in the alkaline hydrolysis of RNA and is utilized by many degradative ribonucleases. Nucleophilic attack of the 2´ oxygen on phosphate occurs through a transition state or short-lived intermediate in which the phosphorus atom is penta-coordinated with five oxygen atoms in a trigonal-bipyramidal arrangement (Fig. 4). The 2´-oxygen nucleophile is axial to the 5´-oxygen leaving group, resulting in inversion of stereochemical configuration about the central phosphorus atom. Since the reaction is a transesterification from a 3´,5´-diester linkage to a 2´,3´-diester linkage, no net change in the number of phosphodiester

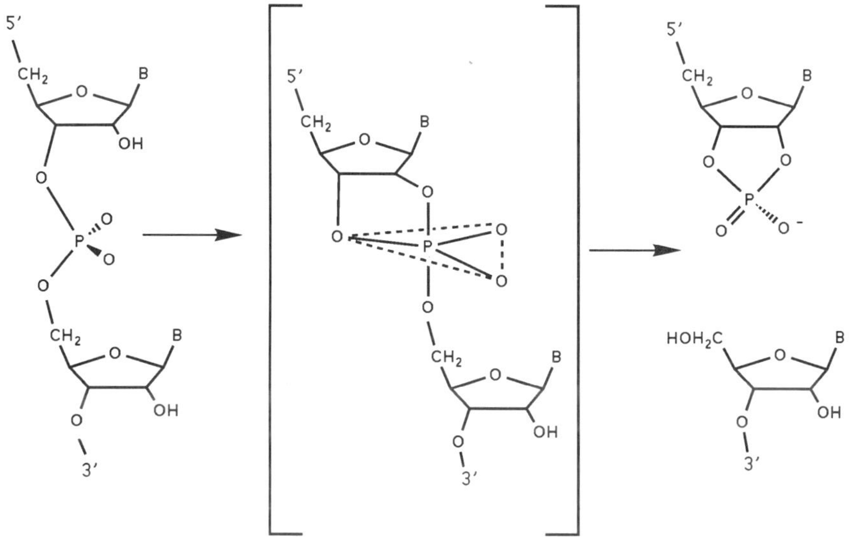

Figure 4 Putative transition state of RNA cleavage reactions that give products containing 2′,3′-cyclic phosphate and 5′-hydroxyl termini.

bonds occurs. Consequently, the principal driving force of this reaction is not a favorable change in enthalpy.

The rate of chain cleavage in the reaction shown in Figure 4 can be accelerated by a number of factors. These include (1) a basic group to aid in deprotonation of the 2′-hydroxyl, in effect enhancing the nucleophilicity of the 2′ oxygen; (2) an acidic group that can stabilize and possibly protonate the 5′ oxyanion; and (3) any environment that can stabilize the pentavalent species, which is either a transition state or a short-lived intermediate. Divalent metal ions can provide one or more of these functions. A metal ion-bound water molecule can lose a proton and potentially act as general base by removing the proton from the 2′-hydroxyl group at the cleavage site. The positive charge on a metal ion could stabilize the attacking or leaving oxyanion or, by binding to negatively charged nonbridging oxygens, stabilize the transition state. The positive charge can also polarize electron density, thereby increasing the susceptibility of the phosphorus atom to nucleophilic attack (electrophilic catalysis).

It has been known for more than 30 years that divalent metal ions stimulate the cleavage of RNA (Butzow and Eichorn 1965; Farkas 1968; Eichhorn et al. 1971). In these early studies, the depolymerization of homopolymers such as poly(U) and poly(A) was measured at various concentrations of divalent metal ions, including Mg^{++}, Ba^{++}, Mn^{++}, Co^{++}, Ni^{++}, Cu^{++}, Zn^{++}, Sn^{++}, and Pb^{++}. Although cleavage rates of individual phosphodiester bonds were not determined directly, it was clear that rates of cleavage varied by more than 1000-fold. The order of reactivity correlated reasonably well with the pK_a of hydrated metal ions, with the fastest rate obtained with Pb^{++} followed by Zn^{++} (Farkas 1968). This suggests that a deprotonated form of the hydrated metal ion is an important component of the reaction. Analysis of reaction products showed both 2′,3′-cyclic phosphate and 2′- and/or 3′-monophosphate termini, suggesting that chain cleavage could be followed by hydrolysis of the 2′,3′-cyclic phosphate. Cleavage rates were also shown to be affected by secondary structure and nucleotide composition of the RNAs. Breakdown of single-stranded poly(A), poly(U), poly(C), and poly(I) with Pb^{++} was several times faster than double-stranded poly(A)-poly(U) or poly(C)-poly(I) (Butzow and Eichorn 1965; Farkas 1968). On the other hand, cleavages of poly(A), poly(C), and poly(U) with Zn^{++} were equally fast, whereas poly(I) was decomposed more slowly (Butzow and Eichorn 1965). These early studies clearly indicated that divalent metal ions can accelerate RNA cleavage and prompted the study of the reaction of divalent metal ions with natural RNAs.

"SPECIFIC" CLEAVAGE OF RNA BY METAL IONS

When divalent metal ions are incubated with heteromeric RNAs, the rate of chain cleavage at certain phosphodiester bonds is often much greater than at others. These cleavage sites are referred to as "specific" if the "background" cleavage at other sites is much slower. A number of reports of RNAs specifically cleaved by different divalent and trivalent cations are summarized in Table 5. In some cases, the products of cleavage have been shown to give 2′,3′-cyclic phosphate and 5′-hydroxyl termini. In a few cases, it has also been shown that specific cleavage requires a folded RNA structure. In this section, we focus on the specific cleavage of RNA by Pb^{++}, since these reactions are the best characterized.

The specific cleavage of yeast $tRNA^{Phe}$ by Pb^{++} was discovered in 1972 (Dirheimer et al. 1972). Although high Pb^{++} concentrations will promote extensive degradation of any RNA, it is striking that incubation of $tRNA^{Phe}$ at 50–200 μM Pb^{++} results in the rapid cleavage of the phosphodiester bond between U17 and G18 to give 2′,3′-cyclic phosphate and 5′-hydroxyl products (Dirheimer et al. 1972; Werner et al. 1976). The reaction requires the presence of sufficient Mg^{++} or other counterion to ensure folding of the tRNA. However, higher Mg^{++} concentrations inhibit the reaction. The reaction rate increases rapidly with pH from 5.5 to 7.2 and then decreases, presumably due to Pb^{++} oxide or

Table 5 Examples of "specific" cleavage of RNAs by divalent and trivalent metal ions

Metal ion	RNA	pH	Reference
Eu^{+++}	Yeast, *E. coli*, Lupin tRNA^{Phe}	7.2	Marciniec et al. (1989)
Mn^{++}	5′ end of *Tetrahymena* group I intron	7.5	Dange et al. (1990); Kazakov and Altman (1992)
Zn^{++}	M1 RNA	7.5	Kazakov and Altman (1991)
Pb^{++} (other than tRNAs)	3′ terminal domain of *E. coli* 16S rRNA	7.5	Gornicki et al. (1989)
	β-tropomyosin pre-mRNA	7.5	Clouet d'Orval et al. (1991)
Mg^{++} (other than viroid RNAs)	tRNAs: cleavage at UpA, CpA	7.5	Dock-Bregeon and Moras (1987)
	Yeast, *E. coli*, Lupin tRNA^{Phe}: cleavage at D loop	8.5	Wintermeyer and Zachau (1973); Marciniec et al. (1989)
	M1 RNA	9.5	Kazakov and Altman (1991)

polyhydrate formation. Other ions, including Eu^{+++} and Mg^{++} have shown slower rates of cleavage at or near the same site (Wintermeyer and Zachau 1973; Marciniec et al. 1989).

An extensive crystallographic study of Pb^{++} binding to tRNA^{Phe} has provided considerable insight into this Pb^{++} cleavage reaction (Rubin and Sundaralingam 1983; Brown et al. 1985). The coordinates of Pb^{++} binding sites determined from crystals soaked with Pb^{++} at low pH indicate that one of the three bound Pb^{++} ions, Pb(1), is in close proximity to the cleavage site at ribose 18. As shown in Figure 3, Pb(1) is directly coordinated to U59 and C60 in the T loop and through a bridging water molecule to phosphate 20 in the D loop. When the pH of the Pb^{++}-containing crystals is raised to 7.0, the positions of Pb(1) and the surrounding RNA atoms are scarcely altered, but analysis of the RNA clearly indicates that cleavage has occurred at the same site as was observed in solution. This provides strong evidence that Pb(1) in the crystal structure is involved in the reaction.

On the basis of structural data, Brown et al. (1985) propose a reaction mechanism that features nucleophilic activation of the 2′ oxygen of ribose 17 by a Pb^{++}-bound hydroxide. The poor cleavage at pH 5 and rapid cleavage at pH 7 are consistent with the pK_a of about 7.2–7.7 for a Pb^{++}-bound water molecule in solution. Pb(1) is about 6.0 Å from the 2′ oxygen, which is rather distant for the mechanism proposed. However, the crystallographic temperature factor is relatively large in this region (especially for ribose 17) so that substantial molecular motion may be required for the reaction to occur. A bridging water molecule has also been proposed (Sundaralingam et al. 1984). Since Pb(2) and Pb(3) are much further from any 2′ oxygen, the absence of cleavage at other sites can be rationalized. Mg(2) binds close enough to Pb(1) so that competition of Pb^{++} cleavage by high concentrations of Mg^{++} can be understood as simple competition for the binding site. Since the Eu^{+++} and Mg^{++} binding sites are close to Pb(1), they may cleave by a similar mechanism. These metal ions are less acidic than Pb^{++}; consequently, they will have a lower concentration of metal ion hydroxide at pH 7 and, therefore, are expected to cleave more slowly.

Other features of the mechanism of Pb^{++} cleavage of $tRNA^{Phe}$ are less clear. Since Pb(1) is too distant from both nonbridging oxygens (>7 Å) or the leaving 5′ oxygen of ribose 18 (5.9 Å) for direct coordination, it is unlikely that it participates in electrophilic activation through coordination to these oxygens. However, it is possible that unlocalized Mg^{++} ions could be present in solution to play this role. Brown et al. (1985) suggest that structural changes in the crystal would be minimized if the reaction proceeded by equatorial attack of the 2′ oxygen or the phosphate followed by pseudorotation and release of the 5′ oxygen. This mechanism has not been observed with protein enzymes and would require retention of stereochemical configuration about the phosphorus. Although the stereochemical aspects of the reaction have not been determined, the known flexibility of this region makes their arguments for this unconventional mechanism seem less persuasive.

Analysis of Pb^{++} cleavage of a variety of mutants of $tRNA^{Phe}$ supports the above model (Behlen et al. 1990). The single-stranded nucleotides U59 and C60 coordinate Pb(1) directly but are not involved in tertiary hydrogen bonds. Thus, the substantial reduction in Pb^{++} cleavage that is observed when these two nucleotides are mutated can be interpreted as disruption of Pb^{++} binding and not a change in the folding of $tRNA^{Phe}$. When U59 and C60 are introduced into several tRNAs that do not cleave with Pb^{++}, they become susceptible to cleavage with Pb^{++} at phosphate 18 (Behlen et al. 1990). The Pb^{++} cleavage properties of a number of other $tRNA^{Phe}$ mutants have also been determined. U16, U17, and G20 are residues near the cleavage site that protrude into the solvent and, as expected, can be mutated without altering Pb^{++} cleavage. However, when nucleotides involved in the tertiary folding of $tRNA^{Phe}$ and the maintenance of the Pb(1) pocket are mutated, Pb^{++} cleavage is substantially reduced. In a few cases, Pb^{++} cleavage can be restored by introducing a second mutation in the $tRNA^{Phe}$ that reforms the tertiary structure. In general, the mutagenic data support the view that Pb^{++} cleavage of $tRNA^{Phe}$ in solution is due to Pb^{++} binding at the Pb(1) site in the folded $tRNA^{Phe}$, as suggested by the crystallographic experiments.

The Pb^{++} cleavage of tRNA^{Phe} has been used as an "assay" to monitor tRNA folding. Mutations in yeast (Behlen et al. 1990; Sampson et al. 1990) and *Escherichia coli* (Peterson and Uhlenbeck 1992) tRNA^{Phe} that were aminoacylated poorly by their cognate synthetase were tested for Pb^{++} cleavage. Mutants cleaved normally by Pb^{++} were assumed to be folded properly, and the mutation was assumed to have disrupted the interaction with the synthetase enzyme. Mutants that did not cleave well with Pb^{++} were assumed to be misfolded, so it was not possible to make conclusions concerning their interaction with the synthetase. Such independent "folding assays" are invaluable in structure-function studies.

The Pb^{++} cleavage of tRNA^{Phe} has also been exploited to determine which circularly permuted isomers of yeast tRNA^{Phe} fold normally (Pan et al. 1991). The experiment involved joining the 3′ and 5′ termini of tRNA^{Phe} to make a circular tRNA, performing partial alkaline hydrolysis to prepare a collection of circularly permuted isomers and then cleaving them with Pb^{++}. It was found that circularly permuted tRNAs with termini at a minimum of 54 sites were able to fold normally. Only 14 circularly permuted tRNA^{Phe} with termini in highly constrained regions of the structure did not cleave with Pb^{++}. Since tRNA can tolerate breaks in the phosphodiester backbone without disrupting folding, tRNAs can be prepared from more than one RNA fragment. One such example uses two tRNA fragments containing residues 1–35 and 34–76. The tRNA^{Phe} tertiary structure and Pb^{++} cleavage are restored when the two are combined (Sampson et al. 1987). By introducing an excess of the 1–35 fragment, catalytic Pb^{++} cleavage by the 34–76 fragment can be shown under conditions where the cleavage products can dissociate.

Further insight into the cleavage of RNA by Pb^{++} was obtained in an in vitro selection experiment designed to identify novel tRNA tertiary interactions (Pan and Uhlenbeck 1992a). A library of tRNA^{Phe} molecules with 9–10 randomized positions was prepared and converted into circular RNAs. Although most of the more than 10^6 molecules in the library did not cleave with Pb^{++}, the small proportion that did were linear and therefore could be purified from the circular RNA on a denaturing gel. The cleaved RNAs were recircularized, converted back into DNA templates by reverse transcription and PCR, and then transcribed to prepare RNAs enriched for cleavage by Pb^{++}. After six rounds of this procedure, the population of enriched circular RNAs cleaved with Pb^{++} at a rate comparable to native circular tRNA^{Phe}. After cloning, individual variants were sequenced and their Pb^{++} cleavage properties were determined. Interestingly, among a large number of RNAs that cleaved rapidly and specifically with Pb^{++}, very few cleaved at the same site as tRNA^{Phe}. Indeed, many of the other cleavage sites were sufficiently far from the three Pb[++] sites in tRNA[Phe] to suggest that new Pb[++] binding sites are formed. This implies that specific Pb[++] cleavage sites are quite common, a view supported by a substantial number of Pb[++] cleavage sites in other RNAs (Table 5). The only two variants found in the in vitro selection experiment that cleaved at ribose 17 lacked several highly conserved tertiary interactions of tRNA[Phe] and more closely resembled a class of mitochondrial tRNAs. Model building suggested formation of an alternate U-U tertiary interaction for these two variants (Pan and Uhlenbeck 1992a).

Further analysis of one of the Pb[++] cleavage motifs isolated from the in vitro selection experiment revealed that it was quite small (Pan and Uhlenbeck 1992b). A motif of two helices surrounding an asymmetric internal loop of six nucleotides cleaved with Pb[++] more than ten times faster than tRNA[Phe] under comparable conditions. More surprisingly, the initial cleavage to produce a $2',3'$-cyclic phosphate was followed by hydrolysis to produce $3'$-monophosphate specifically (Fig. 5). Such a two-step mechanism is common among protein ribonucleases but has not been observed for catalytic RNA reactions. By comparing the rate of each step of the RNA-catalyzed reaction with the corresponding uncatalyzed reaction caused by Pb[++] alone, an estimate of the reaction rate enhancement by this RNA can be made. An 1100-fold rate enhancement

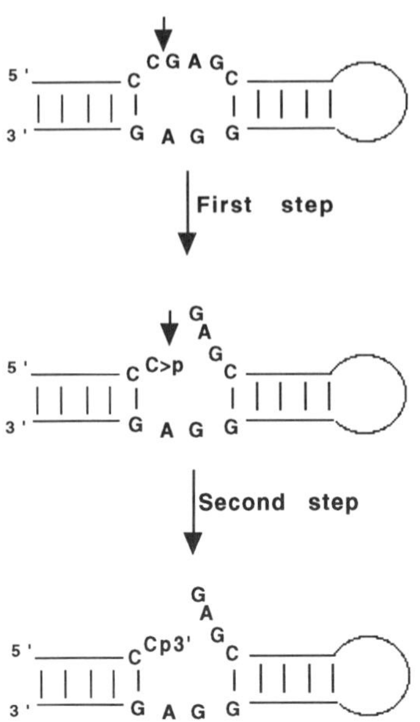

Figure 5 The Pb[++] cleavage motif that carries out cleavage and hydrolysis reactions in a two-step mechanism (Pan and Uhlenbeck 1992b). Only the essential nucleotides are shown.

of the first step and a 40–50-fold rate enhancement of the second step are observed. Although these enhancements are much less than the natural self-cleaving motifs discussed in the next section, they easily explain the very clear specificity that is observed in these cleavage reactions.

In summary, we can understand the highly specific cleavage of RNA by Pb^{++} as the result of a Pb^{++} hydroxide binding to RNA at a site in such a way that the hydroxide ion can perform alkaline hydrolysis of RNA. It is important to realize, however, that this mechanism may not be the only explanation for RNA cleavage reactions catalyzed by metal ions. For example, some metal-dependent cleavage specificity may be achieved by intrinsic differences in the cleavage rate of phosphodiester bonds due to local RNA structure. If the 2′ and 5′ hydroxyl groups are constrained such that in-line attack is stereochemically favored, the cleavage rate will be enhanced. Such sites should show cleavage specificity in solutions containing species that can act as general bases, including metal-bound hydroxide, that do not denature the RNA. As discussed in the previous section, metal ions can also enhance cleavage by electrophilic activation or by stabilizing the transition state. Thus, the importance of these effects on each metal-ion-catalyzed cleavage reaction is expected to vary.

RNA SELF-CLEAVAGE REACTIONS

Under conditions where RNA is normally quite stable (neutral pH and low Mg^{++} concentrations), certain RNA molecules have the capacity to cleave rapidly to yield 2′,3′-cyclic phosphate and 5′-hydroxyl termini. This reaction was first encountered by Bruening and co-workers, who found that the highly purified dimeric form of the satellite RNA of tobacco ringspot virus (sTobRV) spontaneously cleaves into monomeric RNAs of equal length (359 nucleotides) and identical sequence (Prody et al. 1986). cDNA clones of sTobRV transcribed in vitro using SP6 RNA polymerase cleave at a rate and specificity comparable to sTobRV RNA isolated from tobacco, indicating that no accessory proteins or unidentified covalent modifications of the RNA are responsible for the observed cleavage (Buzayan et al. 1986a). When two monomeric RNAs are incubated together under favorable reaction conditions, the 5′-OH and 2′,3′-cyclic phosphate termini are joined, albeit inefficiently, to reform dimeric RNA, demonstrating that the reaction is fully reversible. Since the reaction produces the same termini as found in vivo, it is likely to constitute an integral step in the rolling circle mechanism of sTobRV replication (Keese and Symons 1987).

Self-cleavage has also been found in the RNA genome of avocado sunblotch viroid (ASBV). Unlike sTobRV, ASBV is primarily found as a circular RNA that does not self-cleave. However, when linear dimers of the ASBV genome are prepared by in vitro transcription, self-cleavage at a unique site is observed. Although some cleavage occurs during the transcription reaction, even more occurs after the dimer RNA is heat-denatured and rapidly cooled. This suggests that only a fraction of the molecules form a specific structure that is required for cleavage. Comparison of the sequences of sTobRV, ASBV, lucerne transient streak virus (LSTV), and several other related plant pathogenic RNAs have revealed a common RNA secondary structure that could form in the neighborhood of the cleavage site. This secondary structure motif, called the hammerhead, consists of three RNA helices of variable sequence that are joined by a core of 11 conserved nucleotides (Fig. 6). By truncating the sTobRV and LSTV genomes, it was shown that the hammerhead motif alone was sufficient for cleavage in the presence of Mg^{++} (Buzayan et al. 1986b). These smaller RNAs cleaved much more efficiently than the full-length genomic RNAs.

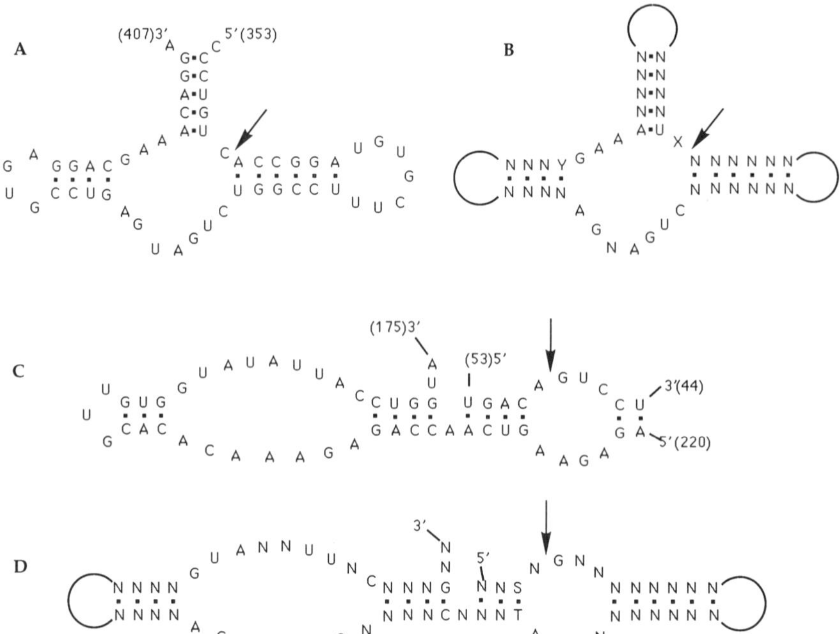

Figure 6 Self-cleaving motifs from plant viroids. (*A*) The "hammerhead" motif from (+)sTobRV. (*B*) Consensus sequence of the hammerhead ribozyme; Y = C, U; X = A, C, U. (*C*) The "hairpin" motif from (–)sTobRV. (*D*) Consensus sequence of the hairpin ribozyme determined by a combination of mutagenesis (R. Tritz et al., in prep.) and in vitro selection (Berzal-Herranz et al. 1992).

The hammerhead is not the only self-cleaving RNA motif. The antigenomic strand of sTobRV contains a different motif, often called the hairpin, that consists of a minimum of 65 nucleotides (Fig. 6). The hairpin cleaves at a rate comparable to the hammerhead, but certain RNAs containing the hairpin also carry out the reverse ligation reaction very efficiently. A self-cleaving motif of about 100 nucleotides has been found as part of the genomic and antigenomic strands of hepatitis δ RNA, a satellite of hepatitis B virus (Sharmeen et al. 1988). Like the hairpin and hammerhead motifs, the δ motif is thought to play a critical role in the RNA replication cycle. Finally, a motif of unknown function was found in a transcript of a satellite DNA in the mitochondria of *Neurospora* (Saville and Collins 1990).

The structure and function of RNA self-cleaving motifs have been conveniently studied by preparing them from two separate RNA molecules. These RNAs can be combined under the desired reaction conditions, and cleavage occurs after the motif has assembled. The RNA that contains the cleavage site is called the substrate and the unaltered RNA is called the ribozyme. By measuring the rate of cleavage for various combinations of mutant ribozymes and mutant substrates, the sequence requirements of the hammerhead were established. These are summarized in Figure 6 and generally support a model proposed by Forster and Symons (1987a,b). Although the lengths and base compositions of the helices can affect the assembly of the hammerhead, only one base pair is strictly required for cleavage. However, mutation of most of the single-stranded nucleotides in the central core results in a substantial reduction in the cleavage rate (Ruffner et al. 1990). Chemical synthesis methods have been used to show that even the removal of functional groups of certain essential nucleotides are deleterious to the reaction (Fu and McLaughlin 1992; Slim and Gait 1992). A number of experiments show that specific parts of the ribose-phosphate backbone in the hammerhead are also needed for optimal activity. Introduction of phosphorothioates into four positions in the hammerhead core reduces cleavage substantially (Ruffner and Uhlenbeck 1990). Substitution of the 2′-OH with 2′deoxy, 2′fluoro, or 2′ amino residues can have substantial effects on cleavage rates at five particular locations (Pieken et al. 1991; Yang et al. 1992).

The rather precise structural requirements for hammerhead cleavage suggest that, like tRNA, the hammerhead folds into a precise three-dimensional structure that is essential for function. Like the central core of tRNA, the central core of the hammerhead probably contains a large number of specific hydrogen bonds between bases and between bases and the backbone. Unlike tRNA, no alternate combinations of bases have been found so far in the hammerhead that can form equivalent tertiary base-base interactions. As was the case for Pb^{++} cleavage of tRNA^{Phe}, the three-dimensional structure of the hammerhead will be illuminating.

The secondary structure and sequence requirements of the hairpin motif have recently been established through a combination of mutagenesis experiments on a ribozyme-substrate complex (R. Tritz et al., in prep.) and in vitro selection experiments (Berzal-Herranz et al. 1992). The secondary structure of the hairpin consists of four helices with the cleavage site placed within an internal loop (Fig. 6). The relatively extensive sequence requirements for hairpin cleavage suggest that, like the hammerhead, a complex tertiary structure must form in order for cleavage to occur. The secondary structures and sequence requirements of the δ and *Neurospora* self-cleaving motifs are not yet fully established. Successful assembly of the δ motif from two RNA fragments has recently been achieved, and several models have been proposed that contain up to four helices (Branch and Robertson 1991; Perrotta and Been 1991; Wu et al. 1992).

An additional consequence of preparing the catalytic motifs from two RNA fragments is that if the helices joining ribozyme to substrate are short, the products will dissociate from the ribozyme. This permits binding of a second substrate molecule to the ribozyme, and a multiple turnover reaction is observed. By measuring the cleavage rate as a function of substrate concentration, the steady-state rate constants of reaction can be obtained. K_M values as low as 40 nM and k_{cat} values as high as 1.5 min^{-1} have been reported for the hammerhead at 25°C and pH 7.5, although analysis is often complicated by the presence of inactive conformational forms of ribozyme and/or substrate. Under similar conditions, similar values of K_M and k_{cat} have been obtained for the hairpin (Hampel and Tritz 1989) and δ (Perrotta and Been 1992) ribozymes. The hammerhead value of k_{cat}/K_M of about 10^7 M^{-1}min^{-1} is similar to protein ribonucleases, although the protein ribonucleases have faster k_{cat} values and higher K_M values.

The minimal kinetic scheme of the two-fragment cleavage reactions involves three steps: substrate binding, cleavage chemistry, and release of each of the cleavage products (Fig. 7). A variety of kinetic methods have been used to determine the elemental rate constants for each of the steps of the hammerhead cleavage reaction (Fedor and Uhlenbeck 1993). The rates of substrate binding, product binding, and product release reflect the helix-coil transition of a standard RNA helix. Measurement of the substrate dissociation rate suggests that it is faster than expected from the length of the helices. Consequently, the lengths of the two helices that form the intermolecular complex greatly affect the overall reaction

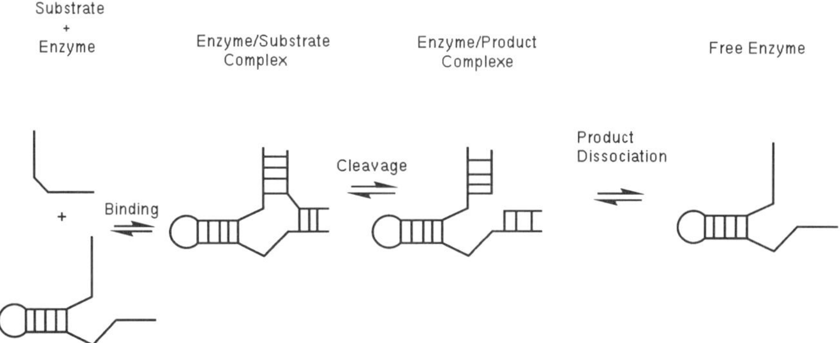

Figure 7 Minimal kinetic scheme of the hammerhead ribozyme. The smaller RNA is the substrate and the larger RNA is the enzyme of the reaction.

rate. In a multiple turnover reaction when the helices are short, the chemical step is rate-limiting at saturating substrate concentration. As the helices become longer, product release becomes slower. At a specific and critical length, product release is slower than the chemical step and so is rate-limiting.

A major benefit of performing the kinetic experiments which employ intermolecular cleavage is that conditions can be found where reaction rates reflect k_2, the rate constant of the cleavage step of the reaction. At pH 7.5 and 25°C, $k_2 = 1$–2 min^{-1} for six hammerheads with different helices. Estimates of the noncatalyzed or background rate of cleavage in the same buffer are in the order of 10^{-6} min^{-1} (K.J. Hertel, pers. comm.). Thus, the folded RNA enhances the rate of cleavage at a unique site by about six orders of magnitude.

Several lines of evidence suggest that divalent metal ions are involved in the mechanism of hammerhead cleavage. Although the observed requirement for divalent ions in the cleavage reaction could simply be to promote the folding of the hammerhead structure, virtually no cleavage occurs when spermine or NaCl is used in place of Mg^{++}. A very low residual cleavage rate ($<10^{-4}$ min^{-1}) in the absence of divalent metal ions may reflect a metal-independent component of the reaction or may be the result of small amounts of residual divalent metal ions, despite considerable efforts to remove impurities. In the presence of spermine, hammerhead cleavage occurs with a large number of divalent ions including Mg^{++}, Mn^{++}, Co^{++}, Ca^{++}, Cd^{++}, Zn^{++}, and Sr^{++} (Dahm and Uhlenbeck 1991). However, other divalent metal ions such as Pb^{++} and Ba^{++} do not stimulate cleavage.

An experiment that clearly implicates a divalent metal ion in the hammerhead cleavage mechanism employs a hammerhead containing a sin-

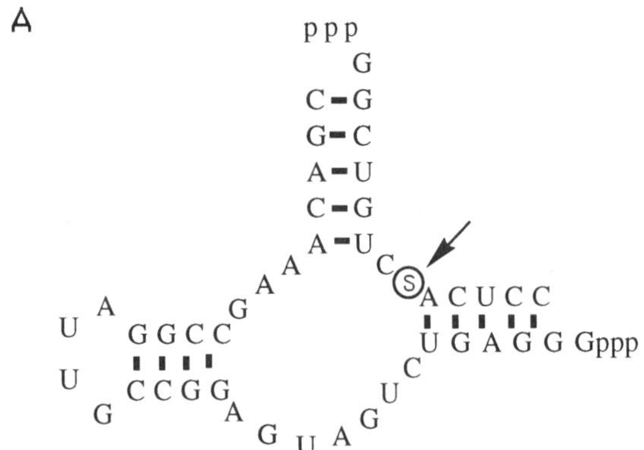

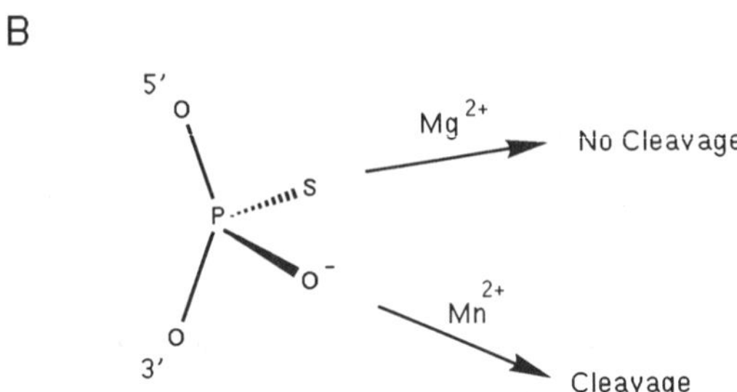

Figure 8 Mn++ "rescue" experiment. (*A*) Sequence of the hammerhead ribozyme that was used for this experiment (Dahm and Uhlenbeck 1991). The *pro*-Rp phosphorothioate incorporated by transcription is shown as a circle. (*B*) Effect of the sulfur substitution on the cleavage reaction. Substantial cleavage occurs only in the presence of Mn++.

gle phosphorothioate at the cleavage site. When the *pro*-Rp oxygen is replaced with a sulfur, the hammerhead loses its ability to cleave in 10 mM MgCl$_2$, whereas it remains active in 10 mM MnCl$_2$. Since Mn++ is much more thiophilic than Mg++ (Jaffe and Cohn 1979), this change in divalent metal ion specificity is consistent with metal ion coordination to the *pro*-Rp oxygen at the cleavage site (Fig. 8). Substitution of the *pro*-Sp oxygen shows no such effect (Koizumi and Ohtsuka 1991; Slim and Gait 1991).

The pH dependence of the rate of hammerhead cleavage also clearly suggests a role of metal ions in the mechanism (S.C. Dahm and O.C. Uhlenbeck, in prep.). Between pH 5.5 and 8.5 where RNA nucleotides generally do not titrate, the rate of the chemical step of cleavage increases with pH in a manner consistent with the titration of a basic group. When the pH rate profile is determined in the presence of different metal ions, the profiles shift in accordance with the pK_a of the metal ion. Thus, at a given pH in the linear range, the rate of cleavage in Ca^{++} (pK_a = 12.9) is about 30-fold less than Mg^{++} (pK_a = 11.4) and 200-fold less than Mn^{++} (pK_a = 10.6). This suggests that the fraction of hammerheads with a bound metal hydroxide at neutral pH are the active species in the cleavage reaction. As would be expected, Co^{++} (pK_a = 10.2) has a similar pH profile as Mn^{++}. However, although Cd^{++} (pK_a = 9.0) cleaves well, it does not cleave better than Co^{++} or Mn^{++} as would be expected. This suggests that differences in cleavage rate may also be dependent on other properties of metal ions, such as the precise location in the RNA structure.

In summary, at this stage of knowledge, the mechanism of hammerhead cleavage appears somewhat similar to the cleavage of yeast tRNA[Phe] by Pb^{++}, in which the precisely folded RNA structure positions a metal hydroxide near the attacking $2'$ oxygen to increase its nucleophilicity and probably abstract a proton. The hammerhead contains a metal ion that directly coordinates to one, but not to the other, nonbridging oxygen at the cleavage site. Although this coordination is required for catalysis, it is not known how it participates in the reaction. Additional metal ions could be involved in transition-state stabilization or stabilizing the leaving group. Although Pb^{++} does not appear to coordinate either of the nonbridging oxygens in tRNA[Phe], it is possible that a Mg^{++} provides this function in solution.

It is instructive to compare the absolute cleavage rates of the hammerhead and tRNA[Phe] by Pb^{++}. If we assume that Pb^{++} binds to tRNA[Phe] just as well as Mg^{++} and that half the bound Pb^{++} is in the hydroxide form, the cleavage rate of tRNA with a bound Pb^{++} hydroxide is 9 min^{-1}. If we assume that the Mg^{++} hydroxide concentration in the hammerhead can be calculated from the pK_a of Mg^{++} in water, the cleavage rate of the hammerhead with a bound Mg^{++} hydroxide is 8000 min^{-1}. Thus, the hammerhead is intrinsically about 900 times better at stimulating the transesterification reaction. Although many factors may be responsible for the difference, it is clear from the crystal structures that the structure of the Pb^{++} cleavage site is not optimal for cleavage. The bound Pb^{++} is too far away from the $2'$ oxygen, and the ribose is not in a configuration for an in-line attack. The hammerhead may have a tertiary structure that places the active Mg^{++} close to a critical phosphodiester bond and has a configuration where in-line attack is more favorable.

CONCLUSIONS AND PERSPECTIVES

Since divalent metal ions were undoubtedly present in the prebiotic solutions of the RNA world, RNA has evolved to employ metal ions in a number of interesting ways. By being able to interact with more than one phosphate simultaneously, divalent metal ions are more effective than monovalent cations in neutralizing the negative charges on the RNA backbone and thereby promoting folding of the RNA chain. Whereas most divalent metal ions bind to RNA with similar affinity and are not well localized, certain divalent ions bind with higher affinity in "pockets" where the folded RNA chain is constrained to allow direct coordination with phosphate oxygens or other functional groups. On the basis of available X-ray data, it is clear that even an RNA as small as tRNA can bind divalent ions in an extraordinary diversity of sites. Thus, we can expect larger RNAs to bind a large number of metal ions at well-defined sites. Unfortunately, it has been very difficult to locate these sites, since no X-ray crystal structure is available. Methods such as mapping metal ion cleavage sites or using the metal ion specificity of phosphorothioates have shown promise; they cannot, however, reliably locate all binding sites.

It is clear that divalent metal ions play an essential role in RNA catalysis. We reviewed a number of biological and nonbiological reactions where a bound metal ion promotes cleavage of the RNA phosphodiester bond at a specific site by transesterification to give a $2',3'$-cyclic phosphate. Cech (this volume) reviews evidence that bound metal ions promote RNA transesterification and hydrolysis reactions by the group I, group II, and RNase P ribozymes. Although the evidence is incomplete, it is possible that one or more of the RNA enzymes stimulate catalysis by a two-metal-ion mechanism that resembles the exonuclease activity of DNA polymerase I. In this type of mechanism, one metal ion enhances the nucleophilicity of water or a ribose hydroxyl, and the other acts as an electrophile to stabilize the oxyanion on the leaving ribose hydroxyl (Fig. 9A). Both metal ions can be coordinated to one of the nonbridging oxygens of the phosphate. Similar mechanisms have been proposed for alkaline phosphatases. Thus, it appears that a common mechanism of protein enzymes is shared by RNA enzymes.

A substantial number of protein enzymes use precisely placed metal ions to promote catalysis. Figure 9 shows examples of postulated reaction mechanisms involving metal ions for a variety of reactions catalyzed by protein enzymes. Although the amino acid functional groups participate directly in the catalysis in some cases, in other cases they simply position the substrate in proximity to the metal ion. This suggests that if specific substrate binding by RNA can be achieved, a substantial number of biochemical transformations could be accomplished using RNA catalysts. Recent success in selecting RNAs that bind a variety of small ligands (Ellington and Szostak 1990) suggests that selective substrate

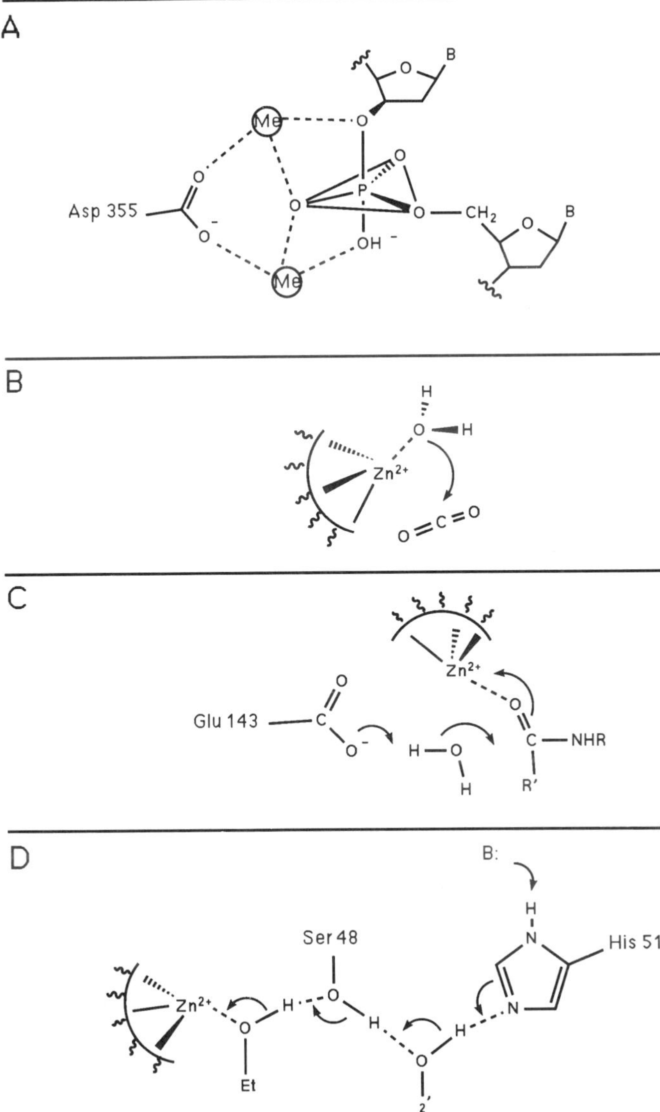

Figure 9 Reaction mechanisms of various metalloproteins. (*A*) 3'-5'-Exonuclease activity of *E. coli* DNA polymerase I by electrophilic and nucleophilic catalysis (Beese and Steitz 1991). (*B*) Carbonic anhydrase by a Zn^{++}-coordinated water or hydroxide (Kannan et al. 1975). (*C*) Thermolysin by electrophilic polarization of the substrate (Holmes et al. 1983). (*D*) Alcohol dehydrogenase by a Zn^{++}-coordinated ethoxide (Andersson et al. 1981).

binding by RNA is clearly possible. Thus, it is reasonable to postulate that the catalytic repertoire of RNA is much broader than various phosphate transfer reactions catalyzed by the ribozymes discovered to date. We can anticipate a vigorous search for new RNA catalysts in biological and nonbiological systems in the coming years.

ACKNOWLEDGMENTS

This work was sponsored by grants from the National Institutes of Health (O.C.U). T.P. acknowledges a postdoctoral fellowship from the Damon Runyon-Walter Winchell Cancer Research Fund (DRG-1103). D.M.L. is supported by a postdoctoral fellowship from the National Institutes of Health (GM-14471). We also thank the W.M. Keck Foundation for their generous support of RNA science on the Boulder campus.

REFERENCES

Andersson, P., J. Kvassman, A. Lindström, B. Olden, and G. Pettersson. 1981. Effect of NADH on the pK_a of zinc-bound water in liver alcohol dehydrogenase. *Eur. J. Biochem.* **113:** 425–433.

Beese, L.S. and T.A. Steitz. 1991. Structural basis for the 3'-5' exonuclease activity of *Escherichia coli* DNA polymerase I: A two metal ion mechanism. *EMBO J.* **10:** 25–33.

Behlen, L.S., J.R. Sampson, A.B. DiRenzo, and O.C. Uhlenbeck. 1990. Lead catalyzed cleavage of yeast tRNA^{phe} mutants. *Biochemistry* **29:** 2515–2523.

Belaich, J.P. and J.C. Sari. 1969. Microcalorimetric studies on the formation of magnesium complexes of adenine nucleotides. *Proc. Natl. Acad. Sci.* **64:** 763–770.

Berzal-Herranz, A., S. Joseph, and J.M. Burke. 1992. In vitro selection of active hairpin ribozymes by sequential RNA-catalyzed cleavage and ligation reactions. *Genes Dev.* **6:** 129–134.

Bina-Stein, M. and A. Stein. 1976. Allosteric interpretation of Mg^{2+} binding to the denaturable *Escherichia coli* tRNA$^{Glu}/_{2+}$. *Biochemistry* **15:** 3912–3917.

Branch, A.D. and H.D. Robertson. 1991. Efficient *trans* cleavage and a common structural motif for the ribozymes of the human hepatitis delta agent. *Proc. Natl. Acad. Sci.* **88:** 10163–10167.

Brown, R.S., J.C. Dewan, and A. Klug. 1985. Crystallographic and biochemical investigation of the lead(II)-catalyzed hydrolysis of yeast phenylalanine tRNA. *Biochemistry* **24:** 4785–4801.

Butzow, J.J. and G.L. Eichorn. 1965. Interactions of metal ions with polynucleotides and related compounds. LV Degradation of polyribonucleotides by zinc and other divalent metal ions. *Biopolymers* **3:** 95–107.

Buzayan, J.M., W.L. Gerlach, and G. Bruening. 1986a. Non-enzymatic cleavage and ligation of RNAs complementary to a plant virus satellite RNA. *Nature* **323:** 349–353.

———. 1986b. Satellite tobacco ringspot virus RNA: A subset of the RNA sequence is sufficient for autolytic processing. *Proc. Natl. Acad. Sci. USA* **83:** 8859–8862.

Celander, D.W. and T.R. Cech. 1991. Visualizing the higher order folding of a catalytic RNA molecule. *Science* **251:** 401–407.

Chaffee, E., T.P. Dasgupta, and G.M. Harris. 1973. Kinetics and mechanism of aquation and formation of carbonato complexes. V. Carbon dioxide uptake by hydroxopentaminecobalt (III) ion to form carbonatopentaamminecobalt (III) ion. *J. Am. Chem. Soc.* **95:** 4169–4173.

Clouet d'Orval, B., Y. d'Aubenton-Carafa, J.M. Brody, and E. Brody. 1991. Determination of an RNA structure involved in splicing inhibition of a muscle-specific exon. *J. Mol. Biol.* **221:** 837–856.

Cohn, M., A. Danchin, and M. Grunberg-Manago. 1969. Proton magnetic relaxation studies of manganous complexes of transfer RNA and related compounds. *J. Mol. Biol.* **39:** 199–217.

Crothers, D.M. and P.E. Cole. 1978. Conformational changes of tRNA. In *Transfer RNA* (ed. S.A. Hman), pp. 196–247. MIT Press, Cambridge, Massachusetts.

Dahm, S.C. and O.C. Uhlenbeck. 1991. Role of divalent metal ions in the hammerhead RNA cleavage reaction. *Biochemistry* **30:** 9464–9469.

Danchin, A. 1972. tRNA structure and binding sites for cations. *Biopolymers* **11:** 1317–1333.

Dange, V., R.B. Van Atta, and S.M. Hecht. 1990. A $Mn^{2(+)}$-dependent ribozyme. *Science* **248:** 585–588.

Dao, V., R.H. Guenther, and P.F. Agris. 1992. The role of 5-methylcytidine in the anticodon arm of yeast $tRNA^{Phe}$: Site-specific Mg^{2+} binding and coupled conformational transite in DNA analysis. *Biochemistry* **31:** 11012–11020.

Dirheimer, G., J.P. Ebel, J. Bonnet, J. Gangloff, G. Keith, B. Krebs, B. Kuntzel, A. Roy, J. Weissenbach, and C. Werner. 1972. Primary structure of transfer RNA. *Biochimie* **54:** 127–144.

Dock-Bregeon, A.C. and D. Moras. 1987. Conformational changes and dynamics of tRNAs: Evidence from hydrolysis patterns. *Cold Spring Harbor Symp. Quant. Biol.* **52:** 113–121.

Ehrenberg, M., R. Rigler, and W. Wintermeyer. 1979. On the structure and conformational dynamics of yeast phenylalanine-accepting transfer ribonucleic acid in solution. *Biochemistry* **18:** 4588–4599.

Eichhorn, G.L., E. Tarien, and J.J. Butzow. 1971. Interaction of metal ions with nucleic acids and related compounds. XVI. Specific cleavage effects in the depolymerization of ribonucleic acids by zinc(II) ions. *Biochemistry* **10:** 2014–2019.

Ellington, A.D. and J.W. Szostak. 1990. *In vitro* selection of RNA molecules that bind specific ligands. *Nature* **346:** 818–822.

Farkas, W.R. 1968. Depolymerization of ribonucleic acid by plumbous ion. *Biochim. Biophys. Acta* **155:** 401–409.

Fedor, M.J. and O.C. Uhlenbeck. 1993. Kinetics of intermolecular cleavage by hammerhead ribozymes. *Biochemistry* **31:** 12042–12054.

Flanagan, J.M. and K.B. Jacobson. 1988. Effect of zinc ions on tRNA structure: Imino proton NMR spectroscopy. *Biochemistry* **27:** 5778–5785.

Forster, A.C. and R.H. Symons. 1987a. Self-cleavage of plus and minus RNAs of a virusoid and a structural model for the active sites. *Cell* **49:** 211–220.

———. 1987b. Self-cleavage of virusoid RNA is performed by the proposed 55-nt active site. *Cell* **50:** 9–16.

Fu, D.-J. and L.W. McLaughlin. 1992. Importance of specific purine amino hydroxyl groups for efficient cleavage by a hammerhead ribozyme. *Proc. Natl. Acad. Sci.* **89:** 3985–3989.

Gornicki, P., F. Baudin, P. Romby, M. Wiewiorowski, W. Kryzosiak, J.P. Ebel, C. Ehresmann, and B. Ehresmann. 1989. Use of lead(II) to probe the structure of large RNA's.

Conformation of the 3′ terminal domain of *E. coli* 16S rRNA and its involvement in building the tRNA binding sites. *J. Biomol. Struct. Dyn.* **6**: 971–984.

Hampel, A. and R. Tritz. 1989. RNA catalytic properties of the minimum (–) sTRSV sequence. *Biochemistry* **28**: 4929–4933.

Hingerty, B., R.S. Brown, and A. Jack. 1978. Further refinement of the structure of yeast tRNA^{Phe}. J. Mol. Biol. **124**: 523–534.

Holbrook, S.R., J.L. Sussman, R.W. Warrant, G.M. Church, and S.-H. Kim. 1977. RNA-ligand interactions: (1) Magnesium binding sites in yeast tRNA^{Phe}. *Nucleic Acids Res.* **4**: 2811–2820.

Holmes, M.A., D.E. Tronrud, and B.W. Matthews. 1983. Structural analysis of the inhibition of thermolysin by an active-site-directed irreversible inhibitor. *Biochemistry* **22**: 236–240.

Huheey, J.E. 1983. *Inorganic chemistry*, 3rd edition, p. 312. Harper and Row, New York.

Jack, A., J.E. Ladner, D. Rhodes, R.S. Brown, and A. Kug. 1977. A crystallographic study of metal-binding to yeast tRNA^{Phe}. *J. Mol. Biol.* **111**: 315–318.

Jaffe, E.K. and M. Cohn. 1979. Diastereomers of the nucleoside phosphorothioates as probes of the structure of the metal nucleotide substrates and of the nucleotide binding site of yeast hexokinase. *J. Biol. Chem.* **254**: 10839–10845.

Kannan, K.K., B. Notstrand, K. Fridborg, S. Lovgren, A. Ohlsson, and M. Petef. 1975. Crystal structure of human erythrocyte carbonic anhydrase B. Three-dimensional structure at a nominal 2.2-Å resolution. *Proc. Natl. Acad. Sci.* **72**: 51–55.

Kazakov, S. and S. Altman. 1991. Site-specific cleavage by metal ion cofactors and inhibitors of M1 RNA, the catalytic subunit of RNase P from *Escherichia coli. Proc. Natl. Acad. Sci.* **88**: 9193–9197.

––––––––. 1992. A trinucleotide can promote metal ion-dependent specific cleavage of RNA. *Proc. Natl. Acad. Sci.* **89**: 7939–7943.

Keese, P. and R.H. Symons. 1987. The structure of viroids and virusoids. In *Viroids and viroid-like pathogens* (ed. J.S. Semancik), pp. 1–47. CRC Press, Boca Raton, Florida.

Koizumi, M. and E. Ohtsuka. 1991. Effects of phosphorothioate and 2-amino groups in hammerhead ribozymes on cleavage rates and Mg^{2+} binding. *Biochemistry* **30**: 5145–5150.

Krakaver, H. 1971. The binding of Mg++ ions to polyadenylate, polyuridylate, and their complexes. *Biopolymers* **10**: 2459–2490.

Labuda, D. and D. Pörschke. 1982. Magnesium ion inner sphere complex in the anticodon loop of phenylalanine transfer ribonucleic acid. *Biochemistry* **21**: 49–53.

Marciniec, T., J. Ciesiolka, J. Wrzesinski, and W.J. Krzyzosiak. 1989. Identification of the magnesium, europium and lead binding sites in *E. coli* and lupine tRNA^{Phe} by specific metal ion-induced cleavages. *FEBS Lett.* **243**: 293–298.

Pan, T. and O.C. Uhlenbeck. 1992a. In vitro selection of RNAs that undergo autolytic cleavage with PB^{2+}. *Biochemistry* **31**: 3887–3895.

––––––––. 1992b. A small metalloribozyme with a two-step mechanism. *Nature* **358**: 560–563.

Pan, T., R.R. Gutell, and O.C. Uhlenbeck. 1991. Folding of circularly permuted transfer RNAs. *Science* **254**: 1361–1364.

Perrotta, A.T. and M.D. Been. 1991. A pseudoknot-like structure required for efficient self-cleavage of hepatitis delta virus RNA. *Nature* **350**: 434–436.

––––––––. 1992. Cleavage of oligoribonucleotides by a ribozyme derived from the hepatitis delta virus RNA sequence. *Biochemistry* **31**: 16–21.

Peterson, E.T. and O.C. Uhlenbeck. 1992. Determination of recognition nucleotides for *E. coli* phenylalanyl-tRNA synthetase. *Biochemistry* **31**: 10380–10389.

Pieken, W.A., D.B. Olsen, F. Benseler, H. Aurup, and F. Eckstein. 1991. Kinetic characterization of ribonuclease-resistant 2′-modified hammerhead ribozymes. *Science* **253:** 314–317.

Porterfield, W.W. 1984. *Inorganic chemistry*, p. 484. Addison-Wesley, London.

Prody, G.A., J.T. Bakos, J.M. Buzayan, I.R. Schneider, and G. Bruening. 1986. Autolytic processing of dimeric plant virus satellite RNA. *Science* **231:** 1577–1580.

Quigley, G.J., M.M. Tetter, and A. Rich. 1978. Structural analysis of spermine and magnesium ion binding in yeast tRNA^Phe. *Proc. Natl. Acad. Sci.* **75:** 64–68.

Reid, S.S. and J.A. Cowan. 1990. Biostructural chemistry of magnesium ion: Characterization of the weak binding sites on tRNA^(Phe)(yeast). Implications for conformational change and activity. *Biochemistry* **29:** 6025–6032.

Rialdi, G., J. Levy, and R. Biltonen. 1972. Thermodynamic studies of transfer ribonucleic acids. I. Magnesium binding to yeast phenylalanine transfer ribonucleic acid. *Biochemistry* **11:** 2472–2479.

Römer, R. and R. Hach. 1975. tRNA conformation and magnesium binding. A study of a yeast phenylalanine-specific tRNA by a fluorescent indicator and differential melting curves. *Eur. J. Biochem.* **55:** 271–284.

Rubin, J.R. and M. Sundaralingam. 1983. Lead ion binding and RNA chain hydrolysis in phenylalimine tRNA. *J. Biomol. Struct. Dyn.* **1:** 639–646.

Rubin, J.R., J. Wang, and M. Sundaralingam. 1983. X-ray diffraction study of the zinc(II) binding sites in yeast tRNA^Phe. *Biochim. Biophys. Acta* **756:** 111–118.

Ruffner, D.E. and O.C. Uhlenbeck. 1990. Thiophosphate interference experiments locate phosphates important for the hammerhead RNA self-cleavage reaction. *Nucleic Acids Res.* **18:** 6025–6029.

Ruffner, D.E., G.D. Stormo, and O.C. Uhlenbeck. 1990. Sequence requirements of the hammerhead RNA self-cleavage reaction. *Biochemistry* **29:** 10695–10702.

Sampson, J.R., A.B. DiRenzo, L.S. Behlen, and O.C. Uhlenbeck. 1990. Role of the tertiary nucleotides in the interaction of yeast phenylalanine tRNA with its cognate synthetase. *Biochemistry* **29:** 2523–2532.

Sampson, J.R., F.X. Sullivan, L.S. Behlen, A.B. DiRenzo, and O.C. Uhlenbeck. 1987. Characterization of two RNA-catalyzed RNA cleavage reactions. *Cold Spring Harbor Symp. Quant. Biol.* **52:** 267–275.

Saville, B.J. and R.A. Collins. 1990. A site-specific self-cleavage reaction performed by a novel RNA in Neurospora mitochondria. *Cell* **61:** 685–696.

Schreier, A.A. and P.R. Schimmel. 1974. Interaction of manganese with fragments, complementary fragment recombinations, and whole molecules of yeast phenylalanine specific. *J. Mol. Biol.* **86:** 601–620.

Sharmeen, L., M.T.P. Kuo, G. Dinter-Gottlieb, and J. Taylor. 1988. Antigenomic RNA of human hepatitis delta virus can undergo self-cleavage. *J. Virol.* **62:** 2674–2679.

Slim, G. and M.J. Gait. 1991. Configurationally defined phosphorothioate-containing oligoribonucleotides in the study of the mechanism of cleavage of hammerhead ribozymes. *Nucleic Acids Res.* **19:** 1183–1188.

———. 1992. The role of the exocyclic amino groups of conserved purines in hammerhead ribozyme cleavage. *Biochem. Biophys. Res. Commun.* **183:** 605–609.

Stein, A. and D.M. Crothers. 1976. Equilibrium binding of magnesium(II) by *Escherichia coli* tRNA^fMet. *Biochemistry* **15:** 157–160.

Sundaralingam, M., J.R. Rubin, and J.F. Cannon. 1984. Nonenzymatic hydrolysis of RNA: Pb(II)-catalyzed site specific hydrolysis of tRNA. The role of the tertiary folding of the polyneucleotide chain. *Int. J. Quant. Chem.* **11:** 355–366.

Werner, C., B. Krebs, G. Keith, and G. Dirheimer. 1976. Specific cleavages of pure tRNAs by plumbous ions. *Biochim. Biophys. Acta* **432:** 161–175.

Westhof, E. and M. Sundaralingam. 1986. Restrained refinement of the monoclinic form of yeast phenylalanine transfer RNA. Temperature factors and dynamics, coordinated waters, and base-pair propeller twist angles. *Biochemistry* **25:** 4868–4878.

Wintermeyer, W. and H.G. Zachau. 1973. Mg2+-katalysierte, spezifische Spaltung von tRN A. *Biochim. Biophys. Acta* **299:** 82–90.

Wu, H.-N., Y.-J. Wang, C.-F. Hung, H.-J. Lee, and M.M.C. Lai. 1992. Sequence and structure of the catalytic RNA of hepatitis delta virus genomic RNA. *J. Mol. Biol.* **223:** 233–245.

Yang, J.-H., N. Usman, P. Chartrand, and R. Cedergren. 1992. Minimum ribonucleotide requirement for catalysis by the RNA hammerhead domain. *Biochemistry* **31:** 5005–5009.

13

Splicing of Precursors to mRNA by the Spliceosome

Melissa J. Moore, Charles C. Query, and Phillip A. Sharp
Center for Cancer Research and Department of Biology
Massachusetts Institute of Technology
Cambridge, Massachusetts 02139-4307

I. INTRODUCTION

The RNA world pictures a time when RNA structures catalyzed many reactions, almost certainly including RNA splicing. The self-splicing reactions of group I and group II introns may be direct remnants of the RNA world. Both of these catalytic structures can facilitate *cis*-splicing, where the two exons are initially part of a contiguous RNA, and *trans*-splicing, where the two exons are initially on two separate RNAs. In the latter case, the partial intron sequences flanking the two exons pair through sequence complementarity to form a catalytic structure joining the associated exons.

Many have speculated that the exon-intron structure was common in the genetic material in the RNA world. In fact, exons have been hypothesized as having originally evolved as units of sequences which encoded functional protein domains that were assembled by RNA splicing into large multidomain proteins with more complex functions. In this period, it is highly likely that RNA splicing was the product of the action of group I- and group II-like catalytic structures, and peptide-bond formation was catalyzed by progenitors of ribosomal RNA. As the potential to generate specific proteins evolved, some of these proteins probably bound to sequences within the catalytic introns to facilitate the rate of splicing. Eventually, the situation must have evolved to a stage where a fragment of RNA derived from a catalytic group I- or group II-like intron was transcribed in abundance and began to associate in *trans* with many different self-splicing introns to facilitate splicing. Perhaps the RNA fragment promoted splicing even more efficiently when associated with a specific protein. Subsequently, introns would have lost the duplicated function, becoming totally dependent on the *trans*-acting small RNA-

The RNA World
© 1993 Cold Spring Harbor Laboratory Press 0-87969-380-0/93 $5 + .00

protein complex. This phenomenon would have been repeated again and again until the only conserved intron sequences left in a *cis* orientation to the exons were those at the intron-exon boundaries, which directed the assembly of the small RNA-protein complexes and specified the exact sites for the transesterification reactions. An additional set of conserved sequences was necessary for the transesterification reaction that formed an RNA branch. Finally, the small RNA-protein complexes matured into the small nuclear ribonucleoprotein particles, U1, U2, U4/U6, and U5, and the complex formed by these particles during excision of an intron became the present-day spliceosome.

The obvious similarity between the intermediates and products of group II self-splicing introns and the nuclear spliceosome, responsible for excision of introns from almost all nuclear genes, provides strong indirect evidence that the above evolutionary hypothesis is correct. However, direct evidence relating the spliceosomal process to the group II intron self-splicing process has not as yet been reported. In fact, it has been remarkably difficult to convincingly demonstrate structure or sequence homology between group II introns and the nuclear snRNAs. Yet, this remains the most likely source of further evidence relating the two processes. Somewhat surprisingly, even though the self-splicing group II intron reaction was discovered more than 8 years ago, little is known about the specific functions of RNA sequences within its catalytic core. On the other hand, the spliceosome has now begun to yield its secrets concerning RNA structure to a combination of deft biochemistry and sophisticated genetics.

This review concentrates on recent advances in the study of the nuclear spliceosomal process. It first focuses on the chemistry of the splicing reactions within the spliceosome and evidence for important snRNA-snRNA interactions. An important concept emerging from this analysis is that the spliceosome (and probably group II self-splicing introns) utilizes two different catalytic sites during the two steps of splicing. The roles of proteins in the spliceosome are reviewed, with emphasis on both factors identified in vertebrate systems by biochemical assays as well as factors identified by genetic analysis in yeast. The phenotypes of temperature-sensitive mutants of *Saccharomyces cerevisiae* confirm the nature of intermediate complexes of snRNP particles and pre-mRNA in the spliceosome cycle first characterized on a kinetic basis. It is speculated that the repetitive arginine/serine sequences found in many proteins important in splicing could be stabilizing unique RNA secondary structures, similar to the function of arginine in the binding of the HIV TAT protein to a bulged loop of RNA. The *cis*-splicing process is contrasted with the closely related *trans*-splicing process. The latter is thought to be not dependent on U1 snRNA-pre-mRNA interaction and may not require sequences equivalent to those of U5 snRNA. However, the *trans*-splicing process is dependent on a unique interaction between the splice leader RNA and U6 snRNA. Recent evidence suggests that the nuclear spliceosomal process may be error-prone by (inappropriately) joining exons out of their functional order. A proofreading system is described that would eliminate these nonfunctional RNAs. Finally, interactions of splicing components that define sequences to be processed as exons and introns are briefly considered, along with the current understanding of regulation of splicing in sex determination in *Drosophila*.

This review draws heavily on recent reviews by Ruby and Abelson (1991), Guthrie (1991), and Green (1991). Material in reviews by Padgett et al. (1986) and Sharp (1987) is considered background for this review and is not specifically referenced.

II. THE CHEMISTRY OF SPLICING

From a chemical standpoint, the reactions involved in pre-mRNA splicing are quite simple: the exchange of one substituent on a phosphodiester bond for another. Yet, uncertainties abound when the problem is considered in slightly more detail. Although the intermediates and products of splicing were characterized some years ago, only recently have the reaction mechanisms been established with certainty, and the pre-mRNA structural features that are important for the transesterification reactions are just now being investigated. Whether the reactions are protein- or RNA-catalyzed also remains an open question. If the spliceosome is indeed a ribozyme, as has been proposed, what is its internal structure, and to what extent is that structure related to the known self-splicing catalytic introns? In addition, are the two steps of splicing catalyzed by a single reactive center or by two separate active sites?

A. A Two-step Mechanism

All introns within nuclear mRNA precursors are thought to be removed by the same two-step mechanism (Fig. 1A). The first step entails cleavage at the 5′ splice site to yield a 5′ exon intermediate (exon 1) with a free 3′-OH terminus. Concurrently, the 5′ end of the intron (IVS) is joined, via a 2′-5′ phosphodiester bond, to an adenosine residue within the intron (the branch site) generally located within 100 nucleotides of the 3′ splice site. This forms the so-called lariat intermediate RNA con-

taining a branched intron with attached 3′ exon (exon 2). The exon 1 and
IVS-exon 2 intermediates are subsequently resolved by cleavage at the
3′ splice site and ligation of the two exons via a 3′-5′ phosphodiester
bond. This yields two products: spliced exons and an excised lariat intron
with a 3′-OH terminus.

In contrast to intramolecular *cis*-splicing, *trans*-splicing involves the
joining of exons from two independent transcripts. In trypanosomes and
nematodes, a short transcript, the spliced-leader (SL) RNA, is *trans*-

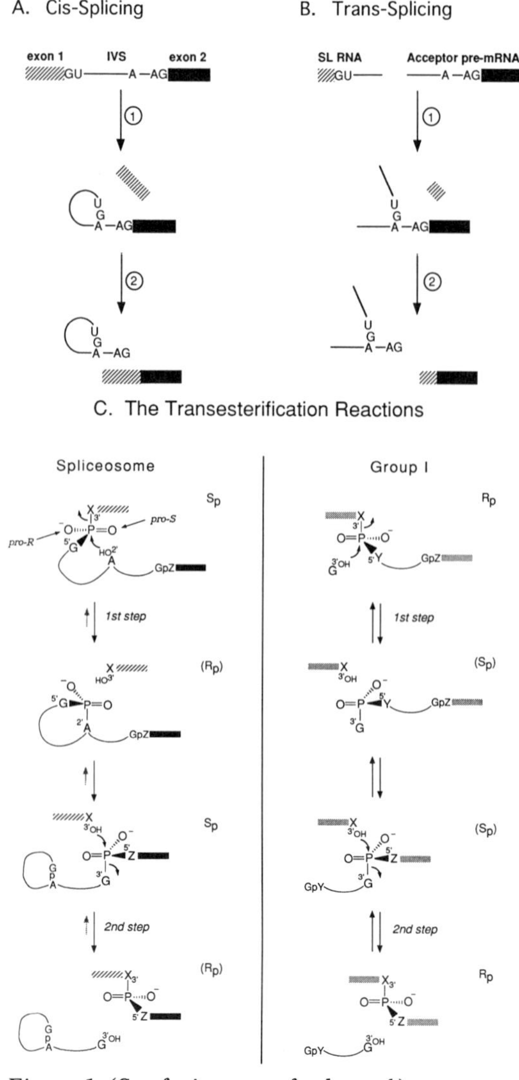

Figure 1 (See facing page for legend.)

spliced to acceptor pre-mRNAs. Except for its intermolecular nature, *trans*-splicing exactly parallels the two steps of *cis*-splicing. Thus, in the first step (Fig. 1B), cleavage at the SL donor site yields free 5′ SL exon; the remainder of the SL RNA is joined to a branch site adenosine in the acceptor pre-mRNA to form a Y-intermediate. In the second step, cleavage at the 3′ splice site and exon ligation yield the spliced exons and an excised Y-intron. In trypanosomes, this reaction results in a common 39-nucleotide sequence at the 5′ end of all mRNAs, whereas in nematodes, both *cis*- and *trans*-splicing occur (discussed further in Section IV.C).

The process by which pre-mRNA introns are removed is thought to be evolutionarily related to that of group II self-splicing introns (see Cech, this volume), because excision of the latter yields exactly the same types of intermediates and products as the former. This observation initially fueled the widely held belief that the two steps of pre-mRNA splicing are RNA-catalyzed (Cech 1985; Sharp 1985). Group I self-splicing proceeds by a different two-step mechanism in which no lariat is involved (Fig. 1C, right). Instead, 5′ exon cleavage in the first step is accompanied by attachment of a free guanylyl nucleotide to the first nucleotide of the intron via a 3′-5′ phosphodiester bond. The second step of group I intron excision is essentially the reverse of the prior step: A conserved guanosine at the 3′ splice site is recognized by the "G-binding site" of the intron and defines the site of intron excision and exon ligation.

Figure 1 Two-step chemical mechanism for pre-mRNA splicing. (*A,B*) Overall schematics of *cis*-splicing and *trans*-splicing, respectively, showing the highly conserved nucleotides and structures of the intermediates and products. (*C*) Differences and similarities between the transesterification reactions involved in pre-mRNA splicing (*left*) and those involved in excision of group I self-splicing introns (*right*). First step differences include opposite phosphorothioate diastereomer preferences, different reactive nucleophiles (2′-OH vs. 3′-OH), and differential placement of the conserved guanosine (5′ vs. 3′) with respect to the reactive phosphate. Second step similarities include preference for the S_p phosphorothioate diastereomer, use of a 3′-OH as the nucleophile and a conserved guanosine attached to the reactive phosphate at the 3′-position. The phosphorothioate diastereomers (R_p or S_p) which do not significantly inhibit each step are shown, with the resultant diastereomer shown in parentheses. (For the group I case, the stereochemical outcome for the second step has only been determined in the reverse direction.) The pro-R and pro-S oxygens are indicated for one phosphate (upper left corner). The potential (but not yet observed) reversibility of each step in pre-mRNA splicing is indicated by the small stippled arrows. X, Y, and Z represent nucleotide positions that are not highly conserved.

B. Transesterification Reactions

Although ATP hydrolysis is required for intron excision from pre-mRNAs (see Section III), none of the phosphates in the spliced products is derived from exogenous ATP. Rather, the splice site phosphates are conserved in the products (Padgett et al. 1984; Konarska et al. 1985; Lin et al. 1985). Additionally, it had been found that neither step of splicing could be separated into bond-breaking and bond-forming reactions. Consistent with these observations, it has recently been shown that both steps proceed through direct "in-line" S_N2 displacement mechanisms (Fig. 1C, left). This was determined by examining the stereochemical outcome of each step using in-vitro-transcribed substrate RNAs into which chiral phosphorothioate analogs (in which one of the non-bridging oxygens, pro-R or pro-S [indicated in Fig. 1C], has been replaced by sulfur to yield the R_p or S_p phosphorothioate diastereomer, respectively) had been specifically incorporated at the splice sites (M.J. Moore and P.A. Sharp, in prep.; R.A. Padgett, pers. comm.). Thus, the first step involves direct nucleophilic attack by the 2'OH of the branch site adenosine on the phosphate at the 5'splice site. This displaces the 3'oxygen of the 5'exon by way of a pentacoordinate transition state. In the second step, the same 3'OH of the 5'exon, which had been the leaving group in the first reaction, reverses its role and acts as a nucleophile at the 3'splice site phosphate to similarly displace the 3'end of the intron. Interestingly, these direct displacement reactions parallel the known reaction mechanism of group I self-splicing introns (McSwiggen and Cech 1989; Rajagopal et al. 1989) and are therefore consistent with the notion that pre-mRNA splicing is an RNA-catalyzed process.

C. Splice Site Structural Requirements

1. Sequence Elements

The exact sites for the transesterification reactions are defined by consensus sequences, primarily within the intron, around both the 5'and 3'splice sites. These sequences are highly conserved in yeast, but much less so in mammals. The 5'splice site consensus is relatively short: AG/<u>GU</u>AUGU in yeast and AG/<u>GU</u>RAGU in mammals (where R=purine, Y=pyrimidine, N=any nucleotide, and / indicates a splice site). In contrast, 3'splice sites are defined by three separate sequence elements: the branch site, the polypyrimidine tract, and the 3'splice site consensus. Together, these elements make up a loosely defined 3'splice site region, which may extend up to 100 nucleotides into the intron upstream of the actual 3'splice site. The branch site consensus in yeast and mammals, respectively, is UACUA<u>A</u>C and YNYUR<u>A</u>C (where <u>A</u> is the site of branch formation), whereas that for the 3′ splice site is C<u>AG</u>/G and Y<u>AG</u>/G, respectively. A polypyrimidine tract is usually found between the branch site and 3′ splice site, but this sequence is more pronounced in mammals than in yeast. Interestingly, introns can be subclassified on the basis of long and short polypyrimidine tracts, and these subclasses have distinguishable properties in splicing (see Section III.C.1). In all the consensus sequences, only the five underlined nucleotides are ubiquitous to (almost) all pre-mRNA introns (<u>GC</u> rather than <u>GU</u> at the 5′ splice site is a relatively common exception to this rule; Jackson 1991). Similar sequence elements have been defined in other systems, such as plants and *Drosophila* (for review, see Senapathy et al. 1990; Goodall et al. 1991; Mount et al. 1992).

2. Sugars and Phosphates

For group I introns, as well as many other catalytic RNAs, the contributions of 2′-OH groups have been widely investigated (see Cech; Pan et al.; both this volume). For pre-mRNA introns, however, only three positions have been similarly examined. Not surprisingly, incorporation of 2′-deoxyadenosine at the branch site blocks lariat formation to that position. Yet, in contrast to the effects of mutations in the branch site consensus sequence, this modification does not promote the use of cryptic sites elsewhere in the intron. Instead, lariat intermediates accumulate that contain branches to non-adenosine residues at positions adjacent to the 2′-deoxyadenosine (C.C. Query and P.A. Sharp, unpubl.). Additionally, specific incorporation of single 2′-deoxy and 2′-O-methyl nucleotides adjacent to either splice site has yielded evidence that the two steps of splicing may be catalyzed by distinct active sites (see discussion in Section I.D; Moore and Sharp 1992).

Recognition of the pre-mRNA phosphate backbone has been probed by the random and site-specific incorporation of stereospecific phosphorothioates. Using a random substitution/interference assay, Maschloff and Padgett (1992) showed that mammalian spliceosome assembly is sensitive to the presence of Rp phosphorothioate diastereomers at several positions in the polypyrimidine tract. Additionally, introduction of the Rp diastereomer at either splice site effectively blocks transesterification at that site (Maschloff and Padgett 1992; M.J. Moore and P.A. Sharp, in prep.). Whether this inhibition results from disruption of important spliceosomal contacts or critical metal-binding sites is not known, although it was reported that attempts to rescue splicing of phosphorothioate containing pre-mRNAs with Mn^{++} (which accepts sulfur as a ligand much more readily than does Mg^{++}; see Cech; Pan et al.; both this volume) were unsuccessful (Maschloff and Padgett 1992). In contrast, an Sp phosphorothioate linkage at the 5′ or 3′ splice site has no detectable effect on the overall rate of either step of splicing (M.J. Moore and P.A. Sharp, in prep.; R.A. Padgett, pers. comm.); this allowed the determination of the stereochemical outcome of the two reactions described above.

D. A Network of RNA:RNA Interactions

The initial proposal for snRNA involvement in splicing stemmed from the observed complementarity between U1 snRNA and the 5′ splice site consensus (Lerner et al. 1980; Rogers and Wall 1980). Subsequent demonstration that U2, U4/U6, and U5 snRNPs were required for splicing, in addition to U1 snRNP, raised the possibility that the other snRNAs also assisted in defining the splice sites by interacting with pre-mRNA substrates through sequence complementarity. The proposal that the transesterification reactions are actually RNA-catalyzed has given further importance to the snRNAs. Therefore, much effort has been devoted to characterizing the skeleton of RNA:RNA interactions within the individual snRNP particles and within the spliceosome. These interactions can be divided into four categories: (1) intramolecular structures formed between different segments of single snRNAs, (2) intermolecular interactions between an snRNA and the pre-mRNA substrate, (3) intermolecular interactions between one snRNA and another snRNA, and (4) intramolecular interactions between specific nucleotides in the pre-mRNA. The first category comprises snRNA secondary structures within the individual snRNP particles, most of which are conserved evolutionarily and are thought to remain unchanged throughout the spliceosome cycle. These have been well reviewed elsewhere (Guthrie and Patterson 1988; Steitz et al. 1988; see also Baserga and Steitz, this volume), and for the most part, they are not considered here. Rather, the main focus is on the dynamic interactions that comprise the latter three categories. Many of these interactions have only recently been defined, and their functions within the spliceosome are controversial. Their possible temporal sequence and their relevance to particular spliceosomal complexes are further discussed throughout Section III.

The 5′ end of U1 snRNA has been shown to base-pair with nucleotides at both the 5′ and 3′ splice sites (Zhuang and Weiner 1986; Siliciano and Guthrie 1988; Reich et al. 1992). Interaction of U1 with the 5′ splice site is required for usage of that site, but it alone is not sufficient to define the exact cleavage site (Seraphin et al. 1988; Siliciano and

Guthrie 1988; Seraphin and Rosbash 1990). Rather, the cleavage site may be defined by interactions between exon nucleotides and a conserved loop sequence in U5 snRNA (k in Fig. 2B; see below) or by other interactions. Genetic experiments in *Schizosaccharomyces pombe* have demonstrated that nucleotides C9 and U10 in U1 snRNA additionally base-pair with the AG dinucleotide at the 3′splice site (f in Fig. 2B) (Reich et al. 1992). Compensatory mutations in C9 and U10 can complement mutations in the AG dinucleotide of an AG-dependent intron (for definition see Section III.C.1), but only for the first step of splicing. Therefore, the 3′splice site AG must be additionally recognized by a different mechanism for the second step to proceed.

U2 snRNA base-pairs with the branch site consensus to form a short helix which presumably bulges out the branch site A (b in Fig. 2A,B). Like the U1 snRNA-5′splice site interaction, the U2 snRNA-branch site helix has been proven by compensatory mutations in both yeast and mammals (Parker et al. 1987; Wu and Manley 1989; Zhuang and Weiner 1989).

Genetic experiments in yeast have shown that U5 snRNA interacts with exon nucleotides adjacent to both splice sites (k in Fig. 2B). In companion studies, different mutations in a conserved loop sequence of U5 snRNA were shown to have one of two effects: (1) In a substrate in which the normal 5′splice site had been mutated, some U5 snRNA mutations could activate cleavage at aberrant 5′splice sites, and (2) in a substrate in which the normal 3′splice site had been mutated, different U5 snRNA mutations could promote the processing of otherwise dead-end lariat intermediates. In both cases, the activating mutations in U5 snRNA increased the complementarity between the loop sequence and exon nucleotides adjacent to the splice site that was being activated (Newman and Norman 1991, 1992). This same U5 snRNA loop sequence has been shown, by specific photocrosslinking, to interact with exon nucleotides adjacent to the 5′splice site of a mammalian intron (Wyatt et al. 1992). It has been proposed that, by forming base pairs (in some cases non-Watson-Crick) with the nucleotides adjacent to each splice site, the loop sequence in U5 snRNA functions both to define the site of 5′exon cleavage, analogous to EBS1 (*exon binding site* 1) of group II introns (Jacquier and Jacquesson-Breuleux 1991), and to orient correctly the 5′and 3′exons prior to the ligation reaction (Newman and Norman 1992).

An additional snRNA/pre-mRNA interaction probably occurs between sequences in U6 snRNA and the pre-mRNA 5′splice site region (g in Fig. 2B). In yeast extracts, Sawa and Abelson (1992) were able to obtain UV-induced cross-links that mapped to the sequence GAAACA in

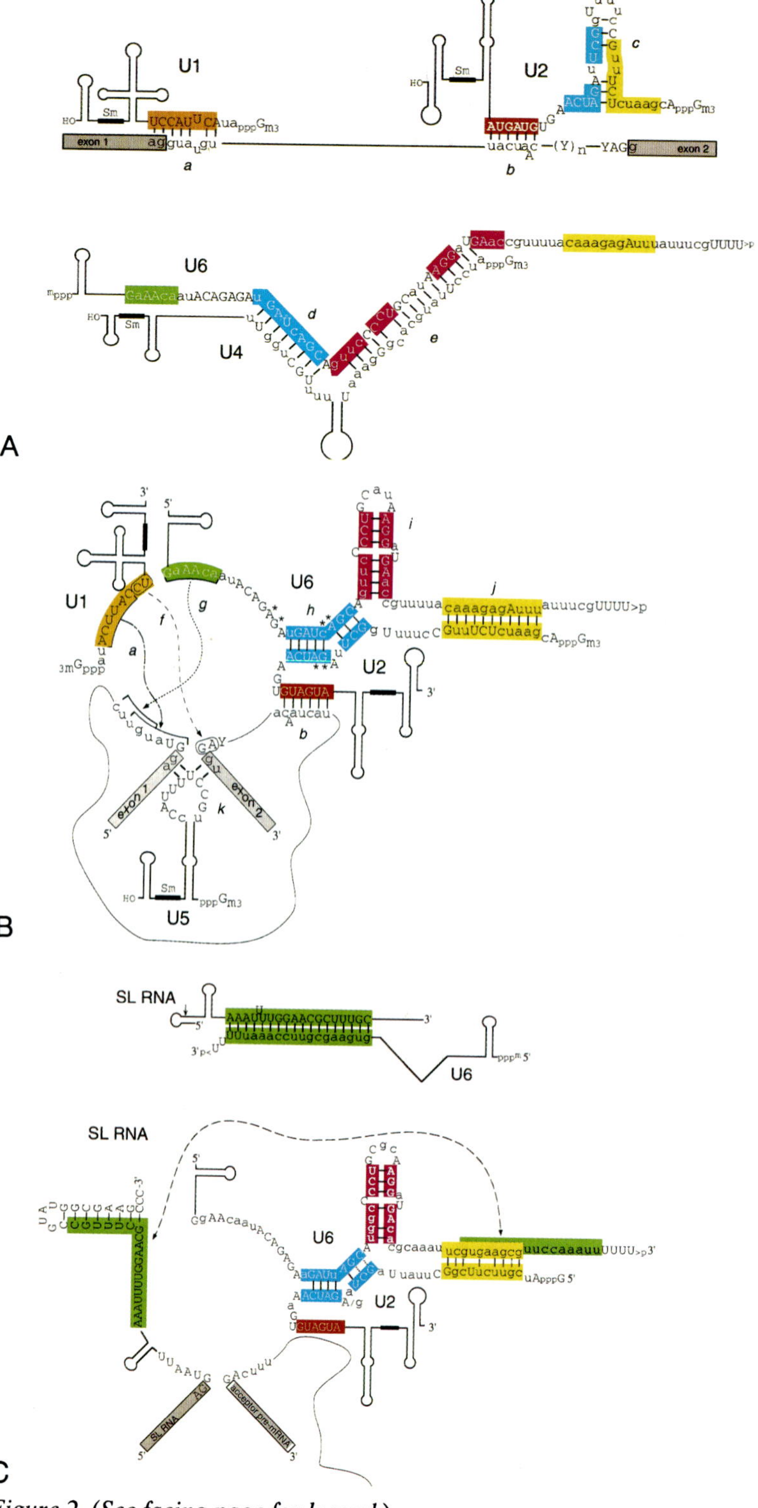

Figure 2 (See facing page for legend.)
312

U6 snRNA and nucleotides downstream from the 5′ splice site, partially overlapping the region of U1 snRNA interaction. Similar cross-links have been obtained in mammalian extracts (Sawa and Shimura 1992; Wassarman and Steitz 1992). A different sequence in nematode U6 snRNA has been shown to interact with SL RNA sequences downstream from its 5′ splice site (Fig. 2C; see Section IV.C) (Hannon et al. 1992).

The best-characterized intermolecular snRNA-snRNA interaction is the extensive base-pairing between U4 and U6 snRNAs (*d* and *e* in Fig. 2A) (Madhani et al. 1990 and references therein). This base-pairing is thought to occur in the U4/U6 and U4/U6.U5 snRNPs but to be disrupted within the context of the spliceosome (see Section III.D.2). The interaction is strongly supported by phylogenetic comparison (Brow and Guthrie 1988), but the deleterious effects of some mutations in U6 snRNA cannot be complemented by predicted compensatory mutations in U4 snRNA (Madhani et al. 1990; Vankan et al. 1990). This suggests additional functions for nucleotides in U6 snRNA. It has been proposed that U6 snRNA comprises part of the catalytic core in the spliceosome and that the U4/U6 base-pairing serves to prevent undesired catalysis by U6 snRNA at inappropriate times (Guthrie and Patterson 1988).

Two separate interactions between U2 and U6 snRNAs have been characterized. The first is a helix between sequences at the 3′ end of U6

Figure 2 Currently defined network of RNA interactions occurring between the spliceosomal snRNAs and pre-mRNA substrates. (*A*) (*Top*) Known base-pairing interactions between U1 and U2 snRNAs and pre-mRNA (orange and red bars, respectively). (*Bottom*) Extensive base-pairing between U4 and U6 snRNAs. (*B*) Known interactions between U1, U2, U5, and U6 snRNAs and pre-mRNA. In both *A* and *B*, pre-mRNA consensus sequences and snRNA sequences are those of *S. cerevisiae*; uppercase nucleotides are absolutely conserved between *S. cerevisiae* and the known sequences of other organisms (excluding trypanosomes, which do not have a GUAGUA sequence in U2 snRNA; Guthrie and Patterson 1988; Madhani and Guthrie 1992). Blue, pink, and yellow bars highlight sequences in U2 and U6 snRNAs that change base-pairing partners during the spliceosome cycle. Internal snRNA secondary structures that do not change between *A* and *B* are shown as stylized stems and loops. Asterisks indicate snRNA positions at which mutations specifically block the second step of splicing. (*C*) (*Top*) A recently described potential base-pairing interaction between nematode splice leader RNA (SL RNA) and the 3′ end of U6 snRNA. (*Bottom*) Possible interactions between nematode U2 snRNA, U6 snRNA, SL RNA, and pre-mRNA, analogous to the interactions shown in *B*. Sequences shown for SL RNA and U6 snRNA are those of *A. lumbricoides* (Hannon et al. 1992); the sequence shown for U2 snRNA is that of *C. elegans* (Thomas et al. 1990).

snRNA and the 5′ end of U2 snRNA (*j* in Fig. 2B). This interaction is required in mammalian cells (Hausner et al. 1990; Datta and Weiner 1991; Wu and Manley 1991) but not in yeast (Bordonné and Guthrie 1992; McPheeters and Abelson 1992). Interestingly, in nematodes the analogous U6 sequence is that which has been shown to base-pair with the Sm-binding site in the SL RNA (Fig. 2C) (Hannon et al. 1992). A second region of U2-U6 snRNA interaction (*h* in Fig. 2B) has only recently been demonstrated (Madhani and Guthrie 1992). Base-pairing in this bulged helix involves U6 snRNA nucleotides that were originally in U4/U6 snRNA helix *d* and U2 snRNA nucleotides that formed stem *c* in U2 snRNA alone (*c* in Fig. 2A) (McPheeters and Abelson 1992). Interestingly, two potential base pairs in helix *h* (dashed bonds) could not be proven by compensatory mutations (Madhani and Guthrie 1992). It is likely that these two U6 nucleotides have additional roles, whereas the two nucleotides in U2 snRNA may remain base-paired with their original partners in the upper half of stem *c* (McPheeters and Abelson 1992).

The sequences in U6 snRNA that pair with U2 snRNA to form helix *h* are immediately adjacent to the U6 snRNA sequence which is thought to pair with nucleotides downstream from the 5′ splice site. Likewise, the sequences in U2 snRNA that pair with U6 snRNA are immediately adjacent to the U2 snRNA sequence that binds to the branch site. The configuration shown in Figure 2B would position the U2-U6 structure near both the 5′ splice site and the branch site, i.e., near the initial catalytic event. It is interesting to note that two different similarities between helix *h* and specific secondary structures in the self-splicing introns have been suggested (Fig. 3). Madhani and Guthrie (1992) have drawn parallels between U2-U6 helix *h* and domain 5 of group II introns (Fig. 3B), whereas McPheeters and Abelson (1992) have argued that this same helix is related to P7 in group I introns (Fig. 3C). Group II domain 5 has been shown to be necessary for 5′ splice site cleavage in those introns (Jarrell et al. 1988; Koch et al. 1992), suggesting a role for helix *h* in at least the first step of pre-mRNA splicing under the Madhani and Guthrie (1992) model. P7 forms the so-called "G-binding site" in group I introns (Michel et al. 1989b; Been and Perrotta 1991), and so McPheeters and Abelson (1992) have proposed that U2-U6 helix *h* functions in binding the 3′ splice site G̲ of pre-mRNA introns, thereby functioning in the second step. Although the two models are not mutually exclusive, neither has yet been tested by doing the appropriate swapping experiments.

The final category of possible RNA-RNA interactions are those between different nucleotides within the pre-mRNA substrate itself. The interactions in this category raise the interesting mechanistic possibility of direct communication between nucleotides in the branch structure and

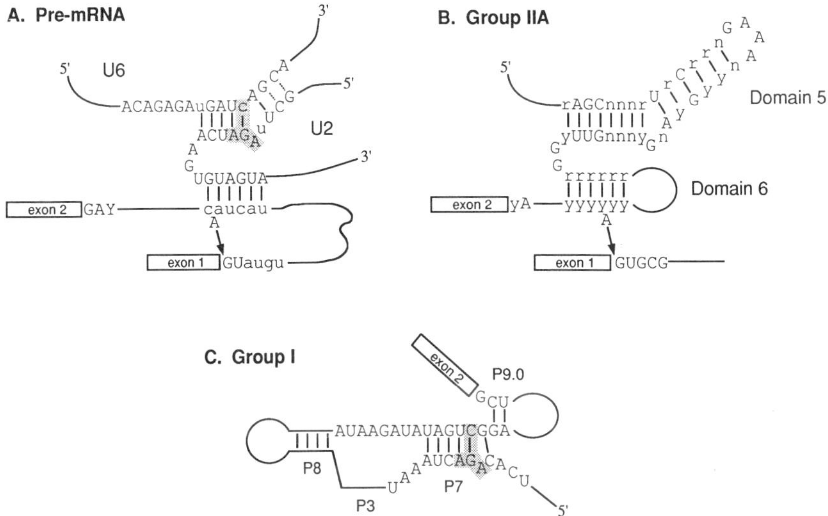

Figure 3 Proposed structural similarities between the spliceosome and self-splicing introns. (*A*) A subset of the interactions between U2 and U6 snRNAs and pre-mRNA (adapted from Madhani and Guthrie 1992). The intron consensus and snRNA sequences are those for yeast; uppercase letters indicate nucleotides that are conserved between yeast and known sequences from other organisms (excluding trypanosomes). (*B*) The consensus sequence for domains 5 and 6 of Group IIA introns (adapted from Michel et al. 1989a; Madhani and Guthrie 1992). Highly conserved nucleotides are shown as uppercase, whereas r, y, and n denote conserved purines, conserved pyrimidines, and variable nucleotides, respectively. (*C*) P7 and associated sequences from the *Tetrahymena* group I intron (adapted from Michel and Westhof 1990; McPheeters and Abelson 1992). The nucleotides that comprise the guanosine binding site (Michel et al. 1989b; Been and Perrotta 1991; Yarus et al. 1991a,b) and their proposed counterparts in U2 and U6 snRNAs (*A*) are shaded.

those at the 3′ splice site. Such direct communication might be partially responsible for the previously recognized ability of the spliceosome to proofread the branch structure in the lariat intermediate, prior to the second step of splicing. That is, although the 5′ splice site /GU and branch site A are required for the production of spliced products, their identity is not critical for the first step of splicing. Rather, /GU→/GA, /GG, or /AU mutations allow lariat formation to proceed (albeit at reduced levels) but effect a total block of the second step of splicing (Aebi et al. 1987 and references therein; Lamond et al. 1987). Changing the branch site A to a different nucleotide can have similar effects (Hornig et al. 1986). This suggests that the specific nucleotides in the branch structure are critically recognized prior to, or during, the second step of splicing. Examination of intron sequences that do not conform to the rule of five invariant nucleotides has revealed that, in addition to the class of introns that begins with /GC instead of /GU, the only other naturally occurring pre-mRNA introns that violate this rule begin with /AU. Most significantly, all of these introns also end with AC/ instead of AG/ (Jackson 1991). This suggests that the two conserved guanosine residues at the 5′ and 3′ splice sites interact in some way that can be mimicked by an interaction between A and C. Recent experiments by Parker and Siliciano (1993) have borne this out: they showed that whereas the single mutations, /GU→/AU or AG/→AC/, blocked splicing of a yeast intron, the two mutations together restored splicing to 10% of the wild-type level.

E. Evidence for Two Catalytic Sites

For group I introns, the two steps of splicing can be explained as nearly identical forward and reverse reactions (Fig. 1C, right) that are catalyzed within a single active site (see Cech, this volume). However, growing evidence supports a hypothesis that the active sites for the two steps of pre-mRNA splicing are different (see also discussion in Section III.E). The first indication that this might be the case derived from the structures of the reactants and products in each step; namely, the transesterification reaction at the 5′ splice site involves an incoming 2′ oxygen atom to form a 2′-5′ phosphodiester bond, whereas that at the 3′ splice site utilizes a 3′ oxygen to form a 3′-5′ phosphodiester. Recognition of the highly conserved guanosines (G) at the splice sites (Fig. 1A) must also differ between the two steps. In the first step, the reactive phosphate is linked to the 5′ position of the 5′ splice site G; in the second step, the reactive phosphate is linked to the 3′ position of the 3′ splice site G (Fig. 1C, left). In group I introns, however, the reactive phosphates are always 3′ to the required guanyl nucleotide (Fig. 1C, right).

Direct experimental evidence against the two steps of pre-mRNA splicing being forward and reverse reactions catalyzed within a single active site has come from the introduction of chemical modifications at the splice sites. First, although specific incorporation of a 2′-O-methyl group at the end of the 5′ exon had little effect on the first step of splicing, it caused a 15-fold reduction in the overall rate of the second step (Moore and Sharp 1992). This suggests that the steric environment around this ribose, which is the leaving group in the first reaction and the nucleophile in the second reaction, differs between the two steps. Even stronger evidence has come from the effects of phosphorothioates on the two steps (see Section II.C.2) (Maschloff and Padgett 1992; M.J. Moore and P.A. Sharp, in prep.). If the two steps were opposite reactions catalyzed by a single reactive center, the principle of microreversibility would require that they be inhibited by opposite phosphorothioate diastereomers. That is, since direct "S_N2" transesterification reactions result in inversion of stereochemistry at the reactive phosphate, the oxygen equivalent to the pro-R oxygen prior to the first reaction would be in the pro-S position prior to the second reaction (Fig. 1C). It has recently been shown that this principle does in fact hold for the two steps of group I splicing: Although the first step is not significantly inhibited by the R_p diastereomer (McSwiggen and Cech 1989; Herschlag et al. 1991), the second step is (Suh and Waring 1992). In contrast to the group I results, the two steps of pre-mRNA splicing are both strongly inhibited by R_p, but not S_p, phosphorothioate linkages at the splice sites. Thus, it would seem clear that the two steps of pre-mRNA splicing are not catalyzed by a single active site as forward and reverse reactions.

A second formal possibility is that the two steps of pre-mRNA splicing are catalyzed by a single active site, but as parallel reactions (i.e., both in the same direction). However, this scenario also seems unlikely, given the substrate structural differences and the effects produced upon removal of the 2′ hydroxyls from the ribose leaving groups in each reaction. Whereas incorporation of a 2′-deoxyribose at the 5′ splice site had no detectable effect on the first step of splicing, a 2′-deoxyribose at the 3′ splice site reduced the overall rate of the second step by 19-fold (Moore and Sharp 1992). However, in a situation where a single active site catalyzes two parallel reactions, the 2′ hydroxyls at the two splice sites would sequentially occupy the same position within that site, so that identical modification of each would be expected to have similar effects on the two steps.

Yet another indication of active-site differences between the first and second steps of pre-mRNA splicing is that mutations in U2 and U6 snRNAs can uncouple the two steps. Changes at either of two positions in U2 snRNA or four positions in U6 snRNA (denoted in Fig. 2B) (Fabrizio and Abelson 1990; McPheeters and Abelson 1992) permit the first step to proceed but arrest splicing before the second step. This set of bases may play a direct role in the reaction at the 3′ splice site or may be important for a transition between two active sites. Thus, taken together, the available evidence is most consistent with a model in which the two steps of pre-mRNA splicing are catalyzed by distinguishable active sites. This does not necessitate, however, that they be entirely different; it is possible that they are largely overlapping, with each containing a subset of components specific to that site.

If the two steps of pre-mRNA splicing are RNA-catalyzed, it is interesting to speculate that the two active sites may be homologous to different known RNA catalysts. The first step may reflect the activity of a catalytic site common to pre-mRNA splicing and group II introns. The characteristics of such a site would be a conserved guanosine linked to the reactive phosphate at the 5′ position, a nucleophilic 2′ OH on an adenosine in a bulged helix, and the ability to accommodate an S_p phosphorothioate diastereomer in the forward reaction. The spliceosomal active site for the second step might be similar to that of the group I introns. The shared characteristics of this site would be a conserved guanosine linked to the reactive phosphate at the 3′ position and the accommodation of an S_p phosphorothioate diastereomer in the forward reaction (as defined by the second-step reaction of group I introns). The active site for the second step of group II introns may, or may not, be similar to that of group I introns, since group II introns have a conserved pyrimidine and not a guanosine at the 3′ splice site. Furthermore, the effects of phosphorothioates in this system have yet to be examined.

III. THE SPLICEOSOME CYCLE

The spliceosome cycle (Fig. 4) was originally described on the basis of the kinetics of appearance of different complexes in splicing reactions in vitro and the splicing intermediates and snRNAs that they contained (Konarska and Sharp 1986, 1987; Cheng and Abelson 1987). The validity of the cycle has been confirmed and extended subsequently by studying the nature of the complexes that accumulate in reactions deficient in particular snRNPs and proteins (see below). A cycle implies both a precise precursor/product relationship between splicing complexes and a conservation of stable constituents, such as snRNPs, and both of these criteria are supported by the biochemical and genetic data. It should be noted that the number of intermediates in the cycle is only constrained by the resolution of future studies. Since many proteins and RNAs participate in splicing, the activity of each could potentially define an intermediate complex. The cycle as currently characterized illustrates only the order of appearance of the major distinct complexes that have been defined by RNA intermediate and snRNP composition.

A. Early Recognition of Pre-mRNA by hnRNP Proteins

Nascent pre-mRNA is almost certainly bound by a multitude of hnRNP proteins immediately after transcription in vivo (for review, see Dreyfuss et al. 1990). The organization of these proteins in recognition of precursor sequences may influence the earliest events in assembly of other spliceosomal components. However, although many hnRNP proteins bind RNA with distinct sequence specificity (Table 1) and associate with different pre-mRNAs in vitro (Bennett et al. 1992), it remains to be determined whether they are uniquely positioned on nascent RNAs in vivo. The best that can be stated is that the roles of hnRNP proteins in splicing are poorly understood.

B. Commitment Complexes

The earliest event that specifically targets pre-mRNA molecules to the splicing pathway in vitro is the formation of a commitment complex. Commitment complexes were originally defined in yeast nuclear extracts by preincubating one substrate RNA with a subset of cellular components and then testing whether this RNA was preferentially processed over a second RNA that was added with the remaining components necessary for splicing. If the first RNA was spliced more rapidly than the second, then the cellular components in the preincubation mixture were capable of forming a stable complex that committed the first RNA to spliceosome assembly (Legrain et al. 1988; Seraphin and Rosbash 1989).

Although not generally stable to nondenaturing gel electrophoresis (Konarska and Sharp 1986, 1987), commitment complexes have been isolated from yeast nuclear extracts under special electrophoresis conditions (Seraphin and Rosbash 1989). Two distinct complexes have been resolved: CC1, formation of which requires only the 5′ splice site sequence, and CC2, for which both the 5′ splice site and branch site sequences are required (Seraphin and Rosbash 1991). Both CC1 and CC2 form in the absence of ATP, and both contain U1 snRNP. Recognition of the 5′ splice site is mediated through base-pairing between the 5′ end of U1 snRNA and the 5′ splice site consensus (Fig. 2B). Recognition of both the branch site and the 5′ splice site in the formation of CC2 also requires U1 snRNP. This initial branch site recognition is independent of U2 snRNP, however, and is thought to be mediated by a non-snRNP splicing factor (Ruby and Abelson 1988; Rosbash and Seraphin 1991).

A similar ATP-independent complex (E or early) has been purified from mammalian extracts by gel filtration chromatography and shown to be commitment-competent (Michaud and Reed 1991). In contrast to CC1 and CC2, however, complex E contains both U1 and U2 snRNPs, although U2 snRNP is not bound as stably as U1 snRNP. Affinity purification of an ATP-independent complex that contained only U1 snRNP was reported by Bindereif and Green (1987), but this species was not tested

Table 1 Pre-mRNA-associated proteins

Pre-mRNA-associated protein	MW (kD)	Binding site	Activity	Motifs			References
				RRM	RS	RGG	
Constitutive factors							
ASF/SF2	33		proximal 5′SS selections; complements S100 strand-annealing	2	+	–	Ge and Manley (1990); Krainer et al. (1990a,b, 1991); Ge et al. (1991)
hnRNP A1 (SF5, UP1, SSP-1, SSP-2)	32	5′SS	distal 5′SS selection; strand-annealing	2	–	+	for review, see Merrill and Williams (1990); Kumar and Wilson (1990) (1990); Stolow and Berget (1990); Mayeda and Krainer (1992)
hnRNP A2/B1	34			2	–	–	Burd et al. (1989); Barnett et al. (1991)
hnRNP C1/C2	41/43	p(U)		2	–	–	Choi et al. (1986); Burd et al. (1989)
hnRNP K	66	p(C)		0	–	+	Matunis et al. (1992)
hnRNP L	68			4	–	–	Piñol-Roma et al. (1989)
hnRNP U	120			0	–	+	Kiledjian and Dreyfuss (1992)
PTB/hnRNP I	62	p(Y)		4	–	–	Garcia-Blanco et al. (1989); Patton et al. (1991); Ghetti et al. (1992); Gil et al. (1992)

	220/280	$5' + 3'$ SS	associated with U5 snRNP				
p220/PRP8	220/280	$5' + 3'$ SS	associated with U5 snRNP				Garcia-Blanco et al. (1990); Whitaker and Beggs (1991); (see also references in Table 2)
SC35	35		association with U1 U2, and 3' SS	1	+	–	Fu and Maniatis (1992a,b)
U2AF(65)	65	p(Y)	U2 recruitment to branch site	3	+	–	Ruskin et al. (1988); Zamore and Green (1991); Zamore et al. (1992)
U2AF(35)	34		U2 recruitment to branch site	0	+	–	Ruskin et al. (1988); Zhang et al. (1992)
Alternative splicing factors							
Sex-lethal (Sxl)	39	*tra* intron	alt. splicing of *tra*	2	–	–	Bell et al. (1988); Inoue et al. (1990)
Transformer (Tra)	29		alt. splicing of *dsx*	0	+	–	McKeown et al. (1987); Hedley and Maniatis (1991)
Transformer-2 (Tra-2)	20	six 13-nt repeats in *dsx*	alt. splicing of *dsx*	1	+	–	Amrein et al. (1988); Gorlaski et al. (1989); Hedley and Maniatis (1991)

Molecular weights are given in kD according to conceptual translations from cDNA sequences, except for p220/PRP8. Binding sites are 5' and 3' SS, 5' and 3' splice sites; p(U), polyuridylate; p(C), polycytidylate; p(Y), polypyrimidine. Motifs are: RRM, RNA recognition motif (for review, see Kenan et al. 1991); RS, arginine-serine dipeptide-rich repeats (see Zahler et al. 1992 and references therein); RGG, arginine-glycine-glycine repeats (Kiledjian and Dreyfuss 1992).

in a commitment assay. An additional ATP-independent complex that contained only bound U2 snRNP and was associated with commitment activity was recently reported by Jamison and Garcia-Blanco (1992).

The commitment reaction in vitro is related, but may not be identical, to the in vivo process by which pre-mRNAs are recognized and retained within the nucleus. In mammalian cells, if a nascent RNA polymerase II transcript contains active 5′ and 3′ splice sites, it is committed to the splicing pathway and is subsequently transported out of the nucleus as spliced mRNA. However, if an RNA contains a single splice site (5′ or 3′), the bulk of the pre-mRNA is retained in the nucleus and then either degraded or processed through the use of a cryptic splice site (Chang and Sharp 1989). This suggests that the 5′ and 3′ splice site regions are each sufficient for initial commitment to the splicing pathway in vivo. Genetic studies in yeast also have suggested that the signals influencing the commitment assay in vitro are related to commitment to the splicing pathway in vivo (Legrain and Rosbash 1989).

Controlling the commitment process may be an important regulatory event in the synthesis of cytoplasmic mRNA. For example, it has been proposed that the human immunodeficiency virus (HIV) protein Rev regulates cytoplasmic composition of specific spliced forms of the viral RNA in this fashion (Chang and Sharp 1989). Rev protein specifically recognizes a cluster of high-affinity binding sites in the unspliced viral RNA and mediates transport of this pre-mRNA to the cytoplasm. Another interesting observation in this system is that the nuclear stability of a family of mutant RNAs from the HIV genome is dependent on the binding of U1 snRNP to a 5′ splice site sequence (Lu et al. 1990). More importantly, in the presence of viral Rev protein, this nuclear RNA, which has interacted with U1 snRNP, is efficiently transported to the cytoplasm in an unspliced state. Thus, in this case, Rev is regulating the transport of RNA that has interacted with U1 snRNP.

C. Complex A

1. Stable Association of U2 snRNP

Subsequent to formation of a commitment complex, U1 snRNP promotes the binding of U2 snRNP to form complex A (Fig. 4). The precise mechanism by which this occurs has not been established. Although commitment complex formation is relatively slow and is ATP-independent, the transition to complex A is rapid and requires ATP hydrolysis (Liao et al. 1992). Additionally, in complex A the pre-mRNA/U2 snRNP interaction is much more stable than the U1 snRNP association. In initial splicing complex analyses carried out by nondenaturing gel electrophoresis,

U1 snRNP did not appear to be associated with any complex (Konarska and Sharp 1986, 1987). However, experiments using different electrophoresis conditions (Zillmann et al. 1988) or gel filtration to separate splicing complexes (Michaud and Reed 1991) have demonstrated U1 snRNP is present in complex A, as well as in later complexes.

In yeast, the primary and critical signal for the initial U2 snRNP/pre-mRNA interaction is the highly conserved branch site sequence, UACUAAC. As discussed in Section II.D, a short bulged helix is formed between this sequence and the GUAGUA sequence in U2 snRNA (Fig. 2). The presence of a polypyrimidine tract adjacent to a branch site sequence can enhance, but is not essential for, the utilization of that site in yeast (Patterson and Guthrie 1991). In contrast, the polypyrimidine tract is much more important in mammalian systems, where branch sequences are less well conserved. In both systems, evidence suggests that the first AG dinucleotide downstream from the branch site and polypyrimidine tract is that most commonly used as the 3′ splice site (see Reed 1989; Smith et al. 1989b; Zhuang and Weiner 1990 and references therein).

The chemical interactions between U1 snRNP and U2 snRNP that promote the stable association of the latter with the 3′ splice site region are not dependent on the RNA duplex formed between the 5′ splice site and U1 snRNA. Degradation or antisense "masking" of the 5′ end of U1 snRNA does not inactivate its potential to affect stable association of U2 snRNP with pre-mRNA; however, depletion of the U1 snRNP particle does (Barabino et al. 1990 and references therein). In mammalian introns with short polypyrimidine tracts (less than 14 nucleotides), the AG at the 3′ splice site is requisite for formation of the lariat intermediate and, thus, probably for the binding of U2 snRNP (Reed 1989). However, introns with longer polypyrimidine tracts can proceed through the first step of splicing without a downstream AG (Smith et al. 1989b). These two intron types are said to be AG-dependent and AG-independent, respectively (Reed 1989). In yeast, Reich et al. (1992) have shown that this early recognition of the 3′ splice site in AG-dependent introns consists of a base pairing interaction between nucleotides C9 and U10 of U1 snRNA and the 3′ splice site AG (see Fig. 2B). This interaction between U1 snRNA and the 3′ splice site could be important for the formation of complex A, but it could potentially occur at a later stage.

2. Protein Factors Required for Complex A Assembly:
The Emerging Importance of Arginine/Serine Domains

Several mammalian factors have been resolved by chromatography on the basis of a reconstitution assay for formation of complex A (as in

Krämer and Utans 1991) or splicing (as in Krainer and Maniatis 1985; Krämer et al. 1987). Additionally, temperature-sensitive mutations in either of two yeast genes, PRP5 and PRP9 (see Table 2), inactivate formation of complex A. However, neither PRP5 nor PRP9 is related to any of the mammalian factors that have been purified and cloned. The best-characterized of these mammalian proteins are U2AF (*U2 auxiliary factor*), ASF/SF2 (*alternative splicing factor/splicing factor 2*), and SC35 (*35-kD spliceosomal component*). Surprisingly, all three of these factors share basic structural similarities and can be divided into two functional subdomains—a region of one or more consensus RNA-recognition motifs (RRM; for review, see Kenan et al. 1991) and a region of arginine/serine repeats (discussed below for each protein). There is, in fact, an entire family of arginine/serine proteins, the SR proteins, that have been shown to complement S100 cytoplasmic extracts for splicing activity (Mayeda et al. 1992; Zahler et al. 1992). Like U2AF, ASF/SF2, and SC35, other SR proteins also contain RRMs. The family is highly conserved throughout evolution, as both *Drosophila* and humans synthesize a similar set of six proteins with similar structures, all reactive with a single monoclonal antibody (Zahler et al. 1992). These were designated (with their names from other systems) SRp20 (X16), SRp30a (ASF/SF2), SRp30b (SC35), SRp40, SRp55 (B52), and SRp75.

U2AF binds the polypyrimidine tract, and perhaps the 3′ splice site <u>AG</u>, independent of any other spliceosome components (Zamore et al. 1992). It was originally identified by its requirement to complement U2 snRNP binding to branch sites in fractionated extracts (Ruskin et al. 1988) and was subsequently shown by biochemical complementation to be required for splicing (Zamore et al. 1992 and references therein). Surprisingly, the sequence specificity of U2AF for binding the polypyrimidine tract is very similar to that of an abundant protein, the polypyrimidine tract-binding protein (PTB) (Garcia-Blanco et al. 1989; Gil et al. 1991; Patton et al. 1991) or hnRNP I (Ghetti et al. 1992). The role of PTB/hnRNP I in splicing is as yet unclear, although a specific antiserum to this protein blocks splicing in vitro at a step before formation of complex A (A. Gil and P.A. Sharp, unpubl.). The hnRNP C protein has binding specificity for polypyrimidine tracts as well, and specific antibodies against this protein can also inhibit splicing in vitro (Choi et al. 1986). Again, a specific role for this protein in splicing has not been determined. It is possible that all three proteins compete for the same site or that they bind sequentially during the spliceosome cycle. Interestingly, recent evidence suggests that different pre-mRNA substrates may bind distinct subsets of hnRNP proteins, such as PTB/hnRNP I versus hnRNP C (Bennett et al. 1992).

U2AF is a heterodimer with subunit molecular masses of 65 kD and 35 kD, both of which have been cloned (Zamore et al. 1992; Zhang et al. 1992). Only the larger subunit is required for reconstitution of splicing; the function of the smaller subunit is currently unknown. The U2AF(65) can be divided into two functional subdomains: a sequence-specific RNA-binding region that contains three prototypical RRM domains and a region of arginine/serine repeats. The former domain is essential for binding to the polypyrimidine tract, whereas the latter is additionally required to effect binding of U2 snRNP to the pre-mRNA (Zamore et al. 1992). Another arginine/serine subdomain is found in the sequence of the smaller subunit, U2AF(35), and it is possible that the two subunits act via their similar sequences to produce a synergistic effect on splicing complex assembly (Zhang et al. 1992; see also Section IV.B).

ASF/SF2 was identified by two different assays: (1) as a factor that facilitates the use of a proximal 5'splice site when presented with a substrate containing competing 5'splice sites (Ge and Manley 1990) and (2) as a factor that complements S100 extracts for splicing activity (Krainer et al. 1990a). It has also been shown to promote RNA:RNA annealing (Krainer et al. 1990b). Recent cloning of this 35-kD protein has revealed that it too contains a consensus RRM and an extensive tract of arginine/serine repeats (Ge et al. 1991). It is of interest to note that an antagonist to ASF/SF2 has been identified: the abundant hnRNP A1 protein can suppress the splice site selection activity of ASF/SF2 by promoting the use of the distal 5'splice site of an alternative set (Mayeda and Krainer 1992). The hnRNP A1 protein is an RNA-binding protein (see Table 1), and it too has RNA:RNA annealing activity (Kumar and Wilson 1990). Whether the yin-yang relationship between ASF/SF2 and hnRNP A1 is functionally important in vivo remains to be determined.

SC35, originally identified as an antigenic determinant in purified spliceosomes (Fu and Maniatis 1990), is another SR protein that has been shown to be critical for splicing in vitro (Fu and Maniatis 1992b). Monoclonal antibodies specific for SC35 will deplete an extract of splicing activity, and this depleted extract can be complemented with recombinant SC35 protein expressed from baculovirus. This protein is required for complex A formation and is present in this complex, as well as in later spliceosomal complexes containing splicing intermediates. In fact, SC35 appears to be specifically required for a binding interaction of U1 and U2 snRNPs with the 3'splice site region (see Fig. 5B) (Fu and Maniatis 1992a), but it is not required for binding of U1 snRNP to the 5'splice site in the absence of ATP. SC35 contains both an arginine/serine domain and an RRM, suggesting sequence-specific binding to RNA.

Table 2 PRP genes

Gene	MW (kD)	Splicing step blocked	Particle association/other activity	Sequence motif	References
PRP2	100	1	spliceosome	RNA helicase (DEAH); zinc finger-like	Lee et al. (1986); Lustig et al. (1986); Lin et al. (1987); Chen and Lin (1990); King and Beggs (1990); Yean and Lin (1991); Kim et al. (1992)
PRP3	56	1	U4/U6 snRNP		Ruby and Abelson (1991)
PRP4	52	1	U4/U6 snRNP; required for U5 snRNP association with U4/U6	β-subunit of G protein	Banroques and Abelson (1989); Dalrymple et al. (1989); Petersen-Bjorn et al. (1989); Bordonné et al. (1990); Xu et al. (1990)
PRP5	96	1	required for U2 addition	RNA helicase (DEAD)	Lustig et al. (1986); Dalbadie-McFarland and Abelson (1990)
PRP6	104	1	U4/U6 snRNP	zinc finger-like; leucine repeat	Abovich et al. (1990); Legrain and Choulika (1990)
PRP7	—	1			Ruby and Abelson (1991)
PRP8	280	1	U5 snRNP		Lossky et al. (1987); Jackson et al. (1988); Anderson et al. (1989); Garcia-Blanco et al. (1990); Whittaker et al. (1990); Whittaker and Beggs (1991); Brown and Beggs (1992)
PRP9	63	1	required for U2 addition	zinc finger-like	Abovich et al. (1990); Legrain and Choulika (1990)
PRP11	30	1	spliceosome	zinc finger-like	Chang et al. (1988); Legrain and Choulika (1990); Schappert and Friesen (1991)
PRP16	120	2	spliceosome	RNA helicase (DEAH)	Couto et al. (1987); Burgess et al. (1990); Schwer and Guthrie (1991, 1992)

PRP17/ SLU4	52	2		β-subunit of G protein	Vijayraghavan et al. (1989); Ruby and Abelson (1991); Frank and Guthrie (1992); Frank et al. (1992)
PRP18	28	2			Vijayraghavan and Abelson (1990)
PRP19	49	1			Ruby and Abelson (1991)
PRP21	33	1			Ruby and Abelson (1991)
PRP22	130	intron turnover and mRNA release		RNA helicase (DEAH); ribosomal protein S1-like	Company et al. (1991)
PRP24	51	1	U2 snRNP; required for assoc. of U4 snRNP with U6	RNA recognition motif	Shannon and Guthrie (1991); Strauss and Guthrie (1991)
PRP25	—	1			Ruby and Abelson (1991)
PRP26	48	intron turnover	debranching activity		Chapman and Boeke (1991)
PRP27	—	intron turnover			Ruby and Abelson (1991)
PRP28	67	1		RNA helicase (DEAD)	Strauss and Guthrie (1991)
PRP29	—	2			Ruby and Abelson (1991)
PRP30	—	1			Ruby and Abelson (1991)
PRP38	28	1			Blanton et al. (1992)
SPP2	21	1			Ruby and Abelson (1991)
SPP41	157	—	suppresses mutations in PRP3, 4, 11		Ruby and Abelson (1991)
SPP81/ DED1	65	—	suppresses mutations in PRP8		Jamieson et al. (1991)
SLU7	44	2		zinc finger-like	Frank and Guthrie (1992); Frank et al. (1992)

In addition to the ability of many SR proteins to complement S100 extracts for splicing activity, it seems that some SR proteins can cross-complement. In side-by-side comparisons of recombinant ASF/SF2 and SC35 proteins in in vitro splice site selection assays, their activities are so far indistinguishable (Fu et al. 1992). The activities of the two proteins in vivo, however, remain unknown.

Given their prevalence in proteins required for spliceosome assembly, can a general role be assigned to arginine/serine sequences? Some arginine/serine domains contain subnuclear targeting signals (Li and Bingham 1991), and these may contribute to the proper localization of splicing factors. Arginine/serine repeats can be modified by phosphorylation on the serine residue (Roth et al. 1990); however, the significance of this modification with regard to splicing remains unclear. Upon phosphorylation, the serine residue in this repeat becomes the specific antigen for a monoclonal antibody that binds sites of active RNA polymerase II transcription on lampbrush chromosomes (Roth et al. 1991). Thus, these proteins are presumably bound to nascent RNA chains and obviously could influence the binding of snRNPs and other proteins. Given that arginine/serine domains are critical for spliceosome formation in mammalian extracts, it is surprising that proteins containing this motif have not been identified among the PRP proteins in yeast (see Table 2). If arginine/serine domains are involved in stabilizing RNA interactions, the flexibility demanded of the mammalian spliceosomal process may make it more dependent on such motifs than the yeast system, which is more constrained in sequence recognition.

Recently, a single arginine residue was shown to be the major determinant for the sequence-specific binding of HIV TAT protein to a bulge loop on HIV TAR RNA (Calnan et al. 1991), and three-dimensional structural analysis of a model complex has shown that a single arginine group can specifically interact with RNA and effect large conformational changes (Puglisi et al. 1992). In an analogous manner, it is possible that arginine/serine domains stabilize RNA:RNA interactions or induce RNA conformational changes that are critical for formation of splicing complexes, such as complex A. The RRM domain(s) in a particular protein could recognize a specific RNA sequence, thereby localizing its arginine/serine domain, which could then function in stabilizing another RNA structure nearby. A sufficiently high concentration of an arginine/serine protein might bypass this requirement for specific localization for function, explaining the ability of many arginine/serine proteins to cross-complement. It is likely that U1 and U2 snRNP interact with an arginine/serine domain, perhaps in SC35 protein, in the generation of complex A. It is also possible that the arginine/serine domain pro-

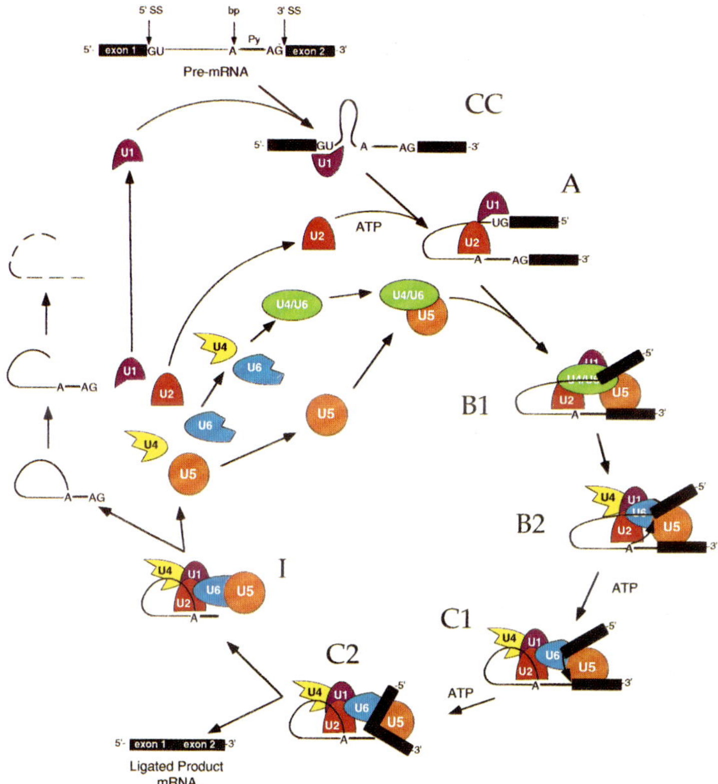

Figure 4 Schematic representation of the involvement of small nuclear ribonucleoprotein (snRNP) particles in pre-mRNA splicing. Other, non-snRNP factors are required for spliceosome formation, but have been omitted for simplicity. Pre-mRNA (top line), containing two exons separated by an intron, enters splicing complexes with snRNPs and exits as ligated mRNA (bottom line) and excised lariat intron (left border). CC, A, B1, B2, C1, C2, and I represent macromolecular complexes within the splicing pathway that have been distinguished biochemically and/or genetically (described in Section III). 5'SS, 3'SS, bp, and Py indicate 5' and 3' splice sites, branch site, and polypyrimidine tract, respectively. The individual snRNPs indicated are U1, U2, U4, U5, and U6. (Modified from Konarska and Sharp 1987; Lamond et al. 1988; Ruby and Abelson 1991.)

vided by U2AF(65) is critical for stabilization of the U2 snRNA/branch site interaction. In considering the role of arginine/serine proteins in splicing complex assembly, the snRNP-specific proteins should also be kept in mind: U1 snRNP itself contains a protein, U1-70K, that has extensive arginine/serine domains in metazoans (Query et al. 1989; Mancebo et al. 1990 and references therein). Although no specific role in

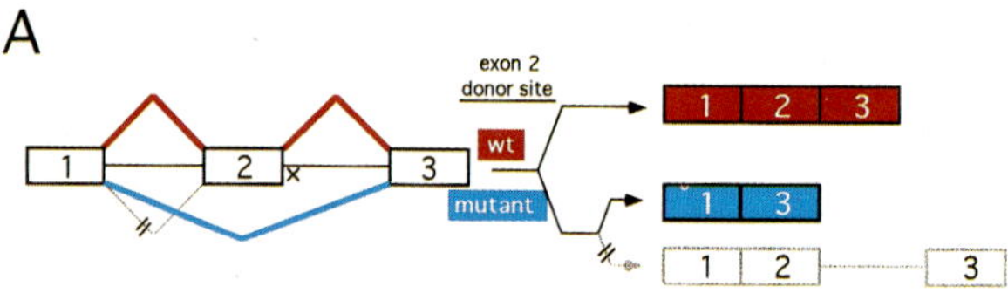
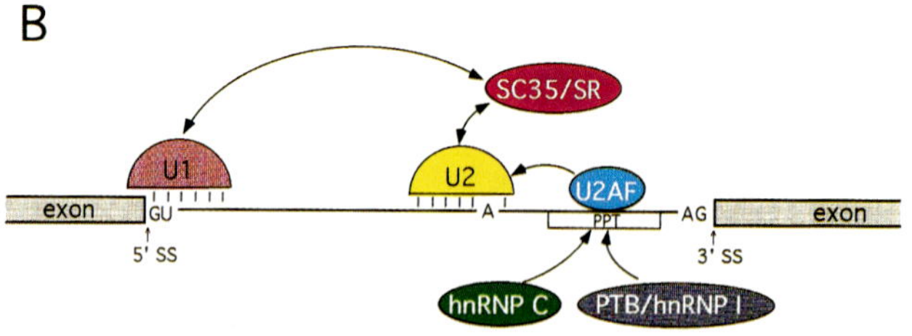
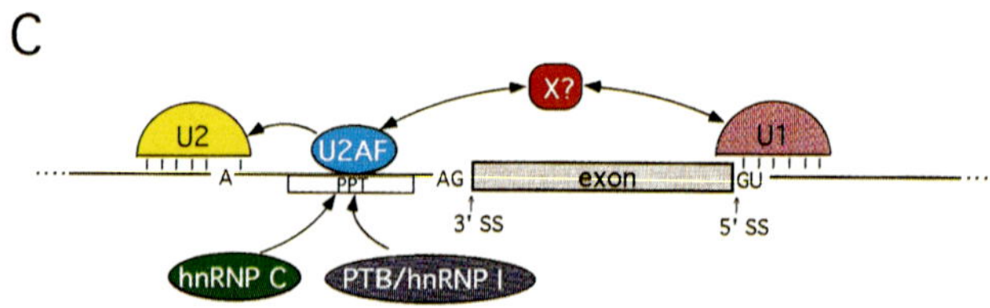

Figure 5 (See facing page for legend.)

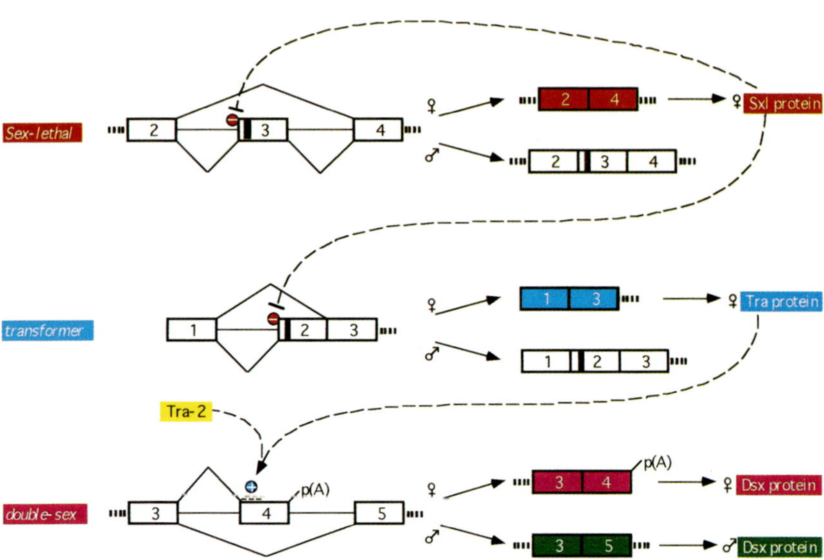

Figure 6 (See facing page for legend.)

Figure 5 (*A*) Mutations at downstream 5′ splice sites in multi-intron-containing pre-mRNAs result in exon-skipping. Mutations that decrease complementarity of 5′ splice sites to U1 RNA (X at the 5′ splice site following exon 2) result in skipping of the upstream exon 2, not intron inclusion (Talerico and Berget 1990). Numbered boxes represent exons and connecting lines represent introns. Solid red slanted lines connecting exons above the introns represent splices occurring in the wild-type pre-mRNA; solid blue slanted lines below the introns represent splices observed in pre-mRNAs with mutant downstream 5′ splice sites; grey slanted lines below the intron represent splices that would be predicted by intron-only definition, but which are not observed. (*B*) Model for interactions in single-intron pre-mRNAs, depicting U1 snRNP base-pairing with the 5′ splice site (Zhuang and Weiner 1986; Siliciano and Guthrie 1988), U2 snRNP base-pairing with the branch site (Parker et al. 1987; Wu and Manley 1989; Zhuang and Weiner 1989), U2AF bound to the polypyrimidine tract, recruiting U2 snRNP to the branch site (Ruskin et al. 1988; Zamore et al. 1992), and interactions between SC35, U1 and U2 snRNPs (Fu and Maniatis 1992a). (*C*) Model for interactions at internal 3′ splice sites of multi-intron-containing pre-mRNAs, depicting exon definition (Robberson et al. 1990) and interactions between U2AF, U1 and U2 snRNPs, and an additional factor (X), dependent on a downstream 5′ splice site (Hoffman and Grabowski 1992). Additional protein factors that interact with polypyrimidine tracts (PTB/hnRNP I, hnRNP C) are shown. SS, splice site; PPT, polypyrimidine tract; SR, proteins with arginine-serine-rich repeats (see Table 2 and Section III.C.2).

Figure 6 Cascade of regulated alternative splicing events in the sex determination pathway of *Drosophila*. (*Upper*) The *Sex-lethal* (*Sxl*) pre-mRNA is alternatively spliced in females (♀) to exclude exon 3 (red), but in males (♂) to include exon 3 (white). The binding of Sxl protein in females to sequences 5′ to exon 3 blocks exon inclusion. (*Middle*) The *transformer* (*tra*) pre-mRNA is spliced to exclude exon 2 (blue), but in males to include exon 2 (white). As with the *Sxl* pre-mRNA, binding of Sxl protein in females to sequences 5′ to the *tra* exon 2 blocks inclusion of exon 2. Both *Sxl* exon 3 and *tra* exon 2 contain stop codons (shaded bars). (*Lower*) The *double-sex* (*dsx*) pre-mRNA is spliced to include exon 4 and an exon 4-specific polyadenylation site [p(A)] in females (pink), but not in males (green). The Tra-2 protein (yellow) binds specifically to six 13-nucleotide repeats (six repeated bars over exon 4) and, along with Tra-2 protein, activates the utilization of the 3′ splice site of exon 4. See text for mechanistic descriptions. Numbered boxes represent exons and connecting lines represent introns. For clarity, only those exons and introns involved in the regulated splicing events are depicted. Slanted lines connecting exons above the introns represent splices occurring in females; slanted lines below the introns represent splices occurring in males. Shaded bars in *Sxl* exon 3 and in *tra* exon 2 represent stop codons. ⊖ represents sites of binding and negative regulation by Sxl protein to the *Sxl* and *tra* pre-mRNAs, resulting in blocking of the adjacent 3′ splice site and subsequent utilization of a distal 3′ splice site. ⊕ represents activation of the *dsx* exon 4 3′ splice site by binding of Tra and Tra-2 proteins.

spliceosome assembly has been defined for U1-70K, its similarity to the other known splicing factors, coupled with its proximity to the 5′ splice site in early complexes, is noteworthy.

D. Complexes B1 and B2

1. Addition of U4/U6.U5 (Complex B1)

Several lines of evidence suggest that it is a tri-snRNP complex, U4/U6.U5, rather than the individual snRNPs, which enters the spliceosome to form complex B1. First, yeast cells that have been depleted of U5 snRNA do not generate complexes of U4/U6 bound to other spliceosomal components (Seraphin et al. 1991). The same effect is observed in vitro when mammalian extracts are depleted of U5 snRNP (Lamm et al. 1991). A second indication that the tri-snRNP complex is functional was provided by complementation of heat-inactivated extracts upon addition of the purified U4/U6.U5 complex (Utans et al. 1992). Additionally, three yeast genes for which temperature-sensitive (ts) mutations have been isolated, PRP4, PRP6, and PRP28 (see Table 2), are defective for formation of the B1 complex. Specific antisera to the PRP4 protein block spliceosome assembly at the same step, preventing splicing, and cause the U4/U6.U5 tri-snRNP complex to accumulate (Banroques and Abelson 1989).

The U4/U6.U5 tri-snRNP has a biochemical complexity worthy of its functional complexity; it sediments at 25S and contains at least 20 distinct polypeptides (Behrens and Lührmann 1991). One protein associated with the tri-snRNP is PRP8 (Lossky et al. 1987), and antisera to this protein block formation of complex B1 (Brown and Beggs 1992). The PRP8 protein, which is also found associated with U5 snRNP alone, is probably important for the stability of the U4/U6.U5 tri-snRNP, because all three snRNAs are more rapidly degraded in cells deficient for this protein (Brown and Beggs 1992). Interestingly, after entering the spliceosome as part of the U4/U6.U5 tri-snRNP, PRP8 and its mammalian homolog p220 ultimately become so closely associated with the pre-mRNA substrate that UV irradiation induces cross-links between the protein and RNA (Garcia-Blanco et al. 1990; Whittaker and Beggs 1991). RNA substrates lacking a 5′ splice site cannot be cross-linked to PRP8/p220; RNAs containing sequences spanning only the 3′ splice site region can form a complex equivalent to A, and associate with the U4/U6.U5 tri-snRNP complex, but do not yield PRP8/p220-specific cross-links (Whittaker and Beggs 1991; M.J. Moore and P.A. Sharp, unpubl.). However, in an intact pre-mRNA, p220 can become cross-linked to both the 5′ and 3′ splice site regions (Wyatt et al. 1992; M.J. Moore and P.A. Sharp, unpubl.). This suggests that components in the U4/U6.U5 tri-snRNP significantly rearrange after binding to complex A formed on an intact pre-mRNA.

Several RNA:RNA interactions are likely to occur upon binding of U4/U6.U5 to complex A. One of these is probably formation of the duplex region between the 3′ end of U6 snRNA and the 5′ end of U2 snRNA (*j* in Fig. 2D; see Section II.D). Another interaction is probably mediated through limited complementarity between the conserved loop in U5 snRNA and exon sequences flanking both the 5′ and 3′ splice sites (*k* in Fig. 2B; see Section II.D). This interaction could persist through step one, and it may be important for specifying the actual 5′ cleavage site.

As discussed below, there can be dramatic changes in the status of base-pairing of the snRNAs during the progression from one splice-osomal complex to another. In this regard, it has been proposed that U1 and U5 snRNAs might simultaneously pair with sequences at the 5′ and 3′ splice sites in a four-stranded structure similar to that diagramed originally by Holliday for branch migration in DNA (Steitz 1992). However, because of the reduced stability of the U1 snRNP association with the spliceosome, it is more likely that base-pairing between U1 snRNA and the 5′ splice site is partially or totally displaced in the later complexes. Interestingly, it appears that both U1 and U5 snRNAs are dispensable in *trans*-splicing systems (see Section IV.C).

2. Structural Rearrangements of U4, U6, and U2 snRNAs (Complex B2)

As shown in Figure 2A, U4 and U6 snRNAs are extensively base-paired in the U4/U6 particle and, presumably, in the U4/U6.U5 particle. Shortly after entering the spliceosome, the RNA duplex between U4 and U6 snRNAs is probably dissociated, resulting in the loose binding of U4 snRNA to the B2 form of the spliceosome (Blencowe et al. 1989). Using a PRP2 ts mutant of yeast, which is defective for the first step in splicing, it was possible to isolate a spliceosomal complex equivalent to B2 that lacked U4 snRNA (Yean and Lin 1991). The isolated complex could be made competent for the generation of splicing intermediates by the addition of PRP2 protein in the absence of U4 snRNA. This result strongly suggests that the latter snRNA is not paired to U6 snRNA in complex B2 and is not essential for catalysis.

Freed from its presumably inhibitory base-pairing with U4 snRNA, this same region of U6 snRNA is then available to pair with a complementary region in U2 snRNA (*h* in Fig. 2B). This interaction could occur prior to either the first or second transesterification reaction. In fact, mutational changes in both U6 and U2 snRNAs in this region can inactivate either or both steps of splicing (Fabrizio and Abelson 1990; Madhani and Guthrie 1992; McPheeters and Abelson 1992). As noted in Section II.D and Figure 3, the secondary structure formed by the pairing of U2 and U6 snRNAs in this region has similarities to conserved regions in both group I and group II introns.

E. Complexes C1 and C2

Cleavage at the 5′ splice site and formation of the lariat intermediate denote formation of complex C1, whereas complex C2 is the form of the spliceosome immediately following exon ligation. A second structural rearrangement, in addition to that described above for the B1 to B2 transition, must occur between C1 and C2; i.e., between the first and second transesterification reactions. One argument for such a rearrangement stems from the evidence that the first and second reactions are catalyzed by different reactive centers (discussed in Section II.E). In addition, there are mutations that uncouple the first and second steps, suggesting different functional and structural requirements for the two active complexes. These second-step mutations have been found in both snRNAs (U6 and U2; see above and Fig. 2B) and proteins (PRP16, PRP17/SLU4, PRP18, PRP29, and SLU7; see Table 2).

Selection for temperature-sensitive (ts) or cold sensitive (cs) alleles of genes by an exacerbation of the growth defects of mutants carrying lesions in the conserved loop of U5 snRNA (synthetic *l*ethal *U*5 mutations; Frank and Guthrie 1992; Frank et al. 1992) yielded two proteins, SLU4 and SLU7, that affect the second step of splicing. SLU4 is an allele of the ts mutation, PRP17 (see Table 2), which had previously been shown to be defective for the second step of splicing. SLU7 appears to have a role in the selection of 3′ splice sites. When the 3′ splice site AG is greater than 22 nucleotides from the branch site in yeast introns, the first step of splicing is independent of recognition of the AG dinucleotide (see Section III.C.1). Yeast strains normally show a 20-fold preference for an AG with a spacing greater than 22 nucleotides, as compared to an AG with a spacing of 15 or fewer nucleotides. In strains deficient in SLU7, this 20-fold preference is not observed, suggesting that the SLU7 protein is particularly important in selecting the 3′ AG with the long spacing. Since SLU7 protein probably interacts with U5 snRNA and functions primarily in the second step, it could be important for U5 snRNA's recognition of exon nucleotides at the 3′ splice site (Newman and Norman 1992). The PRP16 and PRP18 proteins (Table 2) probably interact with SLU7 protein in this process, since the SLU7 mutant was lethal in combination with mutations in either PRP16 or PRP18 (Frank et al. 1992). PRP16 has a sequence motif indicative of an RNA helicase (DEAH; Burgess et al. 1990; see Section III.H) and, thus, could be required to rearrange RNA duplexes in the process of selection of the 3′ AG and the shift from complex C1 to complex C2. SLU4 was also lethal when crossed with either PRP16 or PRP18, suggesting that the four proteins (PRP17/SLU4, SLU7, PRP16, and PRP18) participate in a common structure or step (Frank et al. 1992).

F. Product Release, Complex I, and Intron Degradation

After formation of complex C2 containing the two products, release of the spliced exons from the spliceosome is an active process. Mutations in the PRP22 gene of yeast block this release, resulting in spliced exons remaining bound to the spliceosome (Company et al. 1991). Again, it is possible that this transition requires rearrangements of RNA duplexes, given that PRP22 is probably an RNA helicase (see Section III.H). In vitro, the lariat intron product accumulates and remains associated with the spliceosomal snRNPs to form complex I (Konarska and Sharp 1987). In vivo, however, the mature lariat intron is probably released rapidly and immediately degraded; the half-lives of most excised introns in vivo are a only few seconds. A debranching activity has been described and purified from mammalian extracts (Ruskin and Green 1985; Arenas and Hurwitz 1987). Surprisingly, the gene encoding this activity, PRP26, is not essential in yeast, and excised introns can accumulate with little effect (Chapman and Boeke 1991). Excised introns also accumulate in a yeast strain with a ts mutation in PRP27 (Vijayraghavan et al. 1989); this protein may be essential for dissociation of complex I. The mutant alleles of PRP22 and PRP27 could thus arrest splicing by not regenerating an adequate pool of snRNPs for further rounds of processing at the nonpermissive temperature.

G. Recycling of the snRNPs and Pseudospliceosomes

The half-lives of all the individual spliceosomal snRNAs are very long (~20 hr); therefore, each snRNA must leave and reenter spliceosomes repeatedly. It is highly probable that after the snRNPs are released from spliceosomes in vivo, they are dissociated to single particles, and the prespliceosomal multi-snRNP complexes are reassembled (Fig. 4). A similar dynamic process can be observed in reactions in vitro, where, as conditions change, the distributions of snRNAs in individual complexes, for example, U4/U6 snRNP versus the U4/U6.U5 tri-snRNP, are shifted (Konarska and Sharp 1987). The reassembly of the snRNP complexes U4/U6 and U4/U6.U5 may involve PRP3, PRP4, PRP6, PRP8, and PRP24 (see Table 2). These proteins are bound to snRNP complexes prior to formation of the spliceosome and, at least in the case of PRP8, become incorporated into the spliceosome.

All multi-snRNP complexes may not be intermediates in the spliceosome cycle. A complex containing U2, U4/U6, and U5 snRNPs, a pseudospliceosome, can be generated by incubating a nuclear extract under moderate salt conditions (Konarska and Sharp 1988). Since formation of this multi-snRNP complex does not require pre-mRNA, a backbone of specific snRNA-snRNA interactions must be available to stabilize such pseudospliceosome assembly. The above-mentioned complementarity of U2 and U6 snRNAs could be important in this regard. More interesting, recent evidence suggests that incubation of a splicing reaction with short oligonucleotides containing 5′ splice site sequences stimulates formation of a similar U2.U4/U6.U5 snRNP complex (Hall and Konarska 1992). Surprisingly, although the added ribo-oligonucleotide must contain a 5′ splice site sequence to induce this assembly, the complementary sequences at the 5′ terminus of U1 snRNA are not required and their specific degradation does not inactivate pseudospliceosome assembly. The U2.U4/U6.U5 complex is stable, and formation of an active spliceosome on a subsequently added pre-mRNA is suppressed in reactions that were pretreated with the 5′ splice site ribo-oligonucleotide. This suppression probably reflects depletion of free snRNPs and/or associated non-snRNP factors. It is also possible that the presence of a 5′ splice site sequence in a nuclear RNA in the absence of other splicing signals could trigger the formation of a locally suppressive pseudospliceosome.

H. The PRP genes

Throughout the preceding sections, the effects of various yeast PRP (*precursor RNA processing*) mutant alleles on the formation of particular spliceosomal complexes have been discussed. In this section, the original identification of these genes is described, and several implications relating their primary structures to possible functions are considered.

Most of the PRP genes were isolated from large screens for general temperature-sensitive mutations in yeast. Under restrictive conditions, PRP mutant alleles can have one or more of the following phenotypes: The mutation can (1) cause loss of viability, (2) result in the in vivo accumulation of either pre-mRNAs or splicing intermediates or excised introns, or (3) cause defects for splicing in vitro. In all, at least 30 genes have been identified whose products affect pre-mRNA splicing (Table 2) (for review, see Guthrie 1991; Ruby and Abelson 1991; Rymond and Rosbash 1992). However, these screens have by no means identified all of the yeast genes important for splicing, since most PRP genes are represented by only a single mutant allele. Furthermore, except for PRP8, a U5 snRNP protein, and PRP4 and PRP6, which are associated with the U4/U6 snRNP, mutations have not been isolated in the genes encoding integral snRNP proteins (for review, see Baserga and Steitz, this volume). The total number of yeast genes encoding proteins important for splicing has been estimated to be as large as 100, indicating that approximately 1% of the yeast genome is dedicated to such proteins (Vijayraghavan et al. 1989).

Although specific biochemical functions have not been assigned to most of the PRP proteins, intriguing possibilities are apparent from sequence homologies. Several of the proteins contain zinc finger-like motifs or RRMs, both of which are common to other RNA-binding proteins. Given the presence of snRNAs and pre-mRNA in the spliceosome, it would not be surprising if several of the PRP proteins bind RNA in a sequence-specific fashion. Other motifs common to several PRP proteins are the so-called "DEAD" and "DEAH" boxes conserved among known ATP-dependent RNA helicases (for review, see Wassarmann and Steitz 1991; Schmid and Linder 1992). PRP proteins with such activities could be required to disassociate and rearrange the various snRNA-snRNA and snRNA-pre-mRNA duplex structures. However, it should be kept in mind that no RNA helicase activity has yet been directly demonstrated for any of the PRP proteins. Purified PRP16 protein, a DEAH family member, has been shown to have an RNA-dependent ATPase activity (Schwer and Guthrie 1991). A PRP16 allele was originally isolated as a dominant suppressor of the deficiency of splicing of a mutant gene containing a branch site alteration (A to C) that causes the accumulation of unspliced pre-mRNA (Couto et al. 1987). The mutational change in the suppressor allele altered sequences anticipated to be in the ATP-binding site (Schwer and Guthrie 1991). Thus, ATP hydrolysis is probably related to PRP16 function. It would be anticipated that each individual RNA helicase would be substrate-specific in its activity. For example, each helicase might specifically recognize and dissociate only one duplex region. The recognition could be dependent on either sequence-specific binding or the secondary structure of a particular substrate. However, these duplexes probably occur naturally only in the heart of a spliceosome, making characterization of the various transitions difficult to study.

The third type of consensus sequence common to some PRP proteins is that conserved among GTP-binding proteins. Proteins of this family are allosteric regulators that probably bind to target proteins when charged with GTP, but dissociate when the bound GTP is hydrolyzed to GDP. Thus, these proteins may be pictured as shuttle activities in formation of protein complexes in the spliceosome cycle. As with other GTP-binding proteins (for review, see Bollag and McCormick 1991), it is possible that associated proteins that stimulate GDP←→GTP exchange and GTP hydrolysis may regulate the activity of these factors. The two PRP proteins (PRP4 and 17) with a G-protein motif probably have distinct functions, as they are required for the first and second step of splicing, respectively.

IV. SPECIFICATION OF EXONS AND INTRONS

In addition to the problem of how the precise sites of the chemical events of splicing are determined (Section II), there is the problem of how splice sites are accurately selected amid the large pool of potential sites. The consensus sequences for 5′ and 3′ splice sites do not contain enough information to specify them unambiguously. For example, mammalian genes often contain dozens of introns that may be up to 100,000 nucleotides in length and include many sequences within both introns and exons that mimic authentic splice site sequences. The accurate 5′ and 3′ splice sites must, nevertheless, be distinguished. Even more recognition and specificity must be imposed for the control of alternative splicing. In this section, several levels of pre-mRNA recognition are discussed that may contribute to the sorting of exons and introns. These include exon definition, the best-understood examples of regulated alternative splicing, *trans*-splicing, and proofreading of processed mRNA.

A. A Network of Interactions: Intron and Exon Definition

Most studies of splicing in vitro have utilized single intron-containing RNAs and have characterized interactions and complexes that form across introns (Figs. 4 and 5B). However, studies of multi-intron-containing RNAs demonstrate interactions across exons (Fig. 5C). Most probably, interactions across both introns and exons combine to specify accurate splice sites in long RNAs.

In vertebrate cells, where most exons are less than 300 nucleotides long, the initial unit of exon/intron distinction at internal sites may be the exon; this definition would occur by virtue of interactions between factors bound at a 3′ splice site and a downstream 5′ splice site (Robberson et al. 1990). Evidence for such interactions stems from several observations: (1) Introns preceding artificially long exons are not spliced; (2) U1 snRNP is required for initial recognition of 3 ′ splice sites (Zillman et al. 1987, 1988; see also Section III.C.1); (3) the presence of a 5 ′ splice site within an exon affects splicing of the upstream intron (Robberson et al. 1990); and (4) mutations in 5 ′ splice sites that decrease complementarity to U1 RNA result in skipping of the preceding exon (Talerico and Berget 1990; Grabowski et al. 1991). For example, inclusion of an alternatively spliced exon of the neural cell adhesion molecule (n-CAM) pre-mRNA is altered by the strength of the downstream 5 ′ splice site (Tacke and Goridis 1991). In a compilation of the known in vivo effects of mutations at internal 5 ′ splice sites, 15 of 18 exhibited skipping of the preceding exons (Fig. 5A); in contrast, a simple intron definition model would only predict intron inclusion (Talerico and Berget 1990 and references therein). Furthermore, recent evidence demonstrates specific interactions that span an exon (Hoffman and Grabowski 1992): A protein was identified that was recruited to a poor 3 ′ splice site of an alternatively spliced exon only if a strong 5 ′ splice site was downstream, and this interaction required the presence of U1 snRNP. The protein that bound to the 3 ′ splice site under these conditions was identified as U2AF and was shown to interact with the downstream U1 snRNP via at least one other protein factor.

Together, these data support the model of Berget and colleagues that the exon is a unit of spliceosomal assembly in vertebrates and that multiple factors contribute to communication across exons (Fig. 5C). Following initial exon definition, a rearrangement would be required to make contacts across introns prior to any chemical events. This model would help to explain the limit on exon length of internal vertebrate exons, the avoidance of cryptic 5 ′ splice sites within introns, and the inhibition of splicing of introns by a downstream 5 ′ splice site mutation. Additional mechanisms, however, must contribute to the recognition of first and last introns (Inoue et al. 1989; Niwa and Berget 1991), as well as the processing of very long exons. Finally, even invoking exon definition that couples recognition of sequences at a 5 ′ splice site with sequences at an adjacent 3 ′ splice site, the specificity necessary for accurately processing very long introns cannot be explained.

B. Regulated Splicing: *Drosophila* Sex Determination

Pre-mRNAs of many tissue-specific and developmentally regulated genes are subject to alternative splicing, providing an additional level of genetic control. Such regulated splicing can lead to the production of different proteins or can control expression, for example by including or excluding a stop codon. Whereas most imaginable patterns of alternative splicing have been detected (e.g., skipped exons, included introns, alternative 5'splice sites, alternative 3'splice sites, and mutually exclusive exons [for review, see Smith et al. 1989a]), relatively little is known about the molecular mechanisms involved. The signals responsible for regulating alternative RNA splicing are currently being studied and must be understood in the context of constitutive signals that specify the formation of a spliceosome on an intron. One system, sex determination in *Drosophila*, has seen significant advances in recent years, and this is discussed as an example of regulation of splicing by factors that bind directly to the regulated pre-mRNA. Other notable concepts in regulation that are emerging from other systems include recognition that levels of constitutive factors can modulate alternative splicing choices (see discussion of ASF/SF2 and hnRNP A1 in Section III.C.2 [Ge et al. 1991; Mayeda and Krainer 1992]); a potential role for U1 snRNP and other factors binding to pseudo-5'splice sites in *Drosophila* P elements to regulate authentic 5'splice site usage (Tseng et al. 1991; Siebel et al. 1992); elements involved in neuron-specific splicing of c-*src* (Black 1992); competition between branch sites/polypyrimidine tracts in β-tropomyosin (Smith and Nadal-Ginard 1989; Guo et al. 1991); and the effects of formation of secondary structures on the utilization of splice sites (Chebli et al. 1989; d'Orval et al. 1991; Eng and Warner 1991; Libri et al. 1991; Estes et al. 1992).

In *Drosophila*, sex determination involves a cascade of alternative RNA splicing events, which include examples of both activation and inactivation of 3'splice sites. The primary determinant of sex in *Drosophila* is the ratio of X chromosomes to sets of autosomes. One of the first results of a female ratio (2X:2A) in embryos is activation of the early promoter of the *Sex-lethal* (*Sxl*) gene (Keyes et al. 1992). Once made, the Sxl protein affects the splicing of at least two RNAs that are transcribed throughout the life cycle (Bell et al. 1988; Salz et al. 1989): (1) Sxl autoregulates its own expression from a late promoter by controlling the exclusion of exon 3, resulting in maintenance of the female Sxl protein (Bell et al. 1988); and (2) Sxl mediates exclusion of exon 2 from the *transformer* (*tra*) mRNA, resulting in expression of the female-specific Tra protein (see Fig. 6, upper and middle) (Boggs et al. 1987). In males, where Sxl protein is not expressed, both *sxl* exon 3 and *tra* exon 2 are included (the "default" splicing pattern). In both genes, the male-specific alternative exons contain stop codons, resulting in production of truncated, nonfunctional proteins in males. It is thought that, in females, Sxl protein binds to intron sequences immediately upstream of the excluded

3' splice site on the *tra* RNA (Sosnowski et al. 1989; Inoue et al. 1990), thereby "blocking" the utilization of that 3' splice site and promoting the preferential use of the distal 3' splice site. More recent evidence suggests that this "block" is due to Sxl binding to the *Sxl-* or *tra*-specific polypyrimidine tract and competing with the binding of U2AF (Valcárcel et al. 1993). Interestingly, the Sxl protein sequence contains two RRMs (see Table 1), but lacks an arginine/serine domain (see discussion in Section III.C.2). It seems that the absence of an arginine/serine domain associated with the polypyrimidine tract is what blocks splicing, as Valcárcel et al. were able to restore splicing of the *tra* intron in vitro if they included a hybrid protein in which the RRM domains of Sxl had been fused to the arginine/serine domain of U2AF.

Once expressed in females, the Tra protein acts along with the Transformer-2 (Tra-2) protein to activate the 3' splice site of *double-sex* (*dsx*) exon 4, which is not utilized in males (Fig. 6, lower; Table 1). Tra-2 protein has been shown to bind directly to six 13-nucleotide repeats in exon 4 of *dsx* RNA (Hedley and Maniatis 1991), and both proteins activate exon 4 inclusion in vitro. Although the molecular mechanisms by which these proteins stimulate the formation of splicing complexes are not known, Tra-2 contains an RRM which likely binds to the 13-nucleotide repeat sequences. In contrast, Tra protein does not contain sequences indicating sequence-specific RNA binding, although it can be cross-linked to the 13-nucleotide repeats in the presence of other cellular factors (Tian and Maniatis 1992). Both Tra and Tra-2 contain arginine/serine domains; the placement of these arginine/serine domains near the 3' splice site of *dsx* exon 4 is likely involved in the activation of that splice site.

The alternative splicing of *dsx* pre-mRNA results in female and male sex-specific *dsx* mRNAs that share their first three exons but have different 3' exons and polyadenylation sites. *Dsx* is thought to be the final regulatory gene in this cascade. It binds directly to gene enhancer sequences and is proposed to be a transcriptional repressor (Burtis et al. 1991). Biologically, the female Dsx protein (along with the Intersex protein) represses somatic male differentiation, and the male Dsx protein represses female differentiation.

The regulatory effects of Sxl and the Tra/Tra-2 proteins on splicing exemplify several concepts that apply both to the control of alternative splicing and to constitutive splicing. First, regulation may be accomplished by interfering with factors required for constitutive splicing, such as U2AF. Second, interactions of proteins containing arginine/serine domains are probably essential for splicing (see Section III.C.2), and splicing can be either blocked, by interfering with the normal binding of arginine/serine proteins, or redirected, by the specific placement of new arginine/serine-containing proteins. Finally, it is possible for proteins that bind RNA in a sequence-specific fashion to control splicing of introns located at some distance from the binding site. Thus, the activities of many factors could be necessary for the physiological control of splicing of a particular intron.

C. *trans*-Splicing

The mRNAs from most vertebrate genes are synthesized by the *cis*-splicing cycle described in Section III. A few mRNAs in mammalian cells may also be generated by *trans*-splicing, but these potential events have not been studied in vitro (Vellard et al. 1992). In trypanosomes, all mRNAs are synthesized by *trans*-splicing of a short splice leader (SL) RNA to the 5′ end of precursor RNAs (for review, see Agabian 1990). The situation in the nematodes *Caenorhabditis elegans* and *Ascaris lumbricoides* is mixed, where *cis*-splicing of short introns is the most common, but *trans*-splicing of SL RNAs is not uncommon (Krause and Hirsh 1987; Hannon et al. 1990; for review, see Blumenthal and Thomas 1988; Nilsen 1989). The *trans*-splicing reaction in *A. lumbricoides* has been well characterized in vitro, and the process is similar to the *cis*-reaction (see Fig. 1B) (Hannon et al. 1990).

The two surprising aspects of the *trans*-splicing processes are (1) the reaction does not apparently require U1 snRNP (Bruzik and Steitz 1990), nor does the SL RNA contain sequences equivalent to those U1 RNA sequences that are complementary to 5′ splice site sequences (Maroney et al. 1991) and (2) trypanosomes, which utilize *trans*-splicing exclusively, do not appear to encode an equivalent to U5 snRNA (Tschudi and Ullu 1990). *trans*-Splicing does require analogs of U2, U4, and U6 snRNAs (Tschudi and Ullu 1990). Given the similarity in chemistry between *cis*- and *trans*-splicing, this suggests that only these snRNAs are required for catalysis in either system. Furthermore, if a U4 snRNA's primary function is to shield U6 snRNA until it is assembled in the spliceosome (see Sections II.D and III.D), then the chemistry can probably be assigned to U2 and U6 snRNAs.

Characterization of the *trans*-splicing reaction of *A. lumbricoides* has revealed a functionally important complementarity between the intron sequences in the SL RNA and U6 snRNA (Fig. 2C). This complementarity covers the Sm-binding site and adjacent nucleotides in the SL RNA and sequences immediately adjacent to those in U6 snRNA that are complementary to U2 snRNA (Hannon et al. 1992). An equivalent complementarity has not been identified in the *cis*-splicing process; the presence of this extensive complementarity probably represents a *trans*-splicing-specific modification that allows dispensing with the U1 and U5 snRNA interactions in the localization of the 5′ exon and is only possible when the splicing process utilizes a single leader-intron RNA sequence for all reactions. In *cis*-splicing, many different 5′ exon-intron sequences must be processed by the spliceosome.

D. Proofreading of Spliced RNA

Since introns can be up to 100,000 nucleotides in length, spliceosomes must be able to form on 5′ and 3′ splice sites that are widely separated. However, it would not be surprising if an occasional splicing event combined exons in an incorrect order or *trans*-spliced two inappropriate exons. In fact, direct evidence has been obtained for such errors in the splicing of nuclear RNA from the large genes *DCC* and *ETS-1* (Nigro et al. 1991; Cocquerelle et al. 1992). It is likely that these incorrect RNAs are normally destroyed by a system that monitors open reading frame intactness in newly synthesized mRNAs.

The existence of such a system was first postulated from the decrease in levels of mRNAs expressed from mutant genes containing nonsense codons. This effect was originally noted for thalassemia mutations where β-globin genes with nonsense mutations generate low levels of cytoplasmic mRNA (Maquat et al. 1981; Humphries et al. 1984). For example, mRNAs from genes with nonsense codons near the 5′ end of the protein-coding sequence are degraded to one-tenth the wild-type levels; however, mRNAs from genes with nonsense codons near the 3′ end of the protein-coding sequences accumulate to near wild-type levels (Losson and Lacroute 1979; Baumann et al. 1985; Urlaub et al. 1989; Cheng et al. 1990; R. Pulak and P. Anderson, in prep.). It is thought that this proofreading process precludes the synthesis of protein fragments from mutant genes, which could interfere in a dominant fashion with the function of the wild-type protein. A similar phenomenon has been extensively studied for nonsense mutations in genes in mammalian cells (Daar and Maquat 1988; Urlaub et al. 1989) and has also been described in nematodes (see below) and yeast (Losson and Lacroute 1979; Leeds et al. 1991).

In *C. elegans*, there is a family of six genes (*SMG1-6*) which, when mutated, relieve the decrease in mutant RNA levels (Hodgkin et al. 1989; R. Pulak and P. Anderson, in prep.). Genetic analysis suggests that all six genes contribute to a common process. The proteins encoded by *SMG1-6* have not been well characterized to date, but it would not be surprising if they were closely associated with the nuclear pore complex where they could proofread each newly transported mRNA. Since errors in RNA splicing in the nucleus would probably generate mRNAs with inappropriately positioned nonsense codons, these would be accordingly destroyed by the system.

V. CONCLUSIONS AND PERSPECTIVES

The proposition that the nuclear spliceosomal process for splicing pre-RNAs is evolutionarily related to the group II self-splicing introns has been strengthened by recent data on snRNA-snRNA interactions. The conserved network of complementarity between the branch site, U2 snRNA, U6 snRNA, and the 5′ splice site juxtaposes conserved snRNA structures and the reactive phosphates in the pre-mRNA. Although similarities between the snRNA-snRNA structures formed in the spliceosome and structures within group II and group I RNAs have been suggested, the extent of these similarities is so limited that it does not establish the relationship convincingly. Further experiments will determine the roles of other sequences within the snRNAs and, thus, either extend these preliminary hints of functional similarities or disprove their validity. One current limitation in establishing the functional relationship between group II self-splicing introns and the spliceosome is the paucity of information about the group II reactions. Group II self-splicing introns must be fascinating RNA machines that probably shift between two catalytic sites. Characterization of these sites could reveal interesting correlations to the sites that catalyze the two reactions in the spliceosome.

In considering the evolution of pre-mRNA splicing, it is useful to ask, Why is the spliceosomal process more group II-like than group I-like? A possible answer is that the group II reactions can be designed to be almost irreversible, and thus, introns would tend to be excised, not inserted. In contrast, group I splicing is, by its very nature, reversible. Probably both the group II and spliceosomal processes utilize two reactive sites for the two steps. Thus, completion of the first step could trigger a structural rearrangement of the first catalytic site to make the reverse reaction unfavorable and to form the catalytic site for the second step. In a group II intron, this rearrangement could involve the formation of a stable secondary or tertiary structure upon cleavage at the 5′ splice site and lariat formation. An analogous rearrangement in the spliceosome may be driven by proteins, such as ATP-dependent helicases. After the reaction at the second catalytic site in both systems, the two product RNAs separate, making complete reversal even less likely. In the spliceosome, this step is also promoted by the action of proteins. In the *Tetrahymena* group I intron, where the two steps are catalyzed as forward and reverse reactions at the same site, the predominant free energy driving the process is formation of a stabilized base-pairing interaction between the intron and newly joined exons. Similar complementarity could not be tolerated in the spliceosome, however, since it would constrain the sequences of the flanking exons.

Should it be surprising that the spliceosomal process may now be dependent on the activities of 100 different proteins? Probably not. The complexity of the spliceosome in terms of RNA sequence recognition, the nature of the chemical reactions, the potential for regulation, and its relationship to nucleo-cytoplasmic transport is comparable to that of the ribosome. In fact, the analogy has become more striking with direct evidence that translation is RNA-catalyzed (see Noller, this volume). It is interesting to note that eukaryotic cells are compartmentalized with pre-mRNA splicing in the nucleus and translation in the cytoplasm. Both processes could be equally old in an evolutionary sense; it is likely that they commingled in early organisms in a single compartment. Although no example of coexistence of the ribosomal and spliceosomal processes in a single subcellular compartment has yet been described, group II splicing and translation can clearly coexist in mitochondria. Thus, it is likely that, as the nucleus and cytoplasm separated, these two processes segregrated to their current domains. In the nucleus the genetic code in pre-mRNAs is recognized and processed by the spliceosome, a potential RNA machine, whereas in the cytoplasm the mature mRNA is recognized and decoded into protein by the ribosome, a proven RNA machine.

ACKNOWLEDGMENTS

We thank Margarita Siafaca for help with preparation of this manuscript, and we thank our many colleagues who shared their results with us prior to publication. M.J.M. and C.C.Q. are supported by American Cancer Society and Damon Runyon-Walter Winchell Cancer Research Fund postdoctoral fellowships, respectively. This work was supported by National Institutes of Health grant RO1-GM-34277 (P.A.S.) and National Cancer Institute core grant P30-CA-14051 (Center for Cancer Research).

REFERENCES

Abovich, N., P. Legrain, and M. Rosbash. 1990. The yeast *prp6* gene encodes a U4/U6 small nuclear ribonucleoprotein particle (snRNP) protein, and the *prp9* gene encodes a protein required for U2 snRNP binding. *Mol. Cell. Biol.* **10:** 6417-6425.
Aebi, M., H. Hornig, and C. Weissmann. 1987. 5′ cleavage site in eukaryotic pre-mRNA splicing is determined by the overall 5′ splice region, not by the conserved 5′ GU. *Cell* **50:** 237–246.

Agabian, N. 1990. *Trans*-splicing of nuclear pre-mRNAs. *Cell* **61:** 1157–1160.

Amrein, H., M. Gorman, and R. Nöthiger. 1988. The sex-determining gene *tra-2* of *Drosophila* encodes a putative RNA binding protein. *Cell* **55:** 1025–1035.

Anderson, G.J., M. Bach, R. Lührmann, and J.D. Beggs. 1989. Conservation between yeast and man of a protein associated with U5 small nuclear ribonucleoprotein. *Nature* **342:** 819–821.

Arenas, J. and J. Hurwitz. 1987. Purification of an RNA debranching activity from HeLa cells. *J. Biol. Chem.* **262:** 4274–4279.

Banroques, J. and J.N. Abelson. 1989. PRP4: A protein of the yeast U4/U6 small nuclear ribonucleoprotein particle. *Mol. Cell. Biol.* **9:** 3710–3719.

Barabino, S.M.L., B.J. Blencowe, U. Ryder, B.S. Sproat, and A.I. Lamond. 1990. Targeted snRNP depletion reveals an additional role for mammalian U1 snRNP in spliceosome assembly. *Cell* **63:** 293–302.

Barnett, S.F., T.A. Theiry, and W.M. LeStourgeon. 1991. The core proteins A2 and B1 exist as $(A2)_3B1$ tetramers in 40S nuclear ribonucleoprotein particles. *Mol. Cell. Biol.* **11:** 864–871.

Baumann, B., M.J. Potash, and G. Kohler. 1985. Consequences of frameshift mutations at the immunoglobulin heavy chain locus of the mouse. *EMBO J.* **4:** 351–359.

Been, M.D. and A.R. Perrotta. 1991. Group I intron self-splicing with adenosine: Evidence for a single nucleoside-binding site. *Science* **252:** 434–437.

Behrens, S.-E. and R. Lührmann. 1991. Immunoaffinity purification of a [U4/U6.U5] tri-snRNP from human cells. *Genes Dev.* **5:** 1439–1452.

Bell, L.R., E.M. Maine, P. Schedl, and T.W. Cline. 1988. Sex-lethal, a *Drosophila* sex determination switch gene, exhibits sex-specific RNA splicing and sequence similarity to RNA binding proteins. *Cell* **55:** 1037–1046.

Bennett, M., S. Piñol-Roma, D. Stakins, G. Dreyfuss, and R. Reed. 1992. Differential binding of heterogeneous nuclear ribonucleoproteins to mRNA precursors prior to spliceosome assembly in vitro. *Mol. Cell. Biol.* **12:** 3165–3175.

Bindereif, A. and M.R. Green. 1987. An ordered pathway of snRNP binding during mammalian pre-mRNA splicing complex assembly. *EMBO J.* **6:** 2415–2424.

Black, D.L. 1992. Activation of c-*src* neuron-specific splicing by an unusual RNA element in vivo and in vitro. *Cell* **69:** 795–807.

Blanton, S., A. Spinivasan, and B.C. Rymond. 1992. PRP38 encodes a yeast protein required for pre-mRNA splicing and maintenance of stable U6 small nuclear RNA levels. *Mol. Cell. Biol.* **12:** 3939–3947.

Blencowe, B.J., B.S. Sproat, U. Ryder, S. Barabino, and A.I. Lamond. 1989. Antisense probing of the human U4/U6 snRNP with biotinylated 2′-OMe RNA oligonucleotides. *Cell* **59:** 531–539.

Blumenthal, T. and J. Thomas. 1988. *Cis* and *trans* mRNA splicing in *C. elegans*. *Trends Genet.* **4:** 305–308.

Boggs, R.T., P. Gregor, S. Idriss, M.J. Belote, and M. McKeown. 1987. Regulation of sexual differentiation in *D. melanogaster* via alternative splicing of RNA from the *transformer* gene. *Cell* **50:** 739–747.

Bollas, G. and F. McCormick. 1991. Regulators and effectors of *ras* proteins. *Annu. Rev. Cell Biol.* **7:** 601–632.

Bordonné, R. and C. Guthrie. 1992. Human and human-yeast chimeric U6 snRNA genes identify structural elements required for expression in yeast. *Nucleic Acids Res.* **20:** 479–485.

Bordonné, R., J. Banroques, J. Abelson, and C. Guthrie. 1990. Domains of yeast U4 spliceosomal RNA required for PRP4 protein binding, snRNP-snRNP interactions and pre-mRNA splicing in vivo. *Genes Dev.* **4:** 1185–1196.

Brow, D.A. and C. Guthrie. 1988. Spliceosomal RNA U6 is remarkably conserved from yeast to mammals. *Nature* **334:** 213–218.

Brown, J.D. and J.D. Beggs. 1992. Roles of PRP8 protein in the assembly of splicing complexes. *EMBO J.* **11:** 3721–3729.

Bruzik, J.P. and J.A. Steitz. 1990. Spliced leader RNA sequences can substitute for the essential 5′ end of U1 RNA during splicing in a mammalian in vitro system. *Cell* **62:** 889–899.

Burd, C.G., M.S. Swanson, M. Görlach, and G. Dreyfuss. 1989. Primary structures of the heterogeneous nuclear ribonucleoprotein A2, B1, and C2 proteins: A diversity of RNA binding proteins is generated by small peptide inserts. *Proc. Natl. Acad. Sci.* **86:** 9788–9792.

Burgess, S., J.R. Couto, and C. Guthrie. 1990. A putative ATP-binding protein influences the fidelity of branchpoint recognition in yeast splicing. *Cell* **60:** 705–717.

Burtis, K.C., K.T. Coschigano, B.S. Baker, and P.C. Wensink. 1991. The doublesex proteins of *Drosophila melanogaster* bind directly to a sex-specific yolk protein gene enhancer. *EMBO J.* **10:** 2577–2582.

Calnan, B.J., B. Tidor, S. Biancalana, D. Hudson, and A.D. Frankel. 1991. Arginine-mediated RNA recognition: The arginine fork. *Science* **252:** 1167–1171.

Cech, T.R. 1985. RNA splicing: Three themes with variations. *Cell* **43:** 713–716.

Chang, D.D. and P.A. Sharp. 1989. Regulation by HIV Rev depends upon recognition of splice sites. *Cell* **59:** 789–795.

Chang, T.-H., M.W. Clark, A.J. Lustig, M.E. Cusick, and J. Abelson. 1988. RNA11 protein is associated with the yeast spliceosome and is localized in the periphery of the cell nucleus. *Mol. Cell. Biol.* **8:** 2379–2393.

Chapman, K.B. and J.D. Boeke. 1991. Isolation and characterization of the gene encoding yeast debranching enzyme. *Cell* **65:** 483–492.

Chebli, K., R. Gattoni, P. Schmitt, G. Hildwein, and J. Stevenin. 1989. The 216-nucleotide intron of the E1A pre-mRNA contains a hairpin structure that permits utilization of unusually distant branch acceptors. *Mol. Cell. Biol.* **9:** 4852–4861.

Chen, J.-H. and R.-J. Lin. 1990. The yeast PRP2 protein, a putative RNA-dependent ATPase, shares extensive sequence homology with two other pre-mRNA splicing factors. *Nucleic Acids Res.* **18:** 6447.

Cheng, J., M. Fogel-Petrovic, and L.E. Maquat. 1990. Translation to near the distal end of the penultimate exon is required for normal levels of spliced triosephosphate isomerase mRNA. *Mol. Cell. Biol.* **10:** 5215–5225.

Cheng, S.C. and J. Abelson. 1987. Spliceosome assembly in yeast. *Genes Dev* **1:** 1014–1027.

Choi, Y.D., P.J. Grabowski, P.A. Sharp, and G. Dreyfuss. 1986. Heterogeneous nuclear ribonucleoproteins: Role in RNA splicing. *Science* **231:** 1534–1539.

Cocquerelle, C., P. Daubersies, M.-A. Majérus, J.-P. Kerckaert, and B. Bailleul. 1992. Splicing with inverted order of exons occurs proximal to large introns. *EMBO J.* **11:** 1095–1098.

Company, M., J. Arenas, and J. Abelson. 1991. Requirement of the RNA helicase-like protein PRP22 for release of messenger RNA from spliceosomes. *Nature* **349:** 487–493.

Couto, J.R., J. Tamm, R. Parker, and C. Guthrie. 1987. A *trans*-acting suppressor restores splicing of a yeast intron with a branch point mutation. *Genes Dev.* **1:** 445–455.

Daar, I.O. and L.E. Maquat. 1988. Premature translation termination mediates triosephosphate isomerase mRNA degradation. *Mol. Cell. Biol.* **8:** 802–813.

Dalbadie-McFarland, G. and J. Abelson. 1990. PRP5: A helicase-like protein required for mRNA splicing in yeast. *Proc. Natl. Acad. Sci.* **87:** 4236–4240.

Dalrymple, M.A., S. Petersen-Bjorn, J.D. Friesen, and J.D. Beggs. 1989. The product of the PRP4 gene of *S. cerevisiae* shows homology to β subunits of G proteins. *Cell* **58:** 811–812.

Datta, B. and A.M. Weiner. 1991. Genetic evidence for base pairing between U2 and U6 snRNAs in mammalian mRNA splicing. *Nature* **352:** 821–824.

D'Orval, B.C., Y.D. Carafa, P. Sirand-Pugnet, M. Gallego, E. Brody, and J. Marie. 1991. RNA secondary structure repression of a muscle-specific exon in HeLa cell nuclear extracts. *Science* **252:** 1823–1828.

Dreyfuss, G., M.S. Swanson, and S. Piñol-Roma. 1990. The composition, structure, and organization of proteins in heterogeneous nuclear ribonucleoprotein complexes. In *The eukaryotic nucleus: Structure and function* (ed. P. Strauss and S. Wilson), pp. 503–517. Telford Press, Caldwell, New Jersey.

Eng, F.J. and J.R. Warner. 1991. Structural basis for the regulation of splicing of a yeast messenger RNA. *Cell* **65:** 797–804.

Estes, P.A., N.E. Cooke, and S.A. Liebhaber. 1992. A native RNA secondary structure controls alternative splice-site selection and generates two human growth hormone isoforms. *J. Biol. Chem.* **257:** 14902–14908.

Fabrizio, P. and J. Abelson. 1990. Two domains of yeast U6 small nuclear RNA required for both steps of nuclear precursor messenger RNA splicing. *Science* **250:** 404–409.

Frank, D. and C. Guthrie. 1992. An essential splicing factor, SLU7, mediates 3′ splice site choice in yeast. *Genes Dev.* **6:** 2112–2124.

Frank, D., B. Patterson, and C. Guthrie. 1992. Synthetic lethal mutations suggest interactions between U5 small nuclear RNA and four proteins required for the second step of splicing. *Mol. Cell. Biol.* **12:** 5197–5205.

Fu, X.-D. and T. Maniatis. 1990. Factor required for mammalian spliceosome assembly is localized to discrete regions in the nucleus. *Nature* **343:** 437–441.

———. 1992a. The 35-kDa mammalian splicing factor SC35 mediates specific interactions between U1 and U2 small nuclear ribonucleoprotein particles at the 3′ splice site. *Proc. Natl. Acad. Sci.* **89:** 1725–1729.

———. 1992b. Isolation of a complementary DNA that encodes the mammalian splicing factor SC35. *Science* **256:** 535–538.

Fu, X.-D., A. Mayeda, T. Maniatis, and A.R. Krainer. 1992. General splicing factors SF2 and SC35 have equivalent activities *in vitro*, and both affect alternative 5′ and 3′ splice site selection. *Proc. Natl. Acad. Sci.* **89:** 11224–11228.

Garcia-Blanco, M.A., S. Jamison, and P.A. Sharp. 1989. Identification and purification of a 62,000 dalton protein which binds specifically to the polypyrimidine tract of introns. *Genes Dev.* **3:** 1874–1886.

Garcia-Blanco, M.A., G.J. Anderson, J. Beggs, and P.A. Sharp. 1990. A mammalian protein of 220 kDa binds pre-mRNA in the spliceosome: A potential homologue of the yeast PRP9 protein. *Proc. Natl. Acad. Sci.* **87:** 3082–3086.

Ge, H. and J.L. Manley. 1990. A protein factor, ASF, controls cell specific alternative splicing of SV40 early pre-mRNA in vitro. *Cell* **62:** 25–34.

Ge, H., P. Zuo, and J.L. Manley. 1991. Primary structure of the human splicing factor ASF reveals similarities with *Drosophila* regulators. *Cell* **66:** 373–382.

Ghetti, A., S. Piñol-Roma, W.M. Michael, C. Morandi, and G. Dreyfuss. 1992. hnRNP I, the polypyrimidine tract-binding protein: Distinct nuclear localization and association with hnRNAs. *Nucleic Acids Res.* **20:** 3671–3678.

Gil, A., P.A. Sharp, S.F. Jamison, and M.A. Garcia-Blanco. 1991. Characterization of cDNAs encoding the polypyrimidine tract binding protein. *Genes Dev.* **5:** 1224–1236.

Goodall, G.J., T. Kiss, and W. Filipowicz. 1991. Nuclear RNA splicing and small nuclear RNAs and their genes in higher plants. *Oxford Surv. Plant Mol. Cell. Biol.* **7:** 255–259.

Goralski, T.J., J.E. Edström, and B.S. Baker. 1989. The sex determination locus transformer-2 of *Drosophila* encodes a polypeptide with similarity to RNA binding proteins. *Cell* **56:** 1011–1118.

Grabowski, P.J., F.H. Nasim, and H.-K. Kuo. 1991. Combinational splicing of exon pairs by two-site binding of U1 small nuclear ribonucleoprotein particle. *Mol. Cell. Biol.* **11:** 5919–5928.

Green, M.R. 1991. Biochemical mechanisms of constitutive and regulated pre-mRNA splicing. *Annu. Rev. Cell Biol.* **7:** 559–599.

Guo, W., G.J. Mulligan, S. Wormsley, and D.M. Helfman. 1991. Alternative splicing of b-tropomyosin pre-mRNA: *cis*-acting elements and cellular factors that block the use of a skeletal muscle exon in nonmuscle cells. *Genes Dev.* **5:** 2096–2107.

Guthrie, C. 1991. Messenger RNA splicing in yeast: Clues to why the spliceosome is a ribonucleoprotein. *Science* **253:** 157–163.

Guthrie, C. and B. Patterson. 1988. Spliceosomal snRNA. *Annu. Rev. Genet.* **22:** 387–419.

Hall, K.B. and M.M. Konarska. 1992. The 5′ splice site consensus RNA oligonucleotide induces assembly of U2/U4/U5/U6 small nuclear ribonucleoprotein complexes. *Proc. Natl. Acad. Sci.* **89:** 10969–10973.

Hannon, G.J., P.A. Maroney, J.A. Denker, and T.W. Nilsen. 1990. *Trans*-splicing of Nematode pre-messenger RNA in vitro. *Cell* **61:** 1247–1255.

Hannon, G.J., P.A. Maroney, Y.-T. Yu, G.E. Hannon, and T.W. Nilsen. 1992. Interaction of U6 snRNA with a sequence required for function of the nematode SL RNA in *trans*-splicing. *Science* **258:** 1775–1780.

Hausner, T.-P., L.M. Giglio, and A.M. Weiner. 1990. Evidence for base-pairing between mammalian U2 and U6 small nuclear ribonucleoprotein particles. *Genes Dev.* **4:** 2146–2156.

Hedley, M.L. and T. Maniatis. 1991. Sex-specific splicing and polyadenylation of *dsx* pre-mRNA requires a sequence that binds specifically to *tra-2* protein in vitro. *Cell* **65:** 579–586.

Herschlag, D., J.A. Piccirilli, and T.R. Cech. 1991. Ribozyme-catalyzed and non-enzymatic reactions of phosphate diesters: Rate effects upon substitution of sulfur for a nonbridging phosphoryl oxygen atom. *Biochemistry* **30:** 4844–4854.

Hodgkin, A., A. Papp, R. Pulak, V. Ambros, and P. Anderson. 1989. A new kind of informational suppression in the nematode *Caenorhabitis elegans*. *Genetics* **123:** 301–313.

Hoffman, B.E. and P.J. Grabowski. 1992. U1 snRNP targets an essential splicing factor, U2AF65, to the 3′ splice site by a network of interactions spanning the exon. *Genes Dev.* **6:** 2554–2568.

Hornig, H., M. Aebi, and C. Weissmann. 1986. Effect of mutations at the lariat branch acceptor site on β-globin pre-mRNA splicing in vitro. *Nature* **324:** 589–591.

Humphries, R.K., T.J. Ley, N.P. Anagnou, A.W. Baur, and A.W. Nienhuis. 1984. β⁰-thalassemia gene: A premature termination codon causes β-mRNA deficiency without affecting cytoplasmic b-mRNA stability. *Blood* **64:** 23–32.

Inoue, K., K. Hoshijima, H. Sakamoto, and Y. Shimura. 1990. Binding of the *Drosophila* sex-lethal gene product to the alternative splice site of transformer primary transcript. *Nature* **344:** 461–463.

Inoue, K., M. Ohno, H. Sakamoto, and Y. Shimura. 1989. Effect of the cap structure on pre-mRNA splicing in *Xenopus* oocyte nuclei. *Genes Dev.* **3:** 1472–1479.

Jackson, I.J. 1991. A reappraisal of non-consensus mRNA splice sites. *Nucleic Acids Res.* **19:** 3795–3798.

Jackson, S.P., M. Lossky, and J.D. Beggs. 1988. Cloning of the RNA8 gene of *Saccharomyces cerevisiae*, detection of the RNA8 protein, and demonstration that it is essential for nuclear pre-mRNA splicing. *Mol. Cell. Biol.* **8:** 1067–1075.

Jacquier, A. and N. Jacquesson-Breuleux. 1991. Splice site selection and role of the lariat in a group II intron. *J. Mol. Biol.* **219:** 415–428.

Jamieson, D.J., B. Rahe, J. Pringle, and J.D. Beggs. 1991. A suppressor of a yeast splicing mutation (prp8-1) encodes a putative ATP-dependent RNA helicase. *Nature* **349:** 715–717.

Jamison, S.F. and M.A. Garcia-Blanco. 1992. An ATP-independent U2 small nuclear ribonucleoprotein particle/precursor mRNA complex requires both splice sites and the polypyrimidine tract. *Proc. Natl. Acad. Sci.* **89:** 5482–5486.

Jarrell, K., R.C. Dietrich, and P.A. Perlman. 1988. Group II intron domain 5 facilitates a *trans*-splicing reaction. *Mol. Cell. Biol.* **8:** 2361–2366.

Kenan, D.J., C.C. Query, and J.D. Keene. 1991. RNA recognition: Towards identifying determinants of specificity. *Trends Biol. Sci.* **16:** 214–220.

Keyes, L.N., T.W. Cline, and P. Schedl. 1992. The primary sex determination signal of *Drosophila* acts at the level of transcription. *Cell* **68:** 933–943.

Kiledjian, M. and G. Dreyfuss. 1992. Primary structure and binding activity of the hnRNP U protein: Binding RNA through RGG box. *EMBO J.* **11:** 2655–2664.

Kim, S.-H., J. Smith, A. Claude, and R.-J. Lin. 1992. The purified yeast pre-mRNA splicing factor PRP2 is an RNA-dependent NTPase. *EMBO J.* **11:** 2319–2326.

King, D.S. and J.D. Beggs. 1990. Interaction of PRP2 protein with pre-mRNA splicing complexes in *Saccharomyces cerevisiae*. *Nucleic Acids Res.* **18:** 6559–6564.

Koch, J.L., S.C. Boulanger, S.D. Dib-Hajj, S.K. Hebbar, and P.S. Perlman. 1992. Group II introns deleted for multiple substructures retain self-splicing activity. *Mol. Cell. Biol.* **12:** 1950–1958.

Konarska, M.M. and P.A. Sharp. 1986. Electrophoretic separation of complexes involved in the splicing of precursors to mRNAs. *Cell* **46:** 845–855.

———. 1987. Interactions between small nuclear ribonucleoprotein particles in formation of spliceosomes. *Cell* **49:** 763–774.

———. 1988. Association of U2, U4, U5 and U6 small nuclear ribonucleoproteins in a spliceosome-type complex in absence of precursor RNA. *Proc. Natl. Acad. Sci.* **85:** 5459–5462.

Konarska, M.M., P.J. Grabowski, R.A. Padgett, and P.A. Sharp. 1985. Characterization of the branch site in lariat RNAs produced by splicing of mRNA precursors. *Nature* **313:** 552–557.

Krainer, A.R. and T. Maniatis. 1985. Multiple factors including the small nuclear ribonucleoproteins U1 and U2 are necessary for pre-mRNA splicing in vitro. *Cell* **42:** 725–736.

Krainer, A.R., G.C. Conway, and D. Kozak. 1990a. Purification and characterization of pre-mRNA splicing factor SF2 from HeLa cells. *Genes Dev.* **4:** 1158–1171.

———. 1990b. The essential pre-mRNA splicing factor SF2 influences 5′ splice site selection by activating proximal sites. *Cell* **62:** 35–42.

Krainer, A.R., A. Mayeda, D. Kozak, and G. Binns. 1991. Functional expression of cloned human splicing factor SF2: Homology to RNA binding proteins U1 70K and *Drosophila* splicing regulators. *Cell* **66:** 383–394.

Krämer, A. and U. Utans. 1991. Three protein factors (SF1, SF3 and U2AF) function in presplicing complex formation in addition to snRNPs. *EMBO J.* **10:** 1503–1509.

Krämer, A., M. Frick, and W. Keller. 1987. Separation of multiple components of HeLa cell nuclear extracts required for pre-messenger RNA splicing. *J. Biol. Chem.* **262:** 17630–17640.

Krause, M. and D. Hirsh. 1987. A *trans*-spliced leader sequence on actin mRNA in *C. elegans*. *Cell* **49:** 753–761.

Kumar, A. and S.H. Wilson. 1990. Studies of the strand-annealing activity of mammalian hnRNP complex protein A1. *Biochemistry* **29:** 10717–10722.

Lamm, G.M., B.J. Blencowe, B.S. Sproat, A.M. Iribarren, U. Ryder, and A.I. Lamond. 1991. Antisense probes containing 2-aminoadenosine allow efficient depletion of U5 snRNP from HeLa splicing extracts. *Nucleic Acids Res.* **19:** 3193–3198.

Lamond, A.I., M.M. Konarska, and P.A. Sharp. 1987. A mutational analysis of spliceosome assembly: Evidence for splice site collaboration during spliceosome formation. *Genes Dev.* **1:** 532–543.

Lamond, A.I., M.M. Konarska, P.J. Grabowski, and P.A. Sharp. 1988. Spliceosome assembly involves binding and release of U4 small nuclear ribonucleoprotein. *Proc. Natl. Acad. Sci.* **85:** 411–415.

Lee, M.G., D.P. Lane, and J.D. Beggs. 1986. Identification of the RNA2 protein of *Saccharomyces cerevisiae*. *Yeast* **2:** 59–67.

Leeds, P., S.W. Peltz, A. Jacobson, and M.R. Culbertson. 1991. The product of the yeast *UPF1* gene is required for rapid turnover of mRNAs containing a premature translational termination codon. *Genes Dev.* **5:** 2303–2314.

Legrain, P. and A. Choulika. 1990. The molecular characterization of *PRP6* and *PRP9* yeast genes reveals a new cysteine/histidine motif common to several splicing factors. *EMBO J.* **9:** 2775–2781.

Legrain, P. and M. Rosbash. 1989. Some *cis*- and *trans*-acting mutants for splicing target pre-mRNA to the cytoplasm. *Cell* **57:** 573–583.

Legrain, P., B. Seraphin, and M. Rosbash. 1988. Commitment of yeast pre-mRNA to the spliceosome pathway does not require U2 snRNP. *Mol. Cell. Biol.* **8:** 3755–3760.

Lerner, M.R., J.A. Boyle, S.M. Mount, S.L. Wolin, and J.A. Steitz. 1980. Are snRNPs involved in splicing? *Nature* **283:** 220–224.

Li, H. and P.M. Bingham. 1991. Arginine/serine-rich domains of the *su(w^a)* and *tra* RNA processing regulators target proteins to a subnuclear compartment implicated in splicing. *Cell* **67:** 335–342.

Liao, X.C., H.V. Colot, Y. Wang, and M. Rosbash. 1992. Requirements for U2 snRNP addition to yeast pre-mRNA. *Nucleic Acids Res.* **20:** 4237–4245.

Libri, D., A. Piseri, and M.Y. Fiszman. 1991. Tissue-specific splicing in vivo of the β-tropomyosin gene: Dependence on an RNA secondary structure. *Science* **252:** 1842–1845.

Lin, R.J., A.J. Lustig, and J. Abelson. 1987. Splicing of yeast nuclear pre-mRNA in vitro requires a functional 40S spliceosome and several extrinsic factors. *Genes Dev.* **1:** 7–18.

Lin, R.-J., A.J. Newman, S.-C. Cheng, and J. Abelson. 1985. Yeast mRNA splicing in vitro. *J. Biol. Chem.* **260:** 14780–14792.

Lossky, M., G.J. Anderson, S.P. Jackson, and J. Beggs. 1987. Identification of a yeast snRNP protein and detection of snRNP-snRNP interactions. *Cell* **51:** 1019–1026.

Losson, R. and F. Lacroute. 1979. Interference of nonsense mutations with eukaryotic messenger RNA stability. *Proc. Natl. Acad. Sci.* **76:** 5134–5137.

Lu, X., J. Heimer, D. Rekosh, and M.-L. Hammarskjöld. 1990. U1 small nuclear RNA plays a direct role in the formation of a rev-regulated human immunodeficiency virus env mRNA that remains unspliced. *Proc. Natl. Acad. Sci.* **87:** 7598–7602.

Lustig, A.J., R.J. Lin, and J. Abelson. 1986. The yeast RNA gene products are essential for mRNA splicing in vitro. *Cell* **47:** 953–963.

Madhani, H.D. and C. Guthrie. 1992. A novel base-pairing interaction between U2 and U6 snRNAs suggests a mechanism for catalytic activation of the spliceosome. *Cell* **71:** 803–817.

Madhani, H.D., R. Bordonné, and C. Guthrie. 1990. Multiple roles for U6 snRNA in the splicing pathway. *Genes Dev.* **4:** 2264–2277.

Mancebo, R., P.C.H. Lo, and S.M. Mount. 1990. Structure and expression of the *Drosophila melanogaster* gene for the U1 small nuclear ribonucleoprotein particle 70K protein. *Mol. Cell. Biol.* **10:** 2492–2502.

Maquat, L.E., A.J. Kinniburgh, E.A. Rachmilewitz, and J. Ross. 1981. Unstable β-globin mRNA in mRNA-deficient β⁰-thalassemia. *Cell* **27:** 543–553.

Maroney, P.A., G.J. Hannon, J.D. Shambaugh, and T.W. Nilsen. 1991. Intramolecular base pairing between the nematode spliced leader and its 5′ splice site is not essential for *trans*-splicing in vitro. *EMBO J.* **10:** 3869–3875.

Maschloff, K.L. and R.A. Padgett. 1992. Phosphorothioate substitution identifies phosphate groups important for pre-mRNA splicing. *Nucleic Acids Res.* **20:** 1949–1957.

Matunis, M.J., W.M. Michael, and G. Dreyfuss. 1992. Characterization and primary structure of the poly(C)-binding heterogeneous nuclear ribonucleoprotein complex K protein. *Mol. Cell. Biol.* **12:** 164–171.

Mayeda, A. and A.R. Krainer. 1992. Regulation of alternative pre-mRNA splicing by hnRNP A1 and splicing factor SF2. *Cell* **69:** 365–375.

Mayeda, A., A.M. Zahler, A.R. Krainer, and M.B. Roth. 1992. Two members of a conserved family of nuclear phosphoproteins are involved in pre-mRNA splicing. *Proc. Natl. Acad. Sci.* **89:** 1301–1304.

McKeown, M., J.M. Belote, and B.S. Baker. 1987. A molecular analysis of *transformer*, a gene in *Drosophila melanogaster* that controls female sexual differentiation. *Cell* **48:** 489–499.

McPheeters, D.S. and J. Abelson. 1992. Mutational analysis of the yeast U2 snRNA suggests a structural similarity to the catalytic core of Group I introns. *Cell* **71:** 819–831.

McSwiggen, J.A. and T.R. Cech. 1989. Stereochemistry of RNA cleavage by the Tetrahymena ribozyme and evidence that the chemical step is not rate-limiting. *Science* **244:** 679–683.

Merrill, B.M. and K.R. Williams. 1990. Structure/function relationships in hnRNP proteins. In *The eukaryotic nucleus: Molecular biochemistry and macromolecular assemblies* (Ed. P. Strauss and S. Wilson), pp. 579–604. Telford Press, Caldwell, New Jersey.

Michaud, S. and R. Reed. 1991. An ATP-independent complex commits pre-mRNA to the mammalian spliceosome assembly pathway. *Genes Dev.* **5:** 2534–2546.

Michel, F. and E. Westhof. 1990. Modelling the three-dimensional architecture of group I catalytic introns based on comparative sequence analysis. *J. Mol. Biol.* **216:** 585–610.

Michel, F., R. Umesono, and H. Ozeki. 1989a. Comparative and functional anatomy of group II catalytic introns—A review. *Gene* **82:** 5–30.

Michel, F., M. Hanna, R. Green, D.P. Bartel, and J.W. Szostak. 1989b. The guanosine binding site of the *Tetrahymena* ribozyme. *Nature* **342:** 391–395.

Moore, M.J. and P.A. Sharp. 1992. Site-specific modification of pre-mRNA: The 2′ hydroxyl groups at the splice sites. *Science* **256:** 992–997.

Mount, S.M., C. Burks, G. Hertz, G.D. Stormo, O. White, and C. Fields. 1992. Splicing signals in *Drosophila:* Intron size, information content, and consensus sequences.

Nucleic Acids Res. **20:** 4255–4262.

Newman, A. and C. Norman. 1991. Mutations in yeast U5 snRNA alter the specificity of 5′splice-site cleavage. *Cell* **65:** 115–123.

————. 1992. U5 snRNA interacts with exon sequences at 5′and 3′splice sites. *Cell* **68:** 743–754.

Nigro, J.M., K.R. Cho, E.R. Fearon, S.E. Kern, J.M. Ruppert, J.D. Oliner, K.W. Kinzler, and B. Vogelstein. 1991. Scrambled exons. *Cell* **64:** 607–613.

Nilsen, T.W. 1989. *Trans*-splicing in nematodes. *Exp. Parasitol.* **69:** 413–416.

Niwa, M. and S.M. Berget. 1991. Mutation of the AAUAAA polyadenylation signal depresses in vitro splicing of proximal but not distal introns. *Genes Dev.* **5:** 2086–2095.

Padgett, R.A., P.J. Grabowski, M.M. Konarska, S. Seiler, and P.A. Sharp. 1986. Splicing of messenger RNA precursors. *Annu. Rev. Biochem.* **55:** 1119–1150.

Padgett, R.A., M.M. Konarska, P.J. Grabowski, S.F. Hardy, and P.A. Sharp. 1984. Lariat RNA's as intermediates and products in the splicing of messenger RNA precursors. *Science* **225:** 898–903.

Parker, R. and P.G. Siliciano. 1993. Evidence for an essential non-Watson-Crick interaction between the first and last nucleotides of a nuclear pre-mRNA intron. *Nature* (in press).

Parker, R., P.G. Siliciano, and C. Guthrie. 1987. Recognition of the TACTAAC box during mRNA splicing in yeast involves base pairing to the U2-like snRNA. *Cell* **49:** 229–239.

Patterson, B. and C. Guthrie. 1991. A U-rich tract enhances usage of an alternative 3′ splice site in yeast. *Cell* **64:** 181–187.

Patton, J.G., S.A. Mayer, P. Tempst, and B. Nadal-Ginard. 1991. Characterization and molecular cloning of a polypyrimidine tract-binding protein: A component of a complex necessary for pre-mRNA splicing. *Genes Dev.* **5:** 1224–1236.

Petersen-Bjorn, S., A. Soltyk, J.D. Beggs, and J.D. Friesen. 1989. Characterization of RNA4/PRP4 from *Saccharomyces cerevisiae:* Its gene product is associated with the U4/U6 snRNP. *Mol. Cell. Biol.* **9:** 3698–3709.

Piñol-Roma, S., M.S. Swanson, J.G. Gall, and G. Dreyfuss. 1989. A novel heterogeneous nuclear RNP protein with a unique distribution on nascent transcripts. *J. Cell Biol.* **109:** 2575–2587.

Puglisi, J.D., R. Ton, B.J. Calnan, A.D. Frankel, and J.R. Williamson. 1992. Conformation of the TAR RNA-arginine complex by NMR spectroscopy. *Science* **257:** 76–80.

Query, C.C., R.C. Bentley, and J.D. Keene. 1989. A common RNA recognition motif identified within a defined U1 RNA binding domain of the 70K U1 snRNP protein. *Cell* **57:** 89–101.

Rajagopal, J., J.A. Doudna, and J.W. Szostak. 1989. Stereochemical course of catalysis by the *Tetrahymena* ribozyme. *Science* **244:** 692–694.

Reed, R. 1989. The organization of 3′splice-site sequences in mammalian introns. *Genes Dev.* **3:** 2113–2123.

Reich, C.I., R.W.V. Hoy, G.L. Porter, and J.A. Wise. 1992. Mutations at the 3′splice site can be suppressed by compensatory base changes in U1 snRNA in fission yeast. *Cell* **69:** 1159–1169.

Robberson, B.L., G.J. Cote, and S.M. Berget. 1990. Exon definition may facilitate splice site selection in RNAs with multiple exons. *Mol. Cell. Biol.* **10:** 84–94.

Rogers, J. and R. Wall. 1980. A mechanism for RNA splicing. *Proc. Natl. Acad. Sci.* **77:** 1877–1879.

Rosbash, M. and B. Seraphin. 1991. Who's on first? The U1 snRNP-5′splice site interaction and splicing. *TIBS* 187–190.

Roth, M.B., C. Murphy, and J.G. Gall. 1990. A monoclonal antibody that recognizes a phosphorylated epitope stains lampbrush chromosome loops and small granules in the amphibian germinal vesicle. *J. Cell. Biol.* **111:** 2217–2223.

Roth, M.B., A.M. Zahler, and J.A. Stolk. 1991. A conserved family of nuclear phosphoproteins localized to sites of polymerase II transcription. *J. Cell Biol.* **115:** 587–596.

Ruby, S.W. and J. Abelson. 1988. An early hierarchic role of U1 small nuclear ribonucleoprotein in spliceosome assembly. *Science* **242:** 1028–1035.

———. 1991. Pre-mRNA splicing in yeast. *Trends Genet.* **7:** 79–85.

Ruskin, B. and M.R. Green. 1985. An RNA processing activity that debranches RNA lariats. *Science* **229:** 135–140.

Ruskin, B., P.D. Zamore, and M.R. Green. 1988. A factor, U2AF, is required for U2 snRNP binding and splicing complex assembly. *Cell* **52:** 207–219.

Rymond, B.C. and M. Rosbash. 1992. Yeast pre-mRNA splicing. In *Molecular and cellular biology of the yeast* Saccharomyces: *Gene expression* (ed. E.W. Jones et al.), vol. 2, pp. 143–192. Cold Spring Harbor Laboratory Press, Cold Spring Harbor, New York.

Salz, H.K., E.M. Maine, L.N. Keyes, M.E. Samuels, T.W. Cline, and P. Schedl. 1989. The *Drosophila* female-specific sex-determination gene, *sex-lethal*, has stage-, tissue-, and sex-specific RNA suggesting multiple modes of regulation. *Genes Dev.* **3:** 708–719.

Sawa, H. and J. Abelson. 1992. Evidence for a base-pairing interaction between U6 small nuclear RNA and the 5′ splice site during the splicing reaction in yeast. *Proc. Natl. Acad. Sci.* **89:** 11269–11273.

Sawa, H. and Y. Shimura. 1992. Association of U6 snRNA with the 5′-splice site region of pre-mRNA in the spliceosome. *Genes Dev.* **6:** 244–254.

Schappert, K. and J.D. Friesen. 1991. Genetic studies of the PRP11 gene of *Saccharomyces cerevisiae. Mol. Gen. Genet.* **226:** 277–282.

Schmid, S.R. and P. Linder. 1992. D-E-A-D protein family of putative RNA helicases. *Mol. Microbiol.* **6:** 283–292.

Schwer, B. and C. Guthrie. 1991. PRP16 is an RNA-dependent ATPase that interacts transiently with the spliceosome. *Nature* **349:** 494–499.

———. 1992. A conformational rearrangement in the spliceosome is dependent on PRP16 and ATP hydrolysis. *EMBO J.* **11:** 5033–5039.

Senapathy, P., M.B. Shapiro, and N.L. Harris. 1990. Splice junctions, branch point sites, and exons: Sequence statistics, identification, and applications to genome project. *Methods Enzymol.* **183:** 252–278.

Seraphin, B. and M. Rosbash. 1989. Identification of functional U1 snRNP-premRNA complexes committed to spliceosome assembly and splicing. *Cell* **59:** 349–358.

———. 1990. Exon mutations uncouple 5′ splice site selection from U1 snRNA pairing. *Cell* **63:** 619–629.

———. 1991. The yeast branchpoint sequence is not required for the formation of a stable U1 snRNP pre-mRNA complex and is recognized in the absence of U2 snRNA. *EMBO J.* **10:** 1209–1216.

Seraphin, B., N. Abovich, and M. Rosbash. 1991. Genetic depletion indicates a late role for U5 snRNP during in vitro spliceosome assembly. *Nucleic Acids Res.* **19:** 3857–3860.

Seraphin, B., L. Kretzner, and M. Rosbash. 1988. A U1 snRNA-premRNA base pairing interaction is required early in spliceosome assembly but does not uniquely define the 5′ splice site. *EMBO J.* **7:** 2533–2538.

Shannon, K.W. and C. Guthrie. 1991. Suppressors of a U4 snRNA mutation define a novel U6 snRNP protein with RNA binding motifs. *Genes Dev.* **5:** 773–785.

Sharp, P.A. 1985. On the origins of RNA splicing and introns. *Cell* **42:** 397–400.

———. 1987. Splicing of messenger RNA precursors. *Science* **23:** 766–771.

Siebel, C.W., L.D. Fresco, and D.C. Rio. 1992. The mechanism of somatic inhibition of *Drosophila* P-element pre-mRNA splicing: Multiprotein complexes at an exon pseudo-5′ splice site control U1 snRNP binding. *Genes Dev.* **6:** 1386–1401.

Siliciano, P.G. and C. Guthrie. 1988. 5′ splice site selection in yeast: Genetic alterations in base-pairing with U1 reveal additional requirements. *Genes Dev.* **2:** 1258–1267.

Smith, C.W.J. and B. Nadal-Ginard. 1989. Mutually exclusive splicing of α-tropomyosin exons enforced by an unusual lariat branch point location: Implications for constitutive splicing. *Cell* **56:** 749–758.

Smith, C.W., J.G. Patton, and B. Nadal-Ginard. 1989a. Alternative splicing in the control of gene expression. *Annu. Rev. Genet.* **23:** 527–577.

Smith, C.W.J., E.B. Porro, J.G. Patton, and B. Nadal-Ginard. 1989b. Scanning from an independently specified branch point defines the 3′ splice site of mammalian introns. *Nature* **342:** 243–247.

Sosnowski, B.A., J.M. Belote, and M. McKeown. 1989. Sex-specific alternative splicing of RNA from the *transformer* gene results from sequence-dependent splice site blockage. *Cell* **58:** 449–459.

Steitz, J.A. 1992. Splicing takes a Holliday. *Science* **257:** 888–889.

Steitz, J.A., D.L. Black, V. Gerke, K.A. Parker, A. Krämer, D. Frendeway, and W. Keller. 1988. Functions of the abundant U-snRNPs. In *Structure and function of major and minor small nuclear ribonucleoprotein particles* (ed. M.L. Birnstiel), pp.115–154. Springer-Verlag, New York.

Stolow, D.T. and S.M. Berget. 1990. UV cross-linking of polypeptides associated with 3′ terminal exons. *Mol. Cell. Biol.* **10:** 5937–5944.

Strauss, E.J. and C. Guthrie. 1991. A cold-sensitive mRNA splicing mutant is a member of the RNA helicase gene family. *Genes Dev.* **5:** 629–641.

Suh, E.-R. and R.B. Waring. 1992. A phosphorothioate at the 3′ splice-site inhibits the second splicing step in a group I intron. *Nucleic Acids Res.* **20:** 6303–6309.

Tacke, R. and C. Goridis. 1991. Alternative splicing in the neural cell adhesion molecule pre-mRNA: Regulation of exon 18 skipping depends on the 5′ splice site. *Genes Dev.* **5:** 1416–1429.

Talerico, M. and S.M. Berget. 1990. Effect of 5′ splice site mutations on splicing of the preceding intron. *Mol. Cell. Biol.* **10:** 6299–6305.

Thomas, J., K. Lea, E. Zucker-Aprison, and T. Blumenthal. 1990. The spliceosomal snRNAs of *Caenorhabditis elegans*. *Nucleic Acids Res.* **18:** 2633–2642.

Tian, M. and T. Maniatis. 1992. Positive control of pre-mRNA splicing in vitro. *Science* **256:** 237–240.

Tschudi, C. and E. Ullu. 1990. Destruction of U2, U4, or U6 small nuclear RNA blocks *trans*-splicing in *Trypanosome* cells. *Cell* **61:** 459–466.

Tseng, J.C., S. Zollman, A.C. Chain, and F.A. Laski. 1991. Splicing of the *Drosophila* P element ORF2-ORF3 intron is inhibited in a human cell extract. *Mech. Dev.* **35:** 65–72.

Urlaub, G., P.J. Mitchell, C.J. Giudad, and L.A. Chasin. 1989. Nonsense mutations in the dihydrofolate reductase gene affect RNA processing. *Mol. Cell. Biol.* **9:** 2868–2880.

Utans, U., S.-E. Behrens, R. Lührmann, R. Kole, and A. Krämer. 1992. A splicing factor that is inactivated during in vivo heat shock is functionally equivalent to the [U4/U6.U5] triple snRNP-specific proteins. *Genes Dev.* **6:** 631–641.

Valcárcel, J., R. Singh, P.D. Zamore, and M.R. Green. 1993. Sex-lethal antagonizes the splicing factor U2AF to regulate *tra* alternative splicing in vitro. *Nature* (in press).

Vankan, P., C. McGuigan, and I.W. Mattaj. 1990. Domains of U4 and U6 snRNAs required for snRNP assembly and splicing complementation in *Xenopus* oocytes. *EMBO J.* **9:** 3397–3404.

Vellard, M., A. Sureau, J. Soret, C. Martinerie, and B. Perbol. 1992. A potential splicing factor is encoded by the opposite strand of the *trans*-spliced c-myb exon. *Proc. Natl. Acad. Sci.* **89:** 2511–2515.

Vijayraghavan, U. and J. Abelson. 1990. PRP18, a protein required for the second reaction in pre-mRNA splicing. *Mol. Cell. Biol.* **10:** 324–332.

Vijayraghavan, U., M. Company, and J. Abelson. 1989. Isolation and characterization of pre-mRNA splicing mutants of *Saccharomyces cerevisiae*. *Genes Dev.* **3:** 1206–1216.

Wassarman, D.A. and J.A. Steitz. 1991. Alive with DEAD proteins. *Nature* **349:** 463–464.

———. 1992. Interactions of small nuclear RNA's with precursor messenger RNA during in vitro splicing. *Science* **257:** 1918–1925.

Whittaker, E. and J.D. Beggs. 1991. The yeast PRP8 protein interacts directly with pre-mRNA. *Nucleic Acids Res.* **19:** 5483–5489.

Whittaker, E., M. Lossky, and J.D. Beggs. 1990. Affinity purification of spliceosomes reveals that the precursor RNA processing protein PRP8, a protein in the U5 small nuclear ribonucleoprotein particle, is a component of yeast spliceosomes. *Proc. Natl. Acad. Sci.* **87:** 2216–2219.

Wu, J. and J. Manley. 1989. Mammalian pre-mRNA branch site selection by U2 snRNP involves base pairing. *Genes Dev.* **3:** 1553–1561.

———. 1991. Base-pairing between U2 and U6 snRNAs is necessary for splicing of a mammalian pre-mRNA. *Nature* **352:** 818–821.

Wyatt, J.R., E.J. Sontheimer, and J.A. Steitz. 1992. Site-specific cross-linking of mammalian U5 snRNP to the 5′ splice site before the first step of pre-mRNA splicing. *Genes Dev.* **6:** 2542–2553.

Xu, Y., S. Petersen-Bjorn, and J.D. Friesen. 1990. The PRP4 (RNA4) protein of *Saccharomyces cerevisiae* is associated with the 5′ portion of the U4 small nuclear RNA. *Mol. Cell. Biol.* **10:** 1217–1225.

Yarus, M., M. Illangesekare, and E. Christian. 1991a. Selection of small molecules by the *Tetrahymena* catalytic center. *Nucleic Acids Res.* **19:** 1297–1304.

———. 1991b. An axial binding site in the *Tetrahymena* precursor RNA. *J. Mol. Biol.* **222:** 995–1012.

Yean, S.-L. and R.-J. Lin. 1991. U4 small nuclear RNA dissociates from a yeast spliceosome and does not participate in the subsequent splicing reaction. *Mol. Cell. Biol.* **11:** 5571–5577.

Zahler, A.M., W.S. Lone, J.A. Stalk, and M.B. Roth. 1992. SR proteins: A conserved family of pre-mRNA splicing factors. *Genes Dev.* **6:** 837–847.

Zamore, P.D. and M.R. Green. 1991. Biochemical characterization of U2 snRNP auxiliary factor: An essential pre-mRNA splicing factor with a novel intranuclear distribution. *EMBO J.* **10:** 207–214.

Zamore, P.D., J.G. Patton, and M.R. Green. 1992. Cloning and domain structure of the mammalian splicing factor U2AF. *Nature* **355:** 609–614.

Zhang, M., P.D. Zamore, M. Carmo-Fonseca, A.I. Lamond, and M.R. Green. 1992. Cloning and intracellular localization of the U2 small nuclear ribonucleoprotein auxiliary factor small subunit. *Proc. Natl. Acad. Sci.* **89:** 8769–8773.

Zhuang, Y. and A.M. Weiner. 1986. A compensatory base change in U1 snRNA suppresses a 5′ splice site mutation. *Cell* **46:** 827–35.

———. 1989. A compensatory base change in human U2 snRNA can suppress a branch site mutation. *Genes Dev.* **3:** 1545–1552.

———. 1990. The conserved dinucleotide AG of the 3′ splice site may be recognized twice during in vitro splicing of mammalian mRNA precursors. *Gene* **90:** 263–269.

Zillman, M., S.D. Rose, and S.M. Berget. 1987. U1 small nuclear ribonucleoproteins are required early during spliceosome assembly. *Mol. Cell. Biol.* **7:** 2877–2883.

Zillmann, M., M.L. Zapp, and S.M. Berget. 1988. Gel electrophoretic isolation of splicing complexes containing U1 small nuclear ribonucleoprotein particles. *Mol. Cell. Biol.* **8:** 814–821.

14

The Diverse World of
Small Ribonucleoproteins

Susan J. Baserga and Joan A. Steitz
Department of Molecular Biophysics and Biochemistry
Howard Hughes Medical Institute
Yale University School of Medicine
New Haven, Connecticut 06536-0812

Among the major players in the functioning of present-day eukaryotic
cells are small complexes consisting of RNA and protein. Small ribonucleoprotein (RNP) particles, defined as tight complexes between one
or more proteins and a small RNA molecule (chain length ≤ ~300 nucleotides), come in a variety of shapes and sizes. Since they can be very
abundant (up to 10^7 copies per mammalian cell, as many as ribosomes)
and are often highly conserved from yeast to man, deciphering their cellular roles has been an important challenge for molecular biologists.

Small RNPs inhabit all cellular compartments that have been examined: the nucleoplasm, the nucleoli, the cytoplasm, and the mitochondria.
Further classification of small RNPs (Table 1) has been made possible by
the discovery that they are often targeted by autoantibodies found in the
sera of patients with rheumatic disease, such as systemic lupus erythema-
tosus. Usually the autoepitopes reside on the protein rather than the RNA
moieties of small RNPs, meaning that particles of the same class possess
common polypeptide constituents. The autoantibodies also provide
potent tools for probing the structures and functions of small RNPs.

The vast majority of small RNPs whose functions have been
deciphered play roles somewhere along the pathway of gene expression
(see Table 1). For example, *small nuclear RNPs* (snRNPs) of the nucleoplasm are involved in pre-mRNA processing or tRNA biogenesis. Small
nucleolar RNPs contribute to the maturation of rRNA. *Small RNPs* of the
*c*ytoplasm (scRNPs) function in the control of translation or disposition
of newly synthesized proteins. The only current exceptions to this generalization are the telomerase RNP (see Blackburn, this volume) and
RNase MRP (*m*itochondrial *R*NA *p*rocessing), both of which play roles
in genome maintenance (DNA replication). Even some viral genomes encode small RNAs, which assemble together with host proteins and aid

The RNA World
© 1993 Cold Spring Harbor Laboratory Press 0-87969-380-0/93 $5 + .00

Table 1 Small ribonucleoproteins

		(a) Human small RNPs			
RNP	Function	Abundance (copies/cell)	RNA polymerase	Cap	Common antigen(s)
Nucleoplasm					
U1	pre-mRNA splicing	1×10^6	II	m_3GpppA	Sm
U2	pre-mRNA splicing	5×10^5	II	m_3GpppA	Sm
U4	pre-mRNA splicing	2×10^5	II	m_3GpppA	Sm
U5	pre-mRNA splicing	2×10^5	II	m_3GpppA	Sm
U6	pre-mRNA splicing	4×10^5	III	$mpppG$	none
U7	histone pre-mRNA splicing	5×10^3	II	m_3GpppN	Sm
U11	unknown	1×10^4	II	m_3GpppA	Sm
U12	unknown	5×10^3	II	m_3GpppN	Sm
7SK	unknown	2×10^5	III	$mpppG$	none
8-2	pre-tRNA processing(RNase)	1×10^5	III	$pppG$	Th
Nucleolus					
U3	pre-rRNA processing	2×10^5	II [a]	m_3GpppA [a]	Fb
U8	unknown	4×10^4	II	m_3GpppA	Fb
U13	unknown	1×10^4	II	m_3GpppN	Fb
7-2[b]	unknown	1×10^5	III	$pppG$	Th
Cytoplasm					
7SL	protein translocation (SRP)	1×10^6	III	$pppG$	none
Y1-5	unknown	1×10^5	III	$pppG$ or $pppA$	Ro

		(b) Viral small RNPs			
RNP	Virus	Localization	Abundance (copies/cell)	Polymerase	Common antigen(s)
VAI,II	adeno	cytoplasm	1×10^8	III	La
EBER1,2	Epstein-Barr	nucleoplasm	5×10^6	III	La
HSUR1-7	Herpes saimiri	nucleoplasm	10^3–10^4	II	Sm

[a]U3 is synthesized by RNA polymerase III in plants and has a mpppG cap (Shimba et al. 1992).
[b]7-2 is also part of RNase MRP, which is located in mitochondria (Clayton 1991).

the virus in subversion of cellular functions to allow viral replication or oncogenic transformation.

As in the case of research on ribosomes (a large RNP), a fundamental question is the contribution of the RNA component of each small RNP to whatever cellular function it fulfills. Here, several scenarios can be envisioned: (1) Base-pairing or other interactions of the RNA with the substrate in the vicinity of reacting nucleotides may be earmarks of RNA catalysis, reminiscent of self-splicing or ribozyme-induced cleavage. (2) The small RNA may base-pair with another nucleic acid simply to orient a catalytic polypeptide (either an integral RNP component or associated factor) correctly on the substrate. (3) The RNA may be only a rack on which to hang proteins, conferring an organization essential for both substrate recognition and catalysis by protein enzymes. The mechanisms of small RNP action, as currently known, seem to include examples of all three alternatives. However, further probing may uncover aspects of RNA catalysis inherent to the functioning of every small RNP.

mRNAs ARE FASHIONED BY snRNPs OF THE NUCLEOPLASM

A principal activity within the nucleoplasm of eukaryotic cells is the synthesis, processing, and export to the cytoplasm of mature mRNAs. All small RNPs known to participate in these events belong to the Sm class; that is, they possess a set of small polypeptides (11–29 kD) that are recognized by anti-Sm autoantibodies (see Lührmann 1988). As structural studies of these particles have intensified, more and more specific proteins unique to each particular snRNP have been identified. Thus, particles of the Sm class are quite protein-rich, at least 80% by mass.

The RNA components of Sm snRNPs fold into conserved secondary structures (Fig. 1), even though their sizes can vary widely (e.g., U2 in *Saccharomyces cerevisiae* is eight times the length of trypanosome U2) (Guthrie and Patterson 1988). These folded structures expose highly conserved regions that interact with other RNAs during functioning (for review, see Reddy and Busch 1988; Guthrie 1991). The RNAs are transcribed by RNA polymerase II (for review, see Dahlberg and Lund 1988) and their 5′ caps become hypermethylated to 2,2,7-trimethylguanosine (m_3G); the exception is U6, which is synthesized by RNA polymerase III and acquires a γ-Me cap. All Sm RNAs (except U6) possess a single-stranded region with the sequence ($PuAU_{n\geq3}GPu$), which dictates the binding of the common Sm proteins. The RNAs are encoded by multiple gene copies in vertebrates; single-copy genes in yeast have facilitated genetic analyses.

There are only three known snRNPs of the nucleoplasm that do not belong to the Sm class. The highly abundant 7SK RNP of vertebrate cells, whose 330-nucleotide RNA component is synthesized by RNA polymerase III and becomes γ-Me capped, is of unknown function (Wassarman and Steitz 1991). The vertebrate RNase P snRNP (Bartkiewicz et al. 1989; Doria et al. 1991), which matures the 5′ end of tRNAs, has many more protein components than its bacterial counterpart (for review, see Cech, this volume), at least one of which appears to be shared with the Th RNP (Gold et al. 1988) of the nucleolus (also called 7-2 or MRP; for review, see Clayton 1991; see below); conserved elements in the folding pattern of these RNA components are evident (Forster and Altman 1990), but it is not known whether the RNA alone is catalytic.

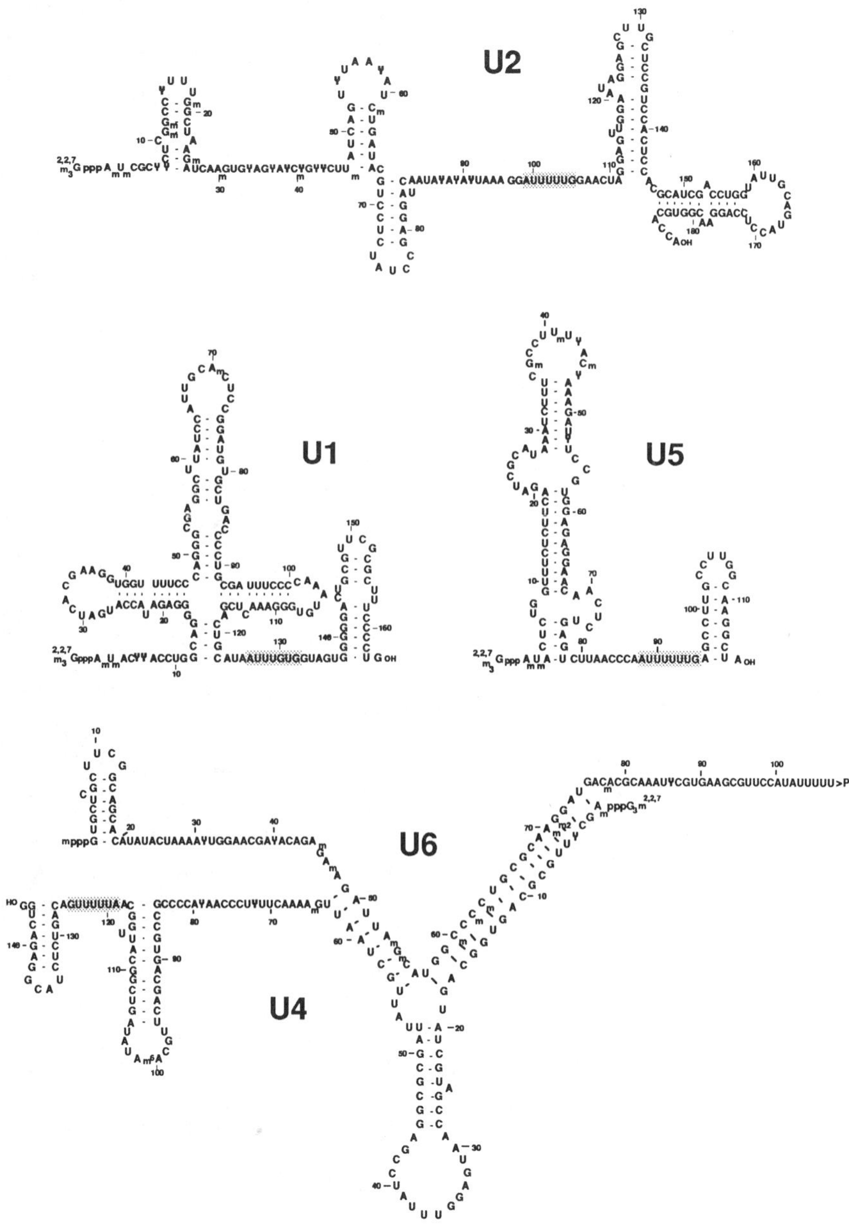

Figure 1 Sequences and conserved secondary structures of the human splicco-somal snRNAs (Guthrie and Patterson 1988; Reddy and Busch 1988). These RNAs, with the exception of U6, contain a binding site for the Sm proteins, shown here by shading (see Lührmann 1988).

Finally, the telomerase RNP has been structurally characterized only from unicellular organisms (see Blackburn, this volume); mammalian telomerase has properties of an RNP, but its RNA moiety has not yet been identified (Morin 1989).

The Spliceosomal snRNPs: An Intron in Pieces?

It has been known for nearly a decade that the removal of introns from pre-messenger RNA (pre-mRNA) proceeds by a chemical mechanism identical to that of group II self-splicing introns (see Cech; Moore et al.; both this volume). Thus, the five highly abundant snRNPs — U1, U2, U4, U5, and U6 — that assemble to form the active spliceosome have quite reasonably been viewed as descendants of an autocatalytic progenitor, carved into pieces and further fashioned by evolution to act on introns with a variety of lengths and sequences and to be susceptible to a multitude of regulatory pressures. If this view is true, one would then expect important interactions between the snRNPs to be largely RNA-based and snRNP interactions with the pre-mRNA substrate to mimic critical RNA·RNA contacts in group II introns.

Indeed, the catalog of RNA·RNA interactions in the spliceosome is growing (for reviews, see Steitz et al. 1988; Guthrie 1991). They are documented in most cases by both biochemical data and genetic suppression experiments. The latter studies (see, e.g., Zhuang and Weiner 1986), showing that deleterious mutations in one of the interacting RNAs can be alleviated by introducing compensatory changes into the other, provide powerful evidence for Watson-Crick base-pairing.

1. The evolutionarily invariant 5′ end of U1 RNA base-pairs with consensus intron sequences at both the 5′ and 3′ splice sites (Reich et al. 1992), probably acting to align these two regions very early in the spliceosome.

2. An internal invariant loop sequence in U5 RNA contacts 5′ exon sequences (Newman and Norman 1991; Wyatt et al. 1992) and, probably later in the reaction, the 3′ exon (Newman and Norman 1992). Both these base-pairing interactions have direct analogies in the group II intron (Jacquier 1990; Jacquier and Jacquesson-Breuleux 1991). Together with the above U1 contacts with intron sequences at the splice sites, these U5 interactions predict formation of a crossover or Holliday-like structure in the newly formed spliceosome (Steitz 1992).

3. A highly conserved region of U2 RNA base-pairs with the branch point region to bulge the A residue whose 2′ OH will attack the 5′

splice site during the first step of the reaction. This interaction precisely mimics the structure of an important domain of group II introns that contains the branch point A residue.

4. Two prominent helical stems (see Fig. 1) hold the U4 and U6 snRNPs together when they join the spliceosome. Subsequent dissociation of U4 (which is then not required for the splicing reaction) is believed to expose regions or allow conformational changes in U6 critical to catalysis. Because it is the most highly conserved RNA in the spliceosome, U6 is believed to be central to the catalytic mechanism.

5. The 5′ end of U2 RNA base-pairs with the 3′ end of U6 RNA, apparently to orient these two snRNPs correctly relative to one another in the spliceosome.

6. Recent psoralen cross-linking data suggest base-pairing of intron sequences at the 5′ splice site with both the invariant loop of U5 and an essential region of U6 RNA (D. Wassarman and Steitz 1992).

7. Additional interactions of U2 and U6 RNA may mimic other domains of group II introns (Madhani et al. 1992).

What emerges is a dynamic series of RNA·RNA interactions (Fig. 2), where partners are exchanged and conformational changes expose new snRNA regions as the two-step splicing reaction proceeds. This scenario explains the need for ATP hydrolysis and for the participation of splicing factors with RNA helicase-like activities in the spliceosome (see Guthrie 1991). Currently, more precise kinetics to allow sequential ordering of the above interactions, as well as characterization of any additional interactions, are needed. Nonetheless, the parallels between nuclear pre-mRNA splicing and group II self-splicing are becoming clear-cut, reinforcing the underlying ribozyme nature of the spliceosome.

Spliced Leader RNPs: Denizens of the Trans Spliceosome

In trypanosomes and their relatives, all mRNAs are fashioned by an unusual RNA processing event called trans-splicing. Regions from two separate transcripts (instead of from a single pre-mRNA) are joined such that every mature message begins with the same (~ 40-nucleotide) sequence, called a spliced leader (SL) (for reviews, see Nilsen 1989; Agabian 1990). The SL sequence is donated by an RNP particle that contains a snRNP-like domain, including an Sm-binding site (although the SL RNP is not recognized by human anti-Sm autoantibodies). Even more curious, certain other classes of organisms (e.g., nematodes) engage in both normal and trans-splicing. A nematode pre-mRNA can have internal introns that are excised in the usual manner, but then acquire a new 5′ end sequence from an SL RNP by trans-splicing.

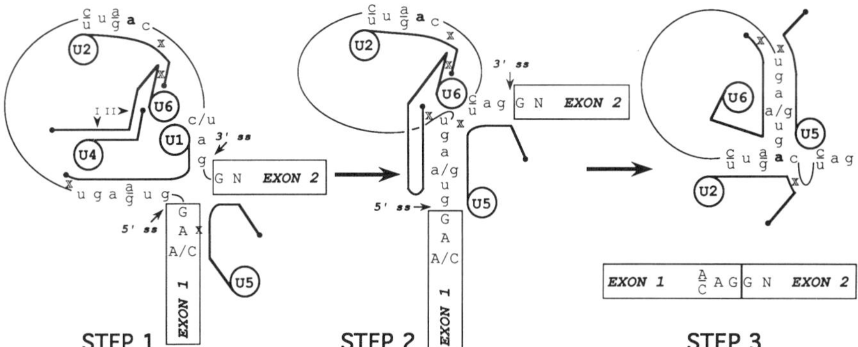

Figure 2 A model for snRNA/pre-mRNA interactions in the spliceosome (courtesy of David Wassarman). Consensus pre-mRNA sequences are written out with the 5′ exon in uppercase and intron in lowercase letters. RNA·RNA cross-links were generated using either psoralen (denoted by open xs; D. Wassarman and Steitz 1992) or a photoactivable nucleotide (denoted by a filled **x**; Wyatt et al. 1992). A more detailed account of the interactions can be found in these references and in Guthrie (1991).

The challenge presented by the trans spliceosome is to decipher how SL RNA sequences functionally substitute for missing snRNAs. Trypanosomes possess U2, U4, and U6 snRNPs, which are required for trans-splicing (Tschudi and Ullu 1990), but so far no U1 or U5 has been detected. Likewise in nematode trans-splicing, U1 appears not to be essential, whereas U2, U4, and U6 are (Hannon et al. 1991). The SL RNAs of trypanosomes are distinctively longer than those of nematodes (~22 nucleotides), arguing that essential U5 functions may be performed by trypanosome but not by nematode SL RNAs (Steitz 1992). In both cases, folded RNA structures can be drawn that suggest how U1 function is supplied (Bruzik et al. 1988). Further exploration of these intriguing variant spliceosomes is needed to define all the similarities and differences. Understanding trans-splicing is sure to provide insights into the evolutionary progression from group II self-splicing to the modern day spliceosome.

Additional Role(s) for the U1 snRNP?

The U1 snRNP in metazoan cells is two- to threefold more abundant than the other spliceosomal snRNPs. Moreover, its cellular distribution is different. In mammalian cells, it is more widely dispersed in the nucleoplasm than U2, U4, U5, and U6, which are concentrated in speckles and coiled bodies (Carmo-Fonseca et al. 1991; Matera and Ward 1993). In

Xenopus oocytes, only U1 snRNPs are present in distinctive small bodies called A snurposomes, whereas all the splicing snRNPs are detected in the morphologically similar B snurposomes (Wu et al. 1991).

Purification and in vitro reconstitution of the components required for cleavage and polyadenylation of pre-mRNA substrates have demonstrated that no snRNP is essential (Wahle and Keller 1992). However, several lines of evidence suggest that the U1 snRNP may serve to couple splicing and polyadenylation (Niwa et al. 1990; Niwa and Berget 1991) and perhaps then hand over the matured mRNA to the export apparatus:

1. Immunoprecipitation studies have detected associations between poly(A) polymerase and the U1 snRNP (Raju and Jacob 1988), as well as between polyadenylation substrates and Sm- and m_3G-bearing components (Hashimoto and Steitz 1986).
2. Psoralen cross-linking has revealed base-pairing interactions between U1 and several polyadenylation substrates; these cross-links are enhanced in the presence of upstream splicing signals, as is polyadenylation itself (Wassarman and Steitz 1993).
3. Genetic suppression experiments have uncovered a requirement for U1 snRNP binding to unspliced mRNA that is exported under the control of the HIV Rev protein (Lu et al. 1990).

Thus, perhaps the promiscuity of the U1 snRNP betrays an in vivo role at the interface of pre-mRNA processing and export from the nucleus. Definitive evidence has yet to be obtained.

The U7 snRNP Base-pairs at a Distance from the Cleavage Site in Histone Pre-mRNAs

The mRNAs for most histones in metazoan cells are neither spliced nor polyadenylated. Rather, their 3′ ends are generated by a nuclear RNA processing reaction that cleaves the pre-mRNA between two conserved signals: an upstream hairpin loop element and a downstream purine-rich sequence (for reviews, see Birnstiel and Schaufele 1988; Mowry and Steitz 1988). A low-abundance snRNP of the Sm class called U7 (Fig. 3) is required and participates by base-pairing its 5′ end with the downstream element (Fig. 4). Genetic suppression experiments reveal that even drastic changes in the sequences of the interacting regions (swaps of six purines for six pyrimidines) allow processing in the mammalian system (Bond et al. 1991). This hints that the U7 snRNP may serve only to orient protein enzymes for action at the somewhat distant cleavage site. Indeed, polypeptide(s) that bind the upstream hairpin loop element and an essential heat-labile factor have been described, but more studies

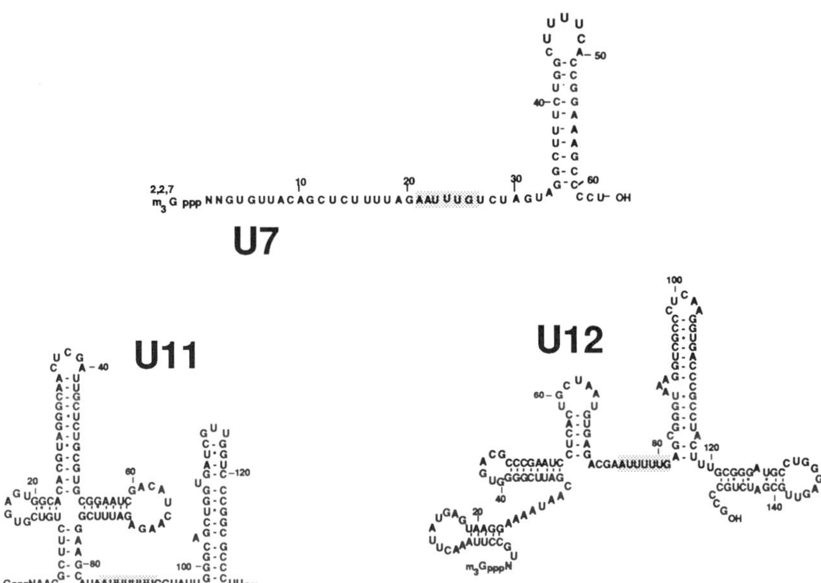

Figure 3 Low-abundance Sm snRNAs. Shown are the sequences and structures for human U7 (Mowry and Steitz 1988), U11, and U12 (Wassarman and Steitz 1993) snRNAs. The Sm protein-binding sites are indicated by shading.

are needed to define both the components of this unique pre-mRNA processing apparatus and its interface with the nuclear export machinery.

U11 and U12 snRNPs Are Implicated in Optional Intron Removal

Intermediate in abundance between the spliceosomal snRNPs and U7 are U11 and U12, also Sm snRNPs (see Fig. 3). U11, which is present in

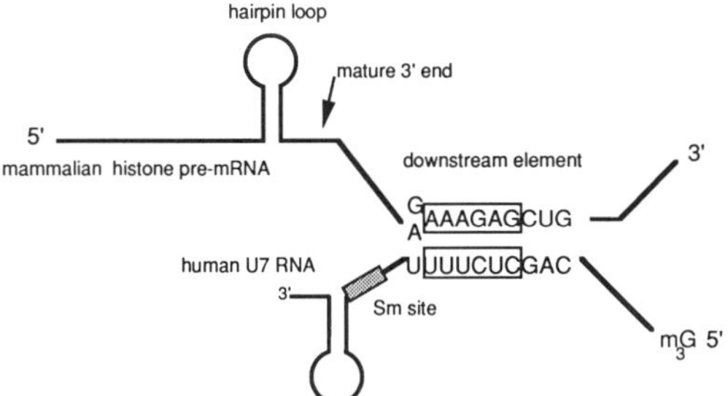

Figure 4 3′ end processing of histone pre-mRNA. Base-pairing between the 5′ end of U7 snRNA and the downstream element of the histone pre-mRNA is shown. The boxed sequences indicate base pairs that can be swapped with no adverse effect on processing (Bond et al. 1991).

slightly higher copy number, exists either as a mono-snRNP or as a (presumably dynamic) two-snRNP complex with U12 in mammalian cell extracts (K. Wassarman and Steitz 1992). Recent elegant studies of the in vitro splicing of a Rous sarcoma virus (RSV) intron whose removal is regulated as a function of the viral life cycle (McNally and Beemon 1992) suggest that the U11 and U12 snRNPs bind to an element in the intron and prevent assembly of the nascent spliceosome (R.R. Gontarek et al., in prep.). Splicing is thereby prevented. Whether base-pairing contributes to U11 and U12 interactions with each other, with the target intron, or with the U1 and U2 snRNPs is not yet known. Also tantalizing is the possibility that U11/12 may interact with the mRNA export apparatus.

NUCLEOLAR snRNPs: COGS IN THE RIBOSOME SYNTHESIS MACHINE

Most small nucleolar RNPs in mammals (Table 1), including U3, U8, U13, U14, X, and Y, belong to a group of particles that contain the protein fibrillarin (Fb). Studies of the assembly and function of these snRNPs have been aided by the discovery that sera from patients with the autoimmune disease scleroderma contain antibodies to Fb. Another abundant nucleolar snRNP called 7-2 contains a different protein autoantigen known as Th (Gold et al. 1988). Since the nucleolus is the site of rRNA transcription, processing, and ribosome assembly, nucleolar snRNPs are expected to participate in one or more of these events. Moreover, the concentration of Fb in the fibrillar regions of the nucleolus, where rRNA transcription and early stages of processing occur, assigns the Fb snRNPs to roles in these early steps of ribosome biogenesis.

Fb snRNPs Process Nascent rRNA Transcripts

The U3 snRNP is the most abundant nucleolar snRNP (for review, see Steitz et al. 1988). U3 RNAs from plants to humans exhibit remarkable sequence conservation over four short regions (Boxes A, B, C, and D; Fig. 5); an intact Box C sequence is required for binding to the Fb protein in vitro (Baserga et al. 1991). The structure of the U3 RNA is less well conserved, particularly the first stem-loop containing Box A. Like the spliceosomal snRNAs, mammalian U3 RNAs are transcribed by RNA polymerase II from multigene families. Little is known about the multiple proteins other than Fb in the U3 snRNP.

The U3 snRNP contributes to the processing of the nascent rRNA transcript, which in mammals contains 18S, 5.8S, and 28S sequences and

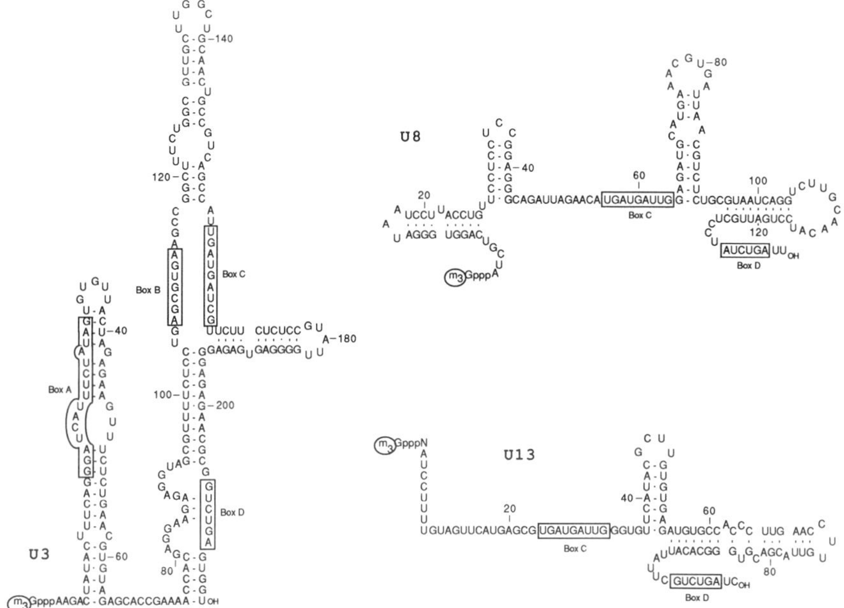

Figure 5 Sequences and structures of human U3, U8, and U13 nucleolar snRNAs. The conserved Box A–D sequences are indicated (for review, see Sollner-Webb et al. 1993).

must therefore be cut on either side of these regions before the rRNAs can be assembled into ribosomes. U3 is required for the earliest cleavage, which occurs within the 5′-most spacer (for review, see Sollner-Webb et al. 1993). The U3 snRNP then curiously stays bound to the pre-rRNA. It can be visualized as a "ball" at the end of an rRNA transcript "branch," suggesting a Christmas tree (Fig. 6). In rRNA transcripts lacking U3-binding sites, no "balls" are visible and only the "branches" can be seen, producing an undecorated pine tree (see Fig. 6). Evidence that the bound U3 snRNP then participates in several subsequent rRNA processing steps comes from studies where U3 has been disabled in oocytes or deleted in yeast (Savino and Gerbi 1990; Hughes and Ares 1991). Interestingly, mutations in U3, Fb, or the U3-binding site near the 5′ end of the rRNA all seem to affect the appearance of 18S rRNA most drastically.

It is not yet known whether U3 base-pairs directly with the rRNA transcript at the cleavage site. However, abundant evidence suggests that the U3 snRNA does associate directly with the rRNA transcript at other sites. Psoralen cross-linking studies both in vivo and in vitro in several mammalian systems have demonstrated that sequences in or near the Box A sequence in U3 can base-pair to the nascent rRNA transcript (Maser

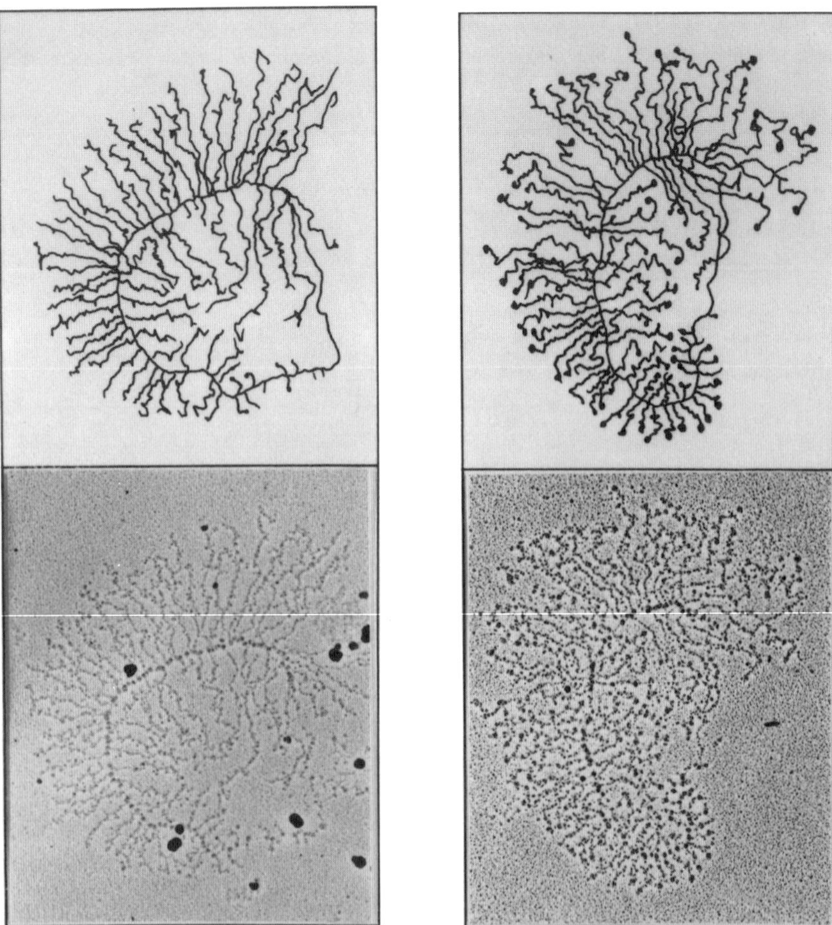

Figure 6 Electron microscopic spreads of rRNA transcription complexes from microinjected *Xenopus* oocytes. (*Right*) rRNA minigene bearing a 250-nucleotide region with the ETS processing site forms "Miller Christmas trees," with terminal knobs composed of processing components, as occurs on the intact rRNA genes of the cell. (*Left*) An analogous rRNA minigene with a small deletion of the essential residues of the processing signal forms "Miller pine trees." (Micrographs courtesy of B. Sollner-Webb and M. O'Reilly, from E.B. Mougey et al., in prep.).

and Calvet 1989; Stroke and Weiner 1989; Tyc and Steitz 1992). The contribution of U3 snRNA to the processing of the rRNA transcript may be an example in which base-pairing orients the snRNP and organizes the substrate such that catalytic polypeptides are directed to the correct processing sites.

The U8 and U13 snRNAs share some features with U3: All bind Fb, all are transcribed by RNA polymerase II, all acquire the m₃G cap structure, and all share two short sequences, Boxes C and D (see Fig. 4), located in similar structural contexts. The U8 and U13 snRNPs, however, are much less abundant (Table 1) than the U3 snRNP. Their roles in rRNA processing or ribosome assembly remain to be determined, although recent evidence suggests that U8 may in some way enable the maturation of 28S rRNA (B.A. Peculis, pers. comm.).

Two variants of RNA X (146 and 148 nucleotides), Y (125 nucleotides), and U14 (87 nucleotides in mouse), which also bind Fb and contain the Box C and D sequences, are distinct from the other Fb snRNPs in several respects. Unlike U3, U8, and U13, they do not have an m₃G cap. Studies in yeast have shown that the U14 snRNA is required for formation of the 18S rRNA (Li et al. 1990). In mammals, the U14 snRNA gene unexpectedly resides within an intron of an unrelated gene, the hsc70 gene (Liu and Maxwell 1990). It therefore does not appear to have its own transcription unit. The gene for at least one variant of human RNA X (now called U15) is also located in the intron of a protein-coding gene (for ribosomal protein S3), suggesting that these snRNAs are also generated by processing of intron sequences (K.T. Tycowski et al., in prep.).

The yeast *S. cerevisiae* possesses several other nucleolar snRNAs besides U3 and U14 (snR128): snR3, snR4, snR5, snR8, snR9, snR10, snR30, snR189, and snR190 (for review, see Sollner-Webb et al. 1992). Of these, only U3, U14, and snR30 are absolutely required for growth: In vivo deletion of snR10 causes some growth impairment. All of the required yeast nucleolar snRNAs so far studied (U3, U14, and snR10), as well as Fb (Tollervey et al. 1991), are involved in similar steps in the production of the yeast 18S rRNA (Hughes and Ares 1991).

Other Nucleolar snRNPs

The 7-2 RNA, although found in the nucleolus, is not associated with Fb. Its bound autoantigen (representing one or more proteins) is known as Th and is shared with another non-nucleolar RNP, RNase P (8-2 RNA; Gold et al. 1988). The human 7-2 is 268 nucleotides long and is transcribed by RNA polymerase III from a single-copy gene (Table 1). The localization of Th to the granular component of the nucleolus, demonstrated by immunoelectron microscopy (Reimer et al. 1988), suggests that the 7-2 snRNP may play a role in ribosome assembly.

Surprisingly, the 7-2 RNA is identical to the RNA of an RNP involved in the in vitro processing of a primer from the leading-strand origin of mitochondrial DNA replication, known as RNase MRP (Gold et al. 1989; for review, see Clayton 1991). Only about 1% of all the RNase MRP in a mammalian cell copurifies with mitochondria; other snRNAs also purify to this extent with mitochondria (Kiss and Filipowicz 1992; Topper et al. 1992). The apparent dual specificity of this RNP, both in location and in activity, remains to be resolved.

CYTOPLASMIC RNPs: PLAYERS IN THE FINAL STAGES OF GENE EXPRESSION

Signal Recognition Particle Aids in Protein Translocation during Translation

The 7SL RNA (~300 nucleotides) is part of a scRNP known as the signal recognition particle, or SRP (for review, see Wolin and Walter 1991; Rapoport 1992). In addition to the RNA component, SRP consists of six distinct polypeptides, several of which form heterodimers. Neither the protein nor RNA components can function alone. SRP binds to secretory proteins as their amino-terminal "signal sequences" emerge from the ribosome and targets nascent chain/ribosome complexes to the SRP receptor in the rough endoplasmic reticulum (RER). Subsequently, these proteins are translocated into the lumen of the RER. The binding of SRP to the SRP receptor appears to involve the two largest polypeptides of the SRP (SRP68 and SRP72). The release of SRP from the signal sequence/ribosome complex, accomplished before protein translocation, requires GTP hydrolysis and is facilitated by the 54-kD protein of SRP (SRP54). Most of our knowledge of SRP comes from in vitro mammalian systems, where the role of the 7SL RNA component remains obscure. Recently, however, SRP has been identified and studied in the yeast, *S. cerevisiae* (Hann and Walter 1991), demonstrating that SRP functions in protein translocation in vivo.

Cytoplasmic RNPs of Unknown Function

The Ro RNPs comprise another class of cytoplasmic RNPs, with function(s) yet unknown. In humans, each Ro particle consists of one of five RNAs (4 unique), called the Y RNAs (80–110 nucleotides), complexed with the 60-kD Ro autoantigen and the 50-kD La protein (Wolin and Steitz 1984). Interestingly, the mouse has only two Ro RNAs, whereas the rat has three. Although previously controversial, Ro RNA localization to the cytoplasm has recently been confirmed by advanced techniques of cell fractionation (C.A. O'Brien et al., in prep.).

Vaults are cytoplasmic RNPs of uniquely large size (35 x 65 nm, larger than ribosomes). In the rat, each vault contains at least 9 copies of a single, novel small RNA (141 nucleotides; Kickhoefer et al. 1993) and at least 55 copies of the major vault polypeptide (104,000 kD). Each half-vault has a structure that resembles an eight-petaled flower. Recently, some intriguing similarities between the plugs in nuclear pore complexes and vaults have been noted, including identical mass and subcellular location (Kedersha et al. 1991), suggesting that vaults may play a role in nuclear/cytoplasmic communication.

SUBVERSION OF CELLULAR FUNCTIONS BY VIRAL SMALL RNPs

Some mammalian viruses carry genes for small RNAs that are abundantly expressed at certain stages of infection. So far the well-documented cases all involve viruses with double-stranded DNA genomes; likewise, the protein constituents of viral RNPs are always host- rather than virus-encoded. The existence of viral small RNPs raises two intriguing questions. First, what important contribution to the viral life cycle makes it worthwhile for a highly compressed viral genome to encode small RNAs? Second, are there host RNAs with analogous roles, whose genes were at some stage co-opted by the virus and tailored to its own ends? Our current incomplete knowledge of the functions of viral small RNAs argues that they all exert their influence by competitively binding a host protein important to a critical cellular pathway.

Adenovirus Uses VA RNA to Circumvent the Interferon Response of the Host

Humans and other mammals combat infection by adenoviruses and other viruses by producing interferon. Interferon then binds to uninfected cells of the host and triggers a cascade of gene expression events that includes the shutdown of protein synthesis should the cell become infected and thereby exposed to double-stranded RNA (for review, see Sen and Lengyel 1992). Inhibition of translation occurs at the initiation stage. An interferon-induced protein kinase (p68), when activated by double-stranded RNA, phosphorylates an initiation factor (eIF2α) and blocks its ability to recycle after depositing the initiator tRNA$^{Met}_i$ on the ribosome.

Yet, protein synthesis does not halt late in the adenovirus life cycle because of the action of VAI RNA (Mathews and Shenk 1991). It and VAII (of unknown function) are transcribed from the adenovirus genome by RNA polymerase III and therefore associate with the host La protein, an autoantigen which binds all nascent transcripts of this class via their U-rich 3′ termini. VAI (about 160 nucleotides) is produced in such great abundance in infected cells (see Table 1) that it effectively competes with double-stranded RNA for binding to the p68 kinase (Mellits et al. 1990; Ghadge et al. 1991). A particular stem-loop structure of VAI has been pinpointed as essential for this competition; another domain of the RNA may then occlude the active site of the enzyme. These beautiful studies leading to a detailed picture of the functioning of a viral small RNA have set an example that is being emulated in investigations of less well understood viral RNPs.

Epstein-Barr Virus EBER Binds a Ribosomal Protein

Epstein-Barr virus (EBV) infection of human B lymphocytes often produces a transformed state (for review, see Miller 1990; Kieff and Liebowitz 1990). During latency, most viral genes are not expressed. Among the exceptions are those for two abundant small RNAs, EBER1 and -2 (for *E*pstein *B*arr-encoded *R*NA). The EBERs (each about 170 nucleotides) are transcribed by RNA polymerase III, are localized in the nucleus, and quantitatively bind the host La protein. Homologous small RNAs are found in cells infected by the related primate viruses *Herpesvirus papio* and *pan*.

The EBER1 RNP was found to contain an additional host protein, initially called EAP (for *E*BER-*a*ssociated *p*rotein) (Toczyski and Steitz 1991). The sequence of the 15-kD human polypeptide revealed that a very close homolog (~80% identical) had been identified as expressed at high levels during sea urchin development. Findings that EAP is abundant in all cells examined, that it is localized in the nucleolus and cytoplasm, and that it binds a G·C-rich stem-loop of EBER1 (Toczyski and Steitz 1993) led to the realization that EAP is ribosomal protein L22 (I. Wool and D. Toczyski, pers. comm.).

Why has EBV designed a small RNA to bind up approximately one half of a particular ribosomal protein in latently infected host cells? Among the most interesting possibilities is that L22 plays some pivotal role in the regulation of ribosome biogenesis and that its sequestration leads to overproduction of ribosomes and elevated protein synthetic activity needed for cellular transformation by EBV. Whether this is indeed the case, as well as a function for EBER2, remains to be established.

Do Herpesvirus saimiri U RNAs Modulate mRNA Stability in Infected Cells?

Herpesvirus saimiri targets the T lymphocytes of new world primates to produce either lytic infection or malignant transformation. In transformed T cells, the most abundant viral transcripts are small RNAs

(75–143 nucleotides) (see Albrecht and Fleckenstein 1992) that assemble into particles of the Sm class; that is, they bind host Sm proteins and acquire m_3G caps (Lee et al. 1988). There are at least seven different HSURs (for *H*erpesvirus *s*aimiri *U R*NAs), whose abundances range from moderate to low compared to the splicing snRNPs (see Table 1). Like cellular U RNAs, HSURs are synthesized by RNA polymerase II in response to unusual promoter and termination signals.

The sequences at the 5′ ends of three of the HSURs mimic mRNA destabilization signals found in the 3′-untranslated regions of messages encoding cellular oncoproteins, cytokines, and lymphokines (Shaw and Kamen 1986). The idea that HSURs might therefore act to manipulate the host's mRNA degradation machinery has some experimental support. The same 32-kD host protein that selectively binds the AU-rich destabilization signal in the c-*fos* mRNA (Vakalopoulou et al. 1991) can be cross-linked to the 5′-end sequences of the relevant HSURs (Myer et al. 1992). The model (Fig. 7) therefore suggests that these HSURs aid in viral transformation by competitively binding the 32-kD protein (presumably a factor in mRNA degradation); this would lengthen the half-lives of unstable mRNAs, leading to slightly increased production of a number of growth-related cellular proteins. Direct proof of this hypothesis is still lacking, as are ideas for the function of the remaining HSUR snRNPs.

snRNP BIOGENESIS FOLLOWS AN UNUSUAL ROUTE

Since snRNAs are transcribed in the nucleoplasm and participate in RNA processing events there, one might suppose that they never leave the nucleus. This is not so. Instead, each of the snRNAs synthesized by RNA polymerase II that have been studied—that is, four of the spliceosomal RNAs (U1, U2, U4, and U5) and one nucleolar RNA (U3)—has a cytoplasmic phase in its biogenesis (for review, see Izaurralde and Mattaj 1992). These RNAs all rapidly exit the nucleus and then bind their common autoantigenic proteins and undergo cap hypermethylation in the cytoplasm. Yet, their requirements for import back into the nucleus differ.

Binding of the common autoantigenic proteins, the Sm proteins, is required for the spliceosomal snRNAs to become localized in the nucleoplasm. Intriguingly, neither the Sm proteins nor the spliceosomal snRNAs can enter the nucleus alone, but together they form a unique karyophilic signal. U1 and U2 further require their Me_3G cap for nuclear import.

In contrast, neither binding of the major autoantigenic protein, Fb, to the U3 snRNP nor cap trimethylation is needed for its nuclear localization (Baserga et al. 1992). Instead, import depends on an intact structural

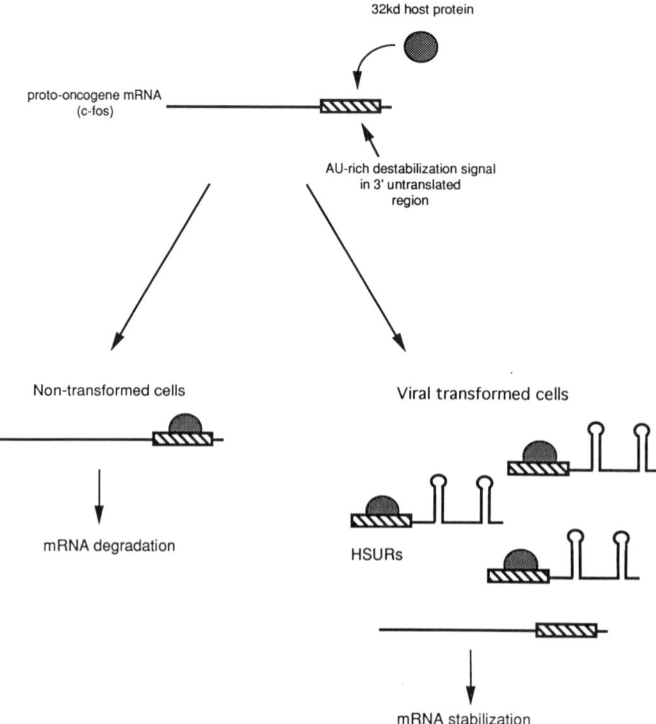

Figure 7 Proposed model for the role of HSURs in viral transformation. In a nontransformed cell, a 32-kD host protein binds to the AU-rich mRNA destabilization signal in proto-oncogene mRNAs, leading to their degradation (Vakalopoulou et al. 1991). After viral infection, binding of the 32-kD host protein is competitively inhibited by the HSUR snRNAs, and stabilization of proto-oncogene mRNA results (Myer et al. 1992). An analogous mechanism involving competitive binding of viral small RNAs to a cellular protein occurs for the VAI RNA of adenovirus (Mathews and Shenk 1991) and may explain EBER function in EBV-transformed cells (Toczyski and Steitz 1993).

element near the 3′ end of the U3 RNA. This 5-nucleotide stem may provide the correct structure or sequence for binding of a protein necessary to import. Once inside the nucleus, the nucleolar snRNPs must be further concentrated in a particular nuclear subregion, the nucleolus. The requirements for snRNP localization in the nucleolus have not yet been elucidated.

PERSPECTIVES

Ribonucleoprotein particles participate in a diverse array of cellular processes in both the nucleus and cytoplasm of eukaryotic cells. Most small

RNPs contribute to the maturation of other RNA molecules (mRNA, rRNA, and tRNA), but some are required for genome replication or for the final stages of gene expression: the targeting of proteins to the RER. Viral RNPs manipulate the normal pathway of gene expression to facilitate viral replication or oncogenic transformation of their host cell. The modes of action of small RNPs range from direct contact between conserved nucleotides and the substrate at the site of catalysis to indirect tethering of important proteins. Further scrutiny of the small RNP habitat is sure to uncover more surprises and valuable insights into the RNA world.

ACKNOWLEDGMENTS

We have drawn heavily on the generosity of many colleagues who have shared ideas prior to publication. We thank Brenda Peculis, Kazimierz Tyc, and Gregory Matera for critical reading of the manuscript; Lynda Stevens and Deborah Boyne for expert typing; and Barbara Sollner-Webb for graciously providing Figure 6. Work from our laboratory was supported by grants GM-26154 and CA-16038 from the National Institutes of Health.

REFERENCES

Agabian, N. 1990. *Trans* splicing of nuclear pre-mRNAs. *Cell* **61:** 1157–1160.

Albrecht, J.C. and B. Fleckenstein. 1992. Nucleotide sequence of HSUR 6 and HSUR 7, two small RNAs of herpesvirus saimiri. *Nucleic Acids Res.* **20:** 1810.

Bartkiewicz, M., H. Gold, and S. Altman. 1989. Identification and characterization of an RNA molecule that copurifies with RNase P activity from HeLa cells. *Genes Dev.* **3:** 488–499.

Baserga, S.J., M. Gilmore-Hebert, and X.W. Yang. 1992. Distinct molecular signals for nuclear import of the nucleolar snRNA, U3. *Genes Dev.* **6:** 1120–1130.

Baserga, S.J., X. Yang, and J.A. Steitz. 1991. An intact box C sequence in the U3 snRNA is required for binding of fibrillarin, the protein common to the major family of nucleolar snRNPs. *EMBO J.* **10:** 2645–2651.

Birnstiel, M.L. and F.J. Schaufele. 1988. Structure and function of minor snRNPs. In *Small nuclear ribonucleoprotein particles* (ed. M.L. Birnstiel), pp. 155–182. Springer Verlag, Berlin.

Bond, U.M., T.A. Yario, and J.A. Steitz. 1991. Multiple processing-defective mutations in a mammalian histone premessenger RNA are suppressed by compensatory changes in U7 RNA both *in vivo* and *in vitro. Genes Dev.* **5:** 1709–1722.

Bruzik, J.P., K. Van Doren, D. Hirsh, and J.A. Steitz. 1988. *Trans* splicing involves a novel form of small nuclear ribonucleoprotein particles. *Nature* **335:** 559–562.

Carmo-Fonseca, M., D. Tollervey, R. Pepperkok, S.M.L. Barabino, A. Merdes, C. Brunner, P.D. Zamore, M.R. Green, E. Hurt, and A.I. Lamond. 1991. Mammalian nuclei contain foci which are highly enriched in components of the pre-mRNA splicing machinery. *EMBO J.* **10:** 195–206.

Clayton, D.A. 1991. Nuclear gadgets in mitochondrial DNA replication and transcription. *Trends Biochem. Sci.* **16:** 107–111.

Dahlberg, J.E. and E. Lund. 1988. The genes and transcription of the major small nuclear RNAs. In *Small nuclear ribonucleoprotein particles* (ed. M.L. Birnstiel), pp. 38–70. Springer-Verlag, Berlin.

Doria, M., G. Carrara, P. Calandra, and G.P. Tocchini-Valentini. 1991. An RNA molecule copurifies with RNase P activity from *Xenopus laevis* oocytes. *Nucleic Acids Res.* **19:** 2315–2320.

Forster, A.C. and S. Altman. 1990. Similar cage-shaped structures for the RNA components of all ribonuclease P and ribonuclease MRP enzyme. *Cell* **62:** 407–409.

Ghadge, G.D., S. Swaminathan, M.G. Katze, and B. Thimmapaya. 1991. Binding of the adenovirus VAI RNA to the interferon-induced 68-kDa protein kinase correlates with function. *Proc. Natl. Acad. Sci.* **88:** 7140–7144.

Gold, H.A., J.N. Topper, D.A. Clayton, and J. Craft. 1989. The RNA processing enzyme RNase MRP is identical to the Th RNP and related to RNase P. *Science* **245:** 1377–1380.

Gold, H.A., J. Craft, J.A. Hardin, M. Bartkiewicz, and S. Altman. 1988. Antibodies in human serum that precipitate ribonuclease P. *Proc. Natl. Acad. Sci.* **85:** 5483–5487.

Guthrie, C. 1991. Messenger RNA splicing in yeast: Clues to why the spliceosome is a ribonucleoprotein. *Science* **253:** 157–163.

Guthrie, C. and B. Patterson. 1988. Spliceosomal snRNAs. *Annu. Rev. Genet.* **22:** 387–419.

Hann, B.C. and P. Walter. 1991. The signal recognition particle in *S. cerevisiae. Cell* **67:** 131–144.

Hannon, G.J., P.A. Maroney, and T.W. Nilsen. 1991. U small nuclear ribonucleoprotein requirements for nematode *cis-* and *trans*-splicing *in vitro. J. Biol. Chem.* **266:** 22792–22795.

Hashimoto, C. and J.A. Steitz. 1986. A small nuclear ribonucleoprotein associates with the AAUAAA polyadenylation signal in vitro. *Cell* **45:** 581–591.

Hughes, J.M.X. and M. Ares, Jr. 1991. Depletion of U3 small nucleolar RNA inhibits cleavage in the 5 ′ external transcribed spacer of yeast pre-ribosomal RNA and impairs formation of 18S ribosomal RNA. *EMBO J.* **10:** 4231–4239.

Izaurralde, E. and I.W. Mattaj. 1992. Transport of RNA between nucleus and cytoplasm. *Semin. Cell Biol.* **3:** 279–288.

Jacquier, A. 1990. Self-splicing group II and nuclear pre-mRNA introns: How similar are they? *Trends Biochem. Sci.* **15:** 351–354.

Jacquier, A. and N. Jacquesson-Breuleux. 1991. Splice site selection and role of the lariat in a group II intron. *J. Mol. Biol.* **219:** 415–428.

Kedersha, N.L., J.E. Heuser, D.C. Chugani, and L.H. Rome. 1991. Vaults. III. Vault ribonucleoprotein particles open into flower-like structures with octagonal symmetry. *J. Cell Biol.* **112:** 225–235.

Kickhoefer, V.A., R.P. Searles, N.L. Kedersha, M.E. Garber, D.L. Johnson, and L.H. Rome. 1993. Vault RNP particles from rat and bullfrog contain a related small RNA that is transcribed by RNA polymerase III. *J. Biol. Chem.* **268:** (in press).

Kieff, E. and D. Licbowitz. 1990. Epstein-Barr virus and its replication. In *Virology*, 2nd edition (ed. B.N. Fields, et al.) vol. 2, pp. 1889–1920. Raven Press, New York.

Kiss, T. and W. Filipowicz. 1992. Evidence against a mitochondrial location of the 7-2/MRP RNA in mammalian cells. *Cell* **70:** 11–16.

Lee, S.I., S.C.S. Murthy, J.J. Trimble, R.C. Desrosiers, and J.A. Steitz. 1988. Four novel

U RNAs are encoded by a herpesvirus. *Cell* **54:** 599–607.

Li, H.V., J. Zagorski, and M.J. Fournier. 1990. Depletion of U14 small nuclear RNA (snR128) disrupts production of 18S rRNA in *Saccharomyces cerevisiae. Mol. Cell. Biol.* **10:** 1145–1152.

Liu, J. and E.S. Maxwell. 1990. Mouse U14 snRNA is encoded in an intron of the mouse cognate hsc70 heat shock gene. *Nucleic Acids Res.* **18:** 6565–6571.

Lu, X., J. Heimer, D. Rekosh, and M.-L. Hammarskjöld. 1990. U1 small nuclear RNA plays a direct role in the formation of a rev-regulated human immunodeficiency virus *env* mRNA that remains unspliced. *Proc. Natl. Acad. Sci.* **87:** 7598–7602.

Lührmann, R. 1988. snRNP proteins. In *Small nuclear ribonucleoprotein particles* (ed. M.L. Birnstiel), pp. 71–99. Springer-Verlag, Berlin.

Madhani, H.D. and C. Guthrie. 1992. A novel base-pairing interaction between U2 and U6 snRNAs suggests a mechanism for the catalytic activation of the spliceosome. *Cell* **71:** 803–817.

Maser, R.L. and J.P. Calvet. 1989. U3 small nuclear RNA can be psoralen cross-linked *in vivo* to the 5′-external transcribed spacer of pre-ribosomal RNA. *Proc. Natl. Acad. Sci.* **86:** 6523-6527.

Matera, A.G. and D.C. Ward. 1993. Nucleoplasmic organization of small nuclear ribonucleoproteins in cultured human cells. *J. Cell. Biol.* (in press).

Mathews, M.B. and T. Shenk. 1991. Adenovirus virus-associated RNA and translation control. *J. Virol.* **65:** 5657–5662.

McNally, M.T. and K. Beemon. 1992. Intronic sequences and 3′ splice sites control Rous sarcoma virus RNA splicing. *J. Virol.* **66:** 6–11.

Mellits, K.H., M. Kostura, and M.B. Mathews. 1990. Interaction of adenovirus VA RNA$_1$ with the protein kinase DAI: Nonequivalence of binding and function. *Cell* **61:** 843–852.

Miller, G. 1990. Biology, pathogenesis, and medical aspects. In *Virology*, 2nd edition (ed. B.N. Fields, et al.), vol. 2, pp. 1921–1958. Raven Press, New York.

Morin, G. 1989. The human telomere terminal transferase enzyme is a ribonucleoprotein that synthesizes TTAGGG repeats. *Cell* **59:** 521–529.

Mowry, K.L. and J.A. Steitz. 1988. snRNP mediators of 3′ end processing: Functional fossils? *Trends Biochem. Sci.* **13:** 445–447.

Myer, V., S.I. Lee, and J.A. Steitz. 1992. Viral snRNPs bind a protein implicated in messenger RNA destabilization. *Proc. Natl. Acad. Sci.* **89:** 1296–1300.

Newman, A. and C. Norman. 1991. Mutations in yeast U5 snRNA alter the specificity of 5′ splice-site cleavage. *Cell* **65:** 115–123.

———. 1992. U5 snRNA interacts with exon sequences at 5′ and 3′ splice sites. *Cell* **68:** 1–20.

Nilsen, T.W. 1989. *Trans*-splicing in nematodes. *Exp. Parasitol.* **69:** 413–416.

Niwa, M., and S.M. Berget. 1991. Mutation of the AAUAAA polyadenylation signal depresses in vitro splicing of proximal but not distal introns. *Genes and Dev.* **5:** 2086–2095.

Niwa, M., S.D. Rose, and S.M. Berget. 1990. In vitro polyadenylation is stimulated by the presence of an upstream intron. *Genes Dev.* **4:** 1552–1559.

Raju, V.S., and S.T. Jacob. 1988. Association of poly(A) polymerase with U1 RNA. *J. Biol. Chem.* **263:** 11067–11070.

Rapoport, T.A. 1992. Transport of proteins across the endoplasmic reticulum membrane. *Science* **258:** 931–936.

Reddy, R. and H. Busch. 1988. Small nuclear RNAs: RNA sequences, structure and modifications. In *Small nuclear ribonucleoprotein particles* (ed. M.L. Birnstiel), pp. 1–37.

Springer-Verlag, Berlin.

Reich, C.I., R.W. VanHoy, G.L. Porter, and J.A. Wise. 1992. Mutations at the 3′ splice site can be suppressed by compensatory base changes in U1 snRNA in fission yeast. *Cell* **69:** 1159-1169.

Reimer, G., I. Raska, U. Scheer, and E.M. Tan. 1988. Immunolocalization of 7-2-ribonucleoprotein in the granular component of the nucleolus. *Exp. Cell Res.* **176:** 117–128.

Savino, R. and S.A. Gerbi. 1990. *In vivo* disruption of *Xenopus* U3 snRNA affects ribosomal RNA processing. *EMBO J.* **9:** 2299–2308.

Sen, G.C. and P. Lengyel. 1992. The interferon system. *J. Biol. Chem.* **267:** 5017–5020.

Shaw, G. and R. Kamen. 1986. A conserved AU sequence from the 3′ untranslated region of GM-CSF mRNA mediates selective mRNA degradation. *Cell* **46:** 659–667.

Shimba, S., B. Buckley, and R. Reddy. 1992. Cap structure of U3 small nucleolar RNA in animal and plant cells is different. *J. Biol. Chem.* **267:** 13772–13777.

Sollner-Webb, B., K. Tyc, and J.A. Steitz. 1993. Ribosomal RNA processing in eukaryotes. In *Ribosomal RNA: Structure, evolution, processing and function in protein synthesis* (ed. A.E. Dahlberg and R.A. Zimmermann). CRC Press, Boca Raton, Florida. (In press.)

Steitz, J.A. 1992. Splicing takes a Holliday. *Science* **257:** 888–889.

Steitz, J.A., D.L. Black, V. Gerke, K.A. Parker, A. Krämer, D. Frendewey, and W. Keller. 1988. Functions of the abundant U-snRNPs. In *Small nuclear ribonucleoprotein particles* (ed. M.L. Birnstiel), pp. 115–154. Springer-Verlag, Berlin.

Stroke, I.L. and A.M. Weiner. 1989. The 5′ end of U3 snRNA can be crosslinked *in vivo* to the external transcribed spacer of rat ribosomal RNA precursors. *J. Mol. Biol.* **210:** 497–512.

Toczyski, D.P.W. and J.A. Steitz. 1991. EAP, a highly conserved cellular protein associated with Epstein-Barr virus small RNAs (EBERs). *EMBO J.* **10:** 459–466.

———. 1993. The cellular RNA binding protein EAP recognizes a conserved stem-loop in the Epstein-Barr virus small RNA EBER1. *Mol. Cell. Biol.* **13:** 703–710.

Tollervey, D., H. Lehtonen, M. Carmo-Fonseca, and E.C. Hurt. 1991. The small nucleolar RNP protein NOP1 (fibrillarin) is required for pre-rRNA processing in yeast. *EMBO J.* **10:** 573-583.

Topper, J.N., J.L. Bennett, and D.A. Clayton. 1992. A role for RNAase MRP in mitochondrial RNA processing. *Cell* **70:** 16–20.

Tschudi, C. and E. Ullu. 1990. Destruction of U2, U4, or U6 small nuclear RNA blocks *trans* splicing in trypanosome cells. *Cell* **61:** 459–466.

Tyc, K. and J.A. Steitz. 1992. A new interaction between the mouse 5′ external transcribed spacer of pre-rRNA and U3 snRNA detected by psoralen crosslinking. *Nucleic Acids Res.* **20:** 5375–5382.

Vakalopoulou, E., J. Schaack, and T. Shenk. 1991. A 32-kilodalton protein binds to AU-rich domains in the 3′ untranslated regions of rapidly degraded mRNAs. *Mol. Cell. Biol.* **11:** 3355–3364.

Wahle, E. and W. Keller. 1992. The biochemistry of 3′-end cleavage and polyadenylation of messenger RNA precursors. *Annu. Rev. Biochem.* **61:** 419–440.

Wassarman, D.A. and J.A. Steitz. 1991. Structural analyses of the 7SK ribonucleoprotein: The most abundant human small RNP of unknown function. *Mol. Cell. Biol.* **11:** 3432–3445.

———. 1992. Psoralen crosslinking detects specific interactions of U1, U2, U5, and U6 small nuclear RNAs with precursor messenger RNA during *in vitro* splicing. *Science* **257:** 1918-1925.

Wassarman K.M. and J.A. Steitz. 1992. The low abundance U11 and U12 snRNPs interact to form a two-snRNP complex. *Mol. Cell. Biol.* **12:** 1276–1285.

———. 1993. Association with terminal exons in premessenger RNAs: A new role for
the U1 snRNP. *Genes Dev.* **7:** (in press).

Wolin, S.L. and J.A. Steitz. 1984. The Ro cytoplasmic ribonucleoproteins: Identification
of the antigenic protein and its binding site on the Ro RNAs. *Proc. Natl. Acad. Sci.* **81:**
1996–2000.

Wolin, S.L. and P. Walter. 1991. Small ribonucleoproteins. *Current Opin. Struct. Biol.* **1:**
251–257.

Wu, Z., C. Murphy, H.G. Callan, and J.G. Gall. 1991. Small nuclear ribonucleoproteins
and heterogenous nuclear ribonucleoproteins in the amphibian germinal vesicle: Loops,
spheres, and snurposomes. *J. Cell Biol.* **113:** 465–483.

Wyatt, J.R., E.J. Sontheimer, and J.A. Steitz. 1992. Site-specific crosslinking of mammalian U5 snRNP to the 5′ splice site prior to the first step of premessenger RNA
splicing. *Genes Dev.* **6:** 2554–2568.

Zhuang, Y. and A.M. Weiner. 1986. A compensatory base change in U1 snRNA suppresses a 5′ splice site mutation. *Cell* **46:** 827–835.

15

RNA Editing: New Uses for Old Players in the RNA World

Brenda L. Bass
Department of Biochemistry
University of Utah
Salt Lake City, Utah 84132

The sequence of a cellular RNA cannot always be inferred from the gene that encodes it. Subsequent to transcription, certain RNAs are edited. RNA editing can involve the insertion or deletion of nucleotides, as well as the modification of encoded nucleotides. RNA editing is required in certain mRNAs to create initiation and termination codons, as well as to eliminate internal frameshifts and change the identity of internal amino acids. Recently, RNA editing has also been observed in certain tRNAs and rRNAs; here the edited residues appear to be necessary to maintain the three-dimensional structure that is crucial to the function of these RNAs.

At present, RNA editing has only been observed in eukaryotes, but in this kingdom it is widespread and found in mitochondrion-encoded RNAs of kinetoplastids, slime molds, and plants; chloroplast-encoded RNAs of plants; and nucleus-encoded RNAs of mammals (Table 1). Although once thought to be limited to the insertion and deletion of uridine residues, editing reactions that produce non-encoded adenosines, guanosines, cytidines, and possibly inosines, have now been described.

The discovery of each new example of editing begins with the observation that a genomic sequence does not correlate with the corresponding cDNA sequence. The discovery of so many examples of RNA editing in recent years is the result of technological advances in sequencing and the polymerase chain reaction (PCR). This review does not focus on the methodology behind the identification of edited transcripts, but rather on the subsequent studies performed to elucidate the mechanisms involved in the editing process. Progress in this area has proceeded rapidly because of the ability of researchers in this field to turn aside dogma and to approach the task with creativity and a willingness to expect the unexpected. Yet, as the mechanisms responsible for RNA editing become more clear, it appears that RNA editing in a certain sense is nothing new

The RNA World
© 1993 Cold Spring Harbor Laboratory Press 0-87969-380-0/93 $5 + .00

Table 1 Summary of observed editing events

Organism	Organelle	Editing type[a]	RNA
Kinetoplastids	mitochondrion	U ; ΔU	mRNA
Slime molds	mitochondrion	C ; U ; G; A	mRNA, rRNA, tRNA
Plants	mitochondrion[b]	C$\rightarrow$U ; U$\rightarrow$C	mRNA, rRNA
	chloroplast	C$\rightarrow$U	mRNA
Mammals	nucleus	C$\rightarrow$U ; A$\rightarrow$G(I[c])	mRNA

[a]Insertion/deletion events are indicated by listing the inserted base or the deleted (Δ) base. Substitution events are indicated with an arrow from the encoded base to the base that is produced by editing.

[b]Although as yet uncharacterized, deletions have recently been observed in plant mitochondrial RNAs; see text.

[c]See text with reference to glutamate receptor mRNAs.

and relies on principles and biochemical reactions already established as important to RNA metabolism.

For the purposes of this review, the various types of editing are placed into one of two categories, depending on whether edited transcripts exhibit the insertion or deletion of nucleotides (Insertion/Deletion Editing) or whether an encoded residue has been substituted by another (Substitution Editing). Whereas the former process changes the overall length of the transcript, in the latter, the length of the transcript is maintained. From what is currently known about the mechanisms involved in the editing process, it appears that these categories are mechanistically distinct. However, for most edited transcripts this remains to be proven. RNA editing as discussed in this chapter is limited to those examples that are thought to occur posttranscriptionally. Cotranscriptional editing as observed in the paramyxoviruses is not discussed but is included in a recent review (Cattaneo 1991).

INSERTION/DELETION EDITING

Insertion/deletion editing has been detected in mitochondrion-encoded transcripts of kinetoplastid protozoa and the slime mold *Physarum polycephalum* (Fig. 1). Nucleotide deletions have also been detected in certain plant mitochondrial RNAs. However, the majority of the editing events in plants involve substitutions, and this type of editing is discussed in a later section. Since insertion/deletion editing is a posttranscriptional event, it must involve a mechanism by which the RNA is cleaved and then religated. As discussed below, cleavage-ligation events have precedence in mechanisms known to be operative in various forms of RNA splicing.

A.

Trypanosoma brucei cytochrome c oxidase III:

UUAUGUGAUUAUGGUUUUGUUUUUUAUUGGUAUUUUUUAGAUUUA

Δ

B.

Physarum polycephalum ATP Synthase, α subunit:

GUCAAUCGGUCAAAUUAUUUCUGUCAAAGAUGGUGUUGCUUUUGUUACAGGACUU

Figure 1 Comparison of editing events in two RNAs of the insertion/deletion category. Only a portion of the published sequence for the cytochrome *c* oxidase III mRNA (Feagin et al. 1988) and ATP synthase mRNA (Mahendran et al. 1991) is shown. Large, bold letters indicate inserted residues; Δ indicates a uridine deletion. (Modified, with permission, from Bass 1991.)

Kinetoplastids

Background

The first example of RNA editing was found in a mitochondrion-encoded mRNA of a kinetoplastid protozoan, or trypanosome (Benne et al. 1986). Subsequent studies have revealed that many mitochondrion-encoded RNAs are edited in kinetoplastids (for review, see Sollner-Webb 1991; Stuart 1991a,b; Hajduk et al. 1993). More is known about the mechanism of editing in kinetoplastids than about editing in other organisms, and thus kinetoplastid editing has become the prototype with which all other types of editing are compared. However, kinetoplastid editing exclusively involves the insertion and deletion of uridine residues, and as yet this particular editing pattern has not been observed in any other organism.

Kinetoplastid RNA editing has been extensively studied in three species: *Trypanosoma brucei, Leishmania tarentolae*, and *Crithidia fasciculata*. Many, but not all, mRNAs are edited in these species, and a recent review provides a tabulation of the number of insertions and deletions observed in each edited RNA (Stuart 1991a). Uridine insertions (≤ 553/transcript) are much more frequent than uridine deletions (≤ 89/transcript). In certain mRNAs, uridine insertions are so extensive that the length of the transcript almost doubles in size (Feagin et al. 1988; Bhat et al. 1990; Koslowsky et al. 1990; Read et al. 1992). Other transcripts have intermediate numbers of uridine insertions or deletions, with the first example reported exhibiting very minor editing, the insertion of only four uridines to correct a frameshift encoded by the cytochrome oxidase II gene (Benne et al. 1986). RNA editing is crucial to the production of translatable mRNAs in kinetoplastids and serves to create initiation and termination codons, as well as to correct internal frameshifts. RNA editing has not been observed in kinetoplastid rRNAs (as cited in Stuart 1991b). Editing in kinetoplastids is a posttranscriptional process

(Harris et al. 1990), and analyses of polycistronic transcripts suggest that (at least in some cases) editing may precede processing and polyadenylation (Read et al. 1992). Other studies also indicate that editing can occur prior to polyadenylation (Decker and Sollner-Webb 1990). However, uridines are sometimes found within poly(A) tails, and partially edited transcripts are frequently polyadenylated, suggesting that editing can also occur subsequent to polyadenylation (Decker and Sollner-Webb 1990; Van der Spek et al. 1990).

Editing Site Recognition: Guide RNAs

A glance at a portion of the edited cytochrome C oxidase III sequence from *T. brucei* is sufficient to recreate the awe and confusion that surrounded initial reports of RNA editing (Fig. 1A). The source of the non-encoded uridines found in kinetoplastid mRNAs, as well as the process that allowed the insertion and deletion of uridines at precise sites, were at first complete mysteries. These mysteries were partially solved by the discovery of small (~45–70 nucleotides) mitochondrion-encoded RNA molecules called guide RNAs (gRNAs; Blum et al. 1990). Numerous gRNAs have now been detected (Blum and Simpson 1990; Pollard et al. 1990; Sturm and Simpson 1990b; Pollard and Hajduk 1991; Van der Spek et al. 1991; Maslov and Simpson 1992; Koslowsky et al. 1992b), and as diagramed in Figure 2A, each gRNA is thought to be involved in "guiding" the editing of a region of a specific message. The 5′ region of a gRNA is complementary to the pre-edited mRNA and serves to anchor the gRNA. The more 3′ region of the gRNA is complementary to the final edited product, rather than to the pre-edited RNA, and serves to direct the production of the fully edited message. Each position within the pre-edited RNA requiring the insertion or deletion of uridines is called an editing site and presumably is recognized as such by its inability to base-pair with the gRNA. In addition to normal Watson-Crick base pairs, the gRNA-mRNA interaction utilizes numerous GU wobble base pairs and (in a few rare instances) AC base pairs (Van der Spek et al. 1991; Maslov and Simpson 1992). Although the interaction of gRNAs with mRNAs has been assumed from their complementarity, recent experiments place this interaction on a firm experimental basis. Specifically, mutations in the mRNA anchor sequence that destroy a step of the editing process in vitro (chimera formation, see below) can be rescued by mutations in the gRNA that restore base-pairing (Blum and Simpson 1992).

Although some transcripts require minimal editing and can be directed by a single gRNA, more extensively edited transcripts require multi-

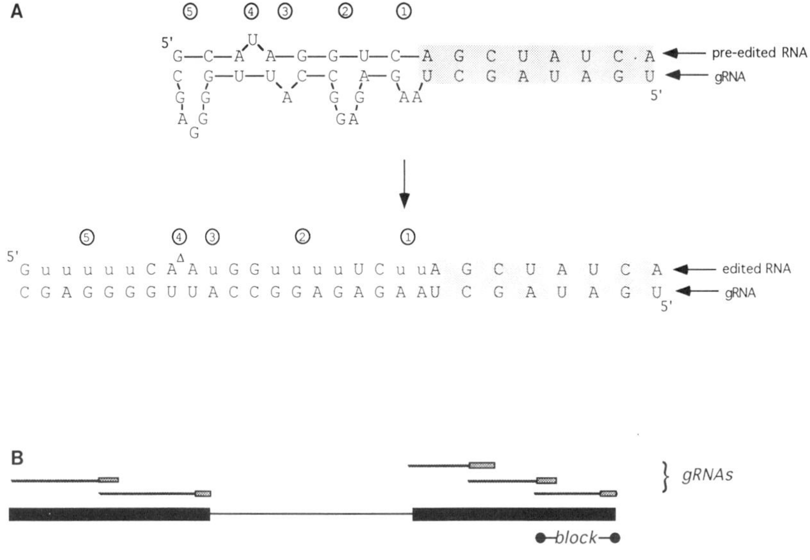

Figure 2 (*A*) Schematic representation of how gRNAs direct kinetoplastid editing. A hypothetical mRNA sequence is shown interacting with a hypothetical gRNA. Anchor sequences are shaded, and five editing sites are indicated with circled numbers. Inserted uridines are shown in lowercase letters, and a uridine deletion is marked with a triangle. (*B*) Five different gRNAs are shown guiding the editing of two domains (solid black bars). The unedited sequence between the two domains is represented with a line. The anchor sequence of each gRNA is indicated by a shaded rectangle.

ple gRNAs to produce the mature mRNA sequence (Fig. 2B) (Blum et al. 1990; Maslov and Simpson 1992). The region of a pre-edited RNA that is edited by a single gRNA is called an editing block, and a region requiring multiple gRNAs for complete editing is called a domain. Editing never occurs uniformly throughout the entire length of an RNA, but a single RNA can have multiple editing domains. Analyses of partially edited transcripts isolated from kinetoplastid mitochondria indicate that editing proceeds in an overall 3′ to 5′ direction (Abraham et al. 1988; Decker and Sollner-Webb 1990; Sturm and Simpson 1990b; Koslowsky et al. 1991). The gRNA complementary to the most 3′ end of an editing domain begins the process, and a 3′ to 5′ polarity is maintained because the anchor region of each successive gRNA is created by editing directed by the last (Fig. 2B) (Maslov and Simpson 1992; Koslowsky et al. 1992b). A comparison of the thermodynamic stability of duplexes predicted to form in anchor regions with those that form between the remainder of the gRNA and the completely edited RNA shows that the base-pairing involving the anchor is consistently more stable; it has been proposed that this enables each successive gRNA to displace the last (Maslov and Simpson 1992).

Although it is generally conceded that editing within a particular domain proceeds with a 3′ to 5′ polarity, the progression of editing within each block is controversial. Analyses of cDNAs made from steady-state mitochondrial RNA reveal molecules containing a fully edited 3′ region linked to a completely unedited 5′ sequence; these molecules would be expected intermediates if editing within each block proceeds with a strict 3′ to 5′ polarity (Fig. 3, expected intermediate) (Koslowsky et al. 1991). However, in many cDNAs the edited and un-

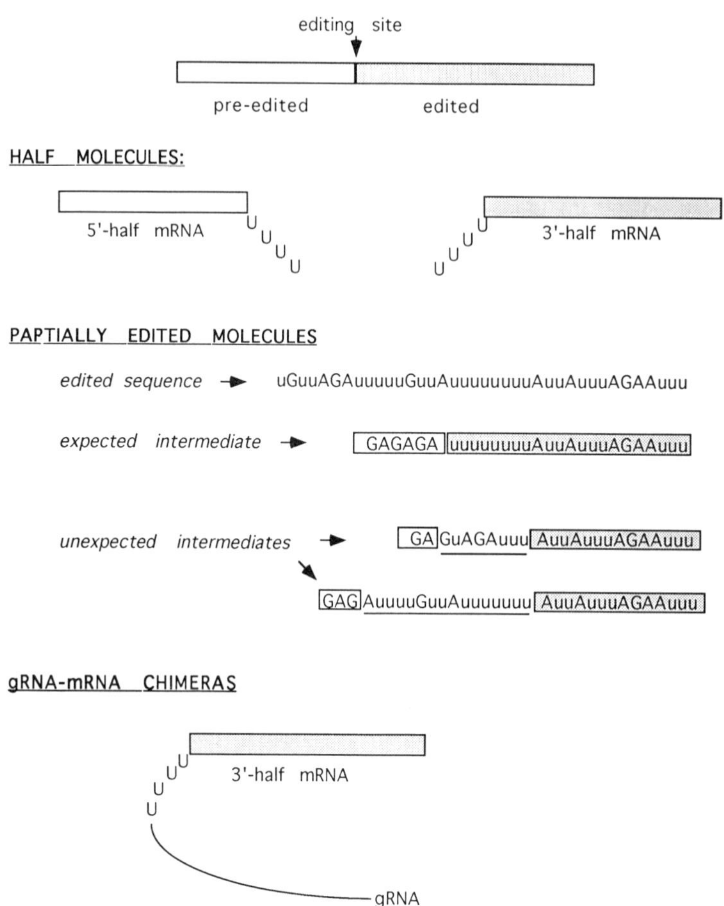

Figure 3 Putative intermediates of kinetoplastid editing. The sequences of the partially edited molecules correspond to actual intermediates observed in editing of *L. tarentolae* cytochrome *b* (Sturm and Simpson 1990a). Junction sequences are underlined.

edited sequences are intervened by a junction sequence (Fig. 3, unexpected intermediates) (Abraham et al. 1988; Decker and Sollner-Webb 1990; Koslowsky et al. 1991). Junction sequences can vary in size from a few to 100 nucleotides and are regions that have been subject to the editing machinery but do not match the fully edited sequence and do not suggest an orderly 3′ to 5′ editing process. Junction sequences have uridines inserted or deleted at sites that are not edited in the mature RNA, as well as aberrant numbers of uridines at editing sites.

A number of hypotheses have been proposed to explain the editing patterns observed in junction sequences. It has been suggested that although editing within a domain proceeds in an overall 3′ to 5′ direction, editing within a designated block is random and any nucleotide within the region is a target (Decker and Sollner-Webb 1990). According to this hypothesis, the ability of a completely and correctly edited block to form a perfectly base-paired duplex with the gRNA signals that editing is finished within this block. Another hypothesis proposes that editing within a block is not completely random but occurs in an order determined by the thermodynamic stability with which different regions of the mRNA can base-pair with the gRNA (Koslowsky et al. 1991). Pairings of lowest thermodynamic stability would be chosen first with successive cycles of editing required to produce the fully edited product, which would form the most stable duplex with the gRNA. Finally, it has been proposed that editing is not random but always occurs at the first mismatched nucleotide adjacent to the anchor and proceeds with a strict 3′ to 5′ polarity (Sturm and Simpson 1990a,b; Sturm et al. 1992). According to this model, the aberrant patterns observed in junction regions are the result of editing by the wrong gRNA or "misediting" by the right gRNA. Editing would thus require successive rounds of re-editing to produce the correctly edited product. At present it is not possible to choose among these models, and whether editing occurs with a strict 3′ to 5′ polarity in editing blocks remains an open question.

Catalytic Mechanism: gRNA-mRNA Chimeras

In addition to their role in guiding the editing process, gRNAs may be directly involved in the series of cleavage and ligation reactions that allow uridine insertion and deletion. gRNAs have non-encoded uridines at their 3′ termini (Blum and Simpson 1990) that are presumably added by a uridylyltransferase activity detected in kinetoplastid mitochondrial extracts (Bakalara et al. 1989; Harris et al. 1990). Most importantly, chimeric molecules (Fig. 3) consisting of the 3′ terminus of the gRNA covalently linked to a 3′-half mRNA have been detected in vivo (Blum et al. 1991; Read et al. 1992b), as well as in vitro (Harris and Hajduk

1992; Koslowsky et al. 1992a). The observed chimeric molecules suggest that the polyuridine tail of the gRNA is the source of the uridines found in edited RNAs, and they also suggest specific mechanisms by which these uridines could be transferred.

Two models have been proposed in regard to the catalytic mechanism of RNA editing in kinetoplastids (Fig. 4). Both models invoke mechanisms that have precedence in RNA splicing reactions. One model proposes that editing is essentially an RNA-catalyzed process involving a series of transesterification reactions (Blum et al. 1991; Cech 1991), whereas the second relies on protein catalysis; in particular, the action of an endonuclease and an RNA ligase (Blum et al. 1990; Sollner-Webb 1991).

The transesterification model derives from considerations of reactions involved in the self-splicing group I and II introns (Cech, this volume), as well as nuclear mRNA splicing (Moore et al., this volume). Like the catalysis observed in these types of splicing, the reaction would proceed through a series of transesterification reactions in which every cleavage

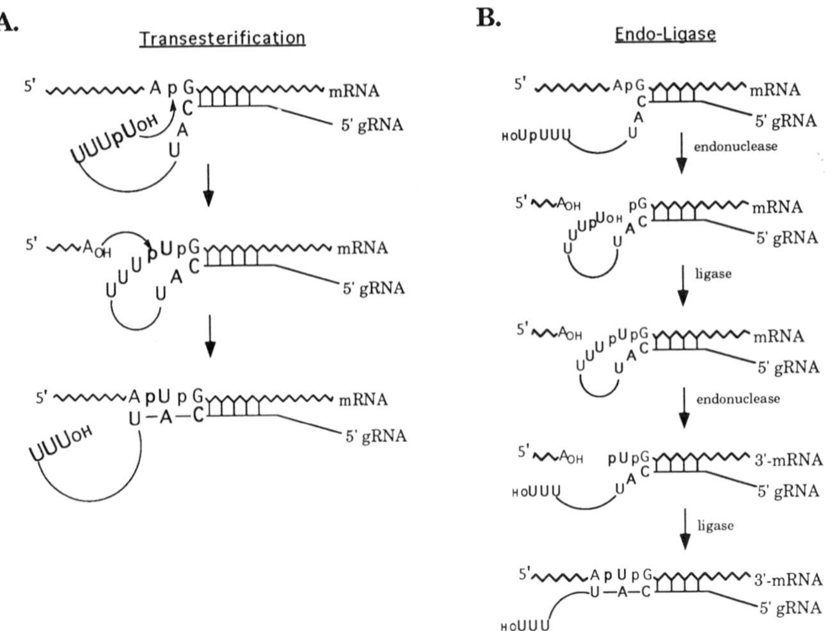

Figure 4 Representation of two models for catalysis of kinetoplastid editing. The editing event shown involves the insertion of a uridine to create an initiation codon. Phosphates (p) are shown only for those phosphodiester bonds directly involved in the reaction. (*A*) Catalysis using a transesterification mechanism; (*B*) catalysis using an endonuclease and an RNA ligase.

is linked to a ligation (Fig. 4A). Two transesterification reactions would be required for each editing event. The 3′ hydroxyl of the terminal uridine of the gRNA would serve as the initial nucleophile, attacking the phosphodiester bond at the editing site. This nucleophilic attack would release the 5′-half mRNA and ligate the 3′ end of the gRNA to the 3′-half mRNA, creating the chimera. The 3′ hydroxyl of the 5′-half mRNA would then initiate a second nucleophilic attack to free the gRNA and insert or delete uridines into the transcript. The first nucleophilic attack would occur 3′ of a mismatched residue for uridine insertions, and 5′ for deletions. The second attack by the 3′ hydroxyl of the mRNA would occur within the polyuridine tail for a uridine insertion or 3′ of the encoded uridine(s) if editing requires a deletion. Although some editing intermediates suggest uridines are transferred one at a time, multiple uridines could be inserted or deleted at a single step by altering the site of the second nucleophilic attack.

Kinetoplastid editing requires proteins and, thus, if editing occurs by transesterification, it will probably be more analogous to nuclear mRNA splicing than to self-splicing. In nuclear mRNA splicing, the RNA is thought to contain the active site involved in catalysis, whereas the proteins aid in formation of the active RNA structure and splicing complex (Moore et al., this volume). In analogy with nuclear mRNA splicing, which occurs in a large complex of proteins and RNA molecules called a spliceosome, similar complexes, or editosomes, have been characterized in kinetoplastid mitochondrial extracts (Pollard et al. 1992; H.U. Göringer et al., in prep.).

The transesterification model is consistent with the recently observed gRNA-mRNA chimeras. It also predicts that a 5′-half mRNA would be a reaction intermediate; such truncated molecules have been detected in cDNA populations (Fig. 3) (Decker and Sollner-Webb 1990; Read et al. 1992a). However, truncated molecules corresponding to the sequence 3′ of an editing site (3′-half mRNA) have also been observed (Fig. 3) (Abraham et al. 1988). Although the detection of 3′-half mRNAs would seem to argue against the transesterification mechanism, such molecules could arise during the transesterification mechanism by aberrant hydrolysis such as occurs in group I introns (Cech 1987). It should be noted that the observed 3′ and 5′-half mRNAs usually contain tracts of uridines or partially edited sequence at their 3′ termini, which is not consistent with a strict 3′ to 5′ polarity. The transesterification model requires a 3′ hydroxyl at the end of the gRNA to serve as the nucleophile in the first cleavage-ligation event. Consistent with this requirement, the addition of a phosphate to the 3′ end of a gRNA, or oxidation of the hydroxyl groups of the terminal gRNA ribose, inhibits chimera formation in vitro (Harris and Hajduk 1992; Koslowsky et al. 1992a).

In contrast to the transesterification model, where every cleavage is linked to a ligation, and each successive cleavage-ligation event could theoretically occur in a single active site (Cech 1991), the second model requires at least two different activities with distinct active sites. This model as originally proposed did not invoke the formation of chimeric molecules (Blum et al. 1990) and now has been modified to include this recent observation (Sollner-Webb 1991). The model is similar to tRNA splicing in that it invokes an endonuclease and an RNA ligase (Fig. 4B). The endonuclease would cleave the pre-edited RNA at an editing site to generate a 5′ as well as a 3′-half mRNA. An RNA ligase activity would then catalyze the ligation of the 3′ hydroxyl of the gRNA to the 3′-half mRNA. Endonuclease as well as RNA ligase activities, the same or different, would then be required a second time to free the gRNA (minus or plus uridines) and religate the two halves of the message.

Like the transesterification mechanism, the endo-ligase mechanism is consistent with certain observations made during studies of kinetoplastid RNA editing. RNA ligase activities have been detected in mitochondrial extracts of *T. brucei* (V.W. Pollard et al., pers. comm.) as well as *L. tarentolae* (Bakalara et al. 1989). An endonuclease activity that specifically cleaves pre-edited transcripts at or near editing sites has also been reported (Harris et al. 1992; Simpson et al. 1992). Since these specific sites are also cleaved by a heterologous single-strand-specific endonuclease, it has been proposed that there is a structural component to recognition by the putative endonuclease (C. Decker et al., pers. comm.). Although the detection of the enzymatic activities required by the endo-ligase mechanism supports this mechanism, it has not been determined whether the observed activities are actually involved in RNA editing or play a role in another aspect of mitochondrial RNA processing.

The endo-ligase model can accommodate chimeric molecules and is consistent with the observed 5′ and 3′-half mRNAs. Although the transesterification model does not require an energy source, since every cleavage is linked to a ligation (Cech 1987 and this volume), some sort of high-energy phosphate-bond cleavage would be predicted for the reaction catalyzed by the RNA ligase. An ATP requirement for uridine incorporation into pre-edited RNA has been observed using crude extracts (A.M. Simpson et al., pers. comm.), but at present this requirement cannot be specifically assigned to a ligase and may reflect the requirement for ATP in editosome assembly. The endo-ligase model invokes enzymes similar to those involved in tRNA splicing, although some differences are apparent. For example, the RNA ligase activity detected in kinetoplastid mitochondrial extracts ligates molecules with 3′ hydroxyls to those containing 5′ phosphates. This type of ligation has precedence in the mechanism used by T4 RNA ligase (Uhlenbeck and Gumport 1982) but differs from that used by the ligases involved in tRNA splicing. The yeast tRNA ligase is a multifunctional enzyme that ultimately ligates molecules containing 2′ phosphate, 3′ hydroxyl termini to those containing a 5′ phosphate (Apostol et al. 1991). Although the type of ligation involved in tRNA splicing in higher eukaryotes is less clear, activities similar to those of yeast, as well as one that ligates 2′, 3′ cyclic phosphates to 5′ hydroxyls, have been detected (for review, see Phizicky and Greer 1993). If a ligase is involved in RNA editing, its substrate requirements may differ slightly from those of the tRNA ligases.

The transesterification model is attractive in its simplicity. However, evidence supporting both mechanisms exists, and at present a definitive choice cannot be made. As discussed, one step in the editing process, chimera formation, has now been reproduced in vitro. The development of an in vitro system that can mimic the complete editing process will no doubt lead to a more complete understanding of the mechanisms required in kinetoplastid RNA editing.

Physarum polycephalum

Background

Outside the kinetoplastids, the only other example of editing that clearly falls into the insertion-deletion category is found in the acellular slime mold *P. polycephalum* (for review, see Miller et al. 1992). Like kinetoplastid editing, *Physarum* editing involves mitochondrion-encoded transcripts. Although very little is known about the mechanism of *Physarum* editing, the pattern of editing observed in *Physarum* RNAs is very different from that of kinetoplastid RNAs. *Physarum* editing involves insertion, but never deletion, of residues and predominantly, but not exclusively, involves the insertion of cytidines. Kinetoplastid editing has only been observed in mRNAs, whereas *Physarum* editing has been observed in mRNAs as well as rRNAs and tRNAs.

Physarum editing was first detected in the α subunit of the mitochondrial ATP synthase mRNA, where the insertion of single cytidines at 54 sites serves to correct frameshifts (Mahendran et al. 1991). Subsequently, editing in *Physarum* has been detected in the small subunit rRNA (45 sites; R. Mahendran et al., in prep.), the large subunit rRNA (57 sites; N. Yang et al., in prep.), several tRNAs (1–2 sites/tRNA; H. Costandy et al., in prep.), and 3 additional mRNAs (9–40 sites/mRNA; M.-L. Ling and D.L. Miller; R. Mahendran et al.; S. Wang et al.; all in prep.; D.L. Miller, pers. comm.). All of these edited RNAs were identified within a 13.8-kb segment of the mitochondrial genome. Since most of the remaining 75% of the genome has not been analyzed for the presence of editing, it seems likely that many *Physarum* RNAs are edited. Among the editing events identified, 90% of the inserted residues are cytidines, 7% are uridines, and 4% are guanosines or adenosines. At most editing sites a single residue is inserted, but 4% of the sites involve dinucleotide insertions.

In addition to providing the first example of insertions involving residues other than uridine, studies of editing in *Physarum* provide the first evidence for insertion/deletion editing in rRNAs and tRNAs. A comparison of the predicted secondary structure of the edited small subunit rRNA with that of *Escherichia coli* 16S rRNA indicates that editing sites occur in variable domains as well as in highly conserved regions of the central core (Miller et al. 1992); thus it is expected that editing is required to produce the functional *Physarum* rRNA. Similarly, the editing observed in tRNAs appears to be necessary to maintain base-pairing within stem regions and to create conserved loop sequences.

Nonrandom Characteristics: Hints of a Mechanism?

In Figure 1, the editing of the *T. brucei* cytochrome oxidase III mRNA is contrasted with the editing observed in the *Physarum* ATP synthase mRNA. Compared to the almost chaotic pattern of editing observed in the kinetoplastids, editing in *Physarum* suggests a more orderly editing process. Although multiple residues are often inserted at a single editing site in the kinetoplastids, all of the insertions in the ATP synthase transcript involve single residues. As discussed, except for rare dinucleotide insertions, this trend continues for the other examples of editing in *Physarum*. Whereas editing sites in kinetoplastid RNAs tend to be localized to particular regions of the transcript, editing sites in *Physarum* RNAs are distributed throughout the transcript, appearing on the average 25 nucleotides apart in mRNAs and 43 nucleotides apart in rRNAs (Miller et al. 1992).

Several additional nonrandom features have been noted among the *Physarum* editing sites characterized (Miller et al. 1992). For example, cytidine and uridine residues, although often inserted as single nucleotides, are never inserted as CC or UU dinucleotides. When present as dinucleotides, C and U residues are inserted as CU, GU, or GC mixed dinucleotides. In contrast, adenosines have not been observed as single insertions but appear as AA dinucleotides. At least 60% of all *Physarum* editing sites are 3′ to a purine-pyrimidine dinucleotide. Finally, for mRNAs, the location of editing sites within particular codons, and the particular codon created, are not random. 50% of the codons consist of AUC, GUC, ACC, or GCC, and at least 68% of all editing sites occur at the wobble position.

Although the nonrandom patterns probably reflect some aspect of how editing sites are chosen, as well as the chemistry involved in nucleotide insertion, the details of these processes are entirely unclear. Molecules analogous to gRNAs have been looked for but not found (Miller et al. 1992). Of course, if gRNAs are involved in *Physarum* editing, somehow they would have to specify the insertion of cytidines, adenosines, and guanosines, as well as uridines. Since editing sites are spaced more widely apart than those in kinetoplastid RNAs, there is more potential for designation of editing sites by intramolecular structure. However, the predicted structure of the small subunit rRNA does not suggest the association of editing sites with any particular structural feature (Miller et al. 1992).

Although partially edited RNAs can be detected in *Physarum* mitochondria, they are much less abundant than those of kinetoplastids, accounting for no more than 5% of the steady-state RNA population (Miller et al. 1992). The detection of partially edited RNAs suggests editing in *Physarum* is a posttranscriptional process and, thus, must involve a cleavage-ligation mechanism. However, truncated half-mRNA molecules have not been observed (D.L. Miller, pers. comm.). The partially edited *Physarum* transcripts show edited regions covalently linked to unedited regions with clear boundaries between the edited and unedited sequences. In contrast to the intermediates observed in kinetoplastids, neither junction regions nor a 3′ to 5′ polarity is observed.

In analogy with the mechanisms proposed for kinetoplastid editing and the fact that most *Physarum* insertions involve cytidine residues, RNAs with oligocytidine tails have been looked for, but also not found. It remains possible that inserted residues derive from nucleotide monomers rather than oligonucleotides. In summary, the mechanism of *Physarum* editing is unclear. Answers will most likely come with the development of an in vitro editing assay.

SUBSTITUTION EDITING

Just as insertion/deletion editing has analogies with RNA splicing, substitution editing has similarities with the RNA modifications known to occur in tRNAs (Björk et al. 1987), rRNAs (Maden 1990), and mRNAs (Singer and Kroger 1979; Tuck 1992). In fact, in certain cases it is hard to know whether recent examples should be referred to as RNA editing or placed in the more traditional category of RNA modification. However, in contrast to the bases generated during RNA modification, which are always non-Watson-Crick bases and often extensively modified, substitution editing appears to simply involve the conversion of one Watson-Crick base to another. In addition, although modifications previously observed in mRNAs do not alter coding capacity, substitution editing clearly alters the protein synthesized from an edited mRNA. All the examples of substitution editing observed to date produce transition, rather than transversion, mutations and furthermore, although not proven, can be explained by a simple deamination or the reverse reaction, an amination.

The discussion of substitution editing begins with examples of nucleus-encoded RNAs that are edited in mammalian cells. The discovery of substitution editing in a mammalian cell occurred soon after the first report of editing in the kinetoplastids. However, in contrast to subsequent studies in kinetoplastids, which led to the rapid identification of many transcripts that required editing, editing in mammalian cells has only been detected in two mRNAs and possibly two tRNAs. Whereas the insertion/deletion editing of mRNA serves to create open reading frames where none exist, editing of mammalian mRNAs merely alters a functional open reading frame to allow the production of two proteins from a single gene. Each of the two edited mRNAs exhibits a different nucleotide transition and thus may reflect a distinct type of editing with different editing machinery. For this reason, each of the mRNAs is discussed separately. The examples of editing of mammalian tRNAs is not discussed, but the reader is referred to the literature (Diamond et al. 1990; Beier et al. 1992).

Mammalian Apolipoprotein B

Background

The discovery of editing within mammalian apolipoprotein B (apoB) mRNA provided the first example of editing in a nucleus-encoded RNA (for review, see Chan 1992; Hodges and Scott 1992). Prior to the discovery that apoB mRNA was edited (Chen et al. 1987; Powell et al. 1987), it was known that two closely related apoB proteins, apoB-100 (512 kD) and apoB-48 (241 kD), existed in mammalian plasma lipoproteins. Comparative sequence analyses of genomic DNA and cDNAs led to the conclusion that the two proteins are products of the same gene. Whereas apoB-100 is the translation product of the mRNA encoded by the genomic sequence, apoB-48 is translated from an edited mRNA. The editing that leads to the production of apoB-48 involves the conversion of a single cytidine (nucleotide 6666 of human cDNA) to a uridine. This substitution editing converts a glutamine codon (CAA) to a stop codon (UAA); thus apoB-48 is a truncated form of apoB-100 and is identical to its amino terminus. Both apoB-100 and apoB-48 have distinct and important roles in lipid metabolism; apoB editing is tissue-specific and under developmental, nutritional, and hormonal control (see reviews cited above). In most mammals, the product of the unedited mRNA, apoB-100, is found in the liver, whereas apoB-48, the protein translated from the edited mRNA, is synthesized in the intestine. An exception to this is found in rodents, where apoB-48 can be detected in liver as well as intestine (Davidson et al. 1988; Tennyson et al. 1989).

Analyses of cDNAs made from nuclear RNA indicate apoB editing is a posttranscriptional event. Edited RNA cannot be detected in populations of unspliced, non-polyadenylated RNA and is low in unspliced polyadenylated RNA (Lau et al. 1991). Furthermore, nuclear RNA that is spliced and polyadenylated is completely edited, suggesting that apoB editing occurs in the nucleus (Lau et al. 1991).

In Vitro Studies

Soon after the discovery of edited apoB transcripts in human and rabbit cells (Chen et al. 1987; Powell et al. 1987), an in vitro assay was developed using cell-free extracts of the rat hepatoma cell line McArdle 7777 (Driscoll et al. 1989). Editing activity has now been studied in vitro using extracts prepared from a variety of cells, including enterocytes of rabbit (Bostrom et al. 1990), baboon (Driscoll and Casanova 1990), and rat (Backus and Smith 1991; Hodges et al. 1991; Navaratnam et al. 1991; Shah et al. 1991), and rat hepatocytes (Lau et al. 1990, 1991; Backus and Smith 1991). As might be expected, a number of tissues and cell lines that do not express apoB or only contain apoB-100 do not show detectable levels of editing activity (Powell et al. 1987; Bostrom et al. 1989; Davies et al. 1989; Driscoll et al. 1989; Chen et al. 1990; Driscoll and Casanova 1990). However, editing activity can sometimes be detected in extracts prepared from cell lines that do not express apoB (Bostrom et al. 1990; Hodges and Scott 1992), and this observation has led to the speculation that the activity that edits apoB may also serve to edit other RNAs.

Although most in vitro studies have used cytoplasmic S100 preparations, as discussed, most evidence suggests that apoB editing occurs in the nucleus. Editing activity can be detected in nuclear as well as cytoplasmic extracts, but the specific activity of nuclear preparations is higher (Lau et al. 1991). It is not known if the presence of the activity in cytosolic extracts is the result of leakage from the nucleus during fractionation or if the editing activity actually resides in both compartments.

Since certain tissues that have editing activity also have unedited apoB mRNA in the cytoplasm, the latter transcripts would have to be inaccessible to or protected from cytoplasmic editing activity.

Optimization of editing in S100 extracts indicates that the activity does not require divalent cations, nucleoside triphosphates, creatine phosphate, or any other cofactors (Driscoll et al. 1989; Driscoll and Casanova 1990; Smith et al. 1991). The activity is abolished by proteinase K treatment, but not by micrococcal nuclease, suggesting that the activity consists of protein rather than RNA components (Driscoll and Casanova 1990; Greeve et al. 1991). The activity has been partially purified from baboon enterocytes, and gel filtration chromatography indicates an apparent molecular mass of 125 kD. Other studies have identified a 40-kD protein (Lau et al. 1990) and a 60-kD protein (Navaratnam et al. 1992) that cross-link to a synthetic apoB RNA substrate. Finally, gel shift assays in combination with glycerol gradient sedimentation have led to the proposal that like nuclear mRNA splicing (Moore et al., this volume), and possibly kinetoplastid editing, apoB editing occurs within a large complex or "editosome." These studies, although controversial, predict that editing occurs in a 27S particle (Smith et al. 1991).

Editing Site Recognition: A Minimal Sequence

A number of studies have focused on determining the minimal apoB sequence required for specific and efficient editing (Bostrom et al. 1989; Davies et al. 1989; Driscoll et al. 1989; Chen et al. 1990; Lau et al. 1990; Backus and Smith 1991; Shah et al. 1991; Smith et al. 1991). It is difficult to draw precise conclusions from these studies, but several generalities can be made. First, it appears that most of the 14-kb apoB mRNA is dispensable with regard to the conversion of C 6666 to U. For example, 26 nucleotides of apoB sequence immediately surrounding the editing site are sufficient to promote precise and specific editing (Fig. 5). When the 26-mer is embedded within heterologous sequences and assayed after transfection, cytidine 6666 is edited with an efficiency only slightly lower than endogenous editing (Davies et al. 1989). Furthermore, a 73-nucleotide RNA containing these 26 residues in addition to polylinker sequences is edited in vitro, although with a reduced efficiency (Shah et al. 1991). The 26-nucleotide sequence is highly conserved among mammalian apoB sequences and is homologous to residues that surround a second apoB editing site that has recently been identified (Navaratnam et al. 1991; Hodges and Scott 1992). A caveat to these observations is that longer segments of apoB are required to observe editing in some cell types (Bostrom et al. 1989; Driscoll et al.

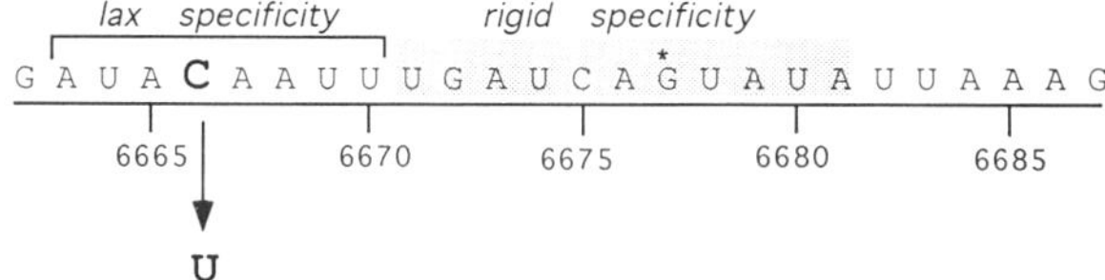

Figure 5 The 26-nucleotide sequence of apolipoprotein B mRNA that is sufficient for editing. The encoded cytidine that is edited to a uridine is shown with a large bold letter. Sequences enclosed in the shaded rectangle are crucial for editing, except G 6677, which is indicated by an asterisk. The sequences are numbered according to the human apoB cDNA sequence.

1989). Furthermore, when full-length human apoB is expressed in transgenic mice, the human sequence is not edited, even though the endogenous mouse apoB exhibits the normal editing pattern (Xiong et al. 1992). These latter results emphasize the obvious: apoB editing is a highly regulated process, and although artificial systems may hint at important aspects of apoB editing, they cannot tell us everything about the process as it occurs in the cell.

The second general principle derived from studies of sequence specificity is that the nucleotides slightly upstream and downstream from the editing site are more important than those that actually span the edited cytidine (Fig. 5). For example, of 22 mutant sequences containing single or multiple base substitutions in the immediate vicinity of the editing site (nucleotides 6663–6670), only two completely abolished editing in vitro (Chen et al. 1990). Other mutagenesis studies support the idea that specificity is lax in the immediate vicinity of the editing site but, in addition, emphasize the importance of nucleotides further downstream (Shah et al. 1991). Specifically, except for position 6677, which is quite tolerant of mutations, mutations within an 11-nucleotide downstream sequence (6671–6681) abolish or drastically affect editing efficiency. The importance of sequences downstream from the editing site is emphasized by the fact that the editing-competent 26-mer contains only 4 residues 5′ of the editing site. Furthermore, it has been observed that sequences 3′ of the editing site can induce editing of heterologous cytidines if translocated to a different region of the apoB sequence (Backus and Smith 1991). Although downstream sequences seem sufficient to allow editing, optimal editing requires additional residues upstream of the editing site (Backus and Smith 1991).

An unresolved aspect of apoB editing is how the residues that seem to be important for editing define the editing site. Several models have been presented based on the importance of distal sequences (Smith et al. 1991; Chan 1992; Hodges and Scott 1992). According to these models a protein, or protein complex, binds to distal sequences and directs editing to nucleotide 6666. All models assume that the active site of the editing enzyme cannot recognize the specific cytidine to be edited unless guided by the initial contact with downstream sequences. In support of this idea is the observation that if residues in close proximity to the editing site are mutated to cytidine, they are sometimes converted to uridines (Chen et al. 1990). It is not clear if the important distal sequences allow recognition by folding into a minimal secondary structure or, alternatively, are recognized in a single-stranded form. Although the mutagenesis studies do not point to a specific structure in an obvious way, regions surrounding the editing site can be folded into a stem-loop structure (Shah et al. 1991). However, this structure is only speculative and is not consistent with the fact that nucleotides 6663–6670 are quite tolerant of mutations, whereas those between 6671–6681 are not.

Catalytic Mechanism

Editing of apoB was first inferred from a cDNA sequence. Formally, the cDNA sequence indicated that a base in the mRNA that could pair with G was converted to a residue that could pair with A. Although a C to U substitution was the most obvious explanation for these results, it was also possible that the C was modified to a different base that could pair with A. However, thin-layer chromatographic analyses of mononucleotides derived from a ^{32}P-C-labeled RNA oligo that was edited in vitro clearly demonstrate that a ^{32}P-labeled U is generated during the reaction (Bostrom et al. 1990; Hodges et al. 1991).

The most direct way to convert a C to a U would be by hydrolytic deamination, the mechanism used by cytidine deaminases (Fig. 6A). However, the mechanism of apoB editing has not been determined, and alternatives such as a ribosyltransferase mechanism (Fig. 6B) or transamination have not yet been ruled out. A transaminase reaction seems unlikely, since C4 of cytidine has partial double-bond character and is sufficiently electrophilic for a direct attack by water. In general, the substrates of transaminase activities are not sufficiently electrophilic for a direct attack. These enzymes use pyridoxal phosphate as a cofactor to form a Schiff base with the leaving group nitrogen that activates the carbon-nitrogen bond for hydrolysis (Braunstein 1973).

Mammalian Glutamate Receptor Channel Subunits

Background

The second example of editing in a mammalian cell is found in nucleus-encoded mRNAs for two classes of brain glutamate receptors (gluR)

	Substitution	Mechanism	Enzymes	References
A.	C → U	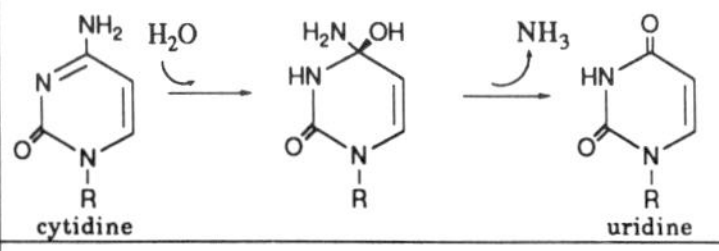 (cytidine → uridine; H_2O, NH_3)	cytidine deaminase	(Frick, Yang et al., 1989)
B.	N' → N''	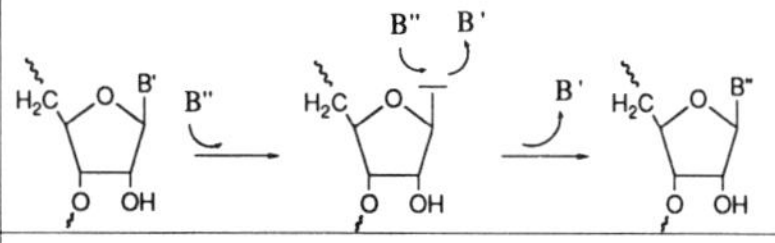 (ribose with B'', B', B''' exchange)	tRNA guanine ribosyl transferase	(Nishimura, 1983)
C.	U → C	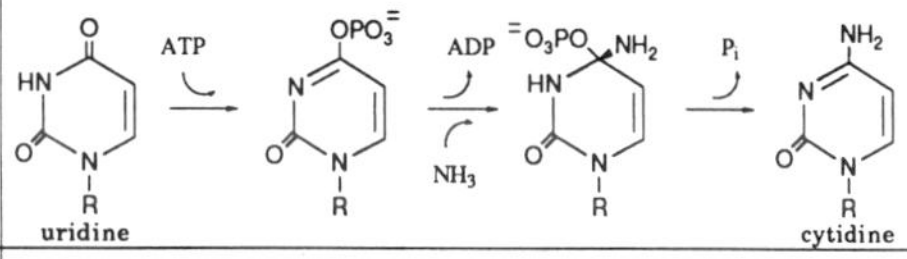 (uridine → cytidine; ATP, ADP, NH_3, P_i)	CTP synthetase	(Lewis and Villafranca, 1989)
D.	A → G	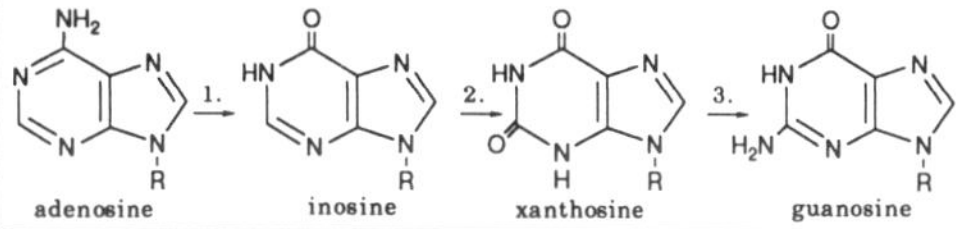 adenosine → inosine → xanthosine → guanosine	1. see E. 2. IMP dehydrogenase 3. GMP synthetase	(Hedstrom and Wang, 1990) (Zalkin and Dixon, 1992)
E.	A → I	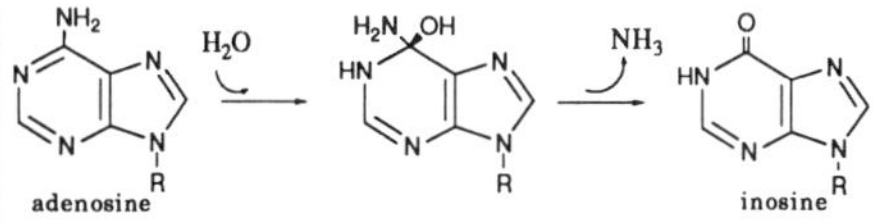 adenosine → inosine; H_2O, NH_3	adenosine deaminase AMP deaminase dsRNA unwinding-modifying activity	(Kurz, Moix et al., 1992) (Merkler, Brenowitz et al., 1990) (Polson, Crain et al., 1991)

Figure 6 Mechanisms used by various cellular enzymes to substitute one nucleic acid base for another. Five examples (*A–E*) are diagramed. N ' and N ' ' are random nucleosides that have different bases, B ' and B ' ', respectively.

(Sommer et al. 1991). Pharmacological studies allow gluRs to be placed in one of three classes according to their agonists: *N*-methyl-D-aspartate (NMDA), α-amino-3-hydroxy-5-methyl-isoxazole-4-propionic acid (AMPA), and kainate (KA). Editing has been demonstrated for mRNAs of the AMPA as well as the KA class. Specifically, cDNAs corresponding to the AMPA selective subunit, gluR-B, as well as the KA selective subunits, gluR-5 and gluR-6, contain a guanosine residue at a position predicted to be an adenosine by the genomic sequence. Although 100% of the cDNAs analyzed for the gluR-B subunit contain the A to G substitution, editing of gluR-5 and gluR-6 is not complete: 40% of the gluR-5, and 80% of the gluR-6, cDNAs exhibit the substitution. The nucleotide change results in the conversion of a glutamine codon (CAG) to an arginine codon (CGG) within the putative channel-forming segment, TMII. This particular amino acid seems to be particularly important in determining the ion transport properties of a channel, and assembly of channels with subunits containing the critical arginine precludes conductance by the divalent cation calcium. It has been proposed that the editing machinery responsible for the sequence changes in the glutamate receptor transcripts may be subject to regulated expression, making the Ca^{++} permeability of channels a controlled event that functions during establishment of excitatory circuits in the developing brain (Sommer et al. 1991).

Editing Site Recognition

Since editing of gluR subunit mRNAs was only recently discovered, investigations into the editing process have not been performed. Sequences surrounding the gluR editing sites are highly conserved among edited and unedited members of the two classes of receptor transcripts, making discrimination based strictly on sequence improbable. Differences in structure are more difficult to discern, and it remains possible that the edited transcripts do indeed assume unique structures. For example, it has been proposed that long-range interactions, perhaps involving intron sequences, are involved (Sommer et al. 1991).

Catalytic Mechanism

Unlike the C to U substitution observed in apoB mRNAs, an A to G substitution would require more than one catalytic event. Using mechanisms with precedence in other enzymes, the conversion of A to G could occur using three different catalytic activities (Fig. 6D). The C6 amino group could be replaced with a ketone by the deamination of adenosine to inosine. This reaction has precedence in the adenosine and AMP deaminases and is also the same mechanism as diagramed for cytidine deaminase. For the amination of C2, two steps would be required, since this carbon does not have a good leaving group. In analogy with mechanisms involved in the biosynthesis of GMP, an enzyme similar to IMP dehydrogenase would first oxidate C2 to produce xanthosine. The C2 ketone of xanthosine could then be aminated with an enzyme such as GTP synthetase, which is thought to use the mechanism diagramed for CTP synthetase (Fig. 6C).

Another less cumbersome mechanism for the gluR editing exists. Specifically, it is possible that the G present in the cDNA actually reflects the presence of an inosine in the mRNA. Inosine, like guanosine, prefers to base-pair with C, and thus an A to G change between a gene and a cDNA is a hallmark of an A to I conversion. The conversion of an adenosine to inosine can occur by a simple deamination (Fig. 6E). Furthermore, an activity that could catalyze an A to I conversion within an mRNA exists in organisms throughout the animal kingdom and is discussed in the next section.

Double-stranded RNA Unwinding/Modifying Activity

Background

Investigations into each of the various types of RNA editing discussed so far began with the identification of an edited RNA and proceeded to a search for the biological activity responsible for the editing event. In contrast, the double-stranded RNA (dsRNA) unwinding/modifying activity was discovered prior to the identification of its biological substrates (for review, see Bass 1992, 1993). In fact, although this activity has been implicated in the editing of several cellular RNAs, as yet, none of these putative substrates has been conclusively demonstrated to be an in vivo substrate. However, the reaction catalyzed by the unwinding/modifying activity suggests that it most likely plays an editing role.

The dsRNA unwinding/modifying activity was discovered when a synthetic dsRNA was microinjected into the cytoplasm of *Xenopus laevis* embryos (Bass and Weintraub 1987; Rebagliati and Melton 1987). After incubation in the cell, the dsRNA exhibited an aberrant mobility on a native gel and an increased sensitivity to single-strand-specific ribonucleases. It was initially proposed that the dsRNA had been subject to an unwinding event, but these phenomena were later understood to result from a covalent modification by an enzyme now known as the dsRNA unwinding/modifying activity (Bass and Weintraub 1988; Wagner et al. 1989). This activity catalyzes the conversion of adenosines to inosines within dsRNA. Since inosine, like guanosine, prefers to base-pair with cytidine, the modification results in the conversion of stable AU base pairs to IU mismatches, producing partially unwound molecules.

Since its discovery in *X. laevis*, the unwinding/modifying activity has been detected in *Caenorhabditis elegans* (M. Krause, unpubl.), *Drosophila melanogaster* (B. Bass, unpubl.), *Bombyx mori* (Skeiky and Iatrou 1991), and numerous mammalian tissues and cultured cells (Wagner et al. 1990). Among organisms of the animal kingdom, RNA editing has been conclusively demonstrated only in mammals. If the unwinding/modifying activity functions in RNA editing, edited transcripts may be widespread within this kingdom.

In most cells, the unwinding/modifying activity is nuclear (Bass and Weintraub 1988; Wagner et al. 1990), but in certain instances, it exists in the cytoplasm (Bass and Weintraub 1987; Skeiky and Iatrou 1991; L. Saccomanno and B. Bass, in prep.). Optimization of the activity in vitro using S100 extracts suggests that like the apoB editing activity, the unwinding/modifying activity does not require cations, nucleoside triphosphates, or any other cofactors (Bass and Weintraub 1987; Wagner and Nishikura 1988). However, more recent studies indicate that the active component has a zinc ion tightly bound at its activity site (see below). Gel filtration chromatography suggests an apparent molecular mass of 210 ± 20 kD (R. Hough and B. Bass, unpubl.).

Substrate Specificity

In vitro studies indicate that the unwinding/modifying activity requires a dsRNA substrate and will not bind to single-stranded RNA, single-stranded DNA, or double-stranded DNA (Bass and Weintraub 1987; Wagner and Nishikura 1988). Although the activity preferentially modifies adenosines in AU-rich regions (Kimelman and Kirschner 1989; A. Polson and B. Bass, unpubl.), very little sequence specificity is observed, and almost any dsRNA molecule can be modified in vitro. Up to 50% of the adenosines in both strands of a given dsRNA can be modified. Although most in vitro studies have used intermolecular dsRNA substrates, intramolecular stem-loop structures can also be modified (Nishikura et al. 1991; Sharmeen et al. 1991). The minimum length of contiguous base pairs necessary for modification is unclear. For intermolecular duplexes, the shortest molecule shown to be a substrate contains 23 base pairs, and a 15-base pair duplex is not modified (Nishikura et al. 1991). However, the same 15 contiguous base pairs are modified in the context of an intramolecular duplex. Since the efficiency of modification depends on the thermodynamic stability of a particular duplex (A. Polson and B. Bass, in prep.), short dsRNA molecules of a variety of sequences must be assayed before a minimum length requirement can be firmly established.

Catalytic Mechanism

Like the C to U conversion observed in apoB mRNA editing, an A to I conversion can occur by hydrolytic deamination, and for the unwinding/modifying activity this mechanism has been proven (Fig. 6E) (Polson et al. 1991). The dsRNA unwinding/modifying activity thus uses the same catalytic mechanism as the adenosine and AMP deaminases (Fig. 6E), suggesting that the active sites of these enzymes may be similar. A recent X-ray crystallographic study of adenosine deaminase complexed with a transition-state analog reveals the presence of a previously undetected zinc cofactor bound at the active site (Wilson et al. 1991). During catalysis by adenosine deaminase, the metal ion is proposed to serve as an electrophile, activating the water for attack at C6. The dsRNA unwinding/modifying activity also appears to require zinc (A. Polson and B. Bass, unpubl.). The zinc atom is apparently tightly bound, since it can be chelated with phenanthroline but not EDTA; the latter characteristics apply to the zinc atom of AMP deaminase as well. Although the unwinding/modifying activity shares mechanistic similarities with the adenosine and AMP deaminases, it is a distinct enzyme. The adenosine and AMP deaminases will not act on dsRNA and, similarly, the unwinding/modifying activity will not deaminate nucleosides or nucleotides (Polson et al. 1991). In addition, the transition-state analog, coformycin, a powerful inhibitor of the adenosine and AMP deaminases, will not inhibit the unwinding/modifying activity (Polson et al. 1991). Since coformycin is a nucleotide analog rather than a polynucleotide analog, the latter merely emphasizes that A to I conversion by the unwinding/modifying activity must occur within the context of dsRNA.

It seems possible that all enzymes that deaminate nucleic acid bases use similar catalytic strategies, perhaps even having similar active sites. The recent cloning of a cytidine deaminase, which like the apoB editing activity converts C to U, indicates that this enzyme may share some active-site amino acids with the adenosine and AMP deaminases (Yang et al. 1992). Furthermore, the enzyme that modifies A to I within the anticodon of certain tRNAs, although originally reported to use a ribosyltransferase mechanism (Elliott and Trewyn 1984), has also been shown to use a hydrolytic deamination mechanism (H. Grosjean et al., pers. comm.).

Biological Substrates

A number of cellular RNAs have been implicated as substrates for the unwinding/modifying activity in vivo (Bass et al. 1989; Kimelman and Kirschner 1989; Sharmeen et al. 1991; Sommer et al. 1991). A recent review evaluates each of these putative substrates with regard to the likelihood that they are indeed true substrates for the unwinding/modifying activity (Bass 1993). Although some putative substrates suggest that the unwinding/modifying activity, like other RNA editing activities, acts to alter the primary sequence of an RNA (Sommer et al. 1991), others suggest an expanded role. Specifically, since the unwinding/modifying activity can disrupt AU base pairs, certain putative substrates invoke the unwinding/modifying activity as a means to change secondary rather than primary structure (Sharmeen et al. 1991). Other studies suggest that the unwinding/modifying activity may act in concert with other cellular factors that serve to provide sequence specificity or, in some cases, afford dsRNA protection from modification (Sharmeen et al.

1991; L. Saccomanno and B. Bass, in prep.). Along these lines, it has also been proposed that the substrate specificity observed in vitro may not be exactly mimicked in vivo (Bass 1993). For example, although some requirement for structure seems probable, it is possible that factors exist in the cell that serve to relax the requirement for a completely base-paired molecule. This idea is relevant to the editing observed in gluR channel transcripts, which, as discussed in the previous section, is consistent with an A to I modification.

Plants

Background

RNA editing has been observed in mitochondrion-encoded RNAs of numerous higher plant species, monocots as well as dicots (for review, see Cattaneo 1991; Walbot 1991). Since the first observations of editing in plant mitochondria (Covello and Gray 1989; Gualberto et al. 1989; Hiesel et al. 1989), every mRNA analyzed has shown signs of editing. The discovery of RNA editing in plant mitochondria explains many anomalies previously observed in these organelles. For example, it had been observed that plant mitochondrial genes sometimes encoded CGG triplets at positions known to code for conserved tryptophans (UGG); this observation led to the suggestion that plant mitochondria used an altered genetic code (Fox and Leaver 1981). In addition, the sequence of certain genes suggested nonconservative amino acid replacements in otherwise evolutionarily conserved proteins (Covello and Gray 1989). It now seems clear that these anomalies were merely evidence of RNA editing events; in contrast to the sequence of the mitochondrial genome, the sequence of edited RNAs predicts a universal codon usage and evolutionarily conserved proteins.

The majority of the editing that occurs in plant mitochondria is like that in apoB mRNA and involves C to U substitutions. Exceptions to this include the reverse substitution (U to C; Gualberto et al. 1990; Schuster et al. 1990a) and, curiously, deletions of genomically encoded Cs (Gualberto et al. 1991). Multiple editing events occur in each edited transcript. For example, editing of NADH dehydrogenase subunit 2 mRNA of *Oenothera* mitochondria involves 39 C to U conversions (Binder et al. 1992) whereas subunit 9 of the ATPase (atp9) mRNA of wheat requires 8 C to U editing events (Nowak and Kuck 1990). Except for a few reports of dinucleotide editing (Lamattina and Grienenberger 1991; Wissinger et al. 1991; Yang and Mulligan 1991), most of the editing sites involve a single nucleotide change. As in *Physarum*, the editing sites are spaced at intervals, predominantly in coding regions, but in a few cases within untranslated regions (Hiesel et al. 1989; Yang and Mulligan 1991) and introns (Conklin et al. 1991; Knoop et al. 1991; Wissinger et al. 1991; Yang and Mulligan 1991; Binder et al. 1992). Most analyses have focused on mRNA, but rRNAs of *Oenothera* have been analyzed. In this species, 18S and 5S rRNA appears to be unedited, whereas the 26S rRNA has one C to U and one U to C change (as cited in Walbot 1991).

Plant editing appears to be a posttranscriptional process, and with some exceptions (Gualberto et al. 1990), steady-state populations of total mitochondrial RNAs contain partially edited transcripts (Schuster et al. 1990b; Gualberto et al. 1991; Yang and Mulligan 1991), whereas polysomal RNAs appear to be fully edited (Schuster et al. 1990b; Yang and Mulligan 1991). The partially edited transcripts do not suggest a polarity to the editing process. However, some sites are more often edited in partially edited transcripts, suggesting variations in the efficiency with which sites are edited (Schuster et al. 1990b; Wissinger et al. 1991; Yang and Mulligan 1991).

Partially edited mRNAs have been detected in populations of unspliced as well as spliced RNA, suggesting that there is not a strict temporal requirement for editing before or after splicing (Sutton et al. 1991). However, a requirement for editing prior to splicing, at least in certain cases, is suggested by studies of trans-spliced genes in mitochondria of *Oenothera* and *Petunia*. Specifically, certain trans-splicing events in these species, involving group II introns, may require editing events to create base-pairing interactions required for splicing (Wissinger et al. 1991; Binder et al. 1992).

Editing Site Recognition

No obvious consensus sequence or structure is apparent among various editing sites. In fact, the editing sites of a particular RNA can differ between different plant species, even when flanking sequences are highly conserved (Covello and Gray 1990). An analysis of nucleotide frequencies surrounding editing sites indicates some element of nonrandomness with 43% of the editing sites considered in one study exhibiting a TCR, where C is the edited nucleotide (Covello and Gray 1990). In another analysis, it was noted that editing sites within the wheat cytochrome oxidase subunit 3 mRNA are always preceded by a pyrimidine (Gualberto et al. 1990). Mechanisms analogous to those of kinetoplastid editing have been considered. By aligning sequences surrounding the editing sites of wheat mitochondrial RNAs, families have been noted that could potentially base-pair with the same antisense or gRNA (Gualberto et al.

1990). However, sequences corresponding to putative gRNAs cannot be detected in mitochondrial DNA.

Catalytic Mechanism

The catalytic mechanism of editing in plant mitochondria has not been investigated. The majority of editing events are C to U conversions, as observed in mammalian apoB mRNA, and thus in theory can be explained using mechanisms similar to those of cytidine deaminase (Fig. 6A) or a ribosyltransferase (Fig. 6B). As for many of the editing examples discussed, the C to U conversions have been inferred from cDNA sequences. Thus, the plant substitutions could in fact involve a hypermodified base that could pair with A, such as lysidine (Weber et al. 1990).

A U to C conversion is not chemically straightforward, since a ketone is not a good leaving group. However, U to C conversions have precedence in the biosynthetic reactions catalyzed by CTP synthetase (Fig. 6C). As shown, the reaction requires ATP, which serves to phosphorylate O4, facilitating the subsequent attack by ammonia and release of inorganic phosphate. A caveat to placing plant mitochondrial editing in the substitution editing category is the recent observation of a deletion event (Gualberto et al. 1991). The observation of deletion events emphasizes how little is understood about certain editing mechanisms and raises the possibility that substitution editing may in some cases involve cleavage-ligation events. Alternatively, plants may employ two distinct editing mechanisms, resulting in insertion/deletion as well as substitution events.

In Vitro Systems

Recently, it was demonstrated that wheat mitochondrial extracts can faithfully edit atp9 mRNA in vitro (Araya et al. 1992). Editing in this in vitro system is inhibited by EDTA, and magnesium but not manganese can fulfill the divalent cation requirement. Monovalent cations are not required, and the addition of ATP does not enhance editing. If U to C substitutions occur using mechanisms similar to those of CTP synthetase, as discussed above, an ATP requirement would be predicted. However, editing of atp9 mRNA requires C to U substitutions but not U to C substitutions. Editing in this in vitro system is inhibited by heat, as well as trypsin and micrococcal nuclease, suggesting a requirement for nucleic acid as well as protein factors. Like the observation of deletions, the nucleic acid requirement suggests that plant editing may have similarities to insertion-deletion editing, requiring molecules analogous to gRNAs. It is hoped that continued studies in vitro will resolve questions concerning the mechanism of plant editing.

Chloroplast Editing

Editing has also been observed in two plant chloroplast mRNAs encoded by the genes rpl2 (Hoch et al. 1991) and *psbL* (Kudla et al. 1992). Like the majority of editing events that occur in plant mitochondria, both of these transcripts exhibit C to U substitutions. Furthermore, in both examples, editing serves to create initiation codons from ACG codons. It is not yet clear how widespread RNA editing is among plant chloroplast RNAs.

CONCLUSIONS AND PERSPECTIVES

The many examples of RNA editing discussed in this review emphasize the importance of this type of posttranscriptional processing to the production of functional cellular RNAs. The identification of edited mRNAs, rRNAs, and tRNAs among diverse eukaryotes and in transcripts encoded by mitochondrial, chloroplast, and nuclear genomes suggests that no particular type of RNA is decidedly exempt from RNA editing. This fact has important implications for all molecular biologists. The primary sequence of a mature RNA can no longer be concluded from a genomic sequence, and analyses of RNA or cDNA must be performed before the sequence of a mature RNA is assumed.

Certainly, additional edited transcripts are yet to be identified, and each new example will be interesting in regard to how editing regulates gene expression. However, one of the most interesting aspects of future studies will be the elucidation of the mechanisms behind the editing process. Although for most examples of editing the mechanism is unclear, studies of editing in kinetoplastid RNAs and mammalian apoB mRNA have revealed two distinct mechanisms by which editing can occur. The mechanisms involved in these types of editing are clearly different from each other and cannot completely explain other examples of editing. Although the possibility that each organism may utilize a different editing mechanism is a bit daunting, the mechanistic studies performed to date suggest that editing is not a mysterious process but utilizes principles involved in other aspects of RNA metabolism. Editing as observed in the kinetoplastids is mechanistically similar to RNA splicing, and apoB editing can be thought of as a form of RNA modification. Thus, future studies can be guided by observations made during studies of previously characterized events in RNA metabolism.

Analyses of the editing events observed to date emphasize that an understanding of editing mechanisms requires answers to two questions: (1) What allows a particular sequence to be recognized as an editing site? (2) What are the catalytic events that lead to the covalent changes required? It seems likely that differences among various forms of editing will be more evident in editing site designation rather than in the actual chemistry involved in catalysis. For example, RNAs that require the insertion or deletion of nucleotides, like the kinetoplastids, most likely utilize cleavage-ligation mechanisms similar to one of the various forms of RNA splicing. Transcripts requiring substitution of an encoded nucleotide, such as apoB mRNA, probably utilize chemistry previously observed among enzymes known to catalyze base modifications.

In contrast, it is difficult to apply what is currently known about editing site designation to all examples of editing. The insertion events observed in *Physarum* RNAs exhibit regularly spaced editing sites of single or dinucleotide insertions, which differs rather dramatically from the editing sites of kinetoplasts that occur one right after another and usually involve multiple insertions or deletions. Similarly, it is conceivable that editing site designation for gluR transcripts is similar to that observed in apoB mRNA, but it is difficult to extrapolate the pattern of apoB editing to that observed in plants, which involves multiple substitutions for each edited transcript.

Regardless of differences in the specific details involved in the designation of editing sites, it seems likely that base-pairing interactions will play an important role. Although RNA aficionados do not need to be told the importance of intramolecular structures to RNA metabolism, kinetoplastid editing emphasizes that intermolecular interactions and non-Watson-Crick base pairs can also be important. This use of intermolecular base-pairing interactions, although surprising when discovered, is not unprecedented in RNA metabolism. Intermolecular interactions occur in ribosomal RNAs (Wool 1986) and in the familiar anticodon-codon interaction. Nuclear mRNA splicing requires the participation of small nuclear RNAs (snRNAs), which like gRNAs direct cleavage-ligation reactions; in this case, to enable precise intron removal and exon ligation (Baserga and Steitz, this volume). Studies of kinetoplastid editing emphasize that even base-pairing of low thermodynamic stability involving GU and AC base pairs can be important, making it probable that almost any RNA can form a structure. Base-pairing interactions may form the basis of editing site designation for all examples of editing, and the different patterns of editing may reflect whether intramolecular or intermolecular interactions are used.

A subtle, but common, theme in editing site designation can be observed among those types of editing for which some aspects of mechanism have been determined. In particular, sequences at a slight distance from the editing site are more important for designating the editing site than are those immediately adjacent. In kinetoplastid editing it is a double-stranded structure slightly downstream from the editing site, formed by interactions of a gRNA with an mRNA, that directs editing of more 5′ nucleotides. In apoB mRNA editing it is also downstream sequences that are most important, but the particular feature of the downstream sequence that allows editing site designation is unclear. It is possible that these important sequences also form a double-stranded structure. In this regard, the in vitro substrate specificity of the dsRNA unwinding/modifying activity suggests that double-stranded structures are crucial for editing site designation within its bona fide in vivo substrates.

Very little is known about the proteins involved in RNA editing, and this will also be an interesting aspect of future studies. As discussed, kinetoplastid and apoB mRNA editing may occur in large protein complexes similar to the spliceosomes of nuclear mRNA splicing. Since the thermodynamic stability of gRNA-mRNA interactions is often low, proteins may serve to stabilize these intermolecular interactions. Furthermore, for a number of the examples of editing discussed, sequences surrounding editing sites are essentially identical to other sequences that are not edited. In these cases, proteins may serve to provide an additional level of discrimination. Given the mechanistic similarities between the various forms of RNA editing and previously characterized biochemical reactions, it is possible that proteins involved in editing may turn out to be previously characterized enzymes that play multiple roles. Regardless, future studies in the new and important field of RNA editing will certainly prove interesting.

ACKNOWLEDGMENTS

I am grateful to B. Blum, L. Chan, J. Grieninberger, H. Grosjean, S. Hajduk, M. Harris, D. Miller, J. Scott, P. Seeburg, L. Simpson, H. Smith, B. Sollner-Webb, and K. Stuart for communicating unpublished results; and to B. Blum, L. Chan, H. Grosjean, S. Hajduk, J. Scott, P. Seeburg, L. Simpson, and B. Sollner-Webb for reading this chapter and offering helpful suggestions.

REFERENCES

Abraham, J.M., J.E. Feagin, and K. Stuart. 1988. Characterization of cytochrome c oxidase III transcripts that are edited only in the 3′ region. *Cell* **55:** 267–272.

Apostol, B.L., S.K. Westaway, J. Abelson, and C.L. Greer. 1991. Deletion analysis of a multifunctional yeast tRNA ligase polypeptide. *J. Biol. Chem.* **266:** 7445−7455.

Araya, A., C. Domec, D. Begu, and S. Litvak. 1992. An in vitro system for the editing of ATP synthase subunit 9 mRNA using wheat mitochondrial extracts. *Proc. Natl. Acad. Sci.* **89:** 1040−1044.

Backus, J.W. and H.C. Smith. 1991. Apolipoprotein B mRNA sequences 3′ of the editing site are necessary and sufficient for editing and editosome assembly. *Nucleic Acids Res.* **19:** 6781−6786.

Bakalara, N., A.M. Simpson, and L. Simpson. 1989. The Leishmania kinetoplast-mitochondrion contains terminal uridylyltransferase and RNA ligase activities. *J. Biol. Chem.* **264:** 18679−18686.

Bass, B.L. 1991. RNA editing. *Physarum*—C the difference? *Nature* **349:** 370−371.

———. 1992. The double-stranded RNA unwinding/modifying activity. In *Antisense RNA and DNA* (ed. J.A.H. Murray), pp. 159−174. Wiley/Liss, New York.

———. 1993. The dsRNA unwinding/modifying activity: Fact and fiction. *Semin. Dev. Biol.* **3:** 425−433.

Bass, B.L. and H. Weintraub. 1987. A developmentally regulated activity that unwinds RNA duplexes. *Cell* **48:** 607−613.

———. 1988. An unwinding activity that covalently modifies its double-stranded RNA substrate. *Cell* **55:** 1089−1098.

Bass, B.L., H. Weintraub, R. Cattaneo, and M.A. Billeter. 1989. Biased hypermutation of viral RNA genomes could be due to unwinding/modification of double-stranded RNA. *Cell* **56:** 331.

Beier, H., M.C. Lee, T. Sekiya, Y. Kuchino, and S. Nishimura. 1992. Two nucleotides next to the anticodon of cytoplasmic rat tRNA^{asp} are likely generated by RNA editing. *Nucleic Acids Res.* **20:** 2679−2683.

Benne, R., J. Van den Burg, J.P. Brakenhoff, P. Sloof, J.H. Van Bloom, and M.C. Tromp. 1986. Major transcript of the frameshifted coxII gene from trypanosome mitochondria contains four nucleotides that are not encoded in the DNA. *Cell* **46:** 819−826.

Bhat, G.J., D.J. Koslowsky, J.E. Feagin, B.L. Smiley, and K. Stuart. 1990. An extensively edited mitochondrial transcript in kinetoplastids encodes a protein homologous to ATPase subunit 6. *Cell* **61:** 885−894.

Binder, S., A. Marchfelder, A. Brennicke, and B. Wissinger. 1992. RNA editing in trans-splicing intron sequences of nad2 mRNAs in Oenothera mitochondria. *J. Biol. Chem.* **267:** 7615−7623.

Björk, G.R., J.U. Ericson, C.E.D. Gustafsson, T.G. Hagervall, Y.H. Jönsson, and P.M. Wikström. 1987. Transfer RNA modification. *Annu. Rev. Biochem.* **56:** 263−287.

Blum, B. and L. Simpson. 1990. Guide RNAs in kinetoplastid mitochondria have a non-encoded 3′ oligo(U) tail involved in recognition of the preedited region. *Cell* **62:** 391−397.

———. 1992. Formation of gRNA/mRNA chimeric molecules in vitro, the initial step of RNA editing, is dependent on an anchor sequence. *Proc. Natl. Acad. Sci.* (in press).

Blum, B., N. Bakalara, and L. Simpson. 1990. A model for RNA editing in kinetoplastid mitochondria: "Guide" RNA molecules transcribed from maxicircle DNA provide the edited information. *Cell* **60:** 189−198.

Blum, B., N.R. Sturm, A.M. Simpson, and L. Simpson. 1991. Chimeric gRNA-mRNA molecules with oligo(U) tails covalently linked at sites of RNA editing suggest that U addition occurs by transesterification. *Cell* **65:** 543−550.

Bostrom, K., Z. Garcia, K.S. Poksay, D.F. Johnson, A.J. Lusis, and T.L. Innerarity. 1990. Apolipoprotein B mRNA editing. Direct determination of the edited base and occurrence in non-apolipoprotein B-producing cell lines. *J. Biol. Chem.* **265:** 22446–22452.

Bostrom, K., S.J. Lauer, K.S. Poksay, Z. Garcia, J.M. Taylor, and T.L. Innerarity. 1989. Apolipoprotein B48 RNA editing in chimeric apolipoprotein EB mRNA. *J. Biol. Chem.* **264:** 15701–15708.

Braunstein, A.E. 1973. Amino group transfer. In *The enzymes* (ed. P.D. Boyer), vol. 9, pp. 379–481. Academic Press, New York.

Cattaneo, R. 1991. Different types of messenger RNA editing. *Annu. Rev. Genet.* **25:** 71–88.

Cech, T.R. 1987. The chemistry of self-splicing RNA and RNA enzymes. *Science* **236:** 1532–1539.

———. 1991. RNA Editing: World's smallest introns? *Cell* **64:** 667–669.

Chan, L. 1992. RNA Editing: Exploring one mode with apolipoprotein B mRNA. *BioEssays* (in press).

Chen, S.-H., X.X. Li, W.S. Liao, J.H. Wu, and L. Chan. 1990. RNA editing of apolipoprotein B mRNA. Sequence specificity determined by in vitro coupled transcription editing. *J. Biol. Chem.* **265:** 6811–6816.

Chen, S.-H., G. Habib, C.-Y. Yang, Z.-W. Gu, B.R. Lee, S.-A. Weng, S.R. Silberman, S.-J. Cai, J.P. Deslypere, and M. Rosseneu. 1987. Apolipoprotein B-48 is the product of a messenger RNA with an organ-specific in-frame stop codon. *Science* **238:** 363–366.

Conklin, P.L., R.K. Wilson, and M.R. Hanson. 1991. Multiple *trans*-splicing events are required to produce a mature nad1 transcript in a plant mitochondrion. *Genes Dev.* **5:** 1407–1415.

Covello, P.S. and M.W. Gray. 1989. RNA editing in plant mitochondria. *Nature* **341:** 662–666.

———. 1990. Differences in editing at homologous sites in messenger RNAs from angiosperm mitochondria. *Nucleic Acids Res.* **18:** 5189–5196.

Davidson, N.O., L.M. Powell, S.C. Wallis, and J. Scott. 1988. Thyroid hormone modulates the introduction of a stop codon in rat liver apolipoprotein B messenger RNA. *J. Biol. Chem.* **263:** 13482–13485.

Davies, M.S., S.C. Wallis, D.M. Driscoll, J.K. Wynne, G.W. Williams, L.M. Powell, and J. Scott. 1989. Sequence requirements for apolipoprotein B RNA editing in transfected rat hepatoma cells. *J. Biol. Chem.* **264:** 13395–13398.

Decker, C.J. and B. Sollner-Webb. 1990. RNA editing involves indiscriminate U changes throughout precisely defined editing domains. *Cell* **61:** 1001–1011.

Diamond, A.M., Y. Montero-Puerner, B.J. Lee, and D. Hatfield. 1990. Selenocysteine inserting tRNAs are likely generated by tRNA editing. *Nucleic Acids Res.* **18:** 6727.

Driscoll, D.M. and E. Casanova. 1990. Characterization of the apolipoprotein B mRNA editing activity in enterocyte extracts. *J. Biol. Chem.* **265:** 21401–21403.

Driscoll, D.M., J.K. Wynne, S.C. Wallis, and J. Scott. 1989. An in vitro system for the editing of apolipoprotein B mRNA. *Cell* **58:** 519–525.

Elliott, M.S. and R.W. Trewyn. 1984. Inosine biosynthesis in transfer RNA by an enzymatic insertion of hypoxanthine. *J. Biol. Chem.* **259:** 2407–2410.

Feagin, J.E., J.M. Abraham, and K. Stuart. 1988. Extensive editing of the cytochrome c oxidase III transcript in *Trypanosoma brucei. Cell* **53:** 413–422.

Fox, T.D. and C.J. Leaver. 1981. The *Zea mays* mitochondrial gene coding cytochrome oxidase subunit II has an intervening sequence and does not contain TGA codons. *Cell* **26:** 315–323.

Frick, L., C. Yang, V.E. Marquez, and R. Wolfenden. 1989. Binding of pyrimidin-2-one ribonucleoside by cytidine deaminase as the transition-state analogue 3,4-dihydrouridine and the contribution of the 4-hydroxyl group to its binding affinity.

Biochemistry **28:** 9423–9430.

Greeve, J., N. Navaratnam, and J. Scott. 1991. Characterization of the apolipoprotein B mRNA editing enzyme: No similarity to the proposed mechanism of RNA editing in kinetoplastid protozoa. *Nucleic Acids Res.* **19:** 3569–3576.

Gualberto, J.M., J.H. Weil, and J.M. Grienenberger. 1990. Editing of the wheat coxIII transcript: Evidence for twelve C to U and one U to C conversions and for sequence similarities around editing sites. *Nucleic Acids Res.* **18:** 3771–3776.

Gualberto, J.M., G. Bonnard, L. Lamattina, and J.M. Grienenberger. 1991. Expression of the wheat mitochondrial nad3-rps12 transcription unit: Correlation between editing and mRNA maturation. *Plant Cell* **3:** 1109–1120.

Gualberto, J.M., L. Lamattina, G. Bonnard, J.H. Weil, and J. M. Grienenberger. 1989. RNA editing in wheat mitochondria results in the conservation of protein sequences. *Nature* **341:** 660–662.

Hajduk, S.L., M.E. Harris, and V.W. Pollard. 1993. RNA editing in protozoan mitochondria. *FASEB J.* (in press).

Harris, M.E. and S.L. Hajduk. 1992. Kinetoplastid RNA editing: In vitro formation of cytochrome b gRNA-mRNA chimeras from synthetic substrate RNAs. *Cell* **68:** 1091–1099.

Harris, M.E., D.R. Moore, and S.L. Hajduk. 1990. Addition of uridines to edited RNAs in trypanosome mitochondria occurs independently of transcription. *J. Biol. Chem.* **265:** 11368–11376.

Harris, M., C. Decker, B. Sollner-Webb, and S. Hajduk. 1992. Specific cleavage of pre-edited mRNAs in trypanosome mitochondrial extracts. *Mol. Cell. Biol.* **12:** 2591–2598.

Hedstrom, L. and C.C. Wang. 1990. Mycophenolic acid and thiazole adenine dinucleotide inhibition of *Tritrichomonas foetus* inosine 5′ monophosphate dehydrogenase: Implications on enzyme mechanism. *Biochemistry* **29:** 849–854.

Hiesel, R., B. Wissinger, W. Schuster, and A. Brennicke. 1989. RNA editing in plant mitochondria. *Science* **246:** 1632–1634.

Hoch, B., R.M. Maier, K. Appel, G.L. Igloi, and H. Kössel. 1991. Editing of a chloroplast mRNA by creation of an initiation codon. *Nature* **353:** 178–180.

Hodges, P. and J. Scott. 1992. Apolipoprotein B mRNA editing: A new tier for the control of gene expression. *Trends Biochem. Sci.* **17:** 77–81.

Hodges, P.E., N. Navaratnam, J.C. Greeve, and J. Scott. 1991. Site-specific creation of uridine from cytidine in apolipoprotein B mRNA editing. *Nucleic Acids Res.* **19:** 1197–1201.

Kimelman, D. and M.W. Kirschner. 1989. An antisense mRNA directs the covalent modification of the transcript encoding fibroblast growth factor in *Xenopus* oocytes. *Cell* **59:** 687–696.

Knoop, V., W. Schuster, B. Wissinger, and A. Brennicke. 1991. Trans splicing integrates an exon of 22 nucleotides into the nad5 mRNA in higher plant mitochondria. *EMBO J.* **10:** 3483–3493.

Koslowsky, D.J., G.J. Bhat, L.K. Read, and K. Stuart. 1991. Cycles of progressive realignment of gRNA with mRNA in RNA editing. *Cell* **67:** 537–546.

Koslowsky, D.J., H.U. Goringer, T.H. Morales, and K. Stuart. 1992a. In vitro guide RNA/mRNA chimaera formation in *Trypanosoma brucei* RNA editing. *Nature* **356:** 807–809.

Koslowsky, D.J., G.R. Riley, J.E. Feagin, and K. Stuart. 1992b. Guide RNAs for transcripts with developmentally regulated RNA editing are present in both life cycle stages of *Trypanosoma brucei. Mol. Cell. Biol.* **12:** 2043–2049.

Koslowsky, D.J., G.J. Bhat, A.L. Perrollaz, J.E. Feagin, and K. Stuart. 1990. The MURF3 gene of *T. brucei* contains multiple domains of extensive editing and is homologous to a subunit of NADH dehydrogenase. *Cell* **62:** 901–911.

Kudla, J., G.L. Igloi, M. Metzlaff, R. Hagemann, and H. Kössel. 1992. RNA editing in tobacco chloroplasts leads to the formation of a translatable psbL mRNA by a C to U substitution within the initiation codon. *EMBO J.* **11:** 1099–1103.

Kurz, L.C., L. Moix, M.C. Riley, and C. Frieden. 1992. The rate of formation of transition-state analogues in the active site of adenosine deaminase is encounter-controlled: Implications for the mechanism. *Biochemistry* **31:** 39–48.

Lamattina, L. and J.M. Grienenberger. 1991. RNA editing of the transcript coding for subunit 4 of NADH dehydrogenase in wheat mitochondria: Uneven distribution of the editing sites among the four exons. *Nucleic Acids Res.* **19:** 3275–3282.

Lau, P.P., S.H. Chen, J.C. Wang, and L. Chan. 1990. A 40 kilodalton rat liver nuclear protein binds specifically to apolipoprotein B mRNA around the RNA editing site. *Nucleic Acids Res.* **18:** 5817–5821.

Lau, P.P., W.J. Xiong, H.J. Zhu, S.H. Chen, and L. Chan. 1991. Apolipoprotein B mRNA editing is an intranuclear event that occurs posttranscriptionally coincident with splicing and polyadenylation. *J. Biol. Chem.* **266:** 20550–20554.

Lewis, D.A. and J.J. Villafranca. 1989. Investigation of the mechanism of CTP synthetase using rapid quench and isotope partitioning methods. *Biochemistry* **28:** 8454–8459.

Maden, B.E.H. 1990. The numerous modified nucleotides in eukaryotic ribosomal RNA. *Prog. Nucleic Acid Res.* **39:** 241–303.

Mahendran, R., M.R. Spottswood, and D.L. Miller. 1991. RNA editing by cytidine insertion in mitochondria of Physarum polycephalum. *Nature* **349:** 434–438.

Maslov, D.A. and L. Simpson. 1992. The polarity of editing within a multiple gRNA-mediated domain is due to formation of anchors for upstream gRNAs by downstream editing. *Cell* **70:** 459–467.

Merkler, D.J., M. Brenowitz, and V.L. Schramm. 1990. The rate constant describing slow-onset inhibition of yeast AMP deaminase by coformycin analogues is independent of inhibitor structure. *Biochemistry* **29:** 8358–8364.

Miller, D.L., R. Mahendran, M. Spottswood, M. Ling, S. Wang, N. Yang, and H. Costandy. 1993. RNA editing in mitochondria of *Physarum polycephalum*. In *RNA editing: The alteration of protein coding sequences of RNA* (ed. R. Benne). Simon & Schuster, New York. (In press.)

Navaratnam, N., R. Shah, D. Patel, V. Fay, and J. Scott. 1992. Apolipoprotein B mRNA editing is associated with UV crosslinking of a 60 kDa protein to the editing site. *Proc. Natl. Acad. Sci.* (in press).

Navaratnam, N., D. Patel, R.R. Shah, J.C. Greeve, L.M. Powell, T.J. Knott, and J. Scott. 1991. An additional editing site is present in apolipoprotein B mRNA. *Nucleic Acids Res.* **19:** 1741–1744.

Nishikura, K., C. Yoo, U. Kim, J.M. Murray, P.A. Estes, F.E. Cash, and S.A. Liebhaber. 1991. Substrate specificity of the dsRNA unwinding/modifying activity. *EMBO J.* **10:** 3523–3532.

Nishimura, S. 1983. Structure, biosynthesis, and function of queuosine in transfer RNA. *Prog. Nucleic Acid Res. Mol. Biol.* **28:** 49–71.

Nowak, C. and U. Kuck. 1990. RNA editing of the mitochondrial atp9 transcript from wheat. *Nucleic Acids Res.* **18:** 7164.

Phizicky, E.M. and C.L. Greer. 1993. Pre-tRNA splicing: Variation on a theme or exception to the rule? *Trends Biochem. Sci.* **18:** 31–34.

Pollard, V.W. and S.L. Hajduk. 1991. *Trypanosoma equiperdum* minicircles encode three distinct primary transcripts which exhibit guide RNA characteristics. *Mol. Cell. Biol.* **11:** 1668–75.

Pollard, V.W., M.E. Harris, and S.L. Hajduk. 1992. Native mRNA editing complexes from *Trypanosoma brucei* mitochondria. *EMBO J.* **11:** (in press).

Pollard, V.W., S.P. Rohrer, E.F. Michelotti, K. Hancock, and S.L. Hajduk. 1990. Organization of minicircle genes for guide RNAs in *Trypanosoma brucei. Cell* **63:** 783–790.

Polson, A.G., P.F. Crain, S.C. Pomerantz, J.A. McCloskey, and B.L. Bass. 1991. The mechanism of adenosine to inosine conversion by the double-stranded RNA unwinding/modifying activity: A high-performance liquid chromatography-mass spectrometry analysis. *Biochemistry* **30:** 11507–11514.

Powell, L.M., S.C. Wallis, R.J. Pease, Y.H. Edwards, T.J. Knott, and J. Scott. 1987. A novel form of tissue-specific RNA processing produces apolipoprotein-B48 in intestine. *Cell* **50:** 831–840.

Read, L.A., R.A. Coreel, and K. Stuart. 1992a. Chimeric and truncated RNAs in *Trypanosoma brucei* suggest transesterification at non-consecutive sites during RNA editing. *Nucleic Acids Res.* **20:** 2341–2347.

Read, L.K., P.J. Myler, and K. Stuart. 1992b. Extensive editing of both processed and preprocessed maxicircle CR6 transcripts in *Trypanosoma brucei. J. Biol. Chem.* **267:** 1123–1128.

Rebagliati, M.R. and D.A. Melton. 1987. Antisense RNA injections in fertilized frog eggs reveal an RNA duplex unwinding activity. *Cell* **48:** 599–605.

Schuster, W., R. Hiesel, B. Wissinger, and A. Brennicke. 1990a. RNA editing in the cytochrome b locus of the higher plant *Oenothera berteriana* includes a U-to-C transition. *Mol. Cell. Biol.* **10:** 2428–2431.

Schuster, W., B. Wissinger, M. Unseld, and A. Brennicke. 1990b. Transcripts of the NADH-dehydrogenase subunit 3 gene are differentially edited in *Oenothera* mitochondria. *EMBO J.* **9:** 263–269.

Shah, R.R., T.J. Knott, J.E. Legros, N. Navaratnam, J.C. Greeve, and J. Scott. 1991. Sequence requirements for the editing of apolipoprotein B mRNA. *J. Biol. Chem.* **266:** 16301–16304.

Sharmeen, L., B. Bass, N. Sonenberg, H. Weintraub, and M. Groudine. 1991. Tat-dependent adenosine-to inosine modification of wild-type transactivation response RNA. *Proc. Natl. Acad. Sci.* **88:** 8096–8100.

Simpson, A.M., N. Bakalara, and L. Simpson. 1992. A ribonuclease activity is activated by heparin or by digestion with proteinase K in mitochondrial extracts of *Leishmania tarentolae. J. Biol. Chem.* **267:** 6782–6788.

Singer, B. and M. Kroger. 1979. Participation of modified nucleosides in translation and transcription. *Prog. Nucleic Acid Res. Mol. Biol.* **23:** 151–194.

Skeiky, Y.A.W. and K. Iatrou. 1991. Developmental regulation of covalent modification of double-stranded RNA during silkmoth oogenesis. *J. Mol. Biol.* **218:** 517–527.

Smith, H.C., S.R. Kuo, J.W. Backus, S.G. Harris, C.E. Sparks, and J.D. Sparks. 1991. In vitro apolipoprotein B mRNA editing: Identification of a 27S editing complex. *Proc. Natl. Acad. Sci.* **88:** 1489–1493.

Sollner-Webb, B. 1991. RNA editing. *Curr. Opin. Cell Biol.* **3:** 1056–1061.

———. 1992. RNA editing. Guides to experiments. *Nature* **356:** 743–744.

Sommer, B., M. Köhler, R. Sprengel, and P.H. Seeburg. 1991. RNA editing in brain controls a determinant of ion flow in glutamate-gated channels. *Cell* **67:** 11–19.

Stuart, K. 1991a. RNA editing in trypanosomatid mitochondria. *Annu. Rev. Microbiol.* **45:** 327–344.

———. 1991b. RNA editing in mitochondrial mRNA of trypanosomatids. *Trends*

Biochem. Sci. **16:** 68−72.

Sturm, N.R. and L. Simpson. 1990a. Partially edited mRNAs for cytochrome b and sub-unit III of cytochrome oxidase from *Leishmania tarentolae* mitochondria: RNA editing intermediates. *Cell* **61:** 871−878.

―――. 1990b. Kinetoplast DNA minicircles encode guide RNAs for editing of cytochrome oxidase subunit III mRNA. *Cell* **61:** 879−884.

Sturm, N.R., D.A. Maslov, B. Blum, and L. Simpson. 1992. Generation of unexpected editing patterns in *Leishmania tarentolae* mitochondrial mRNAs: Misediting produced by misguiding. *Cell* **70:** 469−476.

Sutton, C.A., P.L. Conklin, K.D. Pruitt, and M.R. Hanson. 1991. Editing of pre-mRNAs can occur before *cis-* and *trans*-splicing in Petunia mitochondria. *Mol. Cell. Biol.* **11:** 4274−4277.

Tennyson, G.E., C.A. Sabatos, K. Higuchi, N. Meglin, and H.B.J. Brewer. 1989. Expression of apolipoprotein B mRNAs encoding higher- and lower-molecular weight isoproteins in rat liver and intestine. *Proc. Natl. Acad. Sci.* **86:** 500−504.

Tuck, M.T. 1992. The formation of internal 6-methyladenine residues in eucaryotic messenger RNA. *Int. J. Biochem.* **24:** 379−386.

Uhlenbeck, O.C. and R.I. Gumport. 1982. T4 RNA Ligase. In *The enzymes* (ed. P.D. Boyer), pp. 31−58. Academic Press, New York.

Van der Spek, H., G-J. Arts, R.R. Zwaal, J. Van den Burg, P. Sloof, and R. Benne. 1991. Conserved genes encode guide RNAs in mitochondria of *Crithidia fasciculata*. *EMBO J.* **10:** 1217−1224.

Van der Spek, H., D. Speijer, G-J. Arts, J. Van den Burg, H. Van Steeg, P. Sloof, and R. Benne. 1990. RNA editing in transcripts of the mitochondrial genes of the insect trypanosome *Crithidia fasciculata*. *EMBO J.* **9:** 257−262.

Wagner, R.W. and K. Nishikura. 1988. Cell cycle expression of RNA duplex unwindase activity in mammalian cells. *Mol. Cell. Biol.* **8:** 770−777.

Wagner, R.W., J.E. Smith, B.S. Cooperman, and K. Nishikura. 1989. A double-stranded RNA unwinding activity introduces structural alterations by means of adenosine to inosine conversions in mammalian cells and *Xenopus* eggs. *Proc. Natl. Acad. Sci.* **86:** 2647−2651.

Wagner, R.W., C. Yoo, L. Wrabetz, J. Kamholz, J. Buchhalter, N.F. Hassan, K. Khalili, S.U. Kim, B. Perussia, and F.A. McMorris. 1990. Double-stranded RNA unwinding and modifying activity is detected ubiquitously in primary tissues and cell lines. *Mol. Cell. Biol.* **10:** 5586−5590.

Walbot, V. 1991. RNA editing fixes problems in plant mitochondrial transcripts. *Trends Genet.* **7:** 37−39.

Weber, F., A. Dietrich, J.-H. Weil, and L. Maréchal-Drouard. 1990. A potato mitochondrial isoleucine tRNA is coded for by a mitochondrial gene possessing a methionine anticodon. *Nucleic Acids Res.* **18:** 5027−5030.

Wilson, D.K., F.B. Rudolph, and F.A. Quiocho. 1991. Atomic structure of adenosine deaminase complexed with a transition-state analog: Understanding catalysis and immunodeficiency mutations. *Science* **252:** 1278−1284.

Wissinger, B., W. Schuster, and A. Brennicke. 1991. *Trans* splicing in *Oenothera* mitochondria: Nad1 mRNAs are edited in exon and *trans*-splicing group II intron sequences. *Cell* **65:** 473−482.

Wool, I.G. 1986. Studies of the structure of eukaryotic (mammalian) ribosomes. In *Structure, function and genetics of ribosomes* (ed. B. Hardesty and G. Kramer), pp. 391−411. Springer-Verlag, New York.

Xiong, W., E. Zsigmond, A.M.J. Gotto, L.W. Reneker, and L. Chan. 1992. Transgenic mice expressing full-length human apolipoprotein B-100: Full-length human apolipoprotein B mRNA is essentially not edited in mouse intestine or liver. *J. Biol. Chem.* **267:** 21412–21420.

Yang, A.J. and R.M. Mulligan. 1991. RNA editing intermediates of cox2 transcripts in maize mitochondria. *Mol. Cell. Biol.* **11:** 4278–4281.

Yang, C., D. Carlow, R. Wolfenden, and S.A. Short. 1992. Cloning and nucleotide sequence of the *Escherichia coli* cytidine deaminase (ccd) gene. *Biochemistry* **31:** 4168–4174.

Zalkin, H. and J.E. Dixon. 1992. De novo purine biosynthesis. *Prog. Nucleic Acid Res. Mol. Biol.* **42:** 259–287.

16
Evolution of Functional Structures of RNA

Jun-ichi Tomizawa
National Institute of Genetics, Mishima
Shizuoka-ken 411, Japan

Evolution of new functions for RNA has been crucial to the history of life. New functions could have evolved before the DNA era in RNA molecules capable of replicating or, later, in RNA molecules specified by a DNA genome. Despite the importance of this process during molecular evolution, no one has undertaken a systematic analysis to elucidate the principles and mechanisms involved. Because appearance of a new function would have to be preceded by formation of a new structure, insight can be gained by studying relevant molecular architecture in contemporary RNA. In this chapter, I discuss ways in which certain underlying structures of RNA could evolve new functions.

Many RNA molecules are multifunctional. Any such molecule has probably evolved through sequential addition of domains that have favorable selective effects on the molecule or on the cell carrying it. Each gained function would represent one manifestation of the basic mechanism of evolution within an RNA species. This mechanism can be approached by analysis of multifunctional RNA molecules only, because one cannot imagine the precursors of a monofunctional molecule.

One of the best-studied examples of multifunctional RNA is the primer for DNA replication of plasmid ColE1 in *Escherichia coli*. This 555-nucleotide RNA must form a specific folded structure before it can serve as a primer; it consists of four functional domains, each composed of one, two, or three stem-loops. A number of other plasmid species replicate by means of primer RNAs with nucleotide sequences and structures similar to those in ColE1, so a common ancestor is likely. The evolutionary process leading to the ancestral molecule is deduced from the effects of various mutations on the structure and function of the ColE1 primer RNA and from comparisons of structures among the related primers. The ancestral molecule probably evolved by sequential addition of stem-loop structures on which new functions arose.

The RNA World
© 1993 Cold Spring Harbor Laboratory Press 0-87969-380-0/93 $5 + .00

Formation of a functional primer RNA is regulated by an RNA that is antisense to the primer RNA: The antisense RNA binds to the primer-precursor RNA and prevents primer formation by inhibiting appropriate folding of the molecule. Basic mechanisms of general importance for formation of functional structures in RNA can be deduced by studying this binding reaction. As the antisense RNAs of ColE1-related plasmids are similar but yet specific for each species, comparison of the structures of various antisense RNAs suggests a unique mechanism for diversification. Thus, structural and kinetic analyses of primer formation, and the processes regulating it, could provide a foundation for thinking about how functional structures have evolved in RNA.

First, I describe mechanisms of formation of the ColE1 RNA primer and its regulation by antisense RNA. After presenting general characteristics of formation of folded RNA structures, I discuss evolution of functional domains and propose a scheme for evolution of one of many RNA structures whose formation is triggered by pairing of short inverted repeat sequences. I also suggest a route for diversification of functional structures by a novel mechanism that may assort a stem unit and a loop unit.

OVERVIEW ON ColE1 PRIMER FORMATION AND ITS REGULATION

Closed circular ColE1 DNA can replicate in an extract made from *E. coli* cells (Sakakibara and Tomizawa 1974). The in vitro replication is initiated from one of a few consecutive nucleotides located at a unique position on the DNA (the origin of DNA replication) (Tomizawa et al. 1977) by sequential actions of three enzymes: DNA-dependent RNA polymerase, RNase H, and DNA polymerase I (Itoh and Tomizawa 1979). Transcription that is initiated 555 nucleotides upstream of the origin yields products (RNA II) that provide the primer precursors (Fig. 1, step 1) (Itoh and Tomizawa 1980). Initially, the elongating RNA II transcript separates from the template DNA strand, except for a short region where the synthesis is in progress, as is usually the case in transcription (step 2). However, when RNA polymerase nears the origin, most transcripts start to form persistent hybrids with the template DNA (steps 3 and 4). A minor fraction of the transcripts does not form such a hybrid and continues to separate from the template DNA (step 3, in parentheses). The persistently hybridized transcript is cleaved preferentially at the origin by RNase H (step 5). The cleaved RNA is then used as a primer for DNA synthesis by DNA polymerase I (step 6). Only the nascent transcript that is synthesized on a template DNA molecule can form a persistent hybrid with the DNA molecule and can be processed and used as the primer.

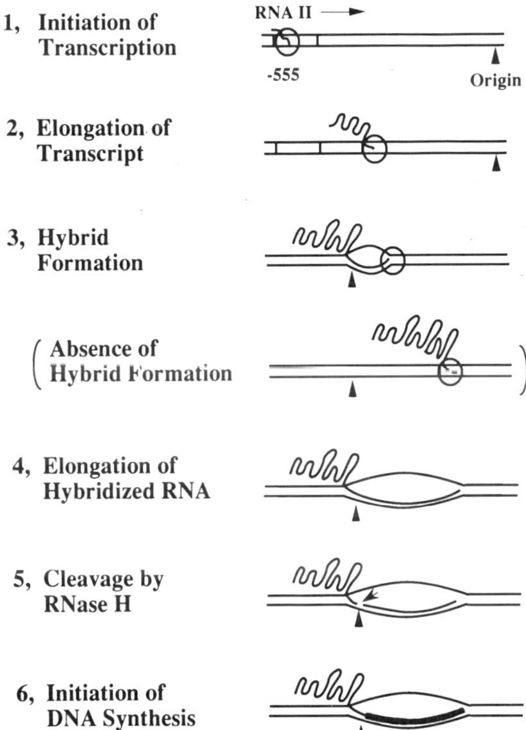

Figure 1 Schematic illustration of the processes of ColE1 primer formation. Straight lines represent double strands of DNA; wavy lines, RNA transcripts; small circles, RNA polymerases. The thick line represents newly synthesized DNA strands.

Once separated from the template DNA, RNA II is inactive as the primer or primer precursor.

Synthesis of the lagging strand is initiated by a replisome, a complex of replication proteins of the host bacteria, on a single-stranded region exposed by the initial DNA synthesis (Minden and Marians 1985; Dasgupta et al. 1987). Subsequent unidirectional DNA synthesis on both strands of the DNA is carried out by a mechanism involving replisomes.

Although normal replication of ColE1 DNA requires the three enzymes noted above, the plasmid can also replicate quite well without RNase H (Naito et al. 1984) or both RNase H and DNA polymerase I (Kogoma 1984). The replication still depends on synthesis of RNA (Naito and Uchida 1986) that hybridizes with the template DNA (Masukata et al. 1987). The single-stranded region exposed by the hybridization of RNA II to the other strand (step 4 of Fig. 1) serves as the template DNA for synthesis of the lagging strand (Masukata et al. 1987). This mechanism of initiation of DNA replication, called "transcriptional activation," is inhibited by RNase H, since it requires the hybridized portion of RNA II that extends beyond the origin.

Studies on the maintenance of various plasmids in bacteria have defined two possibly related regulatory phenomena: copy number control, which fixes the number of plasmid copies per cell, and incompatibility, which prevents stable maintenance of two closely related plasmids. Both of these properties are expressed by a small plasmid that contains a ColE1-derived segment composed essentially of the region which encodes RNA II (Ohmori and Tomizawa 1979; Tomizawa et al. 1981). This region encodes, in addition, an RNA of about 110 nucleotides, named RNA I (Morita and Oka 1979), that is antisense to the 5'-end region of RNA II (Fig. 2). Neither transcript encodes a protein.

Systematic genetic studies showed that mutations in the region that specifies both RNA I and RNA II affect the plasmid copy number and incompatibility (Lacatena and Cesareni 1981; Tomizawa and Itoh 1981). In vitro experiments show that RNA I inhibits formation of a persistent hybrid between RNA II and the template DNA, thus preventing primer formation (Tomizawa and Itoh 1981; Tomizawa et al. 1981). These studies established that RNA I is the specific inhibitor of primer forma-

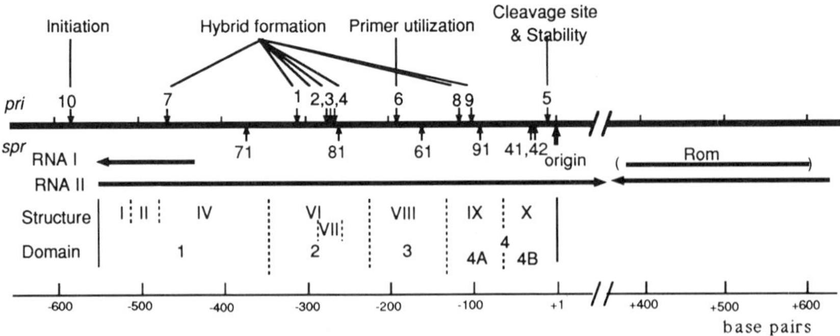

Figure 2 Map of the origin region of ColE1. The thick line indicates the region around the origin of ColE1 DNA. The numbers in the small letters at the bottom are the distances from the origin. The minus sign indicates the upstream of the origin and the plus sign its downstream. Thin arrows are transcripts from the region. The line in brackets is the region that encodes Rom. Positions of relevant *pri* and *spr* mutations are shown. The function affected by each *pri* mutation is presented at the top. The first digit number of the name of an *spr* mutation shows the name of a *pri* mutation that it suppresses. The positions of stem-loop structures and functional domains of primer RNA described below in the text are included.

tion and provided the first evidence for biological regulation by an antisense RNA. To inhibit formation of the hybrid, RNA I must interact with the nascent RNA II transcript well before the RNA polymerase reaches the replication origin (Tomizawa 1986). Therefore, knowledge of the kinetics of this RNA-RNA interaction is critically important for understanding the mechanism of regulation of ColE1 DNA replication. In vivo replication of ColE1 plasmid fits reasonably well with an expectation based on the rates of synthesis and degradation of RNA I and RNA II and the rate constant of their binding (Brenner and Tomizawa 1991; Brendel and Perelson 1993).

ColE1 encodes another factor that reduces the plasmid copy number. A region several hundred base pairs downstream from the origin (Twigg and Sherratt 1980) specifies a transcript (Brenner and Tomizawa 1989) that encodes a protein of 63 amino acids, called Rom (or Rop) (Fig. 2) (Cesareni et al. 1982; Som and Tomizawa 1983). Rom enhances the inhibitory activity of RNA I by stabilizing the product formed by interaction of RNA I with RNA II, as discussed below. Rom is not required for maintenance of ColE1 in bacteria under normal conditions but is required if the rate of binding of RNA I to RNA II is reduced below a certain threshold level by a mutation (Tomizawa and Som 1984).

Various aspects of the mechanisms of formation of the ColE1 RNA primer and its regulation by the antisense RNA have been reviewed (Polisky 1988; Cesareni et al. 1991; Eguchi et al. 1991). Subsequent to our discovery of regulation of ColE1 plasmid replication by an antisense RNA, antisense regulation has been found in many other systems (for reviews, see Simons and Kleckner 1988; Weintraub 1990; Eguchi et al. 1991).

SPECIFIC FOLDING OF PRIMER RNA

RNA II is multifunctional, having several different roles in primer formation and its regulation. Analysis of point mutations, named *pri*, showed that a single base change in RNA II can affect its priming ability (Masukata and Tomizawa 1984, 1986). The positions of some *pri* mutations and their effects on various functions are shown in Figure 2. Many *pri* mutations prevent RNA II from forming a hybrid with the template DNA, whereas others do not affect hybrid formation but instead alter the site of cleavage by RNase H or inhibit utilization of cleaved RNA II. Some *pri* mutations can be suppressed by additional mutations that are named *spr*. The positions of some *pri* mutations and their respective *spr* mutations are separated as far as 240 nucleotides. Of particular interest here is a class of mutations that affects hybrid formation. Members of this class are found in almost any region of RNA II, suggesting that the global structure of the RNA is important for hybrid formation.

The requirement of specific folding of RNA II for its priming activity was revealed by comparison of the structures of active and inactive transcripts. For example, the *pri7* mutation changes the rG residue at position −469 of RNA II to an rA residue and inhibits hybrid formation. The *pri7* mutant transcript has preferential cleavage sites for RNases A, T1, and V1 that differ from the wild-type RNA II. The secondary structures of both wild-type and *pri7* transcripts were predicted by computer analysis that incorporated the biochemical data on RNase sensitivity and genetic data on suppression that suggest pairing of some *pri* sites with the respective *spr* sites (Masukata and Tomizawa 1986). The structures thus predicted for wild-type RNA II and *pri7* RNA II are schematically shown in Figures 3A and 3B, respectively. The direct effect of the *pri7* mutation is inhibition of pairing of the 22-nucleotide-long α region, which includes the *pri7* site, with the complementary β region about 100 nucleotides downstream from the *pri7* site. As a consequence, the structure of an entire region of the RNA is altered. The wild-type structure is recovered by the compensatory *spr71* mutation in the β region. When

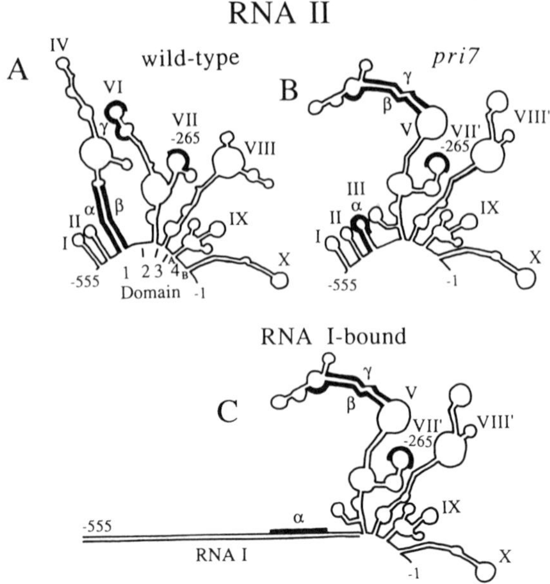

Figure 3 Illustration of secondary structures of various types of RNA II. Roman numerals show the names of the stem-loop structures. The positions of the α, β, and γ regions and of the −265 region are indicated. Domains for the wild-type primer are assigned as indicated.

RNA I binds to wild-type RNA II, the RNA II forms a structure that is analogous to the inactive *pri7* mutant structure (Fig. 3C) (Masukata and Tomizawa 1986).

HYBRID FORMATION OF RNA II WITH THE TEMPLATE DNA

Specific folding of RNA II is required for formation of the persistent hybrid with the template DNA. It is important to know how the folded structure achieves its expected function. To solve this problem, we paid special attention to the observation that the inhibitory effect of the *pri4* mutation (rG to rA change at position −264 in structure VII) on hybrid formation is suppressed by the *spr41* or *spr42* mutation (rG to rA change at position −19 or −18, respectively, in structure X). Because the regions around both the *pri* and *spr* sites in RNA II have stretches of six consecutive rG residues, these regions are unlikely to interact directly. Therefore, we examined the possibility that the region around the *pri4* site (named the −265 region in structure VII) of RNA II might interact with the six consecutive dC residues of the *template DNA* strand in the region around the *spr* sites (named the −20 region) (Masukata and Tomizawa 1990).

The role of the wild-type sequence in the −20 region in hybrid formation was shown by the following experiments. Replacement of the 13 base pairs of ColE1 just upstream of the origin with an unrelated DNA segment does not prevent RNA II from forming a persistent hybrid. However, an additional replacement of the next 7 base pairs farther upstream, which includes a stretch of six dC residues in the template strand, does block hybrid formation. To examine which strand of the DNA in the −20 region is required for hybrid formation, heteroduplex templates were constructed in which the region from 3 nucleotides downstream from the origin to 20 or 60 nucleotides upstream of it was deleted from one or the other DNA strand. In such a template DNA, the 20- or 60-nucleotide segment opposite the deletion formed a bulged loop. Transcription on these heteroduplex templates showed that the −20 region of the template strand cannot be deleted for hybrid formation, but the corresponding region of the nontemplate strand can be deleted.

The transcripts from the *heteroduplex* template deleting the −20 region of the nontranscribed strand were cleaved by RNase H quite efficiently in the −265 region as well as at the origin. This result demonstrates that the rG stretch of RNA II in the −265 region is capable of interaction with the dC stretch of the template strand near the origin. Analysis of the *pri2* mutation demonstrates that this interaction is not fortuitous but is significant for primer formation. The *pri2* mutation changes the rG residue at position −268 to an rA residue, reducing the number of rG residues in the stretch. The *pri2* transcript from the *homoduplex* is not cleaved by RNase H either at the origin or in the −265 region. However, the *pri2* transcript from the *heteroduplex* template is cleaved efficiently in both regions despite the fact that the transcript from the *heteroduplex* template is identical in sequence to that from the *homoduplex* template. Therefore, suppression of the *pri2* mutation by the deletion must be due to exposure of the dC stretch of the template strand, which compensates for the decreased strength of base-pairing between this region and the −265 region of the mutant primer.

Whereas the pairing of the −265 region of RNA II with the dC stretch appears to be essential for cleavage by RNase H at the origin, the wild-type transcript from the *homoduplex* template that is not cleaved detectably in the −265 region can be cleaved at the origin at the standard concentration of RNase H. However, when the RNase H concentration was increased 10,000-fold, a large fraction of the wild-type transcripts from the *homoduplex* template was cleaved in the −265 region. The interaction of the −265 region of the transcript with the −20 region of the *homoduplex* template may be only transient, or the hybridizing −265 region may be in a structure that protects it quite well from the RNase H action. On the other hand, even with the high concentration of RNase H, the *pri2* transcripts from the *homoduplex* template were not cleaved in the −265 region or at the origin. These results indicate that interaction of the −265 region of RNA II with the −20 region of the template strand facilitates formation of the persistent hybrid at the origin. Inhibition of formation of primer by oligocytidylic acid, $(rC)_6$ or $(dC)_6$, which binds to the −265 region, supports this conclusion. The necessity of interaction of the upstream region for formation of the persistent hybrid is shown schematically in Figure 4A.

Structure IX is also involved in hybrid formation. It interacts with structure VII in the regions where compensatory mutations (*pri8* and *spr81*, both in the loop sequences) were found. Together they form a folded structure that may help interaction of the −265 region with the −20 region. Structure X, which has a very GC-rich stem, is also involved in hybrid formation. Very strong pausing of transcription just after synthesis of structure X (Tomizawa and Itoh 1982) should provide a favorable condition for interaction of the −265 region by increasing the time window for the reaction. A mutation that reduces the stability of structure X also reduces the efficiencies both of pausing and of formation of a persistent hybrid (Masukata and Tomizawa 1990).

Normally, the two strands of the template DNA are unpaired locally during transcription, and pairing is restored as transcription advances. This pairing is not restored when RNA II forms a persistent hybrid.

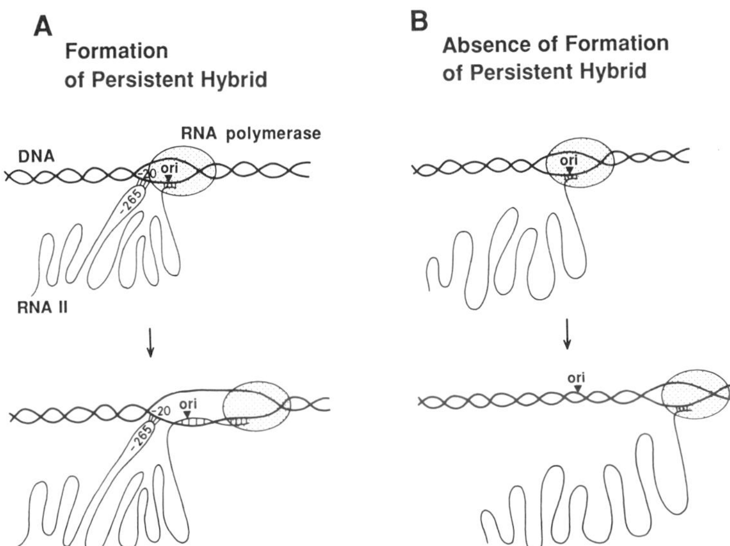

Figure 4 Schematic illustrations of formation and absence of formation of the persistent hybrid. (*A*) The formation of persistent hybrid is facilitated by interaction of the –265 region of RNA II to the –20 region of the template DNA strand. (*B*) Absence of the hybridization is caused by the lack of the interaction. The top diagrams illustrate the states when RNA polymerase is transcribing the origin region, and the bottom diagrams illustrate the states when the polymerase has passed the region.

However, the absence of restoration of pairing per se is not the cause of hybrid formation, because the specific folding of RNA II is required for hybrid formation even on the deletion *heteroduplex* for which the region of the template strand cannot pair anyway. It can therefore be inferred that the interaction of the –265 region, in coordination with formation of structure X, induces formation of a persistent hybrid by somehow modulating the action of RNA polymerase immediately upstream of the origin.

DOMAIN STRUCTURE OF RNA II

Inspection of the structure of the wild-type RNA II shown in Figure 3A suggests that the region upstream of structure IV (inclusive) might not interact with its downstream region. If so, and if the upstream region is not involved in primer formation, the removal of this region in a way that does not disturb the downstream structures might yield a truncated RNA II that is active for the priming function. This is, in fact, the case, because a deletion of the region of 200 nucleotides from the 5′ end of structure I through the 3'end of structure IV has little effect on primer formation (Masukata and Tomizawa 1986). Deletions of 10 nucleotides more or less than the 200 nucleotides have strong detrimental effects, because they affect the structure of the downstream region. Primer formation by the transcript from the deletion template is not susceptible to inhibition by RNA I. These results indicate that the region downstream from structure IV (exclusive) is sufficient for primer formation, whereas the region upstream of structure IV (inclusive) is there solely for regulation of primer formation.

The region of RNA II that forms structure VIII has some function for utilization of the RNase-H-cleaved transcript by DNA polymerase I. This conclusion is obtained by analyses of mutations that affect specifically primer utilization. Such mutations were found in a region between 150 and 200 nucleotides upstream of the origin that is included in structure VIII (Naito and Uchida 1980; Masukata and Tomizawa 1984). Mutants of this group cannot replicate in wild-type bacteria because the RNA II cleaved by RNase H cannot be used as a primer by DNA polymerase I. However, they can replicate in *rnh* bacteria, which are defective in RNase H activity, using the mechanism of transcriptional activation that initiates with lagging-strand synthesis. These mutations prevent formation of structure VIII, and each corresponding suppressor mutation restores the structure. Furthermore, the region of 92 base pairs that specifies structure VIII can be deleted without significantly affecting replication of ColE1 in *rnh* bacteria, although replication of the plasmid in wild-type bacteria is prevented (Masukata and Tomizawa 1990).

Structure X is involved in hybrid formation of RNA II by causing transcriptional pausing, as described above. The region has an additional role in primer formation by determining the sites of cleavage by RNase H. When the position of the bottom of structure X is altered, the cleavage sites are moved to keep the same distances from the bottom of the structure (Masukata and Tomizawa 1984). The cleavage sites are independent of the base sequence around the sites (Masukata and Tomizawa 1984, 1990). Therefore, the distance that is determined by interaction between structure X and RNase H, rather than the base sequence, sets the sites of cleavage. The cleavage of an RNA II molecule at one of a few consecutive sites must be due to flexibility of the structure of the RNA-protein complex. Dependence of the 5'splice-site specificity of a group I intron by the RNA structure rather than the base sequence has been reported (Doudna et al. 1989).

The roles in primer hybridization of the regions other than those mentioned in this section have been described above. Together, these results show that almost the entire structure of the wild-type RNA II is required for formation of a complete primer precursor. The primer RNA is composed of at least four domains, each consisting of one to three stem-loops (Figs. 2 and 3A). Domain 1, which forms structures I, II, and IV, is involved in regulation by RNA I. It is not essential for replication. Domain 2 forms structures VI and VII and is essential for hybrid formation. Domain 3, which contains structure VIII, is essential for replication in the presence of RNase H but not in its absence. Domain 4 is essential for hybrid formation; it may be divided into two subdomains, A and B, consisting of structures IX and X, respectively.

CHARACTERISTICS OF INTERACTIONS INVOLVING LOOP SEQUENCES

The process of binding of RNA I to RNA II is initiated by interaction between complementary loops. Three pairs of complementary loops are involved in the initial phase of the binding reaction. This is followed by a sequence of reactions producing progressively more stable products (Tomizawa 1984, 1990a). Some characteristics of the loop-to-loop interaction were revealed by studies of the interaction between pairs of single stem-loops having various sequences; the stem-loops of each pair consisted of a loop of six to eight bases and a stem of five base pairs (Eguchi and Tomizawa 1990a,b, 1991; Eguchi et al. 1991). In the resulting complex, the complementary loop sequences pair with each other, whereas their stem sequences do not interact. The overall structure of the resulting complex is similar to a linear A-form double-stranded RNA. A model of a complex formed by interaction of complementary single stem-loops is presented in Figure 5.

Binding reactions were compared for the following combinations of complementary sequences: between small complementary linear oligonucleotides, between loops of complementary stem-loops, and between a stem-loop and a linear molecule containing a short sequence that is complementary to the loop sequence (Eguchi and Tomizawa 1991; Eguchi et al. 1991). Binding is initiated by pairing of a few nucleotides (nucleation), and the rate constant of association of the components is relatively independent of their nucleotide sequences and structures. For example, the rate constant of association of a loop sequence with a complementary sequence is similar to that of binding between complementary linear oligonucleotides. On the other hand, the rate constant of dissociation depends very much on the nucleotide sequences and structures in the region of several nucleotides around the nucleation site.

In a more complex process, such as binding of RNA I to RNA II, the initial interaction of a pair of complementary loops is generally followed

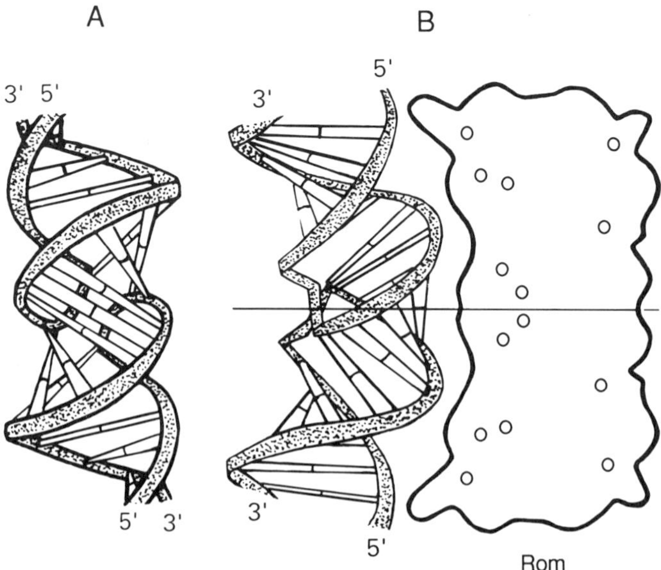

Figure 5 An RNA complex formed by binding of complementary single stem-loops and binding Rom to the complex. The phosphodiester backbones of a complex formed by binding of complementary single stem-loops are shown by ribbons. Each rod indicates a pairing of complementary bases. In the model, all the bases are paired. The overall structure is similar to an A-form double-stranded RNA with a little bend at the center of the complex. The side view of the RNA complex that is shown in *A* is presented in *B* together with the scale model of a Rom dimer interacting with the complex. Circles in the Rom dimer indicate the positions of basic amino acid residues. The horizontal bar indicates an axis of dyad symmetry of the complex. For details, see Eguchi and Tomizawa (1990b, 1991).

by monomolecular reactions that reduce the apparent rate of dissociation of the product of the initial interaction. Therefore, each initial interaction leads with a high probability to irreversible pairing. In such a situation, it is the rate constant of association of the initial interaction that primarily determines the probability of stable binding that is crucial for determination of a biological effect as found, for example, for binding of ColE1 RNAs (Tomizawa 1990a; Eguchi and Tomizawa 1991) and for the copy number control of ColE1 (Brendel and Perelson 1993).

The Rom protein enhances the inhibitory activity of RNA I on primer formation (Lacatena et al. 1984; Tomizawa and Som 1984; Tomizawa 1990b). A single dimer molecule of Rom binds side by side to a complex formed by the loop-to-loop interaction of single stem-loops (Fig. 5B) (Eguchi and Tomizawa 1990a,b, 1991; Eguchi et al. 1991). Binding of

Rom reduces the rate of dissociation of a complex as much as 100-fold but does not affect the rate of association of two complementary single stem-loops. Rom binds with a similar affinity to complexes formed by single stem-loops having various sequences. The protein binds neither to each stem-loop alone nor to A-form double-stranded RNA. In other words, Rom recognizes the structure of a target rather than its exact base sequence. As expected from the fact that dissociation rate constants are less important than association rate constants for the antisense regulation, the effect of Rom on the copy number of ColE1 is moderate.

EVOLUTION OF PRIMER RNA AND ITS ANTISENSE RNA

Phylogenetic Relationships among Domains of RNA II

Several plasmids, including RSF1030, p15A, and CloDF13, replicate using an RNA primer similar to that of ColE1. These plasmids are mutually compatible because replication of each plasmid is regulated by a plasmid-specific RNA I (Tomizawa et al. 1981). Similarity of the base sequences of the RNA primers of these plasmids (Selzer et al. 1983) suggests that they derive from a common ancestor. Nonetheless, the sequences have diverged at a number of sites that are distributed in a highly nonrandom fashion (Fig. 6). Base substitutions in stem regions are infrequent and are often accompanied by complementary base changes in the other strand of the stem. On the other hand, substitutions in the loop regions are quite frequent but occur selectively in certain loops. Divergence of the sequences of stems as well as loops must be strongly constrained by their structures and functions.

On the basis of sequence homologies and similarities of secondary structures, each of the RNA primers of the four plasmids mentioned above can be divided into the same four domains as the ColE1 primer. In addition, presumed RNA primers of two other plasmids, ColA and ColD (Zverev and Khmel 1985), which have sequences homologous to the other four primers, can be similarly divided into four domains. The evolutionary relationships among the four individual domains and whole RNA I and RNA II regions of these plasmids were examined by comparing their phylogenetic trees (Fig. 7). The topology of a phylogenetic tree thus obtained for any particular RNA region represents the evolutionary relationship among the species concerned, if it can be assumed that base changes occurred predominantly by substitutions. Although the topology of a tree that is based on a small number of nucleotides of each component has a limited statistical significance, the following arguments regarding divergence between domains may be permissible, particularly for domains whose trees are of similar topology.

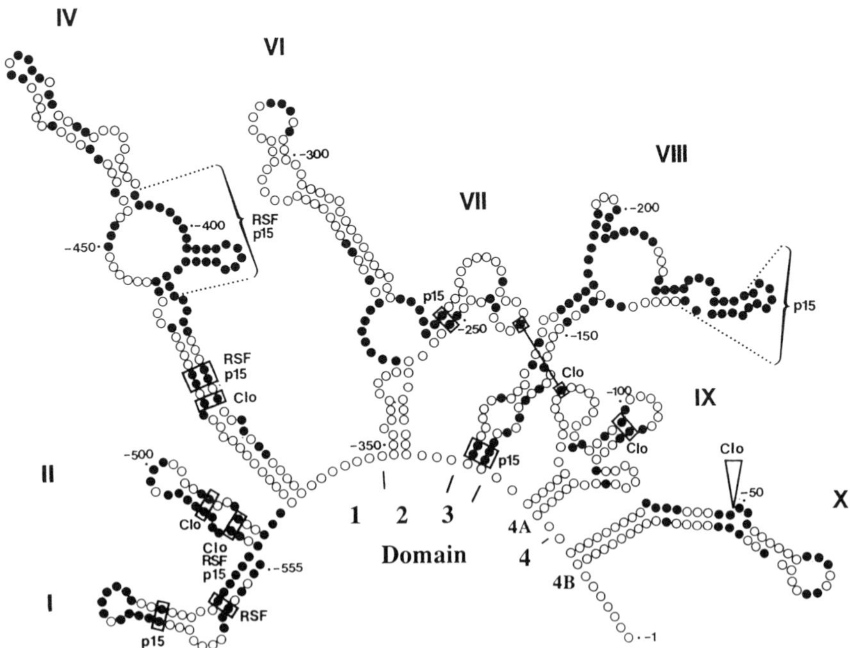

Figure 6 Positions of conserved nucleotides among the ColE1 group plasmids shown in the folded structure of ColE1. Small open circles indicate bases conserved among ColE1, p15A, RSF1030, and CloDF13; solid circles show bases that are different in at least one of the plasmids. Boxes show positions where base-pairing is preserved for the substituted nucleotides in an indicated plasmid. Dotted brackets show locations of deletions, and a triangle indicates a position of a 41-nucleotide insertion. For details, see Masukata and Tomizawa (1986).

When the phylogenetic trees for domains 2 and 4 are compared, ColE1 and RSF1030 and also p15A and ColA are quite similar, whereas CloDF13 and ColD differ significantly, both from each other and from the other four. The similarity in the phylogenetic trees for domains 2 and 4 suggests that these domains evolved together; i.e., that recombination did not occur within and between these domains of different species of plasmid. It could even be possible that recombination has not occurred anywhere within the whole RNA II region. For domain 3, ColE1 and RSF1030 are identical and CloDF13 and ColD are somewhat different from these two. These relations among species are similar to those for domains 2 and 4. In contrast, domain-3 sequences of p15A and ColA are quite different from those of the other plasmids. These differences are probably caused by the deletion of one of the direct repeat sequences and accompanying changes of structure VIII (the position of the change in p15A is shown in Fig. 6) that probably took place after species diversifi-

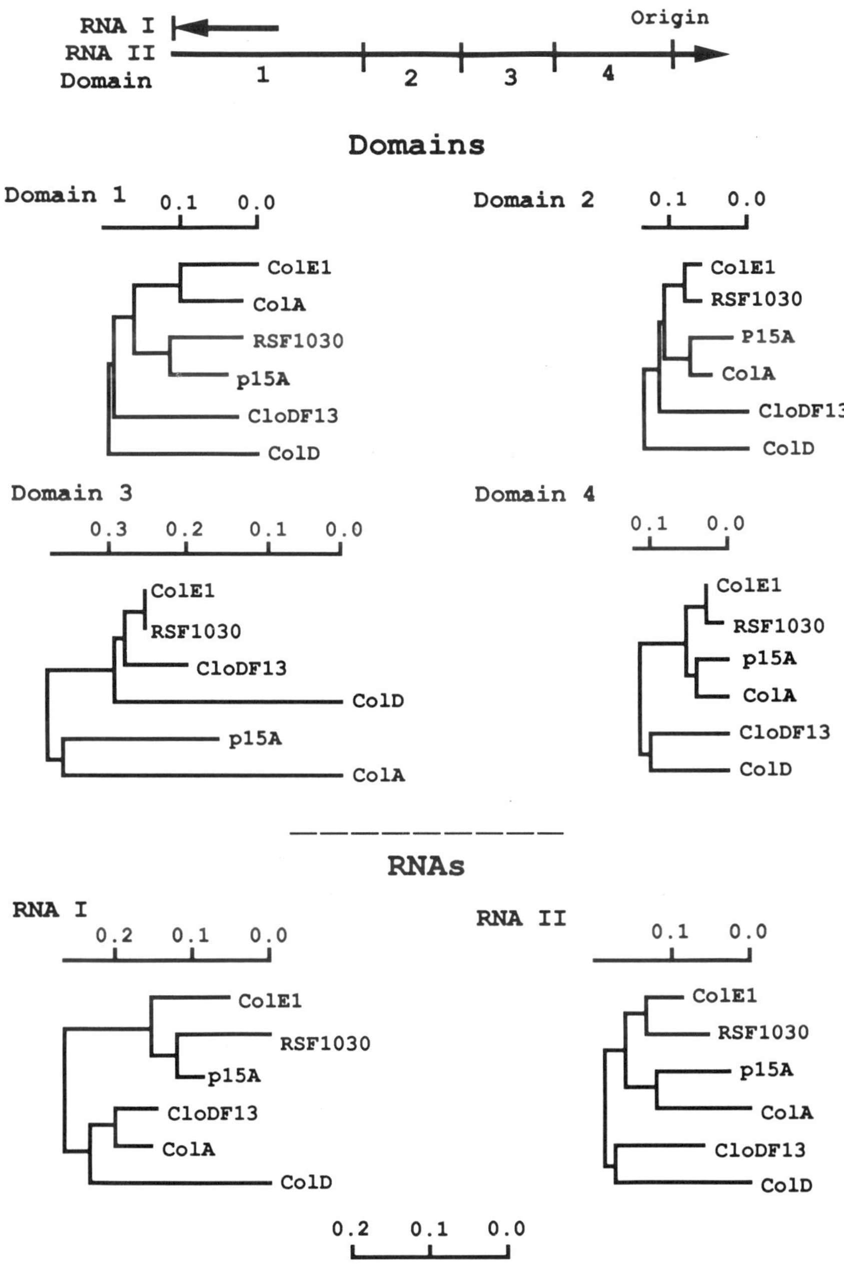

Figure 7 Phylogenetic trees of various domains of the six plasmids. Phylogenetic trees of the RNA I region, primer RNA, and its domains are constructed by the neighbor-joining method (Saitou and Nei 1987).

cation. The similarity of the evolutionary relationships among domains 2, 3, and 4 of various species gives additional support for sharing of a common ancestry by these domains.

The phylogenetic trees for domain 1 and for the RNA I region are both significantly different from those for the other domains. Differences in the mechanisms of divergence probably cause the uniqueness of the topologies of the trees for these regions, as discussed in a later section.

Evolution of Primer Function and Antisense Regulation

If the ancestor of the primer sequences of various plasmids evolved by a successive acquisition of domains, the earliest viable plasmid may have contained only domains 2 and 4 and a promoter for transcription, as these suffice for replication in the absence of RNase H. The ancestor could have evolved by association of domain 4B, which causes transcriptional pausing, and domains 2 and 4A, which further modulate transcription, so that the transcript forms a persistent hybrid. Such a plasmid, which can replicate in *rnh* bacteria by a mechanism involving transcriptional activation, might have evolved in a cell that was lacking or weak in the RNase H activity. Domain 3, which enables use of the RNase-H-cleaved RNA II transcript as a primer, may have evolved in response to increasing RNase H levels. Alternatively, the plasmid may have originated when a sequence containing all these domains was formed in an environment in which the RNase H activity was high.

For stable maintenance of a plasmid in bacteria, replication of the plasmid must be sufficiently efficient and, in addition, must be regulated: If replication is inefficient, the plasmid will be lost by dilution; if it is efficient but unregulated, the host bacteria will be killed by overproduction of the plasmid (Tomizawa 1984). Although regulation of the efficiency of primer transcription could have been used for the regulatory purpose of maintenance of ColE1, there is thus far no evidence for transcriptional regulation of primer formation. Instead, antisense regulation is used for this purpose.

Domain 1 plays a crucial role in regulation of primer formation, because it is the target of regulation by the antisense RNA. Evolution of this domain and the complementary regulatory molecule can be envisioned to have occurred as follows. All species of RNA II have the same sequence in a large part of the α region of this domain (Figs. 3A and 8). For each RNA II species, the α region is complementary to the β-region sequence farther downstream but still in domain 1. In other words, the α region is an inverted repeat of the β region. When a promoter for transcription of the antisense RNA containing the α region evolved, the

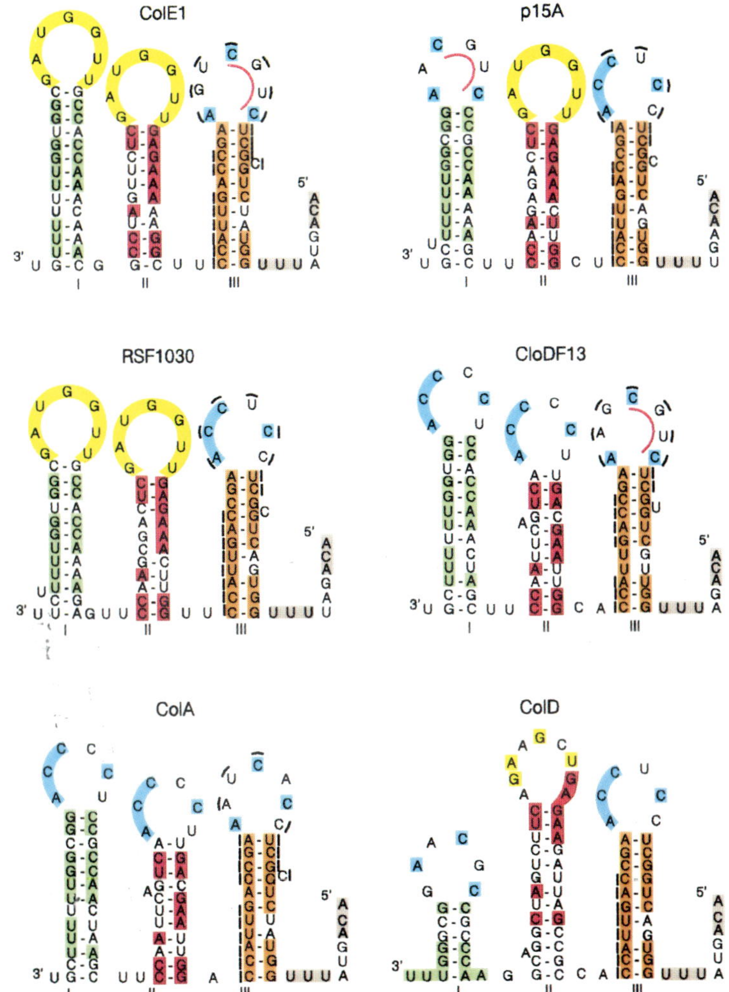

Figure 8 The folded structures of RNA I species of the six plasmids of the ColE1 group. Each identical loop sequence or identical stem sequence is shown by the same color. A broken line around stem-loop III indicates the sequence whose corresponding region in RNA II is complementary to the sequence in the β region of RNA II.

product of transcription would necessarily have some capacity to bind to the complementary RNA II and thus have the potential to affect primer formation. For evolution of this type of regulatory system, an elaborate process of formation of a gene for a regulatory protein and its target sequence is not required. Although any antisense transcript has the potential for regulatory function, its actual use in this capacity requires that binding between sense and antisense RNAs be sufficiently specific and that the binding rate be in a certain suitable range. Interaction between short complementary sequences could satisfy these conditions. The only way to satisfy these conditions by long complementary sequences is by formation of loops of stem-loop structures. In fact, most (if not all) systems of antisense regulation involve folded RNAs (Eguchi et al. 1991). It is important to emphasize that all of the essential features of the inter- and intra-molecular interactions discussed in subsequent sections depend on initial interaction between short single-stranded regions in folded structures.

In antisense regulation, a pair of an effector RNA and its target RNA can be simultaneously altered by a single mutation. If a small number of bases, such as those in complementary loop sequences, are involved in binding, the specificity of inhibition can be changed easily by a mutation. The change would not affect the strength of inhibition very much, because complementarity in the sequences is always maintained. Therefore, use of antisense regulation is particularly advantageous for evolution of a new species of compatible plasmid. In fact, a change of a single base pair in the region that encodes a loop of RNA I can create a new compatible plasmid (Lacatena and Cesareni 1981; Tomizawa and Itoh 1981). However, maintenance of complementarity necessarily limits the range of regulatory activity by antisense regulation. This characteristic might disfavor the use of antisense regulation in a situation where a gene function needs to be regulated over a wide range. However, this limitation could be overcome if the concentration of antisense RNA is subjected to alteration (Simons and Kleckner 1983, 1988).

Having evolved in this manner, the primitive antisense RNA must then have diverged to the several known species of RNA I that form similar folded structures (Fig. 8). The sequences of stem III (in orange) are almost identical for all species of RNA I. They contain α regions that are largely complementary to the downstream β regions. All stem I sequences are very similar, having rG-rC pairs and a run of rUs (in green). These features suggest that stem I probably evolved in an ancestral plasmid for termination of transcription of the antisense RNA. For stem II, the sequences are quite similar for all species of RNA I (in red), except for some differences in the middle regions. Extensive conservation of the sequences for all three stems among all RNA I species indicates that the common ancestral RNA I probably already carried three stem-loops. On the other hand, the three stem sequences differ one from another in any given species of RNA I. This is advantageous, because it should very much reduce the chance of elimination of a stem-loop by intramolecular recombination.

To enhance the inhibitory activity of RNA I, ColE1 specifies the Rom protein. Because the action of Rom is dependent on the presence of RNA I, the participation of Rom in replication regulation must have evolved after establishment of the antisense regulation. The effect of Rom on plasmid replication is small and no effect of Rom on host replication has been detected. In addition, no host sequence homologous to the *rom* gene has been identified. Nonetheless, the finding of the Rom function suggests the presence of a similar host cell protein that could affect kinetic properties of RNA binding. Such a protein, whether or not it is present in the final product, could serve as an auxiliary factor that helps with specific folding of RNA. For such an RNA-chaperoning protein, sequence specificity would not always be essential.

Divergence of Stem-loops of RNA I

The RNA I regions of all six plasmids probably share a common ancestral sequence that includes all three stem-loops. Yet the loop sequences have diverged extensively. Detailed examination of the loop sequences reveals the occurrence of the same or similar loop sequences repeatedly both within a single RNA I species and in various loops of different RNA I species (Fig. 8). The repetitious appearance of the same or similar loops among the different RNA species can be partly accounted for in a general way by point mutations in, or recombination between, genes of established lines of RNA I. For example, RSF1030 could have been made from ColE1, or vice versa, by mutations. The relation between CloDF13 and ColA could be similar. Recombination between ColE1 and p15A could have yielded RSF1030, or recombination between ColA and ColE1 could have given rise to a CloDF13-type plasmid, although compatible plasmids are unlikely to recombine as discussed above. These changes, if they occurred, were probably carried out by the conventional mechanisms of mutation or recombination.

Although a unique combination of three stem-loops for each species of RNA I can be accounted for, in part, by conventional mechanisms, an additional very remarkable feature of the structures of individual stem-loops suggests that a heretofore unrecognized mechanism may be primarily responsible for evolution of these structures. Below, I discuss individual sequences of the stems and loops first, and then I discuss their assembly. As described above, the two major features of the stem sequences are extensive conservation of the sequences for all three stems among all RNA I species and differences of the three stems (shown in green, red, and orange) in any given species of RNA I. I begin the discussion on loop sequences with loop IIIs, which are quite well conserved among all RNA I species. Loop IIIs of three species of RNA I (RSF1030, p15A, and ColD) are identical (in blue), and three other loops share certain similarities (blue spots) with these. Two of these other loop IIIs have the same tetranucleotide sequence (in pink). The following observations suggest that the loop III sequences have diverged from the ancestral sequence by successive point mutations. Among the four base substitutions found in loop III, three have compensatory nucleotide changes in the β region (Fig. 6). Probably, a point mutation that occurred first in the loop III sequence of the ancestral plasmid was selected, as it created a new compatible plasmid, and then the stem IV structure of RNA II was restored by a compensatory mutation in the β region for each species of plasmid.

Comparisons among loops I and II reveal unique distributions of similar loops among different species of RNA I. Six (in blue and in pink) of the twelve loops are very similar to one of the loop III sequences (also in blue and in pink). The other six loops (in yellow) form a distinct set which are identical or nearly identical (for ColD) to each other. The sequences of loops I and II within a given RNA are identical for four species of RNA I, differing only for p15A and ColD.

Stem-loops I and II are made by combinations of one of two types of loops (in yellow and in blue) and one of two types of stems (in green and in red). Although possible formation of these structures by many successive point mutations from an ancestral molecule cannot be dismissed, convergent evolution of one type of loop on two different types of stems and divergent evolution of two different types of loops attached to the same type of stem are unlikely to have occurred simply by such chance events. Note that the sequences of loop III that diverged by point mutations are quite different. Nonetheless, loops whose sequences are almost identical to one of the loop III sequences are present on stems I and II, which are completely different from each other and from stem III. The presence of stem-loops that are composed of different combinations of a loop sequence and a stem sequence suggests that a mechanism exists for assorting a stem unit and a loop unit.

A mechanism that assorts a stem unit and a loop unit must necessarily recognize some aspect of nucleic acid structure reflective of the stem-loop arrangement. Recognition could occur at the DNA level. For example, a mechanism would recognize the structure of DNA regions (in either single- or double-stranded) with inverted repeats carrying a short sequence between them and exchange the sequences. Alternatively, recognition could occur at the RNA level. The structure of a stem-loop RNA would be recognized by a mechanism and then the loop would be replaced by another loop. If replacement covering a loop region were to occur by recombination between RNA stem-loops, the recombined RNA region would be reverse-transcribed into DNA before incorporation of its sequence into the plasmid genome.

It has been reported that the sequence of loop I or II of ColE1 (in yellow) is identical to the D-loops of some *E. coli* tRNAs and very similar to the anticodon-loop of Thr$_3$ tRNA (Yavachev and Ivanov 1988). I found that all the D-loop sequences of the complete set of 46 *E. coli* tRNAs (Komine et al. 1990) are quite similar to the loop sequence (in yellow). In particular, 11 D-loops among them and the anticodon-loop of Thr$_1$ tRNA are identical to the loop sequence (in yellow). The sequences of the 11 D-stems are largely homologous but differ from the Thr$_3$ anticodon-stem and from stems I and II of RNA I. There is some homology between the Thr$_3$ anticodon-stem and stem I of RNA I. It is possible that the loop sequence of an *E. coli* tRNA or its ancestor could have been incorporated into a loop of RNA I during its evolution, probably by either one of the mechanisms discussed above.

The difference in the mechanisms of divergence is the probable cause of the uniqueness of the topology of the phylogenetic tree of domain 1 (and RNA I region). The plasmids are classified by their incompatibility properties, which are determined by their domain 1 sequences and are named accordingly. However, the true phylogenetic relationship between the plasmids is probably better represented by that for domain 2, since base substitutions are the likely major cause of divergence in this domain, and recombination between domains 1 and 2 of compatible plasmids is unlikely to have taken place.

CONCLUSIONS AND PERSPECTIVE

The course of evolution of a functional RNA molecule may be divided into two phases. The first is the process of gaining a new function by a primitive molecule, leading to formation of a prototypic functional molecule. The second is diversification of the prototypic molecule into various descendants with favorable properties for selection under different evolutionary conditions. In either case, formation of some RNA structure must have preceded evolution of a new function for RNA.

Some insight into the general principles for evolution of functional structures of RNA is gained through the studies of formation of the ColE1 primer RNA and its regulation. The primer RNA is composed almost entirely of stem-loops. Each sub-function of the multifunctional RNA probably evolved in a domain consisting of one or more stem-loops, with this structure acting as a part of the integral structure of the complete functional molecule. Formation of an integral structure of a multifunctional RNA may be facilitated by interaction between pairs of single-stranded regions in separate domains of the molecule. In an intramolecular interaction, the rate of interaction between two suitably positioned sequences must be high because of their effective high local concentrations. Therefore, the critical factors that determine the specific folding of RNA must be the potential of prospective sequences to form a stable structure upon their interaction and the suitable positionings of these sequences for efficient interactions.

What is required for a pair of single-stranded regions to form a stable RNA structure through their interaction? I described above that for binding between RNA I and RNA II, the rate constant of association between their loops is crucial in determination of the rate of the reaction that leads to their stable binding. In addition, the rate constant of association of complementary stem-loops is relatively independent of the base sequences of the loops. Therefore, the presence of complementarity between a pair of single-stranded regions is probably sufficient for their interaction to be involved in specific folding of an RNA molecule, and their exact base sequences have relatively minor importance. What about the positioning? It was found that the stem does not directly participate in binding between complementary single stem-loops. Nonetheless, certain stem mutations that disrupt base-pairing in the stem can frequently be suppressed by mutations that restore the base-pairing in the stem. These findings show that the major role of a stem in formation of a specific folding of an RNA is proper positioning of a loop in the molecule.

As for evolution of a new functional structure, I proposed above that the regulatory domain of the primer RNA could have evolved by formation of the α region, which is an inverted repeat of the β region. The β region itself is an inverted repeat of the largely complementary γ region (Fig. 3). One possible mechanism of formation of such an inverted repeat structure is local separation of an end of an elongating strand from the template strand during its synthesis, followed by extension of the end by copying its own sequence before returning to copy the original template strand (Ghosal and Saedler 1978). An alternative way of formation of an inverted repeat structure would be initiated by pairing of short complementary sequences. This pairing would be followed by base changes that could form base-paired regions in many different ways around the initial site of pairing. The number of paired bases in the structure would increase due to selection for more thermodynamically or enzymatically stable structures. Thus, many different structures would be made subsequent to the initial pairing of short complementary sequences. Versatility of structures thus made would provide a chance for evolution of a new function for one of the structures. Presumed error-prone synthesis of polynucleotides in early evolution should have been particularly helpful by contributing versatility of RNA structures and thus increasing the rate of evolution. This mechanism of versatile pairing would represent prevailing evolutionary events, as diversity of the products far exceeds that predicted by the first mechanism. For evolution of a new structure for RNA specified by a DNA genome, the structure needs to express a favorable function for selection. Here, the thermodynamic or enzymatic stability of the RNA would have minor importance. Even when the activity of the structure on which a new function was evolved would not be remarkable, once a new function was evolved there should have been ample time for improvement and refinement of the function of that structure.

In contrast to the probable involvement of formation of inverted repeats in evolution of a new function, formation of direct repeats of short RNA sequences is less likely to be involved in the process, because no obvious mechanism yielding versatility of the resulting structures is envisaged. On the other hand, in eukaryotic genomes, direct repeats that include an entire gene may lead to evolution of a new gene function through diversification of the function of one of repeated genes (Ohno 1970; Ohta 1980). In this case, however, the presence of two or more functional units rather than formation of direct repeats per se is significant for evolution of a new function.

Involvement of tRNA-like structures in evolution of early RNA genomes has been suggested (for reviews, see Weiner and Maizels 1987, 1991; Mans et al. 1991; Okada 1991; Maizels and Weiner, this volume). The presence of such structures at the termini of many contemporary RNA species can be considered as a reflection of their evolutionary histories. The structure of RNA I or of the 5′-end region of RNA II is somewhat similar to that of tRNA, suggesting a possible evolutionary relationship between these RNAs. However, evolution of the primer RNA by addition of domains in linear order starting from domain 1 (or the RNA I region) is unlikely because a functional molecule is complete only when all four domains are assembled, thus eliminating the advantage of sequential selection of more complex functional molecules.

The stem-loop sequences of RNA I have descended probably from a single ancestral sequence. Some other RNA species, such as tRNAs, whose patterns of diversification have some similarity with that of RNA

I, have originated probably from independent ancestors. The presence of some similarity in the patterns of diversification of different RNA species implies that similar mechanisms have operated in the courses of their diversification. It seems probable that a mechanism that specifically assorts a loop unit together with a stem unit could have played important roles in formation of functional structures and in diversification of loop sequences of different RNA species. The specific mechanistic details of such a process remain to be determined.

Studies of the ColE1 RNA primer revealed several reactions that are basic for formation of RNA structures and could have participated in evolution of functional RNAs. Further studies on relationships between structures and functions of various species of RNA should advance our understanding of general principles for evolution of functional structures of RNA.

ACKNOWLEDGMENTS

This chapter is based mostly on our studies carried out at the National Institutes of Health from 1973 through 1989. I am thankful for valuable contributions by my previous associates. Among them, Drs. Michael Brenner, Yutaka Eguchi, Tateo Itoh, Hisao Masukata, Haruo Ohmori, Yoshimasa Sakakibara, and Gerald Selzer gave me useful comments on the manuscript. Comments by my friends, Drs. Tomoko Ohta, Kiyoshi Mizobuchi, Takashi Gojobori, Nancy Kleckner, Martin Gellert, and James F. Crow, are particularly appreciated. I am also grateful for help in the phylogenetic analysis by Dr. Kazuho Ikeo and for secretarial help by Mrs. Sumiko Yamamoto.

REFERENCES

Brendel, V. and A.S. Perelson. 1993. A quantitative model of ColE1 plasmid copy number control. *J. Mol. Biol.* **229:** 860–872.

Brenner, M. and J. Tomizawa. 1989. Rom transcript of plasmid ColE1. *Nucleic Acids Res.* **17:** 4309–4326.

———. 1991. Quantitation of ColE1-encoded replication elements. *Proc. Natl. Acad. Sci.* **88:** 405–409.

Cesareni, G., M.A. Muesing, and B. Polisky. 1982. Control of ColE1 DNA replication: The *rop* gene product negatively affects transcription from the replication primer promoter. *Proc. Natl. Acad. Sci.* **79:** 6313–6317.

Cesareni, G., M. Helmer-Citterich, and L. Castagnoli. 1991. Control of ColE1 plasmid replication by antisense RNA. *Trends Genet.* **7:** 230–235.

Dasgupta, S., H. Masukata, and J. Tomizawa. 1987. Multiple mechanisms for initiation of ColE1 DNA replication: DNA synthesis in the presence and absence of ribonuclease H. *Cell* **51:** 1113–1122.

Doudna, J.A., B.P. Cormack, and J.W. Szostak. 1989. RNA structure, not sequence, determines the 5′ splice-site specificity of a group I intron. *Proc. Natl. Acad. Sci.* **86:** 7402–7406.

Eguchi, Y. and J. Tomizawa. 1990a. Control of ColE1 plasmid replication: Complex formed by RNA I and RNA II and its stabilization by Rom protein. In *Molecular mechanisms in DNA replication and recombination* (ed. C.C. Richardson and I.R. Lehman), pp. 203–214. Wiley-Liss, New York.

―――. 1990b. Complex formed by complementary RNA stem-loops and its stabilization by a protein: Function of ColE1 Rom protein. *Cell* **60:** 199–209.

―――. 1991. Complexes formed by complementary RNA stem-loops: Their formations, structures, and interaction with ColE1 Rom protein. *J. Mol. Biol.* **220:** 831–842.

Eguchi, Y., T. Itoh, and J. Tomizawa. 1991. Antisense RNA. *Annu. Rev. Biochem.* **60:** 631–652.

Ghosal, D. and H. Saedler. 1978. DNA sequence of the mini-insertion IS2-6 and its relation to the sequence of IS2. *Nature* **275:** 611–617.

Itoh, T. and J. Tomizawa. 1979. Initiation of replication of plasmid ColE1 DNA by RNA polymerase ribonuclease H and DNA polymerase I. *Cold Spring Harbor Symp. Quant. Biol.* **43:** 409–418.

―――. 1980. Formation of an RNA primer for initiation of replication of ColE1 DNA by ribonuclease H. *Proc. Natl. Acad. Sci.* **77:** 2450–2454.

Kogoma, Y. 1984. Absence of RNase H allows replication of pBR322 in *Escherichia coli* mutants lacking DNA polymerase I. *Proc. Natl. Acad Sci.* **81:** 7845–7849.

Komine, Y, T. Adachi, H. Inokuchi, and H. Ozeki. 1990. Genomic organization and physical mapping of the transfer RNA genes in *Escherichia coli* K12. *J. Mol. Biol.* **212:** 579–598.

Lacatena, R.M. and G. Cesareni. 1981. Base-pairing of RNA I with its complementary sequence in the primer precursor inhibits ColE1 replication. *Nature* **294:** 623–626.

Lacatena, R.M., D.W. Banner, L. Castagnoli, and G. Cesareni. 1984. Control of initiation of pMB1 replication: Purified Rop protein and RNA 1 affect primer formation in vitro. *Cell* **37:** 1009–1014.

Mans, R.M.W., W.A. Pleij, and L. Bosch. 1991. tRNA-like structures: Structure, function and evolutionary significance. *Eur. J. Biochem.* **201:** 303–324.

Masukata, H. and J. Tomizawa. 1984. Effects of point mutations on formation and structure of the RNA primer for ColE1 DNA replication. *Cell* **36:** 513–522.

―――. 1986. Control of primer formation for ColE1 plasmid replication: Conformational change of the primer transcript. *Cell* **44:** 125–136.

―――. 1990. A mechanism of formation of a persistent hybrid between elongating RNA and template DNA. *Cell* **62:** 331–338.

Masukata, H., S. Dasgupta, and J. Tomizawa. 1987. Transcriptional activation of ColE1 DNA synthesis by displacement of non-transcribed strand. *Cell* **51:** 1123–1130.

Minden, J.S. and K.J. Marians. 1985. Replication of pBR322 DNA *in vitro* with purified proteins. *J. Biol. Chem.* **260:** 9316–9325.

Morita, M. and A. Oka. 1979. The structure of a transcriptional unit on colicin E1 plasmid. *Eur. J. Biochem.* **97:** 435–443.

Naito, S. and H. Uchida. 1980. Initiation of DNA replication in a ColE1-type plasmid: Isolation of mutations in the ori region. *Proc. Natl. Acad Sci.* **77:** 6744–6748.

―――. 1986. RNAase H and replication of ColE1 DNA in *Escherichia coli*. *J. Bacteriol.* **166:** 143–147.

Naito, S., T. Kitani, T. Ogawa, T. Okazaki, and H. Uchida. 1984. *Escherichia coli* mutants suppressing replication-defective mutations of the ColE1 plasmid. *Proc. Natl.*

Acad Sci. **81:** 550–554.

Ohmori, H. and J. Tomizawa. 1979. Nucleotide sequence of the region required for maintenance of colicin E1 plasmid. *Mol. Gen. Genet.* **176:** 161–170.

Ohno, S. 1970. *Evolution by gene duplication.* Springer-Verlag, Berlin.

Ohta, T. 1980. *Evolution and variation of multigene families.* Springer-Verlag, Berlin.

Okada, N. 1991. SINEs. *Curr. Opin. Genet. Dev.* **1:** 498–504.

Polisky, B. 1988. ColE1 replication control circuitry: Sense from antisense. *Cell* **55:** 929–932.

Saitou, N. and M. Nei. 1987. The neighbor-joining method: A new method for reconstructing phylogenetic trees. *Mol. Biol. Evol.* **4:** 406–425.

Sakakibara, Y. and J. Tomizawa. 1974. Replication of colicin E1 plasmid DNA in cell extracts. *Proc. Natl. Acad. Sci.* **71:** 802–806.

Selzer, G., T. Som, T. Itoh, and J. Tomizawa. 1983. The origin of replication of plasmid p15A and comparative studies on the nucleotide sequences around the origin of related plasmids. *Cell* **32:** 119–129.

Simons, R.W. and N. Kleckner. 1983. Translation control of IS10 transposition. *Cell* **34:** 683–691.

———. 1988. Biological regulation by antisense RNA in prokaryotes. *Annu. Rev. Genet.* **22:** 567–600.

Som, T. and J. Tomizawa. 1983. Regulatory regions of ColE1 that are involved in determination of plasmid copy number. *Proc. Natl. Acad. Sci.* **80:** 3232–3236.

Tomizawa, J. 1984. Control of ColE1 plasmid replication: The process of binding of RNA I to the primer transcript. *Cell* **38:** 861–870.

———. 1986. Control of ColE1 plasmid replication: Binding of RNA I to RNA II and inhibition of primer formation. *Cell* **47:** 89–97.

———. 1990a. Control of ColE1 plasmid replication: Intermediates in the binding of RNA I and RNA II. *J. Mol. Biol.* **212:** 683–694.

———. 1990b. Control of ColE1 plasmid replication: Interaction of Rom protein with an unstable complex formed by RNA I and RNA II. *J. Mol. Biol.* **212:** 695–708.

Tomizawa, J. and T. Itoh. 1981. Plasmid ColE1 incompatibility determined by interaction of RNA I with primer transcript. *Proc. Natl. Acad. Sci.* **78:** 6096–6100.

———. 1982. The importance of RNA structure in ColE1 primer formation. *Cell* **31:** 575–583.

Tomizawa, J. and H. Masukata. 1987. Factor-independent termination of transcription in a stretch of deoxyadenosine residues in the template DNA. *Cell* **51:** 623–630.

Tomizawa, J. and T. Som. 1984. Control of plasmid replication: Enhancement of binding of RNA I to the primer transcript by the Rom protein. *Cell* **38:** 871–878.

Tomizawa, J., H. Ohmori, and R.E. Bird. 1977. Origin of replication of colicin E1 plasmid DNA. *Proc. Natl. Acad. Sci.* **74:** 1865–1869.

Tomizawa, J., T. Itoh, G. Selzer, and T. Som. 1981. Inhibition of ColE1 RNA primer formation by a plasmid-specified small RNA. *Proc. Natl. Acad. Sci.* **78:** 1421–1425.

Twigg, A.J. and D. Sherratt. 1980. Trans-complementable copy-number mutants of plasmid ColE1. *Nature* **283:** 216–218.

Weintraub, H.M. 1990. Antisense RNA and DNA. *Sci. Am.* **262:** 34–40.

Weiner, A.M. and N. Maizels. 1987. tRNA-like structures tag the 3′ ends of genomic RNA molecules for replication: Implications for the origin of protein synthesis. *Proc. Natl. Acad. Sci.* **84:** 7383–7387.

———. 1991. The genomic tag model for the origin of protein synthesis: Further evidence from the molecular fossil record. In *Evolution of life* (ed. S. Osawa and T. Honjo), pp. 51–66. Springer-Verlag, Tokyo.

Yavachev, L. and I. Ivanov. 1988. What does the homology between *E. coli* tRNAs and RNAs controlling ColE1 plasmid replication mean? *J. Theor. Biol.* **131:** 235–241.

Zverev, V.V. and I.A. Khmel. 1985. The nucleotide sequences of the replication origin of plasmids ColA and ColD. *Plasmid* **14:** 192–199.

17

Thermodynamic Considerations for Evolution by RNA

Douglas H. Turner and Philip C. Bevilacqua
Department of Chemistry
University of Rochester
Rochester, New York 14627-0216

The only definitive statement that thermodynamics makes about evolution is that ultimately it will result largely in CO_2, H_2O, and N_2. Fortunately, the sun allows us to live in a world that is kinetically controlled due to an input of energy. Nevertheless, since many equilibria are reached rapidly compared with the age of the earth, thermodynamic principles can be used to predict what reactions are possible in this transitory world. In this chapter, we argue that known thermodynamic properties of RNA suggest it is well adapted to play a central role in the evolution of information transfer and catalysis. This is because mononucleotides contain many functional groups and are therefore capable of strong, specific binding interactions. The evidence for the strength of the binding interactions comes largely from optical melting studies of oligonucleotides (Turner et al. 1988). Since evolution presumably started with rather short molecules, oligonucleotides are reasonable model systems for studying the interactions important in early times. In the following, we discuss the interactions thought to be important for organizing RNA, describe some implications for catalysis and information transfer, and speculate on implications for evolution.

DISCUSSION

Fundamental Interactions Determining RNA Structure

Molecular association and organization are essential for information transfer and catalysis. The thermodynamic measure of the strength of an association or the extent of organization is the change in free energy, ΔG^o, since the equilibrium constant for either association or organization into a specific conformation relative to the random coil is given by $K = e^{-\Delta G^{\circ}/RT}$. Here R is 1.987 cal/$M \cdot K$ and T is temperature in kelvins. Thus, the more negative the free-energy change, the more favorable the association or the given conformation. For example, at 37°C (310.15 K),

The RNA World
© 1993 Cold Spring Harbor Laboratory Press 0-87969-380-0/93 $5 + .00

every decrease of 1.4 kcal/mole in free energy, ΔG^{o}_{37}, favors that conformation or association by a factor of 10 increase in the equilibrium constant. As discussed below, individual stacking and hydrogen-bonding interactions can produce changes of this magnitude.

Stacking Interactions

Stacking is the vertical interaction of flat aromatic ring systems, as cartooned in Figure 1. One measure of the strength of stacking interactions is the degree to which homopolymers like polyadenylic acid organize from random coil to stacked, helical conformations. This has been studied by optical, calorimetric, and other methods (Richards et al. 1963; Inners and Felsenfeld 1970; Suurkuusk et al. 1977; Filimonov and Privalov 1978; Freier et al. 1981). The ΔG^{o}_{37} for this intrastrand stacking is both modest and sequence dependent. For polycytidylic, polyadenylic, and polyuridylic acids, the ΔG^{o}_{37} values are −0.3, −0.02, and >+1 kcal/mole, respectively (Richards et al. 1963; Inners and Felsenfeld 1970; Freier et al. 1981). The positive ΔG^{o}_{37} for polyuridylic acid indicates negligible stacking at 37°C.

Another measure of the strength of stacking interactions is the enhanced association constant observed when unpaired nucleotides, dangling ends, are added to the 3′ ends of RNA duplexes (Fig. 1). This can be determined by measuring optical melting curves for duplexes with and without dangling ends (Freier et al. 1983; Petersheim and Turner 1983; Sugimoto et al. 1987; Turner et al. 1988). The interstrand stacking effect can be large. For example, 3′ dangling purines, R, in the sequence $^{CR}_{G}$ make ΔG^{o}_{37} for oligomer association more favorable by 1.7 kcal/mole on average. Thus, a 3′ dangling purine can make a substantial contribution to stability without any requirement for additional complementarity. One trace of an early use of this interaction may be the fact that a 3′ dangling purine typically follows the anticodon in tRNA. The effect is very sequence dependent, however. The ΔG^{o}_{37} increment from a 3′ dangling pyrimidine in the sequence $^{UY}_{A}$ is only −0.1 kcal/mole.

In contrast to 3′ dangling ends, all 5′ dangling ends in RNA add little additional stability, on average making duplex formation more favorable by only 0.2 kcal/mole at 37°C. This is expected from the geometry of an A-form RNA helix, which places a 5′ dangling end away from the opposite strand (Freier et al. 1985).

Hydrogen-bonding Interactions

It is clear that hydrogen bonds are important for providing specificity in the binding and folding of macromolecules. The contributions of hydro-

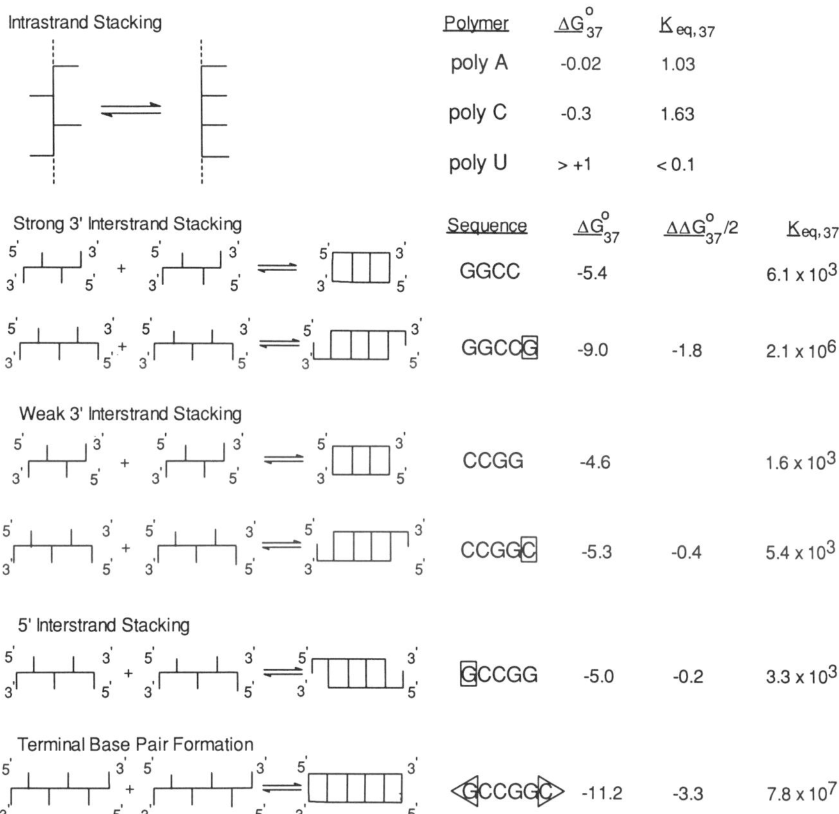

Polymer	ΔG^{o}_{37}	$K_{eq,37}$
poly A	-0.02	1.03
poly C	-0.3	1.63
poly U	> +1	< 0.1

Sequence	ΔG^{o}_{37}	$\Delta\Delta G^{o}_{37}/2$	$K_{eq,37}$
GGCC	-5.4		6.1×10^{3}
GGCCG	-9.0	-1.8	2.1×10^{6}
CCGG	-4.6		1.6×10^{3}
CCGGC	-5.3	-0.4	5.4×10^{3}
GCCGG	-5.0	-0.2	3.3×10^{3}
GCCGGC	-11.2	-3.3	7.8×10^{7}

Figure 1 Some examples of free-energy changes, ΔG^{o}_{37}, and free-energy increments, $\Delta\Delta G^{o}_{37}/2$ for stacking and base-pairing at 37°C. For intrastrand stacking in homopolymers, ΔG^{o}_{37} values come from combining optical and calorimetric data (Freier et al. 1981) or from the absence of any observable stacking in polyuridylic acid (Richards et al. 1963). For interstrand stacking, $\Delta\Delta G^{o}$ values come from subtracting the ΔG^{o}_{37} for duplex formation by a fully Watson-Crick-paired core from the ΔG^{o}_{37} for duplex formation by the core sequence with an added nucleotide that does not pair. For example, $\Delta\Delta G^{o}_{37}$ (2 $^{CG}_{G}$) = ΔG^{o}(GGCCG) – ΔG^{o}(GGCC). The unpaired nucleotide is boxed in the sequence. Also shown is the free-energy increment for adding two terminal CG pairs to a (CCGG)$_2$ core. Note that in all cases, the $\Delta\Delta G^{o}_{37}$ has been divided by two to give the free-energy increment per end.

gen bonds to the free-energy changes for binding and folding have been controversial, however (Abeles et al. 1992). This is because formation of hydrogen bonds in binding and folding must be accompanied by breakage of hydrogen bonds to water. For RNA, thermodynamic studies suggest the net contribution of a hydrogen bond to binding or folding is context-dependent (SantaLucia et al. 1992), but can be significant (Freier et al. 1985, 1986b; Turner et al. 1987). The evidence is discussed below.

The first indication that hydrogen bonds contribute to the ΔG^o values for interactions between RNA oligomers comes from the changes in free energy of duplex formation when an oligomer with a dangling end is extended by one nucleotide to allow the previously unpaired dangling end to form a base pair (Freier et al. 1985, 1986b). One example is illustrated in Figure 1. For this case at 37°C, the duplex with two terminal CG pairs, $(GCCGGC)_2$, is 5.9 kcal/mole more stable than the duplex with two terminal 3′ dangling ends, $(CCGGC)_2$ (Freier et al. 1985). Most of this extra stability depends on the complementarity of the 5′ and 3′ terminal residues, since the duplex with two terminal AC mismatches, $(ACCGGC)_2$, is only 1.4 kcal/mole more stable than $(CCGGC)_2$ (Hickey and Turner 1985). This suggests the hydrogen bonds in the terminal CG pairs are each worth at least (5.9–1.4)/6 = 0.8 kcal/mole. A more refined estimate for the ΔG^o of a hydrogen bond requires a correction for conformational entropy effects. This gives a value of 1.6 kcal/(mole hydrogen bond) for this system (Freier et al. 1986b).

Another approach to estimating the contribution of a hydrogen bond to binding or folding is to measure the free-energy consequences of replacing a single hydrogen-bonding group with a hydrogen (Turner et al. 1987). This "atomic mutation" approach is made possible by advances in the chemical synthesis of RNA (Kierzek et al. 1986; Usman et al. 1987). The results from several studies involving substitutions of single functional groups are illustrated in Figure 2. The changes in free energy upon substituting hydrogen for a single hydrogen-bonding group range from 0 to 1.6 kcal/mole. This range reflects different contexts for the hydrogen bonds. For example, substituting the 5′-terminal Gs of $(GCCGGC)_2$ with inosine replaces the 5′-terminal G amino groups with hydrogens (Fig. 2). This reduces duplex stability by 3.2 kcal/mole or 1.6 kcal/(mole hydrogen bonds), roughly as expected from the comparison of dangling ends and terminal mismatches described above (Turner et al. 1987). Substituting purine for A in GA mismatches replaces the amino group of A with a hydrogen. As shown in Figure 2, this substitution reduces duplex stability by 1.3–1.5 kcal/(mole mismatch) when two GA mismatches are in the center of a helix and by 0.4 kcal/mole for a GA mismatch in a 4-nucleotide hairpin, but it has no effect for a GA mismatch at the end of a helix (SantaLucia et al. 1991a, 1992). The reasons for this range are not clear. One possibility is that loss of a hydrogen bond at the end of a helix is compensated by enhanced conformational freedom. Clearly, however, a single hydrogen bond between bases can contribute on the order of 2 kcal/mole to binding or folding.

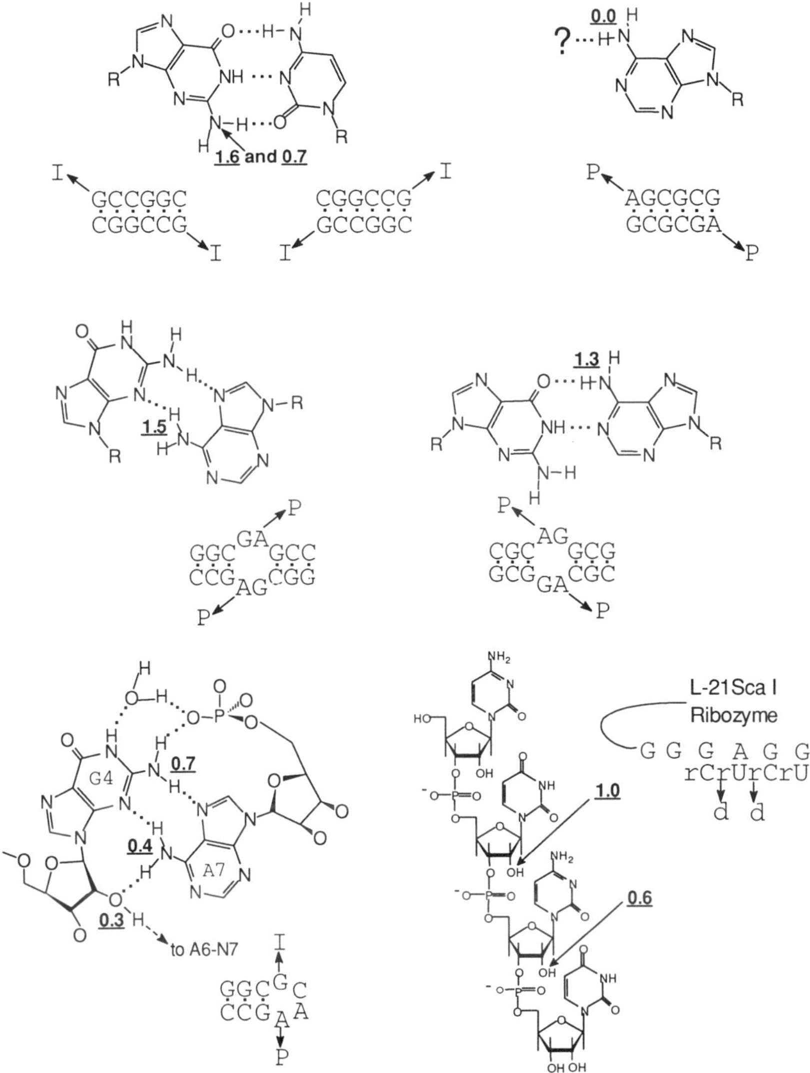

Figure 2 Free-energy increments for hydrogen bonds derived by measuring the thermodynamic effect of substituting H for a functional group. Substitutions are denoted by arrows: I = inosine, P = purine, d = deoxyribose. Changes observed in binding or folding free energy (kcal/mole) are listed in bold, underlined numbers near the functional group that was substituted. Changes in duplex formation (*top* 4 structures) are at 37°C (Turner et al. 1987; SantaLucia et al. 1991a). Changes in hairpin stability (*bottom left*) are at 70°C and are about 0.1 kcal/mole smaller at 37°C (SantaLucia et al. 1992). Changes in ribozyme binding are at 15°C (Bevilacqua and Turner 1991).

Hydrogen-bonded interactions in RNA are not restricted to the bases. It has been shown by kinetic (Sugimoto et al. 1989), gel retardation (Pyle and Cech 1991), and equilibrium dialysis (Bevilacqua and Turner 1991) methods that binding of substrates to a group I ribozyme is partially dependent on particular 2'-OH groups. This provides an advantage for folding and catalytic diversity not possessed by DNA. For example, comparisons of the effects on binding of substrate CUCU to the ribozyme or to an oligomer after replacing substrate 2'-OH groups with hydrogens suggest interactions with two of the 2'-OH groups contribute about 1 kcal/mole each to binding free energy at 15°C (see Fig. 2) (Bevilacqua and Turner 1991). Additionally, 2'-OH group interactions might also help catalyze reactions without affecting binding in the ground state. For example, substitution of H for the 2' OH of the reactive nucleotide of substrates for two different forms of a group I intron hinders reactivity (Sugimoto et al. 1989; Bevilacqua and Turner 1991 and unpubl.) but does not affect ground-state binding. Thus, a 2'-OH can enhance binding in both ground and transition states.

Presumably, hydrogen bonds to phosphate groups are also important for RNA binding and folding (Quigley and Rich 1976). Thus, a single nucleotide has many contact points that can provide significant contributions to free-energy changes for binding and folding.

Counterion Condensation

A unique property of nucleic acids is their high density of negative charge. This leads to condensation of neutralizing counterions into a relatively small volume around the nucleic acid chain, producing a high local concentration of counterions that is independent of the counterion concentration in the bulk solvent (Manning 1978; Record et al. 1978). This concentrating of counterions around the helix is entropically unfavorable, but less so as the counterion concentration in the bulk solvent increases. This underlies the common linear dependence of duplex melting temperature, T_M, as a function of log [Na$^+$]. Another consequence of counterion condensation is that multivalent ions will displace essentially all univalent cations around the chain. Thus, the local volume around a nucleic acid is enriched in multivalent cations. This guarantees a ready source of the metal ions required for catalysis. The details of counterion condensation are relatively well understood for polymeric nucleic acids (Manning 1978; Record et al. 1978). For example, the local concentration of +2 cations near a polymer double helix is on the order of 0.4 M, independent of the bulk concentration of +2 cations (Manning 1978). Studies of the salt dependence of the T_M and the association rate for the hexamer dGCATGC suggest that counterion condensation is also important for oligomers (Williams et al. 1989). The theoretical framework for a quantitative description of the effect with oligomers is currently being developed (Olmsted et al. 1989).

Thermodynamic Advantages for RNA as the First Macromolecular Catalyst and Information Carrier

A prerequisite for catalysis and information transfer is strong binding. For catalysis, simply bringing reactants together into the correct position for reaction can enhance rates by factors of up to 10^8 (Fersht 1985; Jencks 1987; Abeles et al. 1992). We have argued above that individual stacking and hydrogen-bonding interactions in RNA can each contribute on the order of 2 kcal/mole to the ΔG^o for binding at 37°C. This value is even more favorable at low temperature, especially for the stacking component. Thus, the most favorable stacking interaction for a 3′ dangling end contributes –2.6 kcal/mole at 0°C, enhancing a binding constant by a factor of more than 100 (Turner et al. 1988). This interaction would favor chain extension in a 5′ to 3′ direction, since a nucleotide coming onto the 3′ end of an RNA chain would be held tighter than one coming onto the 5′ end where stacking is negligible. In addition to stacking surfaces, of course, a single nucleotide has many hydrogen-bonding groups that can also enhance binding.

In contrast to nucleotides, the number of strong interactions available to a single amino acid is much smaller. This is reflected in the number of monomers required to form ordered structures in oligonucleotides and oligopeptides. Many dinucleotides have equilibrium constants of about 1 for stacking in a helical manner (Powell et al. 1972; Olsthoorn et al. 1981); the equilibrium constant for hairpin formation by $\text{GGC}^{\text{GCAA}}\text{GCC}$ at 0°C is 1×10^5 (SantaLucia et al. 1992). Even for optimized cases, however, an oligopeptide of 14 amino acids is required to get an equilibrium constant of order 1 for α-helix formation at 3°C (Shoemaker et al. 1987). The dissociation constant for duplex formation by GGCC at 0°C is 62 nm (Freier et al. 1983). The dissociation constants for assembly of short peptide strands have not been measured yet, although assembly of covalently linked fragments has been demonstrated (Oas and Kim 1988). Note that formation of GGCC from monomers requires only three coupling reactions; from dimers it requires only one coupling. In a prebiotic world with coupling reactions difficult, nucleic acids have a clear thermodynamic advantage over proteins. Few coupling reactions are required to allow strong associations into ordered assemblies.

Another thermodynamic-based advantage for evolution of nucleic acids is the large temperature dependence for binding constants between nucleic acids. For example, before enzymes evolved, RNA elongation may have been catalyzed by bringing two substrates close via a complementary template. The large temperature dependence of these associations could give rise to a catalytic cycle where low temperatures favored binding for efficient elongation of short oligomers, but high temperatures allowed efficient release of the longer products. Effectively, chain growth becomes a two-step cycle, perhaps regulated by day and night.

Consider the following simple mechanism for the first step of the cycle in which two identical tetramers, GGCC, bind a complementary template, GGCCGGCC, and ligate to form a template/product complex:

$$
\underset{\substack{5'\\ \text{GGCC}}}{P} {}^{3'} +
\underset{\substack{\text{GGCC}\\ \cdots\cdots\\ {}_3\text{CCGGCCGG}_{5'}}}{\overset{P}{\underset{}{}_5} {}^{3'}}
\underset{k_{\text{off}}}{\overset{k_{\text{on}}}{\rightleftharpoons}}
\underset{\substack{\text{GGCCGGCC}\\ \cdots\cdots\cdots\cdots\\ {}_3\text{CCGGCCGG}_{5'}}}{\overset{P}{\underset{}{}_5} {}^{3'}}
\overset{k_{\text{lig}}}{\longrightarrow}
\underset{\substack{{}^{5'}\text{GGCCGGCC}{}^{3'}\\ \cdots\cdots\cdots\cdots\\ {}_3\text{CCGGCCGG}_{5'}}}{}
$$

For short oligomers, K_m ($=(k_{\text{lig}} + k_{\text{off}})/k_{\text{on}}$) is large. Since oligomer abundance is presumably low in early stages of evolution, $K_m \gg$ $[\text{GGCC}]_o \gg [\text{GGCCGGCC}]_o$. The apparent rate, k_{app}, for conversion of GGCC to the template/product complex is then given by $k_{\text{app}} = \{k_{\text{lig}}k_{\text{on}}/(k_{\text{lig}} + k_{\text{off}})\} [\text{GGCC}]_o$. The rate of association for oligomers making GC pairs, k_{on}, is essentially independent of temperature (Pörschke et al. 1973; Williams et al. 1989), so the large temperature dependence of the binding constant is manifested primarily in the rate of dissociation, k_{off}. For example, the half-life, $t_{1/2}$, for dissociation of the first GGCC and template is predicted to be 5 minutes at 0°C, and 9.4 milliseconds at 37°C (Freier et al. 1986a; Turner et al. 1988). Since ligation is likely to be slow, most template-binding events will not result in reaction ($k_{\text{off}} \gg k_{\text{lig}}$). Under these "rapid equilibrium" conditions, the ratio of k_{app} at two temperatures reduces to:

$$
\frac{k_{\text{app}}^{T_{\text{low}}}}{k_{\text{app}}^{T_{\text{high}}}} = \frac{k_{\text{lig}}^{T_{\text{low}}}}{k_{\text{lig}}^{T_{\text{high}}}} \; \frac{k_{\text{off}}^{T_{\text{high}}}}{k_{\text{off}}^{T_{\text{low}}}}
$$

The temperature dependence of each rate can be approximated by the Arrhenius equation, $k = A \exp(-E_a/RT)$. This gives:

$$
\frac{k_{\text{app}}^{T_{\text{low}}}}{k_{\text{app}}^{T_{\text{high}}}} \propto \exp\left[\frac{1}{R}\left(\frac{1}{T_{\text{low}}} - \frac{1}{T_{\text{high}}}\right)(E_{a,\text{off}} - E_{a,\text{lig}})\right]
$$

Thus, if the temperature dependence of k_{off} is larger than k_{lig} (i.e., $E_{a,\text{off}} > E_{a,\text{lig}}$), conversion of GGCC to the template/product complex becomes favored at low temperature by this exponential term. Moreover, under such conditions, the template position of tightest substrate binding is also most efficient, adding a primitive degree of fidelity.

After one or more ligations, no sites would be left on the template for substrate binding, necessitating the second step of the cycle: release of the template/product complex.

$$5'\text{GGCCGGCC}^{3'} \xrightarrow{k_{off}} 2\ 5'\text{GGCCGGCC}\ 3'$$
$$3'\text{CCGGCCGG}_{5'}$$

Kinetic experiments on site-specific cleavage by a group I ribozyme clearly show that product dissociation is rate-limiting under certain conditions (Herschlag and Cech 1990). Presumably, this will be equally or more limiting for the products of ligation reactions on a template. If the length of the product is double that of the substrate as considered above, nearest-neighbor rules indicate ΔG^o will roughly double. As a consequence, $t_{1/2}$ at 0°C for dissociation of the product, GGCCGGCC, from its template becomes 250,000 years! The large temperature dependence of k_{off}, however, permits dissociation by temperature increase; $t_{1/2}$ for dissociation of GGCCGGCC from its template is only 16 hours at 37°C, one warm cycle. In a sense, high temperature could serve as a primitive helicase. Once released, the ligated product could self-assemble and/or undergo further rounds of elongation. Thus, low temperature favors chain elongation of short oligomers, but release of longer products may require high temperature. This is possible because of the large temperature dependence for RNA associations.

It should be noted that template/product formation has an optimum temperature; that is, lower temperature does not *always* favor chain elongation. Temperatures below the optimum disfavor the first step of the cycle for several reasons. Oligomers could bind so tightly to incorrect template positions or positions separated by gaps that a steady-state population of correctly aligned intermediates does not accumulate on a reasonable time scale. Such effects would decrease fidelity and prevent efficient chain elongation. Thus, the optimum temperature for chain elongation depends on both sequence and length. In general, however, release of the elongated product on a reasonable time scale requires an increase above the optimum temperature for chain elongation. It is clear that the strong dependence of k_{off} on both temperature and oligomer length must be considered in discussions of template-mediated coupling of RNA. These considerations suggest that in vitro experiments aimed at simulating evolution might benefit from incorporation of a temperature cycling step.

Thermodynamic Considerations for Evolution of the First Binding Sites and Codes

Even with the many binding contacts available on a nucleotide, initial assembly must have been difficult, given the presumably low concentrations of substrates. This requirement for tight binding may have been one element in the selection of the $3'$-$5'$ sugar-phosphate backbone instead of the $2'$-$5'$ alternative. Short oligomers with $2'$-$5'$ linkages form base pairs, but the ΔG^o_{37} per base pair for duplex formation is typically about 1 kcal/mole less favorable than for the corresponding $3'$-$5'$ linked oligomer (Kierzek et al. 1992). Another factor favoring the $3'$-$5'$ linkage is chemical stability (Usher and McHale 1976).

Given a $3'$-$5'$ backbone, the ΔG^o values for helix propagation by Watson-Crick base pairs (Freier et al. 1986a) suggest that GC pairs would be favored, since on average, adjacent GC pairs contribute 2.2 kcal/mole more than adjacent AU pairs at 0°C. The ΔG^o values for GU pairs are similar to those for AU pairs (He et al. 1991). Thus, there is no thermodynamic requirement for A in early helices. Interestingly, anticodons containing A typically code for amino acids that may not have been necessary at early times: Phe, Tyr, Trp, Cys, Met, Leu, Ile, and Val. Absence of A, however, would give replication in which U could substitute for C, although C would still be favored due to stronger binding.

Formation of simple helices, of course, is probably not sufficient to produce a range of catalytic activities. Presumably, base functional groups must be available to bind substrates in particular orientations and to provide catalysis. This focuses attention on motifs that are not Watson-Crick paired. Unfortunately, most such motifs destabilize a duplex and therefore require more Watson-Crick base pairs to hold it together. The only known exceptions to this general rule involve insertions of single and double GU mismatches into helices (He et al. 1991). Some mismatches are less destabilizing than others, however. For example, recent studies of internal loops containing two mismatches show that insertion of $^{GA}_{AG}$, $^{AG}_{GA}$, and $^{UU}_{UU}$ double mismatches into a helix destabilizes the fully Watson-Crick-paired duplex much less than other non-GU double mismatches (SantaLucia et al. 1991b). Figure 3 shows the change in ΔG^o_{37} measured upon insertion of these and other double mismatches into a helix. The thermodynamic advantage for insertion of the GA and UU double mismatches averages 2.1 kcal/mole at 37°C. NMR studies suggest the enhanced stabilities of loops with GA, UU, and GU mismatches are due to hydrogen bonding (He et al. 1991; SantaLucia et al. 1991b; Nikonowicz and Pardi 1992). If short helices were initially preferred due to difficulty in coupling nucleotides, then the thermo-

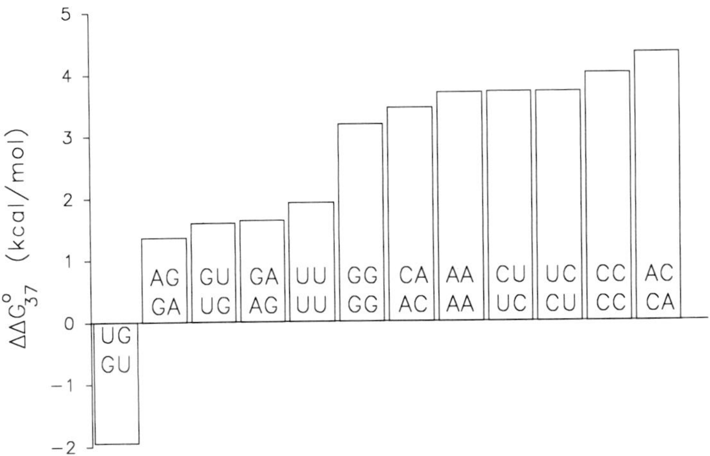

Figure 3 Change in free energy of duplex formation at 37°C upon inserting double mismatches into the middle of a helix. For example, $\Delta\Delta G°_{37}$ ($^{AG}_{GA}$) = $\Delta G°_{37}$(CGC<u>AG</u>GCG) − $\Delta G°_{37}$ (CGCGCG) (SantaLucia et al. 1991b). Positive values of $\Delta\Delta G°_{37}$ are unfavorable for duplex formation.

dynamic results suggest internal loops with GA, UU, and GU mismatches would also be preferred. (Presumably an isolated AU or GC pair in the middle of a loop could also be favorable.)

Modeling of the $^{GA}_{AG}$ mismatch, as shown in Figure 4, suggests even a small internal loop can provide a useful recognition element (SantaLucia 1991). Each GA mismatch is hydrogen-bonded as shown in Figure 2 in a pattern initially seen in DNA by Li et al. (1991) and in RNA by Heus and Pardi (1991). This leaves normal Watson-Crick hydrogen-bonding positions available on G for binding something into the pocket that is formed. Interestingly, the $^{AG}_{GA}$ internal loop has a hydrogen-bonding pattern involving the G imino and carbonyl groups (Fig. 2). Thus, it has fewer functional groups available for further binding. In a survey of 21 16S rRNAs (Gutell et al. 1985) and 38 23S rRNAs (Gutell and Fox 1988), the $^{GA}_{AG}$ motif occurs 45 times, whereas the $^{AG}_{GA}$ motif is absent. Since $^{GA}_{AG}$ and $^{AG}_{GA}$ are thermodynamically similar, this suggests $^{GA}_{AG}$ was selected for its binding possibilities.

Thermodynamic considerations also favor placing mismatches between helices instead of at the ends in order to optimize their effectiveness for binding and catalysis. Mismatches flanked by base pairs have less conformational freedom than mismatches at the ends of helices. This results in greater preorganization of the binding site and hence tighter binding. Moreover, the local Mg^{++} concentration should be greater in the middle than at the end of a helix (Olmsted et al. 1989).

Figure 4 Model of the duplex $\begin{smallmatrix}&\text{GA}&\\\text{GGC}&&\text{GCC}\\\text{CCG}&&\text{CGG}\\&\text{AG}&\end{smallmatrix}$ based on NMR data (SantaLucia 1991).

Base pairs are colored green and the GA mismatches have traditional colors for atoms (O = red, H = white, C = black, N = blue).

Hairpin loops provide another unpaired motif that can be added to an RNA helix while favoring folding. For example, the free-energy penalty for initiating a helix from two separate strands is 3.4 kcal/mole at 37ºC (Freier et al. 1986a). If the helix is initiated intramolecularly, giving rise to a hairpin loop of 4, 5, or 7 Us, then the initiation penalty is 3.0, 3.1, or 3.8 kcal/mole, respectively (Groebe and Uhlenbeck 1988). Some loops of 4, called tetraloops, are even less destabilizing (Tuerk et al. 1988). For example, Antao and Tinoco (1992) report a penalty of only 1.0 kcal/mole for initiating a loop at 37ºC with the sequence UUCG. In addition to the relatively favorable $\Delta G°$ for initiation of a hairpin, hairpin folding is further favored by the transformation of the folding from an intermolecular to an intramolecular process. For the intramolecular folding of a hairpin, half the molecules have a base-paired stem when $\Delta G°_{\text{hairpin}} = 0$ kcal/mole. For helix formation by two self-complementary strands, a $\Delta G°$ of 0 kcal/mole results in half the strands forming duplex helix only when the strand concentration is 1 M. For more realistic strand concentrations, the $\Delta G°$ must be much more favorable. For self-complementary duplexes, the $\Delta G°$ required to drive half the strands into duplex is given by $\Delta G° = RT \ln C_{\text{T}}$, where C_{T} is the total strand concentration. Thus, at 10^{-6} M strand concentration, a ΔG^o of −8.5 kcal/mole is required to drive half the strands into duplex at 37°C. Evidently, there was a strong thermodynamic pressure to select hairpins at an early stage in evolution.

Assembling Motifs

The above discussion focused largely on thermodynamic considerations for the coupling and interaction of short oligonucleotides. These could readily fold into well-known elements of secondary structure such as hairpins and internal loops. At some point, however, it must have been advantageous to assemble two or more motifs into more complicated structures. As with base-pairing, such assembly is facilitated when the process is intramolecular. Moreover, the same stacking and hydrogen-bonding interactions that drive base-pairing can also promote assembly of motifs. This can be seen in the crystal structure of tRNA, where such interactions fix the relative positions of the D and TΨC loops (Kim et al. 1974; Robertus et al. 1974). It is likely, however, that additional interactions are also important in this tertiary folding process.

One tertiary interaction unique to RNA is hydrogen bonding to the 2′-OH group. The role of a 2′-OH group in stabilizing a helix was first suggested by kinetic experiments using chimeric DNA-RNA substrates to open a covalently closed, circular group I intron (Sugimoto et al. 1989). Subsequent gel retardation (Pyle and Cech 1991) and equilibrium dialysis (Bevilacqua and Turner 1991) studies suggest the interaction of a 2′-OH group can contribute on the order of 1 kcal/mole to helix stability. Proposed three-dimensional structures of the group I intron suggest this 2′-OH interaction helps hold and orient the substrate helix in the active site (Kim and Cech 1987; Michel and Westhof 1990). Comparison of the effects of substituting a 2′OH with a 2′H on binding to ribozyme and to a hexamer suggest additional, as yet unidentified interactions are also involved in this helix positioning (Bevilacqua and Turner 1991). It has also been shown that the intramolecular rate for this helix positioning is about 3 s^{-1} at 15°C, much slower than the rate for formation of secondary structure (Bevilacqua et al. 1992). This might be expected if, in fact, secondary structure initially evolved in the absence of tertiary structure.

An electric dichroism study of bundling of DNA helices also suggests there are new interactions involved in helix packing (Mandelkern et al. 1981). This study showed that seven DNA helices, each of about 140 base pairs, can pack together side by side, perhaps with six helices hexagonally surrounding a center helix. Counterintuitively, this association is promoted at low concentrations of monovalent counterion, suggesting it is accompanied by release of Na^+. The assembly is also promoted by +3 ions. Although no equivalent study has been done on RNA, it would not be surprising to find similar effects.

Assembly of helices could have important implications for RNA catalysis by providing a region of low dielectric constant and low water activity. Such a milieu is often important for catalysis by proteins (Abeles et al. 1992). The proposed three-dimensional structures of group I introns contain such a catalytic core (Kim and Cech 1987; Michel and Westhof 1990). Moreover, mapping experiments with hydroxyl radicals provide experimental evidence that substantial regions of the ribozyme are inaccessible to solvent after tertiary folding is induced by Mg^{++} (Celander and Cech 1991).

CONCLUSIONS AND PERSPECTIVES

The many functional groups, high charge density, and strong interactions of RNA provide thermodynamic advantages for RNA in initial stages of evolution. In particular, relatively few coupling reactions are required to form an oligomer capable of tight binding and stable folding. Many of the interactions important for determining secondary and tertiary structure are known, but additional interactions remain to be discovered. The products of evolution were restricted by the building blocks available and their environment. Recent studies of rationally designed, organic replicating systems suggest that similar interactions can result in self-assembly of different building blocks in organic solvents (Tjivikua et al. 1990; Feng et al. 1992). In vitro selection methods have been used to identify nucleic acid sequences capable of specific functions (Ellington and Szostak 1990, 1992; Tuerk and Gold 1990). This suggests similar methods may ultimately be developed for use with different building blocks, thus expanding the possibilities available. Thus, understanding intermolecular interactions may facilitate rational design of a replicating system that can be used to evolve the best structures available for a particular catalytic or other purpose.

ACKNOWLEDGMENTS

This work was supported by National Institutes of Health grant GM-22939.

REFERENCES

Abeles, R.H., P.A. Frey, and W.P. Jencks. 1992. *Biochemistry*. Jones and Bartlett, Boston.

Antao, V.P. and I. Tinoco, Jr. 1992. Thermodynamic parameters for loop formation in RNA and DNA hairpin tetraloops. *Nucleic Acids Res.* **20:** 819–824.

Bevilacqua, P.C. and D.H. Turner. 1991. Comparison of binding of mixed ribose-deoxyribose analogues of CUCU to a ribozyme and to GGAGAA by equilibrium dialysis: Evidence for ribozyme specific interactions with 2′OH groups. *Biochemistry* **30:** 10632–10640.

Bevilacqua, P.C., R. Kierzek, K.A. Johnson, and D.H. Turner. 1992. Dynamics of ribozyme binding of substrate revealed by fluorescence detected stopped-flow. *Science* **258:** 1355–1358.

Celander, D.W. and T.R. Cech. 1991. Visualising the higher order folding of a catalytic RNA molecule. *Science* **251:** 401–407.

Ellington, A.D. and J.W. Szostak. 1990. In vitro selection of RNA molecules that bind specific ligands. *Nature* **346:** 818–822.

———. 1992. Selection in vitro of single-stranded DNA molecules that fold into specific ligand-binding structures. *Nature* **355:** 850–852.

Feng, Q., K. Park, and J. Rebek, Jr. 1992. Crossover reactions between synthetic replicators yield active and inactive recombinants. *Science* **256:** 1179–1180.

Fersht, A. 1985. *Enzyme structure and mechanism*, chapter 2. W.H. Freeman, New York.

Filimonov, V.V. and P.L. Privalov. 1978. Thermodynamics of base interaction in $(A)_n$ and $(A{\cdot}U)_n$. *J. Mol. Biol.* **122:** 465–470.

Freier, S.M., D. Alkema, A. Sinclair, T. Neilson, and D.H. Turner. 1985. Contributions of dangling end stacking and terminal base-pair formation to the stabilities of XGGCCp, XCCGGp, XGGCCYp, and XCCGGYp helixes. *Biochemistry* **24:** 4533–4539.

Freier, S.M., B.J. Burger, D. Alkema, T. Neilson, and D.H. Turner. 1983. Effects of 3′ dangling end stacking on the stability of GGCC and CCGG double helices. *Biochemistry* **22:** 6198–6206.

Freier, S.M., K.O. Hill, T.G. Dewey, L.A. Marky, K.J. Breslauer, and D.H. Turner. 1981. Solvent effects on the kinetics and thermodynamics of stacking in poly(cytidylic acid). *Biochemistry* **20:** 1419–1426.

Freier, S.M., R. Kierzek, J.A. Jaeger, N. Sugimoto, M.H. Caruthers, T. Neilson, and D.H. Turner. 1986a. Improved free-energy parameters for predictions of RNA duplex stability. *Proc. Natl. Acad. Sci.* **83:** 9373–9377.

Freier, S.M., N. Sugimoto, A. Sinclair, D. Alkema, T. Neilson, R. Kierzek, M.H. Caruthers, and D.H. Turner. 1986b. Stability of XGCGCp, GCGCYp, and XGCGCYp helixes: An empirical estimate of the energetics of hydrogen bonds in nucleic acids. *Biochemistry* **25:** 3214–2319.

Groebe, D.R. and O.C. Uhlenbeck. 1988. Characterization of RNA hairpin loop stability. *Nucleic Acids Res.* **16:** 11725–11735.

Gutell, R.R. and G.E. Fox. 1988. Compilation of large subunit RNA sequences presented in a structural format. *Nucleic Acids Res.* (suppl.) **16:** 175–269.

Gutell, R.R., B. Weiser, C.R. Woese, and H.F. Noller. 1985. Comparative anatomy of 16-S-like ribosomal RNA. *Prog. Nucleic Acids Res. Mol. Biol.* **32:** 155–216.

He, L., R. Kierzek, J. SantaLucia, Jr., A.E. Walter, and D.H. Turner. 1991. Nearest-neighbor parameters for G${\cdot}$U mismatches: 5′GU3′/3′UG5′ is destabilizing in the contexts CGUG/GUGC, UGUA/AUGU, and AGUU/UUGA but stabilizing in GGUC/CUGG. *Biochemistry* **30:** 11124–11132.

Herschlag, D. and T.R. Cech. 1990. Catalysis of RNA cleavage by the *Tetrahymena thermophila* ribozyme. 1. Kinetic description of the reaction of an RNA substrate complementary to the active site. *Biochemistry* **29:** 10159–10171.

Heus, H.A. and A. Pardi. 1991. Structural features that give rise to the unusual stability of

RNA hairpins containing GNRA loops. *Science* **253:** 191–194.

Hickey, D.R. and D.H. Turner. 1985. Effects of terminal mismatches on RNA stability: Thermodynamics of duplex formation for ACCGGGp, ACCGGAp, and ACCGGCp. *Biochemistry* **24:** 3987–3991.

Inners, L.D. and G. Felsenfeld. 1970. Conformation of polyribouridylic acid in solution. *J. Mol. Biol.* **50:** 373–389.

Jencks, W.P. 1987. *Catalysis in chemistry and enzymology.* Dover, New York.

Kierzek, R., L. He, and D.H. Turner. 1992. Association of 2′-5′ oligoribonucleotides. *Nucleic Acids Res.* **20:** 1685–1690.

Kierzek, R., M.H. Caruthers, C.E. Longfellow, D. Swinton, D.H. Turner, and S.M. Freier. 1986. Polymer-supported RNA synthesis and its application to test the nearest-neighbor model for duplex stability. *Biochemistry* **25:** 7840–7846.

Kim, S.-H. and T.R. Cech. 1987. Three-dimensional model of the active site of the self-splicing rRNA precursor of *Tetrahymena. Proc. Natl. Acad. Sci.* **84:** 8788-8792.

Kim, S.H., F.L. Suddath, G.J. Quigley, A. McPherson, J.L. Sussman, A.H.J. Wang, N.C. Seeman, and A. Rich. 1974. Three dimensional tertiary structure of yeast phenylalanine transfer RNA. *Science* **185:** 435–440.

Li, Y., G. Zon, and W.D. Wilson. 1991. NMR and molecular modeling evidence for a G•A mismatch base pair in a purine-rich DNA duplex. *Proc. Natl. Acad. Sci.* **88:** 26–30.

Mandelkern, M., N. Dattagupta, and D.M. Crothers. 1981. Conversion of B DNA between solution and fiber conformations. *Proc. Natl. Acad. Sci.* **78:** 4294–4298.

Manning, G.S. 1978. The molecular theory of polyelectrolyte solutions with applications to the electrostatic properties of polynucleotides. *Q. Rev. Biophys.* **11:** 179–246.

Michel, F. and E. Westhof. 1990. Modelling of the three-dimensional architecture of group I catalytic introns based on comparative sequence analysis. *J. Mol. Biol.* **216:** 585–610.

Nikonowicz, E.P. and A. Pardi. 1992. Three-dimensional heteronuclear NMR studies of RNA. *Nature* **355:** 184–186.

Oas, T.G. and P.S. Kim. 1988. A peptide model of a protein folding intermediate. *Nature* **336:** 42–48.

Olmsted, M.C., C.F. Anderson, and M.T. Record, Jr. 1989. Monte Carlo description of oligoelectrolyte properties of DNA oligomers: Range of the end effect and the approach of molecular and thermodynamic properties to the polyelectrolyte limits. *Proc. Natl. Acad. Sci.* **86:** 7766–7770.

Olsthoorn, C.S.M., L.J. Bostelaar, J.F.M. De Rooij, J.H. van Boom, and C. Altona. 1981. Circular dichroism study of stacking properties of oligodeoxyadenylates and poly-deoxyadenylate. *Eur. J. Biochem.* **115:** 309–321.

Petersheim, M. and D.H. Turner. 1983. Base-stacking and base-pairing contributions to helix stability: Thermodynamics of double-helix formation with CCGG, CCGGp, CCGGAp, ACCGGp, CCGGUp, and ACCGGUp. *Biochemistry* **22:** 256–263.

Pörschke, D., O.C. Uhlenbeck, and F.H. Martin. 1973. Thermodynamics and kinetics of the helix-coil transition of oligomers containing GC base pairs. *Biopolymers* **12:** 1313–1335.

Powell, J.T., E.G. Richards, and W.B. Gratzer. 1972. The nature of stacking equilibria in polynucleotides. *Biopolymers* **11:** 235–250.

Pyle, A.M. and T.R. Cech. 1991. Ribozyme recognition of RNA by tertiary interactions with specific ribose 2′-OH groups. *Nature* **350:** 628–631.

Quigley, G.J. and A. Rich. 1976. Structural domains of transfer RNA molecules. *Science* **194:** 796–806.

Record, M.T., Jr., C.F. Anderson, and T.M. Lohman. 1978. Thermodynamic analysis of ion effects on the binding and conformational equilibria of proteins and nucleic acids: The roles of ion association or release, screening, and ion effects on water activity. *Q. Rev. Biophys.* **11:** 103–178.

Richards, E.G., C.P. Flessel, and J.R. Fresco. 1963. Polynucleotides. VI. Molecular properties and conformation of polyribouridylic acid. *Biopolymers* **1:** 431–446.

Robertson, D.L. and G.F. Joyce. 1990. Selection in vitro of an RNA enzyme that specifically cleaves single-stranded DNA. *Nature* **344:** 467–468.

Robertus, J.D., J.E. Ladner, J.T. Finch, D. Rhodes, R.S. Brown, B.F.C. Clark, and A. Klug. 1974. Structure of yeast phenylalanine tRNA at 3Å resolution. *Nature* **250:** 546–551.

SantaLucia, J., Jr. 1991. "The role of hydrogen bonding in the thermodynamics and structure of mismatches in RNA oligonucleotides." Ph.D. thesis. University of Rochester, New York.

SantaLucia, J., Jr., R. Kierzek, and D.H. Turner. 1990. Effects of GA mismatches on the structure and thermodynamics of RNA internal loops. *Biochemistry* **29:** 8813–8819.

————. 1991a. Functional group substitutions as probes of hydrogen bonding between GA mismatches in RNA internal loops. *J. Am. Chem. Soc.* **113:** 4313–4322.

————. 1991b. Stabilities of consecutive A•C, C•C, G•G, U•C, and U•U mismatches in RNA internal loops: Evidence for stable hydrogen-bonded U•U and C•C$^+$ pairs. *Biochemistry* **30:** 8242–8251.

————. 1992. Context dependence of hydrogen bond free energy revealed by substitutions in an RNA hairpin. *Science* **256:** 217–219.

Shoemaker, K.R., P.S. Kim, E.J. York, J.M. Stewart, and R.L. Baldwin. 1987. Tests of the helix dipole model for stabilization of α-helices. *Nature* **326:** 563–567.

Sugimoto, N., R. Kierzek, and D.H. Turner. 1987. Sequence dependence for the energetics of dangling ends and terminal base pairs in ribonucleic acid. *Biochemistry* **26:** 4554–4558.

Sugimoto, N., M. Tomka, R. Kierzek, P.C. Bevilacqua, and D.H. Turner. 1989. Effects of substrate structure on the kinetics of circle opening reactions of the self-splicing intervening sequence from *Tetrahymena thermophila*: Evidence for substrate and Mg^{2+} binding interactions. *Nucleic Acids Res.* **17:** 355–371.

Suurkuusk, J., J. Alvarez, E. Freire, and R. Biltonen. 1977. Calorimetric determination of the heat capacity changes associated with the conformational transitions of polyriboadenylic acid and polyribouridylic acid. *Biopolymers* **16:** 2641–2652.

Tjivikua, T., P. Ballester, and J. Rebek, Jr. 1990. Self-replicating system. *J. Am. Chem. Soc.* **112:** 1249–1250.

Tuerk, C. and L. Gold. 1990. Systematic evolution of ligands by exponential enrichment: RNA ligands to bacteriophage T4 DNA polymerase. *Science* **249:** 505–510.

Tuerk, C., P. Gauss, C. Thermes, D.R. Groebe, M. Gaule, N. Guild, G. Stormo, Y. d'Aubenton-Carafa, O.C. Uhlenbeck, I. Tinoco, Jr., E.N. Brody, and L. Gold. 1988. CUUCGG hairpins: Extraordinarily stable RNA secondary structures associated with various biochemical processes. *Proc. Natl. Acad. Sci.* **85:** 1364–1368.

Turner, D.H., N. Sugimoto, and S.M. Freier. 1988. RNA structure prediction. *Annu. Rev. Biophys. Biophys. Chem.* **17:** 167–192.

Turner, D.H., N. Sugimoto, R. Kierzek, and S.D. Dreiker. 1987. Free energy increments for hydrogen bonds in nucleic acid base pairs. *J. Am. Chem. Soc.* **109:** 3783–3785.

Usher, D.A. and A.H. McHale. 1976. Hydrolytic stability of helical RNA: A selection advantage for the natural 3 ′, 5 ′-bond. *Proc. Natl. Acad. Sci.* **73:** 1149–1153.

Usman, N., K.K. Ogilvie, M.Y. Jiang, and R.L. Cedergren. 1987. Automated chemical synthesis of long oligoribonucleotides using 2′-O-silylated ribonucleoside 3′-O-phosphoramides on a controlled-pore glass support: Synthesis of a 43-nucleotide sequence similar to the 3′-half molecule of an *Escherichia coli* formylmethionine tRNA. *J. Am. Chem. Soc.* **109:** 7845–7854.

Williams, A.P., C.E. Longfellow, S.M. Freier, R. Kierzek, and D.H. Turner. 1989. Laser temperature-jump, spectroscopic, and thermodynamic study of salt effects on duplex formation by dGCATGC. *Biochemistry* **28:** 4283–4291.

18

RNA Structural Elements and RNA Function

Jacqueline R. Wyatt
Isis Pharmaceuticals
Carlsbad, California 92008

Ignacio Tinoco, Jr.
Department of Chemistry and Chemical Biodynamics Laboratory
University of California
Berkeley, California 94720

The structural elements that now exist in RNA must have evolved to provide chemical stability and to facilitate their biological functions. Structural biologists believe that determining the conformation and stability of the structures adopted by RNA will provide a better understanding of their functions. This belief is based mainly on the success of Watson and Crick with double strands for DNA. Even this archetype only shows that complementary base-pairing makes replication easy; the details of helix twist and of A-form, B-form, ... or Z-form geometry are less useful. Nevertheless, we continue to be optimistic and to study RNA structure in the hope that it will provide hints about how RNA does its many jobs. The main value of a structure is to suggest new experiments. The structure allows predictions about the effect of mutations, inhibitors, and enhancers, and about the mechanisms of the reactions. The overall goal is to learn the general features of RNA structure that can be inferred from the sequence and then to relate these structures to biological functions. It is important to recognize that a molecule is not a rigid object; structural elements are dynamic. Single-stranded loops are mobile, base pairs can form and break, and even base-paired helical regions can bend and flex. The presence or absence of structure depends not only on the sequence, but also on the temperature and the environment in general. This means that knowledge of the thermodynamics of structure formation is crucial (for review, see Turner and Bevilacqua, this volume).

Herein we review what is known about the simplest RNA structural elements—double helices, hairpin loops, pseudoknots, internal loops, bulges, base triples, junctions between duplexes—and describe some of

their known roles in function. It is immediately obvious that very little is known about either structure or function and that many simple experiments are waiting to be done to answer the important questions.

DOUBLE-STRANDED HELICES

Structure

An RNA chosen at random from a cell has at least half of its bases in double-stranded helices. The extensive base-pairing in transfer RNAs and ribosomal RNAs can be attributed to their functions requiring a compact structure. However, even messenger RNAs that must be unfolded to be translated, and viral RNAs—such as tobacco mosaic virus—that are unfolded in the virus particle, are over 50% double-stranded in solution (McMullen et al. 1967). It is difficult to devise a sequence containing equal numbers of all four nucleotides that would not pair half of its bases. Of course, a large excess of one type of base, or increasing A·U content, favors single-stranded regions.

Double-stranded regions in RNA molecules adopt an antiparallel, right-handed helical conformation originally observed in fiber diffraction studies and subsequently seen in all crystal structures of RNA molecules that contain base-paired regions (Saenger 1984; Westhof et al. 1985; Dock-Bregeon et al. 1989; Rould et al. 1989; Holbrook et al. 1991; Ruff et al. 1991). Studies of RNA duplexes in solution using nuclear magnetic resonance (NMR) also find A-form geometry (Happ et al. 1988; Chou et al. 1989; White et al. 1992). The NMR data readily distinguish an A-form from a B-form helix. Conformations of the individual nucleotides are accurately obtained, and if a large number of distance and torsion angle constraints are determined, reliable estimates of groove width, displacement, or base tilt can be made (Heus and Pardi 1991; Varani and Tinoco 1991; Varani et al. 1991; White et al. 1992). Sequences of alternating -G-C-G-C- form left-handed Z-form helices (Hall et al. 1984; Davis et al. 1990). Although left-handed helices may occur in natural RNAs (Zarling et al. 1987), the overwhelming majority of helices are A-form. No evidence for B-form helices has been found in RNA.

The $C_{3'}$-*endo* sugar pucker adopted by the nucleotides is one of the distinguishing features of an A-form helix. This sugar pucker is associated with an interphosphate distance of about 5.9 Å. In contrast, the $C_{2'}$-*endo* pucker adopted in B-form helices and in some RNA loop regions results in an interphosphate distance of about 7 Å. The A-form helix is underwound compared to the B-type helix and has a helical repeat of about 11 base pairs per turn. The helical repeat of RNA in solution, measured using gel electrophoresis, is estimated to be between 11 and 12 base pairs per turn (Bhattacharyya et al. 1990; Tang and Draper 1990). The base pairs are displaced away from the helix axis toward the minor groove 4.4 Å and tilted about 18° (Saenger 1984). Thus, the two grooves of the helix are very different; the major groove is deep and narrow, and the minor groove is shallow and wide.

RNA duplexes are not uniform. In tRNA^{Phe}, the number of base pairs per turn of the helices varies from about 10 in the stem of the D loop to 11 in the anticodon stem. The base pairs also differ in their displacement from the helix axis and in the tilt angle (Holbrook et al. 1978). Deviations depend on the sequence and on the structural context of the helix relative to the three-dimensional folding of the molecule. Three-base interactions (base triples) require a widening of the major groove that affects neighboring base pairs (Holbrook et al. 1978). Protein binding may also require a distortion in groove width (Ruff et al. 1991; Weeks and Crothers 1991). Bulges cause a structural distortion in adjacent helices detectable by altered RNase V_1 reactivity and enhanced intercalation (White and Draper 1987, 1989).

Function

Short-range pairings establish the secondary structure of the RNA molecule, creating hairpins, loops, bulges, and junctions. Duplexes formed by long-range pairing serve structural roles in stabilizing the three-dimensional folding of the RNA molecule. These include helices in the conserved core of group I self-splicing introns (Davies et al. 1982; Burke et al. 1987), two long-range pairings suggested by comparison of RNase P RNA from different species (James et al. 1988; Haas et al. 1991), and helices in rRNA that pair bases as far apart as 1990 nucleotides in the primary sequence (Woese et al. 1983; Gutell and Woese 1990). Perrotta and Been (1991) used site-specific mutagenesis to show that a single-stranded region paired to nucleotides at the junction of three stems is required for efficient self-cleavage of hepatitis δ virus RNA.

Some long-range pairings are implicated in critical functional roles. Reactions catalyzed or mediated by RNA allow highly specific and efficient recognition of nucleic acid sequences. In group II self-splicing, 5′ splice site selection is mediated by two exon-binding sites located within the intron; these sites bind the exon through Watson-Crick base-pairing (Jacquier and Michel 1987). In reactions carried out by the *Tetrahymena* ribozyme and the splicing reaction that it performs in vivo, substrate is bound through formation of six base pairs to a region of the intron called the internal guide sequence or 5′ exon-binding site (for review, see Cech 1987). The interaction is additionally stabilized by tertiary interactions involving the 2'-OH of two of the nucleotides of the substrate (Bevilacqua and Turner 1991; Pyle and Cech 1991), but specificity comes from Watson-Crick-type helix formation.

Watson-Crick base-pair formation is also involved in bimolecular recognition events. Several examples come from pre-messenger RNA splicing in eukaryotes. Nuclear pre-mRNA splicing, the removal of non-coding (intron) sequences from mRNA, occurs in the spliceosome, a large complex containing six small nuclear RNA molecules (snRNAs) and at least 30 proteins (for review, see Lührmann et al. 1990; Guthrie 1991; Ruby and Abelson 1991). The interaction between two of the snRNA molecules, U4 and U6, appears to play a regulatory role in the splicing process, as the two enter the spliceosome as an extensively base-paired complex that is disrupted as splicing occurs. The snRNA U1 is involved in recognition of the 5'splice site through base-pair formation. Two other RNA-RNA interactions in the spliceosome are mediated by base-pairing: recognition of the branch point in the mRNA by snRNA U2 and the U2-U6 interaction required for splicing in mammalian cells (for review, see Datta and Weiner 1991; Guthrie 1991; Wu and Manley 1991).

Translation of mRNA into protein also depends on RNA-RNA recognition events. Initiation of translation in prokaryotes requires base-pairing between the Shine-Dalgarno sequence, located about seven bases before the AUG initiation codon of the message, to a hexanucleotide sequence at the 3'end of the 16S rRNA (Lewin 1990). The anticodon-codon interaction is a critical bimolecular duplex-formation reaction, even though only three base pairs are formed. A study of the temperature dependence of the association constant of tetramer binding to the anticodon loop revealed that the interaction is much more stable than expected for formation of a four-base helix (Uhlenbeck 1972). A model for the interaction is provided by the crystal structure of tRNA^{Asp} (Moras et al. 1986). Packing forces in the crystallization resulted in interaction between the 5'-GUC-3' anticodon loops of two tRNAs. The complex is also present in solution (Romby et al. 1985). Comparison of the tRNA^{Asp} structure with the structure of tRNA^{Phe} showed that formation of the short A-form helix between the two anticodon loops resulted in very little change in the conformation of the anticodon loop. The anticodon loop in free tRNA may be in a more favorable conformation for hybridization than a single-stranded oligonucleotide.

Several prokaryotic systems rely on the specificity of RNA duplex formation for antisense regulation of replication, conjugation, and gene expression (for review, see Simons and Kleckner 1988). An example is the inhibition of plasmid ColE1 replication (Eguchi et al. 1991; Tomizawa, this volume). Critical pairing interactions between the antisense RNA and the RNA transcript occur between loop regions, referred to as the "kissing hairpins," and lead to complete hybridization of the two RNAs. The initial interaction is stabilized by a plasmid-encoded protein called Rom. Simons (1988) suggests that loop-loop interactions are typical of antisense mechanisms for several reasons. As is the case for the anticodon-codon interaction, loop regions may be in favorable conformations for hybridization. Loop-loop interactions may increase the specificity of the reaction; if initiation resulted in extensive hybridization, nonspecific interactions might be too stable to permit specific control. Finally, changes in specificity of the interaction require only concerted change at single positions in each RNA rather than changes in the nucleic acid binding domain of a protein.

Current evidence suggests that proteins generally make sequence-specific contacts with single-stranded residues in loops, rather than with base-paired regions. Duplex regions are necessary, however, for presentation of single-stranded nucleotides in the correct orientation. The narrow major groove of the A-form helix may preclude contact between amino acid side chains and the functional groups on the bases that allow discrimination unless structure of loop regions or protein binding itself alters the geometry of the helix. Base-specific contacts are observed between functional groups on bases in stem regions and amino acid residues in the crystal structures of both glutaminyl- and aspartyl-tRNA synthetases complexed with their cognate tRNAs (Rould et al. 1989; Ruff et al. 1991); however, the interactions are at the terminus of the RNA helix where amino acids are not sterically hindered from entering the groove. In the glutaminyl-tRNA the contacts are in the minor groove, and in the aspartyl-tRNA they are in the major groove. Interactions between the phosphate backbone and protein side chains stabilize protein-RNA interactions in a non-sequence-specific fashion as exemplified by the structure of the glutaminyl-tRNA synthetase complex (Rould et al. 1991). The phosphate backbone of helix regions is protected from ethylation by protein binding in the HIV Tat-TAR complex (Calnan et al. 1991) and in tRNA^{Val}-valyl-tRNA synthetase interaction (Florentz and Giegé 1986).

Some proteins function to recognize duplex regions and then disrupt them. A family of helicases, dubbed "DEAD (or DEAH) box" proteins because of a conserved amino acid sequence, includes several of the known nuclear pre-mRNA splicing factors and two eukaryotic translation initiation factors (for review, see Wassarman and Steitz 1991). A different type of protein recognizes double-stranded RNA and irreversibly unwinds it by converting adenosine to inosine. First recognized in *Xeno-*

pus, this unwinding/modifying activity is also present in mammalian cells (Bass and Weintraub 1988; Wagner et al. 1990). The function of the unwinding/modifying enzyme is as yet unknown, although it may help regulate the degradation of mRNA (Kimelman and Kirschner 1989).

HAIRPIN LOOPS

Structure

A hairpin loop results when a single strand doubles back to form a double-stranded region leaving some bases unpaired. The smallest hairpin loop identified so far has two unbonded bases; the largest naturally occurring hairpin loops may contain no more than about ten unbonded bases. Large hairpin loops are shown in secondary structure diagrams, for example, a loop of 16 nucleotides is part of the secondary structure of 16S ribosomal RNA (Noller 1984). As discussed below, non-Watson-Crick base pairs probably form to shorten many of these loops. The phylogenetic arguments used to establish duplex regions give minimum stem lengths and thus maximum loop sizes.

The first RNA hairpin loop characterized was the seven-membered anticodon loop of tRNA[Phe] (Saenger 1984). In the crystal the five bases on the 3′ end of the loop form a stack that continues into the stem. The two bases on the 5′ end of the loop are separated from the other five by a sharp bend accomplished mainly by rotation around the two torsion angles (α, ζ) on either side of the phosphorus atom between U_{33} and G_{m34}. The anticodon loop is stabilized by a hydrogen bond between the imino proton of U_{33} to a phosphate oxygen across the loop. NMR studies of the conformation in solution also show stacking of bases on the 3′ side (Clore et al. 1984). The X-ray as well as the NMR shows that the ribose sugars in the loop and in the stem are in the $C_{3′}$-*endo* conformation. In solution the anticodon loop is definitely more flexible than in the crystal, but it seems to have a very similar structure.

The best-characterized hairpin loops not in tRNA are two extra stable tetraloops; atomic resolution structures in solution have been determined by NMR (Heus and Pardi 1991; Varani et al. 1991). Tetraloops are the most common hairpin loops in bacterial rRNAs, and the most common sequences are GNRA and UNCG (N is any nucleotide, R is G or A) (Woese et al. 1990). The structures of the loops in both these families are remarkably similar (Fig. 1). Both tetraloops are actually biloops with a non-Watson-Crick base pair closing the loop. In the UUCG loop the closing base pair is a reverse wobble G·U pair. In the GCAA loop the closing base pair is an unusual G•A pair. (See Appendix 1 for the structures of these pairs.) Both loops are also stabilized by base stacking in the loop and by base-phosphate or base-ribose hydrogen bonding. It is

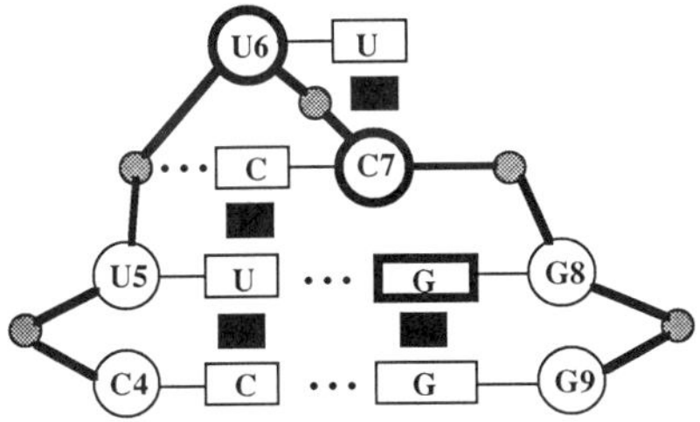

C(UUCG)G tetraloop

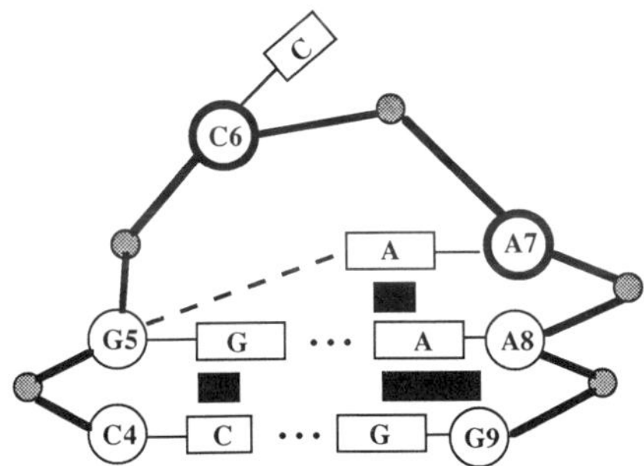

C(GCAA)G tetraloop

Figure 1 The similar conformations of two tetraloops as determined by NMR (Heus and Pardi 1991; Varani et al. 1991). Both contain non-Watson-Crick base pairs and base-sugar-phosphate interactions.

clear that the 2′ hydroxyl of ribose can be a very important part of the loop structure and the loop stabilization. Although a UUCG loop is more stable than a UUUG or UUUU loop in RNA, in DNA, UUCG, TTCG, and TTTT loops are all equally stable thermodynamically (Antao et al. 1991). RNA A-form helices usually have $C_{3'}$-*endo* sugars. In both tetraloops the middle two nucleotides have conformations predominantly $C_{2'}$-*endo*. The $C_{2'}$-*endo* pucker in conjunction with changes in backbone angles relative to those in helical regions allows just two nucleotides to reverse the direction of the strand and form the loop.

A surprising addition to the list of tetraloops is a GAGA tetraloop in the α-sarcin loop — the site of attack by cytotoxic proteins ricin and α-sarcin (Endo et al. 1991; Glück et al. 1992) — of rRNA. The secondary structure diagrams of 28S rRNA show the α-sarcin loop as a loop of 17 non-Watson-Crick-paired nucleotides in the rat 28S rRNA and as 15 nucleotides in *Escherichia coli*. However, NMR (Szewczak et al. 1991) and mutational (Endo et al. 1991) studies indicate that these loops are highly structured. Most of the loop bases are hydrogen-bonded to each other to form a tetraloop (Szewczak et al. 1991; I.G. Wool, pers. comm.). The details of the pairing are still being studied, but model RNA hairpins with a GAGA tetraloop closed by a stem of Watson-Crick base pairs are a good substrate for ricin and α-sarcin (Glück et al. 1992).

A hairpin loop that further illustrates the general feature of structure within large loops is a nominally nine-membered loop with sequence AUUUCUGAC. NMR studies (Puglisi et al. 1990b) show that it is actually a triloop (UCU) with a stem of three extra base pairs — $A^+ \cdot C$, $U \cdot A$, and $U \cdot G$. The protonated $A^+ \cdot C$ pair forms even at pH 7, and the $G \cdot U$ is a normal wobble pair. The three loop nucleotides (UCU) all have $C_{2'}$-*endo* ribose conformations.

The *trans*-activating region (TAR) of human immunodeficiency virus 1 (HIV-1) is a hairpin with a three-base bulge in the stem separated from the loop by four base pairs (Cullen 1991). The loop has the sequence CUGGGA and is closed by a $C \cdot G$ base pair. NMR studies of the loop have been done in several laboratories (G.C. King, Rice University; D. Hoffman, Duke University; J. Jaeger, U. of California, Berkeley). Jaeger finds that the first base (<u>C</u>UGGGA) on the 5′ side of the loop stacks on the stem. The next two nucleotides (C<u>UG</u>GGA) make the bend in the loop; the G of this bend is in a $C_{2'}$-*endo* conformation. The next two Gs are about half $C_{2'}$-*endo*. The last three bases (CUG<u>GGA</u>) tend to stack on the stem, but are not as well stacked as the base on the 5′ side of the loop.

Function

Hairpin loops (at least those studied so far) fall into two structural classes. One class has a definite shape and specific hydrogen-bonding characteristics. Non-Watson-Crick base pairs, base-ribose, and/or base-phosphate interactions produce a compact, stable structure. The stable tetraloops discussed above fall into this class. One of the main roles of this type of loop may be structural. Phylogenetic data suggest that tetraloops and certain three-base loops like UUU are interchangeable in 16S-like rRNA, suggesting that any interactions with proteins are absent or sequence-independent (Wolters 1992). These loops may be able to change conformation on binding, but the change requires an increase in loop free energy and a consequent decrease in the binding constant. The other class of hairpin loops, not characterized in detail yet, is quite flexible. The loop of the TAR element falls into this class. These loops may have many conformations all with similar free energy, so that no one structure dominates. They may undergo conformational changes on binding to a variety of proteins or other RNA molecules with little cost in free energy.

Hairpin loops are involved at every step in the control of mRNA function. Initiation, propagation, and termination of translation, as well as mRNA lifetimes, depend strongly on RNA conformation, in particular, stems and loops (Gold 1988; de Smit and van Duin 1990; Emory et al. 1992). Translation of the mRNA message into protein begins with the binding of the mRNA to a ribosome through base-pairing between the Shine-Dalgarno sequence in the mRNA and the rRNA. A stem and hairpin loop that contains either the Shine-Dalgarno region or the initiator codon greatly reduces the initiation of protein synthesis (see, e.g., McPheeters et al. 1986). Deletion of nucleotides that form the 5′ side of the hairpin (freeing the Shine-Dalgarno sequence and the initiating AUG) increased ribosome binding at least tenfold in bacteriophage f1 gene expression (Blumer et al. 1987). Hairpins do not have to occur at the ribosome-binding site to inhibit binding. In eukaryotes the 40S subunit binds to the 5′ cap of the mRNA and moves in the 3′ direction until it finds the first AUG. Hairpins between the capped site and the first AUG can drastically inhibit translation (Kozak 1991). The position of a hairpin may alter its effect. Hairpins can enhance initiation if they are 12–15 nucleotides downstream from the first AUG. A hairpin may momentarily stall the 40S subunit and facilitate initiation (Kozak 1991).

Several proteins regulate their own synthesis by inhibiting the translation of their mRNAs. T4 DNA polymerase binds to a hairpin a few nucleotides upstream of the Shine-Dalgarno sequence of its mRNA and thus prevents ribosome binding (Andrake et al. 1988). The hairpin loop contains eight nucleotides (AAUAACUC), postulated to be a tetraloop closed by A·C and A·U base pairs (Tuerk and Gold 1990). Both Qβ coat protein (Witherell and Uhlenbeck 1989) and R17 coat protein (Romaniuk et al. 1987) inhibit translation by binding to hairpins that include the initiator codon of the replicase gene in their viral RNA. The hairpins have three (Qβ) or four (R17) nucleotides in their loops. Certain nucleotides in the sequence of each loop are crucial for specific protein binding, but the stem sequence is not (Witherell et al. 1991).

Ribosomal frameshifting causes a change of translational reading frame and allows synthesis of two or more proteins from a single initiation site. In retroviruses a −1 frameshifting event often controls the levels of expression of viral reverse transcriptase relative to viral core proteins. Frameshifting is also known to affect gene expression in a number of systems, including coronaviruses, L-A double-stranded RNA of yeast, bacterial transposons, and even a bacterial gene (see Chamorro et al. 1992 and references therein). In all these cases, frameshifting occurs as the ribosome crosses a seven-nucleotide sequence 5′-X XXY YYZ-3′ (the triplets indicate the original reading frame, and X may be the same as Y). Two of the three base pairs between the anticodons of each of the tRNAs and the mRNA can be maintained after the slip into the −1 reading frame occurs. This slippery sequence is not the only determinant of frameshifting; secondary signals, called "stimulators" by Atkins et al. (1990), are also required. Although these signals are diverse, hairpin loops and pseudoknots immediately downstream from the shift site can cause ribosome pausing and thus lead to increased frameshifting. Hairpin formation immediately downstream from a UGA triplet facilitates its decoding as a selenocysteine codon instead of its normal role as a stop signal (Zinoni et al. 1990); this could also be the result of ribosomal stalling.

As indicated by the above examples, hairpins in the mRNA can affect translation by interacting directly with the rRNAs and proteins. Regulation of translation by binding of non-ribosomal proteins to specific hairpins is also very important. An example is found in the extensively studied system for control of iron metabolism by regulation of the levels of ferritin mRNA and transferrin receptor mRNA (Theil 1990). Ferritin is an iron storage protein, and transferrin receptor is active in the uptake of iron by a cell. The consensus iron-regulatory element or iron-responsive element (IRE) in the mRNAs is a stem-hairpin loop structure (Harrell et al. 1991). The hairpin interacts with different proteins to increase ferritin synthesis and decrease transferrin receptor synthesis as iron concentration increases. Decreasing iron concentration has the opposite effect. The IREs occur at the 5′-untranslated end of ferritin mRNA and at the 3′-untranslated end of transferrin receptor mRNA. Thus, the increase in activity of one mRNA and a concerted decrease in activity of another are regulated by the same RNA structures. A simple explanation follows from the placement of the IREs. A protein that destabilizes or unwinds the IRE hairpin increases the ribosome binding of the ferritin mRNA containing a 5′ IRE, but it increases the rate of degradation by exonucleases of transferrin receptor mRNA with IREs near the 3′ end (Theil 1990). Conversely, a protein that binds to the 5′-IRE hairpin prevents ribosome binding and translation of the ferritin mRNA. The same protein binding to the 3′ IREs may protect the transferrin receptor mRNA from exonuclease degradation. Secondary structure in general decreases the rate of endonuclease and exonuclease cleavage of RNA. However, a comparison of in vitro and in vivo hydrolysis by two 3′ exoribonucleases (McLaren et al. 1991) indicates that the presence of a hairpin near the 3′ end of an RNA is not sufficient to explain the longer lifetime of an mRNA in vivo. Most likely a hairpin-binding protein is necessary to explain the extra stability conferred on an mRNA in vivo by the addition of a hairpin near the 3′ end.

RNA secondary structure is important in all phases of translation; in transcription it is mainly involved in termination. In prokaryotes, a termination signal is an RNA hairpin followed by six to eight uridines; this forms a very unstable dA·rU hybrid with the DNA template facilitating release of the mRNA. An extensive analysis of rho-independent termination hairpins in *E. coli* showed that the majority are tetraloops and the most abundant loop sequences are UUCG and GAAA closed by a C·G pair (Carafa et al. 1990). These loops have similar conformations (Fig.1) and are thermodynamically extra stable (Antao et al. 1991). These loops may initiate the folding of the newly synthesized RNA and cause a pause in transcription. Changes in stem sequence that enhance the thermodynamic stability can also lead to increased terminator efficiency (Cheng et al. 1991). Less is known about the termination of transcription in eukaryotes, but UNCG and GNRA hairpin loops are also abundant in eukaryotes and possibly play the same role.

Attenuation is an important type of termination that controls the transcription of several genes at once in prokaryotes (Watson et al. 1987; Yanofsky 1988). An RNA hairpin (the terminator hairpin) and a run of Us present at a 5′-leader sequence before the beginning of a set of structural genes inhibit the transcription of the downstream genes. Transcription is turned on when an antiterminator hairpin is formed, preventing formation of the terminator hairpin. The equilibrium between terminator and antiterminator hairpins depends on the relative positions of the RNA polymerase and the ribosome. In the attenuator-controlled genes for amino acid biosynthesis, the position of the ribosome is determined by the concentrations of aminoacyl-tRNAs (Watson et al. 1987). Similar mechanisms operate in attenuation in eukaryotes (Resnekov et al. 1989).

The TAR hairpin and bulge region in HIV-1 interacts with a virus-encoded protein, Tat, and the hairpin loop binds to a host protein to greatly enhance the transcription of the mRNA (Cullen 1991; Karn 1991). Both events seem to be necessary for activation of transcription, although the mechanism is not well understood. It is thought that Tat and the host protein collected by the TAR region assemble with RNA polymerase II to form a complex for efficient elongation of the nascent RNA (Cullen 1991; Karn 1991).

MISMATCHES AND INTERNAL LOOPS

Structure

The smallest internal loops are base-base mismatches—non-Watson-Crick base appositions. Every base has both hydrogen-bond donors and acceptors, and thus can form hydrogen bonds with every other base. The two best-characterized non-Watson-Crick base pairs are the G·U wobble and a G·A pair; both occur in tRNA and have two hydrogen bonds each linking them (Saenger 1984). Mismatches distort a double helix and are thus expected to occur most often at the ends of helices; Traub and Sussman (1982) found A·G mismatches mainly at ends of helices in rRNAs. We have already mentioned a reverse wobble G·U and an unusual A·G that occur at the loop end of the stem in extra-stable tetraloops. A number of adjacent non-Watson-Crick base appositions are called a symmetric internal loop. Evidence suggests that in many internal loops, non-Watson-Crick base pairs are formed (SantaLucia et al. 1990, 1991).

A symmetric internal loop of eight nucleotides with four base pairs was serendipitously formed in an attempt to crystallize a UUCG tetraloop. A duplex with two U·C pairs and two wobble G·U pairs formed instead (Holbrook et al. 1991; see also Appendix 3). Both pairs have tightly bound water molecules in the minor groove that help define the structure (Fig. 2a). Although U·C can form a base pair with two direct hydrogen bonds (uracil carbonyl O4 to cytosine amino, and uracil N3 imino to cytosine N3), a water-mediated hydrogen bond between the two N3 nitrogens allows the base pair to fit into a continuous helix with less distortion. The G·U pair has a water-mediated hydrogen bond from the guanine amino to the 2′ hydroxyl of the uridine ribose. In addition to the very well ordered water molecules in the minor groove, each base pair has four more mobile water molecules buttressing the structure in the major groove. The helix is not distorted severely by the run of four non-Watson-Crick base pairs (Fig. 2b). The phosphate displacement from the helix axis varies from 10.9 Å to 8.1 Å, whereas standard A-form has a phosphate displacement of 8.7 Å. The main effect of the replacement of Watson-Crick base pairs is that the major groove becomes significantly wider.

Loop E in eukaryotic 5S RNAs is an asymmetric internal loop of nine nucleotides—four on one strand, five on the other. It forms part of the site for binding transcription factor IIIA (TFIIIA). NMR studies of a 27-nucleotide model of loop E (Wimberly et al. 1993) show that it has the structure schematically shown in Figure 3 (see Appendix 3). There are four non-Watson-Crick base pairs (G·A, Hoogsteen U·A, A·A, and U·U) and a bulged G. The resolution of the structure is not high enough to reveal the hydrogen bonding of the base pairs. However, intrastrand and interstrand contacts of the nucleotides in loop E show that the bases in the loop are stacked and continue the double-helix structure. The step between the G·A and the Hoogsteen A·U is greatly overwound; this is indicated by crossing of the strands in Figure 3. The ribose rings of the

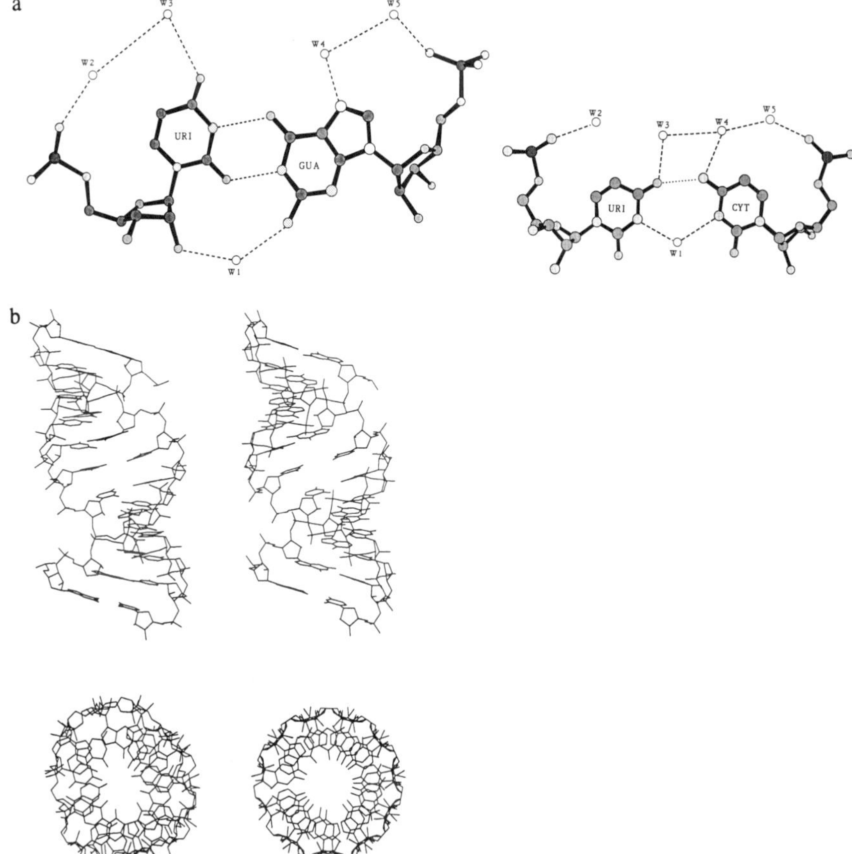

Figure 2 (*a*) Water-mediated non-Watson-Crick base pairs in an internal loop of eight nucleotides. The U•C pair has a water-mediated base•base hydrogen bond; the G•U has a water-mediated base•ribose 2′-hydroxyl hydrogen bond. (*b*) The main effect of these base mismatches is a widening of the major groove (see Holbrook et al. 1991). The loop is on the left; an A-form helix is on the right.

G-bulge and the neighboring A are $C_{2'}$-*endo*; the other nucleotides are mostly $C_{3'}$-*endo*. A model for loop E was studied in which the U underlined in Figure 3 was deleted in an attempt to increase the loop stability and the base-pairing (Varani et al. 1989). The deletion had the opposite effect; the symmetric loop of eight nucleotides had much less base-pairing. The loop bases stacked but were more dynamic than in the native sequence.

Chemical and enzymatic probing of loop E in 5S RNA (Andersen et al. 1984; Westhof et al. 1989) shows that the loop acts like a double-stranded region. It is susceptible to ribonuclease V_1, and its bases are less

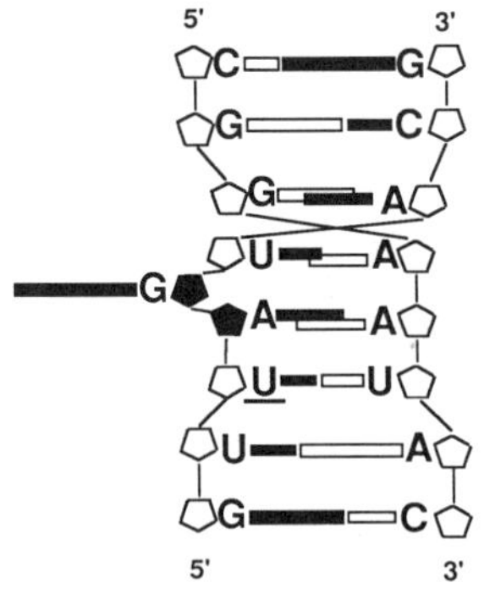

Figure 3 Base-pairing in an internal loop corresponding to loop E of eukaryotic 5S RNA determined by NMR (Wimberly et al. 1993). The loop is an asymmetric loop of nine nucleotides with a bulged G and four non-Watson-Crick base pairs: G•A, Hoogsteen A•U, A•A, U•U. When the underlined U was omitted, the extent of base-pairing decreased significantly (Varani et al. 1989).

reactive to chemical probes. However, there is no agreement about which base is bulged. Andersen et al. (1984) proposed a bulge U; Westhof et al. (1989) proposed a bulge A; NMR data show a bulged G (Wimberly et al. 1993).

On the basis of the limited available data, we think that non-Watson-Crick base-pairing will be maximized and A-form-helix geometry will be observed in most internal loops. Asymmetric internal loops will also form base pairs but will have one or more bulged bases; the presence of a bulge may facilitate non-Watson-Crick base-pairing.

Function

The wider major groove of an internal loop (Holbrook et al. 1991), or its special conformation (Wimberly et al. 1993), should provide a good site for specific protein recognition. The interaction of loop E of 5S RNA with TFIIIA is an example of this. Although TFIIIA protects nearly all 120 nucleotides of 5S from hydroxyl radicals, removing any of the loop-E nucleotides produces the largest decrease in the binding of TFIIIA (Darsillo and Huber 1991). Ribosomal proteins from both the large and small subunits of ribosomes bind to internal loops in rRNA (for review, see Chastain and Tinoco 1991). The binding site of HIV-1 for Rev protein is proposed to be an asymmetric internal loop of 10 nucleotides (Bartel et al. 1991). Internal loops are also important in RNA-RNA interactions. For example, in group I introns, the loop between P4 and P5 stems is involved in the guanosine-binding site (Wang and Cech 1992).

BULGE LOOPS

Structure

Bulge loops interrupt one strand of an otherwise continuous duplex. As with other types of loop regions, the bulge loop may contain any number of unpaired nucleotides. Although bulges are common in the predicted secondary structures of most large RNAs, few high-resolution data about the structure of bulge regions are available. In terms of global structure, it is known that a bulged nucleotide or nucleotides on one strand of a helix cause a bend that can be detected due to anomalous mobility during electrophoresis (Bhattacharyya et al. 1990). Extent of bending is dependent on both bulge size and sequence (Tang and Draper 1990). Lilley and his co-workers have shown that the sequence surrounding the bulge also has an effect on the degree of axial kinking (Riordan et al. 1992).

NMR studies of two RNA duplexes containing single-base bulges suggest that the identity of the bulged nucleotide and surrounding base pairs may influence whether the nucleotide is intercalated into the helix or looped out of the duplex (van den Hoogen et al. 1988; Zhang et al. 1989). Sequence as well as temperature affects the conformation of single-base bulges in DNA duplexes (Joshua-Tor et al. 1992 and references therein). It is not possible yet to predict the conformation of a bulged nucleotide on the basis of the sequence of the bulge and the surrounding base pairs.

The HIV transcriptional activator protein, Tat, binds to an RNA element in the viral mRNA and increases the level of transcription of viral message (for review, see Frankel 1992). The RNA element, called TAR for *trans*-activation response, is a hairpin that contains a three-base bulge. This hairpin has been the focus of intense biochemical and biophysical analysis in the past several years. Mutagenesis studies by a number of groups implicated nucleotides in and around the bulge as critical for specific binding of the Tat protein, as indicated in Figure 4 (Sumner-Smith et al. 1991; Weeks and Crothers 1991). Binding of Tat is mediated by a single arginine within a short stretch of basic amino acids; both a short peptide and free arginine bind specifically to TAR (Tao and Frankel 1992; see also Appendix 3). Intriguingly, an NMR study showed that the 31-nucleo-tide fragment containing the TAR element has different conformations in the presence and absence of argininamide, an arginine analog (Puglisi et al. 1992).

In the absence of argininamide, the NMR data suggest that the duplex

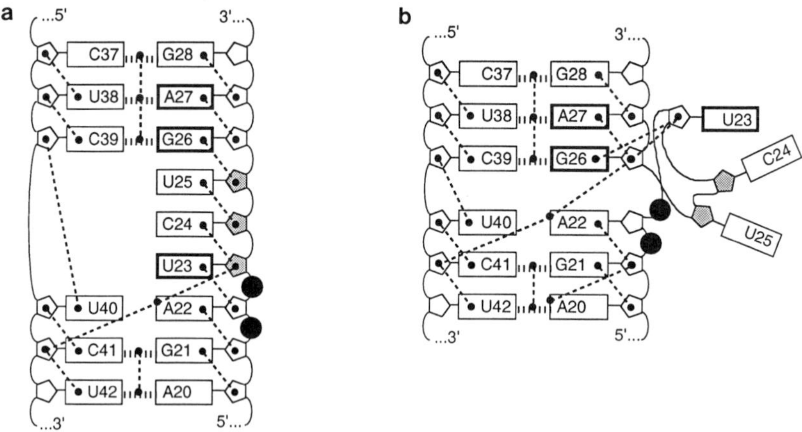

Figure 4 Schematic representation of NMR data on the bulge region of the TAR RNA in (*a*) the absence and (*b*) the presence of argininamide, an arginine analog. The bases implicated by mutagenesis as functionally important are indicated by bold boxes. When either of the two phosphates shown as dark circles is ethylated, Tat does not bind. Observed NOEs are indicated by dashed lines connecting the dots that represent the ribose, base H6/H8, adenine H2, or imino protons. The shaded sugars adopt a $C_{2'}$-*endo* conformation, and all other sugars are primarily $C_{3'}$-*endo*. In the presence of argininamide, U23 forms a triple interaction with the A27·U38 pair. No NOEs are observed to the other two bases in the bulge; these nucleotides are represented as disordered. (Reprinted, with permission, from Puglisi et al. 1992.)

regions above and below the bulge adopt a typical A-form-helix geometry, whereas the bulge region is flexible. Stacking interactions occur between nucleotides in the bulge and the base-paired nucleotides above and below the bulge. Although the NMR data do not provide direct evidence for a bend in the helix, the stacked structure of the bulge could cause a kink between the two helix regions as proposed by Lilley's group (Riordan et al. 1992).

Ligand binding to RNA can result in significant conformational change. When bound to argininamide or to an 11-amino-acid peptide containing a single arginine, the stems above and below the bulge stack coaxially; no stacking is observed between the bulged residues and the stems. Data from nuclear Overhauser effect (NOE) experiments on the TAR-argininamide complex place U_{23} of the bulge in the major groove near the A_{27}·U_{38} base pair above the bulge in what Puglisi et al. (1992) postulate is a base triple (Fig. 4). The other two nucleotides in the bulge are probably swung away from the helix. The argininamide is positioned below the base triple and forms hydrogen bonds to G_{26}. Phosphate contacts (Fig. 4), as proposed by Frankel's group (Tao and Frankel 1992), are consistent with the model.

The sequence specificity of Tat-TAR interaction is due to the ability of the TAR sequence to adopt a particular conformation in addition to a sequence-specific protein contact. The $A_{27} \cdot U_{38}$ base pair above the bulge in the TAR RNA is required for stable triple formation, not because of a specific contact of the argininamide with either base. Nucleotides in the proposed triple of TAR, $(A \cdot U)U$, can be changed to $(G \cdot C)C$ without significantly altering the structure of the RNA oligonucleotide in the presence of argininamide (J. Puglisi, pers. comm.). In another example, the identity of the terminal base pair in the acceptor stem of tRNA[Gln], conserved as a $U \cdot A$ in all isoacceptors, is not important because of specific contacts with glutaminyl-tRNA synthetase (GlnRS), but because it can be easily denatured (Rould et al. 1989). The conformation of the RNA is different in the complex between tRNA[Gln] and GlnRS from what it is free in solution (Rould et al. 1989). The difference observed between the free and bound forms of RNA points to a danger in attempts to study RNA oligonucleotides as models of larger, complex systems.

This study also illustrates the flexibility of short nominally single-stranded regions of RNA. In the TAR-arginine complex, the first nucleotide of the bulge is involved in a tertiary interaction with nucleotides in the adjacent stem. Other interesting tertiary interactions like this one probably await discovery.

Function

The high-resolution structure of the TAR-argininamide complex makes the bulge in TAR the best understood of the protein-binding sites that contain bulges, but other proteins are known to contact bulge regions. A study by Uhlenbeck and his co-workers (Wu and Uhlenbeck 1987; for review, see Witherell et al. 1991) has shown that a bulged A residue is essential for binding of the R17 coat protein to the translational operator region in the RNA; deletion of this nucleotide reduced the equilibrium binding constant to the coat protein 1000-fold. A series of mutations that changed functional groups on the single bulged adenosine indicated that bulky substituents are not tolerated. Mutation to a guanosine was tolerated when surrounding nucleotides were also changed to prevent formation of alternative conformations. The authors speculate that the bulged nucleotide is intercalated into the helix. An NMR study of the RNA uncomplexed to protein indicates that the adenosine is indeed intercalated into the helix (P. Borer, pers. comm.).

Phylogenetic comparison shows that although each of the four bases is about equally represented in 16S rRNA, there are a disproportionate number of unpaired adenosines relative to the other three nucleotides (Gutell et al. 1985). These unpaired adenines may function in protein-binding sites. Chemical probing of the 16S rRNA and the 30S subunit showed that many of the conserved single-stranded adenosines are protected from chemical modification in the 30S subunit, thus suggesting a role in ribosome assembly (Moazed et al. 1986). The binding site of the ribosomal protein L18 on 5S rRNA is a helical region interrupted by a bulged adenosine, and single-base bulges were noted in a number of other putative protein-binding sites (Peattie et al. 1981). Single-base bulges other than adenosine may be important protein contacts as well. A uridine bulge centered in the strictly conserved central domain of plant viroids may interact with a cellular RNA polymerase (Keese and Symons 1985).

A bulged adenosine is the nucleophile in cleavage at the 5′ splice site in both mitochondrial group II and nuclear pre-mRNA splicing. Mechanistically, a 2′-hydroxyl group of an adenosine within the intron attacks the phosphodiester bond at the exon/intron junction, resulting in formation of a lariat intermediate and releasing the 5′ exon (for review, see Sharp 1987; Baserga and Steitz; Moore et al.; both this volume). Phylogenetic comparison of group II introns places the bulged adenosine, the so-called branch point, in the domain VI stem-loop 7 or 8 nucleotides upstream of the 3′ splice site (for review, see Michel et al. 1989). The interaction that bulges the branch-point adenosine in nuclear pre-mRNA splicing is bimolecular. The nucleotides surround the branch point of the intron pair with the small nuclear RNA U2 (for review, see Guthrie 1991). The interactions that bring the 2′ hydroxyl group of the branch point into the orientation required for catalysis of the first transesterification reaction are not yet known. Since group II introns are capable of self-splicing, the three-dimensional structure of the intron itself must be responsible.

PSEUDOKNOTS

Structure

Pseudoknots and knots were originally defined by Studnicka et al. (1978) as base-pairing between nucleotides separated by regions of secondary structure. The terms pseudoknot and knot were defined in order to exclude base-pairing between secondary structure elements from consideration in structure-prediction algorithms. This usage of the term knot is not topologically correct, since formation of a true knot requires linkage between the ends of the strand. By the broad definition, any tertiary interaction involving base-pairing between two regions of unpaired nucleotides is termed a pseudoknot. This pairing can be between any combination of single-stranded regions in hairpin loops, bulges, internal loops, junctions, or unpaired bases between secondary structure motifs. Pleij (1990) points out that this results in 14 different types of pseudoknots (also see Appendix 2). In addition, the RNA chain can trace three different paths in forming the pseudoknot (Fig. 5) (Abrahams et al. 1990). By these broad definitions, virtually any tertiary interaction involving Watson-Crick base-pairing can be called a pseudoknot.

The term pseudoknot was resurrected by Pleij and his co-workers to describe an RNA structure at the 3′ends of certain plant viral RNAs (Pleij et al. 1985). For the purposes of the following discussion, the term pseudoknot is limited to local interactions in which the loop regions of the pseudoknot are unstructured, as described by Pleij et al. (1985). The reasons for this limitation are twofold. First, this definition is in keeping with accepted definitions for other RNA structural motifs. In the secondary structure elements discussed above, the loop regions do not themselves contain structure. If, for example, the loop of a hairpin were able to form additional Watson-Crick base pairs, an internal loop or bulge would be formed. By applying this restriction, only pairings between nucleotides in hairpin loops with adjacent single-stranded regions are considered pseudoknots (Fig. 5a). Second, as discussed by Abrahams et al. (1990), prediction of pseudoknots is much easier if this limitation is applied.

Each pseudoknot has two stem regions, S1 (nearest the 5′end of the molecule) and S2 (nearest the 3′end), and two loop regions. A pseudoknot may also have a single-stranded region (SSR) between the two stems (Fig. 5). The three general types differ in orientation of the potential coaxial stack between the two stem regions. In the best-characterized pseudoknots (Fig. 5a), stem 2 can be thought of as pairing on the 3′side of the hairpin formed by stem 1 (or equivalently, stem 1 pairs on the 5′side of the hairpin formed by stem 2). This results in one loop (L1) bridging the major groove over the 3′stem and the other loop (L2) crossing the minor groove over the 5′stem. The other two types of pseudoknots (Fig. 5b,c) are a result of one loop crossing either the minor or the major groove and the other bridging both stem regions.

The first models for the pseudoknot structure, of the type shown in Figure 5a, were based on biochemical analysis of the 3′end of plant viral RNA from turnip yellow mosaic virus (TYMV) and structural analogy to tRNA (Pleij et al. 1985; Dumas et al. 1987). An NMR study (Puglisi et al. 1990a, 1991) of an oligonucleotide designed to form a pseudoknot revealed a structure similar to that proposed by Pleij et al. (1985). In the

a.

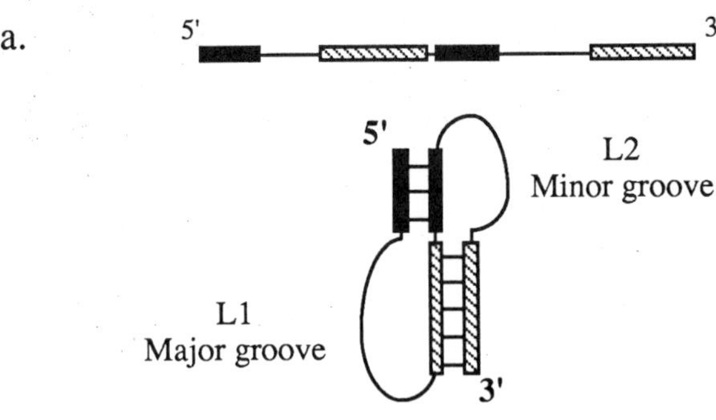

b.

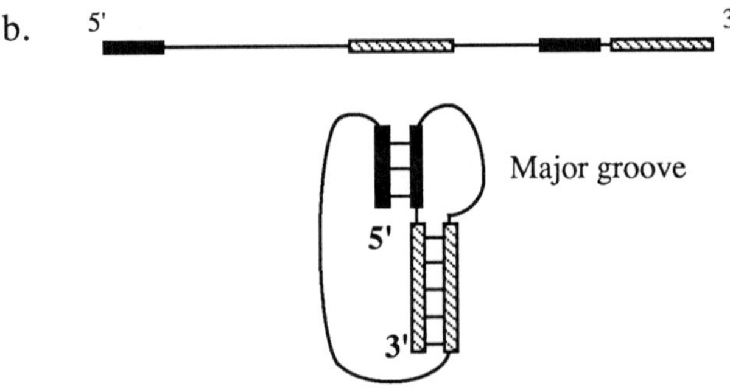

c.

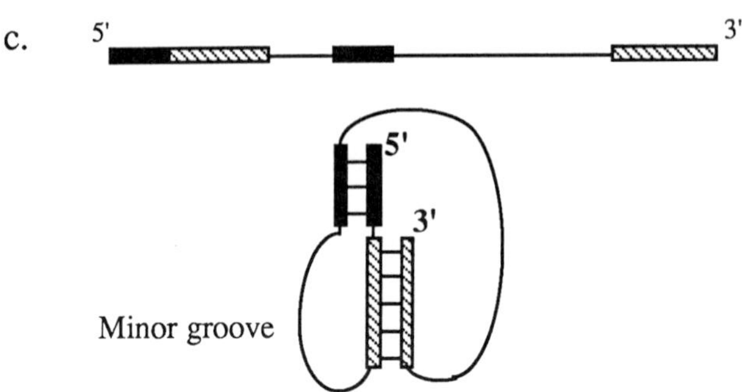

Figure 5 Schematic representation of the three different types of pseudoknots. Each type of pseudoknot has two stems and two loops that cross the stem regions and may have single-stranded nucleotides between the two stems. The best-characterized pseudoknot is shown in *a*; in this structure, L1 crosses stem 2 over its major groove and L2 bridges the minor groove of stem 1. In the other types, one loop crosses either the major (*b*) or minor (*c*) groove and the other bridges both stems.

two stem regions, nucleotide conformations and internucleotide distances are characteristic of A-form geometry. An analysis of the same oligonucleotide by means of circular dichroism was also consistent with A-form geometry (Johnson and Gray 1992). One hallmark of the structural element is coaxial stacking of the two stem regions. This feature was originally inferred by Pleij et al. (1985) to explain how the two stem regions of the pseudoknot in TYMV RNA could mimic the amino acid acceptor stem of tRNA.

There is a minor distortion in stacking at the junctions of the stems and loops (Puglisi et al. 1990a). At this junction, two phosphate groups are in close proximity and the slight unstacking probably reduces unfavorable electrostatic interactions. Either Mg^{++} or a high concentration of Na^+ is necessary for oligonucleotide pseudoknot formation; the cations are probably required to stabilize the junction (Wyatt et al. 1990). More complicated molecules like tRNA (Cole et al. 1972) and the pseudoknot at the 3′ end of TYMV RNA (van Belkum et al. 1988) are also dependent on either Mg^{++} or a high concentration of Na^+ to form native conformations.

The two loops of a pseudoknot are inequivalent. In the oligonucleotide pseudoknot and those in the plant viral RNAs, loop 1 crosses the narrow and deep major groove and loop 2 bridges the shallow and wide minor groove. Experiments with oligonucleotide pseudoknots (Wyatt et al. 1990) suggest that minimum loop sizes can be predicted on the basis of the distance across the grooves of an A-form helix (Pleij et al. 1985). A caveat is that each sequence able to form a pseudoknot also has the potential to form at least two hairpins corresponding to formation of only stem 1 or only stem 2. As the pseudoknot loop size is decreased, so is the loop size of the corresponding hairpin, and equilibrium is shifted in favor of the hairpin conformation for small loop sizes (Wyatt et al. 1990).

Since the pseudoknot is a tertiary interaction, the stability of the structure must be considered relative to the stability of the usual single-stranded reference state and also in comparison to these possible hairpin structures. Studies of oligonucleotides provide evidence that pseudoknots are only marginally more stable than the possible hairpin secondary structures (Wyatt et al. 1990). When the free energies of formation (ΔG^o) of two different oligonucleotides that form pseudoknots were compared to the ΔG^o of oligonucleotides that form each of the possible hairpin structures, the difference in free energy was only about 1.5–2 kcal mole^{-1} at 37°C. Mans et al. (1992) estimated that the free energy gain upon pseudoknot formation is at most 2.0 kcal mole^{-1}, based on aminoacylation efficiency of mutants containing mutations in the pseudoknot regions of the tRNA-like structure of TYMV RNA. It is possible that the conformational lability of the pseudoknot structure may be exploited in systems similar to attenuation and the Tat-TAR complex.

Function

The first discovered role of pseudoknots was in the structural mimicry of tRNA at the 3′ termini of plant viral RNAs. This structure at the 3′ end of brome mosaic virus RNA is required for viral replication (Dreher and Hall 1988). The tRNA-like structures in plant viral RNAs are recognized by a variety of tRNA-processing enzymes, including aminoacyl-tRNA synthetases, tRNA nucleotidyl- and methyl-transferases, RNase P, peptidyl-tRNA hydrolase, and elongation factors (for review, see Haenni et al. 1982); however, most of these reactions do not appear to occur in vivo. Weiner and Maizels (1987) speculated that the tRNA-like structures function as primitive teleomers.

One of the most intriguing functions of the pseudoknot structure is in frameshifting during the translation of certain retroviral mRNAs. As discussed in the hairpin section, frameshifting requires both a slippery sequence and a downstream element of RNA structure (for review, see Atkins et al. 1990). In mouse mammary tumor virus (MMTV) (Chamorro et al. 1992) and in infectious bronchitis virus (IBV) (Brierley et al. 1991), mutational analysis provides strong evidence that the structural element is a pseudoknot. The structures of these two pseudoknots are shown schematically in Figure 6. Jacks et al. (1988) proposed that the structured region may cause the ribosome to pause at the shift site increasing the likelihood of the −1 frameshift. A simple energetic barrier to translation is not sufficient to cause frameshifting, however. In the IBV shift region, an extended stem with the same sequence as the pseudoknot did not promote efficient frameshifting (Brierley et al. 1991). It appears that it is the pseudoknot conformation, rather than sequence, that is important. The sequences of the stems of the pseudoknots are not critical for activity; frameshifting is not sensitive to deletion and/or minimal additions to either loop region of the IBV pseudoknot. The pseudoknot conformation may present an unusually resistant structure to the ribosome helicase activity.

Pseudoknots offer a number of sites for interaction with protein. Florentz and Giegé (1986) used chemical and enzymatic mapping to show that valyl-tRNA synthetase contacts the pseudoknot in the TYMV RNA in the region where the two stems stack coaxially. Mans et al. (1992) speculate that drastic decreases in aminoacylation efficiency due to mutation in loop 2 of the TYMV pseudoknot may be due to a role in aminoacylation of the first nucleotide in the loop. Autoregulation of

a. MMTV

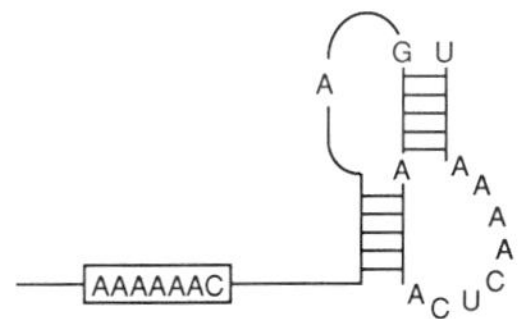

b. IBV

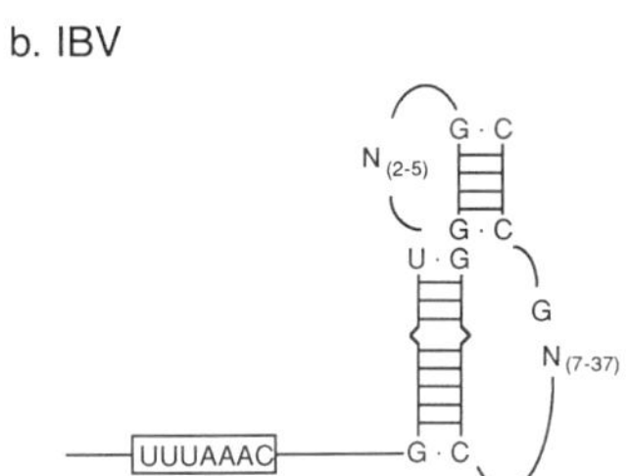

Figure 6 Sequences required for frameshifting in two systems: (*a*) MMTV and (*b*) IBV. The slippery sequence is boxed. The sequences of the stems have been changed without destroying frameshifting. The sequences shown in the loops of MMTV have not been mutagenized, so may or may not be critical. Changes in the bases shown in the IBV pseudoknot affect the efficiency of frameshifting. The number of nucleotides in the IBV loops has been varied within the range indicated with little effect on frameshifting.

translation of the gene 32 mRNA by gene 32 protein involves a pseudoknot. Cooperative binding of the protein to the region around the initiation codon blocks formation of the 30S ribosomal subunit initiation complex. Comparison of operator sequences of a number of bacteriophage led to the proposal that the site for nucleation of binding was a pseudoknot (McPheeters et al. 1988). Recent enzymatic mapping and mutational analysis of the operator sequence has confirmed that a pseudoknot is required for gene 32 protein binding (Y. Shamoo, pers. comm.).

Pseudoknot-like long-range pairings have been observed in several protein-binding sites. These structures are not local interactions, since the pseudoknot loop regions are themselves structured. However, these structures share with local pseudoknots, and with other RNA structural elements like bulges and junctions, the potential for coaxial stacking of two stems. Pairing between a hairpin loop and an internal loop in a group I intron results in an interaction like the one shown in Figure 5b (Jaeger et al. 1991). Ribosomal protein S4 recognizes a structure similar to the structure shown in Figure 5c (Tang and Draper 1989). In this instance, the structure is a complicated "double pseudoknot" due to pairing between the region between the two stems and nucleotides at the 3′ side of the pseudoknot.

TRIPLE-STRAND INTERACTIONS

Structures

A folded RNA molecule consists of double-stranded helices linked by single strands and loops. Whenever a helix and a single-stranded region are near each other, the single-stranded or loop regions can interact with the helices to form triple-strand interactions. Coaxial stacking of stem regions, which can occur at bulges, junctions, and pseudoknots, can be stabilized by triple-strand formation with the single strands, as shown in Figure 7. A-form-helix geometry places the 5′ single strand in the minor groove of the adjacent helix and the 3′ single strand in the major groove (Chastain and Tinoco 1991). The structures of base triples with the third strand in the major groove have been reviewed previously (Saenger 1984; Chastain and Tinoco 1991). Base triples of G·C·A also occur in mixtures of polynucleotide strands of poly(rG)·poly(rC)·poly(rA), but their structure is not known (Chastain and Tinoco 1992a).

Recent studies have provided high-resolution structures of both major and minor groove triples. A major groove base triple is formed in the TAR region of HIV-1 when TAR is bound to argininamide (Puglisi et al. 1992). Chastain and Tinoco (1992b) recently described the structure of a minor groove base triple. The base triple consists of a G·C base pair with an adenine approximately coplanar with the base pair, but hydrogen-bonded to the 2′ hydroxyl of the guanosine ribose (Fig. 7a). The third strand rotates the coaxially stacked helices from the usual A-form value of 33° to about twice that value (Fig. 7b).

Function

Tertiary interactions, including base triples, serve to fix the relative orientations of helices or to allow controlled relative motion of the helices that are necessary for the biological function of the RNA. Some base triples are important in stabilization of the three-dimensional structure of tRNAs. The proposed structure of the catalytic core of the *Tetrahymena* group I intron is determined by several base triples that orient the helices and loops to produce the binding sites for the 5′-splice sequence and the guanosine cofactor (Michel and Westhof 1990; Michel et al. 1990). The three-way junction in the hammerhead self-cleaving RNA must specify the site where cleavage occurs. The single-stranded nucleotides are important (Ruffner et al. 1990), but their conformation is not known.

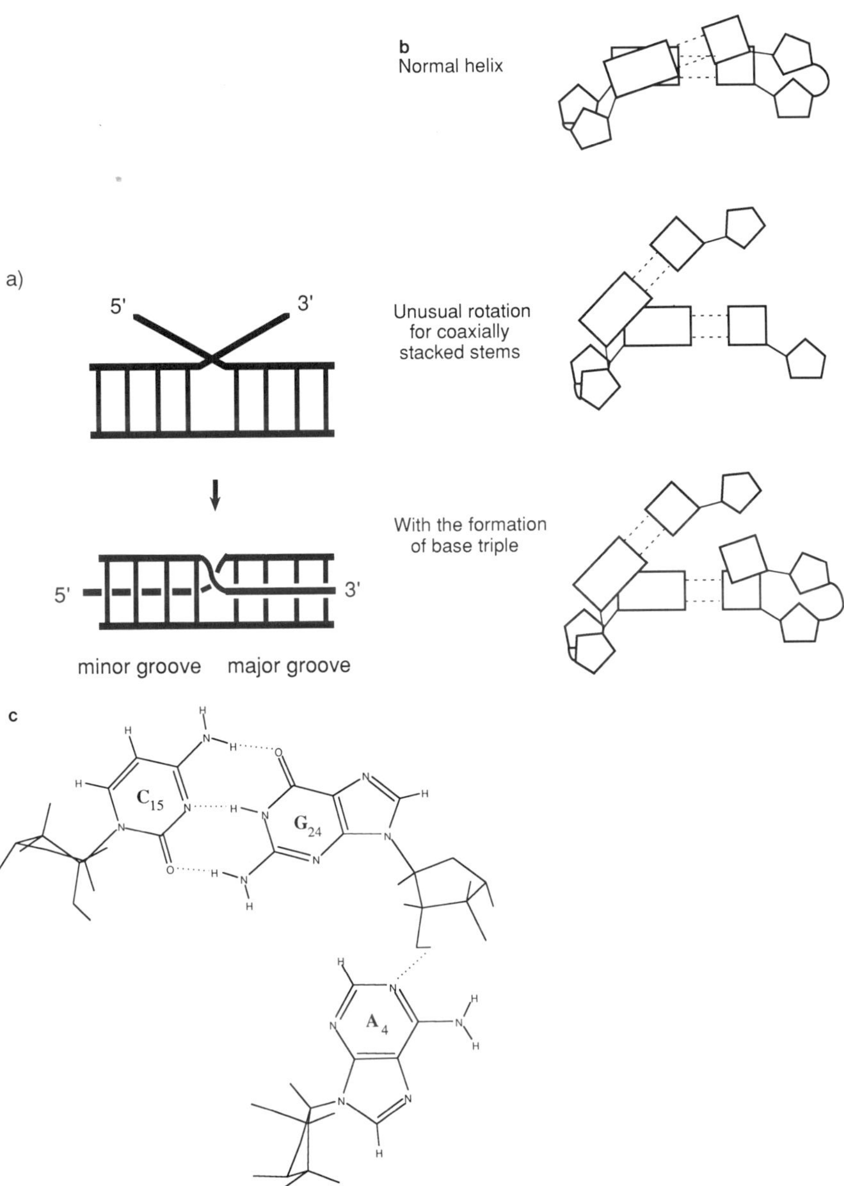

Figure 7 (*a*) Schematic showing that for coaxially stacked stems the 5′ single strand of one stem fits into the minor groove of the other, and the 3′ single strand of one stem fits into the major groove of the other. (*b*) Adjacent base pairs of coaxially stacked stems which do not have the usual 33° rotation per base pair but have about twice that rotation; this seems to be stabilized by three-base interactions. (*c*) A base triple adjacent to coaxially stacked stems as found by NMR. The third strand is in the minor groove and involves a hydrogen bond between an N3 of adenine and a 2′ hydroxyl. These results are from Chastain and Tinoco (1992b).

CONCLUDING REMARKS

Although knowledge of RNA structural motifs is rapidly expanding, prediction of the three-dimensional conformation and the function of an RNA molecule on the basis of primary sequence alone is still in the future. The building blocks of the three-dimensional conformations — stems, loops, simple tertiary interactions — are not yet well understood. RNA is conformationally flexible, and binding of ligands ranging in size from a Mg^{++} ion to a protein may affect the conformation. The relatively simple systems described do provide insight into the biologically relevant structures adopted by RNA. We have reviewed studies that show that even simple loop regions can form unexpected interactions which are critical for the function of the RNA. Continued study of simple RNA structural motifs and simple RNA-protein interactions should facilitate understanding of more complex biological functions of RNA.

ACKNOWLEDGMENTS

We thank Drs. Michael Chastain, Peter Davis, and Susan Freier for carefully reading the manuscript and making useful comments; and Drs. Phillip Borer, Michael Chastain, David Hoffman, John Jaeger, Gary King, Joseph Puglisi, Yusif Shamoo, and Brian Wimberly for sharing results prior to publication. This work was supported in part by National Institutes of Health grant GM-10840 and Department of Energy grant DE-FG03-86ER60406 to I.T. J.W. was supported in part by a Jane Coffin Childs postdoctoral fellowship.

REFERENCES

Abrahams, J.P., M. van den Berg, E. van Batenburg, and C. Pleij. 1990. Prediction of RNA secondary structure, including pseudoknotting, by computer simulation. *Nucleic Acids Res.* **18:** 3035–3044.

Andersen, J., N. Delihas, J.S. Hanas, and C.-W. Wu. 1984. 5S RNA structure and interaction with transcription factor A. 1. Ribonuclease probe of the structure of 5S RNA from *Xenopus laevis* oocytes. *Biochemistry* **23:** 5752–5759.

Andrake, M., N. Guild, T. Hsu, L. Gold, C. Tuerk, and J. Karam. 1988. DNA polymerase of bacteriophage T4 is an autogenous translational repressor. *Proc. Natl. Acad. Sci.* **85:** 7942–7946.

Antao, V.P., S.Y. Lai, and I. Tinoco, Jr. 1991. A thermodynamic study of unusually stable RNA and DNA hairpins. *Nucleic Acids Res.* **19:** 5901–5905.

Atkins, J.F., R.B. Weiss, and R.F. Gesteland. 1990. Ribosome gymnastics — Degree of difficulty 9.5, style 10.0. *Cell* **62:** 413–423.

Bartel, D.P., M.L. Zapp, M.R. Green, and J.W. Szostak. 1991. HIV-1 rev regulation involves recognition of non-Watson-Crick base pairs in viral RNA. *Cell* **67:** 529–536.

Bass, B. and H. Weintraub. 1988. An unwinding activity that covalently modifies its double-stranded RNA substrate. *Cell* **55:** 1089–1098.

Bevilacqua, P.C. and D.H. Turner. 1991. Comparison of binding of mixed ribose-deoxyribose analogues of CUCU to a ribozyme and to GGAGAA by equilibrium dialysis: Evidence for ribozyme specific interactions with 2′OH groups. *Biochemistry* **30:** 10632–10640.

Bhattacharyya, A., A.I.H. Murchie, and D.M.J. Lilley. 1990. RNA bulges and the helical periodicity of double-stranded RNA. *Nature* **343:** 484–487.

Blumer, K.J., M.R. Ivey, and D.A. Steege. 1987. Translational control of phage f1 gene expression by differential activities of the gene V, VII, IX and VIII initiation sites. *J. Mol. Biol.* **197:** 439–451.

Brierley, I., N.J. Rolley, A.J. Jenner, and S.C. Inglis. 1991. Mutational analysis of the RNA pseudoknot component of a coronavirus ribosomal frameshifting signal. *J. Mol. Biol.* **220:** 889–902.

Burke, J.M., M. Belfort, T.R. Cech, R.W. Davies, R.J. Schweyen, D.A. Shub, J.W. Szostak, and H.F. Tabak. 1987. Structural conventions for group I introns. *Nucleic Acids Res.* **15:** 7217–7221.

Calnan, B.J., B. Tidor, S. Biancalana, D. Hudson, and A.D. Frankel. 1991. Arginine-mediated RNA recognition: The arginine fork. *Science* **252:** 1167–1171.

Carafa, Y. d'A., E. Brody, and C. Thermes. 1990. Prediction of rho-independent *Escherichia coli* transcription terminators. *J. Mol. Biol.* **216:** 835–858.

Cech, T.R. 1987. The chemistry of self-splicing RNA and RNA enzymes. *Science* **236:** 1532–1539.

Chamorro, M., N. Parkin, and H.E. Varmus. 1992. An RNA pseudoknot and an optimal heptameric shift site are required for highly efficient ribosomal frameshifting on a retroviral messenger RNA. *Proc. Natl. Acad. Sci.* **89:** 713–717.

Chastain, M. and I. Tinoco, Jr. 1991. Structural elements in RNA. *Prog. Nucleic Acids Res. Mol. Biol.* **41:** 131–177.

––––––. 1992a. Poly(rA) binds poly(rG)•poly(rC) to form a triple helix. *Nucleic Acids Res.* **20:** 315–318.

––––––. 1992b. A base-triple structural element in RNA. *Biochemistry* **31:** 12733–12741.

Cheng, S.-W.C., E.C. Lynch, K.R. Leason, D.L. Court, B.A. Shapiro, and D.I. Friedman. 1991. Functional importance of sequence in the stem-loop of a transcription terminator. *Science* **254:** 1205–1207.

Chou, S.-H., P. Flynn, and B. Reid. 1989. Solid-phase synthesis and high-resolution NMR studies of two synthetic double-helical RNA dodecamers: r(CGCGAAUUCGCG) and r(CGCGUAUACGCG). *Biochemistry* **28:** 2422–2435.

Clore, G.M., A.M. Gronenborn, E.A. Piper, L.W. McLaughlin, E. Graeser, and J.H. van Boom. 1984. The solution structure of a RNA pentadecamer comprising the anticodon loop and stem of yeast tRNA^{Phe}. *Biochem. J.* **221:** 737–751.

Cole, P.E., S.K. Yang, and D.M. Crothers. 1972. Conformational changes of transfer ribonucleic acid. Equilibrium phase diagrams. *Biochemistry* **11:** 4358–4368.

Cullen, B.R. 1991. Regulation of HIV-1 gene expression. *FASEB J.* **5:** 2361–2368.

Darsillo, P. and P.W. Huber. 1991. The use of chemical nucleases to analyze RNA-protein interactions. *J. Biol. Chem.* **266:** 21075–21082.

Datta, B. and A.M. Weiner. 1991. Genetic evidence for base pairing between U2 and U6 snRNA in mammalian mRNA splicing. *Nature* **352:** 821–824.

Davies, R.W., R.B. Waring, J.A. Ray, T.A. Brown, and C. Scazzocchio. 1982. Making ends meet: A model for RNA splicing in fungal mitochondria. *Nature* **300:** 719–724.

Davis, P.W., R.W. Adamiak, and I. Tinoco, Jr. 1990. Z-RNA: The solution NMR structure of r(CGCGCG). *Biopolymers* **29:** 109–122.

de Smit, M.H. and J. van Duin. 1990. Control of prokaryotic translational initiation by mRNA secondary structure. *Prog. Nucleic Acid Res. Mol. Biol.* **38:** 1–35.

Dock-Bregeon, A.C., B. Chevrier, A. Podjarny, J. Johnson, J.S. de Bear, G.R. Gough, P.T. Gilham, and D. Moras. 1989. Crystallographic structure of an RNA helix: [U(UA)$_6$A]$_2$. *J. Mol. Biol.* **209:** 459–474.

Dreher, T.W. and T.C. Hall. 1988. Mutational analysis of the sequence and structural requirements in brome mosaic virus RNA for minus strand promoter activity. *J. Mol. Biol.* **201:** 31–40.

Dumas, P., D. Moras, C. Florentz, R. Giegé, P. Verlaan, A. van Belkum, and C.W.A. Pleij. 1987. 3-D graphics modelling of the tRNA-like 3′-end of turnip yellow mosaic virus RNA: Structural and functional implications. *J. Biomol. Struct. Dyn.* **4:** 707–728.

Eguchi, Y., T. Itoh, and J. Tomizawa. 1991. Antisense RNA. *Annu. Rev. Biochem.* **60:** 631–652.

Emory, S.A , P. Bouvet, and J.G. Belasco. 1992. A 5′-terminal stem loop structure can stabilize messenger RNA in *Escherichia coli. Genes Dev.* **6:** 135–148.

Endo, Y., A. Glück, and I.G. Wool. 1991. Ribosomal RNA identity elements for ricin A-chain recognition and catalysis. *J. Mol. Biol.* **221:** 193–207.

Florentz, C. and R. Giegé. 1986. Contact areas of the turnip yellow mosaic virus tRNA-like structure interacting with yeast valyl-tRNA synthetase. *J. Mol. Biol.* **191:** 117–130.

Frankel, A.D. 1992. Activation of HIV transcription by Tat. *Curr. Opin. Genet. Dev.* **2:,** 293–298.

Glück, A., Y. Endo, and I.G. Wool. 1992. Ribosomal RNA identity elements for ricin A-chain recognition and catalysis: Analysis with tetraloop mutants. *J. Mol. Biol.* **226:** 411–424.

Gold, L. 1988. Posttranscriptional regulatory mechanisms in *Escherichia coli. Annu. Rev. Biochem.* **57:** 199–233.

Gutell, R.R. and C.R. Woese. 1990. Higher order structural elements in ribosomal RNAs: Pseudo-knots and the use of noncanonical pairs. *Proc. Natl. Acad. Sci.* **87:** 663–667.

Gutell, R.R., B. Weiser, C.R. Woese, and H.F. Noller. 1985. Comparative anatomy of 16-S-like ribosomal RNA. *Prog. Nucleic Acid Res.* **32:** 155–216.

Guthrie, C. 1991. Messenger RNA splicing in yeast: Clues to why the spliceosome is a ribonucleoprotein. *Science* **253:** 157–163.

Haas, E.S., D.P. Morse, J.W. Brown, F.J. Schmidt, and N.R. Pace. 1991. Long-range structure in ribonuclease P RNA. *Science* **254:** 853–856.

Haenni, A.-L., S. Joshi, and F. Chapeville. 1982. tRNA-like structures in the genomes of RNA viruses. *Prog. Nucleic Acids Res. Mol. Biol.* **27:** 85–104.

Hall, K., P. Cruz, I. Tinoco, Jr., T.M. Jovin, and J.H. van de Sande. 1984. "Z-RNA" — A left-handed RNA double helix. *Nature* **311:** 584–586.

Happ, C.S., E. Happ, M. Nilges, A.M. Gronenborn, and G.M. Clore. 1988. Refinement of the solution structure of the ribonucleotide 5′ r(GCAUGC)$_2$: Combined use of nuclear magnetic resonance and restrained molecular dynamics. *Biochemistry* **27:** 1735–1743.

Harrell, C.M., A.R. McKenzie, M.M. Patino, W.E. Walden, and E.C. Theil. 1991. Ferritin mRNA: Interactions of iron regulatory element with translational regulator protein P-90 and the effect on base paired flanking regions. *Proc. Natl. Acad. Sci.* **88:** 4166–4170.

Heus, H. and A. Pardi. 1991. Structural features that give rise to the unusual stability for RNA hairpins containing GNRA loops. *Science* **253:** 191–194.

Holbrook, S.R., C. Cheong, I. Tinoco, Jr., and S.-H. Kim. 1991. Crystal structure of an RNA double helix incorporating a track of non-Watson-Crick base pairs. *Nature* **353:**

579–581.

Holbrook, S.R., J.L. Sussman, R.W. Warrant, and S.-H. Kim. 1978. Crystal structures of yeast phenylalanine transfer RNA. II. Structural features and functional implications. *J. Mol. Biol.* **123:** 631–660.

Jacks, T., H.D. Madhani, F.R. Masiarz, and H.E. Varmus. 1988. Signals for ribosomal frameshifting in the Rous sarcoma virus *gag-pol* region. *Cell* **55:** 447–458.

Jacquier, A. and F. Michel. 1987. Multiple exon-binding sites in class II self-splicing introns. *Cell* **50:** 17–29.

Jaeger, L., E. Westhof, and F. Michel. 1991. Function of P11, a tertiary base pairing in self-splicing introns of subgroup IA. *J. Mol. Biol.* **221:** 1153–1164.

James, B.D., G.J. Olsen, J. Liu, and N. Pace. 1988. The secondary structure of ribonuclease P RNA, the catalytic element of a ribonucleoprotein enzyme. *Cell* **52:** 19–26.

Johnson, K.H. and D.M. Gray. 1992. Analysis of an RNA pseudoknot structure by CD spectroscopy. *J. Biomol. Struct. Dyn.* **9:** 733–745.

Joshua-Tor, L., F. Frolow, E. Apella, H. Hope, D. Rabinovich, and J.L. Sussman. 1992. Three-dimensional structures of bulge-containing DNA fragments *J. Mol. Biol.* **225:** 397–414.

Karn, J. 1991. Control of human immunodeficiency virus replication by the *tat, rev, nef* and protease genes. *Curr. Opin. Immunol.* **3:** 526–536.

Keese, P. and R.H. Symons. 1985. Domains in viroids: Evidence of intermolecular RNA rearrangements and their contribution to viroid evolution. *Proc. Natl. Acad. Sci.* **82:** 4582–4586.

Kimelman, D. and M.W. Kirschner. 1989. An antisense mRNA directs the covalent modification of the transcript encoding fibroblast growth factor in *Xenopus* oocytes. *Cell* **59:** 687–696.

Kozak, M. 1991. Structural features in eukaryotic mRNAs that modulate the initiation of translation. *J. Biol. Chem.* **266:** 19867–19870.

Lewin, B. 1990. *Genes 4.* Cell Press, Cambridge, Massachusetts.

Lührmann, R., B. Kastner, and M. Bach. 1990. Structure of spliceosomal snRNPs and their role in pre-mRNA splicing. *Biochim. Biophys. Acta* **1987:** 265–292.

Mans, R.M.W., M.H. van Steeg, P.W.G. Verlaan, C.W.A. Pleij, and L. Bosch. 1992. Mutational analysis of the pseudoknot in the tRNA-like structure of turnip yellow mosaic virus RNA. Aminoacylation efficiency and RNA pseudoknot stability. *J. Mol. Biol.* **223:** 221–232.

McLaren, R.S., S.F. Newbury, G.S.C. Dance, H.C. Causton, and C.F. Higgins. 1991. mRNA degradation by processive 3′-5′ exoribonucleases *in vitro* and the implication for prokaryotic mRNA decay *in vivo. J. Mol. Biol.* **221:** 81–95.

McMullen, D.W., S.R. Jaskunas, and I. Tinoco, Jr. 1967. Application of matrix rank analysis to the optical rotatory dispersion of TMV RNA. *Biopolymers* **5:** 589–613.

McPheeters, D.S., G.D. Stormo, and L. Gold. 1988. Autogenous regulation site on the bacteriophage T4 gene *32* messenger RNA. *J. Mol. Biol.* **201:** 517–535.

McPheeters, D.S., A. Christenson, E.T. Young, G. Stormo, and L. Gold. 1986. Translational regulation of expression of the bacteriophage T4 lysozyme gene. *Nucleic Acids Res.* **14:** 5813–5826.

Michel, F. and E. Westhof. 1990. Modelling of the three-dimensional architecture of group I catalytic introns based on comparative sequence analysis. *J. Mol. Biol.* **216:** 585–610.

Michel, F., K. Umesono, and H. Ozeki. 1989. Comparative and functional anatomy of group II catalytic introns — A review. *Gene* **82:** 5–30.

Michel, F., A.D. Ellington, S. Couture, and J.W. Szostak. 1990. Phylogenetic and genetic evidence for base-triples in the catalytic domain of group I introns. *Nature* **347:** 578–580.

Moazed, D., S. Stern, and H.F. Noller. 1986. Rapid chemical probing of conformation in 16S ribosomal RNA and 30S ribosomal subunits using primer extension. *J. Mol. Biol.* **187:** 399–416.

Moras, D., A.-C. Dock, P. Dumas, E. Westhof, P. Romby, J.-P. Ebel, and R. Giegé. 1986. Anticodon-anticodon interaction induces conformational changes in tRNA: Yeast tRNA^{Asp}, a model for tRNA-mRNA recognition. *Proc. Natl. Acad. Sci.* **83:** 932–936.

Noller, H.F. 1984. Structure of ribosomal RNA. *Annu. Rev. Biochem.* **53:** 119–162.

Peattie, D.A., S. Douthwaite, R.A. Garrett, and H.F. Noller. 1981. A "bulged" double helix in an RNA-protein contact site. *Proc. Natl. Acad. Sci.* **78:** 7331–7335.

Perrotta, A.T. and M.D. Been. 1991. A pseudoknot-like structure required for efficient self-cleavage of hepatitis delta virus RNA. *Nature* **350:** 434–436.

Pleij, C.W.A. 1990. Pseudoknots: A new motif in the RNA game. *Trends Biochem. Sci.* **15:** 143–147.

Pleij, C.W.A., K. Rietveld, and L. Bosch. 1985. A new principle of RNA folding based on pseudoknotting. *Nucleic Acids Res.* **13:** 1717–1731.

Puglisi, J.D., J.R. Wyatt, and I. Tinoco, Jr. 1990a. Conformation of an RNA pseudoknot. *J. Mol. Biol.* **214**; 437–453.

———. 1990b. Solution conformation of an RNA hairpin loop. *Biochemistry* **29:** 4215–4226.

———. 1991. RNA Pseudoknots. *Accts. Chem. Res.* **24:** 152–158.

Puglisi, J.D., R. Tan, B.J. Calnan, A.D. Frankel, and J.R. Williamson. 1992. Conformation of the TAR RNA-arginine complex by NMR. *Science* **257:** 76–80.

Pyle, A.M. and T.R. Cech. 1991. Ribozyme recognition of RNA by tertiary interactions with specific ribose 2'-OH groups. *Nature* **350:** 628–631.

Resnekov, O., M. Kessler, and Y. Aloni. 1989. RNA secondary structure is an integral part of the *in vitro* mechanism of attenuation in simian virus 40. *J.Biol.Chem.* **264:** 9953–9959.

Riordan, F.A., A. Bhattacharyya, S. McAteer, and D.M.J. Lilley. 1992. Kinking of RNA helices by bulged bases, and the structure of the human immunodeficiency virus trans-activator response element. *J. Mol. Biol.***226:** 305–310.

Romaniuk, P.J., P. Lowary, H.-N. Wu, G. Stormo, and O.C. Uhlenbeck. 1987. RNA binding site of R17 coat protein. *Biochemistry* **26:** 1563–1568.

Romby, P., R. Giegé, C. Houssier, and H. Grosjean. 1985. Anticodon-anticodon interactions in solution. Studies of the self-association of yeast or *Escherichia coli* tRNA^{Asp} and of their interactions with *Escherichia coli* tRNA^{Val}. *J. Mol. Biol.* **184:** 107–118.

Rould, M.A., J.J. Perona, and T.A. Steitz. 1991. Structural basis of anticodon loop recognition by glutaminyl-tRNA synthetase. *Nature* **352:** 213–218.

Rould, M.A., J.J. Perona, D. Söll, and T.A. Steitz. 1989. Structure of *E. coli* glutaminyl-tRNA synthetase complexed with tRNA^{Gln} and ATP at 2.8 Å resolution. *Science* **246:** 1135–1142.

Ruby, S.W. and J. Abelson. 1991. Pre-mRNA splicing in yeast. *Trends Genet.* **7:** 79–85.

Ruff, M., S. Krishnaswamy, M. Boeglin, A. Poterszman, A. Mitschler, A. Podjarny, B. Rees, J.C. Thierry, and D. Moras. 1991. Class II aminoacyl transfer RNA synthetases: Crystal structure of yeast aspartyl-tRNA synthetase complexed with tRNA^{Asp}. *Science* **252:** 1682–1689.

Ruffner, D.E., G.D. Stormo, and O.C. Uhlenbeck. 1990. Sequence requirements of the hammerhead RNA self-cleavage reaction. *Biochemistry* **29:** 10695–10702.

Saenger, W. 1984. *Principles of nucleic acid structure.* Springer-Verlag, New York.

SantaLucia, J., Jr., R. Kierzek, and D.H. Turner. 1990. Effects of GA mismatches on the structure and thermodynamics of RNA internal loops. *Biochemistry* **29:** 8813–8819.

————. 1991. Stabilities of consecutive A•C, C•C, G•G, U•C, and U•U mismatches in RNA internal loops: Evidence for stable hydrogen-bonded U•U and C•C$^+$ pairs. *Biochemistry* **30:** 8242–8251.

Sharp, P.A. 1987. Splicing of messenger RNA precursors. *Science* **235:** 766–771.

Simons, R.W. 1988. Naturally occurring antisense RNA control—A brief review. *Gene* **72:** 35–44.

Simons, R.W. and N. Kleckner. 1988. Biological regulation by antisense RNA in prokaryotes. *Annu. Rev. Genet.* **22:** 567–600.

Studnicka, G.M., G.M. Rahn, I.W. Cummings, and W.A. Salser. 1978. Computer method for predicting the secondary structure of single-stranded RNA. *Nucleic Acids Res.* **5:** 3365–3387.

Sumner-Smith, M., S. Roy, R. Barnett, L.S. Reid, R. Kuperman, U. Delling, and N. Sonenberg. 1991. Critical chemical features in *trans*-acting-responsive RNA are required for interaction with human immunodeficiency virus type 1 Tat protein. *J. Virol.* **65:** 5196–5202.

Szewczak, A.A., Y.L. Chan, P.B. Moore, and I.G. Wool. 1991. On the conformation of the alpha sarcin stem-loop of 28S rRNA. *Biochimie* **73:** 871–877.

Tang, C.K. and D.E. Draper. 1989. Unusual mRNA pseudoknot structure is recognized by a protein translational repressor. *Cell* **57:** 531–536.

Tang, R.S. and D.E. Draper. 1990. Bulge loops used to measure the helical twist of RNA in solution. *Biochemistry* **29:** 5232–5237.

Tao, J.S. and A.D. Frankel. 1992. Specific binding of arginine to TAR RNA. *Proc. Natl. Acad. Sci.* **89:** 2723–2726.

Theil, E.C. 1990. Regulation of ferritin and transferrin receptor mRNAs. *J. Biol. Chem.* **265:** 4771–4774.

Traub, W. and J. Sussman, 1982. Adenine-guanine base pairing in ribosomal RNA. *Nucleic Acids Res.* **10:** 2701–2708.

Tuerk, C. and L. Gold. 1990. Systematic evolution of ligands by exponential enrichment: RNA ligands to bacteriophage T4 DNA polymerase. *Science* **149:** 505–510.

Uhlenbeck, O.C. 1972. Complementary oligonucleotide binding to transfer RNA. *J. Mol. Biol.* **65:** 25–41.

van Belkum, A., P. Verlaan, J.B. Kun, C. Pleij, and L. Bosch. 1988. Temperature dependent chemical and enzymatic probing of the tRNA-like structure of TYMV RNA. *Nucleic Acids Res.* **16:** 1931–1950.

van den Hoogen, Y.T., A.A. van Beuzekon, E. de Vroom, G.A. van der Marel, J.H. van Boom, and C. Altona. 1988. Bulge-out structures in the single-stranded trimer AUA and in the duplex (CUGGUGCGG)•(CCGCCCAG). *Nucleic Acids Res.* **16:** 5013–5030.

Varani, G. and I. Tinoco, Jr. 1991. RNA structure and NMR spectroscopy. *Q. Rev. Biophys.* **24:** 479–532.

Varani, G., C. Cheong, and I. Tinoco, Jr. 1991. Structure of an unusually stable RNA hairpin. *Biochemistry* **30:** 3280–3289.

Varani, G., B. Wimberly, and I. Tinoco, Jr. 1989. Conformation and dynamics of an RNA internal loop. *Biochemistry* **28:** 7760–7772.

Wagner, R.W., C. Yoo, L. Wrabetz, J. Kamholz, J. Buchhalter, N.F. Hassan, K. Khalili, S.U. Kim, B. Perussia, F.A. McMorris, and K. Nishikura. 1990. Double-stranded RNA unwinding and modifying activity is detected ubiquitously in primary tissues and cell lines. *Mol. Cell Biol.* **10:** 5586–5590.

Wang, J.-F. and T.R. Cech. 1992. Tertiary structure around the guanosine-binding site of the *Tetrahymena* ribozyme. *Science* **256:** 526–529.

Wassarman, D.A. and J.A. Steitz. 1991. Alive with DEAD proteins. *Nature* **349:** 463–464.

Watson, J.D., N.H. Hopkins, J.W. Roberts, J.A. Steitz, and A.M. Weiner. 1987. *Molecular biology of the gene*, 4th edition, p. 488. Benjamin/Cummins, Menlo Park, California.

Weeks, K.M. and D.M. Crothers. 1991. RNA recognition by Tat-derived peptides: Interaction in the major groove? *Cell* **66:** 577–588.

Weiner, A.M. and N. Maizels. 1987. tRNA-like structures tag the 3′ ends of genomic RNA molecules for replication: Implications for the origin of protein synthesis. *Proc. Natl. Acad. Sci.* **84:** 7383–7387.

Westhof, E., P. Dumas, and D. Moras. 1985. Crystallographic refinement of yeast aspartic acid transfer RNA. *J. Mol. Biol.* **184:** 119–145.

Westhof, E., P. Romby, P.J. Romaniuk, J.-P. Ebel, C. Ehresmann, and B. Ehresmann. 1989. Computer modeling from solution data of spinach chloroplast and of *Xenopus laevis* somatic and oocyte 5 S rRNAs. *J. Mol. Biol.* **207:** 417–431.

White, S.A. and D.E. Draper. 1987. Single base bulges in small RNA hairpins enhance ethidium binding and promote an allosteric transition. *Nucleic Acids Res.* **15:** 4049–4064.

―――――. 1989. Effects of single-base bulges on intercalator binding to small RNA and DNA hairpins and a ribosomal RNA fragment. *Biochemistry* **15:** 4049–4064.

White, S.A., M. Nilges, A. Huang, A.T. Brünger, and P.B. Moore. 1992. NMR analysis of helix I from the 5S RNA of *Escherichia coli. Biochemistry* **31:** 1610–1621.

Wimberly, B., G. Varani, and I. Tinoco, Jr. 1993. Conformation of loop E from eukaryotic 5S RNA. *Biochemistry* (in press).

Witherell, G.W. and O.C. Uhlenbeck. 1989. Specific RNA binding by Qβ coat protein. *Biochemistry* **28:** 71–76.

Witherell, G.W., J.M. Gott, and O.C. Uhlenbeck. 1991. Specific interaction between RNA phage coat proteins and RNA. *Prog. Nucleic Acid Res. Mol. Biol.* **40:** 185–220.

Woese, C.R., S. Winker, and R.R. Gutell. 1990. Architecture of ribosomal RNA— Constraints on the sequence of tetra-loops. *Proc. Natl. Acad. Sci.* **87:** 8467–8471.

Woese, C.R., R. Gutell, R. Gupta, and H.F. Noller. 1983. Detailed analysis of the higher-order structure of 16S-like ribosomal ribonucleic acids. *Microbiol. Rev.* **47:** 621–669.

Wolters, J. 1992. The nature of preferred hairpin structures in 16S-like rRNA variable regions. *Nucleic Acids Res.* **20:** 1843–1850.

Wu, J. and J.L. Manley. 1991. Base pairing between U2 and U6 snRNAs is necessary for splicing of a mammalian pre-mRNA. *Nature* **352:** 818–821.

Wu, H.-N. and O.C. Uhlenbeck. 1987. Role of a bulged A residue in a specific RNA protein interaction. *Biochemistry* **26:** 8221–8227.

Wyatt, J.R., J.D. Puglisi, and I. Tinoco, Jr. 1990. RNA pseudoknots: Stability and loop size requirements. *J. Mol. Biol.* **214:** 455–470.

Yanofsky, C. 1988. Transcription attenuation. *J. Biol. Chem.* **263:** 609–612.

Zarling, D.A., C.J. Calhoun, C.C. Hardin, and A.H. Zarling. 1987. Cytoplasmic Z-RNA. *Proc. Natl. Acad. Sci.* **84:** 6117–6121.

Zhang, P., P. Popieniek, and P.B. Moore. 1989. Physical studies of 5S RNA variants at position 66. *Nucleic Acids Res.* **17:** 8645–8657.

Zinoni, F., J. Heider, and A. Böck. 1990. Features of the formate dehydrogenase mRNA necessary for decoding of the UGA codon as selenocysteine. *Proc. Natl Acad. Sci.* **87:** 4660–4664.

19
RNA: The Shape of Things to Come

Larry Gold, Pat Allen,
Jon Binkley, David Brown,
Dan Schneider, and Sean R. Eddy
Department of Molecular, Cellular and Developmental Biology
University of Colorado, Boulder, Colorado 80309

Craig Tuerk, Louis Green,
Sheela MacDougal, and Diane Tasset
NeXagen, Inc.
Boulder, Colorado 80301

This enormous book and the literature behind it establish a new conventional wisdom. Today it is common to assert that an RNA world once existed and flowed into the present biosphere. Introductory texts and review articles softly (or staunchly [Watson et al. 1987]) proclaim that RNA comprises the collection of genetic and catalytic molecules that began the biosphere and then gracefully stepped into the background as a richer life unfolded. In this volume are papers that define the position RNA might have held. Having read many such papers, we note that by the scientific rules of evidence, the concept of an RNA world may stand only as an attractive idea, an idea that never can be tested adequately. One problem is that known RNA catalysts are few, and thus we have no way to assess the catalytic breadth of RNA by normal observation and experimentation.

Methods now exist to inspect the catalytic and binding properties of small RNA molecules without recourse to the identification of RNAs in the present biosphere. In this paper, we illustrate the chemical and likely catalytic potential of RNA by reference to work done by us using these new methods.The plausibility and limitations of an early RNA world are addressed by the work. We have come to an astonishing and unexpected conclusion. RNA apparently has a huge distribution of sizes and shapes, and a collection of RNA molecules, even a collection of rather short RNA molecules, could have been used for an enormous variety of chemistry prior to the invention of proteins. Proteins seem profoundly better molecules for most biological requirements in the present bio-

The RNA World
© 1993 Cold Spring Harbor Laboratory Press 0-87969-380-0/93 $5 + .00

sphere, yet we now see some advantages for RNA when the degree of polymerization of the two macromolecules under comparison is small.

SELEX: THE FIRST EXPERIMENT

Our laboratory had studied the regulation of the translation of the bacteriophage T4 DNA polymerase (Andrake et al. 1988; Tuerk et al. 1990). In collaboration with Jim Karam's laboratory, we discovered that the biosynthesis of DNA polymerase was autogenously regulated by specific binding of DNA polymerase to the ribosome-binding-site region of its mRNA. The mRNA target included a hairpin with a loop of eight nucleotides. We devised an experiment to test all 65,536 loop sequences for binding to T4 DNA polymerase.

SELEX is an acronym for Systematic Evolution of Ligands by EXponential enrichment. The notion (shown in Fig. 1) is that vast repertoires of nucleic acid sequences, prepared by randomization of DNA during machine synthesis (for use directly as DNA ligands or as templates for transcription), could be bound to a target (in this case T4 DNA polymerase) and that high-affinity ligands could be identified within the repertoire by reiterative partitioning of the target-bound nucleic acids from the nucleic acids of lower affinity. SELEX had been under development for several years in our laboratory, largely because much of our interest had been focused on nucleic-acid-binding proteins and the specific target sequences recognized by those proteins (Winter et al. 1987; McPheeters et al. 1988; Ruckman et al. 1989).

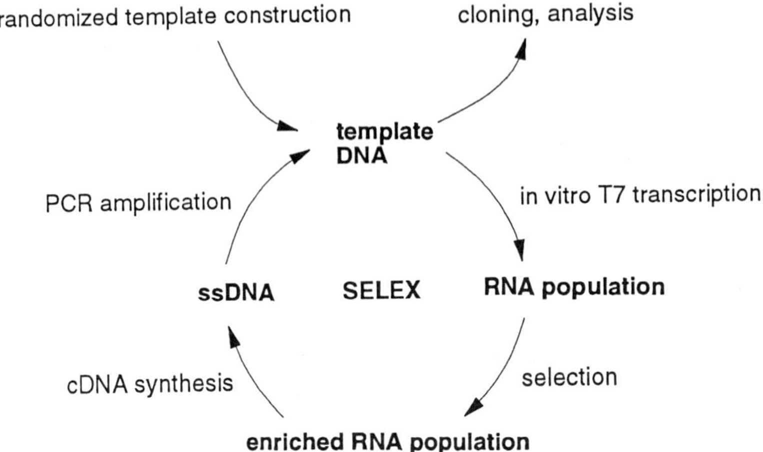

Figure 1 Overview of the SELEX protocol (Tuerk and Gold 1990).

SELEX with T4 DNA polymerase was quite succe
Gold 1990). Within four rounds of selection and enri
tion, the RNA repertoire was reduced to two prima
was the wild-type sequence found in the bacteriophage
merase mRNA, and the other contained four nucleotide subs
within the eight nucleotides that had been randomized (Fig. 2). The two
major sequences had identical dissociation constants and competed for
binding to the same site on the DNA polymerase. In this case, biology
had apparently chosen one best loop sequence and had ignored (at least
in the few phage we inspected) an equivalent ligand four mutations distant from the wild-type sequence.

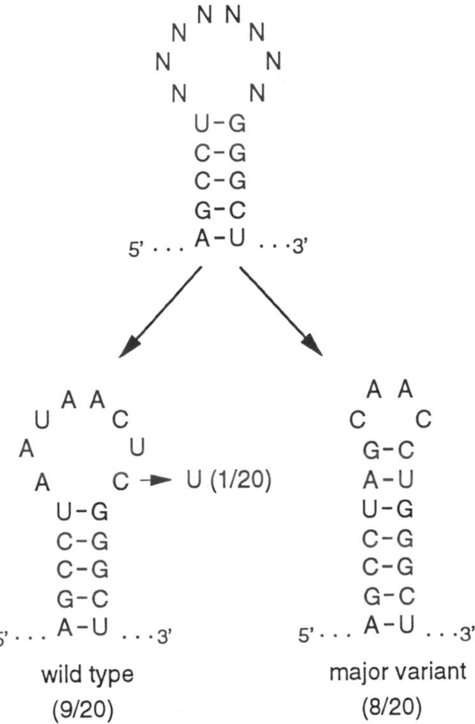

Figure 2 SELEX against the RNA-binding site of T4 DNA polymerase. The
eight positions of the loop were made random by machine DNA oligonucleotide
synthesis. Twenty molecules were isolated and sequenced from the RNA population after four rounds of amplification and selection. Nine molecules had the
wild-type sequence; eight had a different sequence, dubbed the "major variant."
Both molecules bind T4 DNA polymerase with indistinguishable K_d values (4.8
nM). One other sequence from the pool had a point substitution of C to U that
seemed to make the wild-type sequence more like the major variant, but this
molecule bound fourfold less tightly.

We were struck by the fact that the second sequence identified did not seem structurally isomorphic with the wild-type loop, as we drew the usual two-dimensional pictures of the predicted structures (Fig. 2). Moreover, one ligand had a point substitution that seemed to us to make the wild-type loop more like the major variant, yet it was a weaker binder (Tuerk and Gold 1990). We could not easily understand the equivalent binding of the two RNAs in approximately the same target site.

In other words, the major variant sequence had come from a very different part of sequence space than the wild-type. "Sequence space" refers to a useful visualization of the 4^N different sequences of length N as points in a space, each of which has a value (here, binding affinity; more generally, "fitness" for a particular selection or environment). Although formally it is more logical to think of this as an N-dimensional space, it is more intuitive to think of the space as a three-dimensional landscape, where the height of the landscape reflects the fitness, and increasing distance in the XY plane approximately reflects decreasing sequence similarity between two points. A "peak" in sequence space is then, for our purposes, a group of related sequences that have a high binding affinity to the target. Traditional site-directed mutagenesis (or evolution, for that matter) typically explores a single peak in sequence space and would usually start from a peak reached by biological evolution. With SELEX, we realized we had the ability to find and explore peaks that biology may never have explored.

We realized that our views of RNA structure were naive. We had always thought that as an RNA-protein interaction evolved it would become strongly restricted toward a single RNA sequence; we never expected to find another peak in sequence space. We became convinced that RNA (and single-stranded DNA) had, via secondary and tertiary structural motifs, the potential to form unanticipated shapes, and those shapes were well beyond the published imagination of the scientific community. We began to wonder if we might not be able to select and use small RNA molecules much as monoclonal antibodies are used. SELEX was a protocol that could sample vast nucleic acid repertoires with virtually any target molecule, and the number 65,536 was merely the beginning of an entirely new method of ligand discovery from repertoires as vast as the solubility limits of nucleic acids and the volume limitations of a particular partitioning methodology.

SELEX: A VAST REPERTOIRE PROBED WITH A WELL-CHARACTERIZED TARGET

To test the accuracy of SELEX to locate a natural RNA ligand in a large randomized repertoire, we performed an experiment with bacteriophage

R17 coat protein as the target (Schneider et al. 1992). A detailed analysis of important elements of the wild-type RNA ligand was performed by Olke Uhlenbeck's laboratory, by comparing the affinity of many singly and doubly mutant ligands with that of the wild-type (Lowary and Uhlenbeck 1987; Romaniuk et al. 1987; Witherell 1991). Those experiments provided a simple picture of the high-affinity RNA ligand for the coat protein as a hairpin with a tetra loop and a bulged nucleotide in the stem (Fig. 3). Those experiments also showed that the wild-type sequence was not the sequence with the highest binding affinity: a U to C substitution in the loop increased the binding affinity markedly.

SELEX was done for eleven rounds with an RNA randomized at 32 contiguous positions; more than 10^{19} different sequences are in such a repertoire, although, due to volume constraints, about 10^{14} were tested. The resulting collection of high-affinity ligands confirms and extends the findings from the Uhlenbeck laboratory (Fig. 4) (Schneider et al. 1992). The U to C substitution that improves the wild-type ligand is found in all the ligands. Most of the new ligands have C:G base pairs at the top of the

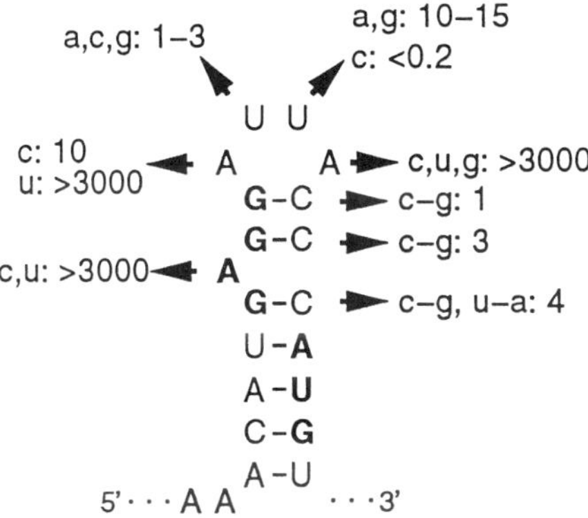

Figure 3 Abridged summary of the Uhlenbeck group's mutational studies of the phage R17 coat protein RNA-binding site (Lowary and Uhlenbeck 1987; Romaniuk et al. 1987; Witherell 1991). The minimal wild-type binding site is shown. In boldface is the Shine-Dalgarno ribosome-binding site (GAGG) and translational initiation codon (AUG), which are additional biological constraints imposed on this sequence. Arrows indicate single-base changes in the loop and bulge or compensatory double-base changes in the stem, and the relative dissociation constant of the mutant compared to the wild-type K_d of 3.3 nM. For our purposes here, note that (1) the stem sequence and the identity of the second base in the AUUA loop are relatively unimportant; (2) a U to C substitution at the third loop position actually improves binding affinity at least fivefold; and (3) the identities of the other loop and bulge positions are important for tight binding.

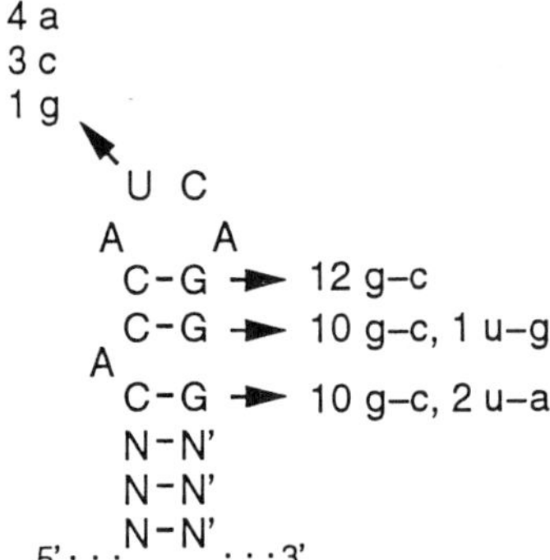

Figure 4 Consensus of the SELEX-generated R17 coat protein ligand sequences of Schneider et al. (1992). 38 individuals were isolated and sequenced; 36 had structures like the known R17 coat protein-binding site (the other two sequences have been ignored). Arrows indicate differences from the consensus at a position or base pair among these 36 sequences; the number of sequences in the 36 that carry a particular substitution is also indicated. Note the variability at the second loop position and in the stem, the C selected at the third loop position, and the conservation of the sequence of the other loop and stem positions.

stem rather than the natural G:C base pairs (in the wild-type sequence the Gs are part of the Shine-Dalgarno sequence of the regulated replicase cistron, a biological constraint not relevant to SELEX in vitro). Most importantly, this SELEX experiment, using perhaps the most thoroughly studied natural RNA:protein interaction, yielded a sensible answer in keeping with our expectations from far smaller repertoires.

Three points should be made here in some detail. Eleven rounds were required to narrow the repertoire to a population from which 36 of 38 cloned molecules formed properly bulged hairpins (for details, see Schneider et al. 1992). We proceed through the mathematics of SELEX to explore the congruence between intuition and reality (Irvine et al. 1991). The mathematics show that more rounds are needed to accomplish SELEX than one might think, largely because the low-affinity ligands in the repertoire are so numerous relative to the high-affinity ligands. Similarly, all partitioning methods have some background in which bulk nucleic acid contaminates the target-bound population; this also increases the number of rounds required to complete SELEX. Finally, Uhlenbeck's work showed that most nonfunctional mutations lowered the affinity for R17 coat protein by roughly 1000-fold, yet our eleven rounds only improved the global dissociation constant about 80-fold. This fact, also observed in our laboratory for several other target proteins, suggests that the nonspecific binding constant may be higher for totally random repertoires than it is for sequences closely related to a functional tight-binding sequence. This implies that one could set out to use SELEX to obtain "anti-binders" with abnormally poor binding affinities to a target protein.

COMPLEX BULGES AND PSEUDOKNOTS: HIV *rev*
AND REVERSE TRANSCRIPTASE

Our first two examples concerned simple hairpins (which are not simple) (Cheong et al. 1990; Heus and Pardi 1991) and a hairpin with a single bulged nucleotide. In two complete SELEX experiments with nucleic-acid-binding proteins encoded by human immunodeficiency virus (HIV), we identified more complicated RNA ligands.

In the first case, we studied the Rev protein (Tuerk et al. 1992a). The resulting high-affinity ligands bind more tightly to the protein than does the viral target (the RRE) (Malim et al. 1990) and are structurally related to that viral target (Fig. 5). The ligands contain asymmetric bulges with paired nucleotides within the bulge.

In the second case, from a repertoire of RNA randomized at 32 positions, high-affinity ligands were found against HIV reverse transcriptase (Fig. 6) (Tuerk et al. 1992b). The ligands bind to the active site of reverse transcriptase and inhibit cDNA synthesis (Tuerk et al. 1992b). In this case, the resultant ligands do not seem to mimic natural ligands, such as the $tRNA^{Lys}_3$ that is utilized for priming replication (unless somehow the mimicry reflects the shape of a primer/template junction, which is not obvious). Most importantly, the reverse transcriptase ligands generated through SELEX are pseudoknots, an RNA motif found increasingly in important RNA molecules and found through SELEX with other targets as well.

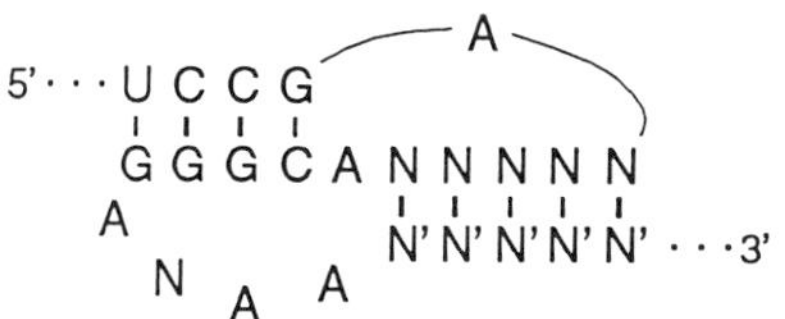

Figure 5 The wild-type sequence of a portion of the HIV *rev*-responsive element (RRE) and the consensus structure of the majority class of SELEX-generated *rev* ligands (dubbed Motif I) (Tuerk et al. 1992a). 53 molecules were isolated and sequenced; 27 had the Motif I structure. The highest-affinity ligand from the Motif I class had a measured K_d of 80 nM, compared to a K_d of 400 nM for a wild-type RRE transcript.

```
                A       G     A
5'···C G C U G    C G    U     C A···3'
     | | | | |    | |    |     | |
3'···G C G A C ‿ G C   G ‿ G U···5'
                     G
```

wild type RRE

```
            U U       A   A
5'···N N N N N    G A G   U   C A N···3'
     | | | | |    | | |   |   | | |
3'···N'N'N'N'N' ‿ C U C   G ‿ G U N'···5'
                       A
```

Motif I REV ligands

Figure 6   Pseudoknotted consensus structure of 112 of the 153 SELEX-generated ligands to HIV reverse transcriptase. The highest-affinity ligands had measured K_d values of approximately 5 nM.

PROTEIN TARGETS THAT ARE NOT NUCLEIC ACID BINDERS

Extrapolating from reverse transcriptase, Rev, the R17 coat protein, and T4 DNA polymerase (which have binding sites for nucleic acids) to all proteins represents the critical challenge for SELEX generalization. We report here one success with nerve growth factor (NGF) (J. Brinkley et al., unpubl.), and note a recent publication (Bock et al. 1992) in which thrombin was the successful target of a SELEX experiment with a single-stranded DNA repertoire. (Diane Tasset [unpubl.] at NeXagen has found a high-affinity RNA ligand aimed at human thrombin; that ligand bears no obvious relationship to the reported DNA ligand.)

In our NGF experiment, 10^{13} RNAs were challenged with protein according to the standard protocol (J. Brinkley et al., unpubl.). After ten rounds, a number of winning ligands were found; many of these sequences had an apparent structure even more complicated than a simple pseudoknot (Fig. 7). Ligands aimed at thrombin and NGF, through the rules of epiphany and logic, suggest that SELEX can work on any protein. Nucleic acids that bind to dyes have been isolated by Ellington and Szostak (1990, 1992); those experiments reinforce our idea that nucleic acids can be used to provide ligands aimed at nearly any target. From the generalization of SELEX comes the idea that novel binding and even altered catalytic properties (Robertson and Joyce 1990) or new catalytic properties can be observed in RNAs identified through SELEX.

RESTATEMENT OF THE DRIVING PRINCIPLE

We restate the startling yet now obvious conclusion. Nucleic acid molecules, containing nothing more chemically profound than the four stan-

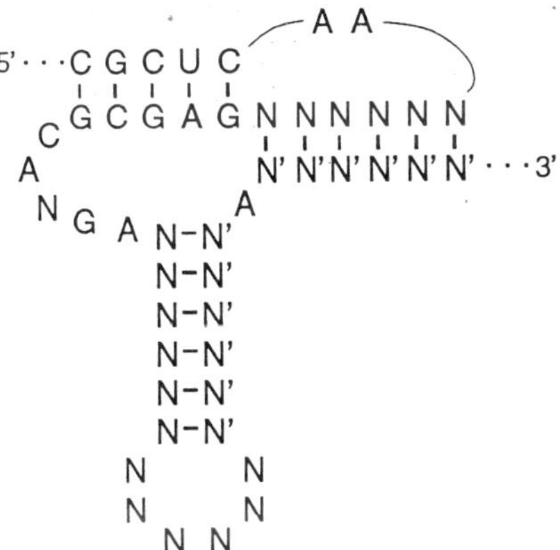

Figure 7   Pseudoknotted (with an elaboration) consensus structure of the SELEX generated ligands to NGF. The 5′ nucleotides CGCUCAA were part of the fixed primer annealing site and were not randomized in the starting population. The K_d of the highest-affinity ligand from this population was approximately 100 nM.

dard nucleotides, have the capacity to acquire substantial intramolecular contacts (standard Watson/Crick base pairs, non-standard base pairs, and novel tertiary interactions). Such structures provide relatively rigid surfaces as potential ligands for specific binding to any target.

In our experience with the targets described in this paper and with other targets (under study both at the University of Colorado and at NeXagen), a common affinity of a winning ligand is in the nanomolar range; these dissociation constants are surprisingly low. We have seen reports for the winning ligands from peptide libraries (Cwirla et al. 1990; Fodor et al. 1991; Houghten et al. 1991; Lam et al. 1991) and expect similar reports for the winning ligands from oligosaccharide libraries; in general, these (very short) polypeptide ligands bind more weakly than do RNAs identified through SELEX. We think that single-stranded nucleic acid libraries have within the repertoires individuals with tight-binding properties, because short oligonucleotides form relatively rigid structures compared to peptides of the same monomer length.

To a first approximation, the structural motifs of RNAs and small regions of protein sequences are similar (Fig. 8). This is not particularly profound; simply put, the first level of description of either a protein or RNA structure is its secondary structure, which describes not only which residues are bound into a particular known conformation (α-helix, β-sheet, or A-form RNA helix), but also what is single-stranded and where the ends of the single-stranded regions are tethered. RNA, through closed hairpin loops and the extended loops of pseudoknots, can tether single

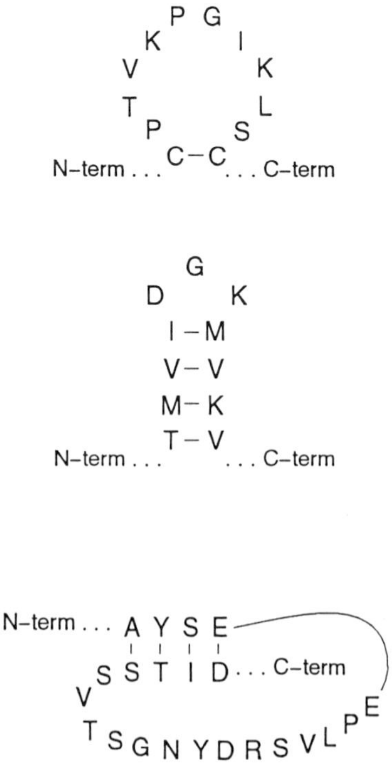

Figure 8 Secondary structure representations of small protein structure motifs. (*Top*) Hairpin loop closed by a cysteine-cysteine disulfide bridge in residues 43–54 of erabutoxin (Brookhaven database entry PDBlNXB). (*Center*) Hairpin loop formed by antiparallel β strands, in residues 297–307 of glyceraldehyde 3-phosphate dehydrogenase (PDB1GD1). (*Bottom*) Long loop (analogous to the extended second loop of an RNA pseudoknot) formed by parallel β strands, in residues 271–293 of glyceraldehyde 3-phosphate dehydrogenase. Similar closed-loop and extended-loop structures are made with parallel and antiparallel α-helix bundles, but we show no examples here.

strands in space with secondary structural motifs much like protein can. The crucial difference, for our purposes, is that whereas random, short RNAs inevitably contain stable base-pairing domains, short peptides usually do not form stable intramolecular contacts.

It may cost several kilocalories to fix a short unstable peptide structure within a binding pocket of a target, thus yielding perhaps the difference between nanomolar and micromolar dissociation constants. (Short peptides containing disulfides can be rigid, and some such peptides do have nanomolar affinities for receptors.) Peptide repertoires for drug discovery and other purposes may require pre-enrichment for rigid structures in order to find molecules with high affinities. Intramolecular structures are found over and over in SELEX-generated ligands because of the intrinsic base-pairing capacity of RNA.

THE EARLY BIOSPHERE, THE RNA WORLD, AND THE GENERALITY OF SELEX

As a result of our SELEX experiments, we feel much more comfortable with the new conventional wisdom presented in this book. The RNA World Hypothesis relies on the assumption that a collection of fairly short RNA molecules could have provided sufficient catalytic variety to initiate life; but the range of activities one has been able to see for RNA has been limited to short walks in the sequence space around extant sequences such as RNase P or the *Tetrahymena* ribozyme. We now know that SELEX can find short RNA sequences with tight binding affinity to nearly anything we have tried so far as a target; because simple binding to a target is not much different from binding and stabilizing the transition state of a reaction pathway, we are sure that we will now be able to isolate molecules with catalytic properties beyond cleavage and ligation of nucleic acids.

We note again that short RNAs tend to form thermodynamically stable structures, whereas short peptides, as a rule, do not. We think that this simple fact may have much to do with the success of SELEX—and perhaps also the RNA world—in finding such a variety of high-affinity ligands in a small sequence space. There are many reasons that proteins may have "won" over RNA; we believe one was that the onset of translation and the resultant availability of larger proteins solved the intramolecular structure problems faced by small peptides and allowed amino acids' superior chemistry to triumph.

As we continue to search our immediately accessible, relatively small RNA sequence spaces for interesting ligands and begin to design experiments to make much larger sequences and to search spaces of stable protein sequences (Tuerk and Gold 1990), we wonder to what extent we are reproducing experiments done billions of years ago.

CONCLUSIONS AND PERSPECTIVES

Most gifted scientists view RNA (or single-stranded DNA) as strings of information rather than sources of interesting globular ligands. A recent example of that thinking is displayed in a proposal for "Encoded Combinatorial Chemistry" (Brenner and Lerner 1992). In that proposal, nucleic acids are used to accompany members of a repertoire encoded by that nucleic acid; the repertoire itself is conceived to be virtually any chemistry. When a ligand is partitioned from the bulk of the repertoire, the accompanying nucleic acid may be used as a coded tag to identify the sequence of the true ligand. A narrow example of the idea would be to isolate peptides attached to their own mRNAs and then to treat the copolymer as both an encoding nucleic acid and a potential ligand (Tuerk and Gold 1990). In this proposal, the encoding nucleic acid is treated as merely the ultimate means to identify (through amplification and sequencing) the true ligand. We predict, for reasons cited above, that this new technology will often yield a copolymer in which the *nucleic acid tag* is the true ligand, instead of the intended polymer.

SELEX provides full access to and exploitation of the sequence space and structure space of short oligonucleotides; those spaces are large enough to provide novel properties not found in the present biosphere. Creative design of the partitioning or otherwise selective step of the SELEX protocol will yield stunning new molecular qualities. Most amazingly, RNA — the very molecules proposed as the transitional enzymes and catalysts billions of years ago prior to the DNA and protein enzyme takeover of the biosphere — will play a significant role in applied health research. The possible uses of "oligonucleotide antibodies" are enticing, providing at a minimum high-affinity binding reagents in a more useful form than monoclonal antibodies. Oligonucleotide ligands from the SELEX procedure can be found quickly, and high binding specificities can be demanded during the selection steps. The next several years of research should provide both useful molecules and clues to events that happened long ago.

ACKNOWLEDGMENTS

This work was supported by NeXagen and by National Institutes of Health grants GM-28685 and GM-19963. We thank the W.M. Keck Foundation for their generous support of RNA science on the Boulder campus.

REFERENCES

Andrake, M., N. Guild, T. Hsu, L. Gold, C. Tuerk, and J. Karam. 1988. DNA polymerase of bacteriophage T4 is an autogenous translational repressor. *Proc. Natl. Acad. Sci.* **85:** 7942–7946.

Bock, L., L. Griffin, J. Latham, E. Vermaas, and J. Toole. 1992. Selection of single-stranded DNA molecules that bind and inhibit human thrombin. *Nature* **355:** 564–566.

Brenner, S. and R. Lerner. 1992. Encoded combinatorial chemistry. *Proc. Natl. Acad. Sci.* **89:** 5381–5383.

Cheong, C., G. Varani, and I. Tinoco, Jr. 1990. Solution structure of an unusually stable RNA hairpin, 5′ GGAC(UUCG)GUCC. *Nature* **346:** 680–682.

Cwirla, S., E. Peters, R. Barrett, and W. Dower. 1990. Peptides on phage: A vast library of peptides for identifying ligands. *Proc. Natl. Acad. Sci.* **81:** 6378–6382.

Ellington, A. and J. Szostak. 1990. In vitro selection of RNA molecules that bind specific ligands. *Nature* **346:** 818–822.

———. 1992. Selection in vitro of single-stranded DNA molecules that fold into specific ligand-binding structures. *Nature* **355:** 850–852.

Fodor, S., J. Read, M. Pirrung, L. Stryer, L.A. Tsai, and D. Solas. 1991. Light-directed, spatially addressable parallel chemical synthesis. *Science* **251:** 767–773.

Heus, H. and A. Pardi. 1991. Structural features that give rise to the unusual stability of RNA hairpins containing GNRA loops. *Science* **253:** 191–194.

Houghten, R., C. Pinilla, S. Blondelle, J. Appel, C. Dooley, and J. Cuervo. 1991. Generation and use of synthetic peptide combinatorial libraries for basic research and drug discovery. *Nature* **354:** 84–86.

Irvine, D., C. Tuerk, and L. Gold. 1991. SELEXION. Systematic evolution of ligands by exponential enrichment with integrated optimization by non-linear analysis. *J. Mol. Biol.* **222:** 739–761.

Lam, K., S. Salmon, E. Hersh, V. Hruby, W. Kazmierski, and R. Knapp. 1991. A new type of synthetic peptide library for identifying ligand binding activity. *Nature* **354:** 82–84.

Lowary, P. and O. Uhlenbeck. 1987. An RNA mutation that increases the affinity of an RNA-protein interaction. *Nucleic Acids Res.* **15:** 10483–10493.

Malim, M., L. Tiley, D. McCarn, J. Rusche, J. Hauber, and B. Cullen. 1990. HIV-1 structural gene expression requires binding of the *rev trans*-activator to its RNA target sequence. *Cell* **60:** 675–683.

McPheeters, D.S., G.D. Stormo, and L. Gold. 1988. Autogenous regulatory site on the bacteriophage T4 gene *32* messenger RNA. *J. Mol. Biol.* **201:** 517–535.

Robertson, D. and G. Joyce. 1990. Selection in vitro of an RNA enzyme that specifically cleaves single-stranded DNA. *Nature* **344:** 467–468.

Romaniuk, P., P. Lowary, H.-N. Wu, G. Stormo, and O. Uhlenbeck. 1987. RNA binding site of R17 coat protein. *Biochemistry* **26:** 1563–1568.

Ruckman, J., D. Parrna, C. Tuerk, D.H. Hall, and L. Gold. 1989. Identification of a T4 gene required for bacteriophage mRNA processing. *New Biol.* **1:** 54–65.

Schneider, D., C. Tuerk, and L. Gold. 1992. The selection of high affinity RNA ligands to the bacteriophage R17 coat protein. *J. Mol. Biol.* **228:** 862–869.

Tuerk, C. and L. Gold. 1990. Systematic evolution of ligands by exponential enrichment. *Science* **249:** 505–510.

Tuerk, C., S. MacDougal, and L. Gold. 1992a. Directed evolution of RNA ligands to HIV-1 *rev* protein. *Proc. Natl. Acad. Sci.* **89:** 6988–6992.

Tuerk, C., S. Eddy, D. Parma, and L. Gold. 1990. Autogenous translational operator recognized by bacteriophage T4 DNA polymerase. *J. Mol. Biol.* **213:** 749–761.

Tuerk, C., S. MacDougal, G. Hertz, and L. Gold. 1992b. In vitro evolution of functional nucleic acids: High affinity RNA ligands to the HIV-1 *rev* protein. In *The polymerase chain reaction* (ed. F. Ferre et al.). Birkhauser, Springer-Verlag, New York. (In press.)

Watson, J., N. Hopkins, J. Roberts, J. Steitz, and A. Weiner. 1987. *Molecular biology of the gene.* Benjamin/Cummings, Menlo Park, California.

Winter, R.B., L. Morrissey, P. Gauss, L. Gold, T. Hsu, and J. Karam. 1987. Bacteriophage T4 *regA* protein binds to mRNAs and prevents translation initiation. *Proc. Natl. Acad. Sci.* **84:** 7822–7826.

Witherell, G. 1991. Specific interaction between RNA phage coat proteins and RNA. *Prog. Nucleic Acid Res. Mol. Biol.* **40:** 185–220.

20

In Vitro Selection of Functional RNA Sequences

Jack W. Szostak
Department of Molecular Biology
Massachusetts General Hospital
Boston, Massachusetts 02114

Andrew D. Ellington
Department of Chemistry
Indiana University
Bloomington, Indiana 47405

A critical but seldom articulated assumption common to models of the evolution of an RNA world is that prebiotically synthesized polynucleotides of random sequence were the source of the first self-replicating molecules. However, the probability of finding functional molecules of any sort in a mixture of random sequences is unknown. A high probability would favor the origin of life, whereas a very low probability would make the origin of life more difficult.

To address this problem, we developed an iterative method for the isolation of nucleic acid sequences with specific ligand-binding or catalytic properties from very large pools of random sequences. The key to this method is the use of cycles of in vitro selection to enrich the pool for RNA species with the desired properties, followed by amplification of the selected molecules. This cycle of in vitro selection and amplification can be repeated as many times as necessary to isolate a population of molecules composed entirely of binding species, which can then be cloned and characterized individually. We used cycles of affinity chromatography followed by polymerase chain reaction (PCR) amplification to isolate ligand-binding RNAs and DNAs. We found that roughly 10^{-10} of the original sequences in our random pools were able to bind to organic dyes and other small ligands, including amino acids. If these results can be extrapolated to the binding of the transition states of chemical transformations, it is likely that a wide range of RNA catalysts might be found in pools of random sequence RNA molecules. Although it is a long way from a ligand-binding RNA sequence to an RNA replicase, our results suggest that the appearance of such a replicase in a pool

The RNA World
© 1993 Cold Spring Harbor Laboratory Press 0-87969-380-0/93 $5 + .00

of random sequence molecules might not have been a limiting step in the origin of life.

Similar in vitro selection methods have been applied to a wide range of problems, including the definition of protein-binding sites on DNA and RNA and the modification of the catalytic properties of ribozymes. We discuss recent advances in these areas and possible future applications of in vitro selection to the isolation of new catalysts.

IN VITRO SELECTION

The essence of in vitro selection is that it is an iterative procedure (Fig. 1). A pool of random or partially random RNA or DNA sequences is subjected to a purification protocol designed to enrich for molecules with some desired property, and the small amount of enriched material that is obtained is then amplified so that the enrichment process can be repeat-

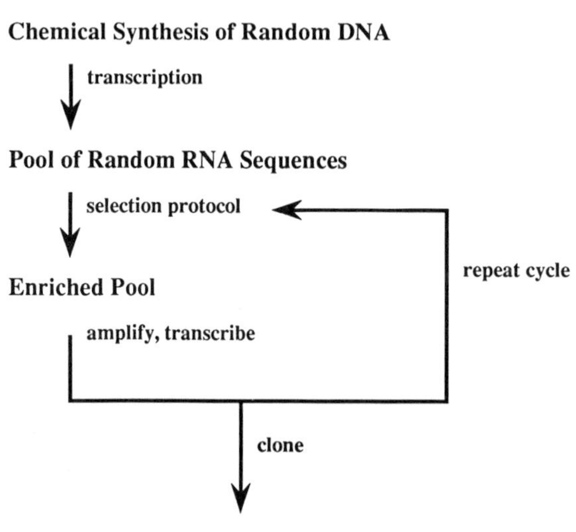

Figure 1 In vitro selection schematic. A chemically synthesized pool of random (or partially random) DNA is transcribed to generate a pool of RNA sequences, which is then subjected to a selection procedure such as affinity chromatography to yield an enriched pool. This enriched pool is amplified, and the cycle of selection and amplification is repeated until the remaining sequences pass the selection procedure. At this point, individual sequences can be cloned and characterized. Aptamers are nucleic acids with specific ligand-binding properties. With appropriate selection procedures, catalytic RNAs or RNAs with other properties can be isolated.

ed. Most purification procedures are limited to a one-step purification factor of 100- to 1000-fold because of an inevitable background of non-specific absorption of unwanted molecules to the purification media and to the glass or plastic walls of the apparatus. For example, we typically obtain a background of 0.1–1.0% of the input RNA from a variety of affinity chromatography purifications. When attempting to purify a nucleic acid with a particular binding or catalytic property from a large pool of sequences, amplification must be applied between purification steps to retain an easily handled amount of material. Repetition of cycles of purification and amplification increases the total purification factor multiplicatively, such that five cycles allows an enrichment of 10^{10}–10^{15}.

Such extreme degrees of purification allow the isolation of fascinating new molecules with unexpected structures and properties, but in order to take full advantage of the power of this method, it is necessary to create very large pools of random or partially random sequences. The synthesis of very large pools pushes the limits of DNA synthesis technology. The power of reiterated in vitro selection may also lead to unexpected problems and artifacts that require more sophisticated approaches to purification and amplification. Some of these issues are addressed in the sections below.

Pool Synthesis

RNAs or DNAs that perform a new function may be selected from a pool of completely random sequences. A major issue in the design of random sequence pools is the length of the random region. Very short regions (say, less than 25 nucleotides in length) have the advantage that a pool can easily be synthesized that will contain every possible sequence in the random region. Thus, in principle, every possible functional sequence can be isolated and characterized; conversely, failure to select any functional sequence implies that no sequence of that length or smaller exists which is capable of carrying out the specified function.

Somewhat longer random regions, of 30–60 nucleotides, allow the formation of more complex structures and motifs that may be essential for function, e.g., multiple stem-loops, bulges, pseudoknots, and combinations thereof. Such pools cannot be fully covered with a practical amount of material, hence the probability of success in the isolation of a rare sequence is related to the total number of sequences in the pool and is thus a function of the amount of material that can be easily synthesized and manipulated. A pool with a 60-nucleotide random region flanked by 20-nucleotide defined sequence primer-binding sites requires the synthesis of a DNA template 100 nucleotides in length. High-quality oligos of this length can be readily synthesized with current DNA synthesis technology on commercially available DNA synthesizers. Pool complexities of 10^{13}–10^{15} different sequences can be obtained. Another advantage of pools of this length is that it is relatively easy to characterize the selected molecules by sequence comparison, chemical protection, etc.

The strategy we have chosen is to synthesize pools with the longest possible length of random sequence. The principal rationale for this approach is the potential advantage derived from the combinatorial effect of sequence interactions within the long sequence. A given 120-nucleotide sequence, of course, contains within itself 60 different (but related) 60-mer sequences in register. However, on top of this linear advantage of length, there is a combinatorial effect that arises from the fact that different parts of the sequence may pair with each other to generate complex secondary and tertiary structures. Furthermore, certain functions, i.e., catalysis of some reactions, may require large structures to provide a framework for the accurate positioning of critical groups and thus might only be isolated from long pools.

Unfortunately, these potential advantages are offset by the technical difficulties associated with the synthesis of very long DNA templates. It appears that very minor side reactions (including depurination, branching reactions, and almost certainly other uncharacterized reactions) that are not a problem during the synthesis of short oligos cause severe problems in the synthesis of oligos greater than 140 nucleotides in length. These side reactions not only decrease the yield of isolated DNA, but also decrease its biological activity, in that a large fraction of the synthetic DNA contains lesions that block the passage of polymerases and thus prevent the synthesis of full-length copies during PCR or transcription. Some modifications to the synthetic protocol, such as the use of aprotic Lewis acids such as BF_3:methanol complex for detritylation (Mitchell et al. 1990) and the use of a step (such as capping) prior to oxidation that will hydrolyze branched molecules (Pon et al. 1986), may help, but there are clearly other problems, and it remains quite difficult to obtain consistent results in the synthesis of very long DNAs. We have recently succeeded in synthesizing a pool with a random region of 220 nucleotides, and a complexity of 2×10^{15} sequences, by first synthesizing three separate shorter pools, amplifying these by PCR, using restriction enzymes to generate sticky ends, and then ligating the three segments together with DNA ligase (D. Bartel and J.W. Szostak, unpubl.). This hybrid approach combines the ease of synthesis of short oligos with the potential advantages of using long sequences for selection.

In view of the difficulties associated with the chemical synthesis of long DNAs, it may seem surprising that enzymatic methods for random DNA synthesis have not been employed. In principle, long random sequences could readily be generated with terminal deoxynucleotidyl transferase or polynucleotide phosphorylase. Two difficulties offset the ease of random synthesis with these enzymes. The first problem is the difficulty of adding a defined sequence tail to the random sequence pool, although a homopolymer tail could be added by removing all nucleotides from the reaction and then restarting the reaction by adding back one nucleotide. A second and more serious problem is that product length cannot be controlled, so that a very heterodisperse population is obtained. As discussed below, it is often convenient to work with a population of molecules that are all at least approximately the same length.

A variation on the synthesis of fully random libraries is to synthesize a library that is mostly random, except for small internal defined sequences which are designed to impose or at least bias the final pool toward a given secondary structure (by analogy with the way that antibody variable regions are presented within the context of a conserved framework). For example, a cloverleaf or pseudoknot library could be constructed with short defined sequences to provide the base-paired framework, and with random loop sequences to confer different potential functions. By ensuring that all members have a minimal level of structure, such pools might allow the isolation of structures that were too complex to be isolated directly from random sequence pools.

In many cases, one may wish to generate a pool biased toward a given initial sequence. Such doped or degenerate pools are useful for the isolation of mutant sequences with altered or optimized properties, and also for defining important positions and for obtaining information such as secondary structure by sequence comparison. In such cases, the major issue is the extent of degeneracy to be introduced during pool synthesis. The optimum level of doping may be difficult to calculate if the number of nucleotides required for function is not known. Too low a level of mutagenesis will make it difficult to identify conserved positions and will require the sequencing of an inordinate number of clones to identify covariations. On the other hand, too much mutagenesis is also bad, as many molecules will be rendered nonfunctional by deleterious mutations. In general, longer sequences require lower levels of mutagenesis for this reason. For example, we used 30% mutagenesis for an analysis of the structure of the Rev-binding site in a 66-nucleotide sequence (of which about 20 were required for function), whereas recent experiments with a 140-nucleotide ribozyme have been done with 5% doping (Bartel et al. 1991; M. Green and J.W. Szostak, unpubl.). A more detailed discussion of this issue may be found in Green et al. (1991).

Very long sequences cannot be synthesized on a single oligo, due to the technical problems alluded to above. However, this upper limit is surprisingly long, and we have been able to generate libraries with sequence complexities of 10^{12}–10^{13} sequences by the synthesis of DNA templates ranging from 140 to 180 nucleotides. Larger libraries may be made by first preparing separate libraries for parts of the sequence, which are then amplified and subsequently joined together by restriction digestion and ligation. Another method that is useful at moderately low levels of mutagenesis (less than 5%) is to synthesize doped oligos that can be annealed to a template strand derived from a single-stranded phage cloning vector (Beaudry and Joyce 1992). Many such oligos can be ligated together on the template strand, and the pool can be amplified either by transformation (which limits the total complexity) or by selective removal of biotinylated template strand on avidin beads followed by PCR amplification of the doped strand.

An alternative strategy that may be useful in focusing on small regions of a large structure is to generate pools in which most of the sequence is defined, but specific regions are fully or partially randomized. This is particularly useful for the identification of all possible functional substitutions (Tuerk and Gold 1990; Berzal-Herranz et al. 1992) or for the identification of covariation between segments of sequence that are suspected to interact (Green et al. 1990).

A major issue in the synthesis of doped libraries, which has not yet been adequately addressed, is the generation of certain classes of mutations, particularly insertions. Substitution mutations are easy to generate simply by preparing mixtures of each nucleotide phosphoramidite with the other three nucleotides. Single-base deletions can be obtained by omitting the capping step in DNA synthesis and occur at a lower level even with the capping step. Single-base insertions and larger insertions and deletions are much harder to generate, and, indeed, no adequate method for random insertion mutagenesis has been published.

A completely different approach to mutagenesis that can be used in place of or in addition to the above methods is to introduce mutations during the amplification process by mutagenic PCR (Leung et al. 1989; Zhou et al. 1991). Although this method yields lower levels of mutagenesis (typically 0.5–2% per position), new mutations are introduced during each round of amplification. Mutations accumulate in noncritical positions but are kept by selection to lower levels in positions of functional importance. Continuous mutagenesis allows one to continue selection/amplification cycles indefinitely and to continue to evolve a sequence to perform its selected function more effectively.

Selection Schemes

Any conventionally used purification scheme can be adapted to repetitive use in cycles of selection and amplification. Affinity chromatography, filter binding, gel mobility shift, and immunoprecipitation have been most widely used in such experiments to date, but other methods will no doubt be applied in the future. The need to guard against the enrichment of undesired classes of RNAs is an unusual aspect of in vitro selection technology and is a direct consequence of the very high purification factors achieved after multiple cycles. A common example is the isolation of RNAs that bind to agarose after a certain number of cycles of affinity chromatography on derivatized agarose supports; similarly, RNAs that bind to nitrocellulose are isolated after repeated selection by nitrocellulose filter binding. The isolation of sequences that bind to a given protein by immunoprecipitation of protein:RNA complexes might instead yield sequences that bind to the antibody.

Such undesired selections can be minimized, if not eliminated, by the use of negative selection steps alternating with the standard positive selections (Fig. 2). In the case of affinity chromatography, the RNA pool may first be passed through a pre-column designed to absorb any species that bind to the matrix itself (Ellington and Szostak 1992; Famulok and Szostak 1992). The pre-column should be as similar as possible to the derivatized column used for selection; i.e., if possible, the pre-column should be subjected to the same activation procedure, should contain any linker molecules used in the attachment of the ligand to the selection column, and should be quenched in the same way as the selection column. The RNA pool should be subjected to an annealing procedure prior to application to the pre-column, and the flowthrough from the pre-column should be applied directly to the positive selection column, to minimize the time available for RNAs to switch between different metastable conformations.

Another way of minimizing the selection of undesired RNAs is to use an elution procedure that is as specific as possible. Early selections using affinity chromatography selected positively for binding of the RNA to the ligand-derivatized column matrix; after washing of the column, all remaining RNA was eluted by denaturation. Such harsh elution procedures have the advantage of eluting even the tightest binders from the column, but have the disadvantage of also eluting RNAs that are binding in undesired ways, e.g., because of affinity for the matrix or a combination of the matrix and the ligand. Milder elution schemes such as ligand elution, in which the desired RNAs are eluted from the column by adding ligand to the elution buffer, are much more specific. This method relies on the RNA dissociating from the matrix with some reasonable rate, and

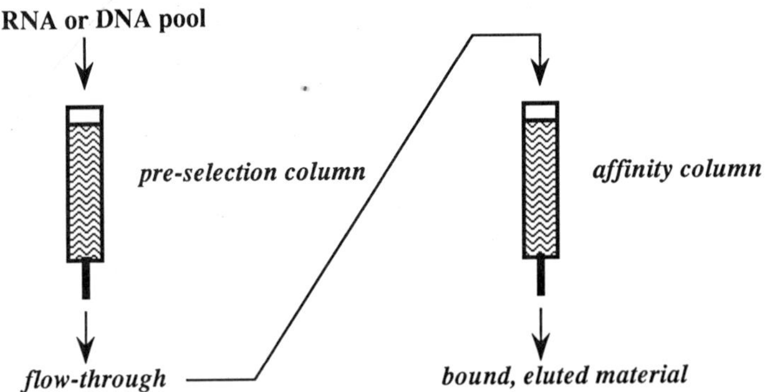

Figure 2 Negative/positive selection. The binding specificity of a pool of aptamers is improved by removing molecules that bind nonspecifically by negative selection on a pre-column. The flowthrough from the pre-column is applied directly to a positive selection affinity column.

then rebinding either to ligand on the column or ligand in solution; clearly, RNAs with very slow off-rates might be lost by this procedure. A variant of this scheme is to attach the ligand to the column with some easily cleaved linker, such as a disulfide bridge. Cleavage of the linker will then result in elution of all ligand-RNA complexes from the column. A similar approach involving binding of a protein to a lectin column was used in the isolation of thrombin-binding DNA sequences (Bock et al. 1992). The DNA-protein complexes were eluted from a con-A agarose column with α-methyl mannoside, leaving behind any sequences that were bound to the lectin or the matrix.

Selection artifacts may also be minimized by alternating between different enrichment protocols. For example, in the selection of ligand-binding RNAs, alternating between different solid supports (e.g., agarose, cellulose, acrylamide) should minimize the significance of interactions between the RNA and the matrix so that interactions with the ligand are emphasized. Similarly, alternation between filter binding and immunoprecipitation might yield more specific RNA-protein interactions.

Wild-type sequences are present in doped pools at much higher levels than any given mutant derivative; selection for a mutant therefore involves competition between the wild type and the mutant. For the selection to succeed, the factor by which the mutant is enriched relative to the wild type in each cycle must be sufficiently large that the mutant becomes more abundant than the wild type within the number of cycles that are carried out over the course of the whole experiment. For example, at 6% mutagenesis, a given quadruple mutant is present at approximately 2 x 10^{-7} of the level of the wild-type sequence in the pool. If the mutant is selected 100-fold more strongly than wild type at each selective step, it can be recovered in four cycles; however, if it is only enriched by 2-fold relative to wild type during each selection, 23 cycles are required. Therefore, it is easier in practice to select for altered function than for marginally improved function.

Amplification Methods

The most commonly used method of amplification for in vitro selection experiments is PCR. The production of very large pools requires large-scale PCR. Since PCR becomes much less efficient as strand concentrations rise above 10 µg/ml, the amplification of 500 µg of synthetic DNA by 10-fold to yield 5 mg of total DNA requires a 500-ml PCR reaction. Such reactions are best carried out in 10–15-ml aliquots, with manual transfer of tubes between water baths; since only 5–6 cycles of PCR are required, this is not too laborious. This initial PCR is highly desirable, since all of the amplified DNA is now enzymatically made and free of chemical lesions; it can be efficiently transcribed by RNA polymerase. Furthermore, the amplified DNA can be stored and aliquots can be used for multiple experiments. This DNA is then converted to RNA by a large-scale in vitro transcription reaction, usually using T7 RNA polymerase.

During the actual cycles of selection and amplification, the selected RNA molecules must first be converted to cDNA with reverse transcriptase or with a DNA polymerase that will also copy RNA templates. It is important to monitor the efficiency of this conversion process, as a failure of the reverse transcription reaction can lead to a catastrophic loss of sequence complexity. This is an especially serious problem in the first round of selection and amplification, when each sequence is represented by only a few copies; molecules that fail to be converted to cDNA at this stage are permanently lost from the pool. At later rounds, each sequence is present in many copies and is less likely to be lost by a single inefficient cDNA synthesis. The efficiency of the reaction can be estimated from the amount of selected RNA and the number of cycles of PCR that are required to amplify the resulting cDNA to a given level.

A number of other amplification systems have been developed that can be used in place of PCR. Isothermal transcription-based amplification (Guatelli et al. 1990) shows particular promise for use in selection and in vitro evolution experiments. In this system, reverse transcriptase,

RNase H, and T7 RNA polymerase are used together to do a one-tube conversion of RNA to cDNA to dsDNA and then to multiple copies of RNA. The rapidity and convenience of this method make it feasible to perform experiments involving many more cycles of selection than with the standard three-step amplification process.

All amplification methods have the potential for giving rise to severe artifacts in the form of molecules with enhanced replication efficiency. Deletions commonly occur, as short molecules often replicate more efficiently. The appearance of such artifacts can be minimized by incorporating a size selection step into each cycle, by gel-purifying either the PCR-amplified DNA or the transcribed RNA. Unfortunately, this slows down the procedure significantly.

In the future, an even more direct amplification process may become practical, namely, the use of Qβ replicase, an RNA-dependent RNA polymerase. Indeed, the first in vitro selection experiments were carried out with Qβ replicase many years ago (Mills et al. 1967; Kramer et al. 1974). Unfortunately, this enzyme has complex template recognition properties and replicates different sequences at very different efficiencies. Sequences to be replicated must be flanked by the appropriate recognition sequences, which are still rather large. Nevertheless, if these problems can be worked out, the system may be very useful.

LIGAND-BINDING NUCLEIC ACIDS

Recognition of Non-protein Ligands by RNA

The large number of different nucleic acid sequences that can be created synthetically represents an equally large number of three-dimensional shapes. Some of the surfaces presented by these nucleic acid structures will be complementary to specific chemical species. It should therefore be possible to use the selection/amplification methods discussed above to select for RNAs which present surfaces that interact specifically with a given compound, and, therefore, to find individual sequences that code for ligand-binding nucleic acids.

The range of molecular shapes and electronic structures that can be recognized by nucleic acids is still unclear. For example, there might be a paucity of RNA sequences that could bind to compounds that are highly negatively charged (because of the repulsion between the sugar-phosphate backbone and the ligand) or that are primarily hydrophobic (because nucleic acids may not be able to form convoluted hydrophobic binding pockets). Prior to the advent of in vitro selection techniques, it was known that nucleic acids could interact semi-specifically with a variety of drugs that either intercalated into or sat in the major or minor grooves of a double helix. In addition, there were a few examples of specific recognition of a substrate, such as guanosine, by a nucleic acid, such as the group I self-splicing intron from *Tetrahymena thermophila* (Bass and Cech 1984).

A variety of selections for ligand-binding nucleic acids have now been carried out, and the surprising result seems to be that there are (so far) few limitations to what nucleic acids can recognize and bind. We initially selected RNA molecules that could bind a number of small organic dyes from a pool of 10^{13} sequences that had been randomized over 100 contiguous bases (Fig. 3) (Ellington and Szostak 1990). We refer to these selected ligand-binding nucleic acids as aptamers, from the Latin *aptus*, to fit. The structural characteristics of the dyes used for selections did not differ greatly from other ligands that were known to interact with nucleic acids: The compounds were all of roughly 600 m.w. and contained conjugated ring structures (such as anthroquinones) and hydrogen bond donor/acceptors (amines and carbonyls). Interestingly, most of the dyes were highly negatively charged: For example, Reactive green 19 contained 6 sulfonates. Nevertheless, under the high ionic strength conditions used for the selections (0.5 M LiCl), many RNA species were able to bind to these dyes.

For RNAs selected against several of the dyes, binding was specific, in that RNAs selected against one dye could not bind to other dyes (Fig. 4). This specificity was especially surprising, given the chemical similarities between the dyes in general and the close relationship between two in particular: Cibacron blue and Reactive blue 4, both of which contained an anthroquinone system coupled via phenylene diamine to a reactive triazine. These results suggested that the selected RNA molecules contained ligand-binding domains that could specifically recognize individual chemical moieties on a molecule. Additional experiments have shown that binding interactions may span large portions of the dye molecules: When Cibacron blue is split into two roughly equal-sized halves, neither half binds well to the RNA aptamers selected against Cibacron blue itself.

The specificity of the dye-nucleic acid interactions could also be inferred from the complexity of the structures that were generated during the selections. Rough calculations suggested that only about one in every 10^{10} RNAs initially present in the pool could bind to a given dye. The fact that there was a relatively small proportion of binding species in the population suggested that the selected interactions were much more complex than simple intercalative binding (since it would be relatively easy to find properly paired helices in a pool of random sequences) or the binding exhibited by drugs such as netropsin, which can recognize A-T-

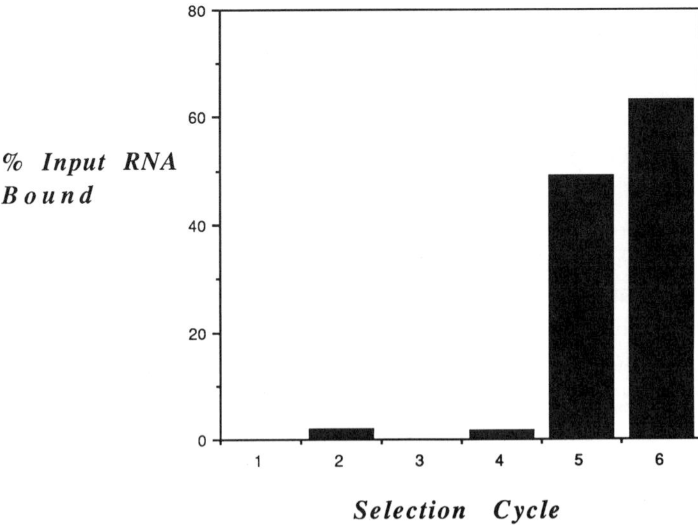

Figure 3 Selection of Cibacron blue-binding RNAs. A pool of 10^{13} different RNA sequences was applied to a Cibacron blue-agarose affinity column in a high-salt buffer containing magnesium, washed extensively, and then eluted in a low-salt no-magnesium buffer. The fraction of the input RNA present in· the eluted material is low for the first four cycles (the variation simply reflects the efficacy of the washing procedure), then jumps in cycle five. At this point, most of the remaining molecules bind to the affinity column (Ellington and Szostak 1990).

rich DNA, but only over short (4 bp) sequences (that, again, would have been relatively abundant in a random pool). This prediction was borne out once the binding site sequences of individual aptamers were determined. In general, from 20 to 30 bases were found to be critical for recognition, and these were organized into recognizable secondary structural elements (stem-loops, stem-bulge-stems). Both the length and structural complexity of the binding sites argued that interactions between aptamers and their ligands involved molecular recognition similar to that seen in enzyme active sites.

We also performed a similar set of experiments with a randomized DNA pool and again found that aptamers could be selected that would bind specifically to dye ligands (Ellington and Szostak 1992). Interestingly, for one of the ligands used, Reactive green 19, there was a consensus 18-base sequence that appeared in a number of aptamers. Given that the pools used for these experiments spanned $>10^{13}$ different sequences, this result suggested that the 18-mer sequence was optimal for binding this particular ligand: That is, no other sequences (of this size or smaller) would be expected to work as well. In addition, the 18-base mo-

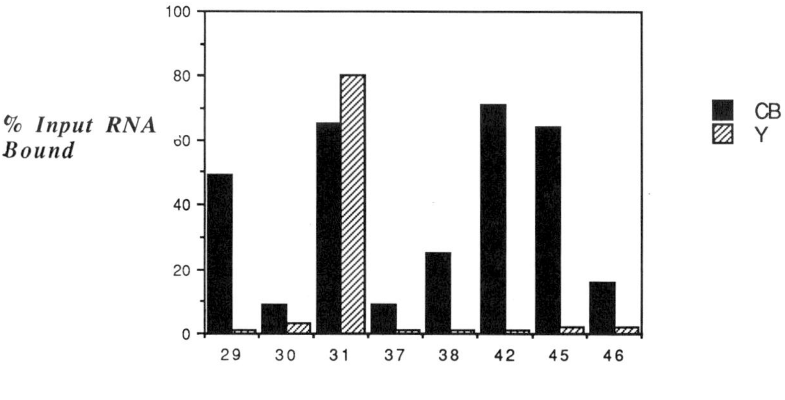

Figure 4 Specificity of cloned Cibacron blue-binding aptamers. Individual molecules from cycle 5 of the selection illustrated in Fig. 3 were cloned, and RNA was prepared from each clone. Some of the clones, such as aptamer 31, bind nonspecifically to both Cibacron blue (B) and reactive yellow (Y) agaroses; others, such as 30 and 37, bind poorly; and others, such as 42 and 45, bind well and specifically (Ellington and Szostak 1990).

tif could itself bind Reactive green 19 outside of the long (174-base) sequences in which it was initially found, but only when presented as a sequence-specific loop on a non-sequence-specific stem. Moreover, when RNA sequences corresponding to the DNA motif were synthesized, they were found not to bind to the dye. Again, this is indicative of specific interactions between the aptamer and its ligand, since DNA and RNA have structures that, although similar, will nevertheless appear quite different to a docking molecule that is searching for a particular pattern of hydrogen bonds, salt bridges, and hydrophobic interactions.

Whereas the dyes were originally chosen as ligands for in vitro selection experiments based on their chemical properties, it became important to know if smaller and more biologically relevant molecules such as amino acids, sugars, and cofactors could also be specifically recognized by nucleic acids. Famulok and Szostak (1992) have selected RNAs that can bind to D-tryptophan immobilized on an agarose column. Binding is both tight ($K_d = 18$ μM) and stereospecific: The discrimination between D- and L-tryptophan is at least 1000-fold at room temperature. The group I intron also exhibits stereospecific discrimination against arginine inhibitors of the splicing reaction, but at only a 2-fold level. The selection experiments with D-tryptophan are also important because they demonstrate that the chemical complexity of a ligand that can be recognized by RNAs of length 20–100 can be as small as 200 m.w. (tryptophan), as opposed to 600 m.w. (dyes), and because tryptophan is a hydrophobic amino acid that is (presumably) being recognized at least in part on the basis of its shape. In line with this, recent work by John Latham (pers. comm.) suggests that selected DNA molecules can bind to prostaglandin F2a, an extremely hydrophobic compound with only three hydrogen bond donors (hydroxyls). Similarly, selection experiments with phosphonate transition-state analogs have demonstrated that structures as small as 2-methylamido-6-carboxy pyridine (m.w. 200) can be specifically recognized by RNA molecules.

Pei et al. (1991) have used in vitro selection technology to explore the recognition of dsDNA sequences by RNA. These authors selected RNA molecules from random sequence pools that could form triplex structures with an immobilized double helix. The target was an oligopurine sequence and, as expected, the selected DNAs contained runs of pyrimidines that corresponded to the (already known) Y.RY motif for triplex formation. This experiment demonstrated a surprising tolerance for structured insertions such as hairpin loops in the third strand.

Recognition Sites of RNA-binding Proteins

In vitro selection methods have been extensively used in the determination of protein-binding sites on double-stranded DNA (for review, see Szostak 1992). More recently, essentially the same approach has been used to examine RNA-protein interactions. This was first done by Tuerk and Gold (1990), who randomized an eight-nucleotide loop sequence in a stem-loop that is a binding site for T4 DNA polymerase. After several rounds of selection by nitrocellulose filter binding of RNA-protein complexes coupled with amplification by reverse transcriptase, PCR, and T7 RNA polymerase transcription, the initial pool of 65,000 sequences had been winnowed to two predominant sequences. Remarkably, whereas one of these was the wild-type sequence, the other was quite different and had extended the stem by two base pairs, leaving only a four-base loop. Since then, the method has been applied to the isolation of RNA sequences that bind to a number of proteins, including proteins that do not normally bind nucleic acids; these experiments are described by Gold et al. (this volume).

Subsequently, we applied a similar approach to the elucidation of the binding site of the human immunodeficiency virus (HIV) Rev protein (Bartel et al. 1991). The location of the binding site had been delimited to a region of roughly 70 nucleotides, but the secondary structure of this sequence was still controversial and the exact location of the binding site was undefined. Since we wanted to define the binding site within a given sequence, and since we did not want to isolate new sequences that might bind in a different way, we prepared a degenerate pool with 30% doping of the wild-type sequence. Sequences capable of binding Rev were isolated by three rounds of filter-binding selection alternating with amplification; individual clones were then sequenced and characterized. The resulting sequences were easily aligned, and two regions of invariant bases were seen, embedded within other sequences that varied at the expected rate. Covariation analysis defined the relevant structure as a stem-bulge-stem, with an unusual and functionally important G:G base pair located within the bulge. Detailed mutational analyses and protection studies have confirmed the location and secondary structure of the binding site (Heaphy et al. 1991; Dayton et al. 1992; Kjems et al. 1992).

U1-snRNP-A protein binds to stem-loop II of U1 RNA; the details of its sequence specificity have been determined by in vitro selection (Tsai et al. 1991). Pools of randomized RNA were generated with either 10 or 13 random nucleotides presented in the context of the loop of a stem-loop, or as a simple linear array of 25 random nucleotides. Selection of protein-RNA complexes was accomplished by coimmunoprecipitation of epitope-tagged U1-A protein. The RNA was extracted from the immunoprecipitate, converted to cDNA, and amplified by PCR. After three rounds of selection and amplification, individual clones were sequenced and aligned. Almost all of the selected clones contained an exact match to the first 7 nucleotides of the loop of stem-loop II. This method should be quite useful in establishing the details of the specificity of related RNA-binding proteins.

The first published example of the selection of a nucleic acid sequence that binds specifically to a protein that is not normally involved in nucleic acid binding was the selection of a DNA sequence that binds to thrombin (Bock et al. 1992). In this case, selection was for a single-stranded DNA that would presumably fold up into a structure with a surface complementary to some part of the protein surface. Selection was accomplished by binding the thrombin DNA complexes to a conA-agarose affinity column, washing, and then eluting the complexes with α-methyl mannoside to compete for binding to the lectin. The recovered DNA was amplified by PCR using one biotinylated primer. Single-stranded DNA was recovered by binding the dsDNA to an avidin column and eluting the non-biotinylated strand with alkali. After several rounds of selection, the remaining pool bound thrombin tightly and specifically; sequencing revealed a 15-nucleotide consensus sequence present in almost every clone. This 15-mer alone binds thrombin with a K_d of approximately 20 nM and inhibits its activity in blood clotting. The isolation of a small DNA molecule that binds to and inhibits such a physiologically important protein highlights the potential therapeutic importance of such molecules.

Catalytic Nucleic Acids

Selection for RNAs with catalytic activity is straightforward if the catalytic activity modifies the RNA itself. We have used in vitro selection to look for interactions between two small regions within the *Tetrahymena* ribozyme (Green et al. 1990). The two regions of interest, totaling nine nucleotides, were randomized using two PCR primers. Catalytically active molecules were selected from this pool by a reversal of the first step in self-splicing. The transcribed RNA was designed to look like the product of splicing, so that upon addition of a synthetic 5′ exon, reversal of splicing led to the ligation of the added exon to active intron sequences. The selection procedure took advantage of the presence of the exon sequence as a "tag" for active molecules. After the exon-ligation reaction, the entire pool was converted to cDNA and then amplified by PCR, with one of the primers being complementary to the exon sequence. Thus, only RNAs that had ligated to an exon could be amplified. After three cycles of selection, the catalytic activity of the pool had increased dramatically, and individual sequences were cloned and analyzed. Alignment of the randomized regions revealed specific covariations between certain positions in the ribozyme. Thus, this sort of selection experiment can be used to generate a sort of instant phylogeny, with the advantage that the activity corresponding to each sequence is known. Of course, the sequences are all equally unrelated and cannot be ordered into a phylogenetic tree.

Selection for ligation of a sequence tag to a population of RNA molecules leads to amplification artifacts that are not observed in simple binding selections. One such artifact is the formation of partial cDNA molecules that loop back and are extended to form a hairpin; the resulting sequence can be amplified during selective PCR with only a single primer and can rapidly take over the population. An additional affinity chromatography selection step using an oligonucleotide-derivatized matrix to enrich for the ligated molecules will overcome this problem.

A similar selection method was used by Robertson and Joyce (1990), in which the 3′ guanosine of the ribozyme itself attacks and cleaves an oligonucleotide bound to the internal guide sequence of the intron; the result is that part of the oligonucleotide becomes covalently linked to the 3′ end of the ribozyme, providing a new primer-binding site so that only catalytically active molecules can be converted into cDNA. The cDNA is then converted into dsDNA, which is transcribed by T7 RNA polymerase to yield multiple copies of RNA. This selection scheme has been used along with a mutagenic amplification procedure to evolve enzyme variants that can efficiently cleave DNA substrate oligonucleotides (Beaudry and Joyce 1992).

In vitro genetic selections may prove especially useful for the optimization of self-replicating ribozymes. The *Tetrahymena* and *sunY* self-splicing introns have been engineered to perform template-directed ligations of short (3–10 bases) oligonucleotide fragments (Doudna and Szostak 1989; Doudna et al. 1991, 1993). These ribozymes evolved to catalyze a single ligation reaction in which one of the pieces is covalently linked (i.e., in *cis*), whereas their engineered counterparts perform multiple ligations using both templates and substrates in *trans*. Since the unnatural *trans*-ligation reactions were relatively inefficient, the overall size of the ribozyme (and thus, of the template to be copied) was reduced by deleting several stem-loops from *sunY* (P7.1, P7.2, P9.1, P9.2). Unfortunately, the 141-nucleotide deletion derivative had only weak catalytic activity. Therefore, Green and Szostak (1992) employed an in vitro genetic selection to generate *sunY* variants that could perform the *cis*-ligation reaction more efficiently. Following selection, a quintuple base substitution was isolated that had 4 times the activity of the wild-type enzyme, and 350 times the activity of the original deletion derivative. This variant could completely (albeit inefficiently) copy itself into cRNA using oligonucleotide substrates of approximately 10 bases in length. The selected base substitutions apparently improved the *trans*- as well as the *cis*-ligation activities of the *sunY* intron, since 20-fold more full-length product was generated than for the wild-type ribozyme.

Ribozymes that catalyze self-cleavage reactions have also been studied by in vitro selection. Yeast tRNA[Phe] will self-cleave in the presence of Pb[++] and, in this context, can be considered a metallo-ribozyme. Pan and Uhlenbeck (1992) isolated new lead-cleaving RNAs by starting with the tRNA[Phe] sequence and preparing RNA pools in which subsets of nine or ten positions had been randomized. The resulting RNA was circularized with T4 RNA ligase, and the circular RNA was incubated with lead. Molecules which had self-cleaved anywhere in the sequence became linear and could be purified from the circular starting material by electrophoresis on a denaturing acrylamide gel. These molecules were then recircularized by treatment with T4 kinase and T4 RNA ligase, and the products were converted to cDNA, amplified by PCR, and retranscribed to generate new RNAs. After six rounds of selection and amplification, all four starting pools showed greatly enhanced cleavage when incubated with lead. Sequence analysis and characterization of clones from each pool showed that a variety of different sequences had been isolated, which resulted in specific lead cleavage at a number of distinct sites.

Functionally important bases in the self-cleaving hairpin ribozyme have been identified by in vitro selection experiments (Berzal-Herranz et al. 1992). In this case, the substrate oligonucleotide was linked to the 3' end of the ribozyme by a short oligo-cytidine linker. After cleavage of the substrate by the ribozyme, the reaction is reversed by the addition of a large excess of an oligonucleotide, which becomes ligated to the ribozyme. The oligonucleotide brings in a new primer-binding site, allowing selective amplification of the active species by conversion to cDNA followed by PCR. Thus, each cycle of selection involves an RNA-catalyzed cleavage and ligation. Berzal-Herranz et al. (1992) identified a single conserved residue in the substrate domain and several conserved residues in an internal loop region of the ribozyme.

FUTURE DIRECTIONS

Strategies for Generating New Nucleic Acid Catalysts

The development of new catalytic nucleic acids is clearly one of the most interesting applications of in vitro selection technology. The naturally occurring ribozymes are all essentially phosphotransferases or phosphohydrolases, although the *Tetrahymena* ribozyme has recently been shown to have a very low but significant activity in the hydrolysis of an ester formed between the 3'-hydroxyl of an oligonucleotide and the carboxyl of an amino acid (Piccirilli et al. 1992), and it now appears that the peptidyl transferase activity of the ribosome is an RNA-catalyzed reaction (Noller et al. 1992). Some models of the RNA world have proposed an extensive metabolism based on RNA catalysis (see, e.g., Benner et al. 1990 and references therein), and the range of reactions that can be catalyzed by RNA enzymes therefore becomes an important question that is now open to experimental attack.

There are several possible approaches to the selection of RNA or DNA molecules with catalytic activity. Perhaps the most obvious of these, and the most straightforward extension of the isolation of ligand-binding RNAs and DNAs, would be to select for sequences that bind to transition-state analogs. Such compounds mimic the geometric and electronic structure of the highest energy intermediates along the reaction coordinate of a given chemical transformation; a good deal of enzymatic catalysis can be described in terms of enzyme stabilization of the transition state. The fascinating work that has been done over the last several years in the isolation of catalytic antibodies provides an excellent precedent for the isolation of novel catalysts by selection for binding of transition-state analogs.

Ester hydrolysis is a facile reaction whose chemical and biochemical mechanisms are well understood; the simplicity of this reaction makes it an ideal test case for determining whether RNA catalysts can be isolated from random sequence pools. In most protein esterases, a polarized water molecule attacks the carbonyl carbon of an ester bond. As the reaction proceeds, the hybridization state of the carbon changes from planar sp2 to tetrahedral sp3; a negative charge develops on the carbonyl oxygen, and a proton, either from solution or from a specific proton donor, stabilizes the newly formed alcohol. Phosphonates are compounds of the form RPO_3R'; except for the presence of phosphorus in place of carbon, their tetrahedral ground state mimics the transition state of base-catalyzed ester hydrolysis. Phosphonate transition-state analogs have been successfully used to elicit antibodies that can act as esterases (Lerner and Tramontano 1987). Catalytic antibody esterases have been generated for a variety of phosphonate compounds (Schultz et al. 1990), have been shown to recognize stereocenters in substrates (Janda et al. 1989), and can catalyze the formation and breakdown of lactones (Napper et al. 1987; Benkovic et al. 1988).

A nucleic acid molecule that bound to a phosphonate transition-state analog might be expected to catalyze the corresponding ester hydrolysis reaction. A comparison of nucleic acid and protein chemistry and structure suggests that nucleic acids should be able to act as esterases. The most basic requirements for an esterase enzyme would be (1) the ability to polarize water for nucleophilic attack on the ester carbonyl and (2) the ability to stabilize the tetrahedral geometry of the transition state. The polarization of water hydroxyls should be within the scope of nucleic acid chemistry. The nucleotide bases contain proton donors/acceptors with pKa values near 4 and 9, and although there is no nucleic acid equivalent of histidine (pKa = 7), the acidic and basic moieties can be polarized and shifted toward neutrality; for example, cytosine is easily protonated at physiological pH when included in a base triple (Rajagopal and Feigon 1989).

The catalytic power of nucleic acids could easily be augmented by making use of bound metal ions. Metal-based esterase mechanisms have precedent among protein enzymes; for example, the protein phospholipase A2 uses a bound calcium ion to polarize the attacking water molecule (Scott et al. 1990). An RNA esterase could also expand its chemical repertoire by using a bound metal ion. For example, the 3' hydroxyl of ribose is transformed into a potent nucleophile to catalyze the cleavage of the phosphodiester backbone in group I self-splicing introns, possibly by interaction with a bound magnesium ion (Grosshans and Cech 1989), and the use of bound Pb^{++} to catalyze site-specific RNA hydrolysis is well documented (Brown et al. 1985; Pan and Uhlenbeck 1992).

Another approach to the isolation of new catalysts is to select for molecules that bind to bisubstrate analogs, which differ from transition-state analogs in that they primarily mimic the distance and orientation between substrates in a biosynthetic reaction. For example, phosphon-acetyl-L-aspartate is a bisubstrate analog for the aspartate transcar-bamoylase reaction. This compound displays chemical features at the same distance and relative orientation as they would be found in adjacent substrates in the active site of an enzyme but does not mimic the structure of any intermediary step in the course of the reaction.

Although the isolation of transition state or bisubstrate analog-binding RNAs may well lead to the identification of new catalysts, it is possible that the necessary structures will prove to be too complex to be isolated from even the largest practical random sequence pools. Moreover, RNAs that have been selected for binding must be subsequently screened to identify the (potentially very small) subset that are indeed catalytic. For these reasons, it is worth considering additional strategies for the isolation of new catalysts. One such approach involves the chemical linkage of a substrate to the 5′(or 3′) end of the pool of RNAs (or DNAs). After a period of incubation to allow any catalytically active molecules to transform their attached substrate into product, the active molecules can be enriched with reagents that have affinity for the product but not the substrate (such as specific antibodies or lectins). This approach has two potential advantages. First, the high local concentration of substrate means that a high-affinity substrate-binding site is not required, because even a weak binding site will be saturated. Second, this is a direct selec-tion for catalysis, so no subsequent screening step is required.

A different way of increasing the likelihood of isolating a catalyst would be to intersperse preselected substrate-binding domains with the random sequences of the pool. This procedure could be used with any form of selection for catalysis, direct or indirect. By dividing the selec-tion for catalysts into two stages, one for substrate binding and a second for actual catalysis, it may be possible to isolate catalytic molecules that are too complex to have a reasonable chance of being present in a pri-mary pool of random sequences.

The chemistries available to the five canonical bases may allow them to carry out a number of simple reactions, but their functionality is still quite limited when compared to the 20 amino acids of proteins. This limitation may have led to the diversification of cofactor chemistries during the evolution of metabolism in an RNA world (Benner et al. 1990). Visser and Kellog (1978) have pointed out that most cofactors contain RNA in one form or another, and they have speculated that this is a vestige of a time when cofactors were used exclusively by ribozymes. If these speculations are correct, then it might be possible to select or construct a variety of RNA enzymes, including dehydrogenases, kinases, and carbon-carbon bond-forming enzymes, by making use of cofactor-binding RNA domains which have themselves been isolated in selection experiments.

SUMMARY

The capacity of nucleic acids to encode information has traditionally attracted more attention than their potential for structural and functional roles. This imbalance was largely restored by the discovery of catalytic RNAs and, more recently, by the finding that both RNAs and DNAs can fold into complex three-dimensional shapes that form an almost infinite variety of highly specific ligand-binding sites. Iterative in vitro selection makes use of the information-encoding ability of RNA and DNA to allow genetic experiments on nucleic acid structure and function to be carried out entirely in vitro. By avoiding transformation and cloning, one gains access to the very large numbers of sequences that must be screened in order to isolate the rare sequences of interest. We expect that this ability to handle very large numbers of sequences in parallel will prove to be quite useful in probing the limits of the structural and functional abilities of nucleic acids.

ACKNOWLEDGMENTS

We thank David Bartel for helpful comments on the manuscript. This work was supported by a grant from Hoechst AG to Massachusetts General Hospital and by a grant from the National Institutes of Health to A.D.E.

REFERENCES

Bartel, D., M. Zapp, M. Green, and J.W. Szostak. 1991. HIV-1 rev regulation involves recognition of non-Watson-Crick base-pairs in viral RNA. *Cell* **67:** 529–536.

Bass, B.L. and T.R. Cech. 1984. Specific interaction between the self-splicing RNA of *Tetrahymena* and its guanosine substrate: Implications for catalysis by RNA. *Nature* **308:** 820–826.

Beaudry, A.A. and G.F. Joyce. 1992. Directed evolution of an RNA enzyme. *Science* **257:** 635–641.

Benkovic, S.J., A.D. Napper, and R.A. Lerner. 1988. Catalysis of a stereospecific bimolecular amide synthesis by an antibody. *Proc. Natl. Acad. Sci.* **85:** 5355–5358.

Benner, S., A.D. Ellington, and A. Tauer. 1990. Modern metabolism as a palimpsest of the RNA world. *Proc. Natl. Acad. Sci.* **86:** 7054–7058.

Berzal-Herranz, A., J. Simpson, and J.M. Burke. 1992. *In vitro* selection of active hairpin ribozymes by sequential RNA-catalyzed cleavage and ligation reactions. *Genes Dev.* **6:** 129–134.

Bock, L.C., L.C. Griffin, J.A. Latham, E.H. Vermaas, and J.J. Toole. 1992. Selection of single-stranded DNA molecules that bind and inhibit human thrombin. *Nature* **355:** 564–566.

Brown, R.S., J.C. Dewan, and A. Klug. 1985. Crystallographic and biochemical investigation of the lead(II)-catalyzed hydrolysis of yeast phenylalanine tRNA. *Biochemistry* **24:** 4785–4801.

Dayton, E.T., D.A.M. Konings, D.M. Powell, B.A. Shapiro, L. Butini, J.V. Maizel, and A.I. Dayton. 1992. Extensive sequence-specific information throughout the CAR-RRE: The target sequence of the human immunodeficiency virus type 1 Rev protein. *J. Virol.* **66:** 1139–1151.

Doudna, J.A. and J.W. Szostak. 1989. RNA-catalyzed synthesis of complementary-strand RNA. *Nature* **339:** 519–524.

Doudna, J.A., S. Couture, and J.W. Szostak. 1991. A multisubunit ribozyme that is a catalyst of and template for complementary strand RNA-synthesis. *Science* **251:** 1605–1608.

Doudna, J.A., N. Usman, and J.W. Szostak. 1993. Ribozyme-catalyzed primer extension by trinucleotides: A model for the RNA-catalyzed replication of RNA. *Biochemistry* **32:** 2111–2115.

Ellington, A.D. and J.W. Szostak. 1990. *In vitro* selection of RNA molecules that bind specific ligands. *Nature* **346:** 818–822.

———. 1992. *In vitro* selection of single-stranded DNA molecules that fold into specific ligand-binding structures. *Nature* **355:** 850–852.

Famulok, M. and J.W. Szostak. 1992. Stereospecific recognition of tryptophan agarose by in vitro selected RNA. *J. Am. Chem. Soc.* **114:** 3990–3991.

Green, R. and J.W. Szostak. 1992. Selection of a ribozyme that functions as a superior template in a self-copying reaction. *Science* **258:** 1910–1915.

Green, R., A.D. Ellington, and J.W. Szostak. 1990. *In vitro* genetic analysis of the *Tetrahymena* self-splicing intron. *Nature* **347:** 406–408.

Green, R., A.D. Ellington, D.P. Bartel, and J.W. Szostak. 1991. *In vitro* genetic analysis: Selection and amplification of rare functional nucleic acids. *Methods Compan. Methods Enzymol.* **2:** 75–86.

Grosshans, C.A. and T.R. Cech. 1989. Metal ion requirements for sequence-specific endonuclease activity of the *Tetrahymena* ribozyme. *Biochemistry* **28:** 6888–6894.

Guatelli, J.C., K.M. Whitfield, D.Y. Kwoh, K.J. Barringer, D.D. Richman, and T.R. Gingeras. 1990. Isothermal, in vitro amplification of nucleic acids by a multienzyme reaction after retroviral replication. *Proc. Natl. Acad. Sci.* **87:** 1874–1878.

Heaphy, S., J.T. Finch, M.J. Gait, J. Karn, and M. Singh. 1991. Human immunodeficiency virus type 1 regulator of virion expression, rev, forms nucleoprotein filaments after binding to a purine-rich "bubble" located within the rev-responsive region of viral mRNAs. *Proc. Natl. Acad. Sci.* **88:** 7366–7370.

Janda, K.D., S.J. Benkovic, and R.A. Lerner. 1989. Catalytic antibodies with lipase activity and R or S substrate specificity. *Science* **244:** 437–440.

Kjems, J., B.J. Calnan, A.D. Frankel, and P.A. Sharp. 1992. Specific binding of a basic peptide from HIV-1 rev. *EMBO J.* **11:** 1119–1129.

Kramer, F.R., D.R. Mills, P.E. Cole, T. Nishihara, and S. Spiegelman. 1974. Evolution *in vitro*: Sequence and phenotype of a mutant RNA resistant to ethidium bromide. *J. Mol. Biol.* **89:** 719–736.

Lerner, R.A. and A. Tramontano. 1987. Antibodies as enzymes. *Trends Biochem. Sci.* **12:** 427–430.

Leung, D.W., E. Chen, and D.V. Goeddel. 1989. A method for random mutagenesis of a defined DNA segment using a modified polymerase chain reaction. *Technique* **1:** 11.

Mills, D.R., R.L. Peterson, and S. Spiegelman. 1967. An extracellular Darwinian experiment with a self-duplicating nucleic acid molecule. *Proc. Natl. Acad. Sci.* **58:** 217–224.

Mitchell, M.J., W. Herschowitz, F. Rastinejad, and P. Lu. 1990. Boron trifluoride-methanol complex as a non-depurinating detritylating reagent in DNA synthesis. *Nucleic Acids Res.* **18:** 5321.

Napper, A.D., S.J. Benkovic, A. Tramontano, and R.A. Lerner. 1987. A stereospecific cyclization catalyzed by an antibody. *Science* **237:** 1041–1043.

Noller, H.F., V. Hoffarth, and L. Zimniak. 1992. Unusual resistance of peptidyl transferase to protein extraction procedures. *Science* **256:** 1420–1424.

Pan, T. and O.C. Uhlenbeck. 1992. *In vitro* selection of RNAs that undergo autolytic cleavage with Pb^{2+}. *Biochemistry* **31:** 3887–3895.

Pei, D., H.D. Ulrich, and P.G. Schultz. 1991. A combinatorial approach toward DNA recognition. *Science* **253:** 1408–1411.

Piccirilli, J.A., T.S. McConnell, A.J. Zaug, H.F. Noller, and T.R. Cech. 1992. Aminoacyl esterase activity of the *Tetrahymena* ribozyme. *Science* **256:** 1416–1420.

Pon, R.T., N. Usman, M.J. Damha, and K.K. Ogilvie. 1986. Prevention of guanine modification and chain cleavage during the solid phase synthesis of oligonucleotides using phosphoramidite derivatives. *Nucleic Acids Res.* **14:** 6453–6470.

Rajagopal, P. and J. Feigon. 1989. Triple-strand formation in the homopurine: homopyrimidine DNA oligonucleotides d(G-A)4 and d(T-C)4. *Nature* **339:** 637–640.

Robertson, D.L. and G.F. Joyce. 1990. Selection *in vitro* of an RNA enzyme that specifically cleaves single-stranded DNA. *Nature* **344:** 467–468.

Schultz, P.G., R.A. Lerner, and S.J. Benkovic. 1990. Catalytic antibodies. *Chem. Eng. News* (5/28/90), pp. 26–40.

Scott, D.L., S.P. White, Z. Otwinowski, W. Yuan, M.H. Gelb, and P.B. Sigler. 1990. Interfacial catalysis: The mechanism of phospholipase A2. *Science* **250:** 1541–1546.

Szostak, J.W. 1992. *In vitro* genetics. *Trends Biochem. Sci.* **17:** 89–93.

Tsai, D.E., D.S. Harper, and J.D. Keene. 1991. U1-snRNP-A protein selects a ten nucleotide consensus sequence from a degenerate RNA pool presented in various structural contexts. *Nucleic Acids Res.* **19:** 4931–4936.

Tuerk, C. and L. Gold. 1990. Systematic evolution of ligands by exponential enrichment: RNA ligands to bacteriophage T4 DNA polymerase. *Science* **249:** 505–510.

Visser, C.M. and R.M. Kellog. 1978. Biotin. Its place in evolution. *J. Mol. Evol.* **11:** 171–187.

Zhou, Y., X. Zhang, and R.H. Ebright. 1991. Random mutagenesis of gene-sized DNA molecules by use of PCR with *Taq* DNA polymerase. *Nucleic Acids Res.* **19:** 6052.

21

Contemporary RNA Genomes

John F. Atkins[1]
Department of Human Genetics
University of Utah
Salt Lake City, Utah 84112

> *So, naturalists observe, a flea*
> *hath smaller fleas that on him prey;*
> *And these have smaller fleas to bite 'em*
> *And so proceed ad infinitum,*
> *Thus every poet [viral evolutionist] in his kind,*
> *Is bit by him that comes behind.*
>
> Jonathan Swift
> 1667–1745

The exclusive existence of RNA genomes is part of the central postulate of the RNA world, a hypothetical pre-DNA and pre-encoded-protein period in early evolution. In contrast, natural extant RNA genomes are almost exclusively composed of infectious RNA agents, viruses, subviruses, and some derivatives. The latter comprise a few degenerate viruses that are known in plants and fungi (one in *Saccharomyces cerevisiae*, L-A, is well characterized) and a "carrier" state entity for some viruses in which the viral RNA is maintained in the cytoplasm for several generations (see Onodera et al. 1992). The known RNA genomes that are not associated with a capsid-like structure, and that are not apparently infectious, are few in number and small in size. For example, a 2800-nucleotide linear single-stranded RNA and a deletion derivative replicate in maize mitochondria and are transmitted with the mitochondria. The RNAs are free in the sense of not being sequestered in nucleocapsids (Zhang and Brown 1993). Detailed characterization of such "RNA plasmids" is clearly worthwhile, but they do not affect the generality of the statement about modern RNA genomes being essentially confined to infectious agents.

All modern RNA genomes rely on the cellular translation apparatus and other components specified by DNA genomes and are very far from being self-sufficient. In reaching their present state of evolution, these

---

[1] On leave of absence from Department of Biochemistry, University College, Cork, Ireland.

The RNA World
© 1993 Cold Spring Harbor Laboratory Press 0-87969-380-0/93 $5 + .00

RNA genomes required either the preexistence or concomitant evolution of their hosts. Whereas some, or all, present-day RNA viruses and subviruses may be descendants of precellular life ("all the way with RNA!"), it is also possible that they are derived from normal constituents of cells (Dinter-Gottlieb 1986; Hadidi 1986). All are highly specialized, and because of the intensity of selection to which they were subjected and their potential for adaptability, attempts to discern which, if any, may be derived from the time at which the common ancestor of all modern forms of life arose are easy to dismiss.

Nevertheless, Maizels and Weiner (this volume) have made provocative suggestions for "molecular fossil" features of contemporary viral replication. The catalytic RNA aspect of viroids is also highly relevant and is discussed by Pan et al. elsewhere in this book. This chapter explores the distinctive features of modern RNA viruses and subviruses without purporting to detail their enormous variety and elegant intricacies. The main themes are the variety of different strategies that coexist in a competitive environment and the potential for genomic plasticity. The small subviruses are considered first.

SUBVIRUSES: CIRCULARITY AND ROLLING CIRCLE REPLICATION

From an evolutionary perspective, the most intriguing self-replicating RNAs are the subviruses, the "ad infinitum" of the opening lines above. There are several types (Table 1). Some subviruses (satellites, defective interfering particles, and the hepatitis delta agent) require a "helper," uncompromised virus for infectivity, and the others (viroids) are infectious as unencapsidated RNA. Viroids are small (311 ± 65 nucleotides), single-stranded, highly base-paired, covalently closed circular RNAs that are infectious for plant cells. They do not encode proteins, and their replication is almost certainly conducted via a rolling circle mechanism (see below), using host enzymes. A variety of viroids are known, and there are major differences in the sequence homology between comparable domains of different viroids. This variability has led to the interpretation that viroid evolution has involved successive RNA rearrangements (Kees and Symons 1985; Rezaian 1990). A more speculative suggestion has been made by Haas et al. (1988) based on their identification of sequence complementarity between viroids and higher plant 7S RNA. They postulated that an RNA·RNA interaction, comparable to antisense mRNA interaction, may contribute to viroid pathogenicity. (Plant 7S RNA closely resembles the RNA in animal signal recognition particles [SRP] and may have a similar function to its SRP counterpart.) The

Table 1 Properties of subviruses

	Coat protein	Helper required	Helper homology	Size	Rolling circle
Viroids	no	no	—	311±65	yes
Satellites	yes	yes	no	200–1500	yes
DI	yes	yes	yes	various	no
Delta agent	yes	yes	no	1700	yes

limited data available, however, do not support the antisense hypothesis (see Hammond 1992 and references therein).

The subviruses that require helper viruses can be considered as parasites of those viruses. There are several types. One type, satellites, range in size from about 200 to 1500 nucleotides (for review, see Roossinck et al. 1992) and have no extensive sequence homology with their helper viruses. In contrast, another type, defective interfering (DI) particles, do have homology with their helper viruses. The third type is represented by human hepatitis delta agent. It requires a helper virus for infectivity, but can replicate in its absence. Like the other subviruses, it is also small (ca. 1700 nucleotides) and has single-stranded covalently closed circular genomic RNA that is highly base-paired. The delta agent is negative-stranded, and RNA copied from it encodes a protein, the 24,000-dalton delta antigen. The delta antigen is found within the virion and is required for replication (for review, see Taylor 1991). A similar-sized, 3,100-nucleotide covalently closed circular RNA, 20S RNA, is found in some strains of yeast (Matsumoto et al. 1990). This is cytoplasmically inherited and synthesizes a protein similar in size (23,000 daltons) to the delta antigen. About 20 copies of the 23,000-dalton protein associate with the 20S RNA to form 32S particles. Further work is required to determine if it can reasonably be grouped with the delta agent.

From an RNA world perspective, the replication strategies of viroids, small satellite RNAs, and the hepatitis delta agent merit consideration. Their small circular genomes replicate by a rolling circle mechanism. There is good evidence for the two variants of rolling circle mechanism depicted in Figure 1 (see Forster and Symons 1987; Branch et al. 1988: Symons 1992). In the first variant (Fig. 1A), the circular plus strand is copied to form a concatameric minus strand (step 1), which is then cleaved to produce a monomer (step 2). The monomer is circularized by host RNA ligase (step 3) and copied by RNA polymerase to yield a con-

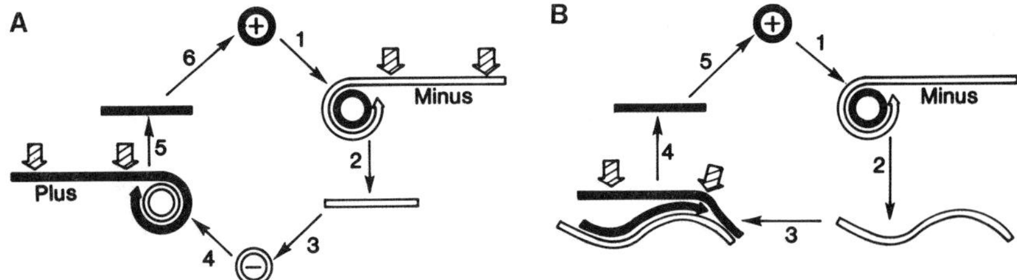

Figure 1 Rolling circle models. In model *A*, the minus-strand concatamer is cleaved before being used to produce the plus strand, whereas in model *B* the minus strand is directly copied. Reproduced, with permission, from Forster and Symons (1987).

catameric plus strand (step 4). Cleavage of the concatamer produces monomers (step 5), which then are circularized (step 6). In the second variation (Fig. 1B), the concatameric minus strand is not cleaved but is copied directly to yield a concatameric plus strand (step 3). Cleavage and circularization then proceed as in the first variant.

Diener (1989) and others have pointed out that a simple rolling circle mechanism could avoid the need for replication initiation and termination signals. At the hypothetical evolutionary stage at which proteins assumed replicase function, this might have been advantageous, but with RNA-based replicase, the RNA structures needed for activity may have had such elaborate structure that initiation was inherent. The multiple-copy product of rolling circle replication, in conjunction with the auto-catalytic cleavage of unit length products, gives transient polyploidy, which may have been advantageous in an error-prone primitive self-replicating RNA system.

Whereas DI particles are obviously deletion variants of their helper viruses, the origins of the other subviruses are far from clear. One of the most intriguing attributes of viroids and satellites is their catalytic RNA capabilities which, of course, are being intensively studied. It would be fitting if such studies of derivatives of the catalytic RNAs found in sub-viruses and *Tetrahymena* lead to self-replicating RNA selection systems, which in turn provide insight to the ways in which viroids and satellites could have arisen.

VIRAL DIVERSITY

There is an enormous diversity of viral properties including genomic composition. However, the coexistence of DNA or RNA—single- or double-stranded; positive, negative, or both; linear or circular; segmented or monopartite—viral genomes of diverse sizes, subject to intense selection means that widely different genomic compositions can give successful viruses. I deal with some properties of viruses with different genomic compositions before considering the basic characteristics of the nucleic acid involved.

Only an ecologically restricted subset of viruses is known. The greatest mass of viruses may infect marine phytoplankton, but their nature is substantially unexplored (Suttle et al. 1990; Olofsson and Kjelleberg 1991). This lack of knowledge also applies to the phages for the vast number of free-living bacteria, including those that live deep in the earth's crust. Neither is dealt with here. Other habitats have been better explored, and in these, RNA viruses are ubiquitous and are perhaps the most abundant cellular parasites known. A high proportion of these have single-stranded RNA (ssRNA) genomes.

WHY SINGLE-STRANDED RNA?

ssRNA viruses are found throughout nature but are especially prevalent among viruses of terrestrial plants; the great majority (more than 75%) of over 600 known plant viruses have ssRNA genomes of the same polarity as mRNA. Why single-stranded and why predominately RNA? Perhaps ssRNA genomes in plant viruses are prevalent because of the requirements for cell-to-cell movement within plants. Whereas some viruses move solely within vascular tissue (mainly phloem, but in some cases probably xylem), short distance cell-to-cell transfer is crucial for productive infection by many plant viruses. Because of tough plant cell walls, all known plant viruses whose infection moves from cell to cell in infected tissues utilize the narrow connections between adjacent cells. These connections, plasmodesmata, are the equivalent of animal cell gap junctions. It has been proposed that all such viruses possess a single-stranded genome or replicate via a ssRNA intermediate (caulimoviruses and badnaviruses). The single strand, encased in "movement" proteins, may be required for plasmodesma transfer. It appears that the complex of ssRNA and movement proteins is longer and thinner than the free nucleic acids and that the movement proteins in the complex also interact with the plasmodesmata (Citovsky et al. 1992). However, there is some evidence that cowpea mosaic virus deals with the plasmodesma transfer problem in a different way. The evidence suggests that alteration of plasmodesma structure permits intact virus particles to pass through (van Lent et al. 1991; for reviews, see Deom et al. 1992; Hull 1992). Consequently, it is premature to draw firm conclusions about the constraints that transfer places on the single-stranded nature of the great majority of plant viruses.

A possible advantage to both plant and animal viruses of having a positive ssRNA genome is the absence of an additional DNA-dependent transcription step (as well as nuclear uptake and splicing). With the exception of poxviruses, viral DNA-dependent transcription occurs in the nucleus, and in some cases, the nucleus is quiescent in infected cells. For replication, RNA viruses use precursors that are abundant in nearly all cells. Some DNA viruses, e.g., the small parvoviruses, are dependent on actively dividing host cells. Others have evolved strategies that stimulate the initiation of DNA synthesis or encode enzymes involved in nucleotide metabolism. Some of the latter category of DNA viruses, e.g., vaccinia, even encode a growth factor, as well as a plethora of factors that influence the host cell (for review, see Moss 1990). On the other hand, the DNA-dependent transcription step permits expression controls not available to positive-stranded RNA genomes. However, many RNA viruses use an equivalent strategy for gene control. They produce subgenomic mRNAs and often have sophisticated controls for highly expressed genes such as those for the coat protein(s).

In contrast to the situation in plant viruses, where only a small minority of the single-stranded viruses have DNA genomes, in bacteria and animals, a larger proportion of single-stranded viruses have DNA genomes. Instead of considering their features, I turn to other types of RNA viruses.

DOUBLE-STRANDED AND NEGATIVE-STRANDED RNA VIRUSES

Not all RNA viruses are positive-stranded; some are negative-stranded and others double-stranded. Negative-stranded viruses have a more host-independent replication strategy than positive-stranded viruses. As their genomic composition implies, negative-stranded genomes are not infectious by themselves. Virus-encoded enzymes, packaged in the virion, perform the necessary first step, transcription. This requires transcription signals that are recognized by the polymerase when in the transcription complex but are ignored when the complex switches to the replicative mode. Transcription initiation and termination signals, which are generally internal, are required for the synthesis of subgenomic mRNAs. Translation to give necessary proteins is required before the switch to replication initiates. As expected, replication produces full-length (positive-stranded) antigenomes. In some viruses (bunyaviruses and arenaviruses), mRNA is also transcribed from the antigenome, so their coding strategy can be considered ambisense. In all, however, the antigenome serves as template for new negative-strand genomes. Viruses that use this general strategy have been found in a wide range of hosts but are not the predominant viruses for any one group of hosts.

In contrast, double-stranded RNA (dsRNA) viruses are the most common fungal viruses, although most, if not all, of these are only transmitted vertically. However, infectious dsRNA viruses are found in a wide variety of hosts, even though the number of different types of virus is not large. All dsRNA viruses have their virus-encoded transcription machinery contained within the virion. The synthesized mRNAs are translated to yield viral proteins that sequester the same mRNAs in subviral particles. Synthesis of the minus strands, and resultant regeneration of progeny dsRNA, occurs within these particles. Replication of negative-stranded RNA viruses is also confined within nucleocapsids. The major distinction in replication strategy from that used by positive-stranded RNA viruses is encapsidation of the equivalent of the replicative form of those viruses and the functioning of the RNA-dependent RNA polymerase, with accessory proteins, in the nucleocapsids. Comparison of the sequence of the RNA polymerase gene from different groups of dsRNA viruses with the equivalent gene in positive-stranded RNA viruses has led to the tentative suggestion that different positive-stranded RNA viruses gave rise to different dsRNA viruses (Koonin 1992). It will be interesting to see if further sequence comparison supports this suggestion.

One expected feature of dsRNA should be its reduced susceptibility, because of the intact complementary strand, to the catastrophic consequences of spontaneous hydrolysis. dsRNA has greater stability than dsDNA due to interstrand stacking, and there is no known biological mechanism for unwinding *long* double-helical RNA. Replication occurs by asymmetric transcription from one strand, with the product serving as template to regenerate a duplex. During transcription, the only duplex is the very short transient duplex at the site of polymerization, and consequently, even in dsRNA viruses, there is no (known) mechanism for error correction. (In some DNA viruses, e.g., adenovirus and the *Bacillus* phage 29, only one strand is made at a time; i.e., replication is continuous and no Okazaki fragments are involved.) The consequence of asymmetric replication of RNA genomes is a limitation on the size of continuous dsRNA due to instability of resulting single strands.

SEGMENTATION

Dividing the genome into several RNA molecules circumvents the severe size-limitation problem for dsRNA genomes and is also extensively used by ssRNA viruses. Segmentation requires strategies to ensure the packaging of (normally) one copy of each segment — the classic example being found in the packaging of each of the ten segments of reovirus. As well as permitting larger dsRNA genome sizes, segmentation may alleviate the lack of ability for error correction. It is well established that when two viruses with different potentially lethal mutations infect the same cell, viable progeny frequently emerge ("multiplicity reactivation"). Although the phenomenon has been observed with undivided viral genomes, segmented genomes seem better positioned to emerge intact due to simple segment reassortment without necessitating recombination (see Reanney 1987).

Segmentation obviously requires replication initiation signals on each segment, and in some cases, the expression of the segments is independently controlled. However, since special signals are required to permit eukaryotic ribosomes to enter mRNAs internally, the use of segmented genomes reduces, or obviates, the need for such signals. Perhaps internal initiation was also a problem for primitive translation systems, and in an organizational sense, segmented genomes may more resemble primitive RNA genomes.

RNA-DNA CHIMERAS

As in much of biology, the boundaries between the different entities are not sharp; there is a continuum between RNA and DNA viruses that extends to host entities such as mobile introns, and plasmids. Well-known examples are retroviruses with genomic RNA but a DNA replicative form, and hepatitis B viruses with genomic DNA but an RNA replicative form. A recently discovered and particularly interesting example is the *Escherichia coli* phage φR73, discovered by the Inouyes and their colleagues (Inouye et al. 1991; Sun et al. 1991; Hsu et al. 1992). φR73 is a dsDNA virus that is normally integrated into the host chromosome. It is the template for an RNA product that is partially reverse-transcribed into a ssDNA molecule that remains covalently linked to part of the RNA (Fig. 2). The DNA branches out from the 15th base, G, of the 75-base RNA chain by a 2′, 5′ phosphodiester linkage rather than by a conventional 3′, 5′ nucleic acid linkage. The chimeric product can be present in numerous copies per cell and so is known as multicopy single-stranded DNA (msDNA). Excision of the double-stranded genomic DNA requires a helper virus (P2); i.e., if φR73 lacked the specific short sequence necessary for packaging by P2 coat proteins, it would be considered host and not virus. (Presumably, it acquired the packaging signals by recombination of a precursor with phage P2, although the exchange could have been in the other direction.) On excision, part of the flanking host DNA remains at the ends of the virus so that when a new cell is in-

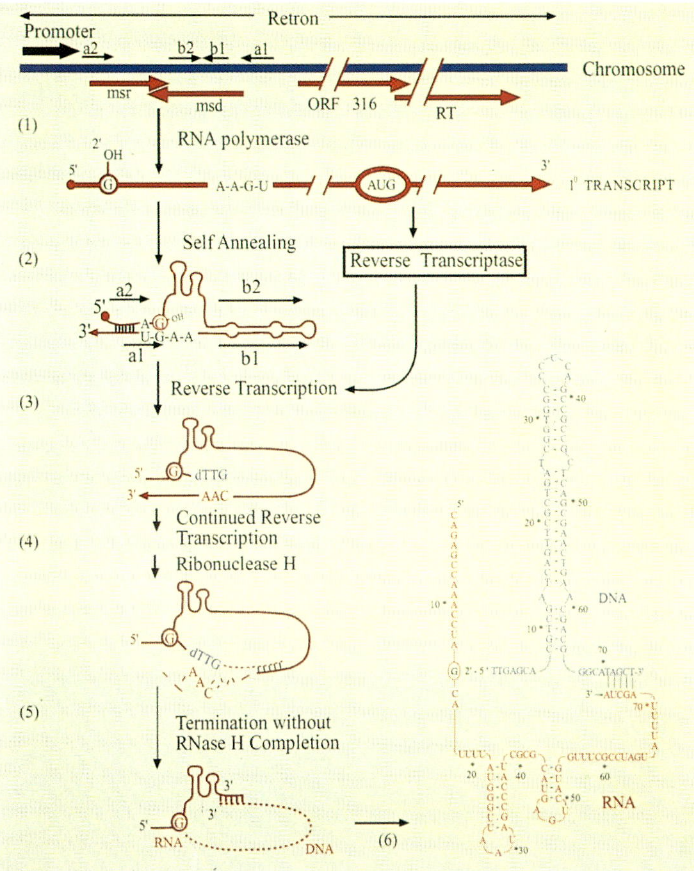

Figure 2 Proposed structure of the msDNA EC73 derived from Retron phage φR73 and mechanism of its synthesis (Sun et al. 1991; Inouye and Inouye 1992). *msr* and *msd* are the regions that encode the RNA and DNA portions of the msDNA molecules (the arrowhead on *msd* indicates polarity and not transcription). In step 1, a long primary transcript is synthesized. In step 2, the 29-base inverted repeats a1 and a2 anneal to form a stable stem structure. The circled rG from which the single-stranded DNA branches is at the very end of this stem. The 5′ end of the DNA synthesized by the reverse transcriptase derived from the RT gene in the primary transcript is complementary to RNA sequence just 5′ of a1 (step 3). Reverse transcription proceeds and ribonuclease H degrades the complementary RNA (step 4), apart from the five 3′ bases (step 5), permitting DNA annealing (step 6).

fected, reintegration occurs into the same matching site within the 3′ end of the gene for selenocystyl-tRNA (*selC*). Other prophages, P22 and P4, also integrate within the 3′ end of tRNA genes (tRNA[Thr] and tRNA[Leu], respectively), which is of evolutionary interest in view of the utilization of the 3′ region of cellular tRNAs by retroviruses and retrotransposons for priming cDNA synthesis (Sun et al. 1991). Retroviruses themselves are clearly very germane to the present discussion, because they have both RNA and DNA phases, but the relevant features are dealt with by Maizels and Weiner (this volume) and are not examined here. Instead, the distinctive properties of DNA and RNA and their relevance for viral genomes will be considered.

VERSATILITY, STABILITY, REPAIR, AND SIZE

An obvious difference between DNA and RNA is potential nucleophilic activity of the 2′ hydroxyl of the pentose of RNA. This might be expected to confer versatility to RNA at the expense of stability. As was first shown for the ssRNA phages (Fiers 1979; Skripkin et al. 1990; Jacobson 1991), probably all ssRNA genomes are highly structured (Fisher and Johnson 1993; Larson et al. 1993). However, based on the limited data available to date (Shen et al. 1979; see McKenna et al. 1992), the ability to form such elaborate secondary and tertiary structure is only slightly more favored for RNA than for ssDNA. Nevertheless, the 2′ hydroxyl in RNA forms additional stabilizing interactions. Just as structure in mRNA affects translational frameshifting, hopping, and readthrough, structures in genomic RNA are likely to affect replicase-associated recombination and probably mutations. Structures also provide functional surfaces for interactions with other macromolecules and relative stability against nuclease attack. In addition, the structures probably help to maintain single-strandedness during replication by sequestering sequences that might otherwise form a template·nascent-strand duplex. (Single-stranded nucleic acids are denser than double-stranded ones due to atoms being brought closer together by van der Waals interactions precluded by hydrogen-bonded base-pairing in duplex molecules, but if this has any relevance for viruses, it is presumably only with respect to packaging.)

Several points can be made concerning differential stability. DNA is subject to depurination at acidic pH. This is unlikely to have relevance to modern genomes, although it may well have been pertinent during very early evolutionary times. In contrast, the deamination of cytosine to uracil is of contemporary relevance. In DNA this problem is corrected by uracil-DNA glycosylase, whereas in RNA it is not corrected. UV irradiation produces a larger number of photochemical changes in RNA than in dsDNA. RNA is hydrolyzed relatively easily compared to DNA, especially in the presence of transition metal ions. However, in the tight virions of some viruses, it is likely that water is excluded, so that the hydrolytic instability of RNA is not relevant. The time window for viral reproduction during which the viral RNA is exposed to water in a cell is short (see Westheimer 1987). In other viruses (influenza ?) it would be surprising if this were true. Other factors, especially the absence, as far as is known, of RNA repair mechanisms, make hazardous relevant simple deductions about the importance of the stability factor from the nature of viral genomes. For viruses, it could also be argued that the other pitfalls and the vast numbers of viral progeny make stability considerations less important.

Is there some limit to RNA length? Viral genomes, especially viral RNA genomes, are short compared to cellular genomes, although still of substantial size. The longest viral RNA (the ssRNA from coronaviruses) is about 30 kb. In contrast, the total transcript length of some cellular genes is much longer (human dystrophin is on the order of 2 Mb and the length of the *Drosophila* cAMP phosphodiesterase transcript is at least 150 kb; see Koenig et al. 1987; Thummel 1992), but the 5′ ends of these transcripts are processed long before transcription is complete. A more relevant size comparison is with the approximately 30-kb late RNA transcript of adenovirus, some of which can be isolated intact prior to processing.

That the type of nucleic acid in viral RNA genomes, and not merely viral genomic size, may be an issue can be inferred from the much larger size of several DNA viral genomes. (For instance, cytomegalovirus has dsDNA of ~225 kb. However, true ssDNA viruses are small; e.g., parvoviruses are ~5 kb and the two molecules in bipartite geminiviruses are ~2.5 kb each.)

When considering the relationship between nucleic acid type and viral genome size and properties, it is natural to think also of the small dsDNA genomes in animal mitochondria. Like RNA genomes, they lack a repair system and also undergo much more rapid evolution than nuclear genomes. The smaller number of encoded products has permitted greater change in the decoding apparatus. Mammalian mitochondria encode the RNAs of their ribosomes, but it appears that on an evolutionary time scale, the role of rRNA is increasingly being taken over by the ribosomal proteins (O'Brien et al. 1990). As is generally the case in mitochondria, these proteins are nuclear-encoded. Mitochondrial rRNAs are evolving 23 times faster than cytoplasmic rRNAs, whereas the mitochondrial ribosomal proteins are changing at the roughly similar rate of 13 times that of their cytoplasmic counterparts (Pietromonaco et al. 1986). Selection constraints on ribosomes, rather than the ability to generate variants, seem more important in limiting change. Similarly, the nucleocapsid-sequestered double-stranded and negative-stranded viruses undergo more rapid evolutionary change than do positive-stranded viruses, whose replication is more interactive with host components (Koonin 1992). This conclusion was reached from an analysis of substitutions in the polymerase genes. Even in positive-stranded viruses, the genes for products that do not interact with host components undergo more rapid change than do polymerase genes (Zimmern 1988). The recurring theme is that selection constraints are predominant despite the plasticity of RNA genomes.

DEFENSE AND ITS CIRCUMVENTION

It is curious that virtually all known animal viruses, both DNA and RNA, generate dsRNA in infected cells. Since dsRNA is either absent in uninfected cells or confined to the nuclei (at least two mammalian genes are transcribed partially off both strands [Viskochil et al. 1991; Colombo et al. 1992]), dsRNA synthesized from viral genomes is the mobilization call for the host defense mediated through the interferon system. Animal viruses have elaborate counter-defense mechanisms that act directly on the products of the various host defense systems. For instance, the adenoviruses' counter-defense involves several virus-encoded proteins that modulate the host tumor necrosis factor or cytotoxic T lymphocytes (Gooding 1992). It also involves the synthesis of a small RNA, known as VA RNA (for *Virus Associated* RNA, named before it was known to be virally encoded). Other viruses, such as human immunodeficiency virus, synthesize a counterpart—the TAR RNA (for review, see Mathews and Shenk 1991). Despite the counter-defenses, it would still seem prudent for viruses to avoid interferon induction, and some, such as vesicular stomatitis virus (VSV), avoid being strong interferon inducers. Having unidirectional transcription, just as the dsDNA phage T7 does, would be a good start to avoid interferon induction. Indeed, adeno-associated virus (AAV) has such a strategy, but it is defective and requires adenovirus as a helper virus to provide the functions it lacks. Adenovirus generates dsRNA by overlapping transcription from both strands, and so productive AAV infection also indirectly generates dsRNA. It seems that viruses have found it easier to evolve counterdefenses than to control tightly their transcription to avoid dsRNA. An alternative possibility is that they are too virulent for their own long-term survival if they do not induce the interferon response and that there is selective value in modulating the host defense instead of avoiding it. Of course, for dsRNA viruses such as reoviruses, the interferon problem is inherent, and this feature presumably explains why they hide their RNA in an inaccessible form in nucleocapsids in infected cells.

The most recent viral counter-defense strategy to come to light is viral acquisition of host genes for cell-surface receptors that help the immune system recognize and fight viruses. The viruses cause the synthesis of large quantities of soluble versions of the receptors, which act as decoys for the immune system (Gooding 1992). Although only large DNA viruses have so far been found to utilize this strategy, it does illustrate the dynamic interplay between viral and host genomes.

GENOME VARIABILITY: POPULATION SIZE

In contrast to viral DNA genomes, viral RNA genomes exhibit great sequence variability. Even in a single virus clone stock, there is a range of related sequences centering on the "mean" sequence ("genetic swarm"). In the small number of cases investigated, 1–10 variants have been found per genome. Potentially, all possible single and double mutants, as well as decreasing proportions of triple, quadruple, etc. mutants, are present in most natural viral populations (Domingo et al. 1978; Domingo and Holland 1988). However, selection acting on populations constrains the rate of divergence, since most variants are deleterious. In addition to this process, selection may sometimes be more directed. For example, it has been suggested, although not established, that each poliovirus replicase molecule acts only once in *cis*, using the RNA from which it was translated as template, and that subsequent reinitiation is prevented by the cleavage of the viral protein genome (VPg) (for review, see Strauss and Strauss 1983). If replicase works only in *cis*, mutants producing nonviable replicase would be instantly aborted. Such options are presumably unreasonably wasteful for a cellular chromosome but would be quite efficient as a feedback control for a molecule devoted largely to coding for its own replication.

Because of selection, the inherent plasticity of the genome often contrasts with the relative stability of particular virus "species." A reasonable way to examine variation without the selection constraint is to follow the fate of extraneous sequences inserted into a virus. This was done in a mutant of tobacco mosaic virus (TMV) with an insertion of 250 "foreign" nucleotides. The insert was found to have accumulated fewer than 3 changes per RNA after continuous replication through 10 passages in different plants (Donson et al. 1991; for review, see Dawson 1992). This frequency translates to less than 1 change per kilobase per replication, and, given the average genome size, corresponds to the range of 1–10 per genome. Since selection was presumably not operating on the TMV insert, one would expect the value to be higher than the variation per genome where selection is operating. Clearly, more studies are needed. Care is needed, however, as other factors, including population size, have to be considered when studying the role of selection.

Selection does not always act on RNA viruses on a large population basis. For instance, some viruses are transmitted in respiratory droplets, and the low virus concentration in these minute droplets means that many of these droplets have only one or a few viruses (Artenstein and Miller 1966; Gerone et al. 1966). Because of the high variability of RNA viruses, most of the particles will have at least one mutation, and most of these will be deleterious. Restoration of greater fitness will require either back, compensatory, or beneficial mutations, and, because of the random nature of mutations and previous optimizations, these secondary mutations will, in general, be less frequent. In the absence of "rescuing" mechanisms, repeated genetic bottlenecks have been proposed to lead to an accumulation of deleterious mutations (Muller 1964). This prediction has been tested following multiple plaque-to-plaque transfers in the segmented dsRNA phage $\phi6$ (Chao 1990) and the nonsegmented ssRNA animal virus, VSV (Clarke et al. 1993). In both cases, decreased fitness was found following the serial passages. In analogy with classical evolution, it is reasonable to regard the isolated populations as viral "genetic islands." Their main consequence is likely to be decreased virulence (see Clarke et al. 1993), but perhaps also a greater chance of a new type arising than in nonisolated large populations.

A major reason for the plasticity of RNA genomes is the apparent lack of proofreading and mismatch repair that are utilized in high-fidelity DNA replication. Several mechanisms generate the variability. Point mutations, mainly caused by misincorporations at the synthetic step, appear to be the most frequent source of variation, at least among viable progeny (for review, see Domingo and Holland 1988). Some variants arise as a consequence of RNA modification giving hypermutation rates at certain positions (see Bass, this volume). The reassortment of segments from different viruses in mixed infections is a minor but interesting source of variation. (An example of the latter is found in pea enation mosaic virus [Demler and de Zoeten 1991]). Finally, an important contributor is recombination, which is dealt with next.

RECOMBINATION

A significant number of RNA viruses are known to exhibit both homologous and nonhomologous recombination. Both processes are fundamentally different from the strand-breaking and rejoining of DNA recombination. Homologous RNA recombination has been studied most with poliovirus, where it was first found (for review, see Lai 1992). During replication, the poliovirus replicase complex can dissociate from one

RNA and reassociate with another. In a mixed infection, reassociation at an equivalent position of another template (template switching or copy choice) can yield homologous recombinants (Kirkegaard and Baltimore 1986). Template switching appears to occur without a consensus sequence at nearly random positions (Jarvis and Kirkegaard 1992), but it is possible that some RNA structures slow the polymerase and locally elevate the frequency. Homologous recombination is only known in some viruses, although the list is growing (for review, see Lai 1992). It now includes the ssRNA phages, where for a long time it was thought to be nonexistent (Palasingam and Shaklee 1992), as well as a dsRNA phage (Mindich et al. 1992). Homologous RNA recombination may occur with all RNA viruses, but if it does, then the frequency in some viruses, e.g. VSV and influenza, is very much lower than in other viruses. Homologous RNA-mediated recombination is not confined to viruses. It has also been detected in yeast chromosomal transcripts, and one type of it has been shown to lead to pseudogene formation (Derr and Strathern 1993).

Nonhomologous RNA recombinations also occur (for review, see Lai 1992). It is possible that some nonhomologous recombination occurrences are due to *trans* splicing. However, it seems clear that detachment and reattachment of the replicase complex at another position, in either the same or a different template, is mainly involved. The "promiscuous" polymerase may not be so promiscuous; it may dissociate from its template and reattach not at random, but instead under the influence of structured regions. This detachment and reattachment in recombination is not conceptually very different from normal transcription in coronaviruses, where there is leader-primed transcription by a nonprocessive RNA polymerase.

It is reasonable to presume that most, if not all, viruses undergo intermolecular arrangements of the nonhomologous type. On an evolutionary time scale, this transfer of RNA genes from virus to virus, or from host cell to virus, by nonhomologous recombination is likely to have been important. Evidence suggestive of such past gene translocations can be seen in the sequence homology between replicase genes of such diverse RNA viruses as Sindbis, brome mosaic virus, and TMV and in the tandem triplication of the VPg (viral protein genome-linked) gene of foot and mouth disease virus (Forss and Schaller 1982). An interesting example of nonhomologous recombination is found in bovine viral diarrhea virus. Normally, it is noncytopathic with a wide distribution. Cytopathic variants arise, and those analyzed have been found to have acquired a host ubiquitin gene, which may be directly related to the cytopathogenicity (for review, see Myers et al. 1990).

A number of the deletion variants formed by nonhomologous recombination, although defective on their own, are viable in the presence of wild-type helper virus. These DI particles have been isolated from almost all animal viruses and, by competing for replication, have major effects on viral propagation and, as a consequence, on virulence. In the presence of DI, the non-defective (helper) virus adopts changes that permit it to cope with, or even be resistant to, the DI (Horodyski et al. 1983; O'Hara et al. 1984). Having been generated by nonhomologous recombination, DIs in turn can also participate in further rounds of nonhomologous recombination. A clear example is a Sindbis DI, which has a new sequence almost identical to a rat tRNA^{Asp} (Monroe and Schlessinger 1983). Intermolecular DI RNAs combining sequences from different genome segments of influenza virus have now also been demonstrated.

In finishing this section, it is worth emphasizing the importance of even a low level of recombination for evolution. In fact, the frequency of recombination almost matches the frequency of RNA reassortment in segmented viruses (Fields 1981).

RELATIONSHIPS OF VIRUSES: SPECIES TO SUPERFAMILIES

The famous quote of Theo Dobzhanski, "Nothing in biology makes sense except in the light of evolution," has been countered by Sydney Brenner with, "Anything that is produced by evolution is bound to be a bit of a mess." In fact, the end products of viral evolution are exquisite. The difficulty, especially with RNA viruses because of their plasticity, is elucidating how they evolved and so determining their present relationships. The accepted groupings of viruses, based on the genome organization and phenotypic properties of species and families, is clearly realistic and useful. Although it is sometimes a question of judgment about the relative importance of particular details, the framework is clear. For instance, alphaviruses, flaviviruses, picornaviruses, and many other RNA viruses have evolved a fairly large number of family members that differ sufficiently to be considered separate viruses, but all have essentially the same structure. These separate viruses are simultaneously present in the field and can have either overlapping or distinct geographical distributions. (Not all have multiple strains present at one time; with influenza viruses only one or a few strains predominate in the human population at any one time.)

From the present perspective, larger-scale groupings are more interesting. As described above, there are ample means for generating diversity, but despite this and the abundance of different RNA viruses, it has proved possible to group viruses into a small number of superfamilies

(Haseloff et al. 1984; Strauss and Strauss 1988; Zimmern 1988; Goldbach 1990). These groupings are based on overall organization, replication, and expression strategies. Two of the supergroups include both plant and animal viruses. The smaller group includes animal picornaviruses and plant viruses that show similarities. The larger, more heterogeneous group includes animal alphaviruses and their plant counterparts.

What are the underlying reasons for the similarities that permit the superfamily groupings? There are three possibilities. One is disparate backgrounds with selective pressure toward convergent evolution to solve common problems in a restricted niche. Another is common ancestry with selection for retention of similar features. A third is exchange of gene modules by recombination. Although the similarities extend from animal to plant viruses, and although their hosts diverged a little more than 10^9 years ago, recombination could, as many authors have suggested, have taken place in insects, which are common hosts for many plant and animal viruses and which harbor a great variety of RNA viruses. However, even if recombination is the basis for the similarities that permit superfamily groupings with plant and animal virus members, it does not mean that the conserved features are not ancient. For instance, if some viruses are derived from cellular components, recombination may have permitted the acquisition of modules from ancient times that have been adapted to the viral life-style. At this stage there is no way of knowing. (From the perspective of an RNA world, the progenitor of RNA viruses may be at least as old as the progenitor of cells.) What can be said, however, is that if recombination played a key role, it is likely to have been RNA recombination rather than recombination via DNA intermediates. It is striking that, as far as is known, chromosomal pseudogenes of RNA viruses, with the exception of retroviruses, do not exist. The concept of module transfer, originally introduced for DNA phages, may be equally valid at the RNA level, with the modification that the modules consist of domains rather than groups of genes. In this it may be more akin to the suggested ultimate advantage of RNA splicing (Gilbert 1987).

A model based on disparate origins channeled by convergent evolution has two facets. One is selection for common expression strategies that may not be accompanied by sequence homology. The second is convergent evolution to common sequences. Are there a significant number of motifs that would be equally good for a given purpose? If so, disparate origins might be expected to lead to different solutions. In contrast, a common ancestry for motifs (and perhaps some elaborate RNA structures), irrespective of whether acquired by recombination "recently" or from an ancient ancestral virus without recombination, might have resulted in only one, or a small number, of the possible motifs. Some conservation of blocks (10–15 nucleotides long) of primary sequence among RNA viral replicases has been noted, as well as similarities to reverse transcriptases (Xiong and Eickbush 1990), which are misnamed if RNA to DNA was the original direction of information flow. (The absence of RNA replicases in contemporary uninfected animal cells precludes determining if host replicases have different motifs.) More data, especially three-dimensional protein determinations, are required. To a greater extent, these data are available for viral capsid proteins rather than for enzymes. However, in either case, the available data indicate that the proteins have diverged from a common ancestral structure (Matthews and Rossmann 1985; McKenna et al. 1992). With RNA viruses, it is not easy to estimate the time elapsed since the existence of such a common ancestor. It seems inherently likely that RNA and DNA viruses have been in existence since DNA became the cellular genetic material (see opening quote from a graduate of my alma mater about 300 years ago!). The temptation to think that the probable ancestral protovirus of modern RNA viruses was ancient rather than recent is considerable.

The above examples show that although major changes do occur, in general, RNA viruses exhibit remarkable stability given the plasticity of their genomes. Their ability to rearrange and mutate is not limiting their evolution. Rather, the rate of change of individual species seems governed by selection pressures operating in the specialized niche they occupy and is presumably influenced by the interdependence of optimized features. The ability to recognize modules among diverse members of viral superfamilies and their plausible ancient origins bodes well for the inferences that Weiner and Maizels have made about the presence of molecular fossils for replication initiation. There may be some method in "viral madness." A deeper understanding of RNA structure is likely to be instructive.

ACKNOWLEDGMENTS

I thank C. Anderson, S. Inglis, S. Lee, A. Weiner, and especially T. Tuohy and R. Gesteland for their comments. I am supported by a grant from the National Institutes of Health.

REFERENCES

Artenstein, M.S. and W.S. Miller. 1966. Air sampling for respiratory disease agents in army recruits. *Bacteriol. Rev.* **30:** 571–572.

Branch, A.D., B.J. Benenfeld, and H.D. Robertson. 1988. Evidence for a single rolling circle in the replication of potato spindle tuber viroid. *Proc. Natl. Acad. Sci.* **85:** 9128–9132.

Chao, L. 1990. Fitness of RNA virus decreased by Muller's ratchet. *Nature* **348:** 454–455.

Citovsky, V., M.L. Wong, A.L. Shaw, B.V.V. Prasad, and P. Zambryski. 1992. Visualization and characterization of tobacco mosaic virus movement protein binding to single-stranded nucleic acids. *Plant Cell* **4:** 397–411.

Clarke, D.K., E.A. Duarte, A. Moya, S.F. Elena, E. Domingo, and J. Holland. 1993. Genetic bottlenecks and population passages cause profound fitness differences in RNA viruses. *J. Virol.* **67:** 222–228.

Colombo, P., J. Yon, K. Garson, and M. Fried. 1992. Conservation of the organization of five tightly clustered genes over 600 million years of divergent evolution. *Proc. Natl. Acad. Sci.* **89:** 6358–6362.

Dawson, W.O. 1992. Tobamovirus-plant interactions. *Virology* **186:** 359–367.

Demler, S.A. and G.A. de Zoeten. 1991. The nucleotide sequence and luteovirus-like nature of RNA 1 of an aphid non-transmissible strain of pea enation mosaic virus. *J. Gen. Virol.* **72:** 1819–1834.

Deom, C.M., M. Lapidot, and R.N. Beachy. 1992. Plant virus movement proteins. *Cell* **69:** 221–224.

Derr, L.K. and J.N. Strathern. 1993. A role for reverse transcripts in gene conversion. *Nature* **361:** 170–173.

Diener, T.O. 1989. Circular RNAs: Relics of precellular evolution. *Proc. Natl. Acad. Sci.* **86:** 9370–9374.

Dinter-Gottlieb, G. 1986. Viroids and virusoids are related to group 1 introns. *Proc. Natl. Acad. Sci.* **83:** 6250–6254.

Domingo, E. and J.J. Holland. 1988. High error rates, population equilibrium, and evolution of RNA replication systems. In *RNA Genetics: Variability of RNA genomes* (ed. E. Domingo et al.), vol.3, pp. 3–36. CRC Press, Boca Raton, Florida.

Domingo, E., D. Sabo, T. Taniguichi, and C. Weissmann. 1978. Nucleotide sequence heterogeneity of an RNA phage population. *Cell* **13:** 735–744.

Donson, J., C.M. Kearney, M.E. Hilf, and W.O. Dawson. 1991. Systemic expression of a bacterial gene by tobacco mosaic virus-based vector. *Proc. Natl. Acad. Sci..* **88:** 7204–7208.

Fields, B.N. 1981. Genetics of reovirus. *Curr. Top. Microbiol. Immunol.* **91:** 1–24.

Fiers, W. 1979. RNA Bacteriophages. In *Comprehensive virology* (eds. H. Fraenkel-Conrat and R.R. Wagner), vol. 13, pp 69–204. Plenum Press, New York.

Fisher, A.J. and J.E. Johnson. 1993. Ordered duplex RNA controls capsid architecture in an icosahedral animal virus. *Nature* **361:** 176–179.

Forss, S. and H. Schaller. 1982. A tandem repeat gene in a picornavirus. *Nucleic Acids Res.* **10:** 6441– 6450.

Forster, A.C. and R.H. Symons. 1987. Self-cleavage of plus and minus RNAs of a virusoid and a structural model for the active sites. *Cell* **49:** 211–220.

Gerone, P.J., R.B. Couch, G.V. Keeter, R.G. Douglas, E.B. Derrenbacher, and V. Knight. 1966. Assessment of experimental and natural aerosols. *Bacteriol. Rev.* **30:** 576–588.

Gilbert, W. 1987. The exon theory of genes. Cold Spring Harbor Symp. Quant. Biol. **52:** 901–905.

Goldbach, R. 1990. Genome similarities between positive-strand RNA viruses from plants and animals. In *New aspects of positive-strand RNA viruses* (ed. M.A. Brinton and F.X. Heinz) pp. 3–11. American Society for Microbiology, Washington, D.C.

Gooding, L.R. 1992. Virus proteins that counteract host immune defenses. *Cell* **71:** 5–7.

Haas, B., A. Klanner, K. Ramm, and H.L. Sänger. 1988. The 7S RNA from tomato leaf tissue resembles a signal recognition particle RNA and exhibits a remarkable sequence complementary to viroids. *EMBO J.* **7:** 4063–4074.

Hadidi, A. 1986. Relationship of viroids and certain other plant pathogenic nucleic acids to group 1 and group 2 introns. *Plant Mol. Biol.* **7:** 129–142.

Hammond, R.W. 1992. Analysis of the virulence modulating region of potato spindle tuber viroid (PSTVd) by site-directed mutagenesis.*Virology* **187:** 654–662.

Haseloff, J., P. Goelet, D. Zimmern, P. Ahlquist, R. Dasgupta, and P. Kaesberg. 1984. Striking similarities in amino acid sequence among nonstructural proteins encoded by RNA viruses that have dissimilar genomic organization. *Proc. Natl. Acad. Sci.* **81:** 4358–4362.

Horodyski, F.M., S.T. Nichol, K.R. Spindler, and J.J. Holland. 1983. Properties of DI particle resistant mutants of vesicular stomatitis virus isolated from persistent infections and from undiluted passages. *Cell* **33:** 801–810.

Hsu, M.Y., S.G. Eagle, M. Inouye, and S. Inouye. 1992. Cell-free synthesis of the branched RNA-linked msDNA from retron-EC67 of *Escherichia coli. J. Biol. Chem.* **267:** 13823–13829.

Hull, R. 1992. Down the tube. *Curr. Biol.* **2:** 224–226.

Inouye, M. and S. Inouye. 1992. Retrons and multicopy single-stranded DNA. *J. Bacteriol.* **174:** 2419–2424.

Inouye, S., M.G. Sunshine, E.W. Six, and M. Inouye. 1991. Retronphage ϕR7 3: An *E. coli* phage that contains a retroelement and integrates into a tRNA gene. *Science* **252:** 969–971.

Jacobson, A.B. 1991. Secondary structure of coliphage Qβ RNA. *J. Mol. Biol.* **221:** 557–570.

Jarvis, T.C. and K. Kirkegaard. 1992. Poliovirus RNA recombination: Mechanistic studies in the absence of selection. *EMBO J.* **11:** 3135–3145.

Kees, P. and R.H. Symons. 1985. Domains in viroids: Evidence of intermolecular RNA rearrangements and their contribution to viroid evolution. *Proc. Natl. Acad. Sci.* **82:** 4582–4586.

Kirkegaard, K. and D. Baltimore 1986. The mechanism of RNA recombination in polio virus. *Cell* **47:** 433–443.

Koenig, M., E.P. Hoffman, C.J. Bertelson, A.P. Monaco, C. Feener, and L.M. Kunkel. 1987. Complete cloning of the Duchenne muscular dystrophy (DMD) cDNA and preliminary genomic organization of the DMD gene in normal and affected individuals. *Cell* **50:** 509–517.

Koonin, E.V. 1992. Evolution of double-stranded RNA viruses: A case for polyphyletic origin from different groups of positive-stranded RNA viruses. *Semin. Virol.* **3:** 327–339.

Lai, M.M.C. 1992. RNA recombination in animal and plant viruses. *Microbiol. Rev.* **56:** 61–79.

Larson, S.B., S. Koszelak, J. Day, A Greenwood, J.A. Dodds, and A. McPherson. 1993. Double-helical RNA in satellite tobacco mosaic virus. *Nature* **361:** 179–182.

Mathews, M.B. and T. Shenk. 1991. Adenovirus virus-associated RNA and translation control. *J. Virol.* **65:** 5657–5662.

Matsumoto, Y., R. Fishel, and R.B. Wickner. 1990. Circular single-stranded RNA replicon in *Saccharomyces cerevisiae. Proc. Natl. Acad. Sci.* **87:** 7628–7632.

Matthews, B.W. and M.G. Rossmann. 1985. Comparison of protein structures. *Methods Enzymol.* **115:** 397–420.

McKenna, R., D. Xia, P. Willingmann, L.L. Ilag, S. Krishnaswamy, M.G. Rossmann, N.H. Olson, T.S. Baker, and N.L. Incardona. 1992. Atomic structure of single-stranded DNA bacteriophage φX17 4 and its functional implications. *Nature* **355**: 137–143.

Mindich, L., X. Qiao, S. Onodera, P. Gottlieb, and J. Strassman. 1992. Heterologous recombination in the double-stranded RNA bacteriophage φ6. *J. Virol.* **66**: 2605–2610.

Monroe, S. and S. Schlessinger. 1983. RNAs from two independently isolated defective interfering particles of Sindbis virus contain a cellular tRNA sequence at their 5′ ends. *Proc. Natl. Acad. Sci.* **80**: 3279–3283.

Moss, B. 1990. Poxviridae and their replication. In *Virology*, 2nd edition (ed. B.N. Fields et al.) pp. 2079–2111. Raven Press, New York.

Muller, H.J. 1964. The relationship of recombination to mutational advance. *Mutat. Res.* **1**: 1–9.

Myers, G., T. Rümenapf, and H.-J. Thiel. 1990. Insertion of ubiquitin-coding sequence identified in the RNA genome of a togavirus. In *New aspects of positive-strand RNA viruses* (ed. M.A. Brinton and F.X. Heinz) pp. 25–30. American Society for Microbiology, Washington, D.C.

O'Brien, T.W., N.D. Denslow, J.C. Anders, and B.C. Courtney. 1990. The translation system of mammalian mitochondria. *Biochim. Biophys. Acta* **1050**: 174–178.

O'Hara, P.J., F.M. Horodyski, S.T. Nichol, and J.J. Holland. 1984. Vesicular stomatitis virus mutants resistant to defective-interfering particles accumulate stable 5′-terminal and fewer 3′-terminal mutations in a stepwise manner. *J. Virol.* **49**: 793–798.

Onodera, S., V.M. Olkkonen, P. Gottlieb, J. Strassman, X. Qiao, D.H. Bamford, and L. Mindich. 1992. Construction of a transducing virus from double-stranded RNA bacteriophage φ6: Establishment of carrier states in host cells. *J. Virol.* **66**: 190–196.

Olofsson, S. and S. Kjelleberg. 1991. Virus ecology. *Nature* **351**: 612.

Palasingam, K. and P.N. Shaklee. 1992. Reversion of Qβ RNA phage mutants by homologous RNA recombination. *J. Virol.* **66**: 2435–2442.

Pietromonaco, S.F., R.A. Hessler, and T.W. O'Brien. 1986. Evolution of proteins in mammalian cytoplasmic and mitochondrial ribosomes. *J. Mol. Evol.* **24**: 110–117.

Reanney, D.C. 1987. Genetic error and genome design. *Cold Spring Harbor Symp. Quant. Biol.* **52**: 751–757.

Rezaian, M.A. 1990. Australian grapevine viroid-evidence for extensive recombination between viroids. *Nucleic Acids Res.* **18**: 1813–1818.

Roossinck, M.J., D. Sleat, and P. Palukaitis. 1992. Satellite RNAs of plant viruses: Structures and biological effects. *Microbiol. Rev.* **56**: 265–279.

Shen, C.-K.J., A. Ikoku, and J.E. Hearst. 1979. A specific DNA orientation in the filamentous bacteriophage fd as probed by psoralen crosslinking and electron microscopy. *J. Mol. Biol.* **127**: 163–175.

Skripkin, E.A., M.R. Adhin, M.H. de Smit, and J. van Duin. 1990. Secondary structure of the central region of bacteriophage MS2 RNA. *J. Mol. Biol.* **211**: 447–463.

Strauss, E.G. and J.H. Strauss. 1983. Replication strategies of single stranded RNA viruses of eukaryotes. *Curr. Top. Micrbiol. Immunol.* **105**: 1–98.

Strauss, J.H. and E.G. Strauss. 1988. Evolution of RNA viruses. *Annu. Rev. Microbiol.* **42**: 657–683.

Sun, J., M. Inouye, and S. Inouye. 1991. Association of a retroelement with a P4-like cryptic prophage (retronphage φR73) integrated into the selenocystyl tRNA gene of *Escherichia coli. J. Bacteriol.* **173**: 4171–4181.

Suttle, C.A., A.M. Chan, and M.T. Cottrell. 1990. Infection of phytoplankton by viruses and reduction of primary productivity. *Nature* **347**: 467–469.

Symons, R.H. 1992. Small catalytic RNAs. *Annu. Rev. Biochem.* **61**: 641–671.

Taylor, J.M. 1991. Human hepatitis delta virus. *Curr. Top. Microbiol. Immunol.* **168:** 141–166.

Thummel, C.S. 1992. Mechanisms of transcriptional timing in *Drosophila. Science* **255:** 39–40.

van Lent, J., M. Storms, F. van der Meer, J. Wellink, and R.Goldbach. 1991. Tubular structures involved in movement of cowpea mosaic virus are also formed in infected cowpea protoplasts. *J. Gen. Virol.* **72:** 2615–2623.

Viskochil, D., R. Cawthon, P. O'Connell, G. Xu, J. Stevens, M. Culver, J. Carey, and R. White. 1991. The gene encoding the oligodendrocyte-mylin glycoprotein is embedded within the neurofibromatosis type 1 gene. *Mol. Cell. Biol.* **11:** 906–912.

Westheimer, F.H. 1987. Why nature chose phosphates. *Science* **235:** 1173–1178.

Xiong, Y. and T.H. Eickbush. 1990. Origin and evolution of retroelements based upon their reverse transcriptase sequences. *EMBO J.* **9:** 3353–3362.

Zhang, M. and G.G. Brown. 1993. Structure of the maize mitochondrial replicon RNA*b* and its relationship with other autonomously replicating RNA species. *J. Mol. Biol.* **230:** (in press).

Zimmern, D. 1988. Evolution of RNA viruses. In *RNA genetics*, vol. 2, *Retroviruses, viroids and RNA recombination* (ed. E. Domingo et al.) pp. 211–240. CRC Press, Boca Raton, Florida.

22
Telomerase

Elizabeth H. Blackburn
Department of Microbiology and Immunology
and Department of Biochemistry and Biophysics
The University of California
San Francisco, California 94143-0414

The view of evolution that asserts that the genetic material was first in the form of RNA raises the question of how the transition to DNA genomes occurred. In the current world, although most of the DNA genome of a eukaryote is replicated by copying the preexisting parental DNA, portions have also resulted from copying RNA into DNA. One of these genomic components is telomeric DNA, which requires for its maintenance the continued action of a ribonucleoprotein enzyme, telomerase. Telomerase contains an essential RNA component. The telomeric DNA sequence is specified by and generated from copying a template sequence within the telomerase RNA moiety. In this chapter, I discuss the properties of telomerase and how they may illuminate the evolutionary transition from RNA-based to DNA-based genomes.

Telomeres were originally defined functionally as the natural ends of eukaryotic chromosomes, without which a chromosome is unstable (for review, see Blackburn and Szostak 1984; Zakian 1989; Blackburn 1991). Telomeric DNA in the form of very simple, tandemly repeated sequences is a conserved feature of widely diverged eukaryotes. However, there are some interesting apparent exceptions in certain species, which are discussed below. Representative examples of the simple type of telomeric repeated sequences are shown in Table 1. A terminal stretch of this simple-sequence DNA, ranging in length from less than fifty base pairs in some ciliated protozoans, through a few hundred base pairs in yeast and some other lower eukaryotes, to thousands of base pairs in human cells, appears to be sufficient to maintain a stable telomere. As shown in Table 1, these sequences usually result in the telomeric DNA having a strand composition bias. At each chromosome end the strand containing the characteristic clusters of G residues (commonly called the G-strand) runs 5′ to 3′ toward the chromosomal terminus. In the various lower eukaryotic species where analysis has been possible, the G-strand protrudes 12–16 nucleotides beyond the complementary C-rich strand

Table 1 Telomeric repeat sequences in eukaryotes

Telomeric repeat (5′ to 3′ toward chromosome end)	Representative species	Reference[a]
AG_{1-8}	*Dictyostelium discoideum*	
TTAGGGGG	*Cryptococcus*	Edman (1992)
$TG_{2-3}(TG)_{1-3}$	*Saccharomyces cerevisiae*	
TTGGGG	*Tetrahymena thermophila*	
TT(T/G)GGG	*Paramecium aurelia*	
TAGGG	*Giardia lamblia*	Adam et al. (1991)
TTAGGG	*Homo sapiens, Neurospora crassa, Trypanosoma brucei*	
TTTAGGG	*Arabidopsis thallium*	
TTTTAGGG	*Chlamydomonas reinhartdii*	
$TTAC(A)AG_{2-7}$	*Schizosaccharomyces pombe*	
TT(T/C)AGGG	*Plasmodium berghei*	
CTTAGG	*Ascaris lumbricoides*	Muller et al. (1991)
ACGGATGTCTAACTTCTTGGTGT	*Candida albicans*	McEachern and Hicks (1993)

[a]For references, see Blackburn (1990) except where shown otherwise.

(Blackburn and Szostak 1984; Zakian 1989; Blackburn 1991).

It is the G-strand of telomeric DNA that is synthesized in an unusual fashion by the ribonucleoprotein enzyme telomerase: As described below, an RNA sequence within the telomerase RNA is used as the template for synthesis of this strand (for review, see Blackburn 1991). Telomerase, then, is a DNA polymerase that can be classified as a reverse transcriptase, because its mechanism of action involves the copying of an RNA template into DNA.

THE TELOMERASE REACTION

Telomerase activity was first identified in vitro in cell-free extracts of the ciliated protozoan *Tetrahymena thermophila* (Greider and Blackburn 1985). A synthetic G-rich telomeric DNA oligonucleotide representing the overhanging G-strand of the telomere was shown to act as a primer for synthesis of TTGGGG repeats, the telomeric sequence of *Tetrahymena* (Greider and Blackburn 1985, 1987; Blackburn et al. 1989). Telomerase elongates this primer by direct polymerization, in the usual 5′ to 3′ direction, of deoxynucleoside triphosphate precursors into tandem repeats of the TTGGGG sequence. Natural telomeres have also been shown to be elongated by telomerase, both in vitro and in vivo (Henderson and Blackburn 1989; Yu et al. 1990; Yu and Blackburn 1991). Although under various circumstances telomerase elongates oligonucleotides with a variety of sequences, significantly, it recognizes the sequence of the 3′ end of the primer, such that when this 3′ end contains G and/or T residues, the correct next nucleotides are added to complete some cyclical permutation of a perfect TTGGGG repeat that includes the 3′ end of the primer (Fig. 1A) (Greider and Blackburn 1985, 1987, 1989). This observation was important in formulating the mechanism of telomerase, as described below.

Telomerase activity was subsequently identified in isolated nuclei of the ciliates *Oxytricha* and *Euplotes* (Zahler and Prescott 1988; Shippen-Lentz and Blackburn 1989, 1990) and in nuclear and S100 extracts of human tissue-culture (HeLa) cells (Morin 1989). Each of these telomerase activities synthesizes its species-specific G-strand telomeric sequence and has primer-recognition properties analogous to those of the *Tetrahymena* activity. Examples of the elongation reactions by these telomerases are shown schematically in Figure 1.

Telomerase synthesizes only the G-strand of telomeres. A scheme for the complete replication of telomeres incorporating the action of telomerase has been proposed and is shown in Figure 2 for *Tetrahymena* telomeres. In this model, it was proposed that synthesis of the com-

A Tetrahymena

5' ....GGGGTTGGGG ttggggttgggg....   3'

    ....GGGGTTGGG gttggggttgggg....

B. Euplotes or Oxytricha

    ...GGGGTTTT ggggttttggggtttt...

    ...GGGGTT ttggggttttggggtttt...

C. Human

    ...AGGGTTAGGG ttagggttaggg...

    ...AGGGTTAG ggttagggtt...

Figure 1 Examples of typical in vitro elongation reactions by telomerases. Shown are reactions carried out by the telomerase from (*A*) *T. thermophila* (Greider and Blackburn 1987), (*B*) *E. crassus* (Shippen-Lentz and Blackburn 1990) or *O. nova* (Zahler and Prescott 1988), and (*C*) human HeLa cells (Morin 1989). The 3'-end region of the primer supplied to the reaction is shown in uppercase letters, with the nucleotide sequence at the extreme 3' end of the primer and the first nucleotides added by each telomerase to make a complete repeat unit underlined. Added nucleotides are shown as lowercase letters.

plementary C-rich strand is carried out by discontinuous synthesis by primase-polymerase, as is typical of lagging-strand synthesis during semi-conservative DNA replication, using the extended G-rich strand as a template (Shampay et al. 1984). Synthesis of the C-rich strand by a primase-like activity has been found using extracts from the ciliate *Oxytricha* (Zahler and Prescott 1989). Because telomerase activity has been found in species as evolutionarily distant as ciliates and humans, and because the dynamic aspects of in vivo telomere behavior shared by many eukaryotes can be explained through the involvement of telomerase activity (for review, see Blackburn 1990, 1991), it is very likely that this enzyme is widespread in eukaryotes.

IDENTIFICATION OF THE RNA MOIETY OF TELOMERASE

Treatment of telomerase either with micrococcal nuclease or RNase A, or with protease, was found to destroy the enzymatic activity. Thus, the enzyme was deduced to contain RNA and protein components, both of which appear to be essential for its activity in vitro (Greider and Blackburn 1987). Similar findings were subsequently made for telomerase ac-

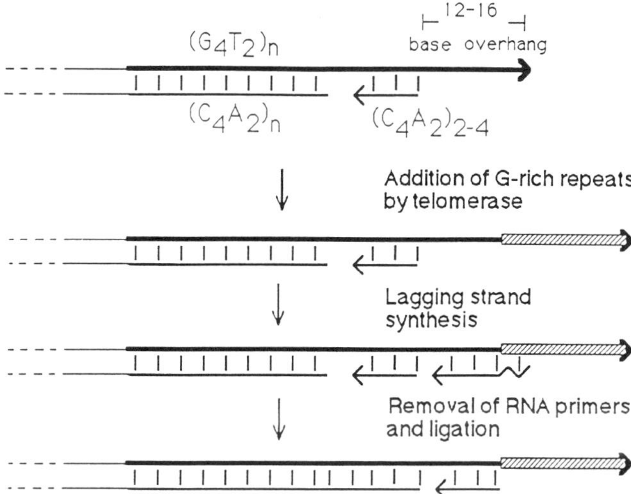

Figure 2 Model for telomere elongation and maintenance by telomerase. The end region of a linear chromosomal DNA molecule of *Tetrahymena* is shown. Telomerase normally adds TTGGGG repeats onto the ends of these chromosomes, counterbalancing the attrition of the ends that is caused by semiconservative replication: Lagging strand synthesis uses an RNA primer (zig-zag) made by primase, which is then elongated by DNA synthesis by DNA polymerase. The lagging strand is synthesized in pieces, which become joined after removal of the RNA primers and DNA polymerization to fill in the resulting internal gap, so that ligation of the resulting internal breaks can occur. However, removal of the 5′-terminal RNA primer leaves a 5′-terminal gap. Thus, without telomerase, repeated rounds of replication would produce increasingly shorter daughter chromosomes.

tivities identified in the ciliates *Oxytricha* and *Euplotes* (Zahler and Prescott 1988; Shippen-Lentz and Blackburn 1989) and in human cells (Morin 1989).

The approximately 160-nucleotide RNA moiety of the telomerase of *T. thermophila* was first identified on the basis of its co-chromatography with telomerase activity and the sensitivity of telomerase activity to ribonuclease H (RNase H) cleavage of a specific region of the RNA (Greider and Blackburn 1987, 1989), as described in more detail below.

The telomerase RNA of the distantly related ciliate *Euplotes crassus* (whose telomeric sequence is T$_4$G$_4$ repeats) was first identified by its sensitivity to cleavage by RNase H directed by a DNA oligonucleotide containing T$_4$G$_4$ repeats (Shippen-Lentz and Blackburn 1990). Additional telomerase RNAs from five other *Tetrahymena* species, and from the related ciliate *Glaucoma*, were identified by cross-hybridization of their single-copy genes to the *T. thermophila* gene (Romero and Blackburn 1991). To date, these ciliate RNAs are the only telomerase RNA sequences that have been identified. However, because the human cell telomerase activity is also ribonuclease sensitive, and its primer recognition properties are very similar to the ciliate telomerase activities, it is almost certain that the human enzyme is also a ribonucleoprotein.

The *Tetrahymena* telomerase RNA, like several other types of small eukaryotic RNAs, was deduced to be an RNA polymerase III transcript on the basis of its lack of a 5′cap, the presence in its gene of a run of T residues encompassing the position of the 3′end of the RNA (Greider and Blackburn 1989), and its sensitivity to α-amanitin in runoff transcription experiments with isolated nuclei (Yu et al. 1990). All the ciliate telomerase RNA genes have a pair of highly conserved upstream sequences and lack a conserved Box A consensus sequence, suggesting that, like U6 RNA, their RNA polymerase III-mediated transcription is regulated by upstream *cis*-acting elements (Shippen-Lentz and Blackburn 1990; Romero and Blackburn 1991). The regulation of telomerase RNA transcription has not been studied. However, when overexpressed in vivo, the steady-state level of accumulated telomerase RNA was unchanged from the situation in untransformed control cells, even though the transcription rate of the overexpressed genes in transformants, as judged by nuclear runoff experiments, was very high. These findings suggested that excess telomerase RNA that was transcribed but not assembled into the telomerase RNA complex was degraded (Yu et al. 1990).

An estimate of the molecular size of the *Tetrahymena* telomerase ribonucleoprotein obtained by gel filtration chromatography was 200–500 kD (Greider and Blackburn 1987). Assuming a stoichiometry of one RNA per telomerase ribonucleoprotein complex, the RNA component accounts for approximately 50 kD. The protein component(s) of telomerase has not yet been identified biochemically. However, a candidate for a telomerase component in yeast is encoded by an essential gene, *EST1* (Lundblad and Szostak 1989). In *est1⁻* mutants, all the telomeres exhibit continuous shortening over the course of several cell generations, leading to senescence. Increased rates of chromosome loss precede cell death. In addition, it is intriguing that the predicted protein sequence of the *EST1* gene product contains four amino acid motifs that are conserved among both viral and nonviral reverse transcriptases, as well as among viral RNA-dependent RNA polymerases (Lundblad and Blackburn 1990). These properties are consistent with the *EST1* product's being an essential telomerase component, because the ciliate telomerases have been shown to be specialized reverse transcriptases, and, as described below, disruption of telomerase RNA function in *Tetrahymena* also leads to telomere loss and cellular senescence.

THE TEMPLATE FUNCTION OF TELOMERASE RNA

In Vitro and Phylogenetic Evidence

A key function for the RNA component of telomerase was suggested by the identification within the *Tetrahymena* telomerase RNA of a 9-nucleotide sequence, 5'-CAACCCCAA-3', which is complementary to the telomeric TTGGGG repeats synthesized by this enzyme (Greider and Blackburn 1989). Evidence suggesting that this sequence acts as the template came from experiments which tested the effect on telomerase activity of RNase H treatment in the presence of DNA oligonucleotides complementary to the RNA sequence. RNase H cleavage directed by a DNA oligonucleotide whose 3'end was complementary to the 5'-CAACCCCAA-3'sequence specifically inactivated telomerase. Since RNase H cleaves RNA where it is base-paired to a complementary DNA sequence, loss of telomerase activity indicated that this sequence was important. Furthermore, this oligonucleotide competed with a $(TTGGGG)_4$ primer for telomerase, suggesting it was blocking access of the primer to a site necessary for telomerase activity. Another antisense DNA oligonucleotide whose 3'end was complementary to the adjacent sequence in telomerase RNA, but ended one nucleotide from the 3'end of the putative template sequence, was also extended by GGGGTT repeat addition, suggesting that base-pairing of a DNA oligonucleotide to the telomerase RNA in the vicinity of the template could allow the oligonucleotide to be utilized as a primer. From these results, it was proposed that the 5'-CAACCCCAA-3'sequence in the *Tetrahymena* telomerase RNA is the template sequence (Greider and Blackburn 1989). Figure 3A shows the model proposed for the mechanism of *Tetrahymena* telomerase that incorporates these findings.

This model rationalized the precise recognition of the 3'end sequence of the substrate primer described above and was further supported by the identification and functional analysis of the RNA moiety of telomerase from the ciliate *E. crassus* (Shippen-Lentz and Blackburn 1990). This enzyme synthesizes T_4G_4 repeats (see Fig. 1). A prediction of the model was that telomerase RNAs from other species would each have a template sequence corresponding to its species-specific telomeric repeat sequence. The *Euplotes* telomerase RNA was identified as an approximately 190-nucleotide nuclear RNA species that was cleaved specifically by RNase H directed by a G-rich telomeric DNA oligonucleotide. It was

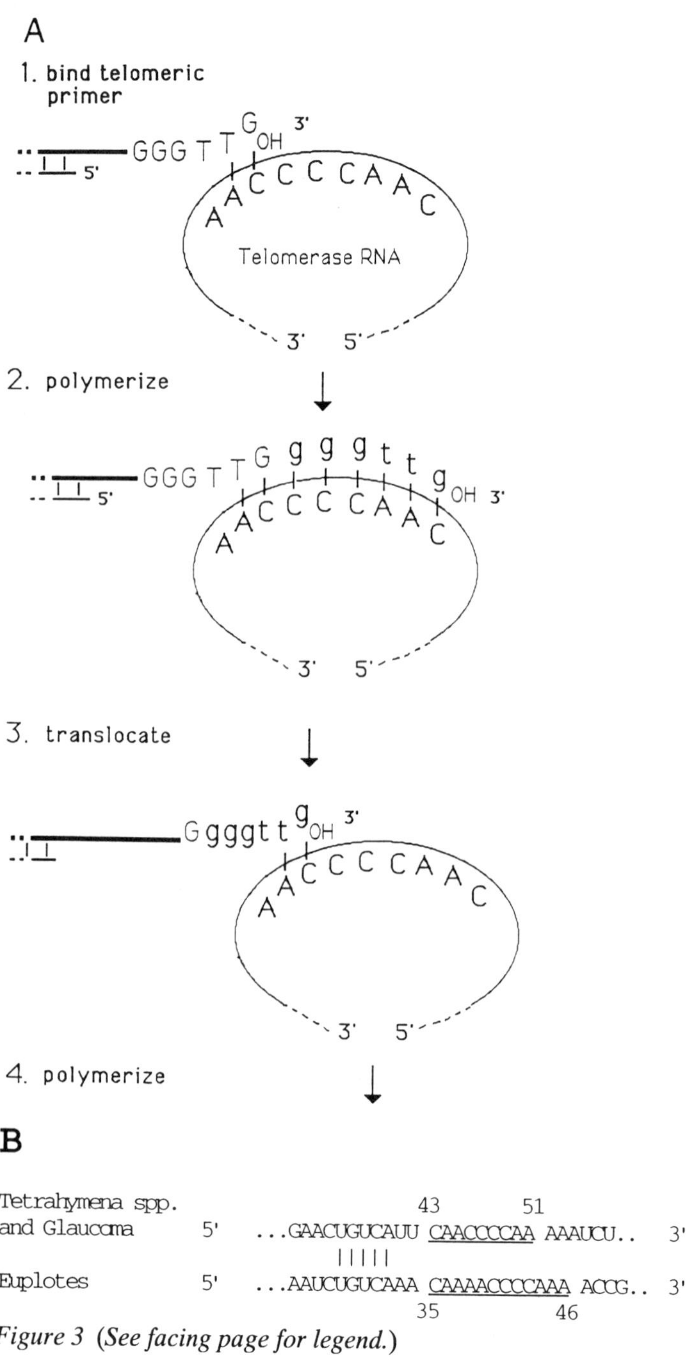

Figure 3 (See facing page for legend.)

shown to contain a sequence, 5'-CAAAACCCCAAAA-3', analogous to the proposed template sequence of the *T. thermophila* RNA. Independent evidence for the templating function of this sequence for synthesis of GGGGTTTT repeats was obtained by in vitro experiments done with the telomerase activity from *Euplotes* (Shippen-Lentz and Blackburn 1990). A series of DNA oligonucleotides complementary to the region of the *Euplotes* telomerase RNA 3' of the putative templating domain, and whose 3' ends extended toward (or varying numbers of nucleotides into) the putative template region, were tested for their ability to prime telomerase. In a manner reminiscent of the situation with the *Tetrahymena* enzyme, with which an oligonucleotide complementary to the region 3' of its template was able to prime repeat addition (Greider and Blackburn 1989), these complementary oligonucleotides competed with a telomeric G-rich DNA primer, suggesting that they interacted with a similar region of the *Euplotes* telomerase. Furthermore, by determining which positions of the 3' ends of these antisense oligonucleotides allowed priming of telomeric repeat addition, the functional template domain of the *Euplotes* RNA was demarcated as shown in the underlined sequence in Figure 3B (Shippen-Lentz and Blackburn 1990). By analogy and extension of the model for telomere synthesis deduced from the ciliate telomerases, a possible template sequence, 5'-CUAACCCUA-3' (Morin 1991), has been suggested for the human telomerase RNA, which synthesizes TTAGGG repeats, although this RNA has not yet been identified.

Figure 3 (*A*) The mechanism of telomeric DNA synthesis by the ribonucleoprotein enzyme telomerase from *Tetrahymena*. (*1*) The 3' few nucleotides of the terminal chromosomal G-rich overhang (see Fig. 1) (thick line; shown arbitrarily as ending in ..TTG 3') base-pair with the telomere-complementary sequence in the telomerase RNA. (*2*) The chromosomal end is extended by polymerization of dGTP and dTTP using the RNA as a template, resulting in the addition of six telomeric nucleotides. (*3*) The extended DNA terminus unpairs from its RNA template, becoming available for (*4*) another round of elongation by telomerase, and/or to primase-polymerase, which uses it in turn as the template for lagging-strand synthesis of the C-rich telomeric strand. (Adapted from Greider and Blackburn 1989.) (*B*) Sequences of the template domain and its surrounding region in the known telomerase RNAs. (*Top*) Telomerase RNAs from *Tetrahymena spp.* and *Glaucoma* (Greider and Blackburn 1989; Romero and Blackburn 1991). (*Bottom*) *Euplotes* telomerase RNA (Shippen-Lentz and Blackburn 1990). Underlined sequences have been shown by in vitro or in vivo studies to be the templating domain. Vertical lines indicate matches between the sequences outside each templating domain, when aligned with reference to the template domain.

Does telomerase have an RNase H activity associated with it, as is typical of conventional reverse transcriptases? During in vitro reactions with the *Tetrahymena* or *Euplotes* enzymes, the overall incorporation rate at saturating primer concentrations typically remains linear for about 30 minutes. Thus, each ribonucleoprotein telomerase molecule carries out many cycles of templated elongation, implying that the template RNA is not destroyed in the course of the extended reaction. Therefore, unlike other reverse transcriptases, telomerase is unlikely to have an associated RNase H activity.

As shown in the model in Figure 3A, up to 9 Watson-Crick base pairs of an RNA-DNA helix could exist at the end of a cycle of templated synthesis on the RNA template of the *Tetrahymena* telomerase. With the *Euplotes* telomerase, from results obtained in vitro using primers complementary to the region 3′ of the template and extending into the template, it was deduced that a minimum of 11 base pairs has to be dissociated after the initial round of elongation of such a primer, and 8 base pairs must be dissociated in each subsequent elongation round, as each GGGGTTTT repeat is added (Shippen-Lentz and Blackburn 1990). These observations raise the question of whether such a long helix is in fact made. If so, unwinding it is expected to require energy. Telomerase activity in vitro has no requirement for deoxynucleoside triphosphates other than those incorporated into the DNA synthesized, or for ribonucleoside triphosphates such as ATP. However, energy could be supplied by one of the deoxynucleoside triphosphate substrates of telomerase, and unwinding could be accomplished by a helicase. The question of whether a helicase activity is either intrinsic to, or associated with, telomerase should be investigated.

In Vivo Demonstration of the Template Function of Telomerase RNA

A prediction of the model shown in Figure 3A is that alteration of the template RNA sequence should result in the synthesis of telomeric DNA with the corresponding altered sequence. This prediction was confirmed by experiments done in vivo (Yu et al. 1990; Yu and Blackburn 1991). The template sequence of the cloned *T. thermophila* telomerase RNA gene was mutated by site-directed mutagenesis to produce different templates: one with an additional C residue, converting the CCCC sequence to CCCCC, which was predicted to specify the synthesis of G_5T_2 repeats, and another with the A at position 44 (see Fig. 3B) substituted by G (predicting G_4TC repeats). Each mutated telomerase RNA gene was expressed in vivo by transformation of *Tetrahymena* with the altered gene. Expression of either of these two altered telomerase RNA genes resulted in the in vivo synthesis of telomeric DNA whose sequence corresponded to that of the altered template. These in vivo results proved that the 5'-CAACCCCAA-3' sequence in the *Tetrahymena* telomerase RNA gene is the template for telomere synthesis.

Primary and Secondary Structures of Telomerase RNA: Other Roles for the RNA?

The findings described above established that telomerase is an unusual reverse transcriptase that carries its own internal RNA template for DNA synthesis. If the telomerase RNA has other functions besides that of acting as a template, it is expected to contain conserved structural and/or sequence features. Telomerase RNAs from five *Tetrahymena* species in addition to *T. thermophila*, and from the related tetrahymenine ciliate *Glaucoma chattoni*, have been identified. In each of these species, the approximately 160-nucleotide telomerase RNA is encoded by a single gene (Romero and Blackburn 1991). Among the *Tetrahymena* group of RNAs, the maximum overall sequence divergence is about 35%, allowing them to be aligned with each other and allowing compensatory base-pair changes in putative secondary structures to be identified. Such comparison of the sequences of these RNAs allowed a phylogenetically supported secondary structure model to be deduced. Although primary sequences have diverged rapidly overall among these RNAs, they appear to share a strikingly conserved secondary structure (Romero and Blackburn 1991), shown schematically in Figure 4. Recently, on the basis of phylogenetic comparison and covariation of these sequenced tetrahymenine telomerase RNAs, a pseudoknot structure has been proposed for them (Ten Dam et al. 1991); this feature is incorporated into Figure 4. The degree of secondary structure conservation in the face of the rapidly diverging primary sequences in these RNAs contrasts with the situation seen with, for example, the sequences of the similarly sized 5.8S RNA in this same group of species, whose maximum differences include only two single-base changes (Preparata et al. 1989).

Excluding the template, no long stretches of primary sequence have been found to be universally conserved in all ciliate telomerase RNAs. The RNAs from *Tetrahymena spp.* and *Glaucoma* all contain a 22-base sequence centered on the template sequence that is absolutely conserved among this group of ciliates (all these tetrahymenine ciliates have the same telomeric repeat sequence, TTGGGG). However, the only sequence that is obviously conserved between the tetrahymenine group and the distantly related ciliate *Euplotes* is a 5-base sequence two nucleotides

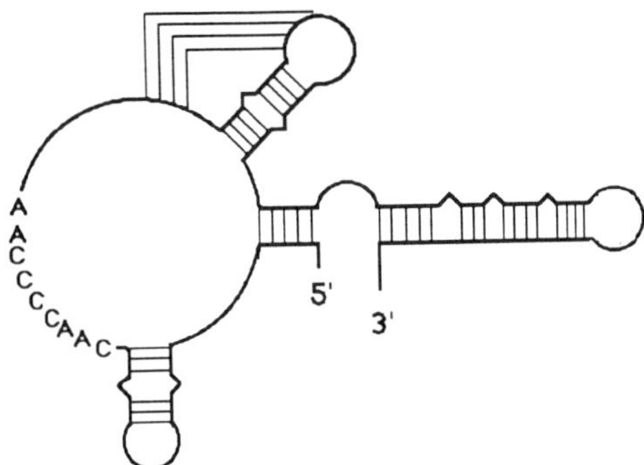

Figure 4 Secondary structure model for telomerase RNAs from *Tetrahymena spp.* and *Glaucoma*. The bases in the template sequence are shown as letters; the rest of the RNA is indicated as the thicker line. Thin lines represent base pairs in secondary structural features whose existence is supported by covariation among these RNAs (Romero and Blackburn 1991; Ten Dam et al. 1991): These include three bulged hairpin stem-loop structures, a long-range pairing close to the 5′ end, and a pseudoknot (indicated by the right-angled lines).

5′ to the template (Fig. 3B). Apart from the template region, however, a reliable alignment cannot be made between the approximately 190-base *Euplotes* and the approximately 160-base *Tetrahymena* telomerase RNAs, because the primary sequences of these more distantly related RNAs are highly diverged.

Whether the telomerase RNA has any catalytic or other function in addition to its known template function has yet to be determined. The strong secondary structural conservation of telomerase RNA, at least among the tetrahymenine ciliates, is one indication that the telomerase RNA has other roles besides simply providing a built-in template. In addition, intriguing results from in vivo analysis of the function of telomerase RNAs altered by site-directed mutations support this possibility. In experiments like those described above which established the template function of the *Tetrahymena* telomerase RNA in vivo, other template mutations, in which the C at position 48 (see Fig. 3B) is substituted by T or G (predicting GAG_2T_2 and GCGGTT repeats, respectively) have been tested. In contrast to the results in which the mutated template directed the synthesis of the expected telomeric repeat sequences, the position-48 mutations each resulted in telomere shortening and loss and a severe cell-senescence phenotype (Yu et al. 1990; J. Bradley and E.H. Blackburn, unpubl.). These phenotypes were not seen in control transformants that received either the wild-type gene or some other telomerase RNA gene mutants with different single-base changes. Furthermore, the expected mutated repeat sequence was not detectable in the telomeric DNA of these cells. These results constituted direct evidence that telomerase is essential for the long-term viability of *Tetrahymena* and that its continued action is necessary to prevent cellular senescence. Moreover, they suggest the possibility that the C residue at position 48, although it is part of the template, also plays a special base-specific role in the telomerase reaction, for which another base cannot substitute. This result is also interesting in light of evidence suggesting a possible base modification of C residues 46, 47, and 48: At these positions, the polynucleotide backbone is resistant to cleavage with RNases, although these positions show normal sensitivity to alkaline hydrolysis (Greider and Blackburn 1989). Thus, some residues in the telomerase RNA may have a function other than templating.

PRIMER SPECIFICITY OF TELOMERASE

Telomeric DNA sequences in most eukaryotic groups examined so far are characterized by a widely conserved feature: clusters of G-residues on the strand that is synthesized and extended by telomerase (see Table 1). The efficient utilization of G-rich telomeric DNA sequences, as opposed to other G+C-rich sequences, was one of the first obvious and striking features of telomerase activity that was noticed. Telomerase can also elongate oligonucleotides with a variety of sequences, with a wide range of efficiencies (Greider and Blackburn 1985, 1987, 1989; Zahler and Prescott 1988; Blackburn et al. 1989; Morin 1989, 1991; Shippen-Lentz and Blackburn 1990; Yu et al. 1990; Harrington and Greider 1991; Yu and Blackburn 1991). The properties of a DNA primer that affect its efficiency as a substrate by telomerase are complex and are still in the process of being dissected. In vitro, the enzyme works most efficiently when provided with an oligonucleotide primer whose 3′ end can at least minimally base-pair with the template region, consistent with the model shown in Figure 3A. However, other DNA oligonucleotides, primarily, but not limited to, those that contain a G-rich sequence somewhere along their length, can also function well as primers in vitro. In vivo as well as in vitro, telomerase recognizes various telomeric sequences that are different from those specified by the enzyme. The addition of yeast telomeric repeats onto *Tetrahymena* telomeres introduced into yeast, one of the lines of evidence that originally suggested the existence of telomerase (Shampay et al. 1984), was proposed to involve such recognition. Such in vivo recognition has been shown directly by experiments in *Tetrahymena* in which telomerases with altered template sequences added the corresponding altered telomeric sequences onto wild-type repeats, and vice versa (Yu et al. 1990; Yu and Blackburn 1991).

In several different eukaryotes, sequence analysis has shown that broken chromosome ends are often healed in vivo by addition of telomeric repeats directly onto nontelomeric DNA sequences (for review, see Blackburn 1991; Muller et al. 1991). In *Tetrahymena*, it has been demonstrated that such in vivo healing is accomplished by telomerase (Yu and Blackburn 1991). In these experiments, a telomerase RNA gene with an altered template was introduced into *Tetrahymena* cells, which, as described above, specifies synthesis in vivo of telomeres containing the corresponding altered sequence. This altered gene can be used as a marker for the action of telomerase in vivo. When cells expressing the altered telomerase RNA gene underwent a developmentally programed genomic rearrangement that involves de novo addition of telomeric DNA to nontelomeric sequences, the altered telomeric repeats were added directly to the nontelomeric end sequences. This result showed that, in this situation, telomerase adds the first as well as subsequent telomeric repeats to the broken end.

TELOMERASE AND G-RICH TELOMERIC DNA

One explanation for the conservation of clusters of G residues on one strand of telomeric DNA in many eukaryotes (see Table 1) is that there are (as yet unknown) specific structural requirements for the functioning of telomeric DNA that are uniquely met by clusters of G or C residues on one or the other strand. However, as seen in Table 1, it is noteworthy that the repeated telomeric sequence in the opportunistic pathogenic yeast *C. albicans* deviates significantly from the short telomeric repeats documented in a great many other eukaryotes (McEachern and Hicks 1993): First, the length of each repeat unit (23 bp) far exceeds that of typical eukaryotic telomeric repeat sequences (9 bp at most), and second, it lacks the prominent bias toward G residues on the DNA strand at each 3′ end of the chromosome characteristic of the other sequences shown in Table 1. The telomeric repeat sequences of several budding yeasts related to *C. albicans* have recently been identified, and their unit lengths vary from 23 bp in *C. albicans* and closely related yeasts and 25 bp in some other budding yeasts, through 16- and 8-bp repeats (M. McEachern and E.H. Blackburn, in prep.), down to the short, irregular G-rich telomeric repeat sequences of *S. cerevisiae* (see Table 1). A second type of unusual telomeric sequence may also occur in the insect *Drosophila melanogaster*. In this organism, broken chromosomes can become healed by the addition of a class of apparently retroelement-derived sequences called HeT-A sequences (Biessmann et al. 1990). HeT-A sequences are clearly derived from a common sequence, but they are complex and bear no obvious relation to the sequences shown in Table 1 (Valgeirsdottir et al. 1990). HeT-A sequences have been identified at the cytological level (by in situ hybridization) at the terminal regions of all *D. melanogaster* chromosomes (Valgeirsdottir et al. 1990). No conventional short-repeat type of telomeric sequence has been identified in this species. The de novo addition of HeT-A sequences to broken chromosome ends has been shown to occur in two distinct situations in *D. melanogaster*: a broken end induced by movement of a transposon (Biessmann et al. 1990) and the new ends created by opening of a ring X chromosome (Traverse and Pardue 1988). These independent events suggest that HeT-A sequences may serve a telomeric role in this species.

Telomeric repeats of the typical very short repeat type are therefore dispensable in the *Candida* yeasts and also perhaps in *Drosophila*. This raises the question of whether the G-richness that is such a widespread feature of telomeric DNAs in other species has been conserved for reasons other than the functioning of telomeric DNA itself. A possible clue comes from comparison of the telomeric sequences of *C. albicans* and the eight other budding yeasts whose telomeric sequences have been determined (M. McEachern and E.H. Blackburn, in prep.). Strikingly, all nine different repeat sequences contain a conserved TGGTGT consensus sequence, imbedded in telomeric unit repeat sequences that are otherwise completely divergent in length, composition, and sequence (M. McEachern and E.H. Blackburn, in prep.). This G-containing motif, together with the observations that mutating the C residue at position 48 in the template of the *Tetrahymena* telomerase RNA appears to inhibit telomere synthesis in vivo (Yu et al. 1990) and that this particular C residue is one of the three C residues that appear to be modified (Greider and Blackburn 1989), hint at a possible mechanistic reason for the occurrence of G-clusters in telomeric DNA: They are simply the consequence of a requirement for one or more C residues in the template region of telomerase RNA.

From RNA Replicase to Telomerase to Retroviruses?

Telomerase is an unusual reverse transcriptase because, unlike the conventional purely protein reverse transcriptases found in systems ranging from retroviruses to prokaryotes, it is a ribonucleoprotein that contains an RNA template as an integral part of the enzyme. The apparent conservation of telomerase in the form of a ribonucleoprotein in widely divergent eukaryotes suggests that this enzyme was already present at the time of the separation of eukaryotes from other kingdoms. This, and the fact that the RNA component acts as a template for DNA synthesis, prompt the speculation that telomerase may be a relic from a time of the evolutionary transition from RNA to DNA genomes. Specifically, the striking ability of the group I self-splicing rRNA intron ribozyme of *Tetrahymena* to polymerize RNA using an internal RNA template (Zaug and Cech 1986) suggests the possibility that the RNA moiety of telomerase is a relic of an RNA replicase ribozyme that acquired the ability to synthesize DNA.

The existence of long telomeric repeats in *C. albicans* and certain other budding yeasts suggests that *Candida* has a telomerase that copies a significantly longer region of its telomerase RNA than the telomerases of those eukaryotes with the more typical short telomeric repeats. Does the *Candida* type of telomeric repeat represent a more primitive state? If an ancestor of telomerase RNA was an RNA replicase ribozyme that could in principle be copied completely, the long *Candida* type of repeat may represent an intermediate stage in the evolution of telomerase toward enzymes that copy only very short regions of their RNA. By this model, it can be imagined that during evolution progressively shorter regions of the RNA were copied, so that most organisms have now evolved to the point where only a minimal region of the telomerase RNA, characterized by one or more Cs, is used as the template for reverse transcription.

Did retroelements and retroviruses evolve from telomerase, or vice versa? It can be envisaged that a self-replicating RNA, the antecedent of telomerase RNAs, became associated with protein. The protein could, in the case of the purely protein reverse transcriptases found in modern eukaryotes and prokaryotes, have taken over the catalytic function of the telomerase RNA. By this model, some or all of the original self-replicating RNA was retained in telomerase. This RNA may, by analogy with group I self-splicing RNAs, have a role in the binding of the primer substrate besides base-pairing, or it may have a catalytic role, either by itself or in conjunction with the protein. For example, telomerase RNA residues may collaborate with amino acid residues of the telomerase protein component in forming an active site. In any case, the telomerase RNA still serves (minimally) a templating role. It is worth noting that the reverse transcriptase of duck hepatitis B virus cannot use an exogenously supplied template RNA in vitro, but can use only the endogenous RNA template within the hepatitis viral core particles (Radziwill et al. 1988). An interesting possibility is that this particular reverse transcriptase represents an intermediate situation between telomerase and more conventional reverse transcriptases, in which the viral template RNA is still closely associated with the protein enzyme.

In *D. melanogaster*, the HeT-A elements that are added to the ends of broken chromosomes have all the hallmarks of having been synthesized by reverse transcription of (often) defective retroelements. Perhaps this situation represents an exploitation of a reverse transcriptase reaction to carry out a role usually accomplished by telomerase: the addition of DNA to the chromosome ends to balance the loss of terminal DNA sequences through incomplete replication, as shown in Figure 2. The elongation of broken (and perhaps all) chromosome ends in *D. melanogaster* is exactly analogous to the action of telomerase. Thus, the essential difference between *D. melanogaster* and more typical eukaryotes may be only that in *D. melanogaster* the added DNA repeats are synthesized by copying a substantial part of the retroelement RNA, so that adding DNA segments to broken or other chromosome ends occurs in large increments, and therefore addition needs to occur only infrequently. On the other hand, in most eukaryotic species the typical telomerase copies only a very short region of RNA, so that lengthening of the chromosome occurs in small increments and thus must occur in most or all cell divisions. It can be imagined that in the presence of a sufficiently efficient HeT-A mechanism, the requirement for telomerase was bypassed in *Drosophila* and telomerase was lost.

It cannot be excluded that telomerase is derived from reverse transcriptase and not vice versa. In this alternative model for telomerase evolution, essentially the converse of the one described above, telomerase evolved from a reverse transcriptase that remained associated with an RNA, part of which is copied into DNA. Synthesis of HeT-A elements onto the ends of *D. melanogaster* chromosomes would then represent an early stage in this evolution, rather than an aberration. Perhaps for reasons of efficiency, in most organisms the trend was toward shorter regions being copied into DNA. By this model, the *Candida* type of telomere would again represent the more primitive state, in which a longer portion of the RNA was copied. It is intriguing to consider the reverse transcriptases that have been found in the prokaryotes *Myxobacter* and *Escherichia coli* in this light. In these cases, the reverse transcriptase gene is tightly linked to a region encoding an RNA (msRNA), which apparently folds so that it can prime synthesis of a DNA copy of a portion of itself by reverse transcription. The 5′ end of the DNA becomes joined, through a 2′-5′ phosphodiester linkage, to a priming rG residue at an internal position of the msRNA (see Fig. 2 in Atkins, this volume). This reverse transcriptase, in common with many others, includes an RNase H activity, which destroys the msRNA template region as the DNA is synthesized. It is possible that current telomerases were derived from such a reverse transcriptase and an ms-like RNA. In telomerase the priming function would now be provided by DNA, and the RNase H activity would be lost. Although either type of model is possible, evolution of telomerase from reverse transcriptase does not suggest an explanation for the conserved clusters of G residues in telomeric repeats.

CONCLUSIONS AND PERSPECTIVES

In conclusion, I return to the question posed at the beginning of this chapter: What was required to make a complete transition to DNA genomes? It is striking that the addition of typical short telomeric repeats and HeT-A sequences involves reverse transcription in both cases. It is tempting to propose that the spectrum of telomeric sequences synthesized by reverse-transcriptase-like activities including telomerase is wider than had been previously supposed and ranges from the very short telomeric repeat units characteristic of most eukaryotes; through the *Candida* telomeric repeat units, representing an intermediate situation; to the *D. melanogaster* HeT-A elements. Once reverse trancriptase activities had evolved, the transition from RNA to DNA genomes may not have required a profound change in terms of enzymatic properties. Reverse transcriptase seems likely to have mediated the evolution from RNA to DNA genomes. The structures of the catalytic DNA polymerase domains of the reverse transcriptase from the HIV retrovirus and of DNA polymerase I from *E. coli* have been determined by X-ray crystallography, revealing a striking structural similarity between these two active sites (Kohlstaedt et al. 1992). The hypothesis that RNA preceded DNA as the genetic material suggests that DNA polymerases have evolved from reverse transcriptases.

ACKNOWLEDGMENTS

I thank Mike McEachern and Renata Gallagher for helpful discussions of the manuscript, the National Institutes of Health for support, and the Markey Charitable Trust for providing laboratory facilities.

REFERENCES

Adam, R.D., T.E. Nash, and T.E. Wellems. 1991. Telomeric location of Giardia rDNA genes. *Mol. Cell. Biol.* **11:** 3326–3330.

Biessmann, H., J.M. Mason, K. Ferry, M. d'Hulst, K. Valgeirsdottir, K.L. Traverse, and M.L. Pardue. 1990. Addition of telomere-associated HeT DNA sequences "heals" broken chromosome ends in *Drosophila. Cell.* **61:** 663-673.

Blackburn, E.H. 1990. Telomeres: Structure and synthesis. *J. Biol. Chem.* **265:** 5919–5921.

———. 1991. Structure and function of telomeres. *Nature* **350:** 569–573.

Blackburn, E.H. and J.W. Szostak. 1984. The molecular structure of centromeres and telomeres. *Annu. Rev. Biochem.* **53:** 163–194.

Blackburn, E.H., C.W. Greider, E. Henderson, M.S. Lee, J. Shampay, and D. Shippen-Lentz. 1989. Recognition and elongation of telomeres by telomerase. *Genome* **31:** 553–560.

Edman, J.C. 1992. Isolation of telomerelike sequences from *Cryptococcus neoformans* and their use in high-efficiency transformation. *Mol. Cell. Biol.* **12:** 2777–2783.

Greider, C.W. and E.H. Blackburn. 1985. Identification of a specific telomere terminal transferase activity in *Tetrahymena thermophila* extracts. *Cell* **43:** 405–413.

———. 1987. The telomere terminal transferase of *Tetrahymena* is a ribonucleoprotein enzyme with two distinct primer specificity components. *Cell* **51:** 887–898.

———. 1989. A telomeric sequence in the RNA of *Tetrahymena* telomerase required for telomere repeat synthesis. *Nature* **337:** 331–337.

Harrington, L.A. and C.W. Greider. 1991. Telomerase primer specificity and chromosome healing. *Nature* **353:** 451–454.

Henderson, E.R. and E.H. Blackburn. 1989. An overhanging 3′ terminus is a conserved feature of telomeres. *Mol. Cell. Biol.* **9:** 345–348.

Kohlstaedt, L.A., J. Wang, J.M. Friedman, P.A. Rice, and T.A. Steitz. 1992. Crystal structure at 3.5 Å resolution of HIV-1 reverse transcriptase complexed with an inhibitor. *Science* **256:** 1783–1790.

Lundblad, V. and E.H. Blackburn. 1990. RNA-dependent polymerase motifs in EST1: Tentative identification of a protein component of an essential yeast telomerase. *Cell* **60:** 529–530.

Lundblad, V. and J.W. Szostak. 1989. A mutant with a defect in telomere elongation leads to senescence in yeast. *Cell* **57:** 633–643.

McEachern, M. and J. Hicks. 1993 Unusually large telomeric repeats in the yeast *Candida albicans*. *Mol. Cell. Biol.* **13:** 551–560.

Morin, G.B. 1989. The human telomere terminal transferase enzyme is a ribonucleoprotein that synthesizes TTAGGG repeats. *Cell* **59:** 521–529.

———. 1991. Recognition of a chromosome truncation site associated with α-thalassaemia by human telomerase. *Nature* **353:** 454–456.

Muller, F., C. Wicky, A. Spicher, and H. Tobler. 1991. New telomere formation after developmentally regulated chromosomal breakage during the process of chromatin diminution in *Ascaris lumbricoides*. *Cell* **67:** 815–822.

Preparata, R.M., F.P. Meyer, E.M. Simon, C.R. Vossbrinck, and D.L. Nanney. 1989. Ciliate evolution: The ribosomal phylogenies of the tetrahymenine ciliates. *J. Mol. Evol.* **28:** 427–441.

Radziwill, G.H., H. Zentgraf, H. Schaller, and V. Bosch. 1988. The duck hepatitis B virus DNA polymerase is tightly associated with the viral core structure and unable to switch to an exogenous template. *Virology* **163:** 123–132.

Romero, D. and E. Blackburn. 1991. A conserved secondary structure for telomerase RNA. *Cell* **67:** 343–353.

Shampay, J., J.W. Szostak, and E.H. Blackburn. 1984. DNA sequences of telomeres maintained in yeast. *Nature* **310:** 154–157.

Shippen-Lentz, D. and E.H. Blackburn. 1989. Telomere terminal transferase activity in the hypotrichous ciliate *Euplotes crassus*. *Mol. Cell. Biol.* **9:** 2761–2764.

———. 1990. Functional evidence for an RNA template in telomerase. *Science* **247:** 546–552.

Ten Dam, E., A. van Belkum, and K. Pleij. 1991. A conserved pseudoknot in telomerase RNA. *Nucleic Acids Res.* **19:** 6951.

Traverse, K.L. and M.L. Pardue. 1988. A spontaneously opened ring chromosome of *Drosophila* has acquired He-T sequences at both new telomeres. *Proc. Natl. Acad. Sci.* **85:** 8116–8120.

Valgeirsdottir, K., K.L. Traverse, and M.L. Pardue. 1990. HeT DNA: A family of mosaic repeated sequences specific for heterochromatin in *Drosophila melanogaster*. *Proc. Natl. Acad. Sci.* **87:** 7998–8002.

Yu, G.-L. and E.H. Blackburn. 1991. Developmentally programmed healing of chromosomes by telomerase in *Tetrahymena*. *Cell* **67:** 823–832.

Yu, G.-L., J.D. Bradley, L.D. Attardi, and E.H. Blackburn. 1990. *In vivo* alteration of telomere sequences and senescence caused by mutated *Tetrahymena* telomerase RNAs. *Nature* **344:** 126–132.

Zahler, A.M. and D.M. Prescott. 1988. Telomere terminal transferase activity in the hypotrichous ciliate *Oxytricha nova* and a model for replication of the ends of linear DNA molecules. *Nucleic Acids Res.* **16:** 6953–6972.

———. 1989. DNA primase and the replication of the telomeres in *Oxytricha nova*. *Nucleic Acids Res.* **17:** 6299–6317.

Zakian, V.A. 1989. Structure and function of telomeres. *Annu. Rev. Genet.* **23:** 579–604.

Zaug, A. and T.R. Cech. 1986. The intervening sequence RNA of *Tetrahymena* is an enzyme. *Science.* **231:** 470–475.

23

The Genomic Tag Hypothesis: Modern Viruses as Molecular Fossils of Ancient Strategies for Genomic Replication

Nancy Maizels and Alan M. Weiner
Department of Molecular Biophysics and Biochemistry
Yale University School of Medicine
New Haven, Connecticut 06510

INTRODUCTION

tRNA Plays a Surprising Number of Roles in Replication

First impressions are often lasting, and it is probably safe to say that most molecular biologists first encounter tRNA as a key component of the translation machinery. This is how tRNA is presented in elementary courses, and this is how the molecule is portrayed in textbooks. Yet because tRNA is commonly introduced as a component of the translation machinery, it is all too easy to come to think of translation as the *primary* or *proper* function of tRNA. In fact, as we discuss in detail, tRNA and tRNA-like molecules also play key roles in a wide variety of replicative processes, including replication of single-stranded RNA viruses of bacteria, plants, and possibly mammals; replication of duplex DNA plasmids of fungal mitochondria; retroviral replication; and replication of modern chromosomal telomeres.

How did tRNA come to have so many different roles in replication? One possibility is that, for reasons which are not yet understood, tRNA or tRNA-like structures have been repeatedly and independently borrowed from translation to serve ad hoc roles in replication. Alternatively, tRNA or tRNA-like structures may be widespread in contemporary replication because tRNA played a central role in the replication of ancient RNA genomes, a role which has been conserved as well as subtly transformed as genomes evolved from RNA to duplex DNA.

We have proposed that tRNA-like structural motifs first evolved as 3′-terminal structures that tagged RNA genomes for replication in the RNA world before the advent of protein synthesis (Weiner and Maizels 1987). This hypothesis provides a natural explanation for the ubiquity of tRNA-like motifs by suggesting that tRNA-like structures arose early and played an essential role in the earliest replicating systems. The central role of the early tRNA-like structures in replication would make it likely that the structural motif was conserved throughout subsequent evolution, and the antiquity of the motif would assure that it was used and reused in many different ways. This simple "genomic tag" hypothesis has surprising power to organize many apparently unrelated aspects of molecular biology into a coherent whole and to suggest new relationships between areas of research that appear superficially to be unconnected.

Molecular Fossils, Coevolution, and Continuity

The genomic tag hypothesis implies that molecular evolution has been so conservative as to preserve a role for tRNA-like structures in replication through 3.5 billion years of genomic evolution. Given the sweep of evolution—from the first hopeful prebiotic smudge to the glories and follies of humankind—it may be tempting to argue that such transformations tend to obliterate the molecular evidence, with the result that the most ancient aspects of cellular structure would be *least* likely to survive in recognizable form. However, the more that is understood about the full complexity of the biological machinery, the clearer it becomes that the components of a living cell interact in so many ways that a change in any key component may require compensatory changes in many others. Thus, most molecules cannot freely evolve to maximize the efficiency of a single function, but must instead *coevolve* with other interacting molecules. When a significant change in one molecule would entail an impossibly large number of simultaneous compensatory changes in others, then the necessity of coevolution can effectively freeze a molecule in time. It was clear even to Lucretius that "Natura non saltus fecit" (Nature never makes a leap), and more recently, Orgel (1968) reformulated this old idea as the principle of continuity.

Skeletons, shells, feathers, leaves, and wood are preserved as fossils by death and mineralization. Molecular structures and functions are preserved because the complexity of the living process retards or prevents further significant evolution. White (1976) appears to have been the first to apply the term "fossil" to biochemical processes. We define a *molecular fossil* as any contemporary structure or function which is ancient in origin and provides us with clues about the history of life. We argue here that the ubiquity and conservation of tRNA-like structures strongly suggest that this motif defines a molecular fossil record of events dating back to the beginnings of life on earth.

The Explanatory Power of the Hypothesis

Molecular biologists are accustomed to experiments that produce instant yes or no answers. The genomic tag hypothesis does not suggest such experiments and thus, like most other ideas in evolutionary biology, it cannot easily be falsified. Nonetheless, during the past 5 years, a remarkable variety of unanticipated experimental results have emerged that directly support, were predicted by, or are consistent with the original genomic tag hypothesis (Weiner and Maizels 1987). Popper (1963) coined the term "explanatory power" to describe the ability of a hypothesis to account for previously unrelated and apparently disparate observations. Not only does the genomic tag hypothesis appear to possess considerable explanatory power, but the hypothesis also appears to be robust, comfortably making sense of new data rather than struggling to accommodate unwelcome experimental results. As discussed below, these new results include cleavage of modern genomic tags by RNase P (Green et al. 1988; Guerrier-Takada et al. 1988; Mans et al. 1990); specific binding of an amino acid by RNA (for review, see Yarus et al. 1991); identification of major tRNA identity elements within the acceptor stem (Musier-Forsyth and Schimmel 1992); division of tRNA synthetases into at least two unrelated classes (Eriani et al. 1990); a role for tRNA as *template* for reverse transcription of a retroplasmid genome (Akins et al. 1989; Saville and Collins 1990); the existence of an internal tRNA-like template in telomerase (for review, see Blackburn 1991); and, most recently, RNA catalysis of reactions at an aminoacylphosphoester center (Piccirilli et al. 1992) and catalysis of the ribosomal peptidyl transferase reaction by ribosomal RNA (Noller et al. 1992). We take the resilience and predictive power of the hypothesis as evidence that it has been fruitful and may be substantially correct.

THE GENOMIC TAG HYPOTHESIS

The Hypothesis in Outline

We proposed that ancient linear RNA genomes possessed 3′-terminal tRNA-like structures, which we called genomic tags (Weiner and Maizels 1987). Figure 1 shows the simplest form of a tRNA-like genomic tag, a stem and loop with a 3′-terminal CCA. This resembles what is sometimes called the top half of tRNA, a coaxial stack of the acceptor stem on the TψC arm (Fig. 2). Like the 3′-terminal tRNA-like motifs of contemporary bacterial and plant RNA viruses (Rao et al. 1989) and also possibly animal picornaviruses (Pilipenko et al. 1992), the genomic tag would have served two main roles, providing an initiation site for replication and functioning as a simple telomere.

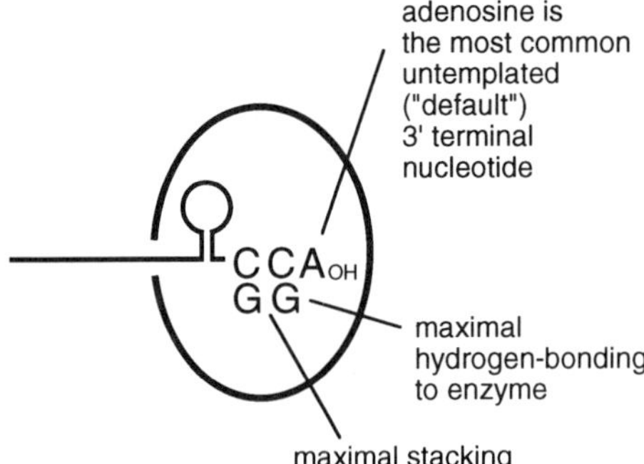

Figure 1 The simplest genomic tag. The tag functions as a simple telomere by sequestering subterminal RNA sequences in secondary structure, thus forcing the replicase to initiate on the 3′-terminal CCA of the genomic RNA. As shown in this figure and described in the text, the 3′-terminal CCA sequence may have been *selected* to facilitate efficient and faithful initiation of replication with the sequence 5′GG. Note that the duplex stem of the simplest tag corresponds to the "top half" of modern tRNA (Fig. 2).

As an initiation site for replication, the tag would bind to the replicase, ensuring replication of genomic (as opposed to nongenomic and random) RNA molecules. In addition to conferring template specificity, the tag would also sequester subterminal RNA sequences in secondary structure, thereby forcing the replicase to initiate on the 3′-terminal CCA of the genomic RNA. Chemical considerations suggest that the 3′-terminal CCA sequence could in fact have been *selected* to facilitate efficient and faithful replication. Initiation with guanosine on the penultimate base of the template might be favored because G:C pairs are stronger than A:U pairs; because G has a greater potential than the other bases for hydrogen bonding with the polymerase; and because strong stacking of G on G in the 5′GG dinucleotide might help to compensate for the absence of a primer (Fig. 1).

As a telomere, the tag would provide a site for untemplated nucleotide addition, ensuring that critical terminal regions of the genome were not lost during replication. If replication initiated on the penultimate G (as Qβ replicase does today) (Blumenthal and Carmichael 1979), then the 3′-terminal A would be the minimal telomeric sequence, and untemplated addition of a 3′-terminal A residue would need to occur in the next round of replication. Addition of an untemplated 3′-terminal nucleotide is a common activity of both RNA and DNA polymerases made of protein and thus could also have been a property of an RNA-RNA replicase. Qβ replicase possesses this terminal transferase activity (Blumenthal and Carmichael 1979); both the bacteriophage T7 (Milligan

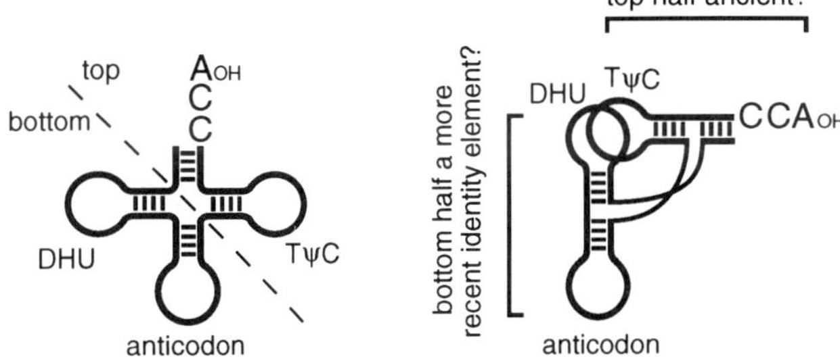

Figure 2 tRNA has two structural domains. The top half of tRNA, a coaxial stack of the acceptor stem on the TψC stem, may be the more ancient domain, with the bottom half, a coaxial stack of the dihydrouracil and anticodon stems, representing a later expansion loop that supplemented the function of the top half without disrupting it.

et al. 1987) and SP6 RNA polymerases will add a single untemplated 3′ nucleotide (Melton et al. 1984); and the ability of *Taq*I DNA polymerase to add an untemplated A to polymerase chain reaction (PCR) products is notorious (see, e.g., Tse and Forget 1990).

Some Implications of the Genomic Tag Hypothesis

The genomic tag hypothesis explains and relates many disparate roles of tRNA motifs in cellular metabolism. In this section, we review some aspects of the hypothesis considered in detail previously (Weiner and Maizels 1987) and discuss data from experiments inspired by, or relevant to, these facets of the hypothesis. In later sections, we develop new aspects of the hypothesis and discuss relevant new data.

RNase P

RNase P is a ribonucleoprotein enzyme that functions as an endoribonuclease to remove the 5′ leader from tRNA precursors. This processing reaction occurs in all contemporary cells and organelles. In some sense this is a rather surprising reaction, since there is no obvious reason why the 5′ end of tRNA could not be generated directly by transcription, as is known to be the case for *Xenopus laevis* selenocysteine tRNA (Lee et al. 1989; Carbon and Krol 1991).

We have suggested that the contemporary RNase P activity derives from an ancient activity that arose to convert genomic RNA molecules into functional subgenomic RNAs by removing the 3′-terminal tRNA-like tag, thereby enhancing the structural and enzymatic versatility of functional RNA molecules. Then, in a reversal of fortunes, an enzyme that had once freed functional RNA from a tRNA-like 3' tag survived as an enzyme for removing the nonfunctional 5' leader from tRNA precursors. Our suggestion that RNase P might have evolved to recognize 3'-terminal tRNA-like genomic tags led directly to experiments showing that *Escherichia coli* RNase P can process the 3'-terminal tRNA-like pseudoknot of turnip yellow mosaic virus (TYMV) at a site corresponding to the 5' leader of a tRNA precursor (Green et al. 1988; Guerrier-Takada et al. 1988; Mans et al. 1990). The fact that a contemporary eubacterial processing enzyme recognizes the 3' structures of plant viruses suggests remarkably conservative coevolution of both the activity itself and the structure of its substrate.

Since the RNA component of RNase P is enzymatically active under appropriate conditions (Guerrier-Takada et al. 1983), it is plausible that early forms of RNase P could have been composed entirely of RNA. Indeed, Alberts (1986) has argued that proteins are such efficient and versatile catalysts that the very presence of RNA in an enzyme may be taken as prima facie evidence that the enzyme is ancient and may predate the advent of templated protein synthesis. Thus, both the function and composition of RNase P are consistent with preservation of this activity since a very early era in molecular evolution. The genomic tag hypothesis plausibly explains why an activity critical for tRNA processing would have arisen so early.

Telomeres and the First Telomerase

While endonucleolytic cleavage by RNase P generates the mature 5' end of tRNA, exonucleolytic processing resects the 3' end of tRNA precursors, and the CCA-adding enzyme then reconstructs the CCA terminus. Once again, this is a rather surprising reaction, because the CCA end is destroyed only to be rebuilt. The purpose of this activity becomes clear, however, when the CCA-adding activity is viewed as a molecular fossil. If the 3' end of a tRNA-like structure served as the initiation site for an early replicase, a CCA-adding activity would have been necessary to restore nucleotides lost as a result of incorrect initiation. The CCA-adding activity therefore would have functioned as the first telomerase, and the modern CCA-adding activity may have evolved from an RNA enzyme into a protein enzyme by stepwise replacement of RNA by protein (White 1976, 1982; Visser and Kellogg 1978a,b). As discussed below, use of a telomeric $C_m A_n$ motif by early RNA genomes may explain why modern DNA telomeres employ very similar sequence motifs.

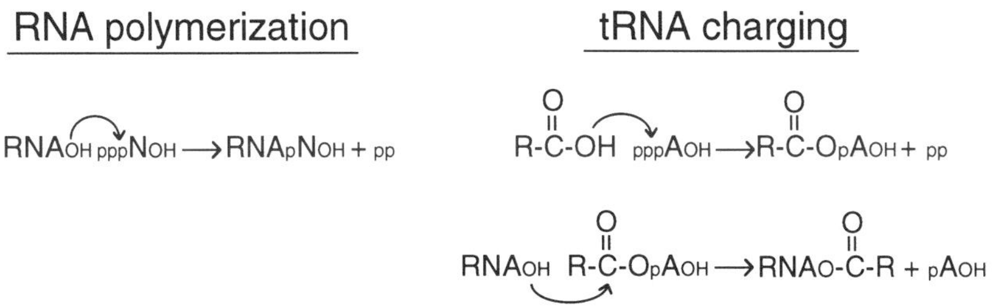

RNA polymerization   ## tRNA charging

$$RNA_{OH}\,_{ppp}N_{OH} \longrightarrow RNA_pN_{OH} + pp$$

$$\underset{\text{O}}{\overset{\text{O}}{\text{R-C-OH}}}\ _{ppp}A_{OH} \longrightarrow \underset{\text{O}}{\overset{\text{O}}{\text{R-C-O}}}_pA_{OH} + pp$$

$$RNA_{OH}\ \ \underset{\text{O}}{\overset{\text{O}}{\text{R-C-O}}}_pA_{OH} \longrightarrow \underset{\text{O}}{\overset{\text{O}}{\text{RNAO-C-R}}} + \,_pA_{OH}$$

Figure 3 tRNA charging resembles RNA polymerization. Note that attack on the phosphoester bond proceeds by an in-line, S_N2 mechanism, resulting in a trigonal bipyrimidal transition state, whereas attack on the aminoacyl-phosphoester bond occurs orthogonal to the π face, resulting in a tetrahedral intermediate. Despite these differences in stereochemistry, a ribozyme that catalyzes phosphoester bond transfer can also catalyze reactions at the carbon center, albeit at vastly reduced efficiency (Piccirilli et al. 1992).

tRNA Synthetase Function and the Origin of Protein Synthesis

We have argued that the series of reactions required for aminoacylation of tRNA chemically resembles RNA polymerization (Fig. 3) (Weiner and Maizels 1987). This prompted us to suggest that 3′-terminal tRNA-like structures of ancient RNA genomes may have been aminoacylated in a reaction resembling the charging of contemporary tRNAs by aminoacyl-tRNA synthetases. In making this suggestion, we explicitly postulated that the enzyme responsible for aminoacylation was an RNA and that this RNA enzyme could bind both an amino acid and a mononucleotide and could catalyze reactions at an aminoacyl-phosphoester center. We also suggested that RNA, as a structured polyanion, would most likely bind basic amino acids and that charging of a genomic tag with a basic amino acid would have the greatest effect on the structure or function of the RNA. The ability of group I introns to bind the mononucleotide guanosine was known at the time (for review, see Michel et al. 1989). It has since been shown that group I introns will bind L-arginine specifically (Hicke et al. 1989; Yarus, this volume); that the group I guanosine binding site can be rationally redesigned to bind other nucleotides (Michel et al. 1989); and that a group I intron can, without significant redesign, function as a credible aminoacylphospho-esterase (Piccirilli et al. 1992).

The genomic tag hypothesis suggests that aminoacylation initially conferred a *replication advantage* on molecules carrying a genomic tag. This could have occurred in any of several ways. Aminoacylation might have facilitated binding of the replicase to the 3′ end of the genome, perhaps simply by countering the net negative charge of the RNA replicase with a positively charged (basic) amino acid. Aminoacylation could also have served as a regulatory mechanism for withdrawing a genomic RNA from the replicative pool by blocking binding of the tag to the replicase. A third possibility, suggested by Wong (1991), is that aminoacylation might be seen as a form of RNA modification (like methylation, thiolation, isopentenylylation, etc.) that would broaden the structural or catalytic range of the RNA bearing it.

Our model for the origin of protein synthesis is unique in postulating that key components of the translation apparatus — tRNA and tRNA aminoacylation activity — first evolved as essential components of the *replication* machinery before the advent of protein synthesis. With these two key components of the translation apparatus in place, the scene was set for the interdependent coevolution of replication and templated protein synthesis. Perhaps at first, random condensation of aminoacylated tRNAs generated short polycations. These simple polymers could have facilitated RNA-catalyzed reactions in a manner analogous to modern polyamines (Jay and Gilbert 1987; Maizels and Weiner 1987; Weiner and Maizels 1987) or they could have stabilized or promoted a particular RNA structure, much as the tract of basic amino acids within the HIV Tat protein shapes the structure of the TAR element (Puglisi et al. 1992). Without several different species of charged tRNA, however, no mRNA would be necessary or (for that matter) useful. We therefore speculated that the original protoribosome evolved to facilitate synthesis of specific peptides by aligning charged tRNAs *before* the advent of mRNA, and that templating of peptide synthesis by mRNA was one of the *last* steps in the evolution of the modern translation apparatus (Maizels and Weiner 1987). This scenario is consistent with recent evidence that contemporary tRNAs comprise two separate structural domains (see below). We also emphasize that borrowing tRNA from replication for protein synthesis might constrain, but would not preclude, further evolution of the role of tRNA in replication.

The Unexpected Diversity of tRNA Synthetases

A further prediction of the genomic tag model for the origin of protein synthesis is that at least some of the tRNA synthetases were originally RNA enzymes (Weiner and Maizels 1987). We originally suggested that this might account for the remarkable diversity in structure of modern synthetases already apparent at the time (Schimmel 1987), because there is unlikely to be a *unique* pathway by which an RNA enzyme would evolve into a protein enzyme by gradual replacement of RNA structures with protein structures (White 1976, 1982; Visser and Kellogg 1978a,b). Thus, we explicitly proposed that the existence of apparently unrelated protein enzymes carrying out the same essential function could be taken as strong evidence for an ancestral RNA enzyme. This notion has been further developed by Benner et al. (1989).

Despite the apparent diversity of tRNA synthetase subunit structure and sequences (Schimmel 1987), the suggestion that there might be more than one class of synthetase was not well received, perhaps because it was so difficult to imagine that 20 tRNA synthetases doing the same enzymatic job could have descended from more than one ancestral form. However, Eriani et al. (1990) subsequently found that tRNA synthetases could be partitioned into at least two classes based on mutually exclusive sets of sequence motifs. Class I has the Rossmann fold for nucleotide binding, class II an antiparallel β sheet (Cusack et al. 1990); class I acylates the 2$'$ hydroxyl, class II the 3$'$ hydroxyl; yet the two classes employ virtually identical reaction mechanisms involving an enzyme-bound aminoacyladenylyl intermediate derived from ATP. From these structural and functional data one must conclude either that (1) tRNA synthetases made of protein evolved twice but independently adopted the same reaction mechanism; or, as we proposed originally, (2) the first tRNA synthetases were made of RNA and evolved stepwise by distinct pathways through RNP intermediates into two ancestral tRNA synthetases made of protein, all the while preserving the aminoacyladenylyl charging mechanism that first evolved in the RNA world.

The Two Structural Domains of Contemporary tRNA May Have Evolved Independently

Contemporary tRNAs are composed of two structural domains, a "top half" consisting of a coaxial stack of the acceptor stem on the TψC arm, and a "bottom half" consisting of a coaxial stack of the dihydrouracil arm on the anticodon arm (Fig. 2). The genomic tag hypothesis suggests the intriguing possibility that these two structural domains evolved in different eras, with the top half of the molecule arising early and the bottom half arising later. Presuming that RNase P is ancient, the ability of RNase P to recognize the top half of tRNA alone (McClain et al. 1987) can be interpreted as evidence that the original genomic tags may have been as simple as the top half of the molecule (Fig. 1). Charging of such a simple genomic tag with different amino acids might have arisen initially to facilitate differential replication, modification, or processing of genomic RNAs; alternatively, differential charging could have arisen to facilitate synthesis of new (or more precise) peptides. In either case, there would have been selective pressure to distinguish different tags from each other, and this is consistent with new functional (Francklyn et al. 1992) and structural studies (Rould et al. 1989) demonstrating that critical tRNA identity elements can lie within the top half of the molecule. In particular, minihelices corresponding to the top half of tRNA are sufficient in some cases for highly specific charging (Francklyn et al. 1992; Musier-Forsyth and Schimmel 1992), and a fragment of the top half is a substrate for the "protein-free" peptidyl transferase activity (Noller et al. 1992; Noller, this volume). Further evidence for the potential complexity of tRNA identity elements within the top half of the molecule comes from crystallographic work showing that critical identity elements are revealed by partial melting of the acceptor stem when *E. coli* glutamine tRNA binds to the cognate synthetase (Rould et al. 1989). The fact that enzymes as different as RNase P (McClain et al. 1987), tRNA synthetases (Rould et al. 1989), ribosomal RNA (Noller et al. 1992), and EF-Tu (Rasmussen et al., 1990) can recognize the top half of tRNA suggests that this is, indeed, the more ancient half of the molecule containing the most essential identity elements.

The bottom half of tRNA appears then to be a more recent addition, used both by synthetases and by mRNAs to distinguish one species of tRNA from another. The bottom half of tRNA can thus be viewed as an expansion "loop," not unlike those found in ribosomal RNA (see Moore; Noller, both this volume). The notion that the bottom half of the tRNA molecule evolved later is also consistent with our suggestion that *templated* protein synthesis, requiring tRNA-mRNA recognition mediated by the anticodon loop, could not have been selected until relatively late in the evolution of the translation apparatus when there were already several different species of charged tRNA to read the mRNA (Maizels and Weiner 1987).

TRACING GENOMIC TAGS FROM RNA PHAGE TO MODERN CHROMOSOMAL TELOMERES: THE NOTION OF "TRANSITIONAL GENOMES"

From the beginning, we suspected that tRNA-like genomic tags survived from an RNA world into a DNA world and were, in the process, transformed into the tRNA primers of retrovirus reverse transcription and the terminal C_mA_n motifs of modern chromosomal telomeres (Weiner and Maizels 1987). Yet we were unable to make these connections explicit because there appeared to be one or more missing links in the molecular fossil record. Recent stunning and completely unanticipated results from the groups of Lambowitz (Kuiper and Lambowitz 1988; Akins et al. 1989), Saville and Collins (1990), and Blackburn (1991) provide molecular fossil evidence of genomes and genomic tags in transition from single-stranded RNA to double-stranded DNA. In this section we discuss the *Neurospora* retroplasmid and the *Tetrahymena* telomerase, two critical links that were missing from the molecular fossil record as it was known in 1987 (Weiner and Maizels 1987).

Neurospora Mitochondrial Retroplasmids: Evidence That Genomic Tags Survived the Transition from RNA to DNA Genomes

Mitochondria of some strains of *Neurospora crassa* contain double-stranded DNA plasmids that were originally detected because their replication competes with replication of mitochondrial DNA and causes a respiratory-deficient phenotype. Most surprisingly, replication of these plasmids requires production of a full-length RNA transcript of genomic DNA (Figs. 4 and 5) (Kuiper and Lambowitz 1988; Akins et al. 1989),

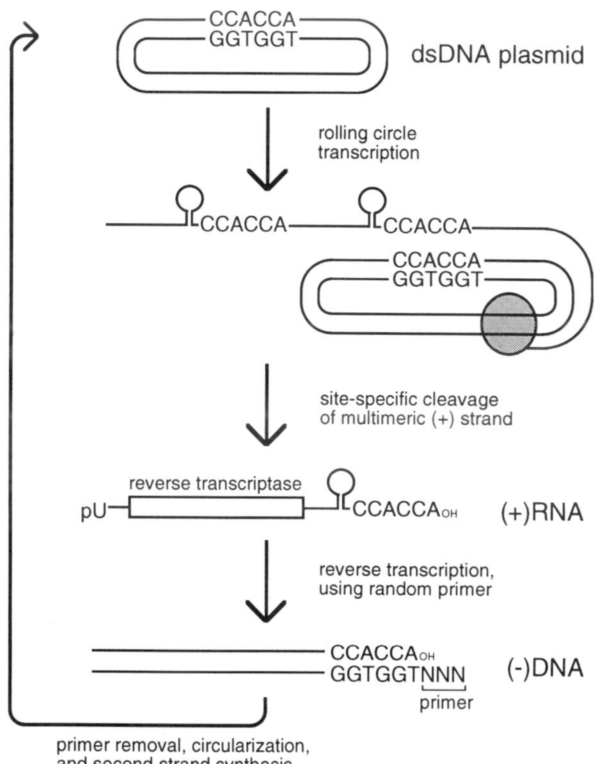

Figure 4 The *Neurospora* mitochondrial Varkud plasmid is a simple retrovirus-like element.

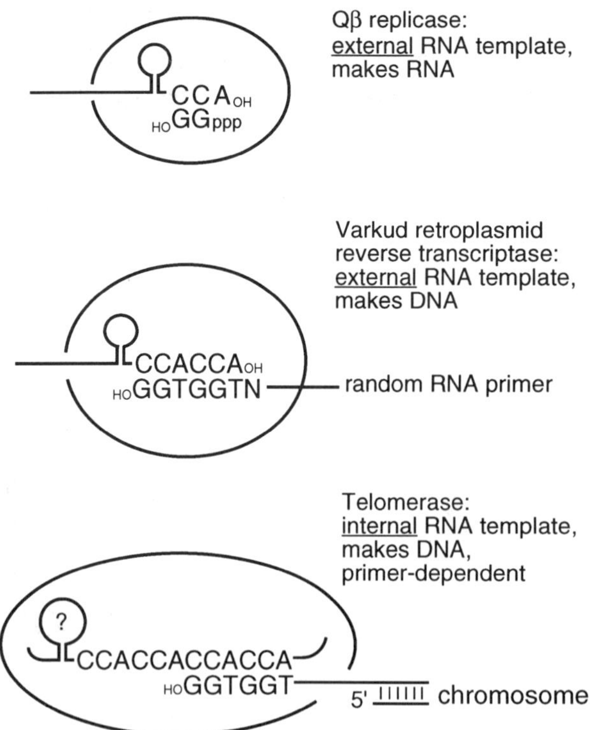

Figure 5   From Qβ replicase to telomerase. Qβ replicase, the *Neurospora* retroplasmid reverse transcriptase, and *Tetrahymena* telomerase all initiate on a tRNA-like template, suggesting that the ancestral forms of these enzymes represent a single line of evolutionary descent (see also Fig. 6).

and this RNA carries a 3′-terminal tRNA-like structure ending in CCACCA, implying that it functions as a tRNA-like genomic tag (Akins et al. 1989). An open reading frame within the genomic RNA encodes a functional reverse transcriptase, and mitochondrial extracts of plasmid-containing strains can produce a full-length cDNA copy of the genomic RNA by initiating cDNA synthesis on the terminal A of the CCACCA. The 3′-terminal genomic tag therefore serves as the initiation site for replication, as in Qβ, but Qβ replicase is replaced by the plasmid reverse transcriptase (Fig. 5). Primers for replication appear to be random RNAs, which are subsequently removed (A.M. Lambowitz, pers. comm.). Whether the full-length RNA/DNA hybrid produced after the first step in plasmid replication circularizes directly or is first converted into double-stranded DNA is not yet known, but it is clear that after circularization, rolling circle *transcription* of the double-stranded DNA replicative inter-mediate yields multimeric plus strands. Monomeric genomic RNAs are thought to be excised from these multimers and, at least in the case of the Varkud-associated VSDNA retroplasmid, a ribozyme contained within the element itself appears to perform this cleavage reaction (Saville and Collins 1990).

The fungal mitochondrial "retroplasmids" can be thought of as linear RNA genomes that replicate via a circular DNA intermediate, or circular DNA genomes that replicate via a linear RNA intermediate. We therefore consider these retroplasmids to be examples of *transitional genomes* from an era in which genomic tags were being transformed from templates for RNA synthesis to templates for DNA synthesis.

Telomerase: A Genomic Tag Carried by the RNA Component of a Reverse Transcriptase

When we first suggested that the $3'$-terminal CCA motif of tRNA-like genomic tags was related to the nearly universal C_mA_n motif of eukaryotic nuclear telomeres (Weiner and Maizels 1987), we were unable to discern any hints in the molecular fossil record regarding the mechanism or chain of events by which a $3'$-terminal CCA motif in RNA could be transformed into a $5'$-terminal motif in DNA. Another missing link in the molecular fossil record emerged from detailed characterization of the *Tetrahymena* telomerase. This enzyme, first described by Greider and Blackburn (1985), possesses an unusual terminal nucleotidyl transferase activity that adds the species-specific T_nG_m repeats, one nucleotide at a time, to an appropriate T_nG_m primer (for review, see Blackburn, 1991). Remarkably, telomerase is a ribonucleoprotein, and the RNA component is in fact an *internal* template for synthesis of the species-specific T_nG_m repeat. (For example, the telomeric repeat in the ciliate *Tetrahymena* is T_2G_4, and the internal template sequence in the *Tetrahymena* telomerase is $5'$-CAACCCCAA-$3'$.) The protein component of telomerase, at least in the yeast *Saccharomyces cerevisiae*, is homologous to retroviral reverse transcriptases (Lundblad and Blackburn 1990). Telomerase can thus be seen as a specialized reverse transcriptase with an internal tRNA-like template (for review, see Blackburn 1991 and this volume).

The Unusual Properties of G and G-rich DNAs

The unusual properties of G and of G-rich DNAs suggest that chemical determinism may explain both the C_nA_m motif in RNA telomeres and use of the T_mG_n motif in modern DNA telomeres (Weiner and Maizels 1987). Telomeric T_nG_m repeats and other G-rich DNAs can spontaneously form unusual four-stranded structures in vitro that contain *intra-*

molecular non-Watson/Crick-base pairs (Henderson et al. 1987). These structures, dubbed "G quartets," may play a role in telomere aggregation or function in vivo (Sen and Gilbert 1988; Sundquist and Klug 1989; Williamson et al. 1989). As detailed in Figure 1, the sequence CCA might have been *selected* in an RNA world as an efficient initiation site because the initiating guanosines stack more strongly and have more potential hydrogen-bonding interactions than any other base. In effect, CCA would have been selected for the unusual properties of its G-rich *complement*, and it is this complementary $T_m G_n$ motif that forms the telomeric repeating unit.

Functional Similarities between Telomerase and the Retroplasmid Reverse Transcriptase

The *Neurospora* retroplasmid reverse transcriptase uses a random RNA as primer to copy a genomic CCACCA template sequence into cDNA (see Fig. 4) (A.M. Lambowitz, pers. comm.). Telomerase uses telomeric DNA as primer to copy a CCACCA-like template within the enzyme itself into a cDNA. This provides a striking link between telomerase and the retroplasmid reverse transcriptase (Fig. 5). Both enzymes copy RNA into DNA and, like Qβ replicase, both initiate on the CCA motif of a tRNA-like template.

Might there be a single line of evolutionary descent for these enzymes and the genomes that they replicate? Figure 5 shows how a tRNA recognition domain critical for initiation may be conserved among these enzymes and how the reverse transcriptase domain may interact with the recognition complex. Figure 5 also suggests a simple pathway by which an enzyme like the retroplasmid reverse transcriptase could evolve into a specialized reverse transcriptase activity like telomerase. The retroplasmid enzyme would acquire an internal template by stably binding a tRNA or tRNA-like structure, and a primer terminus that could pair with this internal template would then allow the reverse transcriptase to enter directly into elongation mode. Although reverse transcriptases require a primer and Qβ replicase does not, the initiating nucleotide can plausibly be regarded as a very short primer because it must occupy the same site on the enzyme as the primer terminus during elongation.

FROM TEMPLATE TO PRIMER

tRNAs Prime Retroviral Reverse Transcription

Initiation of retrovirus replication normally requires a tRNA as primer (see Fig. 8); the hepadnaviruses are the only conspicuous exception

(Ganem and Varmus 1987). tRNA priming of reverse transcription occurs not only in prototypical avian and mammalian retroviruses, but also in such lower eukaryotic retroviral elements as Ty1 in the yeast *S. cerevisiae* (Chapman et al. 1992) and *copia* in the fly *Drosophila melanogaster* (Kikuchi et al. 1986). To function as a primer, the 3′ end of the tRNA becomes partially unfolded so that it can base-pair with the primer-binding site on the genomic RNA (Fig. 6), and due to the requirement for complementary base-pairing, each retroviral element uses only a specific tRNA primer. As discussed below, tRNA also primes reverse transcription of circular extrachromosomal retroviruses such as cauliflower mosaic virus (CaMV) (Hohn et al. 1985; Covey and Turner 1986). When we originally proposed that the use of tRNAs as primers for

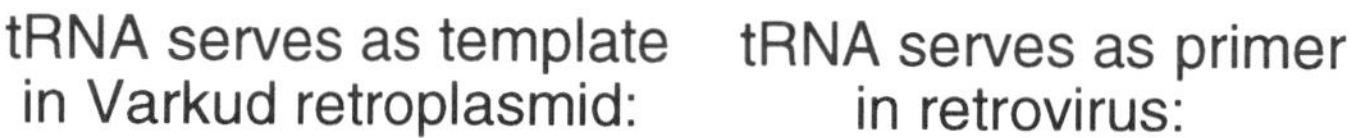
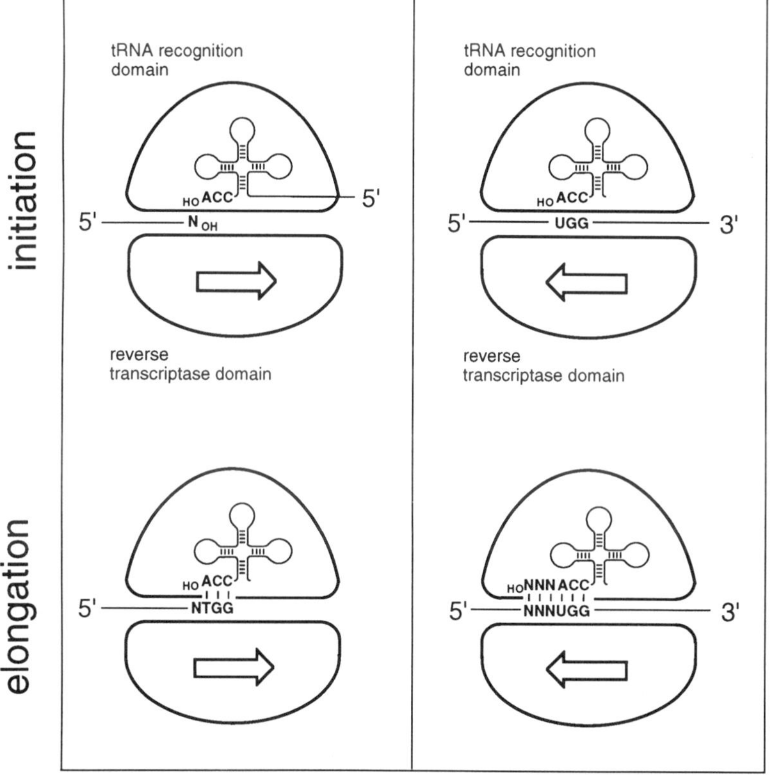

Figure 6 From template to primer. The transformation of tRNA from template for initiation of replication to primer requires that the tRNA-binding domain be separate from, or flexibly tethered to, the catalytic domain.

retroviral replication was derived from the use of tRNA-like structures as *templates* for the initiation of RNA replication (Weiner and Maizels 1987), we were unable to cite transitional forms in the molecular fossil record. The use of tRNA as template for reverse transcription of the retroplasmid RNA genome (Figs. 4 and 5), as described above, now establishes a plausible connection between tRNA and reverse transcription; however, the transformation of tRNA from template to primer remains to be explained. In this section, we discuss evidence suggesting that this transformation is less improbable than it might at first appear.

In Most Polymerases, Distinct Protein Domains Are Responsible for Template Specificity and Catalysis

A wealth of molecular and structural data is consistent with the notion that the two functions of any polymerase—template specificity and catalysis—are typically carried out by distinct protein domains. The classic examples are σ factors, which confer promoter specificity on eubacterial RNA polymerases (for review, see Jaehning 1991). Separability of template specificity and catalysis is also a property of Qβ replicase and of the generic RNA replicase encoded by poliovirus. The Qβ replicase holoenzyme consists of four subunits (Blumenthal and Carmichael 1979): a phage-encoded replicase (subunit II) and three host-encoded subunits (EF-Tu, EF-Ts, and ribosomal protein S1). Subunit II exhibits a generic RNA replicase activity, whereas template specificity lies entirely within the other subunits, with EF-Tu presumably recognizing the 3′-terminal tRNA-like structure just as it recognizes tRNA during protein synthesis (Weiner and Maizels 1987). Similarly, the poliovirus replicase polypeptide 3D[pol] appears to require an additional specificity factor for initiation on the polioviral minus-strand template (Andino et al. 1990; R. Andino, pers. comm.).

Additional evidence that the template specificity and catalytic functions of many polymerases are separable comes from analysis of sequence motifs shared by viral RNA-dependent RNA polymerases (replicases) and retroelement RNA-dependent DNA polymerases (reverse transcriptases). These enzymes display a wide range of template and primer specificity, but nonetheless share essentially invariant "signature" sequence motifs (Poch et al. 1989). Mutations in the signature sequence of DNA-dependent DNA polymerases abolish catalytic function, and these signature residues are therefore thought to be located at the active site rather than in the template recognition domain (Polesky et al. 1992). If template specificity and catalytic function were not at least partially separable, changes in template specificity would obliterate these characteristic signatures. Indeed, we might *expect* that the separation of template specificity and catalysis would be the rule. If the template specificity of a polymerase could not be changed *without* affecting catalysis, the enzyme would be constrained to evolve slowly, whereas a polymerase in which these functions were separate could more easily adapt to new templates, thereby diversifying and ultimately producing more molecular descendants.

Separability of tRNA Recognition and the Reverse Transcriptase Active Site

As shown in Figure 6, the separability of template specificity and catalytic function in many contemporary polymerases suggests a pathway by which tRNA might have evolved from template to primer. A reverse transcriptase that carried a tRNA-binding domain might readily evolve from an enzyme that used the tRNA as template to one that used it as primer, *provided that the tRNA recognition domain was flexibly tethered to the active site.* The ability of HIV reverse transcriptase to form a *binary* complex with primer tRNA (Barat et al. 1989) is consistent with the idea that tRNA recognition can be separated from catalysis.

Use of tRNA as primer presumes the existence of an RNA helicase activity, because the enzyme would need to melt the top half of tRNA (a coaxial stack of the acceptor stem on the TψC arm) (Figs. 1 and 2) in order to allow the primer to base-pair with the genomic primer-binding site (see Fig. 7). However, this helicase activity would *already* be in place, because any enzyme that uses a genomic tag as template must possess an RNA helicase activity that can melt the top half of tRNA once initiation has occurred on the 3′-terminal CCA. Extensive complementarity between the tRNA primer and the template may be a later refinement to assure a unique site of initiation and to stabilize the initiation complex. Use of tRNA as primer also presumes that polymerization can begin in elongation mode, as opposed to initiation mode when tRNA functions as template (Fig. 7). In fact, these two modes may be very similar. Early work on *E. coli* RNA polymerase showed that a primer as short as a dinucleotide dramatically increases the efficiency and specificity of initiation (Downey et al. 1971; Maizels 1973), and recent structural and mechanistic studies on DNA polymerase suggest that the initiating nucleotide occupies the same site on the enzyme as the primer terminus during elongation (see Fig. 7) (Polesky et al. 1992).

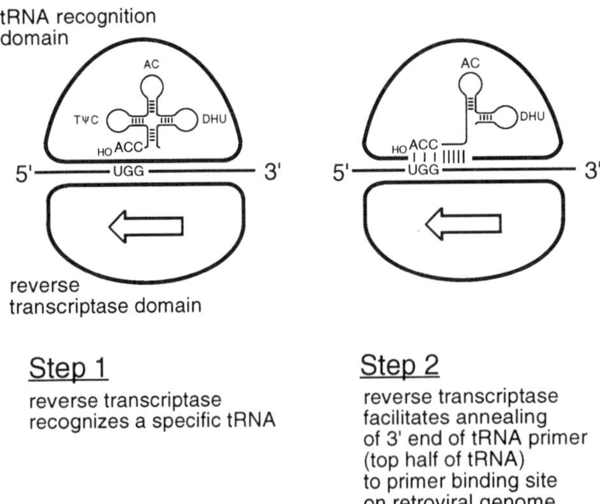

Figure 7 RNA helicase activity is required whether tRNA is template *or* primer. With a genomic tag as template, the top half of tRNA must be melted once initiation has occurred on the 3′ terminal CCA (see Fig. 1); with a genomic tag as primer, the top half of tRNA must be melted in order to adopt the elongation mode at the moment of initiation.

Caulimoviruses are Non-integrating One-LTR Retroviruses

We have discussed several reasons for thinking that mitochondrial retroplasmids may represent the ancestral form of modern retroviruses, but there is yet another missing link between duplex DNA plasmids and the prototypical integrating retrovirus with long terminal direct repeats (LTRs). The plant retrovirus CaMV may provide this connection. The CaMV genome is an extrachromosomal circular duplex DNA that lacks direct repeats and never integrates into chromosomal DNA. Transcription of viral DNA generates a full-length genomic RNA, but because the polyadenylation site for this transcript is located 180 bp downstream from the viral promoter, the transcript contains a 180-nucleotide terminal redundancy (see Fig. 8) (Hohn et al. 1985; Covey and Turner 1986).

Although CaMV differs from prototypical retroviruses in having one rather than two LTRs in the duplex DNA form, CaMV uses exactly the same strategy to generate a terminally redundant genomic RNA. In both cases, a polyadenylation site is ignored when it occurs too near the 5′ end of the RNA, apparently because RNA processing of the nascent transcript is inhibited by proximity to the promoter (Sanfacon and Hohn 1990; but see DeZazzo et al. 1992). As in prototypical two-LTR retroviruses, the tRNA primer-binding site in CaMV genomic RNA lies

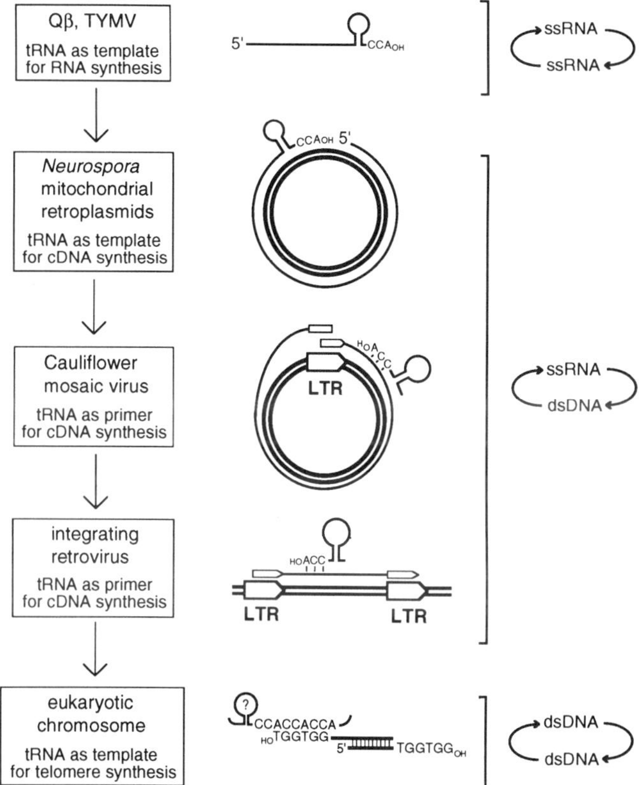

Figure 8 A phylogenetic tree for replication strategies based on tRNA-like structures.

downstream from the 5′-terminal redundancy, so reverse transcriptase can effectively circularize the genome by switching templates from the 5′ end of the genomic RNA to the corresponding position in the 3′-terminal redundancy.

These similarities in genomic structure and replication strategy imply that modern integrating retroviruses descended from an extrachromosomal retroviral element resembling CaMV. The obvious advantage of chromosomal integration is that it assures perpetuation of the element. The existence of *two* LTRs in a prototypical retroviral provirus could then be interpreted as an invention that preserved the established replication strategy by counterfeiting the circular topology of the ancestral genome.

We anticipate that additional examples of transitional genomes will be discovered. One intriguing candidate is 10Sa RNA, a small, stable, highly conserved RNA found in diverse eubacteria (Subbarao and

Apirion 1989; Brown et al. 1990; Tyagi and Kinger 1992). The mature 10Sa molecule is generated by RNA processing from a larger transcript, and the 3'end of 10Sa RNA almost perfectly matches the conserved residues in the top half of tRNA, including the CCA (Tyagi and Kinger 1991). This suggests the intriguing possibility that 10Sa RNA may be the genomic or subgenomic transcript of a new retroelement.

ARE MODERN VIRUSES MOLECULAR FOSSILS OF ANCIENT STRATEGIES FOR GENOMIC REPLICATION?

Most of us have come to have considerable confidence in the validity of sequence-based phylogenies relating organisms and organelles (Pace et al. 1986; Woese and Pace, this volume). Viruses, however, are quite another story. No single molecule, product, or function is common to all viruses, so there is no universal standard for evolutionary comparisons that is useful in quite the same way as small subunit ribosomal RNA sequences have been for organismal phylogenies. Furthermore, the interchangeability of functionally similar viral modules can potentially wreak havoc with attempts to develop viral phylogenies based on a single molecule such as reverse transcriptase (but see Xiong and Eickbusch 1990). Nor can simple assumptions be made about the regularity of viral molecular clocks, because generation times are short, burst sizes are large, and there is no single typical genome, but rather a population of quasispecies and defective interfering particles all propagating simultaneously (Eigen et al. 1981). Compounding these formidable problems, it can be difficult or impossible to trace the source of new viral genetic information, because viruses move horizontally between hosts, exchange genetic information with other viruses by recombination, and acquire new characteristics in each of a succession of hosts.

Although many aspects of viral life cycles violate the assumptions of sequence-based phylogenetic methods, viruses also present a unique opportunity for evolutionary analysis because certain aspects of viral life cycles appear to resist rapid change. Gene order and replication strategy, for example, are remarkably conserved between some plant and animal viruses. The stability of viral strategies for genomic replication and gene expression led Strauss and Strauss (1983) to propose that these properties of the viral life cycle are valuable and appropriate markers for virus evolution.

The attitude that viruses are cellular parasites, assembled in relatively recent evolutionary time from preexisting parts of the cell (Benner and Ellington 1988), is difficult to reconcile with data showing that many aspects of viral life cycles are stable over evolutionary time. In fact, the genomic tag hypothesis suggests an alternative view of viruses, as fossils that reveal the diversity of ancient replication pathways (see also Wintersberger and Wintersberger 1987). Considered in this way, viruses have not devised novel and subversive replication strategies, but rather have conserved the useful features of more ancient forms of cellular chromosomal replication, even as cellular replication strategies continued to evolve. Contemporary viruses may thus preserve a record of the replication strategies used by more ancient chromosomes.

To illustrate how this view of contemporary viruses may be informative about molecular evolution, we show how RNA viruses, transitional genomes, and contemporary DNA genomes might be related using a phylogenetic tree based on the use of tRNA-like structures in replication (Fig. 8). The plus-strand RNA bacteriophages (Qβ) and plant viruses (brome mosaic virus) may represent molecular fossils of an RNA world in which cellular chromosomes were single-stranded RNA molecules with a 3′-terminal genomic tag. The *Neurospora* mitochondrial retroplasmid may represent a transitional stage from RNA to DNA genomes, in which duplex DNA functioned as the storage form for genetic information and genomic RNA with a 3′terminal genomic tag served as a replicative intermediate. CaMV may be a molecular fossil of a more advanced transitional genome, which still replicated through a genomic RNA, but in which the genomic tag had been transformed into a tRNA primer.

This view of viruses further suggests that modern duplex DNA genomes might be usefully regarded as retroviruses writ large. Eigen et al. (1981) originally observed that the role of RNA priming in the initiation of DNA synthesis might be a remnant of earlier RNA genomes. We are intrigued by the more specific possibility that the RNA primer for DNA replication is a degenerate form of the retroviral tRNA primer (see also Wintersberger and Wintersberger 1987). Primases, for example, might turn out to be more closely related to reverse transcriptases than to RNA polymerases. Similarly, as discussed above, the telomerase responsible for completing replication of eukaryotic chromosomes is a specialized form of reverse transcriptase with an internal tRNA-like template (Fig. 5). Taken together, this evidence for the key role of RNA in *both* initiation and completion of chromosomal replication supports the notion that retroelements are the ancestors of modern eukaryotic chromosomes.

In arranging these selected viruses in this way, we in no sense imply that any one of these viruses is unchanged from ancient times. We do wish to point out that essential aspects of a viral replication strategy may remain stable over billions of years and that the genomes of ancient organisms may survive in the form of viruses as the host replication strategy evolves (Weiner 1987). This is, of course, consonant with the genomic tag hypothesis itself, which credits plus-strand RNA genomes similar to Qβ and modern plant viruses with the invention of tRNA and aminoacylation. Viewed as predecessors of modern chromosomes, rather than derivatives of them, viruses may have much more to tell us about molecular evolution than is commonly appreciated.

CONCLUSION

We originally conceived of the genomic tag hypothesis to explain the origin of protein synthesis (Weiner and Maizels 1987), but with time it became clear that this hypothesis had equally distinct implications for the evolution of replicative mechanisms, beginning in an RNA world and continuing to the present day. The new evidence that we have discussed for the central role of genomic tags in the evolution of RNA to DNA genomes strengthens the case that tRNA-like genomic tags arose early in an RNA world.

Translation today is such a complex and sophisticated process that the apparatus must have arisen stepwise, but it has been difficult to imagine how any single component of the apparatus could be useful by itself or how additional components could each individually confer a further selective advantage. Central to the genomic tag hypothesis is the suggestion that the two key components of the translation apparatus — tRNA and tRNA aminoacylation activity — first evolved as essential components of the replication apparatus and were subject to selection *before* the advent of protein synthesis. Once in place, these two key components of the translation apparatus could be co-opted for other purposes, with the result that replication and templated protein synthesis were fated forever after to coevolve.

REFERENCES

Akins, R.A., R.L. Kelley, and A.M. Lambowitz. 1989. Characterization of mutant mitochondrial plasmids of *Neurospora spp.* that have incorporated tRNAs by reverse transcription. *Mol. Cell. Biol.* **9:** 678–691.

Alberts, B.M. 1986. The function of the hereditary materials: Biological catalyses reflect the cell's evolutionary history. *Am. Zool.* **26:** 781–796.

Andino, R., G.E. Rieckhof, and D. Baltimore. 1990. A functional ribonucleoprotein complex forms around the 5′ end of poliovirus RNA. *Cell* **63:** 369–380.

Barat, C., V. Lullien, O. Schatz, G. Keith, M.T. Nugeyre, F. Gruninger-Leitch, F. Barré-Sinoussi, S.F. LeGrice, and J.L. Darlix. 1989. HIV-1 reverse transcriptase specifically interacts with the anticodon domain of its cognate primer tRNA. *EMBO J.*

8:3279–3285.

Benner, S.A. and A.D. Ellington. 1988. Return of the "last ribo-organism." *Nature* **332:** 688–689.

Benner, S.A., A.D. Ellington, and A. Tauer. 1989. Modern metabolism as a palimpsest of the RNA world. *Proc. Natl. Acad. Sci.* **86:** 7054–7058.

Blackburn, E.H. 1991. Structure and function of telomeres. *Nature* **350:** 569–573.

Brown, J.W., D.A. Hunt, and N.R. Pace. 1990. Nucleotide sequence of the 10Sa RNA gene of the β-purple eubacterium *Alcaligenes eutrophus*. *Nucleic Acids Res.* **18:** 2820.

Blumenthal, T. and G.C. Carmichael. 1979. RNA replication: Function and structure of Qß replicase. *Annu. Rev. Biochem.* **48:** 525–548.

Carbon, P. and A. Krol. 1991. Transcription of the *Xenopus laevis* selenocysteine tRNA$^{(Ser)Sec}$ gene: A system that combines an internal B box and upstream elements also found in U6 snRNA genes. *EMBO J.* **10:** 599–606.

Chapman, K.B., A.S. Bystrom, and J.D. Boeke. 1992. Initiator methionine tRNA is essential for Ty1 transposition in yeast. *Proc. Natl. Acad. Sci.* **89:** 3236–3240.

Covey, S.N. and D.S. Turner. 1986. Hairpin DNAs of cauliflower mosaic virus generated by reverse transcription *in vivo*. *EMBO J.* **5:** 2763–2768.

Cusack, S., C. Berthet-Colominas, M. Hartlein, N. Nassar, and R. Leberman. 1990. A second class of synthetase structure revealed by X-ray analysis of *Escherichia coli* seryl-tRNA synthetase at 2.5 Å. *Nature* **347:** 249–255.

DeZazzo, J.D., J.M. Scott, and M.J. Imperiale. 1992. Relative roles of signals upstream of AAUAAAA and promoter proximity in regulation of HIV-1 mRNA 3′ end formation. *Mol. Cell. Biol.* **12:** 5555–5562.

Downey, K.M., B.S. Jurmark, and A.G. So. 1971. Determination of nucleotide sequences at promoter regions by the use of dinucleotides. *Biochemistry* **10:** 4970–4975.

Eigen, M., W. Gardiner, P. Schuster, and H. Winkler-Oswatitsch. 1981. The origin of genetic information. *Sci. Am.* **244:** 88–118.

Eriani, G., M. Delarue, O. Poch, J. Gangloff, and D. Moras. 1990. Partition of tRNA synthetases into two classes based on mutually exclusive sets of sequence motifs. *Nature* **347:** 203–206.

Francklyn, C., J.P. Shi, and P. Schimmel. 1992. Overlapping nucleotide determinants for specific aminoacylation of RNA microhelices. *Science* **255:** 1121–1125.

Ganem, D. and H.E. Varmus. 1987. The molecular biology of the hepatitis B viruses. *Annu. Rev. Biochem.* **56:** 651–693.

Green, C.J., B.S. Vold, M.D. Morch, R.L. Joshi, and A.L. Haenni. 1988. Ionic conditions for the cleavage of the tRNA-like structure of turnip yellow mosaic virus by the catalytic RNA of RNase P. *J. Biol. Chem.* **263:** 11617–11620.

Greider, C.W. and E.H. Blackburn. 1985. Identification of a specific telomere terminal transferase activity in *Tetrahymena* extracts. *Cell* **43:** 405–413.

Guerrier-Takada, C., A. van Belkum, C.W.A. Pleij, and S. Altman. 1988. Novel reactions of RNAase P with a tRNA-like structure in turnip yellow mosaic virus RNA. *Cell* **53:** 267–272.

Guerrier-Takada, C., K. Gardiner, T. Marsh, N. Pace, and S. Altman. 1983. The RNA moiety of ribonuclease P is the catalytic subunit of the enzyme. *Cell* **35:** 849–857.

Henderson, E., C.C. Hardin, S.K. Walk, I. Tinoco, Jr., and E.H. Blackburn. 1987. Telomeric DNA oligonucleotides form novel intramolecular structures containing guanine-guanine base pairs. *Cell* **51:** 899–908.

Hicke, B.J., E.L. Christian, and M. Yarus. 1989. Stereoselective arginine binding is a phylogenetically conserved property of group I self-splicing RNAs. *EMBO J.* **8:** 3843–3851.

Hohn, T., B. Hohn, and P. Pfeiffer. 1985. Reverse transcription in CaMV. *Trends Biochem. Sci.* **10:** 205–209.

Jaehning, J.A. 1991. Sigma factor relatives in eukaryotes. *Science* **253:** 859.

Jay, D.G. and W. Gilbert. 1987. Basic protein enhances the incorporation of DNA into lipid vesicles: A model for the formation of primordial cells. *Proc. Natl. Acad. Sci.* **84:** 1978–1980.

Kikuchi, Y., Y. Ando, and T. Shiba. 1986. Unusual priming mechanism of RNA-directed DNA synthesis in *copia* retrovirus-like particles of *Drosophila*. *Nature* **323:** 824–826.

Kuiper, M.T.R. and A.M. Lambowitz. 1988. A novel reverse transcriptase activity associated with mitochondrial plasmids of *Neurospora*. *Cell* **55:** 693–704.

Lee, B.J., S.G. Kang, and D. Hatfield. 1989. Transcription of *Xenopus* selenocysteine tRNA Ser (formerly designated opal suppressor phosphoserine tRNA) gene is directed by multiple 5 ′ -extragenic regulatory elements. *J. Biol. Chem.* **264:** 9696–9702.

Lundblad, V. and E.H. Blackburn. 1990. RNA-dependent polymerase motifs in EST1: Tentative identification of a protein component of an essential yeast telomerase. *Cell* **60:** 529–530.

Maizels, N. 1973. The nucleotide sequence of the lactose mRNA transcribed from the UV5 promoter mutant of *E. coli*. *Proc. Natl. Acad. Sci.* **70:** 3585–3589.

Maizels, N. and A.M. Weiner. 1987. Peptide-specific ribosomes, genomic tags and the origin of the genetic code. *Cold Spring Harbor Symp. Quant. Biol.* **52:** 743–749.

Mans, R.M., C. Guerrier-Takada, S. Altman, and C.W. Pleij. 1990. Interaction of RNase P from *Escherichia coli* with pseudoknotted structures in viral RNAs. *Nucleic Acids Res.* **18:** 3479–3487.

McClain, W.H., C. Guerrier-Takada, and S. Altman. 1987. Model substrates for an RNA enzyme. *Science* **238:** 527–530.

Melton, D.A., P.A. Krieg, M.R. Rebagliati, T. Maniatis, K. Zinn, and M.R. Green. 1984. Efficient *in vitro* synthesis of biologically active RNA and RNA hybridization probes from plasmids containing a bacteriophage SP6 promoter. *Nucleic Acids Res.* **12:** 7035–7056.

Michel, F., M. Hanna, R. Green, D.P. Bartel, and J.W. Szostak. 1989. The guanosine binding site of the *Tetrahymena* ribozyme. *Nature* **342:** 391–395.

Milligan, J.F., D.R. Groebe, G.W. Witherell, and O.C. Uhlenbeck. 1987. Oligoribonucleotide synthesis using T7 RNA polymerase and synthetic DNA templates. *Nucleic Acids Res.* **15:** 8783–8798.

Musier-Forsyth, K. and P. Schimmel. 1992. Functional contacts of a transfer RNA synthetase with 2 ′ -hydroxyl groups in the RNA minor groove. *Nature* **357:** 513–514.

Noller, H.F., V. Hoffarth, and L. Zimniak. 1992. Unusual resistance of peptidyl transferase to protein extraction procedures. *Science* **256:** 1416–1419.

Orgel, L.E. 1968. Evolution of the genetic apparatus. *J. Mol. Biol.* **38:** 381–393.

Pace, N.R., G.J. Olsen, and C.R. Woese. 1986. Ribosomal RNA phylogeny and the primary lines of evolutionary descent. *Cell* **45:** 325–326.

Piccirilli, J.A., T.S. McConnell, A.J. Zaug, H.F. Noller, and T.R. Cech. 1992. Aminoacyl esterase activity of the *Tetrahymena* ribozyme. *Science* **256:** 1420–1424.

Pilipenko, E.V., S.V. Maslova, A.N. Sinyakov, and V.I. Agol. 1992. Towards identification of *cis*-acting elements involved in the replication of enterovirus and rhinovirus RNAs: A proposal for the existence of tRNA-like terminal structures. *Nucleic Acids Res.* **20:** 1739–1745.

Poch, O., I. Sauvaget, M. Delarue, and N. Tordo. 1989. Identification of four conserved motifs among the RNA-dependent polymerase encoding elements. *EMBO J.* **8:** 3867–3874.

Polesky, A.H., M.E. Dahlberg, S.J. Benkovic, N.D.F. Grindley, and C.M. Joyce. 1992. Side chains involved in catalysis of the polymerase reaction of DNA polymerase I from *Escherichia coli. J. Biol. Chem.* **267:** 8417–8428.

Popper, K.R. 1963. *Conjectures and refutations*, 4th edition. Routledge and Kegan Paul, London.

Puglisi, J.D., R. Tan, B. J. Calnan, A.D. Frankel, and J.R. Williamson. 1992. Conformation of the TAR RNA-arginine complex by NMR spectroscopy. *Science* **257:** 76–80.

Rao, A.L.N., T.W. Dreher, L.E. Marsh, and T.C. Hall. 1989. Telomeric function of the tRNA-like structure of brome mosaic virus RNA. *Proc. Natl. Acad. Sci.* **86:** 5335–5339.

Rasmussen, N.J., F.P. Wikman, and B.F. Clark. 1990. Crosslinking of tRNA containing a long extra arm to elongation factor Tu by *trans*-diamminedichloroplatinum(II). *Nucleic Acids Res.* **18:** 4883–4890.

Rould, M.A., J.J. Perona, D. Söll, and T.A. Steitz. 1989. Structure of *E. coli* glutaminyl-tRNA synthetase complexed with tRNA(Gln) and ATP at 2.8 Å resolution. *Science* **246:** 1135–1142.

Sanfacon, H. and T. Hohn. 1990. Proximity to the promoter inhibits recognition of cauliflower mosaic virus polyadenylation signal. *Nature* **346:** 81–84.

Saville, B.J. and R.A. Collins. 1990. A site-specific self-cleavage reaction performed by a novel RNA in *Neurospora* mitochondria. *Cell* **61:** 685–696.

Schimmel, P. 1987. Aminoacyl tRNA synthetases: General scheme of structure-function relationships in the polypeptides and recognition of transfer RNAs. *Annu. Rev. Biochem.* **56:** 125–158.

Sen, D. and W. Gilbert. 1988. Formation of parallel four-stranded complexes by guanine-rich motifs in DNA and its implications for meiosis. *Nature* **334:** 364–366.

Strauss, E.G. and J.H. Strauss. 1983. Replication strategies of the single stranded RNA viruses of eukaryotes. *Microbiol. Immunol.* **105:** 2–98.

Subbarao, M.N. and D. Apirion. 1989. A precursor for a small stable RNA (10Sa RNA) of *Escherichia coli. Mol. Gen. Genet.* **217:** 499–504.

Sundquist, W.I. and A. Klug. 1989. Telomeric DNA dimerizes by formation of guanine tetrads between hairpin loops. *Nature* **342:** 825–829.

Tse, W.T. and B.G. Forget. 1990. Reverse transcription and direct amplification of cellular RNA transcripts by Taq polymerase. *Gene* **88:** 293–296.

Tyagi, J.S. and A.K. Kinger. 1992. Identification of the 10Sa RNA structural gene of *Mycobacterium tuberculosis. Nucleic Acids Res.* **20:** 138.

Visser, C.M. and R.M. Kellogg. 1978a. Biotin. Its place in evolution. *J. Mol. Evol.* **11:** 171–187.

————. 1978b. Bioorganic chemistry and the origin of life. *J. Mol. Evol.* **11:** 163–169.

Weiner, A.M. 1987. Summary. *Cold Spring Harbor Symp. Quant. Biol.* **52:** 933–941.

Weiner, A.M. and N. Maizels. 1987. 3′ Terminal tRNA-like structures tag genomic RNA molecules for replication: Implications for the origin of protein synthesis. *Proc. Natl. Acad. Sci.* **84:** 7383–7387.

White, H.B., III. 1976. Coenzymes as fossils of an earlier metabolic state. *J. Mol. Evol.* **7:** 101–104.

————. 1982. Evolution of coenzymes and the origin of pyridine nucleotides. In *The pyridine coenzymes* (ed. J. Everse et al.), pp. 1–17. Academic Press, New York.

Williamson, J.R., M.K. Raghuraman, and T.R. Cech. 1989. Monovalent cation-induced structure of telomeric DNA: The G-quartet model. *Cell* **59:** 871–880.

Wintersberger, U. and E. Wintersberger. 1987. RNA makes DNA: A speculative view of the evolution of DNA replication mechanisms. *Trends Genet.* **3:** 198–202.

Wong, J.-T. 1991. Origin of genetically encoded protein synthesis: A model based on selection for RNA peptidation. *Orig. Life Evol. Biosphere* **21:** 165–176.

Xiong, Y. and T.H. Eickbush. 1990. Origin and evolution of retroelements based upon their reverse transcriptase sequences. *EMBO J.* **9:** 3353–3362.

Yarus, M., M. Illangesekare, and E. Christian. 1991. An axial binding site in the *Tetrahymena* precursor RNA. *J. Mol. Biol.* **222:** 995–1012.

APPENDIX 1: Structures of Base Pairs Involving at Least Two Hydrogen Bonds

Provided by Ignacio Tinoco, Jr.
Department of Chemistry and Chemical Biodynamics Laboratory
University of California, Berkeley, California 94720

The structures of the 28 possible base pairs that involve at least two hydrogen bonds are given in Figures 1–4 (for description, see Saenger, in *Principles of nucleic acid structure*, p. 120. Springer-Verlag [1984]). The non-Watson-Crick base pairs are called base-base mismatches, or internal loops of two nucleotides.

The RNA World
© 1993 Cold Spring Harbor Laboratory Press 0-87969-380-0/93 $5 + .00

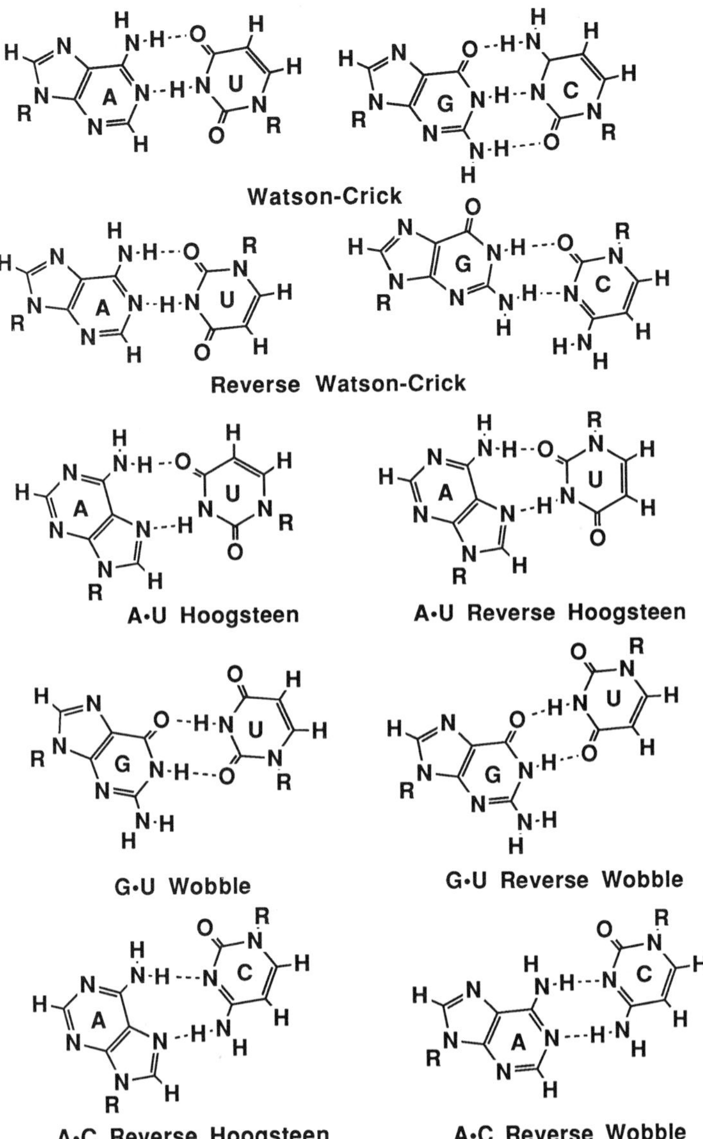

Figure 1 The ten possible purine-pyrimidine base pairs. Yeast tRNA^Phe provides crystal structures for a G·U wobble, an A·U reverse Hoogsteen, and a G·C reverse Watson-Crick (Saenger, in *Principles of nucleic acid structure*, p. 336. Springer Verlag [1984]). A G·U wobble crystal structure is also given by Holbrook et al. (*Nature 353:* 579 [1991]). A G·U reverse wobble was determined by NMR in an extrastable RNA tetraloop (Varani et al., *Biochemistry 30:* 3280 [1991]). Note that the A·C reverse wobble is very similar to the G·U reverse wobble.

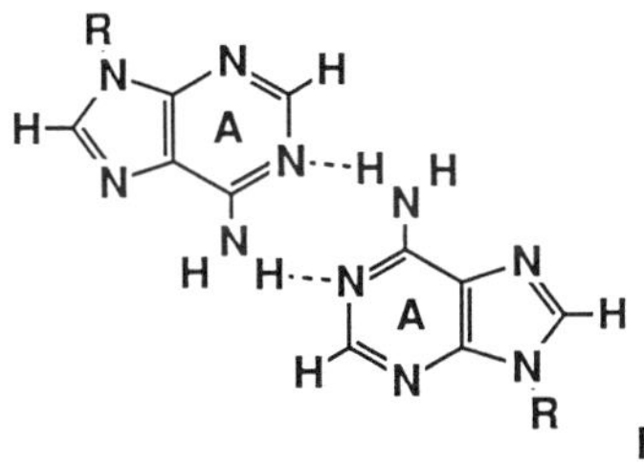

A•A N1-amino, symmetric

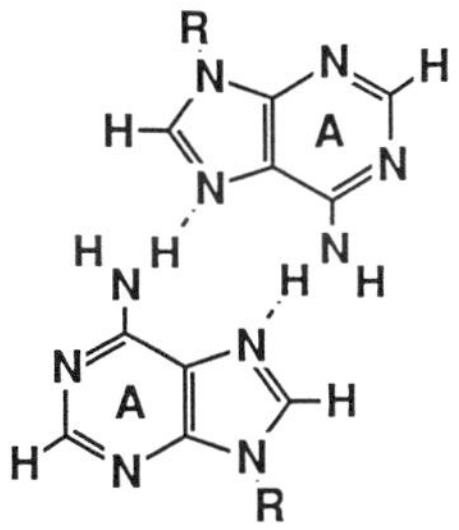

A•A N7-amino, symmetric

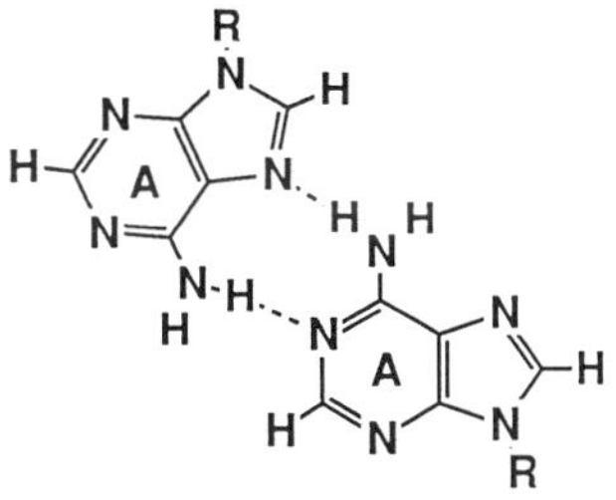

A•A N1-amino, N7-amino

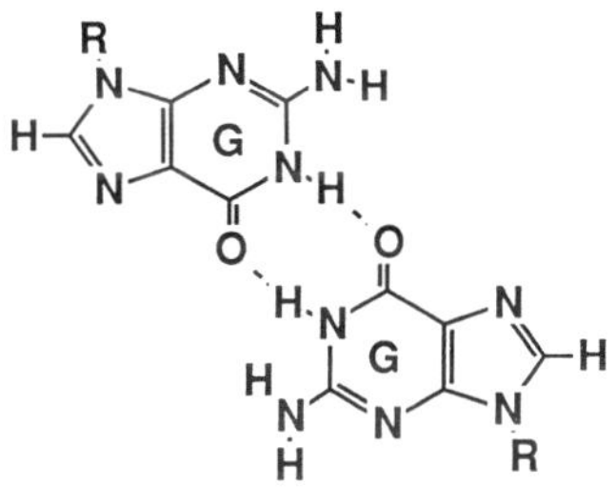

G•G N1-carbonyl, symmetric

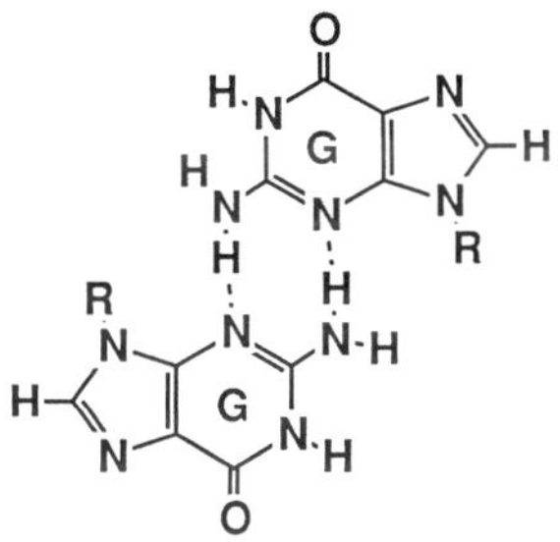

G•G N3-amino, symmetric

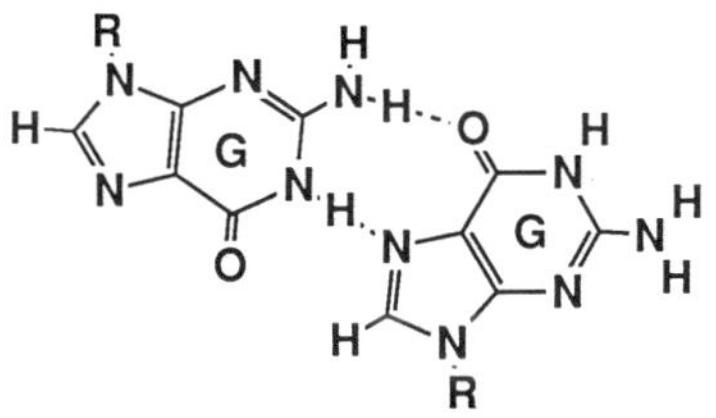

G•G N7-N1, carbonyl-amino

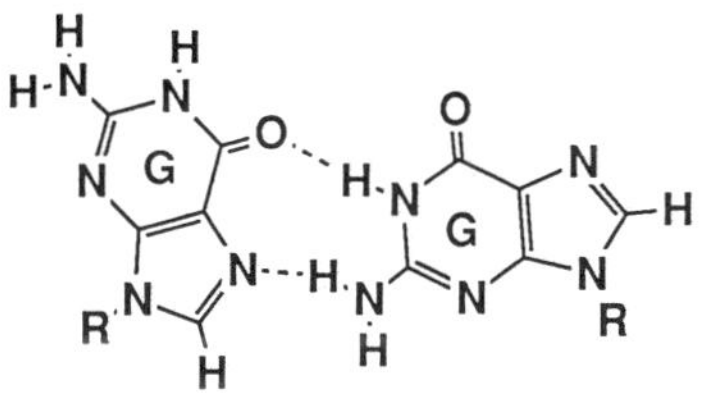

G•G N1-carbonyl, N7-amino

Figure 2 The seven possible homo purine-purine base pairs. Yeast $tRNA^{Phe}$ provides crystal structures for an A•A N7-amino symmetric pair and a G•G N7-N1, carbonyl-amino pair (Saenger, in *Principles of nucleic acid structure*, p. 336. Springer Verlag [1984]). Both pairs are part of base triplets.

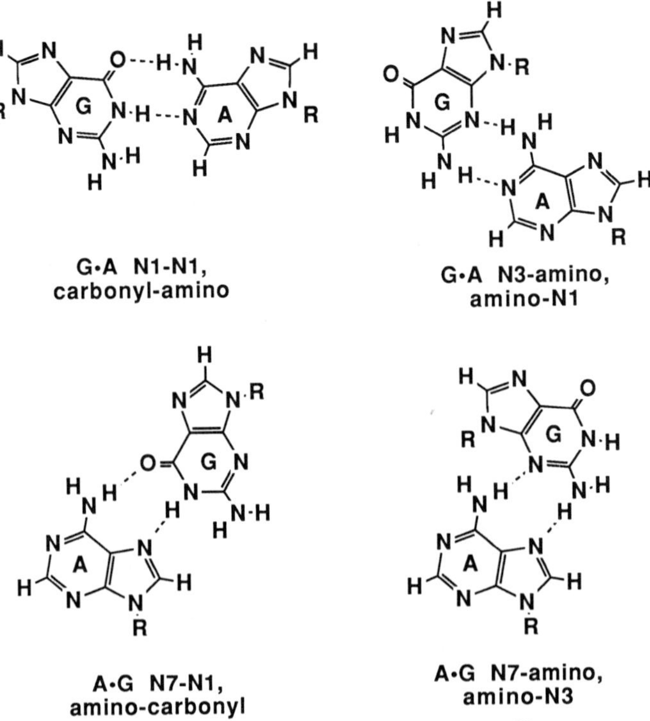

Figure 3 The four possible A·G base pairs. Yeast tRNA^Phe provides a crystal structure for the G·A N1-N1, carbonyl-amino pair (Saenger, in *Principles of nucleic acid structure*, p. 336. Springer Verlag [1984]). The A·G N7-amino, amino-N3 pair has been determined by NMR in an extrastable RNA tetraloop (Heus and Pardi, *Science 253:* 191 [1991]) and in loop E of a 5S rRNA (Wimberly et al., *Biochemistry 32:* 1078 [1993]).

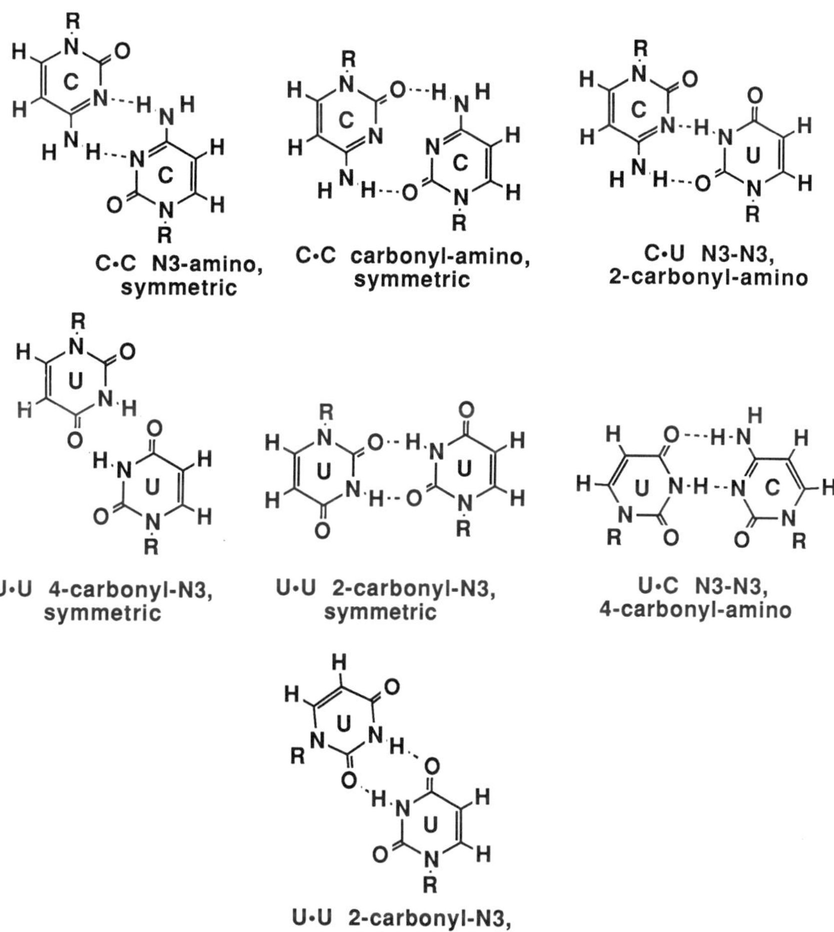

Figure 4 The seven possible pyrimidine-pyrimidine base pairs. The crystal structure of a U·C pair is given by Holbrook et al. (*Nature 353:* 579 [1991]). It is the U·C N3-N3, 4-carbonyl-amino pair, but with a water molecule between the two N3 atoms. The water-mediated hydrogen bond increases the distance between the C1 ′ atoms of the ribose rings (designated as R in the figure) to allow a better fit into a normal double helix.

APPENDIX 2: RNA Pseudoknots

Cornelis W.A. Pleij
Department of Biochemistry, Gorlaeus Laboratories
Leiden, The Netherlands

Understanding the many different biological activities displayed by RNA requires a detailed knowledge of the specific three-dimensional folding of a single-stranded RNA molecule. This folding is dominated by the formation of intramolecular double-helical or stem regions involving the Watson-Crick A·U and G·C base pairs. This gives rise to a collection of stems which is represented in what is called the "classical" or "orthodox" secondary structure of RNA. These structures have in common that loop regions formed (like hairpin [H], bulge [B], or interior [I] loops) do not participate in their turn in base-pairing. In the last decade it has become clear, however, that for the final, biologically relevant, tertiary structure, base-pairing of the loop regions can be essential. It is this kind of base-pairing that leads to the formation of pseudoknotted structures. Some basic features of possible pseudoknot structures are described here. For a more in-depth discussion of structural and functional aspects of RNA pseudoknots, see Westhof and Jaeger (*Curr. Opin. Struct. Biol. 2:* 327 [1992]) and ten Dam et al. (*Biochemistry 31:* 11665 [1992]).

A pseudoknot structure is formed when a loop region in a classical secondary structure is involved in Watson-Crick base-pairing with a complementary region outside this loop. The simplest structural element in RNA is the hairpin and, consequently, base-pairing of the hairpin loop gives rise to the simplest type of pseudoknot (Fig. 1A). Base-pairing can take place with a sequence either downstream or upstream of the pseudoknot. Note that one single pseudoknot can be presented in two different, but in fact equivalent, ways (Fig. 1A). A pseudoknot is always defined by two stems (S1 and S2) and by two or three loop regions. In the generalized form of Figure 1A there are three linkers or connecting loops (L1–L3). If one of the three loop regions is absent, structurally interesting and biologically relevant pseudoknots are formed. In that case, the base-pairing loop region becomes contiguous to the other stem, which enables coaxial stacking of the two stem regions. In Figure 1B, the two resulting connecting loops span the shallow and deep grooves of the quasi-continuous double helix, respectively (see L_S and L_D). The structural requirements for these two loops are reasonably well understood.

The RNA World
© 1993 Cold Spring Harbor Laboratory Press 0-87969-380-0/93 $5 + .00

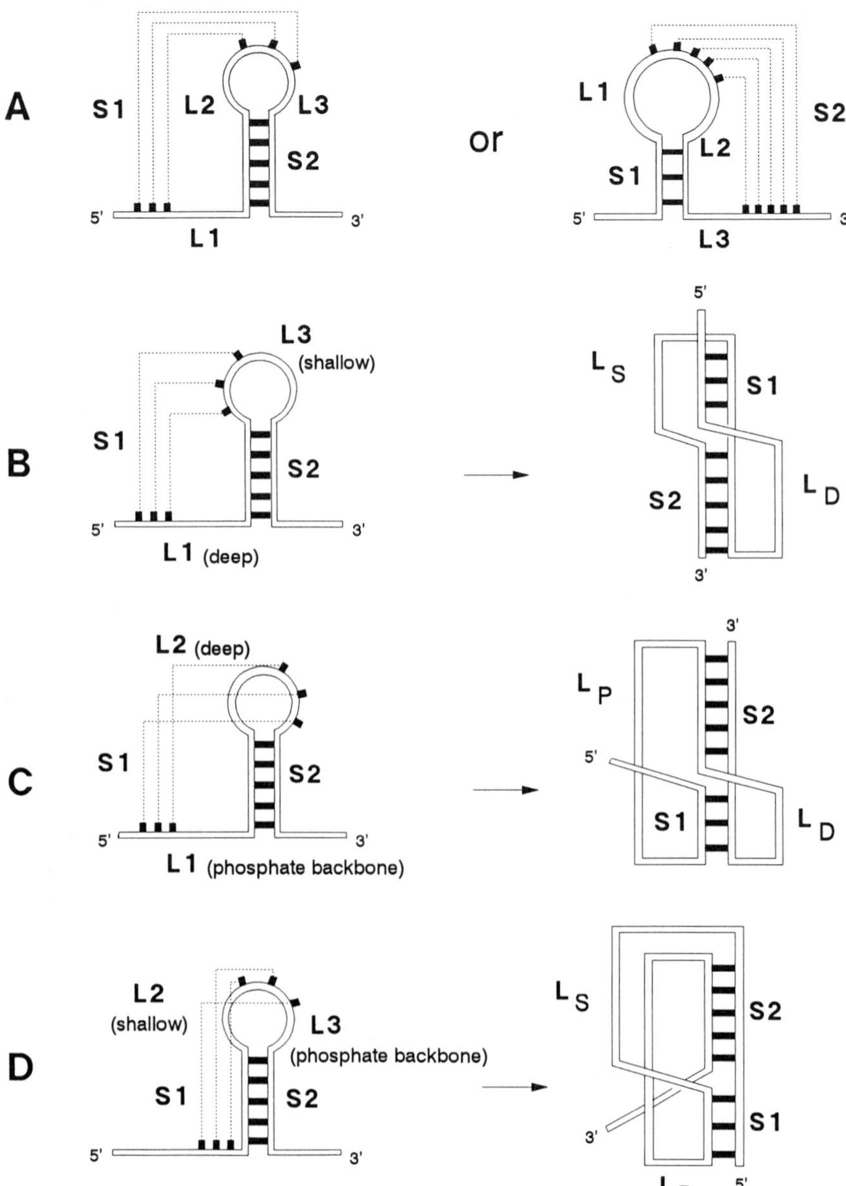

Figure 1 The H-type pseudoknot. (*A*) Two alternative presentations of the generalized H-pseudoknot. (*B–D*) Three classes of pseudoknots with coaxial stacking of the two stems when loop L2 (*B*), L3 (*C*), or L1 (*D*) is zero, respectively. Dashed lines indicate base-pairing. L_D is the loop crossing the deep groove, L_S is the loop crossing the shallow groove, and L_P is the loop spanning the sugar-phosphate backbone.

For instance, the loop crossing the deep groove can be as short as one or two nucleotides when 5–7 base pairs are spanned (Pleij et al., *Nucleic Acids Res. 13:* 1717 [1985]). The pseudoknot shown in Figure 1B is found very often and is also called the classical pseudoknot. It was first described in the tRNA-like structure at the 3′ end of turnip yellow mosaic virus (TYMV) RNA. The coaxial stacking of the two stems was essential in this case to understand the resemblance to tRNA (Rietveld et al., *Nucleic Acids Res. 10:* 1929 [1982]). Figure 2A shows the secondary structure including the pseudoknot at the very 3′ end of TYMV RNA, and Figure 2B presents a stereoscopic view of this pseudoknot as derived from model-building studies using molecular graphics. This classical and relatively simple type of pseudoknot has been found in many different RNAs. It occurs in noncoding regions of plant viral RNAs and animal viral RNAs at both 3′ and 5′ termini, sometimes in tandem, leading to three or more pseudoknots in a row. This type of pseudoknot is also found in coding regions of various viral RNAs, where it plays a role in stimulating ribosomal frameshifting or readthrough. Furthermore, these pseudoknots are also present in leaders of prokaryotic mRNAs, where they are involved in translational autoregulation (for review, see ten Dam et al., *Biochemistry 31:* 11665 [1992]).

Two other possible stem stackings are shown in Figure 1, C and D. As a result, one of the two connecting loops now runs outside the helical cylinder (L_P) and in fact crosses not only grooves, but also the sugar phosphate backbone. So far only a few examples of this type of pseudoknot have been proposed. On the other hand, experimental verification of this possible stacking is urgently needed. If one or more of the three different connecting loops in Figure 1A is replaced partially or entirely by a hairpin or any other elaborated RNA structure, then other types of pseudoknots can be discerned. In Figure 3A, introduction of a stem in loop L3 leads to a bulge loop base-pairing with a region outside this loop. However, this loop can also be seen as a variant of an H-type pseudoknot (Fig. 3A), which may illustrate the pitfalls encountered in proposing a classification of RNA pseudoknots. Another important aspect of this B-type pseudoknot is how the three stem regions are stacked or oriented. There is no information available about the three-dimensional structure of any B-type pseudoknot. A typical example is found in the tRNA-like structure at the end of tobacco mosaic virus (TMV) RNA.

In the same way, a pseudoknot involving an interior loop can be constructed starting from Figure 1A, as illustrated in Figure 3B. Redrawing and reorientation of the stems, however, can transform this I-type pseudoknot into one of the B-type. A well-known example exists in the core structure of group I introns, corresponding to stem region P3.

For reasons outlined above, a classification of pseudoknots involving loop-loop interactions or multibranched loops is complicated and can even become meaningless; the more so because no consequences for the three-dimensional folding are known (Mans, Ph.D. thesis, University of Leiden, The Netherlands [1991]).

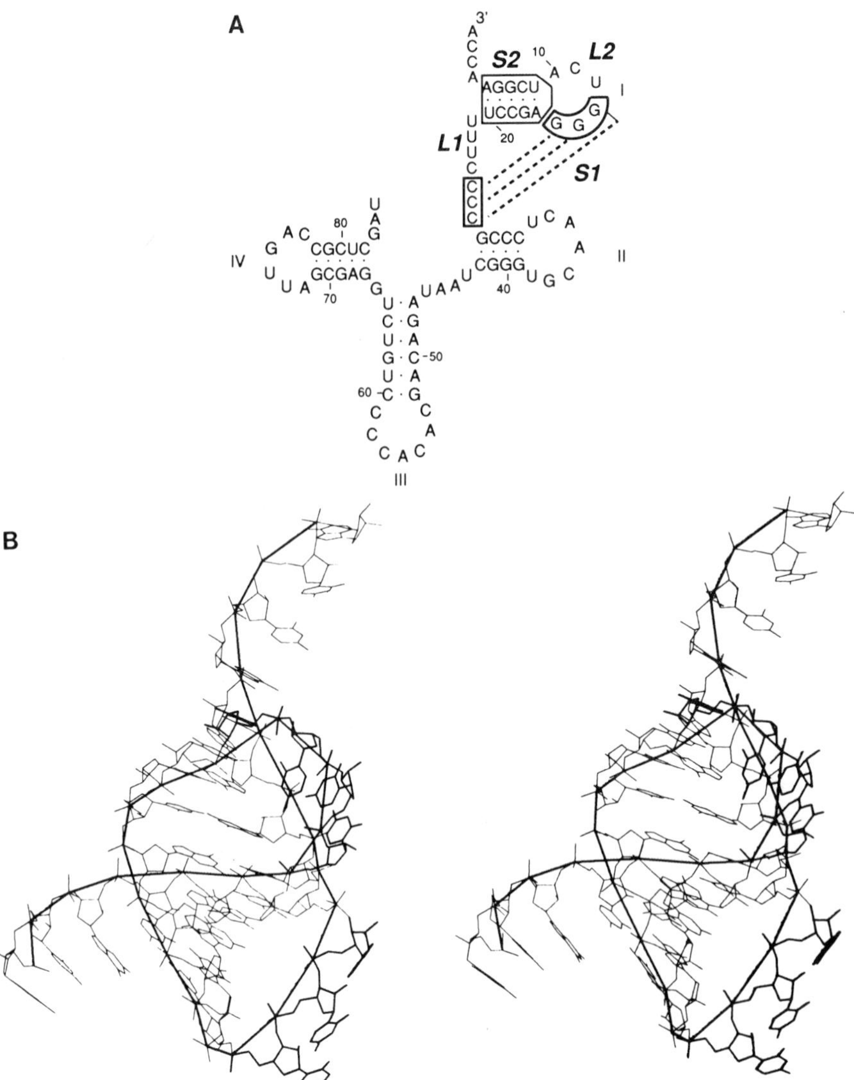

Figure 2 The tRNA-like structure at the 3′ end of TYMV RNA. (*A*) Secondary structure. Dashed lines indicate the pseudoknot-forming base-pairing. Numbering of nucleotides is from the 3′ end. (*B*) Stereoscopic view of the pseudoknot region as obtained by computer modeling.

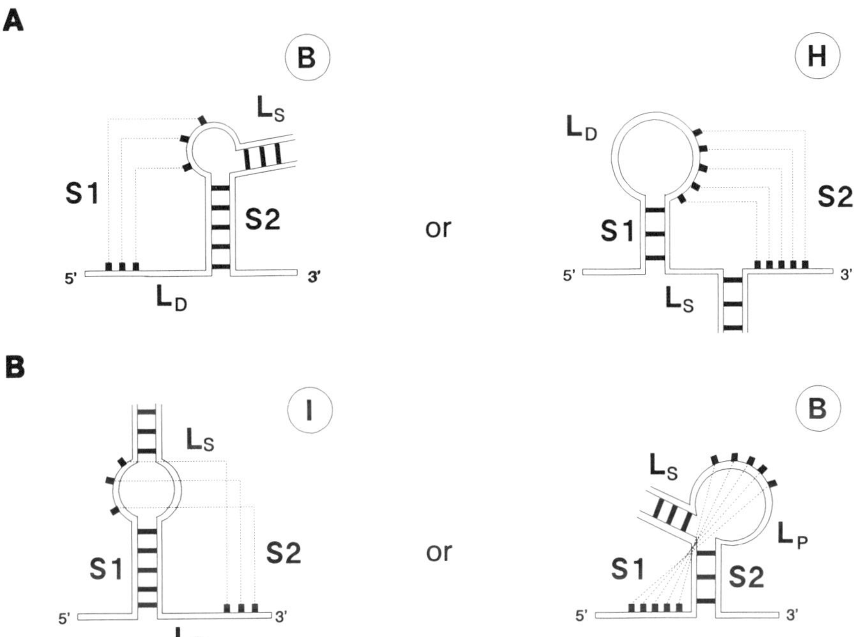

Figure 3 Alternative secondary structure presentations for a B-type pseudoknot (*A*) and an I-type pseudoknot (*B*). For further details, see Fig. 1.

APPENDIX 3: Stereo Pairs of RNA Fragments

Provided by Ignacio Tinoco, Jr.
Department of Chemistry and Chemical Biodynamics Laboratory
University of California, Berkeley, California 94720

Five stereo pairs of RNA fragments are shown here. We thank Stephen Holbrook, Arthur Pardi, and James Williamson for providing their coordinates.

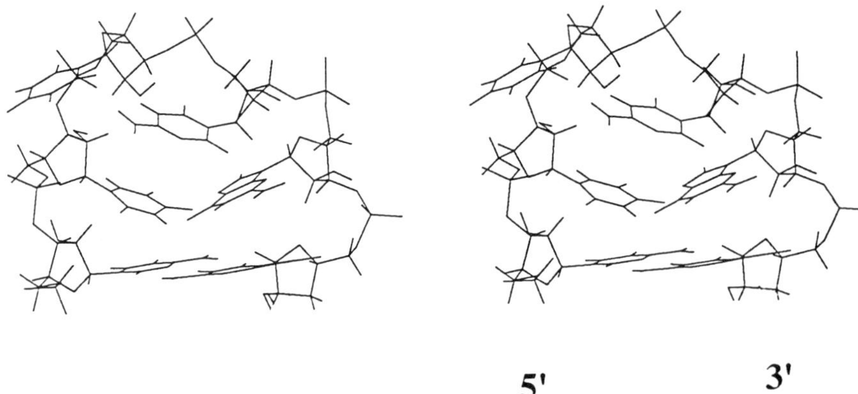

Figure 1 The 5′-C(UUCG)G-3′ extrastable tetraloop and closing base pair as determined by NMR (Varani et al., *Biochemistry 30:* 3280 [1991]). The closing C·G base pair is seen at the bottom of the figure with the buckled U·G pair above it. The second U in the loop is shown out in the solvent; its position is not well defined, and any base can replace the U without perturbing the loop structure. The loop C is seen bridging the loop with its amino group near a phosphate group. Three other Watson-Crick base pairs that continue the stem are not shown.

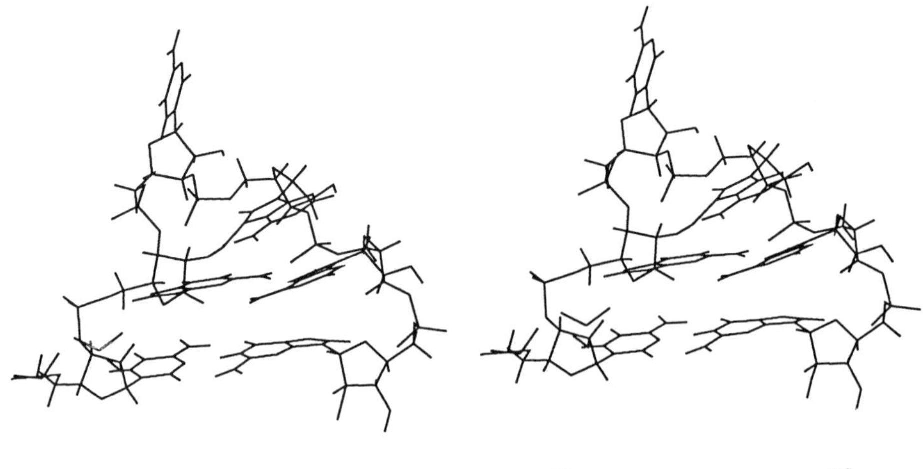

Figure 2 The 5′-C(GCAA)G-3′ extrastable tetraloop and closing base pair as determined by NMR (Heus and Pardi, *Science 253:* 191 [1991]). The closing C·G base pair is seen at the bottom of the figure with the G·A pair above it. The loop C is shown out in the solvent; its position is not well defined, and any base can replace the C without perturbing the loop structure. The next A is shown bridging the loop to interact with a ribose. Three other Watson-Crick base pairs that continue the stem are not shown.

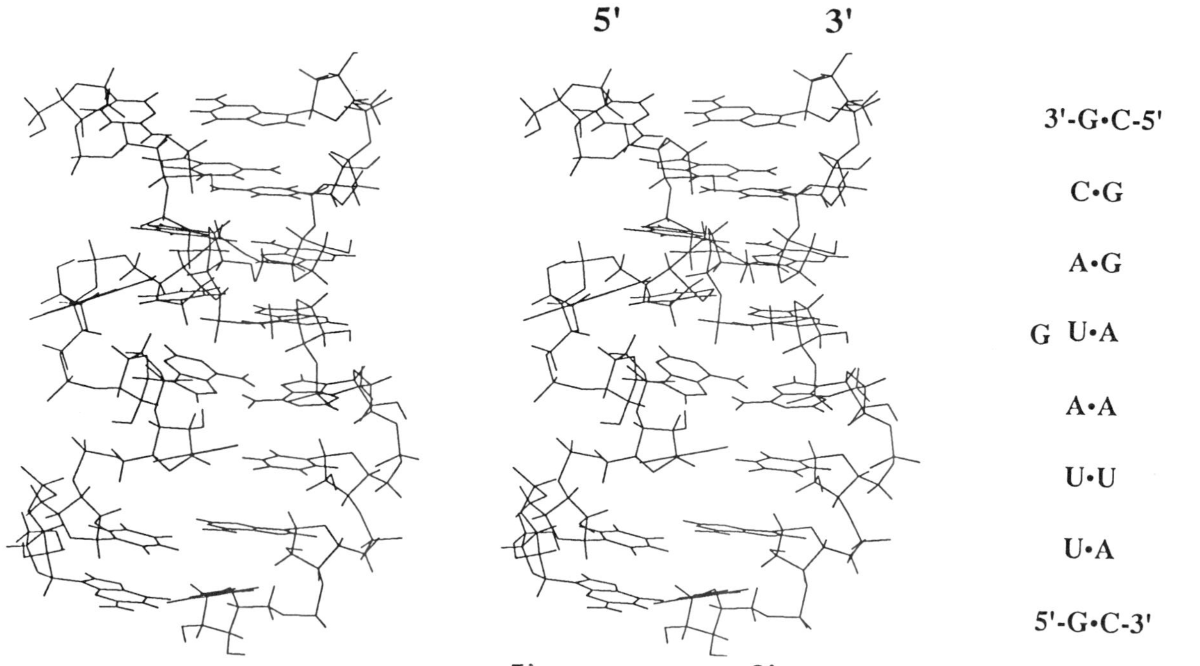

Figure 3 Loop E of eukaryotic 5S RNA as determined by NMR (Wimberly et al., *Biochemistry 32:* 1078 [1993]). Two Watson-Crick base pairs are shown at each end of the asymmetric internal loop of nine nucleotides. The loop consists of a U·U, an A·A, a reverse Hoogsteen U·A, and an A·G pair. A bulge G lies in the major groove interacting with the U·A. The ribose attached to the A on the 5′ side of the bulge G is reversed in direction; thus at the A·A pair, the strands are parallel instead of antiparallel. Two other Watson-Crick base pairs that continue the stems above and below the loop are not shown.

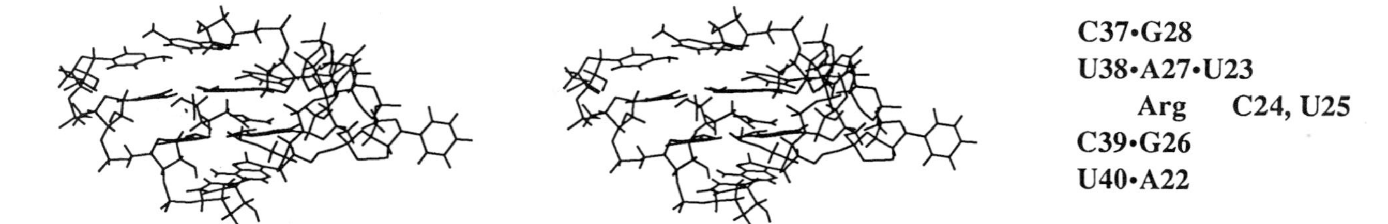

Figure 4 The TAR RNA-arginine complex as modeled by NMR (Puglisi et al., *Science 257:* 76 [1992]). Arginine is in the center of the figure with the base triple formed by the Watson-Crick pair of U38•A27 and the bulge U23 right above it. Thus two base pairs are shown both above and below the arginine; one of the bulge bases is seen on the right of the duplex. Other nucleotides in the oligonucleotide studied are not shown.

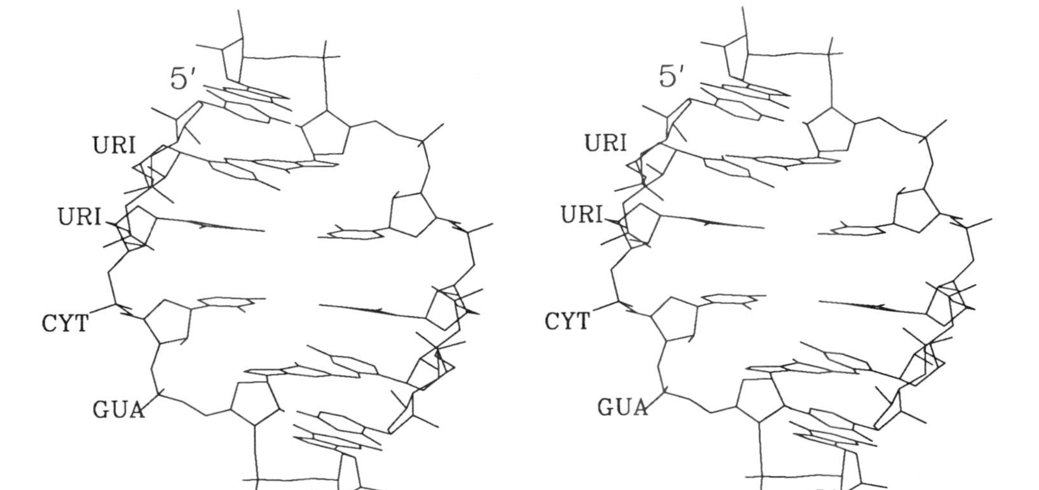

Figure 5 A symmetric internal loop of eight nucleotides formed in a duplex of 5′-C(UUCG)G-3′ tetraloop sequences as determined by X-ray diffraction (Holbrook et al., *Nature 353:* 579 [1991]). The internal loop contains two wobble G·U base pairs and two U·C base pairs which contain water-mediated base-base hydrogen bonds. The loop is very similar to an A-form helix except that the major groove is widened. Three other Watson-Crick base pairs that continue the stems above and below the loop are not shown.

Index